# Bifurcation and Chaos in Engineering

Springer
*London*
*Berlin*
*Heidelberg*
*New York*
*Barcelona*
*Budapest*
*Hong Kong*
*Milan*
*Paris*
*Santa Clara*
*Singapore*
*Tokyo*

Yushu Chen and Andrew Y.T. Leung

# Bifurcation and Chaos in Engineering

With 235 Figures

Springer

Professor Yushu Chen, PhD
Department of Mechanics, Tianjin University, Tianjin, China 300072

Professor Andrew Y.T. Leung, DSc, PhD, CEng, FRAeS, MIStructE, MHKIE
Manchester School of Engineering, Simon Building, University of Manchester, Oxford Road, Manchester M13 9PL, UK

ISBN-13:978-1-4471-1577-9 e-ISBN-13:978-1-4471-1575-5
DOI: 10.1007/978-1-4471-1575-5

British Library Cataloguing in Publication Data
Chen, Yushu
Bifurcation and chaos in engineering
1. Engineering mathematics 2. Differential dynamical systems
3. Bifurcation theory 4. Chaotic behavior in systems
I. Title II. Leung, Andrew
620'.0015'1
ISBN-13:978-1-4471-1577-9

Library of Congress Cataloging-in-Publication Data
A catalog record for this book is available from the Library of Congress



Typesetting: Camera-ready by authors

69/3830-543210 Printed on acid-free paper

# Preface

For the many different deterministic non-linear dynamic systems (physical, mechanical, technical, chemical, ecological, economic, and civil and structural engineering), the discovery of irregular vibrations in addition to periodic and almost periodic vibrations is one of the most significant achievements of modern science. An in-depth study of the theory and application of non-linear science will certainly change one's perception of numerous non-linear phenomena and laws considerably, together with its great effects on many areas of application. As the important subject matter of non-linear science, bifurcation theory, singularity theory and chaos theory have developed rapidly in the past two or three decades. They are now advancing vigorously in their applications to mathematics, physics, mechanics and many technical areas worldwide, and they will be the main subjects of our concern.

This book is concerned with applications of the methods of dynamic systems and subharmonic bifurcation theory in the study of non-linear dynamics in engineering. It has grown out of the class notes for graduate courses on bifurcation theory, chaos and application theory of non-linear dynamic systems, supplemented with our latest results of scientific research and materials from literature in this field. The bifurcation and chaotic vibration of deterministic non-linear dynamic systems are studied from the viewpoint of non-linear vibration. It is advantageous to do so, because on the one hand both have similar concepts and methods, and on the other hand, the explanation of theoretical methods with a view towards application facilitates not only those engaged in the study the work of vibration but also those devoted to studies in other fields. One can use this book as a reference work on dynamical systems. Although this book is far from an exhaustive monograph on non-linear dynamical and chaos in engineering systems, the authors hope that it will serve as a systematic and practical aid to those who carry out teaching and research work on bifurcation and chaos of non-linear vibration systems in the fields of mathematics, physics, mechanics, ecology, engineering and technology with a stress on civil and structural engineering.

This book can be divided into four main parts. The first part describes the theory of dynamic systems, encompassing Chapter 1 — dynamical systems, ordinary differential equations and stability of motion, Chapter 2 — calculations of flow, and Chapter 3 — discrete dynamic systems. The second part presents the main methods of the bifurcation theory. The Liapunov–Schmidt method, centre manifold theorem, vector fields, normal form theory, averaging method, singularity theory, Hopf bifurcation theory and the bifurcation theory of 1/2 subharmonic resonance of non-linear parametrically excited vibration systems, and Euler's buckling problems — are all dealt with in Chapters 4 to 7. The third part consists of Chapter 8, dealing with chaos theory and its application, and Chapter 9, dealing with the construction of chaotic regions. The fourth part consists of Chapter 10, which describes the numerical methods in non-linear dynamics, and Chapter 11, which finally presents some important engineering practice examples of application of bifurcation theory.

We are greatly indebted to Professors William F Langford, Wang Zhaolin and Huang Kelei and Doctors TC Fung, T Ge, SG Mao and SK Chui for their valuable contributions and constructive opinions in the preparation of this book. Thanks are also extended to Wang Deshi, Cao Qingjie, Zhang Weiyi, Hu Jindong, Leung Hau Yan and others in the group of the theory and application of non–linear dynamical systems in the Department of Mechanics, Tianjin University and the Department of Civil Engineering, Hong Kong University for their work and comments in the preparation of manuscripts. The secretary support of Mr Li Weidong and Ms B Knight of the University of Manchester in the final production of the camera ready copy of the book is gratefully acknowledged. We are also thankful to the helping hands of the staff of Springer–Verlag to make the publication of the book possible.

Finally, AYTL is grateful to the support of his family, Anna, Colin, Edwin and Johanna throughout the preparation the book.

# Introduction

Deterministic motion and the transient process of dynamic systems from one deterministic motion to another occur by means of vibrations.

A deterministic motion is one that has repeatability and a certain stability, and the transient process is the whole process that instigates this deterministic motion. The set of transient processes constitutes the attraction domain of the motion. When a physical parameter in the system under study changes to a certain value, the solution curve branches out a family of curves. This is called bifurcation. If the deterministic motion changes rapidly enough (i.e. jumps exist), the new state of motion is considered to be caused by hard elements, otherwise by soft elements. Phenomena arising in such a non-linear system are named non-linear vibrations.

Widely used methods of the theory of non-linear vibration, or methods of finding a periodic solution, are the perturbation method, the averaging method and the numerical method. They have developed rapidly since the 1920s. These methods, however, were used in the past only in studying the periodic solution when the parameters in the vibration system are given constants. But in engineering practice some parameters of a system often may be subjected to small variations (also referred to as perturbation). The speed of a vehicle running on a straight rail, the temperature and density in a chemical process, and the damping coefficient and frequency of the excitation in parametrically excited system are only a few examples, and the perturbation of these parameters often tends to cause the appearance of bifurcation of the periodic solutions.

The bifurcation theory permeates different fields of engineering and natural science. Differential equations describing physical systems often contain parameters whose measured values fluctuate within a very small range. If the differential equation simulating a physical system is structurally unstable when the parameters reach some values, then a change will appear in quality in the behaviour of the solution (i.e. bifurcation) when a small variation takes place the right-hand side of the differential equation. As a result, it is necessary to know how the phase portraits respond to the change of parameters.

It is common to describe a mechanical model by means of a huge group of differential equations, for example, a system with numerous variables. In order to proceed or simplify the analysis, the common practice is to regard those variables which change very slightly in the dynamic process as constants, or to omit some of the variables which affect only minor factors. Yet it is often impossible to evaluate the effects of the dropped terms on the original model according to the simplified equation. In this case, if the omitted terms are taken as perturbations, they can be dealt with by singularity theory.

The parameter family which varies very slowly with time is termed a slowly varying family. The theory of relaxation vibration is closely associated with the bifurcation theory in which parameters do not vary with time. There are slowly varying parameters in the "slow–fast" system of the relaxation vibration. When the rate of change of slowly varying parameters is zero, the "slow–fast" system becomes the bifurcation system mentioned previously; when the rate is not zero, a special phenomenon called "dynamic bifurcation" takes place.

Not only does the bifurcation theory interest researchers with its profound and active topics but also its close ties with practical engineering problems reveal its importance. Complicated theoretical bifurcation problems can be found in the snake motion of high-speed vehicles, heat convection in fluid, vibration of pipes and related problems, oscillations in chemical reactions, dynamic buckling of the Euler rod, oscillation of chain bridges, bifurcation of shock waves, bifurcation in the neural network in human bodies, bifurcation in superconductors, bifurcation of dynamic systems in sociology and economics, and so on.

The singularity theory can be used to simplify the form of the bifurcation equations. It can determine the minimum number of unfolding parameters from among numerous control parameters.

Of the many different deterministic non-linear dynamic systems (physical, mechanical, technical, chemical, biological, economic) the discovery of irregular vibrations in addition to periodic and almost-periodic vibrations is one of the important scientific discoveries in recent years. The reason why irregular vibrations (chaotic) are studied from a viewpoint of the vibration theory is mainly that they have similar concepts and methods.

Chaos is a motion occurring over a limited range in the deterministic system. With no regularity, such a motion is similar to a stochastic one and extremely sensitive to the initial value (that is, a small change in the initial value results in a completely different response — it may fall into Liapunov's instability). In other words, chaotic motion has unpredictability over a long time. The deterministic systems are those that can usually be described by differential equations, partial differential equations, difference equations and even simple iterative equations. The coefficients in these equations are deterministic.

In a word, chaotic motion is an unstable bounded stationary motion (i.e. locally unstable and wholly compressible). This definition unfolds the two aspects of chaotic motion: instability (this character can be made exact when the Liapunov's averaging exponent is greater then zero) and finiteness. Or, chaotic motion is a bounded stationary motion without equilibrium, periodicity and almost-periodicity. Here the bounded stationary motion means that the state of motion does not vary with time in a stochastic sense (from the viewpoint of the finite domain as a whole in phase space).

The appearance of bifurcation in non-linear systems and global bifurcation, in particular, will lead to chaotic motion in many cases. The bifurcation theory is one of the theories which explain how the resolvable system transits from a normal orderly state to a chaotic state. As H. Poincaré well says: the bifurcation theory is a torch, it lights up the road leading from systems which can be studied to those that cannot be studied. Chaos is different from the stochastic motion in that it has definite modes of motion and fine structures, e.g. self-resemblance structure.

Taking advantage of this character of the bifurcation theory, Landau and Hopf gave a picture of the transition from laminar to turbulent flow when the Reynolds number increases. In Landau's description, this transition is realised through the tori of the increased number of dimensions. Since the discovery of the strange attractor by Lorenz in 1963, four typical routes leading to chaos have been known. They are periodic doubling bifurcation, intermittent transition, secondary Hopf bifurcation and the breaking of the KAM (Kolmogorov–Arnold–Mosor) torus. The mechanisms of the first three have all been clear in theory, but that of the fourth remains to be solved. Another important mechanism leading to chaos concern homoclinic and heteroclinic bifurcations. At this point, the stable and unstable manifolds of the saddle points will intersect with each other under disturbances, resulting in a hyperbolic limit set — the Smale horseshoe. This is thought to be one of the basic structures of chaos by mathematicians. In 1963 Milnikov proposed a method of analyzing homoclinic bifurcation. Though important advances have been made in the study on chaos, little has been reported of the chaos in high and infinite dimensional systems. The main difficulty is that it is hard to imagine the visual structure of the stable and unstable manifolds in the high dimensional systems.

The ubiquitous presence of bifurcation and chaos in different fields of engineering and natural sciences gives impetus to the rapid development of the theory of dynamic systems.

At the turn of this century Poincaré and others evolved the concept of dynamic systems from the study of classical mechanics and qualitative theory of differential equations. Instead of integrating the differential equations, he took the right-hand side of the equations as a defined vector field and through the form of the known vector field he explored the solution curve family defined by differential equations. (Taking a curve as the path moved through by a particle, we assume that every particle on the curve moves simultaneously, and in this context we call the solution curve an orbit or a trajectory.) Poincaré introduced such concepts as equilibrium state, periodic solution, recurrence and studied the behaviour of a solution developing with time as a whole. Poincaré 's innovation signifies that he regards the state variable of a system not only as a function of the time, but also as a function of the initial conditions, i.e. the continuously varying structure of flow is the total sum of curves passing through all the points in definition domain to value domain.

Modern research work on the differential dynamic system originated from the work of M. M. Peixoto and others in the 1960s. Thanks to the advocacy and promotion of scholars such as V.I. Arnold, S. Smale and many others, research of basic theories of this discipline has made important progress. However, such new tools were only at the disposal of mathematicians before the mid-1970s. In the past decade the study of differential dynamic systems has spread extensively to many different areas of application.

The two main developing directions of dynamic systems are the study of orderly (or regular) motion and chaotic (or irregular) motion. The orderly and disorderly motions of the discrete system are merely behaviours of time, its disorderly behaviour is only time chaos with its geometrical feature being the strange attractor; the orderly and disorderly motions of the continuous system are behaviours of both time and space. Turbulent flow, being a continuous system, is chaos of space as well as chaos of time (its geometrical feature is a saddle invariant set). They are associated with each other, but one cannot be derived from the other.

This book consists of eleven chapters and an introduction.

Chapter 1 introduces the basic concepts of dynamic systems, the important properties of flow, Poincaré –Bendixon's theorem, a brief description of the main theories of ordinary differential equations and stability of motion.

Chapter 2 treats the calculation of flow: first the divergence of flow, then the calculation of linear flow, non-linear differential equations and the calculation of its flow, and finally the stable manifold theorem.

Chapter 3 discusses discrete dynamic systems, introducing discrete dynamic systems and the linear map, the non-linear map and the stable manifold theorem of the map, the classification of the generic map, the Poincaré map and the structural stability of the vector field.

Chapter 4 deals with the Liapunov–Schmidt reduction, the concepts of bifurcation, the implicit function theorem and singularity theory. Examples include the one-half subharmonic resonance bifurcation of non-linear parametrically excited vibration systems and Hopf bifurcation.

Chapter 5 is focused on the centre manifold theorem and normal form theory of a vector field by the matrix method.

Chapter 6 is devoted to Hopf bifurcation as an example of application of the centre manifold theorem and normal form theory, the complex normal form of Hopf bifurcation, the real normal form, the calculation formulae of the analytical method of the stability of the solution, and lastly, bifurcation problems of systems with double zero eigenvalues.

Chapter 7 introduces the application of the averaging method in bifurcation theory, the averaging method for systems with many degrees of freedom, the geometrical description of the averaging method, averaging method and local bifurcation, and averaging method and global bifurcation.

Chapter 8 presents a brief description of chaos, examples of chaos in non-linear systems, the method of study and the statistical character of chaos.

Chapter 9 deals with the construction of chaotic regions. A new method of numerical simulation is given to describe the characteristics of the solutions of dynamic systems. The characteristics of the solutions, such as the number of solutions, the type and periodicity of the solutions, and, more importantly, the existence of chaotic solutions in a physical parametric space are of interest.

Chapter 10 introduces some numerical methods in common use, such as the construction of normal form, symplectic numerical integration and Toeplitz matrices.

Chapter 11 presents the application of bifurcation theory to non-linear structural dynamics, in which are included the bifurcation analysis of oscillations with piecewise-linear characteristics and other applications of bifurcation theory.

# Contents

# Chapter 1

# Dynamical Systems, Ordinary Differential Equations and Stability of Motion

This chapter presents thè basic concepts and important properties of differential dynamical systems, and introduces the basic concepts of flow and its important properties by means of examples. Sections 1.4 and 1.6 discuss the important properties of the limit set in plane, especially the Poincaré–Bendixson theorem and its application. Also touched on are the basic concepts of initial value problems in ordinary differential equations and the basic concepts of Liapunov's stability of motion.

## 1.1 Concepts of Dynamical Systems

A dynamical system is the extension of a mechanical system as described by Newtonian differential equations. The methods of describing dynamical systems are in constant change and development; among the current ones are ordinary differential equations, algebraic logical functions, diagrams and Markov chains.

### 1.1.1 Definition of Dynamical Systems

The following two-dimensional differential equations can be used as an example to explain some concepts

$$\begin{cases} x_1' = a_1 x_1 \\ x_2' = a_2 x_2 \end{cases} \tag{1.1}$$

where $x_i' = \mathrm{d}x_i/\mathrm{d}t$, $i = 1.2$ and $a_1$ and $a_2$ are constants. The exact solution of the above differential equation can be obtained as follows:

$$x_1(t) = k_1 \exp(a_1 t). \qquad x_2(t) = k_2 \exp(a_2 t),$$

where $k_1$ and $k_2$ are constants determined by the initial conditions. Eq. (1.1) can be analyzed from a geometrical point of view. If $x_1$ and $x_2$ are taken as a curve to be solved on $(x_1 .x_2)$ plane $\mathcal{R}^2$, written as:

$$x(t) = (x_1(t), x_2(t)), \quad x(t) \quad \mathcal{R}^2, \quad t \in \mathcal{R}$$

The left-hand side of eq. (1.1) stands for the tangent vector as follows

$$x'(t) = (x_1'(t), x_2'(t))$$

Rewriting eq. (1.1) in the form of matrix

$$x' = Ax \tag{1.1'}$$

where $x = \begin{bmatrix} x_1 \\ x_2 \end{bmatrix}$, $A = \begin{bmatrix} a_1 & 0 \\ 0 & a_2 \end{bmatrix}$. $Ax$ in eq. (1.1)' represents a vector with its starting point at $x$. The initial value $x(t_0) = u = (u_1, u_2)$ is a given point on $\mathcal{R}^2$, which the solution curve must pass through when $t = t_0$ from the geometrical point of view.

Map $A$: $\mathcal{R}^2 \to \mathcal{R}^2$ (or $x \to Ax$) can be regarded as a vector field on $\mathcal{R}^2$, i.e. a vector $Ax$ is determined by every point $x$ on $\mathcal{R}^2$. Let us draw a vector $Ax$ starting from $x$, that is, draw an oriented segment from $x$ to $x+Ax$. For example, if $a_1 = 2$, $a_2 = -\frac{1}{2}$, and $x = (1,1)$, then the oriented segment from (1,1) to $\left(3, -\frac{1}{2}\right)$ can be determined (shown in Fig. 1.1).

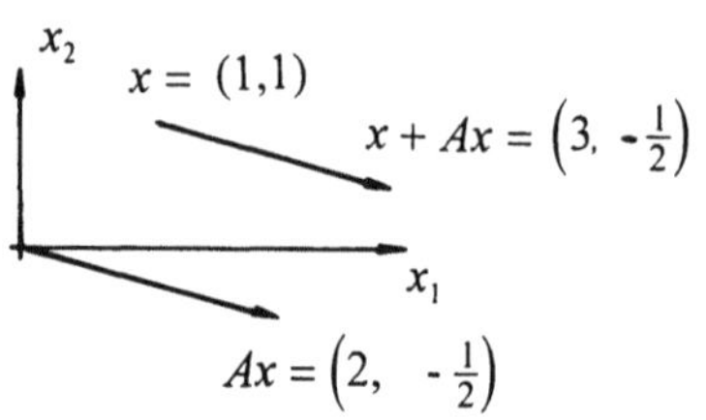


Fig. 1.1

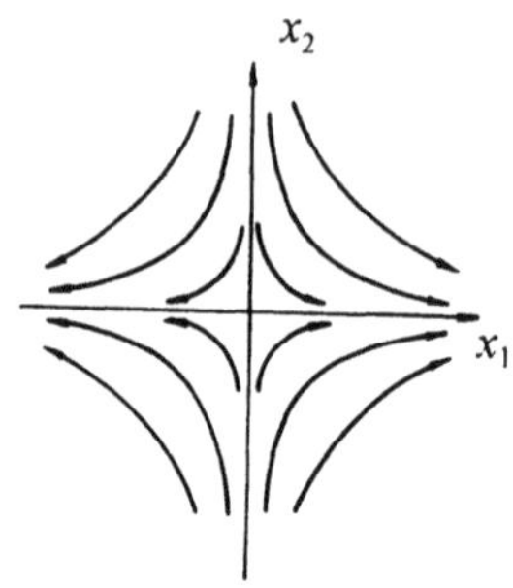


Fig. 1.2

To find the solution $u$ by solving differential eq. (1.1) or (1.1)' to satisfy the initial condition $(u_1, u_2)$ when $t = 0$ is to find the curve on the plane passing through $u = (u_1, u_2)$ when $t = 0$. The trivial solution $(x_1(t), x_2(t)) = (0,0)$ can also be taken as a curve. The family made up of solution curves is a subset on $\mathcal{R}^2$, and is called the phase space diagram of system (1.1) or (1.1)' (shown in Fig. 1.2).

To take eq. (1.1) as a dynamical system is to take the independent variable $t$ as parameter and solution curve $x(t)$ as path of a material point moving on plane $\mathcal{R}^2$. The point, when $t = 0$, can be assumed to be an arbitrary point $u = (u_1, u_2)$ in plane $\mathcal{R}^2$. As the time goes by, the point moves along the solution curve satisfying $x(0) = u$. For any time $t > 0$ (or $t < 0$), the point will be at another place $x(t)$. $g^t(u)$ represents the dependent relation between $t$ and $u$, written as:

$$g^t(u) = (u_1 \exp(a_1 t), u_2 \exp(a_2 t))$$

Supposing all material points on the plane curves move simultaneously (as dust is stirred by steady wind) in this context, then the solution curve is called an orbit or trajectory. For every fixed time $t$ in $\mathcal{R}$ there exists a transformation that makes each point in the plane correspond to another point $g^t(u)$ in the same plane. The transformation written as $g^t: \mathcal{R}^2 \to \mathcal{R}^2$ is obviously a linear one, that is, all vectors $u$, $v$ and any real number $\lambda$ satisfy the following conditions

$$g^t(u+v) = g^t(u) + g^t(v)$$

$$g^t(\lambda u) = \lambda g^t(u)$$

As time passes, every point on the plane moves simultaneously along the trajectory which passes through this point. In this way the whole of map $g^t: \mathcal{R}^2 \to \mathcal{R}^2$ is the one-parameter family of transformation. The parameter is the time $t$. This family is called a flow or a dynamical system, or a vector field determined by the transformation $x \to Ax$. It is also equivalent to eq. (1.1). In other words, a flow is the total of all trajectories which pass through all points in the domain of definition and come within the range. The theory of dynamical systems is concerned with the global behaviour of a flow.

### Mathematical Definitions of Dynamical Systems

Suppose that $\mathcal{M}$ is an arbitrary set in a metric space on which an operation is defined, and $\mathcal{G}$ is a one-parameter family of transformation on metric space $\mathcal{M}$:

$$\mathcal{G} = \{ g^t \mid t \quad \mathcal{R} \}$$

$$g^t : \mathcal{M} \to \mathcal{M}, \text{ for any } t \in \mathcal{R}$$

where $t$ is a parameter.

Suppose that $g^t$ satisfies the properties of the group:

i) $g^0 = \mathrm{I} = \text{Identity}$;

ii) $g^{t+s} = g^t \bullet g^s = g^s \bullet g^t$;

iii) $g^t x$ are continuous for $t$ and $x$;

where "•" denotes the composition of the operations.

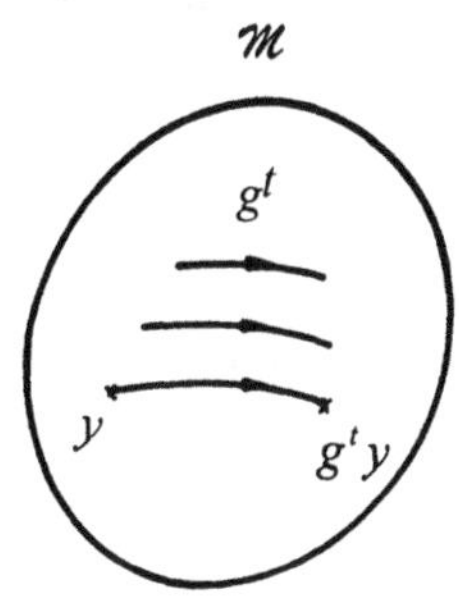


Fig.1.3

The binary group $(\mathcal{M}.\mathcal{G})$ is called a dynamical system or a flow. It fills the real line $\mathcal{R}$ ($t \in \mathcal{R}$) as well as $\mathcal{M}$ ( $y \in \mathcal{M}$). Here $\mathcal{M}$ is called the phase space, or $g^t$ is the $\mathcal{C}^0$ (or $\mathcal{C}^r$) dynamical system on $\mathcal{M}$. Having nothing to do with differentiation, the flow is also called the topological dynamical system. Note since for any given t there exists an inverse map $g^{-t}$ of $g^t$, and $g^{-t}$ belongs to $\mathcal{C}^0$ (or $\mathcal{C}^r$), then $g^t$ is a homomorphism (or a diffeomorphism).

### 1.1.2 Classification of Dynamical Systems

**Continuous dynamical system**: Dynamical system $(\mathcal{M},\mathcal{G})$ is said to be continuous (or phase flow) if $g: \mathcal{R} \times \mathcal{M} \to \mathcal{M}$ is continuous (where $g^t x = g(x,t), t \in \mathcal{R}, x \in \mathcal{M}$).

**Differential dynamical system:** Dynamical system $(\mathcal{M},\mathcal{G})$ is said to be differentiable (or smooth phase flow) if $g: \mathcal{R} \times \mathcal{M} \to \mathcal{M}$ is differentiable.

**Discrete dynamical system:** Suppose $t \in \mathcal{R}$ is replaced by $k \in \mathcal{Z}$ (the set of integers), then $\mathcal{G} = \{g^k | k \in \mathcal{Z}\}$, $\{\mathcal{M}, \mathcal{G}\}$ is said to be a discrete dynamical system. If $g$ satisfies $g^0$=Identity, then $g^{m+n} = g^m g^n = g^n g^m$ requires that $g^k$ be continuous on $\mathcal{M}$ (or differentiable).

**Semi-dynamical system:** If $t, s \geq 0$, then $g^{t+s} = g^t g^s$ is called a semi-dynamical system.

**Note** The set $\mathcal{G} = \{g^t | t \in \mathcal{R}\}$ is an Abelian group.

### 1.1.3 Dynamical Systems and Vector Field

A dynamical system is the description of the total orbits passing through all points in the domain of definition in the given phase space $\mathcal{S}$ within the change of time. Our aim is to find in sequence the state ($t \neq 0$) of the system in $\mathcal{S}$ from the initial state $x_0$ ($t_0 = 0$) according to a certain method of finding solutions for a deterministic physical system. The dynamical system on $\mathcal{S}$ tells us that where $x_1$ and the position of $x$ will be after one unit time, where $x_2$ is after two units times, and so on. It also tells us the state $x_{-1}$ one unit time before the time $t = 0$, etc. If it is extended to the whole real line, the orbits on the phase space can be obtained. The map from $t$ to $x_t (\mathcal{R} \to \mathcal{S})$ is a curve in $\mathcal{S}$. It represents the whole course of state $x$ when the time changes from $-\infty$ to $+\infty$. The map $g^t : \mathcal{S} \to \mathcal{S}$ from $x$ to $x_t$ is defined at any $t$. Since we take $g^t$ as the motion state with the time changing, the inverse $g^{-t}$ exists for all maps $g^t$. Suppose that $A$ is an operator in vector space $\mathcal{S}$, and $g: \mathcal{R} \times \mathcal{S} \to \mathcal{S}$ is determined by $g(t, x) = e^{tA} x$, then $g^t$ can be written as $g^t = e^{tA}$. Clearly, $g^0 = e^0 =$ Identity, again $e^{(t+s)A} = e^{tA} e^{sA}$, and the flow is in accordance with the group properties. Therefore $g^t = e^{tA}$ can determine a dynamical system on $\mathcal{S}$. A dynamical system $g^t$ on $\mathcal{S}$ can also determine a differential equation, that is, a vector field $f: \mathcal{S} \to \mathcal{E}$ can be determined (here $\mathcal{E}$ is assumed to be an open set in metric space). Let

$$f(x) = \frac{\mathrm{d}}{\mathrm{d}t} g^t(x) \Big|_{t=0}$$

then for any point $x$ in $\mathcal{S}$, $f(x)$ is the vector starting from $x$. Each vector is assumed to be a tangent vector corresponding to the orbit $g^t x$ at $t = 0$, and hence the dynamical system determines a differential equation.

In summary, we can give the relationship between a differential equation and a dynamical system:

1) a dynamical system on $\mathcal{S}$ determines a differential equation on $\mathcal{S}$;

2) for any given differential equation defined at any $t$, this is also accompanied by a dynamical system.

### 1.1.4 Phase Space Diagrams

To study the behaviour of the systems is to study the behaviour of the trajectories in the phase space. Unfolding the phase space into the structure of the trajectories, the structure of the trajectories is constructing the phase space diagram of the dynamical systems. From a geometrical point of view this structure is the geometrical pattern of the relative positions of trajectories in the phase space. It should be noted that it is difficult to depict the orbits completely, and this is a problem that needs to be solved. Nevertheless, some of the important properties of the structure have been explored, and the complete description of the structures of dynamical systems is now possible. The separating line through singularity points plays a very important role in the description of the phase structure. The singularity points mentioned here include: the equilibrium states or the stationary state of the system; the isolate closed orbit called limit cycle (corresponding to periodic motion); and the separating curve and surface -- the boundaries of the domain attracted to various special stable trajectories. Some of the special elements can form an integral manifold in phase space which may be of the stable, unstable or saddle type. The stable equilibrium and periodic motions are the simplest examples of deterministic motion. The trajectory structure of various dynamical systems will be discussed in the following sections.

## 1.2 Ordinary Differential Equations

The ordinary differential equation can be written in the following standard form:

$$y' = f(t,y) \tag{1.2}$$

where $y \in \mathcal{R}^n$, $t \in \mathrm{R}$, $f : \mathcal{D} \subseteq \mathcal{R}^n \to \mathcal{R}^n$, and $\mathcal{D}$ is an continuous open subset. For example, a second-order differential equation

$$x'' + x = 0$$

can be reduced to the standard form as follows by the transformation $y_1 = x.\ y_2 = x'$

$$\begin{aligned} y_1' &= y_2 \\ y_2' &= -y_1 \end{aligned}$$

Obviously, if the function $\varphi : \theta \to \mathcal{R}^n$, ($\theta$ is a real interval) is the solution vector of the equation, then $\varphi$ is differentiable and satisfies

(i) $(t, \varphi(t)) \in \mathcal{D}$, for any $t \in \theta$;

(ii) $\dfrac{\mathrm{d}\varphi}{\mathrm{d}t} = f(t, \varphi(t))$, for any $t \in \theta$.

**The initial value problem (IVP):** Solving a set of equations with the initial condition $y(t_0) = y_0$ is called the initial value problem, written as

$$\begin{cases} y' = f(t,y) \\ y(t_0) = y_0 \end{cases} \tag{1.3}$$

where $(t_0, y_0) \in \mathcal{D}$, and $f(t,y)$ is the same as in eq. (1.2). The solution $\varphi$ of the initial value problem is the solution of eq. (1.2) with the condition $\varphi(t_0) = y_0$.

Without loss of generality, $t_0$ is usually taken as $t_0 = 0$ (in general, $t_0$ can be changed to 0), and the solution is dependent on $y_0$. Therfore, we introduce the sign $\varphi(t) \equiv \varphi(t, y_0) \equiv \varphi_t(y_0) \equiv \varphi^t(y_0)$, where $\varphi$ is an advanced map, which represents the value at $t$ starting from $y_0$.

**Definition 1.1** System (1.2) or (1.3) is called autonomous if the vector function is independent of the time variable explicitly, written as

$$y' = f(y), \ y(t_0) = y_0 \tag{1.4}$$

otherwise it is called non-autonomous.

Dynamical systems mainly treat autonomous systems. A non-autonomous system in $\mathcal{R} \times \mathcal{R}^n$ can be regarded as an autonomous system in $\mathcal{R}^{n+1}$ by introducing variable $y_{n+1} = t$ and changing the original equations to the autonomous system in $\mathcal{R}^{n+1}$:

$$\frac{dy_1}{dt} = f_1(t, y), \cdots, \frac{dy_n}{dt} = f_n(t, y), \frac{dy_{n+1}}{dt} = 1$$

$$y = \begin{bmatrix} y_1 \\ y_2 \\ \vdots \\ y_n \end{bmatrix}, f = \begin{bmatrix} f_1 \\ f_2 \\ \vdots \\ f_n \end{bmatrix}, y_{n+1} = t$$

The main problems to be solved in this section are:

(1) Is there a solution $\varphi(t, y_0)$ of initial value problem (1.3)?
(2) If yes, is the solution $\varphi(t, y_0)$ unique?
(3) Is the solution $\varphi(t, y_0)$ continuous?
(4) What is the relation between the solutions and the dynamical system?

The dependent relation between the solutions and $y_0$ and $f$ will be solved by stability theory while that between the solutions and the parameters will be sloved by bifurcation theory. To solve these problems, it is necessary to introduce some frequently used signs and concepts. The norm Euclidean in $\mathcal{R}^n$ is written as $|\bullet$, i.e. $|y| = \sqrt{y_1^2 + \ldots + y_n^2}$, $y \quad \mathcal{R}^n$. $\mathcal{C}_n^k(\mathcal{D})$ is the set of vector functions $\varphi: \mathcal{D} \to \mathcal{R}$ with a $k$-order continuous derivative ($\mathcal{D}$ is a real interval), and forms a vector space in infinite linear space. The norm $||\varphi|| = \sup_{t \in [a,b]} |\varphi(t)|$ is called the supremum norm or uniform norm. Usually the norm $|\bullet$ is for discrete systems and $||\bullet||$ for continuous: $||\varphi||_k = ||\varphi|| + \sum_{l=1}^{k} ||\varphi^l||$ where $\varphi^l$ is the $l$-order derivative of $\varphi$. The vector space is called the normed space if a norm is defined in the space. Thus $\mathcal{C}_n[a,b]$ is a normed space in the meaning of the norm $||\varphi|| = \sup|\varphi(t)|$.

Let $\{x^n | n = 1,2,...\}$ be a sequence in normed space $\mathcal{X}$. The sequence is said to be convergent in $\mathcal{X}$, if there exists $x^* \in \mathcal{X}$ such that $\lim_{n\to\infty} \|x^n - x^*\| = 0$, and the point $x^*$ is called a limit, written as $\lim_{n\to\infty} x^{+n} = x^*$.

**Corollary:** The limit of the sequence in a normed space is unique.

**Proof** Let $\lim_{n\to\infty} x^n = x$, $\lim_{n\to\infty} x^n = y$, then $\|x - y\| \le \lim\|x^n - x\| + \lim\|x^n - y\| \to 0$.

**Definition 1.2** Sequence $\{x^n | n = 1,2,...\} \subseteq \mathcal{X}$ is said to be a Cauchy sequence, if for any $\varepsilon > 0$, there exists an $N = N(\varepsilon)$ such that $\|x^n - x^m\| < \varepsilon$ for any $n, m > N$.

**Definition 1.3** The normed space is said to be complete, if any Cauchy sequence in the space is convergent in the space.

A complete normed space is called a Banach space. For example, $\mathcal{R}, \mathcal{C}, \mathcal{R}^n$ are Banach spaces in the sense of general norms, as is $\mathcal{C}_n[a,b]$ in the norm of the supremum. Suppose $\mathcal{B}$ is a Banach space, then $T: \mathcal{B} \to \mathcal{B}$ is a map on $\mathcal{B}$.

**Definition 1.4** Map $T$ on $\mathcal{D} \subseteq \mathcal{C}$ is said to be compressive, if there exists constant $L$ satisfying $0 < L < 1$ such that $\|Tx - Ty\| \le L\|x - y\|$ for any $x, y \in \mathcal{B}$.

**Corollary:** A compressive map is continuous.

**Definition 1.5** $x \in \mathcal{R}$ is said to be a fixed point of map $\mathcal{T}$, if $x = Tx$.

**Theorem 1.1** (Compressive map theorem) Suppose $\mathcal{B}$ is a Banach space, $\mathcal{M}$ is a closed subset in $\mathcal{B}$, then if map $T: \mathcal{M} \to \mathcal{M}$ is compressive (compressive constant $0 < L < 1$)

(i) there exists a fixed point $x^* \in \mathcal{M}$ for map $T$;

(ii) $x^* \in \mathcal{M}$ is unique;

(iii) for any $x^0 \in \mathcal{M}$, the sequence $\{x^0, Tx^0, T^2x^0, ...\}$ is convergent to $x^*$ ($x^*$ is the limit of the sequence);

(iv) the error in (iii) (or the convergence rate) is

$$\|T^k x^0 - x^*\| \le \left(\frac{\|Tx^0 - x^0\|}{1-\lambda}\right)\lambda^k$$

where the "error" tends to zero when $k$ tends to infinity.

**Proof** First prove that the sequence $\{x^0, Tx^0, T^2x^0, ...\}$ is a Cauchy sequence. In fact:

$$|T^{k+p}x^0 - T^k x^0| \le \lambda^k \| T^p x^0 - x^0 \| = \lambda^k \| T^p x^0 - T^{p-1}x^0 + T^{p-1}x^0 - \ldots + Tx^0 - x^0 \|$$
$$\le \lambda^k \{\| T^p x^0 - T^{p-1}x^0 \| + \ldots + \| Tx^0 - x^0 \|\} \le \lambda^k \{\lambda^k + \ldots + \lambda + 1\} \| Tx^0 - x^0 \|$$

so that $\|T^{k+p}x^0 - T^k x^0\| \le \lambda^k \dfrac{\|Tx^0 - x^0\|}{1-\lambda} \to 0 \ (k \to \infty)$

and hence the sequence $\{x^0, Tx^0, T^2x^0, ...\}$ is a Cauchy sequence.

From the completeness of $\mathcal{B}$, we know that the sequence is convergent in $\mathcal{B}$. Suppose $x^* \in \mathcal{B}$ satisfies $\lim_{k\to\infty} T^k x^0 = x^*$. As $\mathcal{M}$ is closed and $\{T^k x^0\} \subseteq \mathcal{M}$, then we thus have $x^* \in \mathcal{M}$. As $T$ is continuous, then

$$Tx^* = T(\lim_{k\to\infty} T^k x^0) = \lim_{k\to\infty} T(T^k x^0) = \lim_{k\to\infty} T^{k+1} x^0 = x^*$$

This completes the proof of (i).

(ii). Suppose $x^* = Tx^*$ and $y^* = Ty^*$ are two fixed points of map $T$, then it follows that $\|x^* - y^*\| = \|Tx^* - Ty^*\| \le \lambda \|x^* - y^*\| < \|x^* - y^*\|$. This is a contradiction and hence the fixed point of $T$ is unique.

The proof of (iii) and (iv) can be shown by combining the proving process of (i) with the conclusion of (ii).

**Definition 1.6** $f: \mathcal{D} \subseteq \mathcal{R} \times \mathcal{R}^n \to \mathcal{R}$ is said to satisfy Lipschitz condition with respect to $y$ on $\mathcal{D}$, if there exists a constant $L > 0$, such that

$$\|f(t,y) - f(t,z)\| \le L\|y - z\| \text{ for any } (t,z) \text{ and } (t,z) \in \mathcal{R}.$$

**Corollary:** If $f(t,y)$ satisfies the Lipschitz condition on $\mathcal{D} \subseteq \mathcal{R} \times \mathcal{R}^n$, then $f(t,y)$ is continuous on $\mathcal{D}$.

The existence and uniqueness theorem, the Picard theorem of the initial value problem is presented below.

**Theorem 1.2** (The existence and uniqueness theorem) If $f$ satisfies the Lipschitz condition on $\mathcal{D} \subseteq \mathcal{R} \times \mathcal{R}^n$, then there exists a unique solution for the initial value problem (1.3) on the closed interval $[t_0 - k, t_0 + k]$, written as $y = \varphi(t)$, $\varphi(t_0) = y_0$, $k > 0$.

**Proof** As $\mathcal{D}$ is open, there exists a closed rectangular $\bar{\mathcal{R}} \subseteq \mathcal{D}$

$$\bar{\mathcal{R}} = \{(t,y) \mid |t - t_0| \le a, |y - y_0| \le b, a, b > 0\} \qquad \text{(Fig. 1.4).}$$

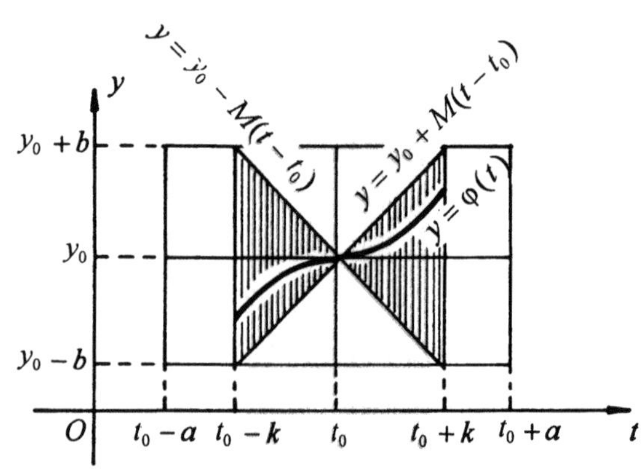


Fig. 1.4

Since $f$ satisfies the Lipschitz condition ($\mathcal{L}>0$ is the Lipchitz constant) on $\bar{\mathcal{R}}$, then $f$ is continuous, and there exists $M > 0$ ($M = \max_{[t,y]\in\mathcal{R}} |f(t,y)|$) such that $|f(t,y)| \le M$, $(t,y)$ $\bar{\mathcal{R}}$. Define a cone: $\mathcal{K} = \{(t,y) \| y - y_0 | \le M|t - t_0| < k\}$, where $k < \min\left\{\frac{1}{L}, a, \frac{b}{M}\right\}$.

The remaining parts of the proof can be obtained by means of the compressive map theorem. Let $\mathcal{B} = \mathcal{C}_n[t_0 - k, t_0 + k]$ be a Banach space, then $\mathcal{M} = \{\varphi \in \mathcal{B} | (t, \varphi(t)) \in \mathcal{K}$, for any $t \in [t_0 - k, t_0 + k]\}$ is the closed subset of $\mathcal{B}$. As the map $T$: $\mathcal{M} \to \mathcal{B}$ is defined by $T\varphi(t) = y_0 + \int_{t_0}^{t} f(s, \varphi(s))ds$, it is evident that $T$ is continuous and hence $T$: $\mathcal{M} \to \mathcal{B}$. Now we prove $T$: $\mathcal{M} \to \mathcal{B}$. Since $\|T\varphi(t) - y_0\| = \|\int_{t_0}^{t} f(s, \varphi(s))ds\| \le M|\int_{t_0}^{t} ds| \le M|t - t_0|$, therefore $T$: $\mathcal{M} \to \mathcal{B}$. Next we prove $T$ is compressive. Take arbitrary $\varphi(t)$ and $\psi(t)$, so that

$$\|T\varphi(t) - T\psi(t)\| = \|\int_{t_0}^{t} [f(s, \varphi(s)) - f(s, \psi(s))]ds\|$$

$$= \sup_{t\in[t_0-k, t_0+k]} |\int_{\varphi}^{t} [f(s, \varphi(s)) - f(s, \psi(s))]ds|$$

$$\le \sup_{s} |\int_{\varphi}^{t} [f(s, \varphi(s)) - f(s, \varphi(s))]ds|$$

$$\le L \sup_{t} |\int_{t_0}^{t} [\varphi(s) - \psi(s)]ds|$$

$$\le L\|\varphi - \psi\| \, |\int_{t_0}^{t} ds| \le KL\|\varphi - \psi\|$$

From the hypothesis of the theorem it can be assumed that $KL < 1$, so according to the compressive theorem there exists a unique solution $\varphi \in \mathcal{M}$ such that $T\varphi = \varphi$, where $\varphi(t)$ satisfies $\varphi(t) = y_0 + \int_{t_0}^{t} f(s.\varphi(s))ds$ and $\varphi'(t) = f(t.\varphi(t))$. . $\varphi(t_0) = y_0$. This completes the proof of the theorem. The unique existence theorem of the initial value problem tells us that the solution appears locally in the neighbourhood of the initial point.

**Definition 1.7** Function $f$ is said to satisfy the local Lipschitz condition in $\mathcal{D} \subseteq \mathcal{R} \times \mathcal{R}^n$, if for any point $(t_0, y_0) \in \mathcal{D}$ there exists a neighborhood $N$ of the point on which $f$ satisfies the Lipschitz condition with respect to $y$, written as $f \in \mathrm{Lip}(\mathcal{D})$.

**Note 1:** The Lipschitz constants can be different in different neighbourhood.

**Note 2:** It is difficult to decide whether $f$ satisfies the Lipschitz condition on $\mathcal{D}$. In practice the condition is replaced by a partial derivative of $f(t,y)$ with respect to $y$.

**Corollary:** If $f(t,y)$ has continuous partial derivatives with respect to every variable, then $f(t,y) \in \mathrm{Lip}(\mathcal{D})$.

For the initial value problem (1.3), where $(t_0, y_0) \in \mathcal{D}$, $f(t,y) \in \mathrm{Lip}(\mathcal{D})$, according to the unique existence theorem there exists a unique solution on the interval $\theta = [t_0 - a_1, t_0 + a_1]$, written as $\varphi_1 : \mathcal{D} \to \mathcal{R}^n$, $\varphi_1(t_0) = y_0, a_1 > 0$. Consider the following initial value problem determined by the right-hand point $P(t_0 + a_1), \varphi(t_0 + a_1)$ of solution $\varphi_1(t_0)$, $y' = f(t,y)$, $y(t_0 + a_1) = \varphi_1(t_0 + a_1)$. Since $\varphi \in \mathcal{D}$, from the unique existence theorem we know that there exists a unique solution $\varphi_2$ on the interval $\theta = [t_0 + a_1 - a_2,\ t_0 + a_1 + a_2]$, with the condition $\varphi_2(t_0 + a_1) = \varphi_1(t_0 + a_1)$. Repeating the above procedure, we arrive at the conclusion that the right-hand end of the interval $[t_0 - a_1, t_0 + a_1 + a_2 + \ldots)$ is open. Similar methods can be applied to the left-hand ends. The interval thus obtained is open, and its end point can be finite or infinite. By fixing the above intervals, we obtain the following solution $y = \varphi(t)$, $t \in [t_0 - a_1, t_0 + a_1 + a_2 + \ldots)$. Obviously, the interval where the solution $\varphi(t_0)$ lies is expanded. $\mathbf{J}(t_0 . y_0) = (\alpha.\beta)$, the union of the intervals on which the solution exists, where $\mathbf{J}$ is called the maximal interval of the solution of the initial value problem (1.3), and is open. $\alpha$ can be $-\infty$ and $\beta$ can be $+\infty$ . Below is a discussion about the global uniqueness determined by the local uniqueness.

**Lemma:** If there are two solutions $U(t)$ and $V(t)$ on $\theta \subseteq \mathbf{J}$ for the initial value problem (1.3), then $U(t) = V(t)$.

**Proof** Let $\theta^*$ be the maximal interval on which $U(t) = V(t)$, and $t^*$ is the right-hand end point of $\theta^*$. If $t^* \in \theta$, we can apply the unique existence theorem at point $(t^*, U(t^*)) = (t^*, V(t^*))$, and then there exists $\delta > 0$, such that $U(t) = V(t)$ on $[t^* - \delta . t^* + \delta]$. It is a contradiction with $t^*$ which is the end point of the maximal interval $\theta^*$. In the same way, the left-hand side can be considered. The maximal $\theta^* = \theta$ is chosen, that is, $U(t) = V(t)$ for any $t \in \theta$. The unique solution on the maximal interval is called the maximum solution. Now let us examine the behaviour of the maximum solution on $\mathbf{J}$ when $t \to \alpha(\beta)$.

**Theorem 1.3** Suppose that $\varphi$ is the maximum solution on $\mathbf{J}$ and $\mathcal{K}$ is the compact set (bounded and closed set), $\mathcal{K} \subseteq \mathcal{D}$ ($\mathcal{D}$ is open), then $(t, \varphi(t))$ leaves $\mathcal{K}$ when $t \to \beta(t \to \alpha)$.

**Proof** Consider $[t_0, \beta) \subseteq (\alpha, \beta) = \mathbf{J}$. Suppose $(t, \varphi(t)) \in \mathcal{K}$ for any $t \in [t_0, \beta)$. From the compactness of $\mathcal{K}$ and the continuity of $f$ we know that there exists $M > 0$, such that $|f(t,y)| \le M$ for any $(t,y) \in \mathcal{K}$. Using the Cauchy criterion, we can prove the existence of $\varphi(\beta^-) = \lim_{t \to \beta} \varphi(t)$. Take $t_1, t_2 \in [t_0, \beta)$, according to the equivalence of the initial value problem and the integral equation, we have

$$\varphi(t_1) = y_0 + \int_{t_0}^{t_1} f(s, \varphi(s)) ds$$

$$\varphi(t_2) = y_0 + \int_{t_0}^{t_2} f(s, \varphi(s)) ds$$

$$\varphi(t_1) - \varphi(t_2) | \le | \int_{t_1}^{t_2} f(s, \varphi(s)) ds \le M | t_1 - t_2 |$$

then follows $|\varphi(t_1)-\varphi(t_2)|\to 0$, (when $|t_1-t_2|\to 0$) and $\varphi(\beta^-)=\lim_{t\to\beta}\varphi(t)$. Besides, $\varphi(\beta^-)=\lim_{t\to\beta^-}\varphi(t)=y_0+\lim_{t\to\beta^-}\int_{t_0}^{t} f(s,\ \varphi(s))ds=y_0+\int_{t_0}^{\beta} f(s,\ \varphi(s))ds$. Therefore, $\varphi(t)$ satisfies the integral equation, and hence $y=\varphi(t)$ is a solution on $[t_0,\beta]$. Applying the unique existence theorem at point $(\beta,\varphi(\beta))$, we can see that there exists $\delta>0$ such that $\varphi(t)$ is the unique solution defined on the interval $[\beta-\delta,\beta+\delta]$. This contradicts that $\varphi(t)$ is the maximum solution. In the same way, we can consider the behaviour when $t\to\alpha$. Therefore, the solution must leave the compact set $\mathcal{K}\subseteq\mathcal{D}$.

**Corollary:** If $\mathbf{J}=(\alpha,\beta)$ the maximal interval, then either $(t,\varphi(t))$ tends to the boundary of $\mathcal{D}$ or $|\varphi(t)|\to+\infty$ when $t\to\infty$ (or both cases occur at the same time).

**Example 1** (Global existence)

Consider the linear initial value problem $y'=Ay,\ \ y(0)=y_0$, where $A=(a_{i,j})_{n\times n}$, a square matrix. The solution of the system can be written as $y=e^{tA}y_0$, for any $t>0$. $e^{tA}$ is the fundamental matrix of the system and the maximal interval is $\mathcal{J}=\mathcal{R}$. Then the solution will leave any compact set which contains the initial point.

**Example 2** Consider the non-linear initial value problem

$$y'=1+y^2,\ \ y(0)=0,$$
$$\mathcal{D}=\mathcal{R}\times\mathcal{R},\ f\in\text{Lip}(\mathcal{D})$$

So far as the initial condition $y(0)=0$ is considered, the solution of the system is that $y(t)=\tan t$. When $t\to\frac{\pi}{2}-$ and $t\to\frac{\pi}{2}+$, then $y(t)\to\pm\infty$.

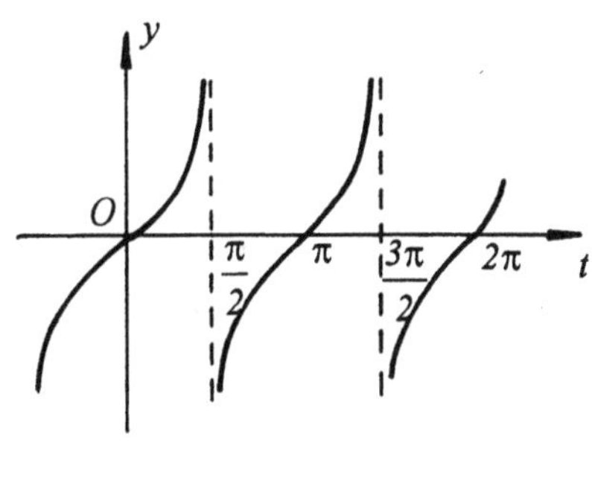


Fig. 1.5

This solution leaves any given bounded set $|y|\le M$ within a finite time which is called the "finite escape time" (Fig. 1.5).

For initial value problem (1.3) there are other forms of the unique existence theorem in addition to the Picard theorem. These theorems will be presented here without proof.

**Theorem 1.4** (Cauchy—Peano) Suppose $f\in\text{Lip}(\mathcal{D})$. and $\mathcal{D}=\mathcal{R}\times\mathcal{R}$, $(t_0,y_0)\in\mathcal{D}$, then there exists a solution $y=\varphi(t),\ \varphi(t_0)=y_0$ on interval $[t_0-h,\ t_0+h],\ h>0$.

**Theorem 1.5** (Cauchy—Lindlof) Suppose $f$ is a real analytic, then there exists a unique analytic solution $y=\varphi(t),\ \varphi(t_0)=y_0$ on interval $[t_0-h,\ t_0+h],\ h>0$, for problem (1.3).

The Cauchy—Lindlof theorem is the basis for the method of power series for the ordinary differential equation. A unique solution is $y=\varphi(t),\ \varphi(t_0)=y_0$ on interval $[t_0-h, t_0+h]$ with $\varphi \in \mathcal{C}^k([t_0-h, t_0+h])$. It is easy to see that $\varphi \in \mathcal{C}^k([t_0-h, t_0+h])$ from $\frac{d\varphi}{dt} = f(t,\varphi(t))$.

So far we have dealt with the unique existence theorem of the initial value problem (1.3). We shall now discuss the continuous dependence of the solution upon the initial values.

**Theorem 1.6** Suppose $y' = f(t,y)$, $f$ Lip($\mathcal{D}$), $\mathcal{D} \subseteq \mathcal{R} \times \mathcal{R}^n$. If $y=\varphi_1(t)$ and $y=\varphi_2(t)$ are solutions satisfying $y(t_0)=y_1$ and $y(t_0)=y_2$ respectively, then the following inequality holds: $|\varphi_1(t)-\varphi_2(t)| \le |y_1-y_2| e^{L|t-t_0|}$, $t \in \theta$, where $L > 0$ is the Lipschitz constant.

In order to prove this theorem we introduce the Bellman—Gronwall inequality.

**Lemma (Bellman—Gronwall):** Suppose that $\varphi(t)$ is a non-negative function on interval $\theta$; if there exist $K>0, \lambda \ge 0, r \ge 0$, such that $\varphi(t) \le \lambda + |\int_{t_0}^t (k\varphi(s)+r)ds|$, $t \in \theta$, then $\varphi(t) \le \lambda e^{K|t-t_0|} + \frac{r}{k}(e^{K|t-t_0|}-1)$

**Proof** We shall the theorem only for $t \ge 0$. According to the hypothesis we have $|\varphi(t)| \le \lambda + k\int_{t_0}^t \left(\varphi(s)+\frac{r}{k}\right)ds$, $t \ge t_0$. Let $U(t) = \int_{t_0}^t \left(\varphi(s)+\frac{r}{k}\right)ds$, then

$$U'(t) = \varphi(t) + \frac{r}{k} \le \left(\lambda+\frac{r}{k}\right) + k\int_{t_0}^t \left(\varphi(s)+\frac{r}{k}\right)ds$$

that is $U'(t) \le \left(\lambda+\frac{r}{k}\right) + kU(t)$. Multiplying both sides by the factor $e^{-k(t-t_0)} \ge 0$, we have $U'(t)e^{-k(t-t_0)} \le \left(\lambda+\frac{r}{k}\right)e^{-k(t-t_0)} + kU(t)e^{-k(t-t_0)}$. After simplifying the equation, we have $\frac{d}{dt}U(t)e^{-k(t-t_0)} \le \left(\lambda+\frac{r}{k}\right)e^{-k(t-t_0)}$. Integrating both sides from $t_0$ to $t$, we get $kU(t)e^{-k(t-t_0)} \le \left(\lambda+\frac{r}{k}\right)[1-e^{-k(t-t_0)}]$. Further

$$\lambda + kU(t) \le \lambda e^{-k(t-t_0)} + \frac{r}{k}(e^{-k(t-t_0)}-1),\ t \ge 0,$$

therefore $\varphi(t) \le \lambda e^{-k(t-t_0)} + \frac{r}{k}(e^{-k(t-t_0)}-1)$.

**Proof of the theorem:** Suppose that $y=\varphi_1(t)$ and $y=\varphi_2(t)$ are the solutions satisfying $y(t_0)=y_1$, $y(t_0)=y_2$ on the common interval $\theta$ respectively, $t \in \theta$, we have

$$\begin{aligned}|\varphi_1(t)-\varphi_2(t)| &= |y_1-y_2+\int_{t_0}^t (f(s,\varphi_1(s))-f(s,\varphi_2(s)))ds| \\ &\le |y_1-y_2| + L|\int_{t_0}^t (\varphi_1(s)-\varphi_2(s))ds|\end{aligned}$$

where $L > 0$ is the Lipschitz constant, $t \in \theta$. Applying the Bellman—Gronwall inequality we have $|\varphi_1(t) - \varphi_2(t)| \le |y_1 - y_2| e^{k|t-t_0|}$, $t \in \theta$.

**Corollary** If the solution of the initial value problem

$y' = f(t,y)$, $y(t_0) = y_0$ is $y(t) = \varphi(t, y_0), \varphi(t_0, y_0) = y_0$, $t \in \theta$, then $\varphi(t, y_0)$ as the function of $y_0$ is continuous.

**Proof** If $y(t) = \varphi(t, y_0)$ is the solution of the initial value problem, we have

$$|\varphi(t, y_0 + \Delta y_0) - \varphi(t, y_0)| \le |y_1 - y_2| e^{L|t-t_0|}, \text{ for any fixed } t \in \theta,$$

so that $|\varphi(t, y_0 + \Delta y_0) - \varphi(t, y_0)| \to 0$, when $\Delta y \to 0$. Therefore, $y = \varphi(t, y_0)$ is continuous for variable $y_0$. Likewise, when $t_0$ changes, the solution of the initial value problem can be written as $y = \varphi(t, t_0, y_0)$, $\varphi(t_0, t_0, y_0) = y_0$. It can also be proved that $\varphi(t, t_0, y_0)$ is continuous as a function of $t_0$.

Bellman—Gronwall inequality can be used to prove the following theorem.

**Theorem 1.7** Suppose that $\varphi_1(t)$ and $\varphi_2(t)$ are solutions of the initial value problems respectively

$$\begin{cases} y' = f(t,y) \\ y(t_0) = y_0 \end{cases} \quad \text{and} \quad \begin{cases} y' = g(t,y) \\ y(t_0) = y_0 \end{cases}$$

If $f, g \in \text{Lip}(\mathcal{D})$, $\mathcal{D} \subseteq \mathcal{R} \times \mathcal{R}^n$, and $|f(t,y) - g(t,y)| \le \varepsilon$, for any $(t,y) \in \mathcal{D}$, then on the same interval $\varphi_1(t)$ and $\varphi_2(t)$ satisfy $|\varphi_1(t) - \varphi_2(t)| \le \frac{\varepsilon}{L}[e^{L|t-t_0|} - 1]$, $t \in \theta$, where $L$ is the Lipschitz constant.

**Note** For an arbitrary fixed $t \in \theta$ there holds $|\varphi_1(t) - \varphi_2(t)| \to 0$, as $\varepsilon \to 0$.

## 1.3 Properties of Flow

### 1.3.1 Dynamical Systems and Initial Value Problems

An initial value problem can be determined by a smooth flow $g^t$ ($g^t$ is continuous and differentiable). In a given dynamical system $(\{g^t\}, \mathcal{S})$, $\mathcal{S} \subseteq \mathcal{R}^n$, $g \in \mathcal{C}^1(\mathcal{R} \times \mathcal{S})$, $\frac{\partial g}{\partial t}$ $\mathcal{C}(\mathcal{R} \times \mathcal{S})$ for every point $x$ in phase space, that is, $x \in \mathcal{S}$ can define a vector velocity $V(X) = \frac{\partial}{\partial t} g^t x|_{t=0}$. From the geometrical point of view, $(x)$ is a vector field that is tangential to the phase curve $\{g^t x | t \in \mathcal{R}\}$ at $x$ when $t = 0$. From the physical point of view, $V(x)$ is the phase velocity.

First let us consider the autonomous initial value problems. Autonomous initial value problems can be written as

$$y' = V(y), \; y(0) = x, \; V\colon \mathcal{S} \to \mathcal{R} \tag{1.4}$$

According to the existence theorem we can see that there exists a solution to the problem $y = g^t x = \varphi(t,x)$. If $V(y)$ Lip($\mathcal{D}$), then $g^t x$ is the unique solution. Substituting $y(t) = g^t x$ to (1.4) we have

$$\frac{\mathrm{d}y}{\mathrm{d}t} = \frac{\mathrm{d}}{\mathrm{d}t} g^t x = \lim_{s \to 0} \frac{g^{t+s}(x) - g^t(x)}{s}$$

$$= \lim_{s \to 0} \frac{g^s(g^t(x)) - g^0(g^t(x))}{s} = \frac{\mathrm{d}}{\mathrm{d}s} g^s(g^t(x)) = V(g^t(x))$$

with the initial condition that $y(0) = g^0 x = x$. Therefore, $y(t) = g^t x$ is the solution of the initial value problem of an autonomous system. Thus, a smooth flow $g^t x$ can determine an autonomous initial value problem.

**Lemma:** Consider system (1.4), let flow $\varphi(t,x)$ be the unique maximal solution, then $\varphi(t+s,x) = \varphi(t,\varphi(s,x))$, for any $s,t,x$, is also the solution of system (1.4).

**Proof** Let $\psi_1(t) = \varphi(t+s,x), \psi_2(t) = \varphi(t,\varphi(s,x))$. Both $\psi_1$ and $\psi_2$ satisfy the same initial value problem $\frac{\mathrm{d}\psi_i(t)}{\mathrm{d}t} = f(\psi_i(t)), i = 1,2$, $\psi_i(0) = \varphi(s,x)$. According to the unique existence theorem on the same time interval, it follows that $\psi_1(t) = \psi_2(t)$, so that $\varphi(t)$ is the unique maximal solution. This completes the proof of the lemma.

The lemma stated above tells us that an autonoumous system can define a dynamical system. But how about is the non-autonomous system? Let us see an example: $y' = t$, $y(0) = x$ whose exact solution is $y(t) = \frac{1}{2}t^2 + x = \varphi(t,x) = g^t x$.

Can the solution define a flow? It is necessary to examine whether it conforms to the property of the group.

$$\varphi(t+s,x) = \frac{1}{2}(t+s)^2 + x = \frac{1}{2}t^2 + ts + \frac{1}{2}s^2 + x \tag{a}$$

$$\varphi(s,\varphi(t,x)) = \frac{1}{2}s^2 + \varphi(t,x) = \frac{1}{2}t^2 + \frac{1}{2}s^2 + x \tag{b}$$

The comparison between (a) and (b) shows that (a) differs from (b). In other words, the exact solution of the non-autonomous system does not conform to the property of group and the system cannot determine a flow. As stated in the preceding section, dynamical systems are concerned with autonomous systems. Non-autonomous systems must be changed to autonomous systems before they are treated. As we know, $t$ in flow $g$ is defined on the complete real axis ($t \in \mathcal{R}$). If $t$ is defined in some subinterval $\mathbf{J}(x)$ of the time axis, then the maximal interval of the solution is contained in $\mathcal{R}$ where the solution lies. To define the initial value problem by means of a flow in this case, we have to introduce the concepts of generalized dynamical systems. Now consider the initial value problem defined on the maximal interval $\mathbf{J}(x)$:

$$y' = f(y),\ y(0) = x,\ f \in \mathrm{Lip}(\mathcal{D}), \text{ or } f \in \mathcal{C}^1(\mathcal{D}) \tag{1.5}$$

Suppose that there exists an unique solution $\varphi(t,x)$ defined on the maximal interval $\mathbf{J}(x)$, where $x \in \mathcal{D}$ , $t$ $\mathbf{J}(x) \subseteq \mathcal{R}$. The properties of a flow for $t \in \mathcal{R}$ proved above are all valid for $t$ $\mathbf{J}(x)$. Define a set $\Omega = \{(t,x) | x \in \mathcal{D}, t \in \mathbf{J}(x)\}$ where $\Omega$ is the domain of definition of flow $g$, that is $g{:}\Omega \to \mathcal{D}$. The binary group $(\{g\},\Omega)$ is called a generalized dynamical systems, where $g$ satisfies the following:

1) $g \in \mathcal{C}^1(\Omega)$ , which can also be written as $g^t x = g(t,x)$;

2) $g^0 x = x$;

3) $g^{t+s} x = g^t \bullet g^s x$

### 1.3.2 Properties of the Flow

The following shows the difference between the method of analyzing initial value problems and the methods of analyzing dynamical systems from the geometrical point of view. The purpose of the method of the initial value problem is to study the existence and the uniqueness of the solution of an ordinary differential equation passing through a point as an initial value. Consider the initial value problem

$$y' = f(y),\ y(0) = x,\ f \in \mathrm{Lip}(\mathcal{D})$$

The trajectory of the solution $y = \varphi(t,x) = g^t x$ of the system passing through the initial point $x$ in phase space is a curve in the phase space which can be obtained as the time changes. In other words, for any fixed point $x$ in phase space the solution defines a curve when time changes. It can be seen in Fig.1.6 that of the solution is periodic, then the curve is closed.

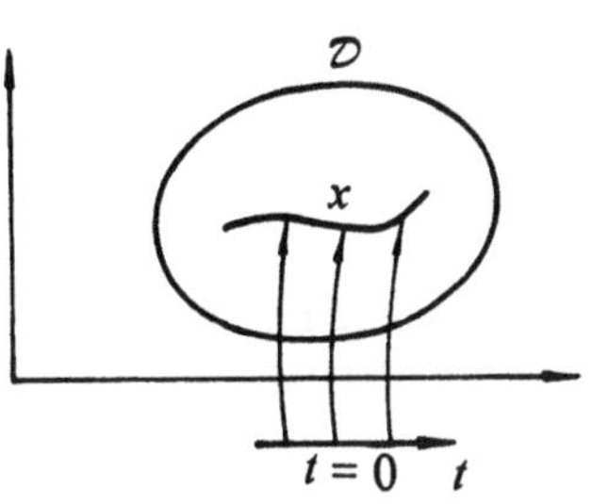


Fig. 1.6

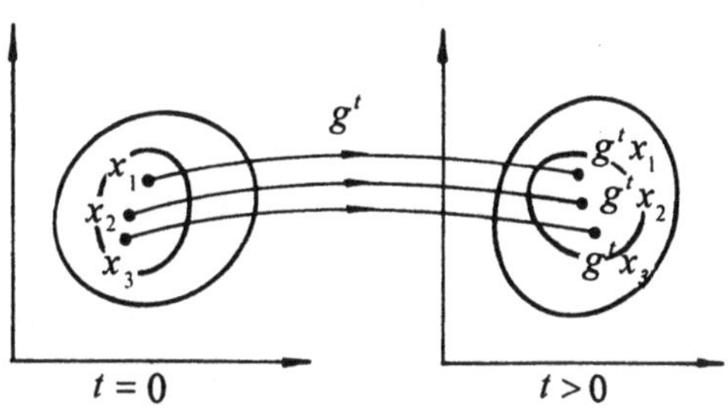


Fig. 1.7

The method of flow is to study the total of all orbits passing through all points in the initial domain and all points on the orbits moving simultaneously. The method of flow is to fix $t$, change $x$ and map the domain $\mathcal{X}$ of the initial values (the set of initial points) to the sequent set of points $g^t\mathcal{X}$ by $g^t$. For any $t > 0$ when $x$ varies $g^t$ maps the initial set $\mathcal{X}$ to the sequent set $g^t\mathcal{X}$. $g^t$ is the totality of the orbits from $\mathcal{X}$ to $g^t\mathcal{X}$. The purpose of the method of flow is to study the properties of flow as shown in Fig. 1.7.

**The properties of a flow**

For a given flow $\varphi: \mathcal{D} \to \mathcal{D}$ defined on the real line, the orbit is defined as

$$r(x) = \{\varphi^t x | t \quad \mathcal{R}\} \subseteq \mathcal{D}$$

and the phase time course (graph) is defined as

$$\mathcal{G}(x) = \{(t, \varphi x) | t \in \mathcal{R}\} \subseteq \mathcal{R} \times \mathcal{D}$$

The orbit is a curve (or the phase portrait) in phase space passing through point $x$, and the phase time course is a motion (integral curve) in generalized space $x_1 - x_2 - t$ as shown in Fig. 1.8. The orbit is unique in the sense of multiparameters.

1) Different initial phases determine the same orbit. Define $\psi^t x = \varphi^{t-c} x = \varphi^t(\varphi^{-c} x)$, then it is clear that $\{\psi^t x | t \in \mathcal{R}\} = \{\varphi^t x | t \in \mathcal{R}\} = r(x)$, that is $\psi$ and $\varphi$ define the same orbit through different initial phases.

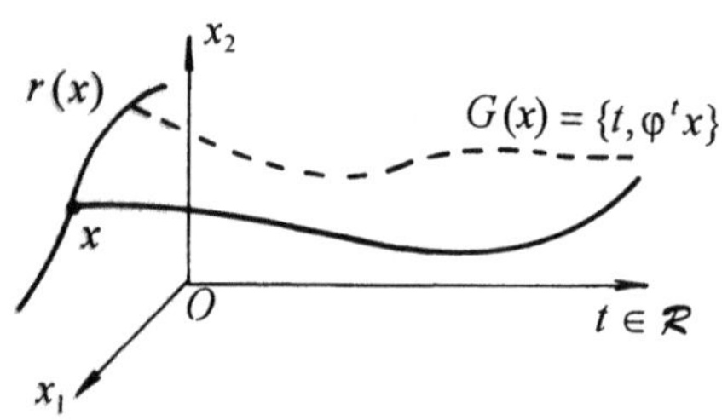


Fig. 1.8

2) Different solutions passing through the same point at different time define the same orbit. Suppose $\psi(t)$ and $\varphi(t)$ are solutions satisfying the conditions $\varphi(0) = x$, $\psi(c) = x$ respectively. Since $\psi(t)$ and $\varphi(t)$ pass the same point at different time, assuming $\psi(t, \psi(0)$ and $\varphi(t-c, x)$ it follows from the unique existence theorem that $\psi(t, \psi(0) = \varphi(t-c, x)$. The orbits $\psi$ and $\varphi$ are shifted by the phase difference.

**The phase time course of the orbits is not unique**

The orbit in the phase space for a flow passing through a fixed point is unique. On the orbit lie numerous phase points, through each of which there exists a phase time course. The difference of all the points on the orbit is the shift of the time.

**Lemma** If there are two orbits of a flow $\varphi^t: \mathcal{D} \to \mathcal{D}$ which intersect at any point, then they are identical.

**Proof** Suppose $p \in r(x) \cap r(y)$, $x \neq y$, then there exist $t_1, t_2 \in \mathcal{R}$ such that $p = \varphi^{t_1} x$ and $p = \varphi^{t_2} y$, that is $\varphi_{t_1} x = \varphi_{t_2} y$ or $y = \varphi_{t_1 - t_2} x$ therefore, $y \in r(x)$. In fact, $\varphi^t y = \varphi^t \varphi^{t_1 - t_2} x = \varphi^{t + t_1 - t_2} x$, so that $r(y) \subseteq r(x)$. Likewise, $r(x) \supseteq r(y)$, then $r(x) = r(y)$.

**Corollary:** If $y \in r(x)$, then $r(y) = r(x)$.

If one chooses an initial point on the same trajectory arbitrarily, one would get the same trajectory, therefore the trajectory is denoted by r only, without specifying the position of the phase point.

### 1.3.3 Definitions of Some Terms

1) **Trajectory** $r(x)$: A trajectory is the union of the positive semi-trajectory and the negative semi-trajectory, written as $r(x) = r^+(x) \cup r^-(x)$ where $r^+(x) = \{\varphi^t(x) \mid t \geq 0\}$ and $r^-(x) = \{\varphi^t(x) \mid t \leq 0\}$.

2) **The closure of** $r(x)$**:** The closure of the trajectory consists of all the points on the trajectory and all its accumulation points, written as

$H(x)$ = the closure of $r(x)$,

the positive closure $H^+(x)$ = the closure of $r^+(x)$,

the negative closure $H^-(x)$ = the closure of $r^-(x)$.

3) **Equilibrium point:** An equilibrium point for a flow is the fixed point. A point for an autonomous system $x' = f(x)$ is called the equilibrium point if $f(x) = 0$. The equilibrium point is also called the critical point or singular point.

4) **Stationary state:** The state of a motion for a system is called the stationary state if it does not change with time *t*.

a) If $\varphi^t$ is a flow corresponding to an ordinary differential equation $y' = f(y)$, the stationary state is $\varphi^t x = x$ or $f(x) = 0$ for any *t*.

b) If *x* is an equilibrium point, then the time course is $r(x) = x$, $H(x) = x$ and the phase time course $\mathcal{G}(x) = \{(x) = \{(t, x) | t \in \mathcal{R}\}$ is a straight line in the generalized phase space $x - t$, that is, $x$ does not change with time.

c) If $x$ is an equilibrium point, $x \neq p$, $x \in \mathcal{H}(p)$, $x \notin r(p)$, then there exists $t_k \in \mathcal{R}$, $t_k \to +\infty$ or $t_k \to -\infty$, as $n \to +\infty$, such that $x = \lim_{t \to \infty} \varphi^{t_k} p$

**Proof by contradiction:** Suppose $x \in r(p)$, an equilibrium point, then $r(x) = r(p) = x$. This implies that $x = p$ and contradicts $x \neq p$. So $x \notin r(p)$.

This completes the proof.

## 1.4 Limit Point Sets

### 1.4.1 $\omega(\alpha)$—limit point set

Suppose a dynamical system $\varphi^t$ corresponds to an *n*-dimensional system. In order to describe the behaviour of $\varphi^t$ when time tends to infinity, we introduce the concept of the limit point set. A point $x^0$ is called an $\omega$-limit point if there exists a sequence $t_n \to \infty$ such that $\lim_{n \to \infty} \varphi^t x = x^0$ (positive limit point). The set of all $\omega$-limit points is called the $\omega$-limit point set. This is written as: $\omega(x) = \{p \in \mathcal{D} \mid$ there exists a sequence $t_1 < t_2 < \ldots < t_n \to +\infty$ such that $\varphi^{t_n} x \to p\}$. In the same way we can define $\alpha$-limit point set as follows: $\alpha(x) = \{p \in \mathcal{D} \mid$ there exists a sequence $t_1 > t_2 > \ldots > t_n \to -\infty$ such that $\varphi^{t_n} x \to p\}$.

**Example 1** Consider the linear system:

$$y' = -y, \quad y(0) = 0$$

The solution of the system is $y(t) = Ce^{-t}$ for the equilibrium point $y = 0$. The orbit passing through the equilibrium point is $r(0) = 0$. The phase space is $\mathcal{D} = \mathcal{R}$ (see Fig. 1.9(a)) and the phase time course is $(t, \varphi' y) = (t, Ce^{-t})$ (see Fig. 1.9(b)).

$\omega(y)$: $y = 0, \omega(y) = 0,$

$y > 0,$

$y < 0, \omega(y) = 0,$

$\alpha(y)$: $y = 0, \alpha(y) = 0,$

$y > 0, \alpha(y) = \phi,$

$y < 0, \alpha(y) = \phi.$

where $\phi$ is null set.

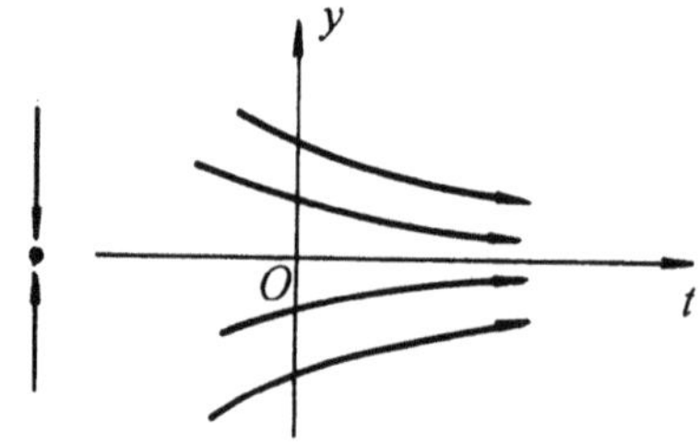


(a) (b)

Fig. 1.9

**Example 2**. Consider the logistic equation: $y' = y(1-y),\ y \in \mathcal{R}$.

The equilibrium point of the system is where $y = 0$ and $y = 1$, and the solution is $y = \varphi(t) = \dfrac{1}{1-\left(1-\dfrac{1}{y_0}\right)e^{-t}}$. The phase diagram and the phase time course are shown in Fig. 1.10.

The limit sets are

$\omega(y) = 0$ for $y > 0$

$\omega(y) = \phi$ for $y < 0$

$\omega(0) = 0 = \alpha(0)$

and $\alpha(y) = 0$ for $y < 1$

$\alpha(y) = \phi$ for $y > 1$

$\alpha(1) = 1 = \omega(1)$

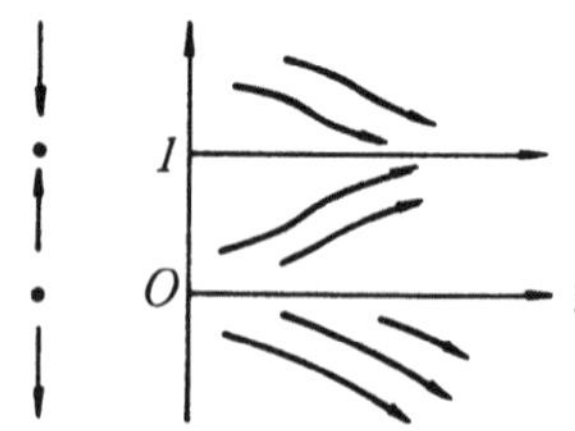


(a) (b)

Fig. 1.10

**Example 3** Consider the two-dimensional system:

$$x' = -y + x(1 - x^2 - y^2)$$
$$y' = x + y(1 - x^2 - y^2)$$

By introducing a transformation $x = r\cos\theta$, $y = r\sin\theta$ we get the following equation $r' = r(1-r^2)$ or $\dfrac{dr^2}{dt} = 2r^2(1-r^2)$. If we let $\mathcal{R} = r^2$, we can get the logistic equation as in example 2.

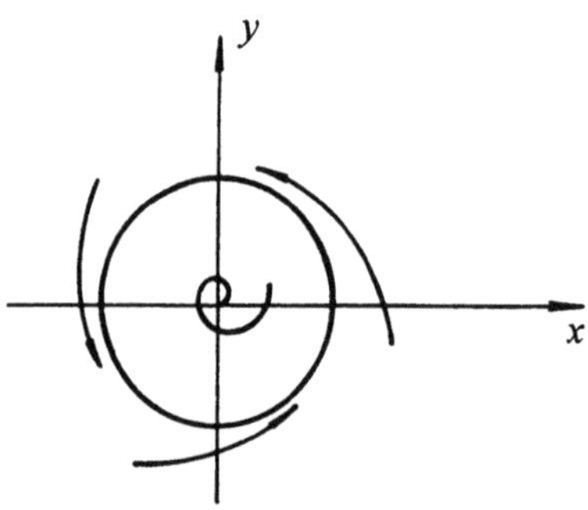


Fig.1.11

The solution for $r$ is similar to that in example 2, and the solution $\theta$ for $\theta = t + \theta$. Equilibrium points of (1) are (0, 0) and of (2) $r = 1, \theta = t + \theta_0$. Limit point sets are $\omega(x,y) = \{\text{unit circle}\}$ $(x,y) \neq (0,0)$, $\omega(0,0) = (0,0)$, $\alpha(x,y) = \phi$, $x^2 + y^2 > 1$, $\alpha(x,y) = (0,0)$, $x^2 + y^2 \leq 1$.

**Proposition 1.1** If $x$ is an equilibrium point of the trajectory passing this point, then $x$ is the $\omega(\alpha)$ -limit point set.

**Proposition 1.2** Every point in a closed trajectory is the $\omega(\alpha)$ -limit point of itself, so the closed orbit is the $\omega(\alpha)$ -limit point set itself.

**Proposition 1.3** If $x$ is an asymptotic stable equilibrium point, then it is the $\omega$ -limit point of all the orbits in phase space tending to $x$ when $t \to \infty$.

**Proposition 1.4** A stable limit circle is the $\omega$ limit point set of all the trajectories in phase space tending to it when $t \to \infty$.

**Proposition 1.5** A unstable limit circle is the $\alpha$-limit point set of all the trajectories in phase space tending to it when $t \to \infty$.

A trajectory is said to be homoclinic, if a point on the trajectory tends to the same point $P$ as $t$ tends to positive and negative infinity $P$ is called a homoclinic point, which is the $\omega(\alpha)$-limit point as shown in Fig. 1.12. A trajectory is said to be heteroclinic, if a point of itself tends to different phase points as $t$ tends to positive and negative infinity respectively as shown in Fig. 1.13. $P$ is the $\alpha$-limit point of the trajectory $l_2$ and the $\omega$-limit point of the trajectory $l_3$. In the same way, $Q$ is the $\alpha$-limit point of the trajectory $l_3$ and the $\omega$-limit point of the trajectory $l_2$.

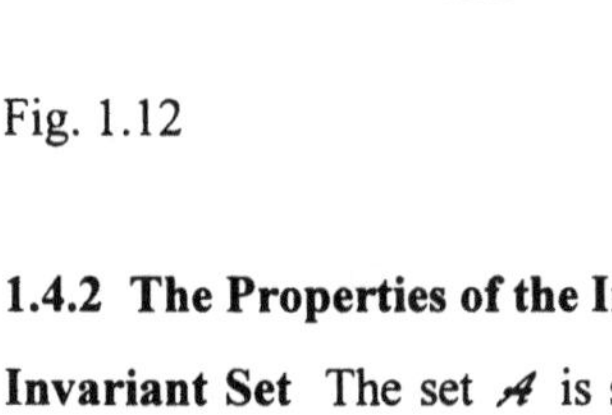


Fig. 1.12

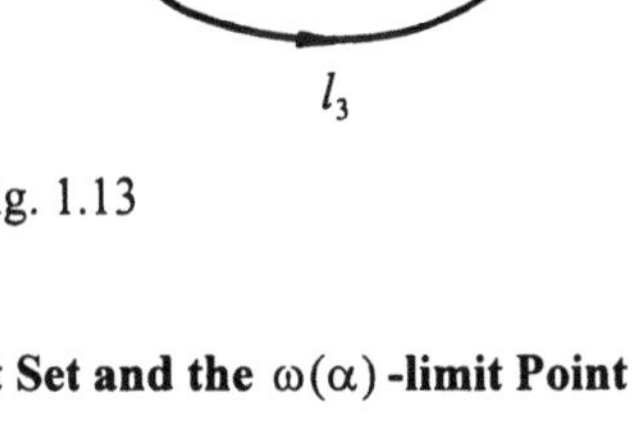


Fig. 1.13

### 1.4.2 The Properties of the Invariant Set and the $\omega(\alpha)$ -limit Point Set

**Invariant Set** The set $\mathcal{A}$ is said to be a positive (negative) invariant for dynamical systems, if for any $x \in \mathcal{A}$ there exists $\varphi' x \in \mathcal{A}$, for any $t$ >0 (<0). If $\mathcal{A}$ is positively and negatively invariant at the same time, then it is called the invariant set.

**Theorem 1.8**

(1) Any whole trajectory is an invariant set; and the semi-positive (negative) trajectory is the positive (negative) invariant set.

(2) Any invariant set is composed of some whole trajectories, and any positive (negative) invariant set is composed of some semi-positive (negative) trajectories.

The following are the properties of the limit set of a given trajectory. Since $p \in \omega(x)$ and $p$ is a limit point, there exists a sequence $\{\varphi x$ such that $\varphi x \to p, (t \to \infty)$.

(1) The $\omega(\alpha)$-limit set is closed and invariant, and it is composed of some whole trajectory. The sufficient and necessary condition for S to be closed is that S is equal to its closure or every point on the boundary which belongs to S.

(2) The $\omega(\alpha)$-limit set is empty if the trajectory $\varphi^t x_0$ tends to infinity as $t \to +\infty$ (or $t \to -\infty$), that is, $\|\varphi^t x_0\| \to \infty$, when $t \to +\infty$ (or $t \to -\infty$).

(3) The $\omega(\alpha)$-limit set is the single point set $\{\bar{x}\}$ if $\lim_{t\to\infty} \varphi^t x_0 = \bar{x}$ or $\lim_{t\to-\infty} \varphi^t x_0 = \bar{x}$.

(4) The $\omega(\alpha)$-limit $\omega(\alpha)$-limit set of a semi-positive (negative) trajectory on a bounded region is connected.

(5) Suppose $\mathcal{A}$ is a closed invariant set, if $x \in \mathcal{A}$ then both $\omega(x)$ and $\alpha(x)$ are in $\mathcal{A}$.

(6) If $x$ and $y$ are in the same trajectory, then $\omega(x) = \omega(y)$.

(7) If $x \in \omega(y)$, then $\omega(x) \subseteq \omega(y)$.

**Proof** First prove that $\omega(x)$ is closed. Suppose that there exists a sequence $\{q_1, q_2, \ldots, q_n, \ldots\} \subseteq \omega(x)$ such that $\lim_{n\to\infty} q_n = q$. As $q_n \in \omega(x)$, there exists a sequence $\varphi^{t_{nk}} x \to q_n$ $(t_{nk} \to \infty)$. In order to prove $\omega(x)$ to be closed, it is necessary to prove that sequence $\{q_1, q_2, \ldots, q_n, \ldots\}$ has a limit $q$ in $\omega(x)$. Now choose a diagonal sequence $\varphi^{t_n k_n}$

$$\begin{vmatrix} \varphi^{t_1 k_1} & \varphi^{t_2 k_1} & \cdots & \varphi^{t_n k_1} \\ \varphi^{t_1 k_2} & \varphi^{t_2 k_2} & \cdots & \varphi^{t_n k_2} \\ & \cdots & \cdots & \\ \varphi^{t_1 k_n} & \varphi^{t_2 k_n} & \cdots & \varphi^{t_n k_n} \\ q_1 & q_2 & \cdots & q_n \end{vmatrix}$$

For any given $\varepsilon > 0$, there exists $N(\varepsilon) > 0$, such that $|q_n - q| < \frac{\varepsilon}{2}$, for any $n > N(\varepsilon)$.

For such $n$ there exists $k_n$ such that $|\varphi^{t_n k_n} - q_n| < \frac{\varepsilon}{2}$. Considering the sequence $\varphi^{t_n k_n}\}$ we can see that $|\varphi^{t_n k_n} - q| \leq |\varphi^{t_n k_n} - q_n| + |q_n - q| < \frac{\varepsilon}{2} + \frac{\varepsilon}{2} = \varepsilon$. It follows that $q \in \omega(x)$ and $\omega(x)$ is closed.

The following is the proof of invariance of $\omega(x)$. For $p \in \omega(x)$, it is necessary to prove $r(p) \subseteq \omega(x)$. For any $t$, then $\lim_{t_k\to\infty} \varphi^{t+t_k} x = \varphi^t (\lim_{t_k\to\infty} \varphi^{t_k} x) = \varphi^t p$.

On the other hand, $\lim_{t+t_k\to\infty}\varphi^{t+t_k}x = p$, so that $\varphi^t p \in \omega(x)$, for any $t \in \mathcal{R}$, that is, $r(p) \subseteq \omega(x)$ or $\omega(x)$ is invariant.

We shall now prove item (5). If for any $p \in \omega(x)$, $x \in \mathcal{K}$ then there exists a sequence $\{\varphi^{t_k}x\}$ such that $\varphi^{t_k}x \to p(t_k \to \infty)$; since $\mathcal{K}$ is invariant, then $\varphi^{t_k}x \in \mathcal{K}$. On the other hand, $\mathcal{K}$ is closed so that $p \in \mathcal{K}$. This proves that $\omega(x) \subseteq \mathcal{K}$.

**Theorem 1.9** If $r^+(x) \subseteq \mathcal{K} \subseteq \mathcal{D}$ where $\mathcal{K}$ is compact and $\mathcal{D}$ is open, then

(1) $\omega(x)$ is non-empty;

(2) $\omega(x)$ is compact;

(3) $\omega(x)$ is connected.

(4) $\text{dist}(\varphi^t x, \omega(x)) \to 0$ as $t \to \infty$, and here $\text{dist}(y, S) = \inf_{s\in S}|y - s|$

**Proof:** (1) Since $\mathcal{K}$ is compact, then for any sequence $\{\varphi^{t_k}x\}(t_k \to \infty)$ there exists an accumulation point $p \in \mathcal{K}$, and there exists a subsequence $\{\varphi^{t_{k_e}}x\}$ such that $\varphi^{t_{k_e}}x \to p\ (t_{k_e} \to \infty)$. From theorem 1.9 we can see that $p \in \omega(x)$. That is $\omega(x) \neq \phi$.

(2) From the boundedness of $\mathcal{K}$ it follows that $\omega(x)$ is bounded. According to theorem 1.9 we can see that $\omega(x)$ is closed. It is thus proved that $\omega(x)$ is compact.

(3) The Proof: of (3) will be dealt later.

(4) Suppose there exists a number $\delta > 0$ and a sequence $\{\varphi^{t_k}x\}(t_k \to \infty)$ , such that $\text{dist}(\varphi^{t_k}x, \omega(x)) > \delta > 0$. As $\{\varphi^{t_k}x\} \subseteq \mathcal{K}$, then there exists a subsequence of $\{\varphi^{t_{k_e}}x\}$ convergent to point $q \in \omega(x)$. That implies

$$0 = \text{dist}(\varphi^{t_{k_e}}x, \omega(x)) > \text{dist}\ (\varphi^{t_k}x, \omega(x)) > \delta > 0$$

Thus is a contradiction which completes the Proof: of (4).

Now let us define the connected set.

**Definition 1.8** A set $\mathcal{S}$ in a metric space $\mathcal{M}$ is said to be unconnected, if there exist two open sets $\mathcal{A}$ and $\mathcal{B}$, such that $\mathcal{A}\cap\mathcal{S} = \mathcal{P} \neq \phi$, $\mathcal{B}\cap\mathcal{S} = \mathcal{Q} \neq \phi$, $\mathcal{S} \subseteq \mathcal{A}\cup\mathcal{B} = \phi$ (see Fig. 1.14). Otherwise it is called a connected set.

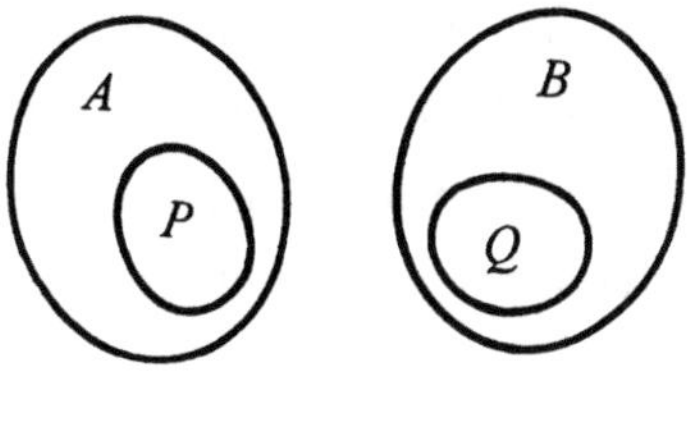


Fig. 1.14

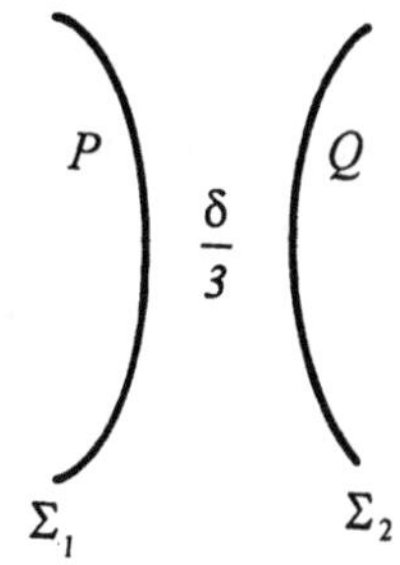


Fig. 1.15

**Corollary:** A set $\mathcal{S}$ is unconnected if there exist two non-empty open sets $\mathcal{A}$ and **B** in $\mathcal{S}$ corresponding to $\mathcal{S}$ satisfying $\mathcal{A} \bigcup \mathcal{B} = \mathcal{S}$, $\mathcal{A} \bigcap \mathcal{B} = \phi$, and if there exist two non-empty closed sets $\mathcal{P}$ and $\mathcal{Q}$ in $\mathcal{S}$ corresponding to $\mathcal{S}$ satisfying $\mathcal{P} \bigcup \mathcal{Q} = \mathcal{S}$, $\mathcal{P} \bigcap \mathcal{Q} = \phi$ (Fig. 1.14).

Now return to Proof: (3) of theorem 1.10. Suppose that $\omega(x)$ is unconnected, then there exist non-empty sets $\mathcal{P}$ and $\mathcal{Q}$ closed in $\omega(x)$ and $\mathcal{P} \bigcup \mathcal{Q} = \omega(x)$, $\mathcal{P} \bigcup \mathcal{Q} = \phi$. Since $\omega(x)$ is compact then $\mathcal{P}$ and $\mathcal{Q}$ are compact in $\omega(x)$. Then there exists a positive number $\delta > 0$, such that $\text{dist}(\mathcal{P}, \mathcal{Q}) > \delta > 0$. Consider semi-positive trajectory $r^+(x)$; it contains two sequences $\{\varphi^{t'_k} x\}$ and $\{\varphi^{t'_e} x\}$ convergent to the points in $\mathcal{P}$ and $\mathcal{Q}$ respectively, where $\mathcal{P}$ and $\mathcal{Q}$ are closed. From the continuity of $r^+(x)$ we can infer that there must be a sequence $\{\varphi^{t'_m} x\}$ such that $\text{dist}(\{\varphi^{t'_m} x\}, \mathcal{P}) \geq \frac{\delta}{3}$ and $\text{dist}(\{\varphi^{t'_m} x\}, \mathcal{Q}) \geq \frac{\delta}{3}$ (Fig. 1.15). Obviously, there is at least one accumulation point $q$ of $\{\varphi^{t'_m} x\}$ assumed that $\{\varphi^{t'_m} x\}$ converges to the point $q$, that is $q \in \omega(x)$, but $q$ is not in where $\mathcal{P}$ and $\mathcal{Q}$. This contradicts $\mathcal{P} \bigcup \mathcal{Q} = \omega(x)$ and completes the Proof.

**Definition 1.9** A point $p$ is called non-wandering for the flow $\varphi^t$, if for any neighbourhood $\mathcal{U}$ of $p$ there exists arbitrarily a large $T$, such that $\varphi^t(\mathcal{U}) \bigcap \mathcal{U} \neq \Phi$ $(t > T)$. The set of all such point $p$ is called the non-wandering set written as $\Omega(\varphi)$, where $\varphi^t(\mathcal{U})$ = $\{x | x = \varphi^t(y), y \in \mathcal{U}\}$. If $q \in \mathcal{E} \backslash \Omega(\varphi)$, then $q$ is called the wandering point.

If $p$ is non-wandering for $\varphi$, then there exists a point $r$ in any neighbourhood $\mathcal{U}$ of $p$ such that $\varphi^t(r)$ returns to $\mathcal{U}$ innumerous times. It is easy to prove that $\Omega(\varphi)$ is invariant and closed. Fixed points and periodic trajectories of $\varphi^t$ are non-wandering.

**Definition 1.10** A closed invariant set $\mathcal{A}$ is called an attracting set, if there is some neighbourhood $\mathcal{U}$ of $\mathcal{A}$, such that $\varphi^t(x) \in \mathcal{U}$ for $t \geq 0$, and $\varphi^t(x) \to \mathcal{A}$, as $t \to \infty$ for all $x \in \mathcal{U}$. The set $\mathcal{B} = \{x | \lim_{t \to \infty} \text{dist } (\varphi^t(x), A) = 0\}$ is the domain of attraction of $\mathcal{A}$.

A repelling set is defined analogously, replacing $t$ by $-t$. The next problem should help us to work out the definition of an attractor as an attracting set which contains a densed trajectory. A repellor is defined analogously. A single point attractor is called a sink and the single point repellor is called a source. Attractors can be divided into two classes: the ordinary, such as the single point attractor, the periodic attractor and the torus attractor; and the "strange attractor" dealing with chaotic problems. The characteristic of an ordinary attractor is the integral dimension and that of a strange atttractor is the fractional dimension [5, 59].

## 1.5 Liapunov Stability of Motion

### 1.5.1 Concepts of Liapunov Stability

Consider the non-autonomous system: $y' = f(t,y)$ where $f(t,y)$ is a vector function defined in region $\mathcal{D}$, satisfying the condition of existence and the uniqueness theorem $\mathcal{D}: t \geq 0$, $|y - y_0| \leq \mathcal{H}$. Suppose $y = \bar{y}(t)$ is the solution of the system (1.3) satisfying $\bar{y}(t_0) = y_0$ defined on $[t_0, T]$, $T > 0$. Then the solution is continuously dependent on the initial value, that is, when the initial value changes slightly, the solution satisfying $y(t_0) = y_0$ exists in the same interval $[t_0, T]$ and changes slightly in $[t_0, T]$ also. Or rather, for any given $\varepsilon > 0$, there exists $\delta > 0$ such that there exists the solution $y(t)$ satisfying $y(t_0)$ defined on the same interval and $|y(t) - \bar{y}(t)| \leq \varepsilon$ for any $y_0$ to satisfy $|y_0 - \bar{y}_0| \leq \delta$, $t_0 \leq t \leq T$.

The continuous dependence on the initial value shows that with small changes of the initial value the motion state of the system is close to the specific motion. As there is always some error in the initial value of practical systems, the continuous dependence on the initial value of the solution ensures that the system can work according to the specific law of motion. However, continuous dependence alone is far from sufficient in many practical problems where the system must work for a long time. For a differential equation the interval where the solution lies must be $[t_0, +\infty)$, and in the whole interval the solution must be kept close to the specific motion state. It is in this situation that continuous dependence cannot be ensured. The introduction of the concept of stability is inevitable.

Consider the non-autonomous system:

$$y' = f(t,y), \qquad f: \mathcal{D} \to \mathcal{R}^n \tag{1.6}$$

where $f \in \mathrm{Lip}(\mathcal{D})$ or $f \in \mathcal{C}^1(\mathcal{D}) \subseteq \mathcal{R}^n$.

Let $y = \psi(t)$ be the solution defined on $t \geq t_0$ satisfying $y_0 = \psi(t_0)$.

**Definition 1.11** The solution $y = \psi(t)$ is called stable, if for any given $\varepsilon > 0$, there exists $\delta = \delta(\varepsilon, t) > 0$ such that $|\varphi(t) - \psi(t)| < \varepsilon$, $t \geq 0$ for any solution $y = \varphi(t)$ satisfying $|\varphi(t_0) - y_0| < \delta$, or else the solution is called unstable.

**Definition 1.12** The solution $y = \psi(t)$ is called uniformly stable, if it is stable and $\delta = \delta(\varepsilon) > 0$ is independent of $t$.

**Note:** For an autonomous system, if $y = \psi(t)$ is stable, then it is also uniformly stable.

**Definition 1.13** The solution $y = \psi(t)$ is called asymptotically stable, if it is stable and there exists $\delta = \delta_1(t_0) > 0$ such that $\lim\limits_{t \to \infty} |\varphi - \psi| = 0$ if the solution $y = \varphi(t)$ satisfies $|\varphi(t_0) - \psi(t_0)| < \delta_1$.

**Example 1** Consider the stability of the zero solution of equation $y' = -y$. The solution of the system is $y = C\exp(-t)$, and obviously the solution $y = 0$ is asymptotically stable.

**Example 2** The exact solution of logistic equation $y' = y(y-1)$ is $y = \frac{C}{(1-c)e^{-t} + C}$. The equilibrium solution $y = 1$ is asymptotically stable, and $y = 0$ is unstable (see Fig. 1.10). Without loss of generality, only the zero solution of eq. (1.6) is considered. For the solution of the system $y = \psi(t)$ (1.6), we can introduce a transformation $y = x + \psi(t)$ and the system can be reduced to

$$x' = f(t, x + \psi(t)) - f(t, \psi(t)) \tag{1.7}$$

The stability of the solution $y = \psi(t)$ of eq. (1.6) is equivalent to that of the trivial solution $x = 0$ of eq. (1.7). For convenience we shall study the stability of another trivial solution $y = 0$ of the following system:

$$y' = f(y), \quad f \in \text{Lip}(\mathcal{D}) \text{ or } f \in \mathcal{C}^1(\mathcal{D}) \tag{1.8}$$

**Definition 1.14** A function $V(y) \in \mathcal{C}^1(\mathcal{D})$ , $\mathcal{D} \subseteq \mathcal{R}^n$ with $V(0) = 0$ is called sign constant definite if $V(y)$ is of the same sign except zero for any $y \in \mathcal{D}$. It is called positive constant if $V(y) \geq 0$ and negative constant if $V(y) \leq 0$.

**Definition 1.15** A function $V(y) \in \mathcal{C}^1(\mathcal{D})$, $\mathcal{D} \subseteq \mathcal{R}^n$ with $V(0) = 0$ is called sign definite if $V(y)$ is of the same sign except zero for any $y \in \mathcal{D}$, $y \neq 0$. It is called positive defined if $V(y) > 0$ and negative defined if $V(y) < 0$.

**Definition 1.16** For a continuously differentiable function $V(y)$ the function

$$\frac{\partial V(y)}{\partial y_1} f_1(y) + \ldots + \frac{\partial V(y)}{\partial y_n} f_n(y)$$

is called the total derivative along system (1.8), written as $\dot{V}(y)$.

### 1.5.2 Liapunov Second Method (Direct Method)

**Theorem 1.10 (Liapunov's first theorem)** If there exists a sign definite function $V(y)$ with the total derivative $\dot{V}(y)$ along eq. (1.8) being sign constant definite opposite to $V(y)$, or identical zero, then the zero solution of eq. (1.8) is stable.

**Proof** Supposing that $V(y) > 0$, the total derivative $\dot{V}(y) \leq 0$. For any given $\varepsilon > 0$ (sufficiently small) such that $|y| \leq \varepsilon\} \subseteq \mathcal{D}$, the given sphere $\mathcal{S} = \{|y| = \varepsilon$ is compact, and there is a constant $K > 0$ satisfying $\min V(y) = K > 0$. From the continuous properties of $V(y)$ we can see that for this $K > 0$, there is a value $\delta > 0$ $(\delta < \varepsilon)$ such that $V(y) < K$ when $|y| < \delta$.

We shall now prove that any solution $y = \varphi(t)$ satisfying $|\varphi(0)| < \delta$ satisfies $|\varphi(t)| < \varepsilon$ for $t \geq 0$. If false, a value $T > 0$ exists such that $|\varphi(t)| = \varepsilon$ and $|\varphi(t)| < \varepsilon$ for $0 \leq t < T$.

From $\dot{V}(y) \leq 0$ it follows that $V(\varphi(T)) - V(\varphi(0)) \leq 0$, and $V(\varphi(T)) - V(\varphi(0)) < K$. It is easily seen that $V(\varphi(T)) \geq K$. This contradicts $V(\varphi(T)) < K$ and completes the proof of the theorem.

**Theorem 1.11 (Liapunov's second theorem)** If there exists a sign definite $V(y)$ with the total derivative $\dot{V}(y)$ along eq. (1.8) being sign definite opposite to $V(y)$, then the zero solution of eq. (1.8) is asymptotically stable.

**Proof** Without loss of generality, supposing $V(y)>0$, the total derivative $\dot{V}(y)<0$. It is easy to see that the solution is stable from theorem 1.11. Then for $\varepsilon = h > 0$, there exists a value $\delta > 0$ such that $|\varphi(t)| < h$, $t \geq 0$ when $|\varphi(0)| < \delta$. Next we shall prove $\lim_{t\to\infty} \varphi(t) = 0$.

Considering function $V(\varphi(t))$, since $\dot{V}(y)<0$, then $V(t) = V(\varphi(t))$ is a monotone decreasing function. It follows that $\lim_{t\to\infty} V(t) = \alpha \geq 0$ from $V(y)>0$.

If $\alpha = 0$, then $0 = \lim_{t\to\infty} V(\varphi(t)) = V(\lim_{t\to\infty} \varphi(t)) = V(0)$, therefore, $\lim_{t\to\infty} \varphi(t) = 0$.

If $\alpha > 0$, there exists $\delta_1 > 0$ such that $\delta_1 \leq |\varphi(t)| \leq \varepsilon$. Let $S = \{y | \delta_1 \leq |y| \leq \varepsilon\}$. Then there exists $m > 0$, such that $\max_{y\in S} \dot{V}(y) \leq m < 0$. By integrating we can show that $V(\varphi(t)) \leq V(\varphi(0)) + mt \to -\infty (t \to \infty)$ which contradicts $V(\varphi(t)) > 0$, that is, $\alpha = 0$.

This completes the proof.

The function $V(y)$ in theorems 1.11 and theorem 1.12 is called Liapunov's function.

**Theorem 1.12 (Liapunov's third theorem)** If there exists a function $V(y)$ such that the total derivative $\dot{V}(y)$ along eq. (1.8) is sign definite, and there exists a point $y_0$ in any neighbourhood of (0, 0) such that align $\dot{V}(y_0)V(y_0) > 0$, then the zero solution of eq. (1.8) is unstable.

**Proof** What we want to prove is that the solution $\varphi(t)$ starting from the vicinity of the origin cannot stay forever at $S = \{y|\ |y| \leq h, h > 0$ . In fact, for any given $\varepsilon > 0$, there exists $y_0$ such that $|y_0| < \varepsilon$ and $\dot{V}(y_0)V(y_0) > 0$, supposing $\dot{V}(y_0) > 0$. $V(y_0) > 0$. If $y = \varphi(t)$ is the solution starting from $y_0$, from $\dot{V}(y_0) > 0$, we can see that $(\varphi(t)) > V(y_0)$. As $(0) = 0$, according to the continuity of $(y)$, there must be $\lambda > 0$, such that $|\varphi(t)| \geq \lambda (t \geq 0)$, that is, the solution $\varphi(t)$ satisfies $\lambda \leq |\varphi(t)| \leq h$. Consider the compact loop $S_1 = \{\lambda \leq |y| \leq h\}$, on which $V(y)$ is bounded, and $\dot{V}(\varphi(t)) \geq l > 0$, then $\dot{V}(\varphi(t)) \geq V(y_0) + lt \to +\infty$, when $t \to +\infty$. This contradicts the fact that $(\varphi(t))$ is bounded. Thus the zero solution is unstable.

**Example:** Consider the stability of the zero solution of the non-linear oscillation system

$$m\ddot{x} = -k(x + x^3) \tag{1.9}$$

and

$$m\ddot{x} = -k(x + x^3) - \alpha x \tag{1.10}$$

where $m$ is the mass and $k$ is the rigidity of the spring, and $\alpha$ is the damping coefficient ($m, k$ and $\alpha$ are positive constants). The total energy of eq. (1.9) is $E = \frac{1}{2}m\dot{x}^2 + k\left(\frac{x^2}{2} + \frac{x^3}{4}\right)$. Let $y = \dot{x}$, then $E(x, y) = \frac{1}{2}my^2 + k\left(\frac{x^2}{2} + \frac{x^4}{4}\right)$ is the positive definite function neighbouring point (0,0). The total derivative of $E(x, y)$ along eq. (1.9) is $\dot{E}(x, y) = ky(x + x^3) - ky(x + x^3) = 0$. Then the zero solution is stable. For system (1.10) the total derivative is $\dot{E}(x, y) = my\left(-\frac{\alpha}{m}y\right) = -\alpha y^2 \leq 0$, therefore the zero solution is stable. But $\dot{E}(x, y)$ is not sign definite, so the solution cannot be judged as asymptotically stable in the vicinity of point (0,0). If,

$$(x, y) = \frac{1}{2}my^2 + \beta xy + k\left(\frac{x^2}{2} + \frac{x^4}{4}\right) \text{ then } (x, y) = \frac{m}{2}\left(y + \frac{\beta}{m}x\right)^2 + \frac{1}{2}\left(k - \frac{\beta}{m}\right)x^2 + \frac{k}{4}x^4$$

when $k - \frac{\beta^2}{2} > 0,\ V(x, y) > 0,\ (x, y) \neq (0,0)$,

and $$\dot{V}(x, y) = \frac{\beta k}{m}x^4 - \left(\alpha - \beta - \frac{\beta\alpha^2}{4km}\right)y^2 - \beta\frac{k}{m}\left(x + \frac{2y}{k}\right)^2$$

when $0 < \beta < \dfrac{\alpha}{1 + \dfrac{\alpha^2}{4km}},\ \dot{V}(x, y) < 0,\ (x, y) \neq (0,0)$.

Then for a sufficiently small $\beta$, $V(x, y)$ is positive and $\dot{V}(x, y)$ is negative definite. Then the zero solution of the system is asymptotically stable.

As a general rule, it is hard to find a proper Liapunov function for a non-linear system. But for a linear stationary system

$$y' = Ay, \quad A = (a_{i,j})_{n\times n}, \quad y = (y_1, y_2, \ldots, y_n)^T \tag{1.11}$$

the stability of the zero solution can be determined by the Liapunov method. The following are the results in this conclusion.

**Theorem 1.13 (Asymptotic stability)** If all the eigenvalues of matrix $A$ in system (1.11) are of negative real parts, then for any negative defined quadratic form $V_1(y) = -y^T Cy < 0$, there exists a positive definite quadratic form $V_2(y) = y^T Py > 0$ such that the total derivative of $V_2(y)$ along (1.11) is $V_1(y)$. That is, the positive definite matrices $C$ and $P$ satisfy the Liapunov matrix equation.

$$A^T P + PA = -C \tag{1.12}$$

therefore the zero solution of (1.11) is asymptotically stable.

**Theorem 1.14 (Stability Theorem)** If all the eigenvalues of matrix $A$ in system (1.11) are of non-positive real parts, and the elementary divisors corresponding to the eigenvalues with zero real parts are simple, then there exist a positive definite quadratic form $V_1(y) = y^T Py > 0$, a negative constant definite quadratic form $V_2(y) = -y^T Cy < 0$, and matrices $C$ and $P$ that satisfy Liapunov's matrix equation.

**Lemma** If there exists a function with the total derivative along (1.8), the total derivative with respect to (1.8) has the following form in $\mathcal{D}$: $\dfrac{dV}{dt} = \lambda V(y) + W(y)$ where $\lambda > 0$, and $W$ is sign constant definite or zero. When $W(y) \neq 0$, there exists $y_0$ at any neighbourhood, of point (0,0), such that $V(y_0)W(y_0) > 0$, and the zero solution is then unstable.

**Theorem 1.15 (Unstable theorem)** If there is at least one eigenvalue with a positive real part or the elementary divisor is complex for a zero real part eigenvalue, even though all the eigenvalues have non-positive real parts, then for any negative definite quadratic form $W(y)$ (not vanishing at the whole trajectory of system (1.11)) there exists $\alpha > 0$ and quadratic form $V(y)$ not being positive, such that $dV/dt = \alpha V + W$ and the zero solution is unstable.

The proofs of the above theorems are omitted. Readers are referred to [39].

### 1.5.3 Floquet Theory in a Periodic System

For subsequent understanding and application, introduction of the Floquet theory is necessary. In order to study the stability of the periodic solution $y = \varphi(t)$ with period $T > 0$ for a given autonomous system:

$$y' = f(y) \qquad f \in \mathcal{C}^1(\mathcal{D}) \quad y \in \mathcal{R} \tag{1.8}$$

We reduce this to the stability of the zero solution of the system by means of the periodic coefficients. Let us introduce transformation $x = y - \varphi(t)$, and take the first approximate equation — the first variation equation or the perturbation equation of $\varphi(t)$). We have

$$x' = A(t)x, \quad A(t+T) = A(t), \quad A(t) = D_x f(\varphi(t)) \tag{1.13}$$

Suppose $\Phi(t)$ is the fundamental matrix of periodic system (1.13), then $\Phi(t)$ satisfies the matrix equation

$$\frac{d\,\Phi(t)}{d\,t} = A(t)\Phi(t) \tag{1.14}$$

Obviously, $\Phi(t+T)$ is also the fundamental matrix of the system replacing $t$ by $t+T$ in $\Phi(t)$, therefore there exists a non-singular matrix $A$ such that

$$\Phi(t+T) = \Phi(t)A$$

Defining the equation $D(\rho) = \det(A - \rho E) = 0$ as the characteristic equation of eq. (1.14), corresponding to period $T$ or simply the characteristic equation. The roots of $D(\rho) = 0$ are the Floquet multipliers of system (1.13).

The characteristic equation is important in the study of the stability of the solutions. For this reason, two properties of the characteristic equation are presented here.

**Property 1** The characteristic equation is independent of the choice of fundamental matrix.

**Proof** Suppose $\Psi(t)$ is another fundamental matrix of the system, then there exist non-singular matrices $C$ and $B$ such that $\Psi(t+T)=\psi(t)C$ and $\Psi(t)=\Phi(t)B$, therefore $\Psi(t+T)=\Phi(t+T)B=\Phi(t)AB=\Psi(t)B^{-1}AB$, and then it follows that $C=B^{-1}AB$. From linear algebra we know that the characteristic equation of matrix $A$ coincides with that of matrix $C$.

**Property 2** The characteristic equation of the system does not change through periodic non-singular transformation.

**Proof**

Suppose $y=Q(t)x$, $Q(t+T)=Q(t)$ then $\dfrac{\mathrm{d}y}{\mathrm{d}t}=[Q(t)A(t)Q^{-1}(t)+\dot{Q}(t)Q^{-1}(t)]y$. Change $\Psi(t)$ to $\Psi(t)=Q(t)\Phi(t)$: $\Psi(t+T)=Q(t+T)\Phi(t+T)=Q(t)\Phi(t)A$, we have, $\Psi(t+T)=\Psi(t)A$.

This completes the proof.

For convenience, we take $\Phi(0)=E$, so that $\Phi(T)=A$.

**Definition 1.17** Matrix $L(t)$ is said to be a Liapunov matrix if it satisfies the following conditions:

(1) the derivative $\dfrac{\mathrm{d}L(t)}{\mathrm{d}t}$ is continuous for $t\geq 0$;

(2) both $L(t)$ and $\dfrac{\mathrm{d}L(t)}{\mathrm{d}t}$ are bounded for $t\geq 0$;

(3) $L^{-1}(t)$ is bounded for $t\geq 0$.

The transformation $y=L(t)x$ is called the Liapunov transformation if $L(t)$ is a Liapunov matrix.

**Corollary 1** The stability of system (1.13) does not change through Liapunov transformation.

**Proof** The proof can be followed from the definition of Liapunov transformation.

**Corollary 2** If $L(t)$ is the Liapunov matrix, then $L^{-1}(t)$ is still the Liapunov matrix.

**Definition 1.18** System (1.13) is called a "reducible" in the sense of Liapunov, if there exists a Liapunov transformation reducing the system to the following system:

$$\frac{\mathrm{d}y}{\mathrm{d}t}=By \tag{1.15}$$

where $B$ is a constant matrix.

**Theorem 1.16 (Reducible theorem)** Linear system (1.13) with periodic coefficients is reducible in the sense of Liapunov.

**Proof:** Suppose $\Phi(t)$ is the fundamental matrix and $A$ is the non-singular matrix such that $\Phi(t+T)=\Phi(t)A$. Let $A=e^{TB}$. Suppose that $\rho_j=\alpha_j+i\beta_j$ are the eigenvalues of $A$, or the Floquet multipliers which are related to the eigenvalues $\lambda_j$ of $B$

$$\lambda_j=\frac{1}{T}\left[\ln\sqrt{\alpha_j^2+\beta_j^2}+i\ \mathrm{arctg}\left(\frac{\beta_j^2}{\alpha_j^2}\right)\right]\quad (j=1,2,\ldots n)$$

$\lambda_j \ (j = 1,2,\ldots,n)$ are called the characteristic exponents corresponding to periodic $T$.

Let $L(t) = \Phi(t)e^{-Bt}$, then $L(t)$ is periodic with $T$. In fact, $L(t+T) = \Phi(t+T)e^{-Bt} = \Phi(t)e^{-Bt} = L(t)$. It is easy to verify that both $L(t)$ and $L^{-1}(t)$ are Liapunov matrices. Let $x = L(t)y$, then

$$\begin{aligned}\dot{y} &= \dot{L}^{-1}(t)x + L^{-1}(t)x \\ &= [-L^{-1}(t)L(t) + L^{-1}(t)A(t)L(t)]y \\ &= -L^{-1}(t)[A(t)L(t) - L(t)B]y + L^{-1}(t)A(t)L(t)y \\ &= By\end{aligned}$$

that is,

$$\frac{\mathrm{d}y}{\mathrm{d}t} = By \tag{1.15}$$

where $B$ is a constant matrix.

The fundamental matrix of system (1.15) is $\Psi(t) = L^{-1}(t)\Phi(t) = e^{Bt}$.

**Theorem 1.17 (Floquet theorem)** If all the moduli of the Floquet multipliers of eq. (1.13) are less than 1, then the zero solution of eq. (1.13) is asymptotically stable. If there is at least one multiplier with the modulus greater than 1, then the solution is unstable. If all the moduli of the multipliers are no more than 1, and elementary divisors of the multiplier with modulus 1 are simple, then the solution is stable. If there exists at least one multiplier with modulus 1 which has a complex elementary divisor, then the solution is unstable.

**Corollary** Periodic linear system (1.13) has a periodic solution with period $T$ if the system has a Floquet multiplier of 1.

**Proof**

**Necessity** Suppose that $y = \varphi(t)$ is a solution with periodic $T$ of eq. (1.13), then

$$\varphi(t) = \varphi(t+T) = \Phi(t+T)v = \Phi(t)Av$$

But $\varphi(t) = \Phi(t)v$, $\Phi(t)$ is the fundamental matrix and $v$ is the non-zero vector, therefore $\Phi(t)(A - E)v = 0$, so that $\rho =$ is the root of the characteristic equation.

**Sufficiency** Suppose that $\rho =$ is the root of the characteristic equation $\det(A - E) = 0$, and $v$ is an eigenvector corresponding to $\rho =$ , then $\varphi(t) = \Phi(t)v$ is the solution with period $T$.

This completes the proof.

## 1.6 Poincaré—Bendixson Theorem and its Applications

This section deals with the important properties of the limit sets of planar dynamical systems, that is, the Poincaré—Bendixson theorem and its applications in the study of limit cycles.

### 1.6.1 Transversality

Non-empty subspaces $V_1$ and $V_2$ in $V$ are said to be transversal if $V_1 \oplus V_2 = V$. The following are some examples.

In a two-dimensional space if span $\{V_1, V_2\} = \mathrm{R}^2$, where $V_1$ and $V_2$ are non-zero vectors, then $V_1$ and $V_2$ are transversal. For a given flow $\varphi^t : \mathcal{D} \to \mathcal{R}^2$, the transversality (or Poincaré section) is an open segment $\mathcal{L}$ which satisfies the following properties. If $x \in \mathcal{L}$, then span$\{V(x).l\} = \mathcal{R}^2$, where

1) $V(x) = \dfrac{\mathrm{d}}{\mathrm{d}t}\varphi^t x\,|_{t=0}$ is the tangent line of the flow at the non-equilibrium point $x$
2) $l$ is a vector in $\mathcal{L}$.

For an $n$-dimenssional space in an ($n$-1)-dimensional hyperplane $\mathcal{H}$ (linear subspace), $\mathcal{L}$ is replaced by a neighbourhood $\mathcal{S}$. If $x \in \mathcal{S}$ and span $V(x),, \mathcal{H}\} = \mathcal{R}^n\}$, then $\mathcal{S} \in \mathcal{H}$ is a Poincaré section.

**Definition 1.19 (Local section)** If $f(x)$ is a continuous $n$-dimensional vector field, $\bar{x}$ is not an equilibrium point, that is, $f(\bar{x}) \neq 0$. Make an ($n$-1)-dimensional hyperplane $\mathcal{H}$ such that $f(\bar{x}) \notin \mathcal{H}$, and $\mathcal{S}$ is called a local section at $\bar{x}$; if there is a neighborhood of $\bar{x}$ on $\mathcal{H}$, then any $x \in \mathcal{S}$ implies $f \notin \mathcal{H}$.

For the $n$-dimensional autonomous system $x' = f(x)$, the trajectory through every $x$ in $\mathcal{S}$ intersects $\mathcal{S}$ transversally. Now let us discuss some lemmas.

The lemmas are limited to the two-dimensional systems shown in eq. (1.16), except for lemma 0 [115].

Consider the two-dimensional autonomous system

$$x' = f(x) \qquad x \in \mathcal{R}^2 \tag{1.16}$$

where $f : \mathcal{R}^2 \to \mathcal{R}^2$ is continuous, satisfying the uniqueness conditions. Suppose every solution exists on $(-\infty, +\infty)$. Let $\varphi^t(x)$ be the flow of system (1.16).

**Lemma 0** Suppose $x_0$ is not a fixed point of the given smooth flow $\{\varphi^t(x)\}$ defined on $\mathcal{D}$ in $\mathcal{R}^\ell$, then:

(1) there is a section $\mathcal{L}$ through $x_0$;

(2) there is a neighbourhood $\eta$ of $x_0$ in $\mathcal{D}$, and for any $x \in \eta$, the $\varphi^t x$ intersects transversality $\mathcal{L}$;

(3) all the trajectories starting from $x \in \eta$ pass through $\mathcal{L}$ in the same sense (in the same direction).

Since $x_0$ is not a fixed point, then $V(x_0) = \varphi^t x|_{t=0} \neq 0$, therefore $V(x) \neq 0$ is in some neighbourhood $\mathcal{S}$ of $x$. Choose vector $l$ to satisfy $l \perp V(x_0)$, then it follows that $l$ is not parallel to $V(x)$ in $\mathcal{S}$. Then lemma 0 is correct.

Define a monotone sequence. For a given orbit $r(x) = \{\varphi^t x | t \in \mathcal{R}\}$, the sequence $y = \{\varphi^{t_n} x\}$ is said to be monotone for time $t$ if $t_1 < t_2 < \ldots < t_n < \ldots$.

For a transversality $\mathcal{L}$ in $\mathcal{R}^2$, if vector $[z_{k-l} - z_k]$ is in the same direction, then the sequence $\{Z_1, Z_2, \ldots, Z_n, \ldots\}$ is called monotone. (See Fig. 1.16.)

**Lemma 1** Suppose $\mathcal{L}$ is a transversality at $x \in r(x)$ and $\mathcal{Y} = \{y_1, y_2, \ldots, y_n, \ldots\}$ is the intersection sequence of $r(x)$ with $\mathcal{L}$, then $\mathcal{Y}$ is monotone on $\mathcal{L}$ if $\mathcal{Y}$ is monotone for time $t$.

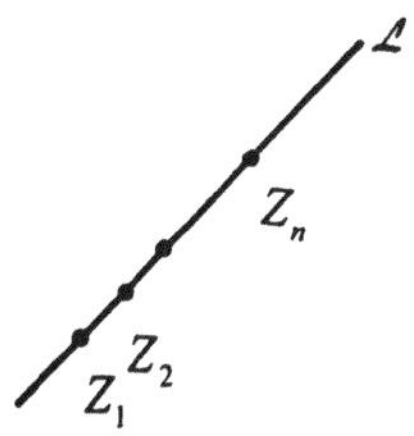


Fig. 1.16

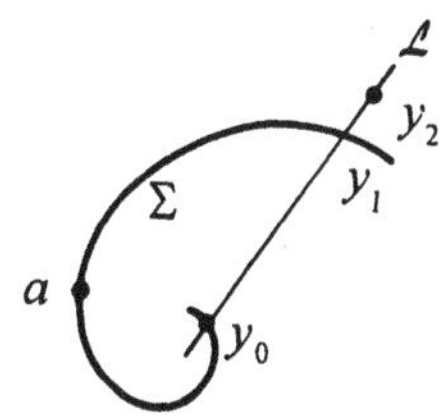


Fig. 1.17

**Proof** It is necessary to prove that $y_2$, the sequent point of $y_0$ and $y_1$, is outside the segment $\overline{y_0 y_1}$ in the direction of $\overline{y_0 y_1}$ . Consider the simple closed curve $\Sigma$ formed by segment $\overline{y_0 y_1}$ and arc $\overline{y_0 a y_1}$ (see Fig. 1.17). For the uniqueness of the orbit and the property of $\mathcal{L}$ (segment $\overline{y_0 y_1}$) we can see that after $y_1$ the orbit cannot go into (or out of) the region enclosed by $\Sigma$. That is, $y_2$ must be outside the segment $\overline{y_0 y_1}$ in the direction of $\overline{y_0 y_1}$.

This completes the proof of the lemma.

**Lemma 2** Suppose $\mathcal{L}$ is an arbitrary transversality of $\omega(x)$, $y \in \omega(x)$, then $r(y)$ cannot intersect $\mathcal{L}$ more than once.

**Proof** Suppose $\mathcal{L}$ is an arbitrary transversality of $\omega(x)$, $y \in \omega(x)$, then $r(y)$ intersects $\mathcal{L}$ at point $y_1$, $y_2$. It is easy to see that $r(y) \subseteq \omega(x)$, and there are non-touched neighbourhoods $\mathcal{S}_1$ and $\mathcal{S}_2$ of $y_1$ and $y_2$ respectively. In the neighbourhoods of $\mathcal{S}_1$ and $\mathcal{S}_2$ there exist sequences (see Fig. 1.18)

$$\varphi^{t_{1k}} x \to y_1 . \text{as } t_{1k} \to \infty; \varphi^{t_{2k}} x \to y_2 . \text{as } t_{2k} \to \infty$$

From the second point of lemma 0, we know that there are innumerable $t$ in the vicinity of $t_{1k}$ and $t_{2k}$ such that $\varphi^t x$ intersects innumerable times in the vicinity $y_1$ and $y_2$ with $\mathcal{L}$, that is $\varphi^t x$ intersects with $\mathcal{L}$ innumerable times. As the relative positions of $y_1$ and $y_2$ are arbitrary, this is a contradiction to the monotoneness of lemma 1. That is, the number of intersections of any transversality with $r(y)$ cannot be more than one.

**Lemma 3** Suppose $\omega(x)$ is an non-empty bounded set with no fixed point in it, then $\omega(x)$ contains at least one periodic orbit $\mathcal{C}$.

**Proof** Let $y \in \omega(x)$, then $r(y) \subseteq \omega(x)$, which implies that $\omega(y) \subseteq \omega(x)$ non-empty. Take $z \in \omega(y)$, and let $\mathcal{L}$ be the transversality through $z$, then there exists a sequence $\{\varphi^{t_k} y\} \subseteq \mathcal{L} \cap r(y)$. It is known from lemma 2 that $\{\varphi^{t_k} y\}$ is a single point sequence. Now let $T = \min|t_i - t_j| > 0$, then $\varphi^T y = y$. For any $t > 0$, this implies that $\varphi^{t+T} y = \varphi^t y$. Therefore $r(y)$ is a periodic orbit in $\omega(x)$ (see Fig. 1.19).

**Lemma 4** If $\omega(x)$ contains one periodic orbit $\mathcal{C}$, then $\mathcal{C} = \omega(x)$

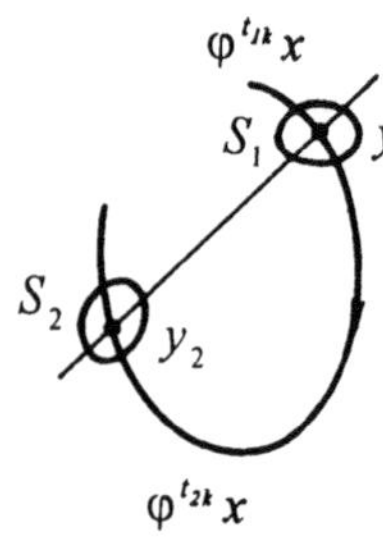


Fig. 1.18

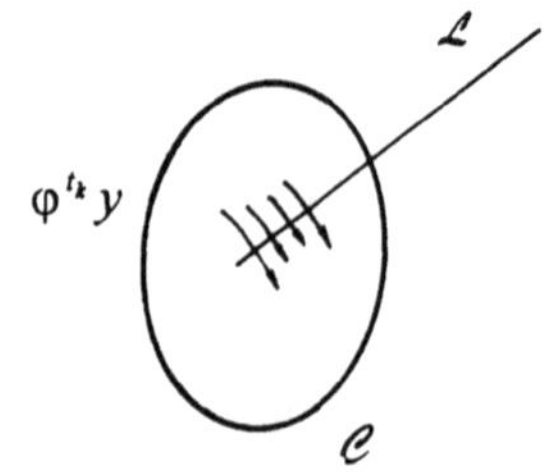


Fig. 1.19

**Proof** Suppose there is a point $z \in$ , then there is a sequence $\varphi^{t_k} x \to z$. If $z$ is not a fixed point, from the second point of lemma 0 we know that with the change of $t_k$, then $\varphi^{t_k} x$ intersects with $\mathcal{L}$ sequently at point $x_k$, that is, $x_k = \varphi^{t_k} x \in \mathcal{L} \cap r(x)$, $x_k$ is monotone on $\mathcal{L}$. From the continuity we know $x_k \to z$.

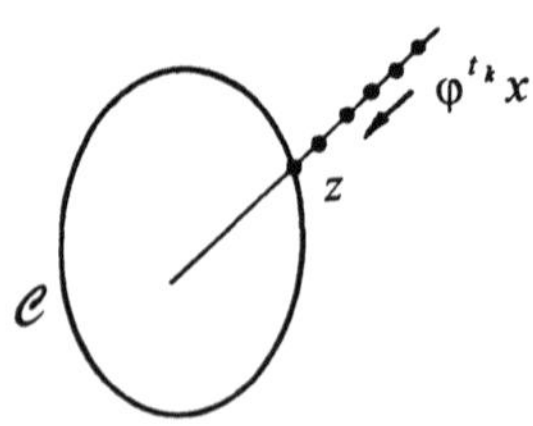


Fig. 1.20

If $|x - z| < \delta$, then $|\varphi^t x - \varphi^t z| < \varepsilon$ for any $t \in [0, T + P$ where $T$ is the period of $\mathcal{C}$. Therefore, when $x_k \to z$ we can take a small quantity of $\delta$ and $\varepsilon$ such that dist $\varphi^t x, \mathcal{C}$ $a \to 0$, as $t \to \infty$. Therefore, arbitrary $y \in \omega(x)$ must be on $\mathcal{C}$ (see Fig. 1.20).

### 1.6.2 Poincaré—Bendixson Theorem and its Applications

**Theorem 1.18 (Jordan curve theorem)** Suppose that $\mathcal{C}$ is a simple curve in $\mathcal{R}^2$, then $\mathcal{R}^2$ can be divided into two mutually disjoint open parts written as $P_i$ and $P_e$: $P_i$ is bounded, called the internal region, and $P_e$ unbounded, called the external region. Any two points in $P_i$ (or in $P_e$) can be connected by a continuous curve located in $P_i$ ($P_e$).

The proof of the theorem is omitted. The following theorems are based on the uniqueness and existence theorem and Jordan theorem. The Poincaré—Bendixson theorem is valid for $\mathcal{R}^2$ owing to the limitation of the Jordan theorem.

**Theorem 1.19 (Poincaré—Bendixson theorem)** Suppose that $r(x)=\{\varphi^t(x)|t\in\mathcal{R}\}$ $x\in\mathcal{D}$ is the orbit of system (1.16), if $\omega(x)\subseteq\mathcal{K}\subseteq\mathcal{D}\subseteq\mathcal{R}^2$, where $\mathcal{K}$ is compact, and $\omega(x)$contains no equilibrium point, then orbit $r(x)$ only satisfies one of the following two cases:

(1) $r(x)=\omega(x)$, where $x$ is in a periodic orbit;

(2) $\omega(x)$ is a limit cycle upon which $r(x)$ winds as $t\to\infty$, or $r(x)$ is not a closed orbit itself but $\omega(x)=\bar{r}(x)\setminus r(x)$, where $\bar{r}(x)$ is the closure of $r(x)$.

**Proof** If $r(x)$ is a closed orbit, then $\omega(x)=r(x)$; if not, then from the boundedness of $r(x)$ we can see that $\omega(x)$ is a bounded invariant non-empty set. There must be a bounded orbit $\Gamma\subseteq\omega(x)$ through point $\eta$ at $t=0$ written as $\Gamma=\{\varphi^t(\eta)|t\in\mathcal{R}\}$. If $\Gamma$ is bounded, then $\omega(\eta)$ is non-empty, therefore $\omega(\eta)\subseteq\omega(x)$. What is needed is to prove that $\Gamma$ is a closed orbit. For this, suppose $\xi\in\omega(\eta)$, and $\mathcal{L}$ is the section through point $\xi$ inferring that $\xi\in\omega(x)\bigcap\mathcal{L}$. There must be a sequence $\{\xi_k\}\in\Gamma\bigcap\mathcal{L}$, $\xi_k\to\xi$ as $k\to\infty$. It is easy to see that $\xi\in\omega(x)\bigcap\mathcal{L}$ from $\Gamma\subseteq\omega(x)$, $k=1,2,\ldots$ . From lemma 2 it follows that $\xi_k=\xi$, $k=1,2,\ldots$ . This proves that $\xi\in\Gamma$. By taking a view of the supposition that $\xi\in\omega(\eta)$ we can see that $\Gamma$ is a closed orbit. From lemma 4 and $\Gamma\subseteq\omega(x)$, it follows that $\Gamma=\omega(x)$. That implies that $\omega(x)$ is a closed orbit.

**Theorem 1.20** Suppose there are at most finite equilibrium points for system (1.16). If the semi-positive orbit of $r(x)$ is bounded, then $\omega(x)$satisfies one of the following:

(1) $\omega(x)$ is an equilibrium point;

(2) $\omega(x)$ is a closed orbit; and

(3) $\omega(x)$ is the union of fixed points and the orbits connecting them. The point on the orbits tends to one of the above fixed points as $t\to+\infty$ and $t\to-\infty$.

The same conclusions can be given for $\alpha(x)$.

**Note 1.** A closed simple curve formed of some fixed points and orbits connecting them is called a singular closed orbit. The path is referred to as a heteroclinic orbit when it connects distinct points and a homoclinic orbit when it connects itself. Closed paths formed of heteroclinic orbits are called heteroclinic cycles, and those formed of one homoclinic orbit are called homoclinic cycles. If a fixed point on the singular closed orbit is a hyperbolic equilibrium point, then it must be a saddle point.

**Note 2.** All possible non-wandering sets fall into three classes: fixed points, closed orbits, and homoclinic or heteroclinic cycles.

For proofs of the above theorems refer to [48].

**Theorem 1.21 (Poincaré loop domain theorem)** Suppose $\mathcal{G}$ is a loop domain enclosed by an inner boundary curve $\Gamma_1$ and an external boundary curve $\Gamma_2$ (see Fig. 1.21). All trajectories through the boundary $\Sigma=\Gamma_1\bigcup\Gamma_2$ go into (out of) $\mathcal{G}$ and there is no singular point in $\mathcal{G}$, thus there exists at least one external stable (unstable) limit cycle and one inner stable (unstable) limit cycle.

This theorem is the important corollary of the Poincaré—Bendixson theorem.

**Proof** Suppose $r_1$ is a trajectory which goes into $\mathcal{G}$ as time increases through $\Gamma_1$ (see Fig. 1.22), then the fact that $\mathcal{G}$ is an invarint bounded set shows $r$ is positively bounded. From the Poincré theorem and as $r_1$ would not be a closed orbit (if it was closed, $r_1$ would go outside $\mathcal{G}$ through $\Gamma_1$, a contradiction of the assumption), we can see that the $\omega$—limit set of $r$ is a closed orbit $\mathcal{C}_1 \subseteq \mathcal{G}$, an inner stable limit cycle. In the same way we can prove that there is an external stable cycle $\mathcal{C}_2 \subseteq \mathcal{G}$. As there is no fixed point in $\mathcal{G}$ and there must be a fixed point enclosed in a closed orbit, then the inner regions of $\mathcal{C}_1$ and $\mathcal{C}_2$ cannot be totally enclosed in $\mathcal{G}$, so that the inner of $\mathcal{C}_1$ and $\mathcal{C}_2$ encloses boundary curve $\Gamma_1$.

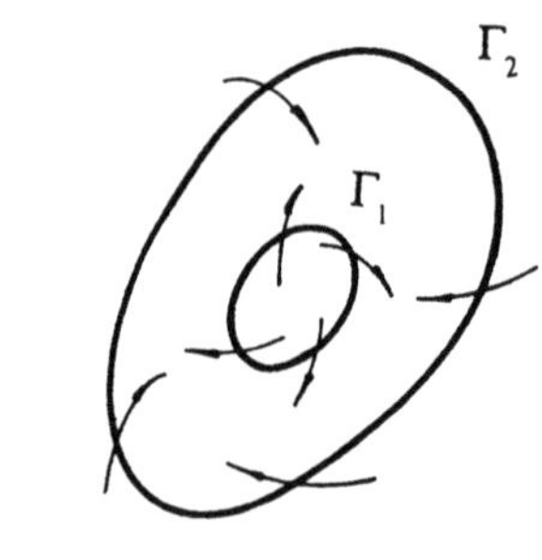


Fig. 1.21

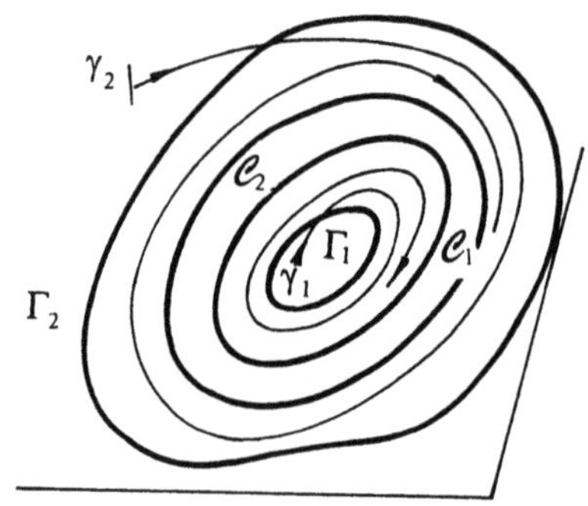


Fig. 1.22

**Example** Consider the two-dimensional system

$$\frac{dy}{dt} = -y - x(x^2 + y^2)$$

$$\frac{dx}{dt} = x - y(x^2 + y^2)$$

or equivalently

$$\frac{dr}{dt} = -r(r-1), \quad \frac{d\theta}{dt} = 1$$

Take $\mathcal{C}_1 : r = 1/2$, and $\mathcal{C}_{r_2} : r = 2$ then

$$\frac{dr}{dt}\Big|_{\mathcal{C}_1} = \frac{2}{2\sqrt{2}} > 0, \quad \frac{dr}{dt}\Big|_{\mathcal{C}_2} = -\sqrt{2} < 0$$

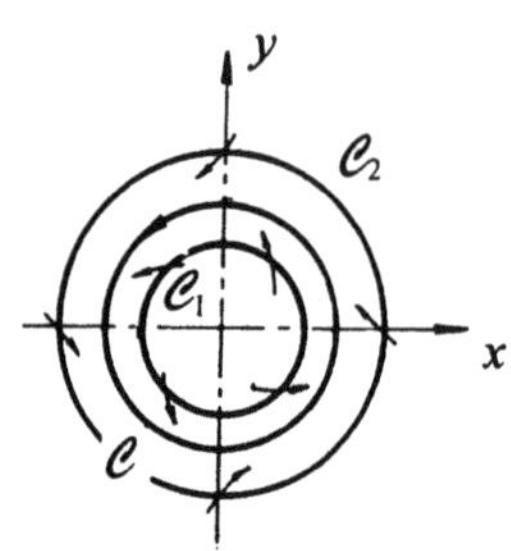


Fig. 1.23

It is easy to see that there is no fixed point in $\mathcal{G}$ formed of $\mathcal{C}_1$ and $\mathcal{C}_2$, thus from the Poincaré loop domain theorem there must be a limit cycle. It is clear that the cycle $r =$ is stable (see Fig. 1.23).

# Chapter 2

# Calculation of Flows

In this chapter, the calculations of flows are discussed. Section 2.1 introduces divergence of flows and Bendixson's negative criterion. We shall deal with the algebric characteristics and calculations of linear flows in section 2.2. Section 2.3 discusses linearly hyperbolic flows and their classification. Section 2.4 deals with calculations of local flows of non-linear system. Finally, we introduce the stable manifold theorem.

## 2.1 Divergence of Flows

### 2.1.1 Divergence of Flows and Bendixson's Negative Criterion

To determine variations of flows with time in terms of differential equations without solving the equations, we have to study the divergence of flows.

Consider flow $\mathbf{J}:[t_0,\beta)\subseteq(\alpha,\beta)^n$. Let $y'=1+y^2, y(0)=0$, $\varphi^t:\mathcal{V}\to\varphi^t\mathcal{V}$, where [illegible] is called volume (Fig. 2.1), then the initial volume is

$$\iiint_{\mathcal{V}} \mathrm{d}\bar{x} = \iint_{\mathcal{V}}\dots\iint \mathrm{d}x_1\mathrm{d}x_2\dots\mathrm{d}x_n \tag{2.1}$$

After time $t$ the volume of $\varphi^t\mathcal{V}$ becomes

$$\varphi^t\mathcal{V} = \iiint_{\varphi^t\mathcal{V}} \mathrm{d}\bar{y} = \iiint_{\mathcal{V}}\left|\frac{\partial y}{\partial x}\right|\mathrm{d}\bar{x} \tag{2.2}$$

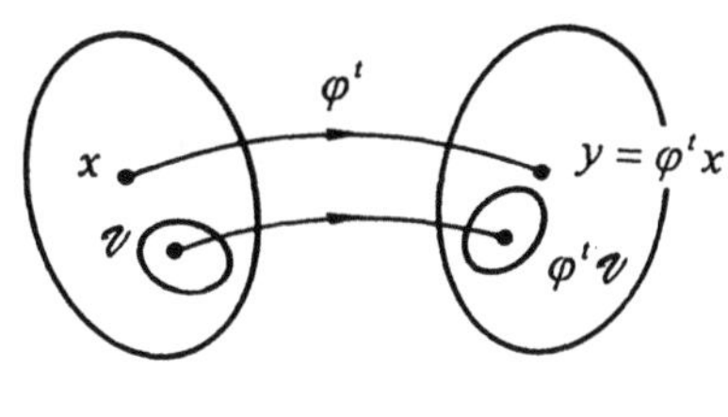


Fig. 2.1

where $\left|\dfrac{\partial y}{\partial x}\right|$ = the Jacobian determinant.

Assuming $\varphi^t x$ is a solution of IVP (initial value problem),

$$\mathcal{D}=\mathcal{R}\times\mathcal{R},\ f\in\mathrm{Lip}(\mathcal{D}) \tag{2.3}$$

After differentiating eq. (2.3) for $x$, we obtain

$$\left(\frac{\partial y}{\partial x}\right)=\left(\frac{\partial f(y)}{\partial y}\right)\left(\frac{\partial y}{\partial x}\right) \tag{2.4}$$

where all three matrices in eq. (2.4) are $n\times n$ Jacobian matrixes. That is

$$\left(\frac{\partial y}{\partial x}\right)=\begin{bmatrix}\dfrac{\partial y_1}{\partial x_1} & \dfrac{\partial y_1}{\partial x_2} & \cdots & \dfrac{\partial y_1}{\partial x_n}\\ & \vdots & \vdots & \\ \dfrac{\partial y_n}{\partial x_1} & \dfrac{\partial y_n}{\partial x_2} & \cdots & \dfrac{\partial y_n}{\partial x_n}\end{bmatrix}=Y \tag{2.5}$$

Further, eq. (2.4) can be rewritten as

$$Y' = A(t)Y \tag{2.6}$$

which is a linear equation of $Y$, where the elements of $A(t)$ are known:

$$A(t) = \left[\frac{\partial f(\varphi' x)}{\partial y}\right] \tag{2.7}$$

If the initial condition of eq. (2.6) is $Y(0) = \left.\frac{\partial y}{\partial x}\right|_{t=0} = \mathbf{I}$, then $\mathrm{Y}(t) = \frac{\partial y}{\partial x}$ is a fundamental matrix of eq. (2.4) and the columns of $\mathrm{Y}(t)$ are $y^{(1)}(t),\ y^{(2)}(t), \ldots,\ y^{(n)}(t)$ respectively, where an arbitrary element of $\mathrm{Y}(t)$

$$y^{(i)}(t) = \left[\frac{\partial y_1}{\partial x_i} \quad \frac{\partial y_2}{\partial x_i} \quad \cdots \quad \frac{\partial y_n}{\partial x_i}\right]^T \tag{2.8}$$

satisfies eq. (2.6)

$$\begin{aligned} &y^{(i)\prime}(t) = A(t)y^{(i)}(t) \\ &y^{(i)}(0) = \{0,0,\ldots,1,0,\ldots,0)^T \end{aligned} \tag{2.9}$$

where the position of 1 is in the position of the $j$th element. So $\det \mathrm{Y}(t) = \left|\frac{\partial y}{\partial x}\right|$ satisfies Abel's formula

$$|\mathrm{Y}(t)| = |\mathrm{Y}(0)|\exp\{\int_0^t \mathrm{trace}\ A(s)\,\mathrm{d}s\} \tag{2.10}$$

where the expression

$$\mathrm{trace}\,A(t) = A_{11} + A_{22} + \ldots + A_{nn} = \frac{\partial f_1(\varphi' x)}{\partial y_1} + \frac{\partial f_2(\varphi' x)}{\partial y_2} + \ldots + \frac{\partial f_n(\varphi' x)}{\partial y_n} \tag{2.11}$$
$$= \mathrm{div}\ f(\varphi' x)$$

is called divergence. From $\left|\frac{\partial y}{\partial x}\right| = \exp\{\int_0^t\ \mathrm{div}\,f(\varphi^s)\,\mathrm{d}s\}$, we obtain

$$\varphi'\upsilon = \iiint_v \left|\frac{\partial y}{\partial x}\right| \mathrm{d}\bar{x} = \iiint_v \exp\{\int_0^t \mathrm{div}\,f(\varphi^s x)\mathrm{d}s\}\mathrm{d}\bar{x} \tag{2.12}$$

If $\mathrm{div}\ f > 0$, flows expand with the increase of volume for $t > 0$ in $\mathcal{D}$. If $\mathrm{div}\ f < 0$, flows contract with the decrease of volume for $t > 0$ in $\mathcal{D}$. If $\mathrm{div}\ f = 0$, the volume remains unchanged for $t > 0$ in $\mathcal{D}$, and flows with unchanged volume.

**Example** Consider "volume" = "area" in phase-plane, then

$$\varphi^t v = \iint_v \exp\{\int_0^t \left[\frac{\partial f_1}{\partial x_1} + \frac{\partial f_2}{\partial x_2}\right](\varphi^s x)\mathrm{d}s\}\mathrm{d}x_1\mathrm{d}x_2$$

**Theorem 2.1 Bendixson's Negative Criterion**

Suppose the definition domain $\mathcal{D} \subseteq \mathcal{R}^2$ is an open simple connected domain

$$y' = f(y), \qquad f \in \mathcal{C}^1(\mathcal{D}) \tag{2.13}$$

If $\operatorname{div} f(y) \neq 0$ for any $y \in \mathcal{D}$, then no periodic orbit exists entirely in $\mathcal{D}$. The expression $\operatorname{div} f(y)$ is not identical zero and does not change its sign.

**Proof** We adopt a disproved method. Suppose there exists a periodic orbit $\mathcal{C}$ in $\mathcal{D}$. Let $\mathcal{S}$ equal the union of the periodic orbit $\mathcal{C}$ and the interior of $\mathcal{C}$, then $\mathcal{S} \subseteq \mathcal{D}$. From $f \in \mathcal{C}^1$, $\operatorname{div} f$ is continuous and $\operatorname{div} f \neq 0$ results in two cases, $\operatorname{div} f > 0$ and $\operatorname{div} f < 0$ for any $y \in \mathcal{S}$.

First, consider $\operatorname{div} f > 0$. As the volume expands with $t > 0$, then the area of $\varphi^t \mathcal{S}$ is larger than that of $\mathcal{S}$. Since $\mathcal{S}$ is a positive invariant set and $\varphi^t y$ cannot cross $\mathcal{C}$ because of the uniqueness of the orbit, then the area of $\varphi^t \mathcal{S}$ should be contained in that of $\mathcal{S}$. The two cases are contradictory. Therefore, the equation has no periodic orbit on $\mathcal{D}$. Similarly, for $\operatorname{div} f < 0$, we arrive at the conclusion that no periodic orbit exists either, according to the negative criterion invariant for $t < 0$.

### 2.1.2 Divergence for Linear Flows

Consider the initial value problem

$$y' = Ay, \qquad y(0) = x \tag{2.14}$$

where $A$ is an $n \times n$ constant matrix whose volume is $\varphi^t v = \iiint \exp(\operatorname{trace} A) \mathrm{d}\bar{x}$. Since trace $A =$ constant, there are three cases:

If trace $A > 0$, then flows expand.

If trace $A < 0$, then flows contract.

If trace $A = 0$, then the volume of flows remains invariant.

## 2.2 Linear Autonomous Systems and Linear Flows and the Calculation of Flows about the IVP

In this section we discuss the relationship between linear flows and autonomous systems as well as their calculation.

### 2.2.1 Linear Flows and Algebra Characteristic of $e^A$

Assume flow $\varphi^t : \mathcal{S} \to \mathcal{S} \subseteq \mathcal{R}^n$. If it satisfies $\varphi^t(\alpha x + \beta y) = \alpha \varphi^t x + \beta \varphi^t y$, for any $x, y \in \mathcal{S}$ and $\alpha,\ \beta \in \mathcal{R}$, then it is called linear.

Consider the linear IVP

$$y' = Ay, \qquad y(0) = x \tag{2.15}$$

where $A: \mathcal{R}^n \to \mathcal{R}^n$ is a linear operator. Equivalent to $A$ is an $n \times n$ matrix with real elements. $Ay$ is called the constant matrix multiplier. Here, on the one hand, after a linear flow $\|A\| = \sup_{|x|=1} |Ax|$, we need to prove that the phase velocity $v(x)$ is linear, where $v(x) = \frac{\partial}{\partial t} \varphi^t x \big|_{t=0}$. On the other hand, we need to prove that the solution of the linear IVP is a linear flow. Assume $y = e^{tA}x$ is a solution of the linear differential equation $y' = Ay$. Define

$$\varphi^t = e^{tA} \tag{2.16}$$

as a flow for $t \in \mathcal{R}$. It is linear since

$$e^{tA}(\alpha x + \beta y) = \alpha e^{tA} x + \beta e^{tA} y \tag{2.17}$$

Since

$$e^{(t+s)A} = e^{tA} e^{sA} \tag{2.18}$$

it satisfies the properties of groups and initial condition $y(0) = e^{0A}x = x$. Substituting $y = e^{tA}x$ into eq. (2.15), we have

$$\frac{\mathrm{d}}{\mathrm{d}t}(e^{tA}x) = \frac{\mathrm{d}}{\mathrm{d}t}\left[ I + tA + \frac{1}{2!}(tA)^2 + \ldots + \frac{t^{n-1}}{(n-1)!} A^{n-1} + \ldots \right] x$$

$$= A\left[ I + tA + \frac{t^{n-2}}{(n-2)!} A^{n-2} + \ldots \right] x = A[e^{tA}]x$$

thus, $y' = Ay$ is verified.

We discuss the extended expression of $e^A$

$$e^A = I + A + \frac{1}{2!}A^2 + \frac{1}{3!}A^3 + \ldots + \frac{1}{n!}A^n + \ldots \tag{2.19}$$

or
$$e^A = \sum_{n=0}^{\infty} \frac{1}{n!} A^n \tag{2.20}$$

where $I$ is an $n \times n$ identical matrix, e.g. $I = \begin{bmatrix} 1 & 0 \\ 0 & 1 \end{bmatrix}$, and $A^0 = I$, $0! = 1$. In order to prove the convergence we introduce an operator norm of $A$ $\|A\| = \sup_{|x|=1} |Ax|$.

It is not difficult to prove that the scalar operational series $e^{\|A\|}$ converges for any real $\|A\|$. The conclusion can be proved by the comparison test principle, but we shall not give the process of proof. Please refer to [40] for details.

**Lemma** Algebric characteristics of $e^A$ are as follows.

Let $A$, $B$ and $P$ be n × n matrices and $P$ be invertible.

(1) If $PAP^{-1} = B$, then $e^B = e^{PAP^{-1}} = Pe^A P^{-1}$

(2) If $AB = BA$ then $e^{A+B} = e^A e^B$

(3) If $e^0 = I$, then $e^{-A} = (e^A)^{-1}$

where $0 = \begin{vmatrix} 0 & \dots & 0 \\ \dots & \dots & \dots \\ 0 & \dots & 0 \end{vmatrix}_{n \times n}$

**Proof**

$$(1)\ Pe^A P^{-1} = P\left(I + A + \frac{1}{2!}A^2 + \ldots + \frac{1}{n!}A^n + \ldots\right)P^{-1}$$

$$= PP^{-1} + PAP^{-1} + \frac{1}{2!}PA^2P^{-1} + \ldots + \frac{1}{n!}PA^nP^{-1} + \ldots$$

$$\because PA^nP^{-1} = PA(P^{-1}P)A(P^{-1}P)\ldots(P^{-1}P)AP^{-1} = B^n$$

$$\therefore Pe^AP^{-1} = I + B + \frac{1}{2!}B^2 + \ldots + \frac{1}{n!}B^n + \ldots = e^B$$

$$(2)\ e^{A+B} = I + (A+B) + \ldots + \frac{1}{n!}(A+B)^n + \ldots$$

By using binomial theorem and $AB = BA$, we have

$$e^{A+B} = I + (A+B) + \ldots + \frac{n!}{n!}\sum_{j+k=n}\frac{A^jB^k}{j!k!} + \ldots = \sum_{n=0}^{\infty}\left(\sum_{j+k=n}\frac{A^jB^k}{j!k!}\right)$$

$$= \left(\sum_{j=0}^{\infty}\frac{A^j}{j!}\right)\left(\sum_{k=0}^{\infty}\frac{B^k}{k!}\right) = e^Ae^B$$

(3) Let $B = -A$ in part (2), then $e^{A-B} = e^0 = I$ and $e^{A-A} = I$.

So $\quad e^{-A} = (e^A)^{-1}$.

**Lemma** can be generalized to $e^{tA}$ for $t \in \mathbb{R}$. In fact, because

$$e^{tA} = I + tA + \frac{1}{2!}(tA)^2 + \ldots + \frac{1}{n!}t^nA^n + \ldots$$

is uniformly convergent on bounded t-intervals, we can differentiate the equation term by term

$$\frac{d}{dt}(e^{tA}) = 0 + A + tA^2 + \ldots + \frac{t^{n-1}}{(n-1)!}A^n + \frac{t^n}{n!}A^{n+1} + \ldots$$

$$= A\left[I + tA + \ldots + \frac{t^{n-1}}{(n-1)!}A^{n-1} + \frac{t^n}{n!}A^n + \ldots\right] = Ae^{tA}$$

So $e^{tA}$ satisfies the ordinary differential equation $Y' = AY$, where $Y = Y(t)$ is a column matrix. If $t = 0$, then $e^{tA} = e^{0A} = e^0 = I$. So $e^{tA}$ is a fundamental matrix.

**Example** Consider IVP $Y' = AY, Y(0) = C \quad \mathbb{R}^n$. Suppose $Y(t) = e^{tA}C$ is a solution of the IVP. For linear differential equation $y' = Ay$, its solution can be expressed in the form of flow $\varphi^t x$. According to the uniqueness theorem we have $\varphi^t x = e^{tA}x$, so the flow $\varphi^t = e^{tA}: \mathcal{R}^n \to \mathcal{R}^n$ for $t \in \mathcal{R}$. This flow satisfies the properties of group.

### 2.2.2 The Calculation of $e^{tA}$

Firstly, we transform $A$ to a Jordan form in terms of a similar transformation. Substituting $Y = TX$ into $Y' = AY$, we obtain

$$Y' = TX' = ATX, \qquad X' = T^{-1}ATX = JX \tag{2.21}$$

where $J = T^{-1}AT$ is the Jordan form, and $T$ is not a singular transformation:

$$T = \begin{bmatrix} v_1^1 & v_1^2 & \cdots & v_1^n \\ v_2^1 & v_2^2 & \cdots & v_2^n \\ \vdots & \vdots & \vdots & \vdots \\ v_n^1 & v_n^2 & \cdots & v_n^n \end{bmatrix} \tag{2.22}$$

and eigenvectors $v^j$ are given by

$$Av^j = \lambda_j v^j \tag{2.23}$$

where $v^j$ is the abbreviation of a column vector defined by

$$v^j = \begin{bmatrix} v_1^j & v_2^j & \cdots & v_n^j \end{bmatrix}^T \tag{2.24}$$

$y$ represents a column vector for $y \in \mathcal{R}^n$

$$y = (y_1, y_2, \ldots, y_n)^T \tag{2.25}$$

So the transformation can be written in full as

$$Y = TX = x_1 v^1 + x_2 v^2 + \ldots + x_n v^n \tag{2.26}$$

where $x_1, x_2, \ldots, x_n$ are the coordinates of $y$ on the basis $v^1, v^2, \ldots, v^n$. Jordan canonical matrix consists of Jordan blocks

$$J = \begin{bmatrix} J_1 & \cdots & 0 \\ 0 & J_2 & \cdots \\ \vdots & \vdots & \vdots \\ 0 & \cdots & J_n \end{bmatrix} \tag{2.27}$$

where $J_1, \ldots, J_n$ associate with the different types of the eigenvalue of $A$. They may take the following forms:

$$J_{k_1} = \begin{bmatrix} \lambda_1 & \cdots & 0 \\ 0 & \lambda_2 & \cdots \\ \cdots & \cdots & \cdots \\ 0 & \cdots & \lambda_l \end{bmatrix} \tag{2.28}$$

with real eigenvalues without multiplicity,

$$J_{k_2} = \begin{bmatrix} \lambda_{l_2} & 1 & \cdots & 0 \\ 0 & \lambda_{l_2} & 1 & \cdots \\ \vdots & \vdots & \vdots & \vdots \\ 0 & 0 & \cdots & \lambda_{l_2} \end{bmatrix} \tag{2.29}$$

with real eigenvalues with multiplicity,

$$J_{k_3} = \begin{bmatrix} \alpha & -\beta \\ \beta & \alpha \end{bmatrix} \tag{2.30}$$

with complex eigenvalue without multiplicity; and

$$J_{k_4} = \begin{bmatrix} \alpha & -\beta & 1 & 0 \\ \beta & \alpha & 0 & 1 \\ 0 & 0 & \alpha & -\beta \\ 0 & 0 & \beta & \alpha \end{bmatrix} \tag{2.31}$$

with complex eigenvalue with multiplicity. For further details of Jordan blocks with multiplicity more than two, [40] is recommended. From the above discussion, we can transform flows into the Jordan form $P^{-1}e^{tA}P = e^{tP^{-1}AP} = e^{tJ}$, and the calculation of $e^{tJ}$ is much easier than that of $e^{tA}$, such as

$$\exp(t\begin{bmatrix} J_1 & \cdots & 0 \\ 0 & J_2 & \cdots \\ \vdots & \vdots & \vdots \\ 0 & \cdots & J_k \end{bmatrix}) = \begin{bmatrix} e^{tJ_1} & \cdots & 0 \\ 0 & e^{tJ_2} & \cdots \\ \vdots & \vdots & \vdots \\ 0 & \cdots & e^{tJ_k} \end{bmatrix} \tag{2.32}$$

This results in reducing a whole system to many small subsystems.

If $J_l = \begin{bmatrix} \lambda_1 & 0 \\ 0 & \lambda_2 \end{bmatrix}$, then

$$e^{tJ_l} = \begin{bmatrix} e^{\lambda_1 t} & 0 \\ 0 & e^{\lambda_2 t} \end{bmatrix} \tag{2.33}$$

If $$J_l = \begin{bmatrix} \lambda_l & 1 & \cdots & 0 \\ 0 & \lambda_l & \cdots & 0 \\ \vdots & \vdots & \vdots & \vdots \\ 0 & \cdots & \lambda_l & 1 \end{bmatrix} \tag{2.34}$$

where $$N = \begin{bmatrix} 0 & 1 & 0 & 0 & \cdots & 0 \\ 0 & 0 & 1 & 0 & \cdots & 0 \\ \vdots & \vdots & \vdots & \vdots & \vdots & \vdots \\ 0 & 0 & 0 & 0 & \cdots & 1 \\ 0 & 0 & 0 & 0 & \cdots & 0 \end{bmatrix} \tag{2.35}$$

then $e^{t(\lambda_i I+N)} = e^{t\lambda_i I}e^{tN}$ (2.36)

$$\text{where } e^{tN} = \begin{bmatrix} 1 & t & \frac{t^2}{2} & \cdots & \frac{t^{n-1}}{(n-1)!} \\ 0 & 1 & t & \cdots & \frac{t^{n-2}}{(n-2)!} \\ \vdots & \vdots & \vdots & \vdots & \vdots \\ 0 & \cdots & \cdots & \cdots & 1 \end{bmatrix} \tag{2.37}$$

If $J_i = \begin{bmatrix} \alpha & \beta \\ \beta & \alpha \end{bmatrix}$,

$$\text{then } e^{tJ} = e^{\alpha t}\begin{bmatrix} \cos\beta t & -\sin\beta t \\ \sin\beta t & \cos\beta t \end{bmatrix} \tag{2.38}$$

*Case 1* Suppose $A$ has $n$ linearly independent eigenvectors $v^{(1)},\ldots,v^{(n)}$ associated with $n$ real eigenvalues $\lambda_1,\ldots,\lambda_n$.

Having transformed, we obtain

$$J = T^{-1}AT = \begin{bmatrix} \lambda_1 & \cdots & 0 \\ \vdots & \vdots & \vdots \\ 0 & \cdots & \lambda_n \end{bmatrix} = \Lambda \tag{2.39}$$

where $AT$ is the same non-singular transformation as in (2.22). Thus

$$e^{t\Lambda} = \begin{bmatrix} e^{t\lambda_1} & \cdots & 0 \\ \vdots & \vdots & \vdots \\ 0 & \cdots & e^{t\lambda_n} \end{bmatrix} \tag{2.40}$$

The original linear differential equation can be decoupled and written as

$$x_1' = \lambda_1 x_1, \cdots, x_n' = \lambda_n x_n \tag{2.41}$$

with its solution

$$x_j(t) = c_j e^{\lambda_j t} \tag{2.42}$$

where the $n$ unknown constants $c_j$ are to be determined by the initial conditions.

Returning to variable $y$, we have

$$y = Tx = [v^1, v^2, \ldots, v^n]\begin{bmatrix} c_1 e^{\lambda_1 t} \\ c_2 e^{\lambda_2 t} \\ \vdots \\ c_n e^{\lambda_n t} \end{bmatrix} = [e^{\lambda_1 t}v^1, e^{\lambda_2 t}v^2, \ldots, e^{\lambda_n t}v^n]\begin{bmatrix} c_1 \\ c_2 \\ \vdots \\ c_n \end{bmatrix} \tag{2.43}$$

**Definition 2.1** Matrix $V(t) = [e^{\lambda_1 t}v^1, e^{\lambda_2 t}v^2, \ldots, e^{\lambda_n t}v^n]$ is a non-singular matrix solution, where $\lambda_e$ are simple real or complex eigenvalues. From the uniqueness theorem of the IVP, then

$$e^{tA} = v(t)[v(0)]^{-1} \tag{2.44}$$

can be obtained.

*Case 2* $A$ has the simply complex eigenvalues.

Since $A$ is a real matrix, $\det(A - \lambda I) = 0$ must be a real characteristic polynomial and hence complex roots must be in conjugate pairs. Take

$$\lambda_j = \alpha_j + i\beta_j, \quad \bar{\lambda}_j = \alpha_j - i\beta_j \tag{2.45}$$

having eigenvectors

$$\begin{aligned} v^j &= v_R^j + iv_I^j \\ \bar{v}^j &= v_R^j - iv_I^j \end{aligned} \tag{2.46}$$

where $v^j$ and $\bar{v}^j$ should satisfy

$$\begin{aligned} Av^j &= \lambda_j v^j \\ A\bar{v}^j &= \bar{\lambda}_j \bar{v}^j \end{aligned} \tag{2.47}$$

In this case, the diagonal matrix of $A$ is $\begin{bmatrix} \lambda_1 & \cdots & \cdots & 0 \\ 0 & \bar{\lambda}_1 & \cdots & 0 \\ \vdots & \vdots & \vdots & \vdots \\ 0 & \cdots & \lambda_m & 0 \\ 0 & \cdots & \cdots & \bar{\lambda}_m \end{bmatrix}$, where $\lambda_j$ and $\bar{\lambda}_j$ belong to diagonal elements.

For complex eigenvectors, if we take linearly independent vectors $\operatorname{Re} v^j$ and $\operatorname{Im} v^j$ instead of $v^j$ and $\bar{v}^j$, it will result in the construction of the real transformation $T$. The Jordan form with a pair of conjugate roots is

$$J = \begin{bmatrix} \alpha & -\beta \\ \beta & \alpha \end{bmatrix} = \alpha I + \beta S \tag{2.48}$$

where $I = \begin{bmatrix} 1 & 0 \\ 0 & 1 \end{bmatrix}, \qquad S = \begin{bmatrix} 0 & -1 \\ 1 & 0 \end{bmatrix}$

Since $\alpha I$ can exchange with any variables, then

$$e^{tJ} = e^{t\alpha I} e^{t\beta S} = \left( I + t\alpha I + \frac{t^2\alpha^2}{2!} I^2 + \ldots \right)\left( I + t\beta S + \frac{t^2\beta^2}{2!} S^2 + \ldots \right)$$

Noticing $I^k = I,\ S^2 = -I,\ S^3 = -S,\ S^4 = I,\ S^5 = S$, we get

$$e^{tJ} = [e^{t\alpha} I]\begin{bmatrix} \cos\beta t & -\sin\beta t \\ \sin\beta t & \cos\beta t \end{bmatrix} = e^{t\alpha}\begin{bmatrix} \cos\beta t & -\sin\beta t \\ \sin\beta t & \cos\beta t \end{bmatrix} \tag{2.49}$$

*Case 3* $A$ has eigenvalues with multiplicity.

(1) $A$ is still diagonal, that is, there exist $n$ linearly independent eigenvectors (which may be complex, too). If $T$ is not a singular matrix, we call this case semi-simple since it is not difficult to solve this type of problem.

(2) $A$ has non-singular real eigenvectors when there exist less than $n$ linearly independent eigenvectors, that is, we can prove that there exist generalized eigenvectors and they can satisfy

$$[A-\lambda_j I]^k v = 0 \tag{2.50}$$

where $1 \le k \le \lambda^j$ is the algebric multiplicity of $\lambda$. When $k=1$, it corresponds to ordinary eigenvectors. It is easy to prove that all of the generalized eigenvectors constitute a basis on $\mathcal{R}^n$.

The Jordan form with multiple real eigenvalues can be written as

$$T^{-1}AT = \begin{bmatrix} J_1 & \dots & 0 \\ \vdots & \vdots & \vdots \\ 0 & \dots & J_k \end{bmatrix} \tag{2.51}$$

here
$$J_l = \begin{bmatrix} \lambda_l & 1 & \dots & 0 \\ \vdots & \vdots & \vdots & \vdots \\ 0 & \dots & .\lambda_l & 1 \\ 0 & \dots & \dots & \lambda_l \end{bmatrix} \tag{2.52}$$

while $\lambda_j$ are the multiple roots whose number is less than the algebric multiplicity of $\lambda_j$. Let $\Sigma$ represent the rank of $J_j$ or the algebric multiplicity of $\lambda_j$. Under conditions of the multiple roots for simplicity of calculation, erase $j$ from $J_j$ to become a simple calculation. Thus $J$ can be written as

$$J = \lambda I + N \tag{2.53}$$

and
$$N = \begin{bmatrix} 0 & 1 & \dots & 0 \\ \vdots & \vdots & \vdots & \vdots \\ 0 & \dots & \dots & 1 \\ 0 & \dots & \dots & 0 \end{bmatrix}_{k\times k} \tag{2.54}$$

$$N^2 = \begin{bmatrix} 0 & 0 & 1 & \dots & 0 \\ \vdots & \vdots & \vdots & \vdots & \vdots \\ 0 & \dots & \dots & \dots & 1 \\ 0 & \dots & \dots & \dots & 0 \\ 0 & \dots & \dots & \dots & 0 \end{bmatrix} \tag{2.55}$$

$$N^{k-1} = \begin{bmatrix} 0 & 0 & \dots & 1 \\ 0 & 0 & \dots & 0 \\ \vdots & \vdots & \vdots & \vdots \\ 0 & 0 & \dots & 0 \end{bmatrix} \tag{2.56}$$

where $N$ is called the nilpotent matrix. Then

$$e^{tJ}=e^{t\lambda I+tN}=e^{\lambda t}I\begin{bmatrix}1 & t & \frac{t^2}{2!} & \cdots & \frac{t^{k-1}}{(k-1)!}\\ \vdots & \vdots & \vdots & \vdots & \vdots\\ 0 & \cdots & \cdots & 1 & t\\ 0 & \cdots & \cdots & 0 & 1\end{bmatrix} \tag{2.57}$$

**Definition 2.2** The definition of the multiplicity of the eigenvalue $\lambda$ of matrix $A$ is that the algebraic multiplicity of $A$ is the multiplicity of the root $\lambda$ of a characteristic polynomial $\det(\lambda I-A)=0$. Geometric multiplicity equals the dimensions of the eigenspace corresponding to $\lambda$.

From linear algebra we know that 1 $\leq$ geometric multiplicity $\leq$ algebraic multiplicity. If algebraic multiplicity is 1, then $\lambda$ is called the simple eigenvalue, or multiple eigenvalue. If geometric multiplicity equals algebraic multiplicity, it is said to be the semi-simple, otherwise, non-semi-simple. If all eigenvalues of $A$ are semi-simple, then $A$ is diagonalizable, otherwise, $A$ can be transformed into the Jordan form. We shall now further discuss multiple complex eigenvalues.

(1) No new problems arise from semi-simple cases.

(2) Jordan blocks with real number forms can be expressed as

$$J_j=\begin{bmatrix}C_2 & I_2 & \cdots & 0\\ 0 & C_2 & \cdots & 0\\ \vdots & \vdots & \vdots & \vdots\\ 0 & \cdots & C_2 & I_2\\ 0 & \cdots & \cdots & C_2\end{bmatrix} \tag{2.58}$$

with non-semi-simple, where $I_2=\begin{bmatrix}1 & 0\\ 0 & 1\end{bmatrix}$, $I_2=\begin{bmatrix}\alpha & -\beta\\ \beta & \alpha\end{bmatrix}$, and the eigenvalue is $\alpha+i\beta$. The multiplicity $\Sigma$ of $C_2$ equals the algebraic multiplicity of $\lambda$.

In conclusion, according to the different eigenvalues of $A$, the solution of linearly IVP may have the following functional forms: $e^{\lambda t}$ with real numbers, $\sin\beta t$, $\cos\beta t$ and $t,t^2,\ldots,t^k$. In general, the general solution is the linear combination of the aforementioned functions.

**Lemma** If $\lambda$ is an eigenvalue of $A$, then $e^{\lambda}$ is an eigenvalue of $e^{A}$. If $v$ is an eigenvector of $A$ corresponding to $\lambda$ , then $v$ is an eigenvector of $e^{A}$ corresponding to $e^{\lambda}$.

If span $\{\mathrm{Re}(v^j),\mathrm{Im}(v^j)\}$ is a subspace corresponding to a pair of conjugate complex eigenvalues $\lambda_j$ and $\bar{\lambda}_j$ , then it belongs to $e^{\lambda_j}$ and $e^{\bar{\lambda}_j}$.

### 2.2.3 Invariant Subspaces of Linear Flows

Assume that linear flow $\varphi^t:\mathcal{R}^n\rightarrow\mathcal{R}^n$ and is a subspace on $\mathcal{R}^n$.

**Theorem 2.2** If any subspace $V$ is invariant for operator $A$, then it is also invariant for flow $\varphi^t = e^{tA}$.

**Proof** Expand the flow into a power series $e^{tA}X = \left(1 + tA + \frac{t^2}{2!} + \ldots + \frac{t^n}{n!}A^n + \ldots\right)X$.

As every term on the right-hand side is in $V$, obviously, $e^{tA}$ must belong to $V$ for finite $t$.

**Corollary** Any eigenspace (or generalised eigenspace) of $A$ is invariant for flow $e^{tA}$.

**Definition 2.3** Following is the definition of stable, unstable and centre subspace:

Let $\lambda_1, \ldots, \lambda_{n_s}$ be the eigenvalues of $A$ with negative real part.

Let $\mu_1, \ldots, \mu_{n_u}$ be the eigenvalues of $A$ with positive real part.

Let $\gamma_1, \ldots, \gamma_{n_c}$ be the eigenvalues of $A$ with zero real part.

Where multiple roots are included, and $n_s + n_u + n_c = n$. Let $V_1, \ldots, V_{n_s}$ be the eigenvectors corresponding to $\lambda_j$ (or generalised eigenvectors). Let $U_1, \ldots, U_{n_u}$ be the eigenvectors corresponding to $\mu_j$ (or generalised eigenvectors). Let $W_1, \ldots, W_{n_c}$ be the eigenvectors corresponding to $\gamma_j$ (or generalised eigenvectors).

We call $\mathcal{E}^s = \text{span}\{V_1, \ldots, V_{n_s}\}$ stable subspace, $\mathcal{E}^u = \text{span}\{U_1, \ldots, U_{n_u}\}$ unstable subspace, and $\mathcal{E}^c = \text{span}\{W_1, \ldots, W_{n_c}\}$ centre subspace. From linear algebra we have $\mathcal{R}^n$ as the direct sum of the above-mentioned subspaces, i.e.

$$\mathcal{R}^n = \mathcal{E}^s \oplus \mathcal{E}^u \oplus \mathcal{E}^c$$

**Theorem 2.3** Subspaces $\mathcal{E}^s, \mathcal{E}^u$ and $\mathcal{E}^c$ are invariant for flow $e^{tA}$. For $x_0 \in \mathcal{E}^s$, $\|e^{tA}x_0\| \to 0$ as $t \to \infty$, or $x_0 \in \mathcal{E}^u, \|e^{tA}x_0\| \to \infty$ (or $0$ ) as $t \to \infty$ (or $-\infty$ ) and for $x_0 \in \mathcal{E}^c$, $e^{tA}$ remains invariant for any $t$. In addition, the flow decays (or grows ) in an exponential way with $t$ in $\mathcal{E}^s$ (or $\mathcal{E}^u$).

**Proof** Firstly we transform $A$ into Jordan canonical form $A = TJT^{-1}$. Hence $e^A = Te^{tJ}T^{-1}$. With semi-simple case, conclusions are obvious for flow $e^{tJ}$. With the non-semi-simple case, $e^{tJ}$ contains elements in the form of $P(t)e^{\alpha t}$ (possibly times $\cos\beta t$, $\sin\beta t$ when there are complex roots). Here $P(t)$ is a polynomial with a degree less than algebraic multiplicity of $\alpha$ or $\alpha + i\beta$.

Suppose $P(t)e^{\alpha t} = [P(t)e^{\alpha_1 t}]e^{\alpha_2 t}$, where $\alpha < 0, \alpha = \alpha_1 + \alpha_2$, so $\alpha_1, \alpha_2 < 0$. When $0 \le t \le \infty$, $P(t)e^{\alpha_1 t}$ is bounded, that is, for $0 \le t \le \infty$, $|P(t)e_1^{\alpha}{}^t| \le K$. Thus $|P(t)e^{\alpha t}| \le Ke^{\alpha_2 t} \to 0$ as $t \to \infty$.

Similarly, if $\alpha < 0$, then $|P(t)e^{\alpha t}| \to 0$ for the flow in $E^u$ when $t \to -\infty$. With the semi-simple case the solution is bounded for the flow in $E^c$ and the solution has a polynomial growth with the non-semi-simple case.

## 2.3 Hyperbolic Operator (or Generality)

### 2.3.1 Generality

Let $\mathcal{V}$ be an $n$-dimensional vector space and $\mathcal{S}$ has the following two properties for subset $\mathcal{S} \subseteq \mathcal{V}$:

(1) $\mathcal{S}$ is open in $\mathcal{V}$, that is, for any $\varepsilon > 0$ and $s \in \mathcal{S}$, $\{x \in \mathcal{V} \| x - s| < \varepsilon\} \subseteq \mathcal{S}$;

(2) $\mathcal{S}$ is dense in $\mathcal{V}$, that is, let $x \in \mathcal{V}$, for any $\varepsilon > 0$, there exists $s \in \mathcal{S}$ so that $|x - s| < \varepsilon$ or any $x \in \mathcal{V}$ can be approached arbitrarily to $s \in \mathcal{S}$.

**Definition 2.4** $\mathcal{S}$ is called generic in $\mathcal{V}$ if and only if there exists $\mathcal{S} \subseteq \mathcal{S}$ and $\mathcal{S}$ is open and dense in $\mathcal{V}$.

**Examples**

(1) If $\mathcal{V} = \mathcal{R}$, then $\mathcal{S} = \{x \in \mathcal{R} | x \neq 0\}$ is open and dense in $\mathcal{V}$. Therefore, $\mathcal{S}$ is generic in $\mathcal{V}$.

(2) If $\mathcal{V} = \mathcal{R}^n$, then $\mathcal{S} = \{(x, y) | x^2 + y^2 \neq k^2, k = 1,2,...\}$ is open and dense in $\mathcal{V}$. Hence $\mathcal{S}$ is generic in $\mathcal{V}$.

**Proposition** Let $X,...,X_m$ be dense and open sets in $F$, then $X = X_1 \cap X_2 ... \cap X_m$ is also dense and open. Let $\mathcal{V} = L(\mathcal{R}^n)$ be composed of all linear operators on $\mathcal{R}^n$. Take $\mathcal{S} \subseteq \mathcal{V}$, $\mathcal{S} = \{x \in \mathcal{V} | x$ satisfies a property **P** given$\}$. If $\mathcal{S}$ contains a dense open subset on $\mathcal{V}$, then $\mathcal{S}$ is generic in $\mathcal{V}$. Further, property **P** is also generic in $\mathcal{V}$.

**Definition 2.5** If all eigenvalues of operator $A$ have a non-zero real part, then $A$ is hyperbolic and the corresponding eigenspace is also generic, otherwise, non-hyperbolic. In such a case the origin is called the hyperbolic equilibrium point of linear system (2.14) and $e^{At}$ is said to be the linear hyperbolic flow. Obviously, the hyperbolic equilibrium point is the unique equilibrium point of system (2.14). The linear hyperbolic flows have only stable and unstable subspaces, but no centre subspace. The hyperbolic equilibrium points can be divided into three types (Fig. 2.2):

(1) If all eigenvalues of $A$ have negative real parts, then the origin is a sink and $e^{At}$ is called a contracting flow (Fig. 2.2(a)).

(2) If all eigenvalues of $A$ have positive real parts, then the origin is a source and $e^{At}$ is called an expanding flow (Fig. 2.2(b))

(3) If all eigenvalues of $A$ have both a positive real part and a negative part, then the origin is a saddle point (Fig. 2.2(c))

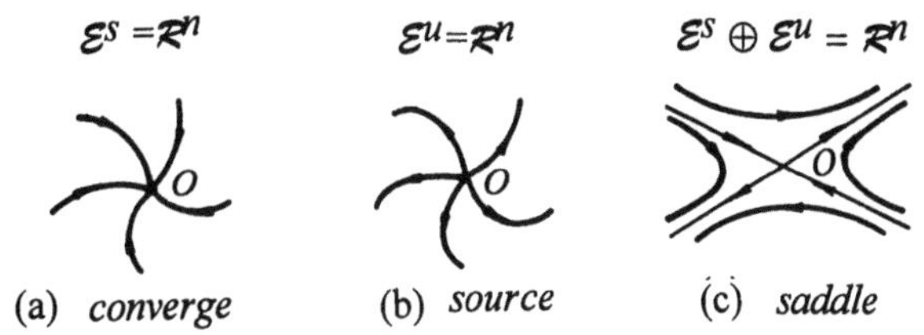


Fig. 2.2

**Theorem 2.4** If $\mathcal{V} = L(\mathcal{R}^n)$ is equivalent to $A: \mathcal{R}^n \to \mathcal{R}^n$, then the following properties are generic:

(1) The set $\mathcal{S}$ of operators on $\mathcal{R}^n$ that have $n$ distinct eigenvalues is dense and open in $L(\mathcal{R}^n)$. On the contrary, if $\mathcal{S}_1$ is a dense and open subset, then $A$ has $n$ distinct eigenvalues and it is called semi-simple operator.

(2) Semi-simplicity is a generic property in $L(\mathcal{R}^n)$.

(3) The set $\mathcal{S}_2 = \{T \in L(\mathcal{R}^n) \mid e^{Tt}$ is a hyperbolic flow$\}$ is open and dense in $L(\mathcal{R}^n)$.

For proof see [40].

### The Significance of Generity

If an operator $A \in L(\mathcal{R}^n)$ is semi-simple, the differential equation $X' = AX$ breaks down into a number of simple uncoupled equations of one or two dimensions. If the eigenvalues of $A$ have non-zero real parts, the differential equation might be complicated from the analytic point of view, but the geometric structure of the hyperbolic flow $e^{At}$ is very simple: it is the direct sum of a contraction and an expansion flow. Operators that generate hyperbolic flows form a dense open set in $L(\mathcal{R}^n)$ while the set of semi-simple operators contains a dense open set. Thus if $A$ generates a hyperbolic flow, so does any operator sufficiently close to $A$.

The significance of generity for linear differential equations is: if there are uncertain elements in matrix $A$ and no reasons to assume the contrary, we might as well assume that the flow $e^{At}$ is hyperbolic. For example, $A$ might be obtained from physical observations and there is a limit to the accuracy of the measuring instruments. If the eigenvalues of $A$ are semi-simple, the accuracy of measurement need not be too exact to reach them, and as the property of motion is unchanged we can save a lot of time.

**Definition 2.6** The property of the subset $\mathcal{S}$ is degenerated if the complementary set of $\mathcal{S}$ is generic in $\mathcal{V}$.

### 2.3.2 Classification of Generic Linear Flows

The behaviours of singular points are determined by the phase locus in its arbitrary small neighbourhood. We are going to discuss the behaviours of the singular points of the generic linear flows of two and three dimensions.

1. Classification of Generic Linear Flows on $\mathcal{R}^n$.

It is known that the linear part of the differential equation in the neighbourhood of the singular points $y_1 = y_2 = 0$ is

$$Y' = AY, \qquad Y \quad \mathcal{R}^2 \tag{2.59}$$

where $A = \begin{bmatrix} a & b \\ c & d \end{bmatrix}$, so the characteristic polynomial is

$$A - \lambda I \mid = \lambda^2 - (\mathrm{tr}A)\lambda + \det A = 0 \tag{2.60}$$

where $\mathrm{tr}\,A = a+d$, $\det A = ad-bc$

The eigenvalues of $A$ are

$$\lambda_{1,2} = \frac{1}{2}[\mathrm{tr}\,A + \sqrt{(\mathrm{tr}\,A)^2 - 4(\det A)}] \tag{2.61}$$

Let $\lambda_j$ be semi-simple. Recall that

$$\mathrm{tr}\,A = \sum_{j=1}^{2} \lambda_j \tag{2.62}$$

$$\det A = \prod_{j=1}^{2} \lambda_j \tag{2.63}$$

$$\mathrm{disc}\,A = (\mathrm{tr}\,A)^2 - 4(\det A) \tag{2.64}$$

We shall study the distribution of the phase portrait near the singular point and determine the various types of singular point. Take (tr $A$) and (det $A$) as two parameters and produce a parameter plane as shown in Fig. 2.3. The degenerate sets on the parameter plane are $\det A = 0$, $\mathrm{tr}\,A = 0$ and $\mathrm{disc}\,A = 0$. These sets divide the parameter plane into five regions in which flows are hyperbolic as shown in table 2.1.

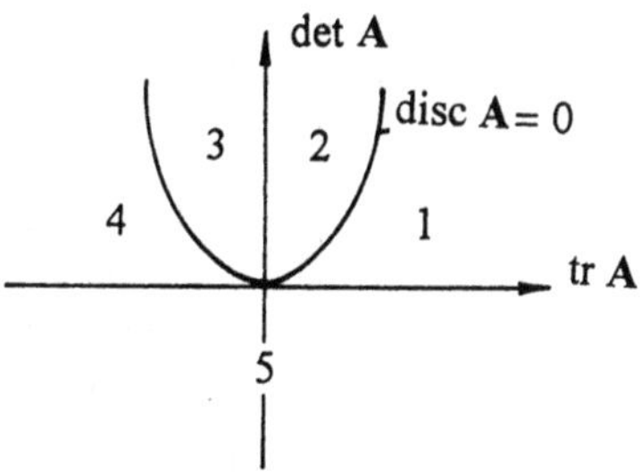


Fig. 2.3

2. Classification of Generic Linear Flows on $\mathcal{R}^3$

Consider a three-dimensional ordinary differential equation on $\mathcal{R}^3$:

$$x' = X(x,y,z), \quad y' = Y(x,y,z), \quad z' = Z(x,y,z) \tag{2.65}$$

where $X, Y, Z$ are the smooth functions of $(x,y,z)$.

Let the right-hand side of eq. (2.65) equal zero and obtain the equilibrium point $M_0(x_0, y_0, z_0)$. The phase portrait is given by the linearized equation of the characteristic equation near equilibrium point $M_0$:

$$\chi(\lambda) = \begin{bmatrix} \frac{\partial X}{\partial x} - \lambda & \frac{\partial X}{\partial y} & \frac{\partial X}{\partial z} \\ \frac{\partial Y}{\partial x} & \frac{\partial Y}{\partial y} - \lambda & \frac{\partial Y}{\partial z} \\ \frac{\partial Z}{\partial x} & \frac{\partial Z}{\partial y} & \frac{\partial Z}{\partial z} - \lambda \end{bmatrix}_{M_0} = 0 \tag{2.66}$$

**Table 2.1**

| Region No. | Parameter | Characteristic value | Phase diag. | Stability |
|---|---|---|---|---|
| 1 | tr $A > 0$<br>det $A > 0$<br>disc $A > 0$ | two positive real character values | 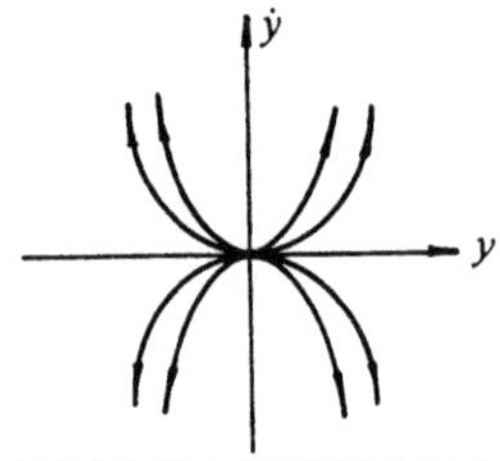 | unstable node |
| 2 | tr $A > 0$<br>det $A > 0$<br>disc $A < 0$ | two imaginary character values with positive real part | 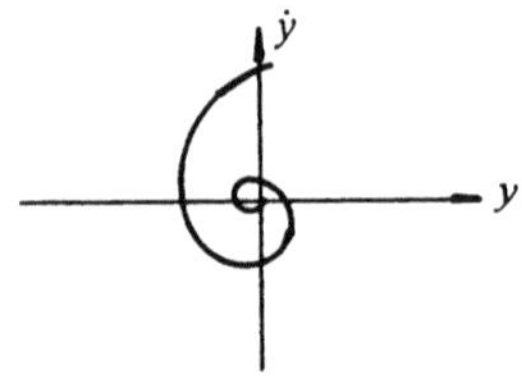 | unstable focus |
| 3 | tr $A < 0$<br>det $A > 0$<br>disc $A < 0$ | two imaginary character values with negative real part | 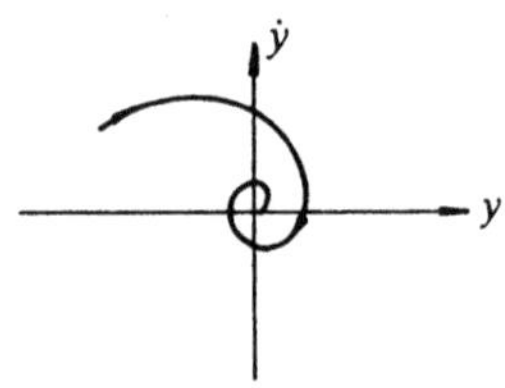 | stable focus |
| 4 | tr $A < 0$<br>det $A > 0$<br>disc $A > 0$ | two negative real character values | 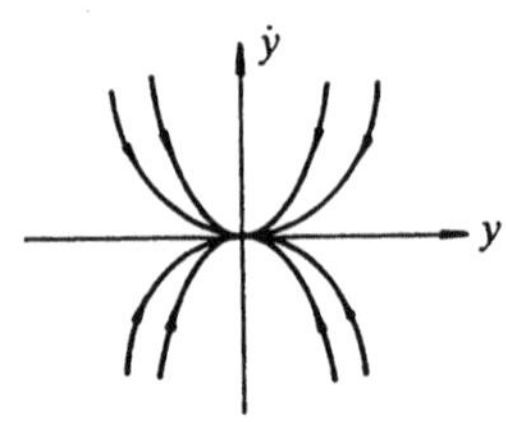 | stable node |
| 5 | tr $A \neq 0$<br>det $A < 0$<br>disc $A > 0$ | two real character values with opposite sign | 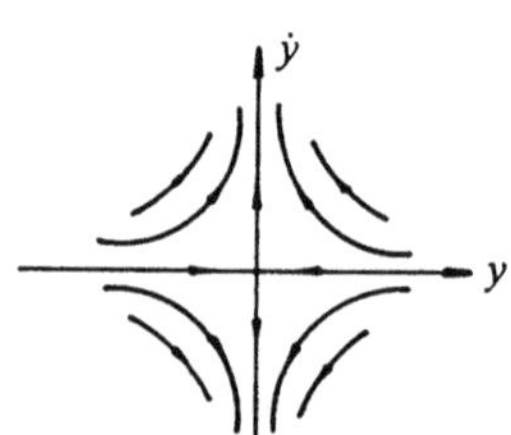 | saddle |

Now let us discuss the following important cases.

1. If equation $\chi(\lambda) = 0$ has negative (or positive) real roots, the equilibrium point is a stable (or unstable) node as shown in Fig. 2.4(a) and (or (b)).

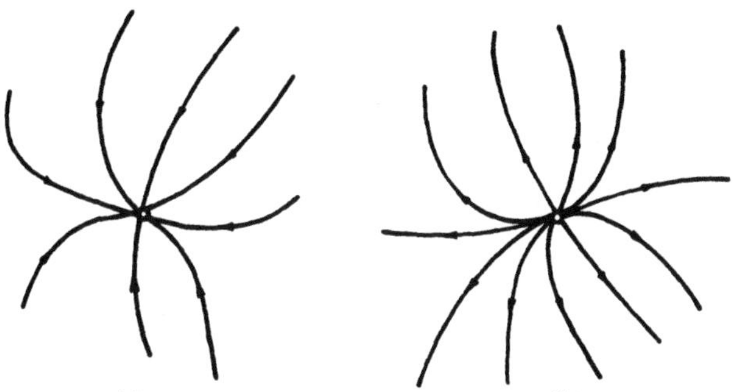

(a) (b)

Fig. 2.4

2. If equation $\chi(\lambda) = 0$ has a real root and two complex roots, and all roots have a negative (or a positive) real part, the equilibrium point is called a stable (or an unstable) focus, as shown in Fig. 2.5(a) and (or (b)).

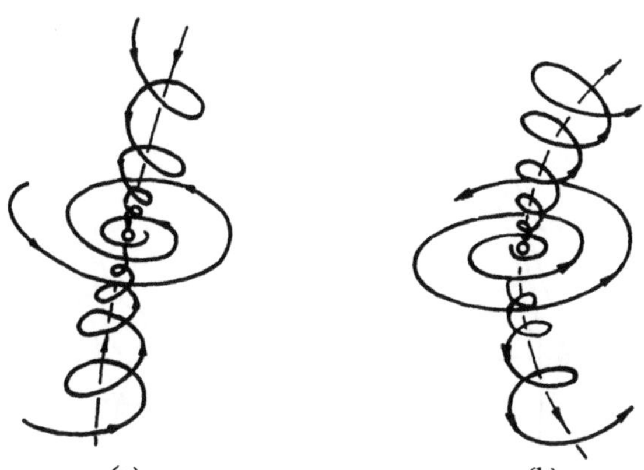

(a) (b)

Fig. 2.5

3. If equation $\chi(\lambda)=0$ has a real root and two complex roots, and the real root and the real part of the conjugate roots have distinct signs, the equilibrium point is a saddle-focus point as shown in Fig. 2.6(a) and (or (b)).

(a) (b)

Fig. 2.6

4. If all roots of equation $\chi(\lambda)=0$ are real but have different signs, the equilibrium point is called a saddle-node point as shown in Fig. 2.7(a) and (or (b)).

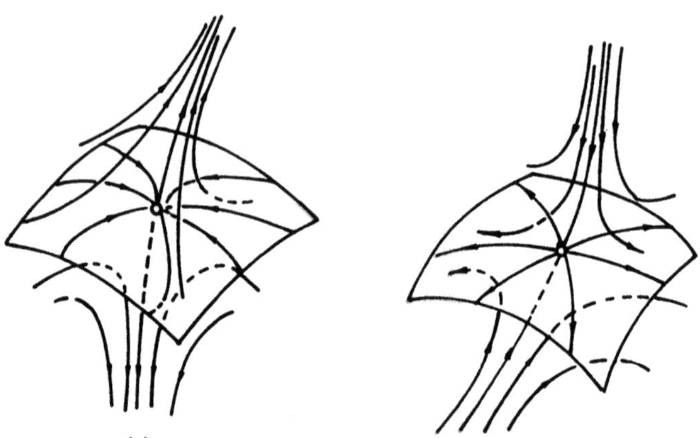

(a) (b)

Fig. 2.7

The concepts of Liapunov stability and orbit stability are different for periodic motion. If there exists a quantity $\delta(\varepsilon)$ for any $\varepsilon > 0$ and a motion $X(t)$ which can satisfy the initial condition $\rho(X_0(t_0), X(t_0)) < \delta$ so that $X(t)$ satisfies $\rho(X_0(t_0), X(t)) < \delta$ for $t > t_0$, then we call the periodic motion $X = X_0(t)$ Liyapunov stable as shown in Fig. 2.8.

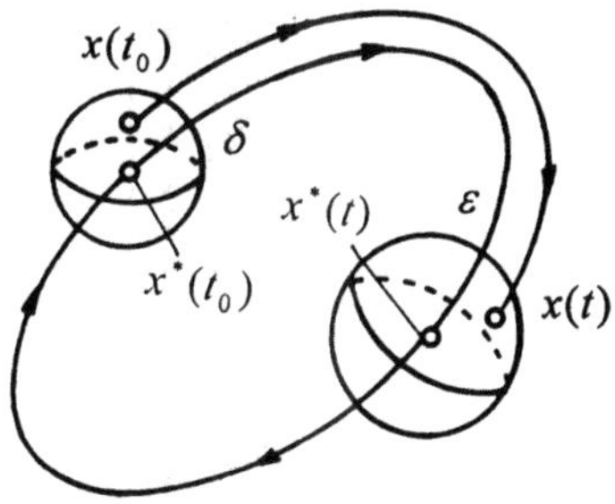


Fig. 2.8

The orbit stability is defined as follows. Let $\gamma$ correspond to the closed orbit of the periodic motion and $X = X(t)(-\infty < t < +\infty)$ be the arbitrary phase locus. If there exists a small quantity $\delta(\varepsilon) > 0$ for any $\varepsilon > 0$ and when $\rho(X(t),\gamma) < \delta$ is satisfied for $t > t_0$, then we call the periodic motion $X = X_0(t)$ orbit stable as shown in Fig. 2.9.

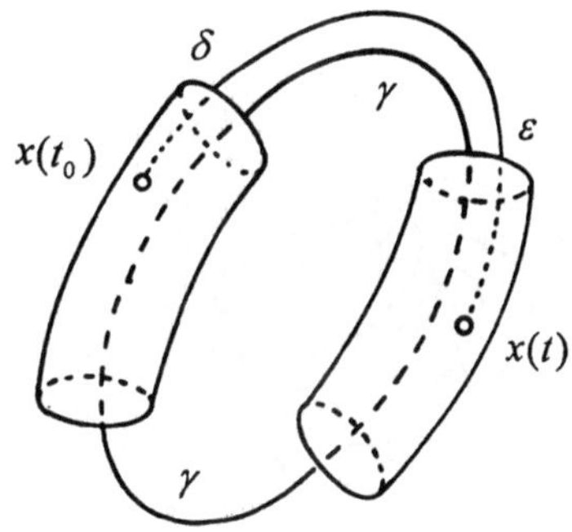


Fig. 2.9

The geometric explanation of the definition is that Liapunov stability requires that the distance between the phase points be less than $\delta$ at the initial time, thus this distance is less than $\varepsilon$, and orbit stability requires that the distance from the phase points to the closed orbit be less than $\delta$ at the initial time, thus this distance is less than $\varepsilon$ after a certain time. Hence from the Liapunov stability of the periodic motion we can get the orbit stability, but the converse is not possible. Taking a three-dimensional space as an example, we shall study the expanded structure of the trajectory on the phase orbit near periodic motion. Let $x = x_0(t)$, $y = y_0(t)$ and $z = z_0(t)$ be the periodic solution with period $\tau$ for three-dimensional systems. After linearizing, we obtain the linearized equation, similar to that of the equilibrium point, where $\left(\frac{\partial X}{\partial x}\right)_{x=x_0(t),\ y=y_0(t),\ z=z_0(t)} \cdots\cdots \left(\frac{\partial Z}{\partial z}\right)_{x=x_0(t),\ y=y_0(t),\ z=z_0(t)}$ are the periodic functions with period $\tau$. Let $\xi^i(t), \eta^i(t), \varsigma^i(t)$ $(i = 1,2,3)$ be the basic solution of the linear differential equations with periodic coefficients, that is, they satisfy the initial conditions $\xi^i(0) = \delta_{1i}$, $\eta^i(0) = \delta_{2i}$, $\varsigma^i(0) = \delta_{3i}$ , where $\delta^{ij} = 1$ when $i = j$; $\delta_{ij} = 0$ when $i \neq j$. For the periodic motion under study, the characteristic equation is

$$\chi(\lambda) = \begin{bmatrix} \xi^1(\tau) - \lambda & \eta^1(\tau) & \varsigma^1(\tau) \\ \xi^2(\tau) & \eta^2(\tau) - \lambda & \varsigma^2(\tau) \\ \xi^3(\tau) & \eta^3(\tau) & \varsigma^3(\tau) - \lambda \end{bmatrix} \tag{2.67}$$

One of its roots is always zero. According to the distribution of the other two roots the cases may appear:

1. Two real roots having different signs, corresponding to the periodic motion of the saddle type as shown in Fig. 2.10. From the closed orbit with saddle-type periodic motion we can obtain two surfaces $S_2^+$ and $S_2^-$ which consist of the phase curves tending to $\gamma$ as $t \to +\infty$ and $t \to -\infty$.

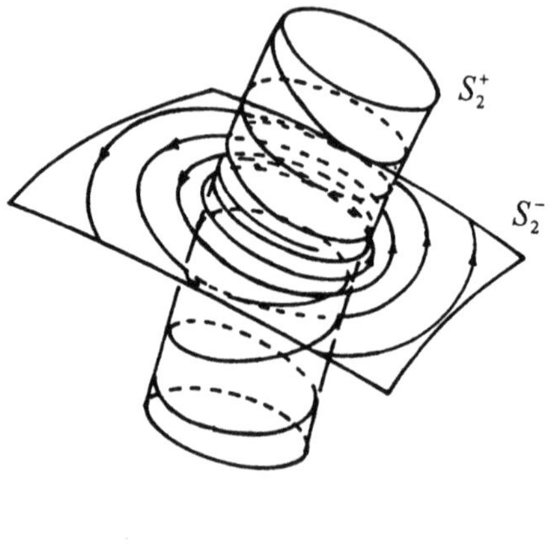


Fig. 2.10

2. Two negative (or positive) real roots corresponding to stable (unstable) node-type periodic motion as shown in Fig. 2.11.

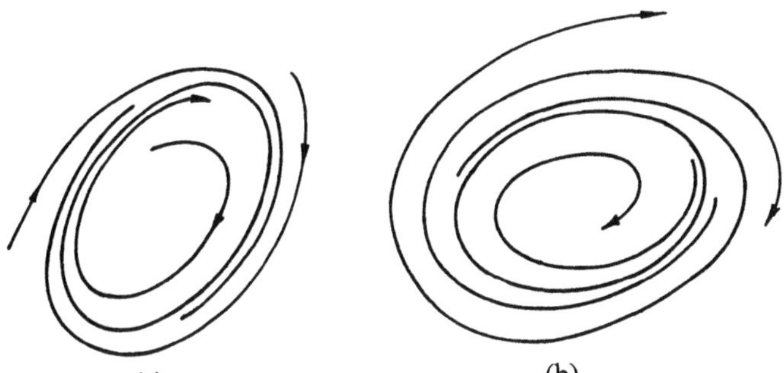

(a) (b)

Fig. 2.11

3. Two complex roots with a negative (positive) part, similar to the spiral line on the closed orbit as shown in Fig. 2.12.

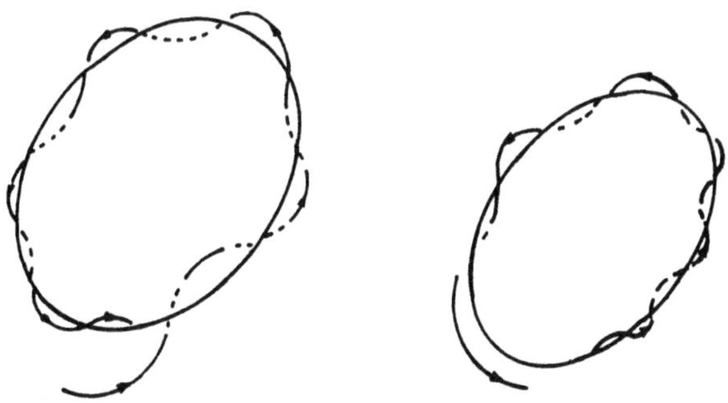

(a) (b)

Fig. 2.12

## 2.4 Non-linear Differential Equations and the Calculation of their Flows

In this section we shall study how to find the flows of non-linear differential equations. Since there is no general method of studying non-linear systems, the ordinary practice is to let the function of vector fields be near $t = 0$ and let the phase point $\bar{x}$ be non-equilibrium, so as to study local flows near $\bar{x}$ .

### 2.4.1 Flow Box Theorem

Consider a non-linear autonomous system

$$X' = f(X) \tag{2.68}$$

where $f(x)$ is a vector field with $n$-dimensions and $\bar{x}$ is not an equilibrium point.

Assume the flow of non-linear system $\varphi^t \in \mathcal{C}^1$. Through $\chi$ a local section (or transverse) $\mathcal{S}$ is made and time interval $(-\varepsilon, \varepsilon) \in \mathcal{R}$ taken. The mapping is defined by $\psi := (-\varepsilon, \varepsilon) \times \mathcal{S} \to \mathcal{R}^n$. Because $\bar{x}$ is not an equilibrium point, $\psi$ is a family of parallel straight lines.

**Theorem 2.5 Flow box theorem**

Take $\psi(t,x) = \varphi^t x$, that is, use a family of parallel lines to represent local flows of non-linear systems. The theorem shows that $\psi$ is a diffeomorphism, so long as positive $\varepsilon > 0$ is sufficiently small. This theorem also points out that in the neighbourhood of the non-equilibrium point (regular point) the trajectory of non-linear systems is diffeomorphical onto a family of parallel lines; thus near the regular point; flows have the same topological properties as the straight lines. The theorem is also known as the linearization theorem.

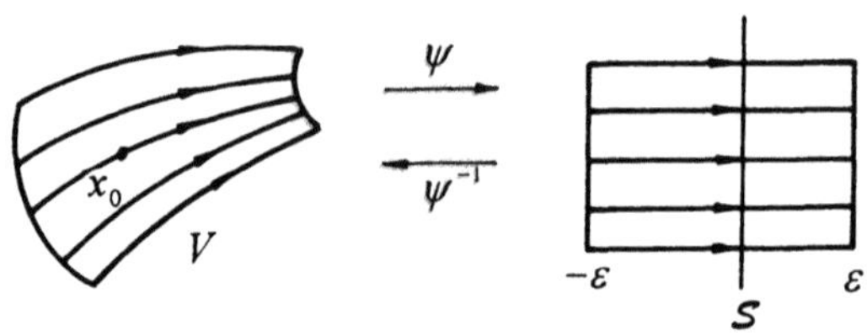


Fig. 2.13

**Definition 2.7** $V_\varepsilon = \psi((-\varepsilon,\varepsilon) \times \mathcal{S})$ is called a flow box at non-equilibrium $\bar{x}$. Properties of flow boxes are:

(1) the trajectory from any point in the box $(-\varepsilon,\varepsilon)$ will reach the section $\mathcal{S}$ in a certain time $t(|t| < \varepsilon)$;

(2) two stable (or unstable) trajectories from two different points in the flow box will never intersect. A stable (or an unstable) trajectory does not intersect by itself, but stable and unstable manifolds starting from different points (or the same point) may intersect.

We are interested in the topological structure of trajectories near equilibrium points.

### 2.4.2 Non-linear Differential Equations and Flows

Now we are going to study the calculation of flows of non-linear differential equations. Consider an $n$-dimensional autonomous system

$$y' = f(y) \qquad f \in \mathcal{C}^1 \tag{2.69}$$

If $x = x_0$ is an equilibrium point, then $f(x_0) = 0$. In this case, the flow box theorem is invalid at $x_0$.

To determine flows near equilibrium point $x_0$, let $x - x_0 = \xi$ be a small vector. Flows of non-linear system can be written as

$$\varphi(t,x) = \varphi^t x \tag{2.70}$$

The flows through equilibrium point $x_0$ satisfy $\varphi^t x_0 = x_0$ for any $t$. Since the flow is a solution, $\varphi'(t,x) = f(\varphi(t,x))$. When $t = 0$, substituting $x = x_0$ into the above equation, we obtain $(\varphi(0,x_0))' = 0 = f(x_0)$, that is, $\varphi(t,x)$ also satisfies a non-linear differential equation when $t = 0$. In order to find the expression of the flow approximately, after $t$ is fixed, expand the flow into a Taylor's series for $x = x_0 + \varsigma$ near $x_0$ in $\mathcal{R}^n$:

$$\begin{aligned}\varphi(t,x) &= \varphi(t,x_0) + D_x\varphi(t,x_0)(x - x_0) + o(\|x - x_0\|^2) \\ &= x_0 + D_x\varphi(t,x_0)\xi + o(\|\xi\|^2)\end{aligned} \tag{2.71}$$

where $D_x\varphi(t,x_0)$ is the Frechet derivative of $\varphi(t,x)$ for $x$ at $x = x_0 + \varsigma$, that is,

$$D_x\varphi(t,x) = \frac{\partial\varphi(t,x)}{\partial x}\Big|_{x=x_0} = \text{Jacobian matrix}$$

By computing, the approximate expression of $\varphi(t,x)$ can be obtained.

**Definition 2.8** The Frechet derivative of a mapping $g: \mathcal{R}^n \to \mathcal{R}^n$ at $x_0$ is a linear operator and satisfies

$$\|g(x_0 + h) - g(x_0) - Ah\| = o(\|h\|^2) \tag{2.72}$$

The problem is whether $D_x\varphi(t,x_0)$ is $A$. Since $\varphi'(t,x) = f'(\varphi(t,x))$, differentiating the two sides for $x$ and interchanging $D_x$ for d/dx , we get

$$\left[D_x\varphi(t,x_0)\right]' = D_\varphi f(\varphi(t,x))D_x\varphi(t,x)|_{x=x_0} = D_\varphi f(x_0)D_x\varphi(t,x_0) \tag{2.73}$$

where $D_\varphi f(x_0)$ is a constant. Hence $D_x\varphi(t,x_0)$ satisfies a linear ordinary differential equation (variational equation) or $D_x\varphi(t,x_0)$ is a solution of the linearized equation. The linearized equation of the autonomous system at equilibrium point $x = x_0$ is as follows

$$y' = Ay \tag{2.74}$$

where $A = D_y f(x_0)$ is the Jacobian matrix at $x_0$. As $\varphi(t,x_0)|_{t=0} = x_0$, we have $D_x\varphi(t,x_0)|_{t=0} = I$. Hence there exists the solution of the linearized equation

$$y = e^{tA} \tag{2.75}$$

Comparing eq. (2.73) with eq. (2.74), we know that $D_x\varphi(t,x_0) = e^{tA}$ by the uniqueness of the solution. Thus, the Taylor's formula of flows for a non-linear differential equation can be expressed as

$$y = \varphi(t,x) = x_0 + e^{tA}\xi + o(\|\xi\|^2) \tag{2.76}$$

### 2.4.3 The Hartman—Grobman's(HG) Theorem

If $Df(x_0)$ has no zero or purely imaginary eigenvalues, that is, $x_0$ is a hyperbolic point of eq. (2.69), then there is a homomorphism $h$ defined on some neighbourhood $u$ of $x_0$ in $\mathcal{R}^n$ locally taking orbits of non-linear flow of eq. (2.69), for those of linear flow. The homomorphism preserves the sense of orbits and can be chosen to preserve parameterization with time as shown in Fig. 2.14.

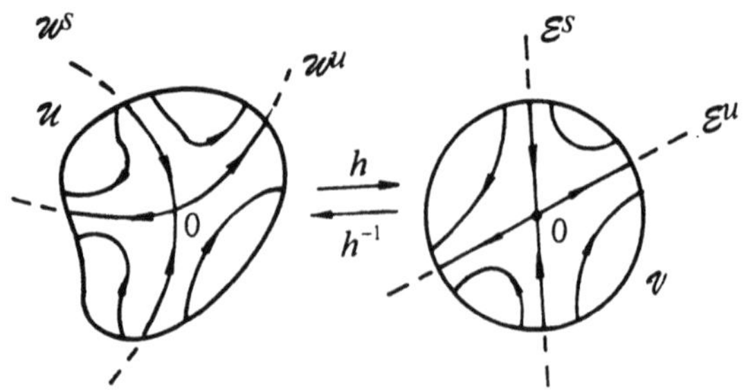


Fig. 2.14

**Topological Equivalence** The topological equivalence between two systems preserves the same orbital construction. Therefore, two systems of topological equivalence can be considered identical for dynamical systems. Consider two maps: $g_1:\mathcal{U}_1 \to \mathcal{U}_1$ and $g_2:\mathcal{U}_2 \to \mathcal{U}_2$, where $\mathcal{U}_1$ and $\mathcal{U}_2$ are open sets and $g_1$ and $g_2$ shows homomorphism. If there exists a homomorphism mapping $h:\mathcal{U}_1 \to \mathcal{U}_2$ satisfying

$$g_1 = h^{-1} g_2 h \tag{2.77}$$

then we say that $g_1$ and $g_2$ are topological equivalent, written as $g_1 \sim g_2$. The diagrams shown in Fig. 2.15 and Fig. 2.16 are exchangeable.

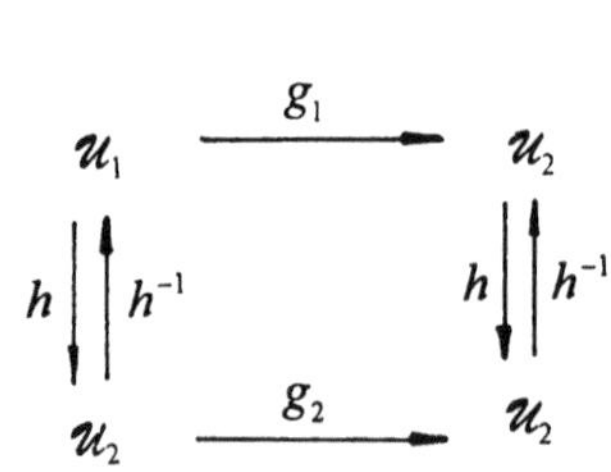


Fig. 2.15

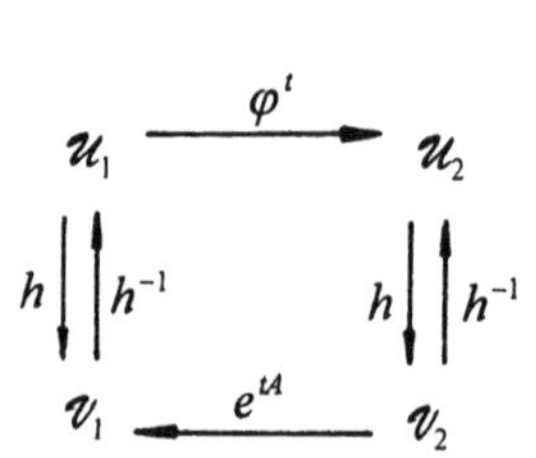


Fig. 2.16

The homomorphism $h$ transforms the trajectory of system $g_1$ across $x$ into that of system $g_2$ across $h(x)$, and changes the ω—limit set (or the α—limit set) of orbit $g_1(x)$ into that of orbit $g_2(h(x))$ and also changes the $n$-period point of $g_1$ into that of $g_2$ and so on.

According to the HG theorem, we have the exchangeable diagram (Fig. 2.16) in which $\mathcal{U}_i \subseteq \mathcal{U}$ is the neighbourhood of $x_0$ and $\mathcal{U}_i \subseteq \mathcal{U}$ is the neighbourhood of $0 \in \mathcal{R}^n$, where $i = 1,2$, $A = Df(x_0)$ and $h: \mathcal{U} \to \mathcal{V}$.

**Note** If points are non-hyperbolic, then the theorem is invalid. The HG theorem can be applied to the saddle point because it is hyperbolic. Hyperbolic assumption is necessary for two topological maps to be equivalent. For example, consider the non-linear system, $x'' + \varepsilon x^2 x' + x = 0$. Let $x = x_1$, and $x_1' = x_2$, then the equation can be shown as

$$\begin{bmatrix} x_1' \\ x_2' \end{bmatrix} = \begin{bmatrix} 0 & 1 \\ -1 & 0 \end{bmatrix} \begin{bmatrix} x_1 \\ x_2 \end{bmatrix} - \begin{bmatrix} 0 \\ \varepsilon x_1^2 x_2 \end{bmatrix}$$

where (0,0) is an equilibrium point, the operator $A = D_x f(0,0) = \begin{bmatrix} 0 & 1 \\ -1 & 0 \end{bmatrix}$ and the eigenvalues of $A$ are $i$ and $-i$. Since the real parts of two eigenvalues of linearized systems are zero, the equilibrium point is non-hyperbolic. Hence the linearized system is not the topologically equivalent system of the non-linear system. The linearized system has only a centre point as shown in Fig. 2.17(a). The non-linear system has a stable or an unstable focus as shown in Fig. 2.17(b) and (c)

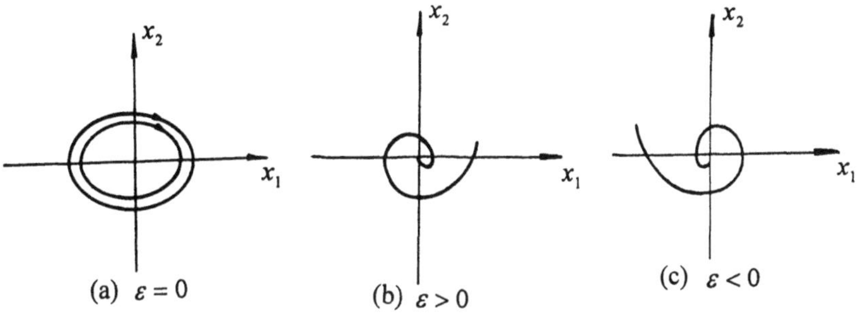


Fig. 2.17

We cannot find a homomorphism $h$ near (0,0) so that the foci of the non-linear system are mapped to the phase-plane of the linear system to be a centre. In other words, the non-linear and linearized system are not topologically equivalent.

**Note** If equilibrium points are hyperbolic, the systems are generic near equilibrium points. In such a case, flows of the non-linear system can be studied through its linearized system. Hence if they are generic, the non-linear system and its linearized system are locally topologically equivalent.

## 2.5 Stable Manifold Theorem

### 2.5.1 Stable Manifold Theorem

To verify the theorem we give the following definition first.

**Definition 2.9** A linear manifold $L$ in $\mathcal{R}^n$ consists of parallel moving trajectories of a subspace and each point on them moves simultaneously as shown in Fig. 2.18. Its expression is given by $L = x_0 + \text{span}\{x_1,...,x_k\}$, where $x_0, x_1,...,x_k \in \mathcal{R}^n$

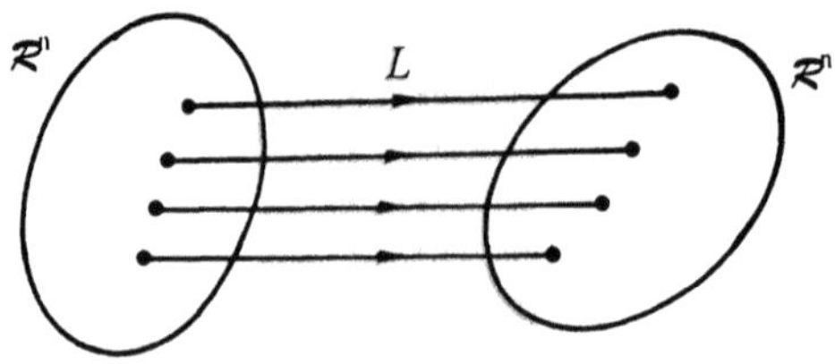


Fig. 2.18

**Definition 2.10** ($k$-dimensional manifold) A set $\mathcal{W} \in \mathcal{R}^n$ is called a $k$-dimensional manifold if every point $x_0 \in \mathcal{R}^k$ has a neighbourhood $\eta_\varepsilon^{\mathcal{W}}(x_0)$ and the neighbourhood is homomorphic to a neighbourhood of $0 \in \mathcal{R}^k$.

**Definition 2.11** The ε—neighbourhood of $x_0 \in \mathcal{W} \subseteq \mathcal{R}^n$ is the set of

$$\eta_\varepsilon^{\mathcal{R}^k}(x_0) = \{y \in \mathcal{R}^k \,||\, y - x_0| < \varepsilon\} \tag{2.78}$$

**Definition 2.12** The ε—neighbourhood of $x_0 \in \mathcal{W} \subseteq \mathcal{R}^n$ is

$$\eta_e^{\mathcal{W}}(x_0) = \mathcal{W} \cap \eta_\varepsilon^{\mathcal{R}^n}(x_0) \tag{2.79}$$

Thus we can find a homomorphism which makes the topological structure in $\mathcal{R}^n$ correspond to that in $\mathcal{W}$ and vice versa.

**Example** On the a sphere (e.g. the surface of the earth):

[neighbourhood of a city on the globe in $\mathcal{R}^3$] $\xrightarrow{morph}$ [one of the pages of atlas in $\mathcal{R}^2$]

**Definition 2.13** If homomorphism $h$ and $h^{-1}$ belong to $\mathcal{C}^k$, then $\mathcal{W}$ is a $\mathcal{C}^k$ manifold.

Let us return to the autonomous system

$$y' = f(y), \quad f \in \mathcal{C}^1, \quad f(x_0) = 0 \tag{2.80}$$

Take $\varphi^t$ as a flow near $x_0$ $\mathcal{R}^n$.

**Definition 2.14** Local stable manifold $\equiv \mathcal{W}^s(x_0) \equiv \{x \in \mathcal{U} | \varphi^t x \to x_0$, as $t \to \infty$ and $\varphi^t x \in \mathcal{U}$, for $t \geq 0\}$ that is, trajectories, starting from any point in the neighbourhood $\mathcal{U}$ of $x$ $\mathcal{R}^n$ as $t \to \infty$, tend to the initial point $x_0$. The sum total of these trajectories is called the local stable manifold. We can define the local unstable manifold similarly:

Local unstable manifold $\equiv \mathcal{W}^u(x_0)$

$\equiv \{x \in \mathcal{U} | \varphi^t x \to x_0$, as $t \to -\infty$ and $\varphi^t x \in \mathcal{U}$, for $t \leq 0\}$

**Example 1** If $x_0$ is asymptotically stable for a certain $\mathcal{U}$, then $\mathcal{W}^s_{loc}(x_0) = \mathcal{U}$.

**Example 2** For the linear ordinary differential equation

$$y' = Ay \tag{2.81}$$

Let $\varphi^t = e^{tA}$, then $\mathcal{W}^s_{loc}(x_0)$ is a neighbourhood of $x_0$ in $\mathcal{E}^s$ as $t \to \infty$ and it is a neighbourhood of $x_0$ in $\mathcal{E}^u$ as $t \to -\infty$, where $\mathcal{E}^s$ and $\mathcal{E}^u$ are present at the stable and unstable subspaces respectively. So in the linear ordinary differential equation we know that there are manifolds and $\mathcal{W}^s_{loc} \subseteq \mathcal{E}^s$, $\mathcal{W}^u_{loc} \subseteq \mathcal{E}^u$.

It is our goal to generalize linear flows to local flows of non-linear systems. To solve this problem, we must answer whether $\mathcal{W}^s$ and $\mathcal{W}^u$ are manifolded or not. If so, how many dimensions? Hence we introduce the stable manifold theorem.

Consider the autonomous system

$$y' = f(y),\ f \in \mathcal{C}^1,\ f{:}\mathcal{R}^n \to \mathcal{R}^n,\ f(x_0) = 0,\ A = D_x f(x_0) \tag{2.82}$$

If the eigenvalues $\lambda_1, \ldots, \lambda_s$ of $A$ have negative real parts, the corresponding stable subspace $\mathcal{E}^s$ is spanned by their eigenvectors. If the eigenvalues $\mu_1, \ldots, \mu_u$ of $A$ have positive real parts, the corresponding unstable subspace $\mathcal{E}^u$ is spanned by their eigenvectors. Where $s + u = n$. There exist stable and unstable manifolds $\mathcal{W}^s$ and $\mathcal{W}^u$ uniquely, satisfying $\dim \mathcal{E}^s = \dim \mathcal{W}^s$, $\dim \mathcal{E}^u = \dim \mathcal{W}^u$, and being tangential to $\mathcal{E}^s, \mathcal{E}^u$ at $x_0$. Further, stable manifold $\mathcal{W}^s_{loc} = \{(x_s, x_u) | x_u = y(x_s)\}$, where $\psi{:}\mathcal{E}^s \to \mathcal{E}^u$.

**Proof** Step (1), separate the stable subspace from the unstable subspace. For

$$y' = f(y) = Ay + F(y) \tag{2.83}$$

where $A \equiv D_y f(y)|_{y = x_0 = 0}$, $F(y) = [f(y) - D_y f(0) y]$

and $F(0) = D_y F(0) = 0$

There are non-singular transformations $y = Tx$ so that

$$\begin{bmatrix} x_1{}' \\ x_2{}' \end{bmatrix} = \begin{bmatrix} B_{s\times s} & 0 \\ 0 & C_{u\times u} \end{bmatrix} \begin{bmatrix} x_1 \\ x_2 \end{bmatrix} + \begin{bmatrix} g_1(x_1, x_2) \\ g_2(x_1, x_2) \end{bmatrix} \tag{2.84}$$

where the eigenvalues of $B$ have negative real parts and those of $C$ positive real parts.

$$g = \begin{bmatrix} g_1 \\ g_2 \end{bmatrix} \in \mathcal{C}^1,\ g = o(\|x\|^2)$$

The basic solution matrix of its linearized system is

$$\Phi(t) = \exp\begin{bmatrix} B & 0 \\ 0 & C \end{bmatrix} t = \begin{bmatrix} e^{tB} & 0 \\ 0 & e^{Ct} \end{bmatrix} \equiv \Phi_s(t) + \Phi_u(t) \tag{2.85}$$

where flows $\Phi_s(t) = \begin{bmatrix} e^{tB} & 0 \\ 0 & 0 \end{bmatrix}$, $\Phi_u(t) = \begin{bmatrix} 0 & 0 \\ 0 & e^{tC} \end{bmatrix}$.

Obviously, there exist constants $K > 0$, $\alpha > 0$ for $\Phi(t)$, having

$$\|\Phi_s(t)\| \le Ke^{-\alpha t}, \quad \text{for } t \ge 0; \quad \|\Phi_u(t)\| \le Ke^{\alpha t}, \quad \text{for } t \le 0 \tag{2.86}$$

Step (2), change eq. (2.84) into an integral calculus equation and find the bounded conditions of solution $X(t)$.

**Lemma** If $X(t)$ is a bounded solution of system eq. (2.84), then $X(t)$ is bounded and satisfies

$$x(t) = \Phi_s(t)x_s(0) + \int_0^t \Phi_s(t-\tau)g(x(\tau))\,\mathrm{d}\tau - \int_t^{+\infty} \Phi_u(t-\tau)g(x(\tau))\,\mathrm{d}\tau \tag{2.87}$$

under the initial condition

$$x_u(0) + \int_0^{\infty} \Phi_u(-\tau)g(x(\tau))\,\mathrm{d}\tau = 0 \tag{2.88}$$

**Proof** If $X(t)$ is a bounded solution of eq.(2.84), then we know, in terms of variation of parameter, that

$$x(t) = \Phi(t)x(0) + \int_0^t \Phi(t-\tau)g(x(\tau))\,\mathrm{d}\tau = \Phi_s(t)x_s(0) + \Phi_u(t)x_u(0) + \int_0^t \Phi_s(t-\tau)g(x(\tau))\,\mathrm{d}\tau$$

$$+\int_0^t \Phi_u(t-\tau)g(x(\tau))\,\mathrm{d}\tau = \Phi_s(t)x_s(0) + \int_0^t \Phi_S(t-\tau)g(x(\tau))\,\mathrm{d}\tau - \int_t^{+\infty} \Phi_u(t-\tau)g(x(\tau))\,\mathrm{d}\tau$$

$$+\Phi_u(t)[x_u(0) + \int_0^{+\infty} \Phi_u(-\tau)g(x(\tau))\,\mathrm{d}\tau] \tag{2.89}$$

From the properties of $g$ and the boundedness of $X(t)$ there exist $N > 0$, so that $\|g(x(t))\| \le N$, therefore

$$\|\int_t^{+\infty} \Phi_u(t-\tau)g(x(\tau))\,\mathrm{d}\tau\| \le \int_0^{+\infty} \|\Phi_u(t-\tau)g(x(\tau))\|\,\mathrm{d}\tau \le KNe^{\alpha t}\int_0^{+\infty} e^{-\alpha\tau}\,\mathrm{d}\tau = \frac{KN}{\alpha} < +\infty \tag{2.90}$$

From the properties of $\Phi(t)$, the first three terms of eq. (2.89) are bounded. Since $\Phi_u(t)$ is unbounded as in the fourth term, we must have

$$x_u(0) + \int_0^{+\infty} \Phi_u(-\tau)g(x(\tau))\,\mathrm{d}\tau = 0$$

in order that $X(t)$ is bounded.

Step (3), construct a stable manifold. A non-linear map is defined by

$T: \mathcal{B} \to \mathcal{B}$, $\quad \mathcal{B} = \{x \mid \|x(t)\| \le \delta\}$

$$Tx = \Phi_s(t)x_s(0) + \int_0^t \Phi_s(t-\tau)g(x(\tau))\,\mathrm{d}\tau - \int_t^{+\infty} \Phi_u(t-\tau)g(x(\tau))\,\mathrm{d}\tau \quad (2.91)$$

It is easy to prove that $T{:}\mathcal{B} \to \mathcal{B}$ is a compressive mapping. Therefore there exists a unique $x^* \in \mathcal{B}$, so that $Tx^* = x^*$, i.e.

$$x^*(t) = \Phi_s(t)\,x_s^*(0) + \int_0^t \Phi_s(t-\tau)g(x^*(\tau))\,\mathrm{d}\tau - \int_t^{+\infty} \Phi_u(t-\tau)g(x^*(\tau))\,\mathrm{d}\tau \quad (2.92)$$

which satisfies

$$x_u^*(0) + \int_0^{+\infty} \Phi_u(-\tau)g(x^*(\tau))\,\mathrm{d}\tau = 0 \quad (2.93)$$

Define a mapping $\psi{:}\mathcal{E}^s \to \mathcal{E}^u$:

$$\psi(x_s) = -\int_0^{+\infty} \Phi_u(-\tau)g(x^*(x_s(\tau)))\,\mathrm{d}\tau \quad (2.94)$$

From the uniqueness of $x^*(t)$, we know that $\psi(x_s)$ is of a single value. The stable manifold is

$$\mathcal{W}_{loc}^s = \{(x_s, x_u) | x_u = \psi(x_s)\} \quad (2.95)$$

From $g(0) = 0$ and $\frac{\partial}{\partial x} g(0) = 0$, $\frac{\partial \psi(0)}{\partial x_s} = 0$ can be obtained. Hence $\mathcal{W}_{loc}^s$ is tangential to $\mathcal{E}^s$ at $x = 0$.

Similarly, we can prove the existence and uniqueness of $\mathcal{W}_{loc}^u$.

### 2.5.2 Phase Portrait

**Definition 2.15** The set of all trajectories is called a phase portrait. This is the mathematical definition of a phase portrait. Practically, the individual diagram of some typical loci is called the phase portrait.

To illustrate the steps in drawing a phase portrait, take a two-dimensional system as an example: $x' = f(x,y)$, $\quad f \in \mathcal{C}^1$

$$y' = g(x,y), \quad g \in \mathcal{C}^1 \quad (2.96)$$

1. Draw Isoclines

$$\frac{\mathrm{d}y}{\mathrm{d}x} = \frac{g(x,y)}{f(x,y)} = \text{const}, \qquad f(x,y) \neq 0 \quad (2.97)$$

The slope of these loci is constant at point $(x,y)$.

If $f(x,y) = 0$, $g(x,y) \neq 0$, then the orbits have a vertical tangent. If $g(x,y) = 0$, $f(x,y) \neq 0$, then the orbits have a horizontal tangent. The point of intersection between $f(x,y) = 0$ and $g(x,y) = 0$ is called an equilibrium point.

2. Applying the stable manifold theorem to the equilibrium point, make the system linearized and find out the stable and unstable subspaces $\mathcal{E}^s$ and $\mathcal{E}^u$ spanned by the eigenvectors. Draw local manifolds tangent $\mathcal{E}^s$ and $\mathcal{E}^u$ at $(x,y)$.

3. In a two-dimensional phase space, the local phase portrait can be expanded to global using information about isoclines. We should pay particular attention to separative orbit. For example, we should first determine those trajectories tending to a saddle point as $t \to \pm\infty$ (equivalent to 1-dimensional $\mathcal{W}^s$ and $\mathcal{W}^u$ in one two-dimensional systems)

4. Find the periodic orbit.

**Example** To illustrate how to draw a phase portrait, consider a model of species competition.

Let $x$ and $y$ represent the populations of a biological species competing for the same resources:

$$\frac{dx}{dt} = x(\alpha_1 - \delta_1 x - \beta_1 y), \quad \frac{dy}{dt} = y(\alpha_2 - \beta_2 x - \delta_2 y) \tag{2.98}$$

where, $\alpha_i, \beta_i, \delta_i > 0$, $i = 1,2$. If $y = 0$ then,

$$\frac{dx}{dt} = x(\alpha_1 - \delta_1 x) \tag{2.99}$$

This is a logistic equation studied previously. $X(t)$ increases with time from 0 toward a saturated state $x_1 = \dfrac{\alpha_1}{\delta_1}$, which the living things of $x$ class may never reach. The same is true of $y$ if $x = 0$. Included in eq. (2.98) are terms other than the logistic equation. For $x \neq 0$ and $y \neq 0$, the point of resources will reduce the carrying capacities to the two classes of biological species, respectively. Fix the coefficient of eq. (2.98) for convenience, so that

$$\frac{dx}{dt} = x(3 - x - 2y) = f(x,y), \quad \frac{dy}{dt} = y(2 - x - y) = g(x,y) \tag{2.100}$$

Sketch the phase portrait of eq. (2.100) in phase-plane $D: x \geq 0,\ y \geq 0$.

1. Sketch Isoclines

If $f(x,y) = 0$, we get

$$x = 0, \text{ and } x + 2y = 3 \tag{2.101}$$

Eq. (2.101) represents two straight lines on which the trajectories have vertical tangent line. If $g(x,y) = 0$, we obtain

$$y = 0 \text{ and } x + y = 2 \tag{2.102}$$

Eq. (2.102) represents two straight lines on which the phase-trajectory has horizontal tangents. Intersecting points of $f(x,y) = 0$ and $g(x,y) = 0$ or equilibrium points $(x_0, y_0)$ are $(0,0), (3,0), (0,2), (1,1)$. Because $\dfrac{dy}{dx} = \dfrac{0}{0}$ at equilibrium points, the slope of the phase portraits is undetermined.

2. Linearization system

Operator $A$ of a linearized system is

$$A = \frac{\partial(f,g)}{\partial(x,y)}|_{(x_0,y_0)} \tag{2.103}$$

(1) At equilibrium point (0,0), $A=\begin{bmatrix}3 & 0\\0 & 2\end{bmatrix}$ and its eigenvalues are 3,2, corresponding to eigenvectors $(1,0)^T,(0,1)^T$. Thus (0,0) is an unstable node.

(2) At equilibrium point (3,0), $A=\begin{bmatrix}-3 & -6\\0 & -1\end{bmatrix}$ and its eigenvalues are $-3,-1$, corresponding to eigenvectors $(1,0)^T,(-3,1)^T$. Thus (3,0) is a stable node.

(3) At equilibrium point (0,2), $A=\begin{bmatrix}-1 & 0\\-2 & -2\end{bmatrix}$ and its eigenvalues are $-1,-2$, corresponding to eigenvectors $(-1,-2)^T,(0,1)^T$. Thus (0,2) is a stable node.

(4) At equilibrium point (1,1), $A=\begin{bmatrix}-1 & -2\\-1 & -1\end{bmatrix}$ and its eigenvalues are $-1+\sqrt{2}$, $-1-\sqrt{2}$, corresponding to eigenvectors $(\sqrt{2},1)^T$, $(\sqrt{2},1)^T$. Thus (1,1) is a saddle point.

3. Sketching

Firstly, we sketch the orbits through saddle point (1,1), and then draw the phase portrait based on information from the isoclines.

4. No closed orbit

From Fig. 2.19, in the course of competing when the numbers of the two classes of living things tend to approximate equality, i.e. reaching (1,1), such a state is unstable. For some reason, when the number of a class of living things is greater than that of the others, then $x$ will continuously increase and $y$ will reduce until it disappears in terms of orbits toward the stable node. Thus $x$ obtains the whole carrying capacity. The stable manifold separates these two later classes of results.

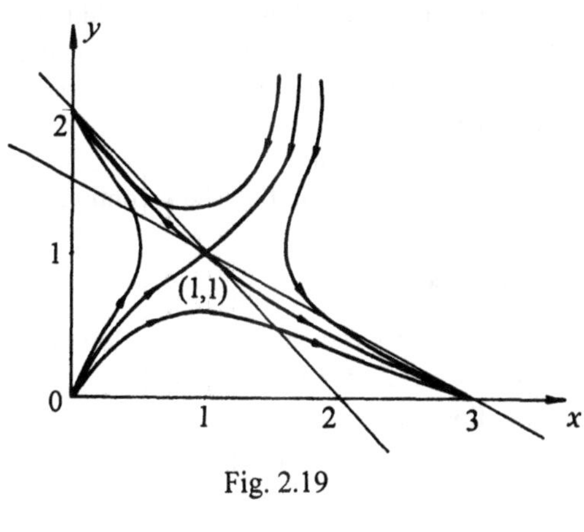


Fig. 2.19

# Chapter 3

# Discrete Dynamical Systems

We have discussed the main properties of continuous dynamical systems (or flows) in preceding chapters. In this chapter, we will study the corresponding properties of discrete systems. Because research on discrete dynamical systems is relatively simple and straightforward, theorems on diffeomorphism are often presented first, followed by the relevant discussion. In addition, flows are sometimes discretized in order to obtain their properties by studying relatively low-dimensional dynamical systems. As preparative knowledge for the bifurcation theory, the concept of structural stability will be explained in the final part of this chapter.

## 3.1 Discrete Dynamical Systems and Linear Maps

### 3.1.1 Discrete Dynamical Systems

Assume an open domain $D \subseteq \mathcal{R}^n$, and suppose that it is a phase space. Define a discrete map $G:\mathcal{D} \to \mathcal{D}$, where $G$ is a diffeomorphism, i.e. $G$ is a biunivocal map from $\mathcal{R}^n$ to $\mathcal{R}^n$, and both $G$ and $G^{-1}$ are continuously differentiable.

If the orbits of $G$ are sets

$$\{x_k | x_k = G^k x_0, k \in Z\} \tag{3.1}$$

then the pair $\{\mathcal{D}, G\}$ defines a differentiable discrete dynamical system (or differentiable discrete flow), where $G^k$ is a bidirectional sequence

$$G^k = G \bullet G \bullet\bullet\bullet G \qquad k > 0$$

$$G^{-k} = [G^{-1}]^k = G^{-1} \bullet G^{-1} \bullet\bullet\bullet G^{-1} \tag{3.2}$$

Obviously, the sequence satisfies that $G^0$ is an identical mapping and $G^k \bullet G^l = G^{k+l}$ for any $k,l \in \mathcal{Z}$. Therefore $\{G^k | k \in \mathcal{Z}\}$ is a simple parameter transformation group from $\mathcal{D}$ to $\mathcal{D}$, and the range of parameter $k$ is an integer additive group $\{\mathcal{Z},+\}$. So we call it a discrete dynamical system. For convenience studying dynamical systems, the term "homomorphism" (or diffeomorphism) is often directly used to designate corresponding discrete dynamical systems. If $G^k$ is defined only for limited $k$ in the orbit set of $G$, then the system is called a local discrete (or generalized discrete) system. The relationship between discrete flows and continuous flows is introduced as follows.

**Example 1 Stroboscopic map**

Define a flow $\{\mathcal{D}, \varphi^t\}$ for $t \in \mathcal{R}$; we can obtain discrete flows by discrete sampling of continuous flows. Fixing time $T$, for example, yields stroboscopic map $G = \varphi^T$, which gives a discrete flow (or orbit) $\{G^k\} = \{\varphi^{kT}, k \in \mathcal{Z}\}$ for a bidirectional sequence. The expression of this orbit is quite effective for the numerical computation.

**Example 2** The discrete dynamical system can also be obtained using the difference method of differential equations

$$x_{k+1} = G(x_k) \tag{3.3}$$

where $x_k$ is preset. The equation approximately describes the proliferation of living things, the rule of economic development and computer programs.

**Example 3** Discrete flow can be obtained through the discrete approximation of solutions to differential equations.

Assume an initial value problem

$$y' = f(t,y), \qquad y(t_0) = y_0 \tag{3.4}$$

Choose $h > 0$, so that its solution can be expressed as

$$y_{k+1} = y_k + hf(t_0 + hk,\, y_k) = G(y_k) \tag{3.5}$$

Such a process of solving equations is called the Euler method.

The foregoing examples demonstrate that discrete flows can be obtained by discrete sampling of continuous flows. On the other hand, for a given homomorphism $G$ in $\mathcal{D}$ we can obtain a continuous flow (called a suspension flow) one-dimension higher in space by the 'suspension' method. Therefore, define a simple flow on $\mathcal{R}\times\mathcal{D}$, $\psi'(p+x) = (t+p,\, x)$ for $p, x \in \mathcal{R}\times\mathcal{D}$ and define an equivalent relationship

$$(p,x) \sim (q,y) <=> p-q = n \in \mathcal{Z}$$

where $y = G^{p-q}(x)$, noting that $\widetilde{\mathcal{D}} = \mathcal{R}\times\mathcal{D}/\sim$. It is easy to verify that if $(p,x) \sim (q,y)$, then

$$\psi'(p,x) \sim \psi'(q,y) \tag{3.6}$$

As a result $\psi'$ induces a flow $\widetilde{\psi}'$ on $\widetilde{\mathcal{D}}$. This is the suspension flow. In order to study the relationship between suspension flow $\widetilde{\psi}'$ and homomorphism $G$, we let $\mathcal{D}$ equal $\Sigma = \{0\}\times\mathcal{D}/\sim$. So the map $G$ can be regarded as a homomorphism on $\Sigma$. The orbital flow $\widetilde{\psi}'$, starting from any point $x$ on $\Sigma$, will return to $\Sigma$. The first returning point is $G(x)$. From section 3.4 we know that $G$ is exactly the Poincaré mapping of flow $\widetilde{\psi}'$ on the transversal plane $\Sigma$.

So there exists a close association between continuous and discrete flows. The theory of dynamical systems includes the study of continuous and discrete flows, that is, the dynamical systems of $\mathcal{C}^r$ include both $\mathcal{C}^r$ flows and discrete diffeomorphism. for a detailed description of discrete dynamical systems refer to references [43, 44].

### 3.1.2 Linear Maps

**Stroboscopic Map of Linear Flows**

Consider the linear differential equation

$$y' = Ay, \qquad y \in \mathcal{R}^n \tag{3.7}$$

corresponding to flow $e^{tA}$ where the flow is linear and non-singular. For a fixed $T$, we have stroboscopic mapping $B = e^{TA}$, which is a diffeomorphism. Assuming $\lambda_i$ is the eigenvalue of $A$, then $e^{TA}$ is the eigenvalue of $B$.

If $\text{Re}\lambda_i > 0$, then $e^{T\lambda_i}|> 1$, and the flow is expansive.

If $\text{Re}\lambda_i < 0$, then $|e^{T\lambda_i}|< 1$, and the flow is contractive.

If $\text{Re}\lambda_i = 0$, then $|e^{T\lambda_i}|= 1$, and the flow is preserved.

**General Linear Diffeomorphism Maps.**

**Definition 3.1** Let $B:\mathcal{R}^n \to \mathcal{R}^n$ be a general linear diffeomorphism mapping, and $\rho$ an eigenvalue of $B$.

(1) When $|\rho|< 1$, $x \in \mathcal{E}^s$, then $x_1 = Bx = \rho x$, $x_2 = Bx_1 = B^2 x = \rho^2 x,\ldots,x_k = B^k x = \rho^k x$, so $||B^k x||=|\rho^k|\bullet||x|| \to 0$ as $t \to \infty$ (or $k \to \infty$).

(2) Similarly, when $|\rho|>$ for $x \in \mathcal{E}^u$, $||B^k x||=|\rho^k|\bullet||x|| \to \infty$ as $t \to \infty$ (or $k \to \infty$)

where $\mathcal{E}^s$ is a stable and an invariant subspace corresponding to the eigenvalue of $B$ with a norm less than one. $\mathcal{E}^u$ is an unstable and invariant subspace corresponding to the eigenvalue of $B$ with a norm more than one. $\mathcal{E}^c$ is a centre and invariant subspace corresponding to the eigenvalue of $B$ with a norm of one. The dimension is $n = n_s + n_u + n_c$.

We shall consider the eigenvalues with multiplicity and complexity.

### 3.1.3 Some Definitions

1. **Theorem 3.1** assume a linear and non-singular map $B:\mathcal{R}^n \to \mathcal{R}^n$. Suppose $k > 0$ and $0 < \alpha <$ , then $\|B^k x_0\| \le K\alpha^K$ for $K \ge 0$ as $x_0 \in \mathcal{E}^s$

and $\|B^k x_0\| \ge K\alpha^K$ for $K \le 0$ as $x_0 \in \mathcal{E}^u$.

2. **The Fixed Points of Diffeomorphism (or Equilibrium Points)**

For linear mapping $B$, it is obvious that $B\underline{O} = \underline{O}$, so $\underline{O}$ is a fixed point (or an Equilibrium Point). If $B$ is non-singular so that there are no other fixed points, then $\underline{O}$ is an isolated fixed point.

3. **Definition 3.2** If $O$ is a hyperbolic point of $B$ (or $B$ is hyperbolic), and if no eigenvalue of $B$ has $|\rho|=$ , then $\mathcal{R}^n = \mathcal{E}^s + \mathcal{E}^u, \mathcal{E}^c = \phi$.

## 3.2 Non-linear Maps and the Stable Manifold Theorem

### 3.2.1 Non-linear Diffeomorphism Map $G$

The fixed point $x^* = G(x^*)$ is hyperbolic if $D_x G(x^*)$ has no eigenvalue of $|\lambda|= 1$ and non-zero eigenvalues, where

$$D_x G(x^*) = \left[\frac{\partial G}{\partial x}\right]_{x=x^*}. \tag{3.8}$$

is a Jacobian matrix.

**Hartman—Grobman's theorem** Given a diffeomorphism $G$ with a hyperbolic fixed point $x^*$, then there exists a homomorphism $h$ in the neighbourhood $\mathcal{U}$ of $x^*$ and the neighbourhood $\mathcal{V}$ of $0 \in \mathcal{R}^n$ so that $h: \mathcal{U} \to \mathcal{V}$ and

$$h(x)G(x) = D_x G(x^*)h(x), \quad x \in \mathcal{U} \qquad (3.9)$$

In this case, $G(x)$ is the local topological equivalent to $D_x G(x^*)$, written as $G(x) \sim D_x G(x)$ (cf. Fig. 3.1).

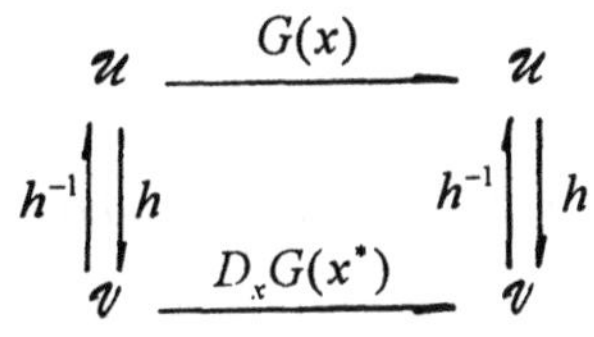


Fig.3.1

### 3.2.2 Stable Manifold Theorem of Discrete Maps

Let $G$ be a diffeomorphism with a hyperbolic fixed point $x^*$, then there exist the local stable and unstable manifolds $\mathcal{W}^s_{loc}(x^*)$ and $\mathcal{W}^u_{loc}(x^*)$. They are tangential to $\mathcal{E}^s$ and $\mathcal{E}^u$ at $x^*$ respectively and of the same dimensions as for $D_x G(x^*)$. For the neighbourhood $\mathcal{U}$ of $x^*$, we have

$$\mathcal{W}^s_{loc}(x^*) = \{x | x \in \mathcal{U},\ G^k x \to x^*, \text{ as } k \to \infty \text{ and } G^k x \in \mathcal{U}\} \qquad (3.10)$$

$$\mathcal{W}^u_{loc}(x^*) = \{x | x \in \mathcal{U},\ G^{-k} x \to x^*, \text{ as } k \to \infty \text{ and } G^{-k} x \in \mathcal{U}\} \qquad (3.11)$$

These manifolds may be continuous to the maximum extension. If $G^L$ and $G^{-L}$ are existent, then the globle stable and unstable manifold can be defined as

$$\mathcal{W}^S(x^*) = \bigcup_{l \ge 0} G^L(\mathcal{W}^s_{loc}(x^*)) \qquad (3.12)$$

$$\mathcal{W}^u(x^*) = \bigcup_{l \ge 0} G^{-L}(\mathcal{W}^u_{loc}(x^*)) \qquad (3.13)$$

The trajectories of the maps are made up of discrete points, and the invariant manifold of the maps is the union of the trajectories of innumerable discrete points. The manifolds are comprised of continuous points, and the invariant manifold of the flow is the union of innumerable manifolds. Therefore, the flow is substantially different from a discrete map.

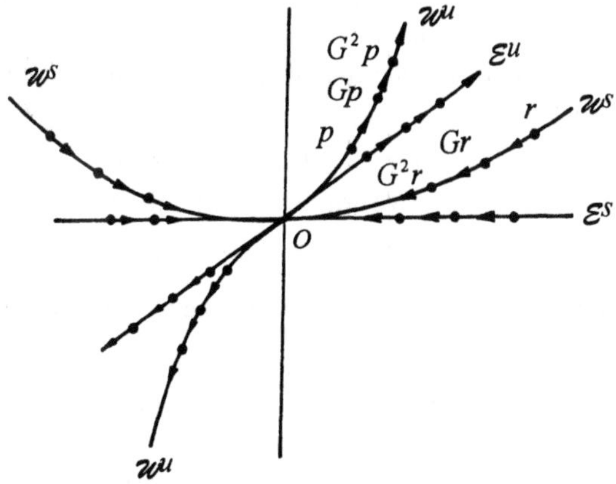


Fig.3.2

**Example The manifold of maps in $\mathcal{D} = \mathcal{R}^2$**

Suppose that (0,0) is a hyperbolic equilibrium point, stable subspace $\mathcal{E}^s$ is the $x$—axis, unstable subspace $\mathcal{E}^u$ has an arbitrary direction and the stable and unstable manifolds, $\mathcal{W}^s$, $\mathcal{W}^u$, are tangential to $\mathcal{E}^s$ and $\mathbf{E}^u$ respectively at (0,0) as shown in Fig. 3.2.

### 3.2.3 Periodic Orbit of Maps

If $x_p = x_0$ or $G^p x_0 = x_0$ for $0 < p \in \mathfrak{Z}$ and $G^L x_0 \neq x_0$, for $< L < p-$ , then orbit

$$\{x_n = G^n x_0 | n \in \mathfrak{Z}\} \tag{3.14}$$

is $p$—periodic or $x_0, x_1, \ldots, x_{p-1}$ are distinct and $x_p = x_0$. By the above definition we can say that if orbits are $p$-periodic, then

$$x_{k+p} = x_k \qquad k \in \mathfrak{Z} \tag{3.15}$$

**Definition 3.3** Set $\{x_0, x_1, \ldots, x_{p-1}, x_p = x_0\}$ is called a cycle of $G$ or the periodic orbit of $G$ if $x_0, x_1, \ldots, x_p$ are $p$—periodic. In addition, if $x_0$ is $p$—periodic for $G$, then $x_0$ is a fixed point for $G^p$.

The chain rule is used to find the derivative of $G^p x_0 = x_0$. The formula is

$$\begin{aligned} DG^p x_0 &= D[G \bullet G \bullet\bullet\bullet G]x_0 = DG \bullet (G^{p-1}) DG \bullet (G^{p-2}) \bullet\bullet\bullet DG(x_0) \\ &= DG(x_{p-1}) DG(x_{p-2}) \bullet\bullet\bullet DG(x_0) \end{aligned} \tag{3.16}$$

**Definition 3.4 The hyperbolic property of** $G$

The $p$—periodic point $x_0$ is called a hyperbolic if and only if the fixed point $x_0$ of $G^p$ is hyperbolic.

**Definition 3.5 The property of fixed points**

Assume that $x^*$ is a hyperbolic fixed point of $G$. If the absolute values of the eigenvalues of $G$ satisfy $|\lambda| <$ and $n_s = n$, then the fixed point $x^*$ is called a sink. If all of the eigenvalues of $G$ satisfy $|\lambda| > 1$ and $n_u = n$, then the fixed point $x^*$ is called a source. If both cases exist at the same time, the fixed point $x^*$ is called a saddle point. Similarly, the sink, the source and the saddle point of $p$—periodic point can be defined.

Given a set $\Lambda$. If $G^k(x) \in \Lambda$ for any $x \in \Lambda$ and $k \in \mathfrak{Z}$, then $\Lambda$ is the invariant of $G$. We can also define the positive and negative invariant sets in the same way. The sets $\omega(x) = \bigcup_{m \in N} \{G^k(x) \mid k \geq m\}$. $\alpha(x) = \bigcup_{m \in N} \{G^{-k}(x) \mid k \geq m\}$ are called, respectively, the $\omega$—limit set and the $\alpha$—limit set of $G$ across trajectories of $x$.

If there is $k \neq 0$ for any neighbourhood of $x$ so that $G^k(U) \bigcap \mathcal{U} \neq \phi$, then $x$ is called a non-wandering point. The set of all non-wandering points of $G$ is called a non-wandering set, written as $\Omega(G)$. Trajectories (including fixed points, periodic orbits) $\omega(x)$, $\alpha(x)$ and $\Omega(G)$ are all invariant sets.

## 3.3 Classification of Generic Systems

### 3.3.1 One-dimensional Systems

Given a one-dimensional linear homomorphism mapping $B:\mathcal{R} \rightarrow \mathcal{R}$, here $B$ is a $\times$ matrix. If $b \in \mathcal{R}$ is an eigenvalue of $B$, then

$$Bx = bx \qquad (3.17)$$

Divide the $x$—axis into four open intervals $I_1, I_2, I_3$ and $I_4$ as shown in Fig. 3.3.

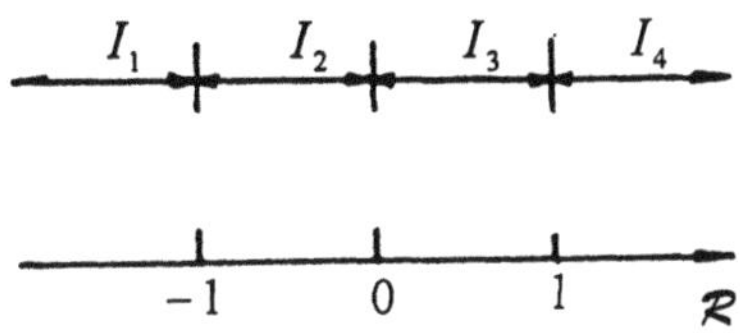


Fig. 3.3

According to the relationship of the mapping $x_k = \lambda^k x_0$, we consider the rule in different intervals as shown in table 3.1. Since $I_1 \cup I_2 \cup I_3 \cup I_4$ is a dense open set in $\mathcal{R}$, mapping $B$ is generic.

Degenerated cases of one-dimensional mapping $B$:

1. $I\eta$ for $b = -1$, so $B$ is two-periodic.
2. Since $B$ is singular for $b = 0$, there is no solution.
3. $Bx = x$ for $b = 1$, so any point $x$ is a fixed point (or periodic point).

**Table 3.1**

| Interval | Action in Phase Space | Phase Portrait |
|---|---|---|
| $I_1$<br>$-\infty < b < -1$ | expanding, reverse sign<br>$x_1 = -\|> 1\| \, x_0$<br>$x_2 = \ \|> 1\|^2 \, x_0$<br>$x_3 = -\|> 1\|^3 \, x_0$<br>... | <br>$-1 \; 0 \; x_0 1$<br>Alte. Source |
| $I_2$<br>$-1 < b < 0$ | contracting, reverse sign | 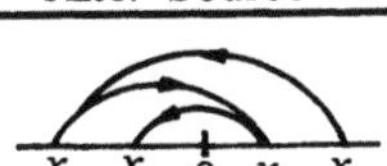<br>$x_1 \; x_3 \; 0 \; x_2 \; x_0$<br>Alte. Sink |
| $I_3$<br>$0 < b < 1$ | contracting, same sign | <br>$0 \; x_2 \; x_1 \; x_0$<br>Monotone Sink |
| $I_4$<br>$< b < \infty$ | expanding, same sign | <br>$0 \; x_0 \; x_1 \; x_2$<br>Monotone Source |

### 3.3.2 Two-dimensional Systems

Consider a two-dimensional linear differential equation:

$$y' = Ay \qquad y \in \mathcal{R}^2 \tag{3.18}$$

and assume $e^{TA} = B: \mathcal{R}^2 \to \mathcal{R}^2 (x_k = B^k x)$ is a linear diffeomorphism mapping and non-singular, where $B = \begin{bmatrix} a & b \\ c & d \end{bmatrix}$.

**Case 1** If $B$ has two distinct real eigenvalues $\lambda_1, \lambda_2$ or $P^{-1}BP = \begin{bmatrix} \lambda_1 & 0 \\ 0 & \lambda_2 \end{bmatrix}$, then the two-dimensional eigenspace can be reduced to a one-dimensional eigenspace. It is the same as the above.

**Case 2** If $\mathcal{B}$ has complex eigenvalues $\lambda = \alpha + i\beta$, $\bar{\lambda} = \alpha - i\beta$, where $\alpha, \beta \in \mathcal{R}$, corresponding to eigenvectors $= a + ib$, $\bar{V} = a - ib$, here $a$, $b \in \mathbf{R}$ and

$$a = \frac{1}{2}(V + \bar{V}) \qquad b = \frac{1}{2i}(V - \bar{V})$$

$$Ba = \frac{1}{2}(BV + B\bar{V}) = \frac{1}{2}(\lambda V + \bar{\lambda}\bar{V}) = \mathrm{Re}(\lambda V) = \alpha a - \beta b$$

$$Bb = \frac{1}{2i}(BV - B\bar{V}) = \frac{1}{2i}(\lambda V + \bar{\lambda}\bar{V}) = \mathrm{Im}(\lambda V) = \alpha b + \beta a$$

Using transformation to change $B$ in $x_k = B^k x$ into a diagonal form

$$T = (a, b) \tag{3.19}$$

where $T$ is a $2 \times 2$ matrix with columns $a$ and $b$, and

$$x = T\eta \tag{3.20}$$

Substituting (3.20) into $x_1 = Bx$, we have

$$\eta_1 = T^{-1}BT\eta \tag{3.21}$$

Suppose

$$T^{-1}BT = \begin{bmatrix} \alpha & \beta \\ -\beta & \alpha \end{bmatrix} = C \tag{3.22}$$

then $C$ can be written as $C = \alpha I + \beta S$, where I is an identical matrix and $S = \begin{bmatrix} 0 & 1 \\ -1 & 0 \end{bmatrix}$. Multiplying the two sides by $\eta$, we have:

$$C\eta = \alpha I\eta + \beta S \tag{3.23}$$

where $I\eta$ represents the pure expansion or contraction in a fixed direction, and $\beta S\eta$ is the length after a rotation through $90^\circ$ for $\alpha I\eta$: $s\eta = s\begin{bmatrix} \eta_1 \\ \eta_2 \end{bmatrix} = \begin{bmatrix} \eta_2 \\ -\eta_1 \end{bmatrix}$. Because $I\eta$ and $[S\eta]^{-1}$ one vertical to each other, we have

$$||C\eta||^2 = ||\alpha\eta||^2 + ||\beta s\eta||^2 = (\alpha^2 + \beta^2)||\eta||^2$$

$$||C\eta|| = \sqrt{\alpha^2 + \beta^2}\,||\eta|| = |\lambda|\,||\eta|| \tag{3.24}$$

The equation is the length of $|\lambda$ after re-scaling. It will contract if $|\lambda|<$ . Now we consider mapping $B^k x = B^k T\eta = TT^{-1}BTT^{-1}...TT^{-1}BT\eta$, as $T^{-1}BT = C$:

$$||B^k x|| \le ||T||\ ||C^k \eta|| = ||T||\ |\lambda^k|\ ||\eta|| = |\lambda|^k\ ||T||\ ||T^{-1}||\ ||x|| \tag{3.25}$$

In general, $||T||\ ||T||>$ . If $|\lambda|<1$ then there certainly exists $K>0$, so that $||B^k x||<||x||$ and when $k\to\infty$, $\|B^k x\|\to 0$. It is remarkable to note that $|\lambda|<$ does not imply $||B^k x||<||x||$ for all $x$. For example, if $a$ and $b$ have different lengths, although flows are eventually contracting, expansion may appear locally as shown in Fig. 3.4.

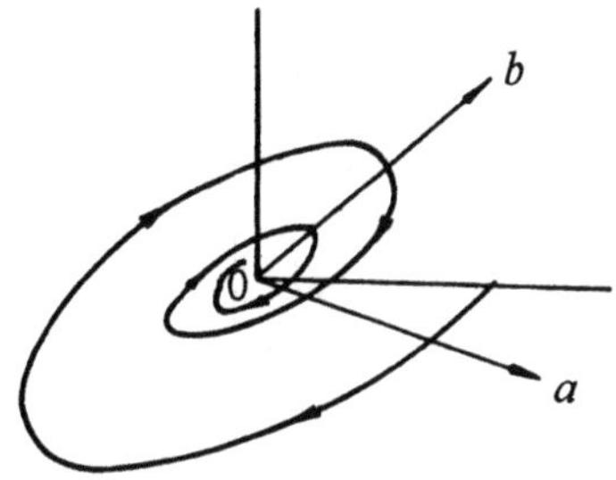

Fig. 3.4

If the eigenvalues of $B$ are simple (real or complex or both) for a given linear homomorphism mapping $B$ on $\mathcal{R}^n$, then the invariant subspace can be divided into a one– or two–dimensional eigensubspace.

**Corollary** If all the eigenvalues of $B$ $|\lambda|<$ , then $B^k$ is contracting for some $k>0$.

## 3.4 Stability of Maps and Poincaré Mapping

### 3.4.1 The Stability of a Diffeomorphic Fixed Point

Let $x^*$ be a fixed point (or equilibrium point) of $G$, i.e. $Gx^* = x^*$.

**Definition 3.6** $x^*$ is stable if there exists $\delta>0$ for any $\varepsilon>0$ so that $|x-x^*|<\delta$, that is $|G^k x - x^*|<\varepsilon$ for $k>0$. Similarly, we can define asymptotically stable and unstable cases. Now we introduce the principle of linearized stability of maps. Let $x^*$ be a fixed point. If all eigenvalues of $G(x^*)$ have $|\lambda|<$ , then $x^*$ is asymptotically stable.

**Proof** This can be proved by using the stable manifold theorem. $\mathcal{W}^s$ is a full neighbourhood of $x^*$ in this case. From $|\lambda|<1$, we know that $x^*$ is a sink by definition.

### 3.4.2 Orbital Stability of Flows in $\mathcal{R}^n$ $(n\ge 2)$

Let $\gamma$ be a periodic orbit of $\varphi^t$ and make $T>0$ so that $\varphi^T x = x$ for any $x\in\gamma$.

**Definition 3.7** $\gamma$ is orbitally stable if there exists $\delta>0$ for any given $\varepsilon>0$ so that when $d(y,\gamma)<\delta$, we have $d(\varphi^t y,\gamma)<\varepsilon$ for $t>0$, where $d(y,\gamma)=\inf||y-x||$. If $\gamma$ is orbitally stable and there exists $\delta>0$ so that when $d(y,\gamma)<\delta$ we have $d(\varphi^t y,\gamma)\to 0$ as $t\to\infty$, then $\gamma$ is asymptotically orbitally stable.

### 3.4.3 Poincaré Mapping

Let $\varphi^t$ be a flow based on $\mathcal{C}^r$ vector field $f(x)$ in $\mathcal{R}^n$. We establish the following general concepts similar to the intersecting lines of flows in the plane.

**Definition 3.8** Let $\Gamma$ be part of a $(n-1)$-dimensional sup-surface $H$ in $\mathcal{R}^n$. If $f(x)$ is not a tangent to $\Sigma$ for any $x \in \Sigma$, i.e. $< f(x), n(x) > \neq 0$, where $n(x)$ is the normal line of $\Sigma$ at $x$, then $\Sigma$ is the transversal section of flow $\varphi^t$. Flow $\varphi^t$ intersects with $\Sigma$ everywhere on $\Sigma$. As long as point $x$ is not an equilibrium point, we can always make a transversal section at $x$.

Let $\Gamma$ be a closed orbit of $\varphi^t$ with a period of $T$. It is possible to make transversal section $\Sigma$ through any point $p \in \Sigma$ by choosing the size of $\Sigma$ so that $\Gamma$ and $\Sigma$ intersect only at one point $p$. Since $\varphi^T(p) = p \in \Sigma$, which means that orbit $\Gamma = q(p)$ starting from $p$ first returns to $\Sigma$ after time $T$ according to the continuity of flows, the neighbourhood $\mathcal{U} \subseteq \Sigma$ of $p$ certainly exists, so the orbit starting from $q \in \mathcal{U}$ can return to $\Sigma$ again. Thus we define the first returning map $p: \mathcal{U} \to \Sigma$ so that for $q \in \mathcal{U}$

$$p(q) = \varphi^{\tau(q)}(q) \in \Sigma, \ \forall q \in U \tag{3.26}$$

where $\tau(q)$ represents the time required for the orbit to return for the first time from $q$ and is a continuous differentiable function. Usually, when $q \neq p$, $\tau(q) \neq \tau(p) = T$, but $\tau(q) \to T$ as $q \to p$. $p$ is usually known as the Poincaré map.

When $\varphi^t$ belongs to $\mathcal{C}^r$, $P$ is a $\mathcal{C}^r$ diffeomorphism on $\mathcal{U}$. Thus we get a discrete flow $\{p^k\}$ near the closed orbit of $\varphi^t$ by using the above method. It is easy to see that $p$ is a fixed point on the Poincaré map. The study of the closed orbit of flows, put in a nutshell, is the study of the fixed point $P$ on the Poincaré map. This is easily done because $P$ is the map in $(n-1)$-dimensional space.

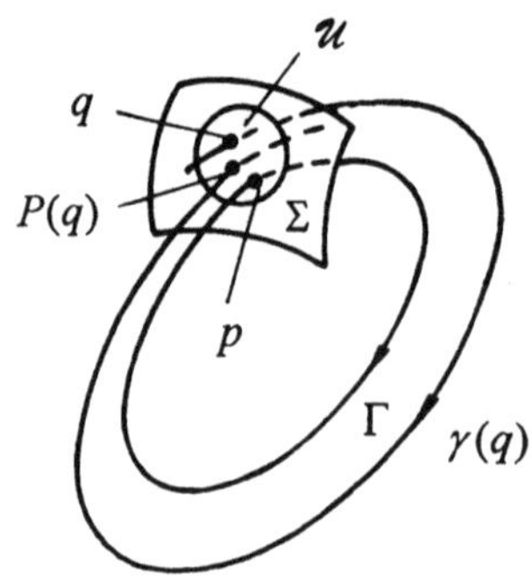

Fig.3.5

**Theorem 3.2** If map $P$ is contracting on the neighbourhood $\mathcal{U}$, then periodic orbit $\gamma$ is asymptotically stable ($|Pu - Pv| \leq \alpha|u - v|$, $u, v \in \mathcal{U}$).

**Proof** Since $p$ is an asymptotically stable fixed point for map $P$, $P^k q \to p$ for $q \in \mathcal{U}$. From the continuity of $\varphi^t$, we have $\varphi^t q \to \gamma$.

Usually, from the stability of $p$ we cannot derive the stability of periodic orbits since the period of the orbits through distinct points in $U$ is distinct or $\tau(q) \neq$ constant. The stability of the periodic orbits can be judged by that of the orbits of flows.

**Theorem 3.3** If all eigenvalues of $DF(p)$ have $|\lambda| < 1$, then $\gamma$ is asymptotically orbital stable.

**Proof** Since $Pq = p + DP(p)(q - p) + O(\|p - q\|^2)$, according to the given conditions of the theorem and the principle of liberalised stability, the theorem can be proved.

Or from the given conditions of the theorem, if $[D(P(p)]^k$ is contracting, then $P$ is contracting. Based on theorem 3.2 $\gamma$ is certainly asymptotically orbital stability.(2) Jordan blocks with real number forms can be expressed

### 3.4.4 Floquet's Theorem—Calculation of Eigenvalues of $DP(p)$

Let $$\vec{v} = \frac{d\varphi' p}{dx}\Big|_{t=0} \tag{3.27}$$

and $\mathcal{V} = \{\alpha v | \alpha \in \mathcal{R}\}$ be a one-dimensional subspace, then

$$\mathcal{R} = \Sigma \oplus \mathcal{V} \tag{3.28}$$

Let $B$ be a sphere around $p$ in $\mathcal{R}^n$, then $\mathcal{U} = \mathcal{B} \cap \Sigma$ is the neighbourhood of $p$. If $x \in \mathcal{B}$, we have

(a) $x = (q, u), \quad q \in \Sigma, \quad u \in \mathcal{V}$;

(b) $\varphi^T x = (\varphi_1' x, \varphi_2' x) \in \Sigma \oplus \mathcal{V}$.

Not losing any generality, suppose $p = 0$. The Frechet derivative $D$ can be written as $D = (D_1, D_2)$ in $\mathcal{R}^n$, where $D_1 = \dfrac{\partial}{\partial q}$ for $q \quad \Sigma$ and $D_2 = \dfrac{\partial}{\partial u}$ for $u \in \mathcal{V}$.

For a diffeomorphism $f$ write $f = \begin{bmatrix} f_1 \\ f_2 \end{bmatrix}$ so the Frechet derivative

$$Df = \begin{bmatrix} D_1 f_1 & D_2 f_1 \\ D_1 f_2 & D_2 f_2 \end{bmatrix} \tag{3.29}$$

where $D_1 f_1$ is an $(n-1) \times (n-1)$ Jacobian matrix.

**Floquet's theorem** Taking $f = \varphi^T$, we have

$$D\varphi'(0) = \begin{bmatrix} D_1 p(0) & 0 \\ * & 1 \end{bmatrix} \tag{3.30}$$

where $*$ is an $\times (n-1)$ matrix, which is usually non–zero.

Since $Pq = \varphi^{\tau(q)} q$, from the chain rule, it is not difficult to get the dependent relation $\tau(q)$.

**Corollary 1** The eigenvalues of $D\varphi^T(p)$ are those of $D_1 P(p)$ with one eigenvalue 1.

**Corollary 2** If all $(n-1)$ eigenvalues of $D\varphi^T(p)$ satisfy $|\lambda| <$ then $\gamma$ is an asymptotically orbital stable.

**Definition 3.9** The eigenvalues of $DP(p)$ are the Floquet multipliers of orbit $\gamma$.

It is not difficult to prove that the results obtained using the Floquet method are the same as those obtained by the Poincaré mapping method.

Since $\varphi^t$ is a flow of (1.4),

$$\varphi^t(x) = x + \int_0^t f(\varphi^s(x))\,\mathrm{d}s \tag{3.31}$$

Deriving for $x$ and assuming $x = 0$, we have

$$D\varphi^t(0) = I + \int_0^t Df(\varphi^s(0))D\varphi^s(0)\,\mathrm{d}s \tag{3.32}$$

Deriving for $t$ again, we have

$$\frac{\mathrm{d}}{\mathrm{d}t}D\varphi^t(0) = Df(\varphi^t(0))D\varphi^t(0) = A(t)D^t(0) \tag{3.33}$$

So $\quad \Phi(t) \overset{def}{=} D\varphi^t(0)$

is a basic matrix of the solution. In eq. (3.32) suppose $t = 0$, then $\Phi(0) = D\varphi^0(0) = I$, and we have

$$C = \Phi^{-1}(0)\Phi(T) = \Phi(T) = D\varphi^T(0) \tag{3.34}$$

Thus the eigenvalues of $D\varphi^T(0)$ are the characteristic exponent of system (1.13). We can judge the stability of the periodic solution according to whether the eigenvalues of $D\varphi^T(0)$ have negative real parts or not. This reinforces the of the stability of the closed orbits, both the methods of Floquet and Poincaré mapping giving the same result.

## 3.5 Structural Stability Theorem

It is necessary to introduce some ideas about the structural stability before discussing dynamical bifurcation theory. Structural stability means that when a dynamical system undergoes a small perturbation and becomes a "nearby" dynamical system, the topological structure of the system remains invariant.

### 3.5.1 Structural Stability

Consider an autonomous system

$$y' = f(y), \qquad f \in \mathcal{C}^1(\mathcal{D}), \qquad \mathcal{D} \subseteq \mathcal{S} \subseteq \mathcal{R}^n \tag{3.35}$$

where the vector field $f$ is connected, compact and positive invariant, i.e. $f$ is continuously differentiable, non-empty, inward and invariant. In other words, if $\partial\mathcal{D}$ (boundary of $\mathcal{D}$) is smooth, then $f$ points to the interior of $\mathcal{D}$ at each point of $\partial\mathcal{D}$ as shown in Fig.3.6.

Suppose set $\mathcal{U} \subseteq \mathcal{R}^n$. For map $f \in \mathcal{C}^1(\mathcal{U} \subseteq \mathcal{R}^n)$ and $\varepsilon > 0$, define the $\varepsilon$-neighbourhood as follows

$$B_\varepsilon(f) \overset{def}{=} \{g \in \mathcal{C}^1(\mathcal{U},\mathcal{R}^n) | \sup_{x \in k}(\|f(x) - g(x)\| + \|Df(x) - Dg(x)\|) < \varepsilon, \text{ for any } k \subset \mathcal{U}\}$$

The concept can be applied to vector fields and all diffeomorphism.

Write the set of all vector fields belonging to $\mathscr{C}^1$ on $\mathscr{U}$ as $\mathscr{H}^1(\mathscr{U})$ and the set of all diffeomorphism as $\mathrm{Diff}^1(\mathscr{U})$. We will give the definition of the structural stability of vector fields or diffeomorphism.

Let $f \in \mathscr{H}^1(\mathscr{U})$ (or $F \subseteq \mathrm{Diff}^1(\mathscr{U})$). If there exists $\varepsilon > 0$ so that for each $g \in B_\varepsilon(f) \subseteq \mathscr{H}^1(\mathscr{U})$, we have $\|f - g\|_1 < \varepsilon$, i.e. $g$ obtained after $f$ undergoes a small perturbation satisfies the requirement that $|f - g|$ and $|Df - Dg$ are small quantities of the same order, then we say $f$ is structurally stable.

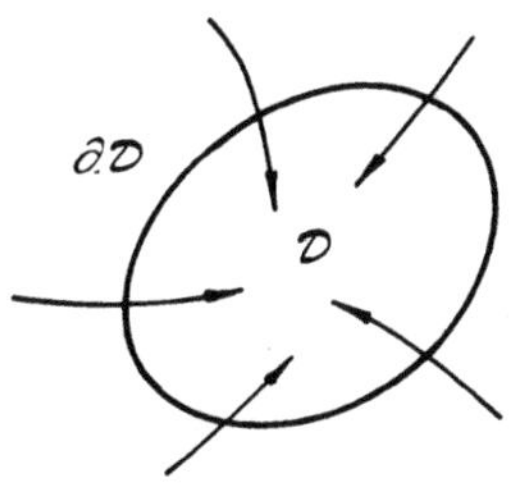


Fig. 3.6

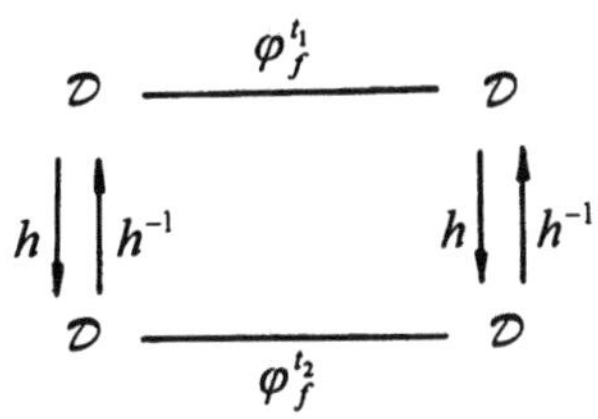


Fig. 3.6a

According to the above definition, if $f$ is structurally stable then $g$ is obtained after $f$ undergoes a small perturbation. $f$ is a $\mathscr{C}^0$ topological conjugate (or the equivalent), i.e. there is homomorphic transformation $h:\mathscr{D} \to \mathscr{D}$, which transfers the equilibrium points of $f$ into those of $g$ and the orbits of $f$ into those of $g$, preserving the direction of the orbits, so $h\varphi_f^{t_1} = \varphi_g^{t_2} h$. This is written as $f \sim g$, where $\varphi_f^{t_1}$ and $\varphi_g^{t_2}$ are flows for $f$ and $g$ respectively. There is an exchange diagram between them (Fig. 3.6a). In general, $t_1 \neq t_2$, but their directions are the same.

**Note 1** Liapunov stability only studies the dynamic properties of single dynamical system undergoing an initial perturbation. Structural stability discusses the relation of topological properties among adjacent dynamical systems in a family. A dynamical system becomes a "close" dynamical system after it undergoes the perturbation. We need to find out under what conditions the topological structure of systems will preserve invariance. This is the problem that structural stability will deal with.

**Note 2** More exactly, the above definition only explains the $\mathscr{C}^1$ structural stability. We can also deal with the corresponding structural stability in some subsets of $\mathscr{H}^1(\mathscr{U})$ (or $\mathrm{Diff}^1(\mathscr{U})$) in a manner analogous to that in the above statement.

### 3.5.2 Smoothness of Homomorphic Transformation *h*

We demand that $f, g \in \mathscr{C}^1$ and $h \in \mathscr{C}^0$ in the definition of topological equivalence. Why is the smoothness of $f$ (or $g$) different from $h$? We will answer this problem as follows.

1. Let $f$ and $h$ belong to $\mathscr{C}^0$. Suppose $f(x)$ is an increasing function as shown in Fig. 3.7(a).

Take a small perturbation $p(x)$ and $p(x) \in \mathcal{C}^0$. So $|p(x)|$ is very small. Since $p(x)$ is not required to belong to $\mathcal{C}^1$, $|p'(0)|$ may be very large, i.e. $p'(0) << 0$ as shown in Fig. 3.7(b). In other words, the slope of $p(x)$ is zero when far from the origin but has very large negative values near the origin. So its slope is not continuous.

Taking $g(x) = f(x) + p(x)$, we can see that $|g - f|_0$ is very small while $|g - f|_1$ is very large, as shown in Fig. 3.7(c). Because $f$ has only one equilibrium point but $g$ has three near zero, i.e. when $f, g \in \mathcal{C}^0$, $f$ and $g$ are not topological equivalent. If $f$ and $g$ belong to $\mathcal{C}^1$, then it is impossible for $g$ to change rapidly near the origin because of the continuity of the derivative of $g$. So the number of equilibrium points of $g$ will be the same as those of $f$, that is, $y = h(x)$.

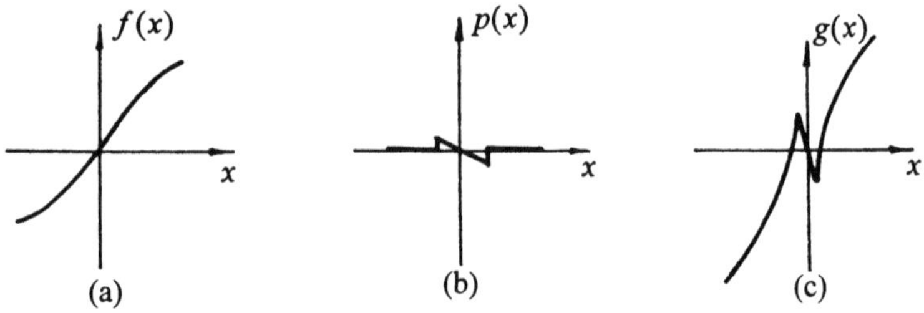


Fig. 3.7

2. Take $f, g \in \mathcal{C}^1$, $h \in \mathcal{C}^1$

$$\frac{dy}{dt_1} = f(g) \qquad \frac{dx}{dt_2} = g(x) \tag{3.36}$$

and $h(x)$. Suppose $t_2 = \alpha(x)t_1$, here, $\alpha(x) > 0$, $t_2 > t_1$. Substituting $t_2$ into eq. (3.36), we have d

$$\frac{dx}{dt_1} = \alpha(x)g(x) \tag{3.37}$$

Also, we obtain

$$\frac{dy}{dt_1} = \frac{dh(x)}{dt_1} = D_x h\left(\frac{dx}{dt_1}\right) = f(y) = f(h(x)) \therefore \frac{dx}{dt_1} = (D_x h)^{-1} f(h(x)) \tag{3.38}$$

From eq. (3.37) and eq. (3.38), we get

$$\alpha(x)g(x) = (D_x h)^{-1} f(h(x)) \tag{3.39}$$

Let $p$ be an equilibrium point of $g(x)$, i.e. $g(p) = 0$. Since $(D_x h(p))^{-1} \neq 0$,

$$f(h(p)) = 0 \tag{3.40}$$

Deriving the linearised equation (3.39) for $x$ at $p$, we get

$$\alpha(p)Dg(p) = (Dh)^{-1} Df(h(p))(Dh) \tag{3.41}$$

From the above equation, $Dg(h)$ is derived from $D(hf(p))$ by a similar transformation. So $Dg$ and $Df$ are similar matrices, and their eigenvalues are proportional. Therefore, $f \sim g$ as $f, g \in \mathcal{C}^1$ and $h \in \mathcal{C}^1$.

3. Demonstrate that $f$ and $g$ are topologically equivalent when $h \in \mathcal{C}^0$. Let us see some examples

**Example 1**

$$\begin{bmatrix} y_1 \\ y_2 \end{bmatrix}' = \begin{bmatrix} 1 & 0 \\ 0 & -2 \end{bmatrix} \begin{bmatrix} y_1 \\ y_2 \end{bmatrix} \tag{3.42}$$

Obviously, the equilibrium point (0, 0) of the system is a saddle point. If we take $y = h(x)$, then

$$\begin{bmatrix} x_1 \\ x_2 \end{bmatrix}' = \begin{bmatrix} 1+\varepsilon & 0 \\ 0 & -2 \end{bmatrix} \begin{bmatrix} x_1 \\ x_2 \end{bmatrix} \tag{3.43}$$

where $\varepsilon <<$ .

The structure of the phase portraits of eq. (3.42) and eq. (3.43) are topologically equivalent at equilibrium point (0,0). Therefore, if $h(x)$ is homomorphic, $\varepsilon$ is a small quantity because $h(x)$ is continuous, and the phase portraits of eq. (3.42) and eq. (3.43) have the same topological structure. It is unnecessary to demand that $h \in \mathcal{C}^1$, that is the eigenvalues of the linearized operator coming from $f$ and $g$ through a homomorphic transformation are in proportion.

**Example 2 A single degree of freedom system without damping**

$$x'' + \omega^2 x = 0 \qquad x \in [-1,1] \tag{3.44}$$

Equation (3.44) can be written as

$$x' = y, \qquad y' = -\omega^2 x, \qquad (x,y) \in \mathcal{K} \tag{3.45}$$

where $\mathcal{K}$ is an elliptic region $x^2 + \left(\frac{y}{\omega}\right)^2 \leq 1$. The origin is the centre of system (3.45), and is full of closed orbits around the origin in $\mathcal{K}$.

Consider its perturbation system

$$x' = y, \qquad y' = -\omega^2 x - \alpha y, \qquad (x,y) \in \mathcal{K} \tag{3.46}$$

When $\alpha > 0$ $(< 0)$, the origin is a stable (unstable) focus and there is no closed orbit in $\mathcal{K}$. Since systems (3.46) and (3.45) are not equivalent, system (3.45) is structurally unstable.

**Note** System (3.45) is structurally unstable in the $\mathcal{C}^1$ vector field. If we concentrate our discussion on the $\mathcal{C}^1$ Hamiltonian vector field, then system (3.45) will preserve the structure after undergoing small perturbations in the set. So we call system (3.45) structurally stable based on Hamiltonian vector fields.

**Example 3** If system

$$x' = f(x) \qquad x \in \mathcal{R}^n \tag{3.47}$$

has a non-hyperbolic equilibrium point $p$, then it is structurally unstable. In this case $Df(p)$ has some eigenvalues with zero real parts. If system (3.47) is perturbed, then the equilibrium point may disappear, or the eigenvalues of the linearised matrix of the system has non-zero real parts, i.e. responding equilibrium points will be hyperbolic. Hence a perturbation system is not topologically equivalent to a non-perturbation system (3.47).

Similarly, if diffeomorphism $x \rightarrow F(x)$ has non-hyperbolic fixed points, then discrete dynamical systems are also structurally unstable. By using the relationship of closed orbits on flows and the Poincaré map, it is known that the system (3.47) will be structurally unstable if the system has non-hyperbolic orbits.

**Example 4** If system

$$x' = f(x) \qquad x \in \mathcal{R}^2 \tag{3.48}$$

has homoclinic or heteroclinic orbits which connect saddle points, then the system is structurally unstable. Let us see how heteroclinic orbits are used to demonstrate this conclusion. Suppose there is a heteroclinic orbit $\gamma = \mathcal{W}^u(p_1) \cap \mathcal{W}^s(p_2)$ (cf. Fig. 3.8(a)) connecting saddle points from $p_1$ to $p_2$ in system (3.48). If (3.48) is perturbed suitably, the saddle connection can be split as shown in Fig. 3.8(b). Obviously, the two systems are not topologically equivalent.

But the conclusion of example 4 cannot be applied directly to $n$-dimensional systems ( $n > 2$ ). To do this we have to employ the concept of transversal section.

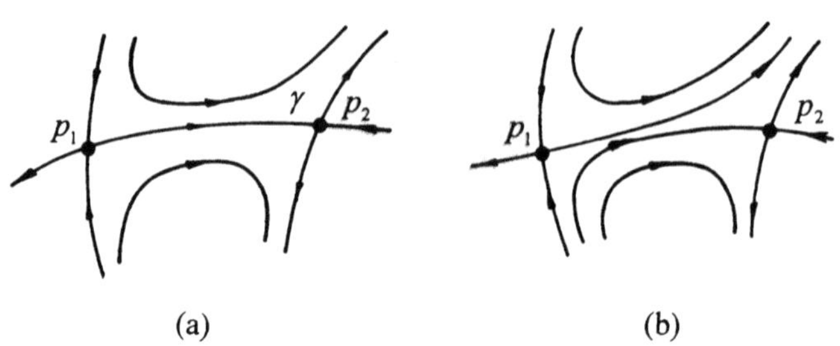


Fig. 3.8

It is easy to understand that if planar system (3.48) has homoclinic or heteroclinic orbits, then the relevant stable and unstable manifolds at saddle points coincide. Therefore, they are unlikely to intersect transversely. If the stable manifolds of system (3.47) are not transversal to the unstable manifolds at equilibrium points, then the system is structurally unstable. The conclusion can be applied to closed orbits of system (3.47), fixed points of discrete dynamical systems, and stable and unstable manifolds of periodic orbits.

From the above conclusion, if system (3.47) is structurally stable, then all eigenvalues and closed orbits of the system are hyperbolic and their stable and unstable manifolds are transversal if intersecting. This is true for discrete dynamical systems. However, as can be seen from some conditions based on the above discussion, we cannot conclude generally that the structural stability of dynamical systems cannot be guaranteed since study of structural stability often deals with a deeper global analysis.

### 3.5.3 Structural Stability of Planar Systems

The important result here is the introduction of the structural stability of planar vector fields. Note $\mathcal{B}^2 = \{(x,y)|x^2+y^2 \le 1\}$. Consider the system

$$x' = P(x,y),\ \ y' = Q(x,y) \qquad (x,y) \in \mathcal{B}^2 \qquad (3.49)$$

Assume function $P,Q \in \mathcal{C}^2(\mathcal{B}^2, \mathcal{R})$, i.e. vector $(P,Q) \in \mathcal{H}^1(\mathcal{B}^2)$. Let vector field $(P,Q)$ not be a tangent to boundary $\partial\mathcal{B}^2$ of $\mathcal{B}^2$.

**Theorem 3.4 Andronov—Pontriagin theorem**

Let $(P,Q) \in \mathcal{H}^1(\mathcal{B}^2)$, then system (3.49) is structurally stable if and only if,

(1) The number of equilibrium points and closed orbits is finite and each is hyperbolic in eq. (3.49).

(2) There are no orbits connecting saddle points in eq. (3.49). Refer to [45] and [46] for more details of the proof. Here only some explanation about the conditions of the theorem will be given. Condition (1) requires that the equilibrium points be only sinks, sources or saddle points; and closed orbits be only stable (or unstable) simple limit cycles and their numbers be finite. Obviously this condition is necessary. Condition (2) requires that there be no homoclinic or heteroclinic orbits. This condition is also necessary. Thus for a structurally stable planar system (3.49), the separatrices of saddle points are not likely to meet. In fact, from condition (2) they can not intersect either non-transversally or transversally as shown in Fig. 3.9(b).

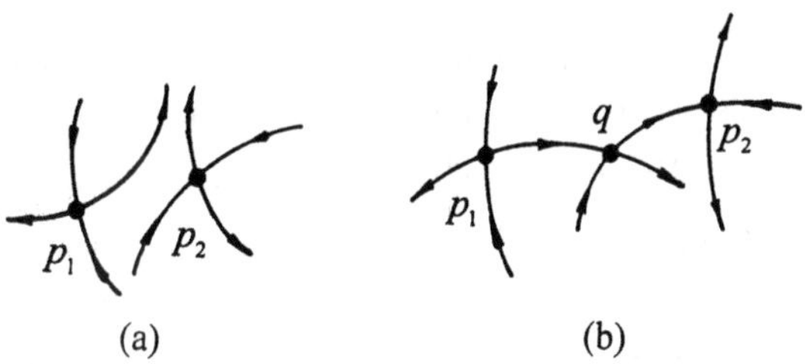

Fig.3.9

For the planar vector fields, if there is a compact set $\mathcal{D} \subseteq \mathcal{R}^2$ so that the vector field is not tangential to $\partial\mathcal{D}$, or trajectories on $\partial\mathcal{D}$ pointing inward (or outward), then Andronov-Pontriagin theorem can be generalised to the system based on a two-dimensional oriented compact manifold $\mathcal{M}^2$.

**Theorem 3.5 Peixoto's theorem**

Let $\mathcal{M}^2$ be a two-dimensional oriented compact manifold and $f \in \mathcal{H}^1(\mathcal{M}^2)$, then vector field $f$ is structurally stable if and only if

(1) the number of equilibrium points and closed orbits is finite and each is hyperbolic;

(2) there are no orbits connecting saddle points.

(3) non-wandering sets are composed of equilibrium points and closed orbits.

For the proof refer to [45]. Peixoto's theorem is an expansion of the Andronov—Pontriagin theorem since condition (3) can be derived from condition (2) for the system on $\mathcal{B}^2$. In fact, non-wandering sets are made up of only equilibrium points, closed and homoclinic orbits for planar autonomous systems. However, it is impossible for homoclinic or heteroclinic orbits to exist in this case from condition (2), so condition (3) can be obtained.

**Theorem 3.6** Let $\mathcal{M}^2$ be a two-dimensional oriented compact manifold. If $\Sigma$ represents a subset made up of all structurally stable vector fields in $\mathcal{H}^1(\mathcal{M}^2)$, then $\Sigma$ is dense and open in $\mathcal{H}^1(\mathcal{M}^2)$.

Readers can find the proof of the theorem in [45]. The theorem indicates that structurally stable systems are universal and structurally unstable systems can be approximated arbitrarily by structurally stable systems in $\mathcal{H}^1(\mathcal{M}^2)$. In order to discuss the subject further, let us introduce some ideas about generic.

**Definition 3.10** Let $\mathcal{V}$ be a metric space and $\mathcal{S}$ be a subset of $\mathcal{V}$. If $\mathcal{S}$ is the intersection of numerable dense open subsets, then $\mathcal{S}$ is called the remainder set.

In particular, $\mathcal{S}$ is a remainder set when $\mathcal{S}$ is a dense open set in $\mathcal{V}$.

From the definition, we know that property $p$ is generic if the set of vector fields, satisfying property $p$, is a remainder set in $\mathcal{H}^1(\mathcal{U})$.

The opposite of generic is the idea of degeneracy.

**Definition 3.11** Property $p$ is called a degenerate if a complement set of vector fields in $\mathcal{H}^1(\mathcal{U})$ (or $\mathrm{Diff}^1(\mathcal{U})$), satisfying property $p$, is a remainder set.

**Example 1** Property $p_1$ in $\mathcal{H}^1(\mathcal{U})$ (or $\mathrm{Diff}^1(\mathcal{U})$) is defined in such a way that each equilibrium point (fixed point) and closed orbit (periodic orbit) is hyperbolic and their stable and unstable manifolds are transversal when intersecting. So property $p$ is generic (cf. [47]).

**Example 2** In $\mathcal{H}^1(\mathcal{U})$ (or $\mathrm{Diff}^1(\mathcal{U})$), property $p_2$ is definite in such a way that a non-wandering set is the closure of equilibrium points and closed orbits (or fixed points and periodic orbits). So property $p_2$ is also generic (cf. [47]).

From Peixoto's theorem, we know that structural stability is generic and structural instability is degenerate in $\mathcal{H}^1(\mathcal{M}^2)$.

So far, most of the important results of two-dimensional structural stability have been based on two-dimensional compact manifolds. However, manifolds that we meet in practice are often not compact, such as phase-plane $\mathcal{R}^2$. So it is necessary to study structural stability on 2-dimensional open manifolds. Here we will not be able to study this problem further since space is limited. But the results of two-dimensional compact manifolds have important functions for the study of two-dimensional open manifolds.

The structural stability of higher dimensional dynamical systems is very complex, so at present we can only discuss some dynamical systems with simple topological characteristics and certain extensions, such as the Morse—Smale system (cf. [43]).

Hyperbolicity of equilibrium points (fixed points) and closed orbits (periodic orbits) and transversal intersection of stable and unstable manifolds are the necessary conditions for structural stability for $n$-dimensional vector fields. It should be noted that stable and unstable manifolds can intersect transversally on homoclinic or heteroclinic orbits for vector fields of dimension greater than two.

Also, all structurally stable systems with vector fields of dimension greater than two make up an open subset, but may not be dense, so structural stability may not be generic.

Finally, we give some simple results.

**Theorem 3.7** Let $\mathcal{U} \subseteq \mathcal{R}^n$ be a compact set and $f \subseteq \mathcal{H}^1(\mathcal{U})$. $f$ is structurally stable if it satisfies

(1) $f$ has only one equilibrium point $p \in \mathcal{U}$ and it is a sink or a source;

(2) $f$ points inwards or outwards on boundary $\partial\mathcal{D}$;

(3) $\lim_{t \to +\infty} \varphi^t(x) = p$ or $\lim_{t \to -\infty} \varphi^t(x) = p$, for any $x \in \mathcal{U}$.

The proof can be found in [40].

**Theorem 3.8** Let $\mathcal{U} \subseteq \mathcal{R}^n$ be a compact set, $f = -\nabla V$ be a gradient vector field and $f$ be non-tangential to $\partial\mathcal{U}$. $f$ is structurally stable in $\mathcal{U}$ if all equilibrium points are hyperbolic and their stable and unstable manifolds are transversal when intersecting. In addition, structurally stable systems make up a dense open subspace in the set of the gradient system, where function $V \in \mathcal{C}^2(\mathcal{U}, \mathcal{R})$

$$\nabla V = \left(\frac{\partial V}{\partial x_1}, \ldots, \frac{\partial V}{\partial x_n}\right)^T$$

The autonomous system defined by this grad vector field

$$x' = \nabla V, \qquad x \in \mathcal{R} \tag{3.50}$$

is called a gradient system. If $\nabla V(x) \neq 0$ at $x$, then is called a regular point, otherwise a critical point. Obviously, $\bar{x}$ is an equilibrium point of a gradient system if and only if $\bar{x}$ is the critical point of the function . It is easy to verify that $\bar{x}$ is an asymptotically stable (or unstable) equilibrium point when $\bar{x}$ is an isolated minimum (or maximum) point.

A non-wandering set is composed of equilibrium points in gradient systems.

There are no sinks and sources in a Hamiltonian system. So hyperbolic equilibrium points are only saddle ones. Obviously, if there are nonhyperbolic equilibrium points in a Hamiltonian, such as centre points, or stable and unstable manifolds intersect through saddle points non-transversally, then the system is structurally unstable.

# Chapter 4

# Liapunov—Schmidt Reduction

The theory and application of bifurcation are presented in this chapter and chapter 5. This chapter describes the basic concepts of bifurcation in ordinary differential equations, Liapunov—Schmidt reduction (LS reduction for short), singularity theory and applications of all these theories. Chapter 5 introduces the centre manifold theorem and the normal form of vector fields. Chapter 6 presents the Hopf bifurcation and double zero eigenvalues. Chapter 7 explains the applications of the averaging method in bifurcation theory.

## 4.1 Basic Concepts of Bifurcation

Bifurcation theory is concerned with the instability of the solution of ordinary differential equations arising from changes in parameters and the consequential variations in a number of solutions. We have discussed the structural stability of dynamical systems in section 3.5 of chapter 3. Bifurcations of dynamical systems (primarily bifurcation in ordinary differential equations) will be introduced in this chapter and discussed further in chapter 5 to 7. If a dynamical system is structurally unstable, a slight proper perturbation would cause a sudden change in quality of its topological structure. Such a change is bifurcation. Bifurcation always occurs in unstable structures and this confirms the close association of bifurcation with structural stability. The study of dynamical systems will be focused on the change of qualitative behaviour caused by structural instability as well as stability. Moreover, a series of bifurcations may give rise to chaotic motion in the dynamical system, and this indicates that bifurcation and chaos are closely related complicated motions. Being a comprehensive subject, the bifurcation of dynamical systems is an integral part of the study of dynamical systems and non-linear differential equations, and has broad prospects of application. Since the 18th century studies on the loss of stability in celestial, elastic and fluid mechanics, and non-linear oscillation, have come within the scope of bifurcation. By the 1970s, with the impetus from dynamical systems, non-linear analysis, non-linear differential equations, and the aid of highly effective numerical calculation methods, the mathematical theory of bifurcation has begun to take shape and find wide uses in mechanics, physics, chemistry, biology, ecology, cybernetics, numerical calculation, engineering, technology, economics and sociology. At present, the study of bifurcation is advancing in-depth and at high speed both in theory and application. In practice, since many systems contain one or more parameter, we determine whether or not the topological structure of a system changes when continuous slight changes in parameters take place. Let us examine a few examples first.

**Example 1 Bifurcation of the buckling of a wire arch**

Imagine an arch made of a metallic wire, Fig. 4.1(a) and (b). The wire arch can be used to demonstrate buckling caused by gravity. The equation of the arch in motion is [124]: $\frac{d\theta}{dt} = F(l,\theta)$ (4.1)

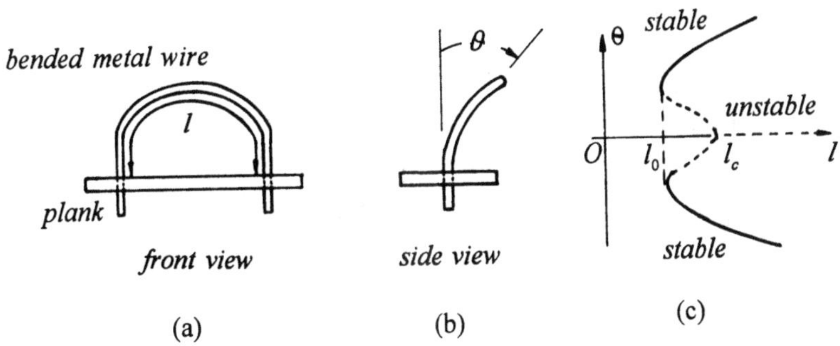


Fig. 4.1

When the length $l$ of the wire arch is very small, eq. (4.1) has only one stable solution $\theta = 0$ (expressing a perpendicular state), Fig. 4.1(b). When $l < l_c$ (the critical value of the loss of stability), the position of leftward or rightward buckling of the arch is stable. When $l_0 < l < l_c$, there may be three stable steady solutions: the upright one $(\theta = 0)$, the leftward buckling or rightward buckling $(\theta \neq 0)$. When $l > l_c$, the arch cannot keep upright but inclines leftward or rightward. The relationship between the change in length of the wire and the number of positions can be expressed by a curve. This is the bifurcation diagram of the wire arch, Fig. 4.1(c).

The $\theta$ in the figure is the state variable, and the length $l$ of the wire is the bifurcating parameter. It can be seen from the bifurcation diagram that with variation of the magnitude of the wire, i.e., from large to small or vice versa, the path in the bifurcation diagram varies. Such a phenomenon is called hysteresis.

**Example 2** Consider a one-dimensional system

$$x' = \mu x - x^3 \qquad x \in \mathcal{R} \tag{4.2}$$

where $\mu \in \mathcal{R}$ is a bifurcating parameter.

From eq. (4.2) we know that when $\mu \leq 0$, eq. (4.2) has an equilibrium branch $x = 0$, which is asymptotically stable. When $\mu > 0$, eq. (4.2) has three equilibrium points, among which the point $x = 0$ is unstable, and the point $x = \pm\sqrt{\mu}$ is asymptotically stable. Fig. 4.2 is the phase portrait of the equilibrium points of system (4.2), when $\mu$ is fixed. In addition, Fig. 4.2 also shows the bifurcation diagram—the change in position of the equilibrium points of system (4.2) with change of $\mu$. The solid line in the figure denotes the stable points (described as S), while the broken line denotes the unstable points (described as u). Obviously, when $\mu \le 0$ and $\mu > 0$, the system has different topological structures. In other words, a sudden change in quality occurs at $\mu = 0$ in the topological structure, and bifurcation of equilibrium branches appears at that point (such a bifurcation is called a pitchfork bifurcation).

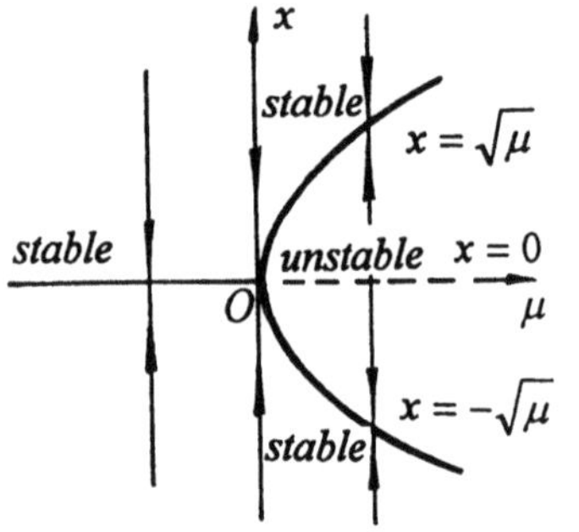


Fig. 4.2

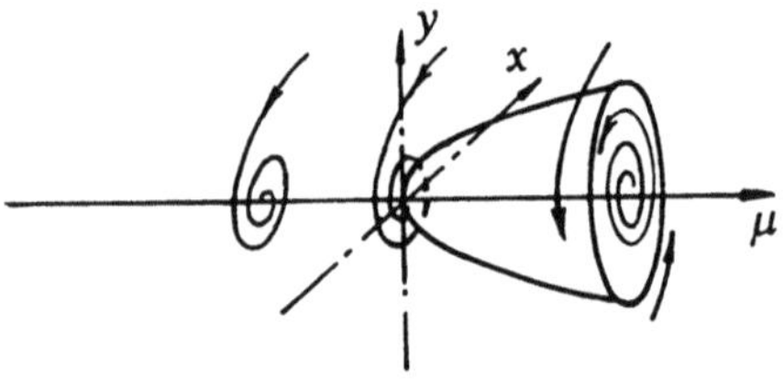


Fig. 4.3

**Example 3** Consider a planar system

$$\begin{cases} x' = -y + x[\mu - (x^2 + y^2)] \\ y' = x + y[\mu - (x^2 + y^2)] \end{cases} \qquad (x,y) \in \mathcal{R}^2 \tag{4.3}$$

where $\mu \in \mathcal{R}$ is a bifurcating parameter.

In polar coordinates eq. (4.3) can be written as

$$\begin{aligned} r' &= r(\mu - r^2) \\ \theta' &= 1 \end{aligned} \tag{4.4}$$

From the above formulae we know that when $\mu \le 0$, eq. (4.3) has a unique asymptotically stable focus (0,0). When $\mu > 0$, the focus (0,0) becomes the unstable focus of eq. (4.3). At this point there is also a stable limit cycle $r = \sqrt{\mu}$. On the transversal section perpendicular to the $\mu$-axis Fig. 4.3 shows the phase portrait of system (4.3) when $\mu$ is fixed. Moreover, Fig. 4.3 also shows the bifurcation diagram—the change in position of stable points and the limit cycle of eq. (4.3) with the change of $\mu$. From the figure, we know that a sudden change occurs at $\mu = 0$ in the topological structure of system (4.3), when a bifurcation appears at this point (such a bifurcation is called a Hopf bifurcation).

The existence of bifurcations implies that a small change in parameter near the bifurcation value will lead to a change in quality of the topological structure of the system. As a result, the structure is not stable.

In this chapter, emphasis will be placed on the bifurcation of dynamical systems described by ordinary differential equation with parameters. Some relevant concepts will be introduced first.

Suppose $\mathcal{U} \subseteq \mathcal{R}^n$, $\mathcal{J} \subseteq \mathcal{R}^m$. Consider the $n$ ordinary differential equations with m parameters

$$x' = f(x, \mu) \tag{4.5}$$

where $x \in \mathcal{U}$ is a state variable, $\mu = (\mu_1, \mu_2, \ldots, \mu_m)^T \subseteq \mathcal{R}^m$ is the vector of bifurcating parameters (also known as the control variable). Suppose for fixed $\mu \in \mathcal{J}$ there is $f(x, \mu) \in \mathcal{H}'(\mathcal{U})$, where $\mathcal{H}'(\mathcal{U})$ is the set of all $\mathcal{C}^1$ vector fields on $\mathcal{U}$.

**Definition 4.1** We say that a bifurcation appears at $\mu = \mu_0$ in system (4.5) (or vector field $(x, \mu)$) if parameter $\mu$ varies continuously, and a sudden change takes place at $\mu \in \mathcal{J}$ in the topological structure of the given system (4.5). $\mu_0$ is a bifurcating value (critical value) and the set composed of bifurcating values in space of parameter $\mu$ is called the bifurcation set.

**Note** The concept of bifurcation of the system with parameter is given by definition 4.1 is for the subset $\{f(x, \mu) \mid \mu \in \mathcal{J}\}$ (given by $f$) of $\mathcal{H}'(\mathcal{U})$, and is different from the general concept of bifurcation discussed in the set of vector fields $\mathcal{H}'(\mathcal{U})$. Obviously, the study area of the latter is much wider than that of the former. However, bifurcation of the system with parameters is often found in practical applications, and each vector field $f(x, \mu)$ corresponds to point $\mu$ in the parameter space $\mathcal{R}^m$, thereby facilitating the study of bifurcation. That is why bifurcation of such a kind becomes the chief concern in the present study.

In order to give a clear picture of bifurcation, we plot in the space $(x, \mu)$ the change of the limit set (equilibrium points, limit cycle, etc.) of system (4.5) with change of parameters. This is called the bifurcation diagram. Figures 4.1(c), 4.2 and 4.3 are all such diagrams.

Now let us consider the subset of all vector fields comprised of $m$ parameters in $\mathcal{H}'(\mathcal{U})$: $D = \{f(x) \mid f \in \mathcal{H}'(\mathcal{U}), \ \mu \in \mathcal{J}\}$.

**Definition 4.2** Suppose $\mu_0 \in \mathcal{J}$ is a bifurcating value of $f \in \mathcal{D}$. If there exists a neighbourhood $\mathcal{W} \subseteq \mathcal{D}$ of $f$ so that there is a homomorphism $h: \mathcal{U} \times \mathcal{J} \to \mathcal{U} \times \mathcal{J}$, $(x, y) \to (y(x, \mu), \gamma(\mu))$ for any $g \in \mathcal{W}$, which maps orbits of vector field $f(x, \mu)$ to that of vector field $g(y, \gamma)$ and preserves time-orientation, then the bifurcation of $f$ at $\mu_0$ is considered generate, otherwise the bifurcation is degenerate.

Obviously, a small perturbation in $\mathcal{D}$ will not change the qualitative behaviour of generic bifurcations. But this will not be so for the degenerate bifurcation. Therefore, we can say that generate bifurcation is structurally stable and degenerate is structurally unstable. We can unfold degenerate bifurcation into generate by adding some parameters.

In general, it is necessary to study the global topological structure to attain a complete analysis of bifurcation. It is a tough and complicated task, often difficult to fulfil. In practice, one investigates only the change in the topological structure of the equilibrium points or the path curve in the neighbourhood of the closed orbit, or rather, the bifurcation of equilibrium points or the vector field in a neighbourhood of the closed orbits. Bifurcations of this kind are called local bifurcations. If the global behaviour of the vector field is to be considered in the analysis of bifurcation, such bifurcations are global bifurcations. Of course, "local" and "global" are only relative. Sometimes local ones can affect the structure of global ones. In this chapter emphasis is placed on the local.

Conventionally, bifurcation falls into two classes: static and dynamical according to the object of study. Static bifurcation is concerned with the change in the number of solutions caused by the change of parameters $\mu$ in equations of the following form:

$$f(x,\mu)=0, \quad x\in\mathcal{U}\subseteq\mathcal{R}^n,\ \mu\in\mathcal{J}\subseteq\mathcal{R}^m \tag{4.6}$$

The dynamical bifurcations are those mentioned in definition 4.1. They induce bifurcation of equilibrium points, closed orbits, homoclinic and heteroclinic orbits, and invariant tori. As the solutions of eq. (4.6) correspond to the equilibrium points of eq. (4.5), static bifurcation comes under bifurcation of the equilibrium points. In other words, static bifurcation is virtually contained in dynamical bifurcation. In practical applications many problems (e.g. the periodic solutions of non-linear vibration (determinate solution)) are static in nature. Static bifurcation is therefore always the important aspect in the study of bifurcation.

To sum up, the main topics of bifurcation are to study:

(1) determination of bifurcating sets, i.e. the necessary and sufficient conditions of bifurcation;

(2) change in topological structure with change of parameters in universal unfolding (i.e. the study of bifurcating behaviour) when bifurcation takes place;

(3) calculation of the bifurcating solution (especially equilibrium points, limit cycles, etc.);

(4) interaction between different bifurcations, and their relationships with other behaviours of dynamical systems, such as lock phase, chaos, etc.

## 4.2 Classification of the Bifurcations of Planar Vector Fields

According to Andronov—Pontriagin's theorem, we can express the general result as follows.

**Theorem 4.1** Bifurcations will appear (i.e. structurally unstable) in $\mathcal{H}^1(\mathcal{B}^2)$ if and only if one of following conditions is satisfied:

(1) there are non-hyperbolic equilibrium points
(2) there are non-hyperbolic closed orbits
(3) there are homoclinic or heteroclinic orbits

where $\mathcal{B}^2=\{(x,y)\,|\,x^2+y^2\le 1\}$.

In order to give a clear idea of the change in topological structure in planar systems when bifurcation appears, let us discuss the planar system autonomous with the simple parameter family

$$\begin{cases} x' = P(x, y, \mu) \\ y' = Q(x, y, \mu) \end{cases} \qquad (x, y) \in \mathcal{B}^2 \tag{4.7}$$

where $P$ and $Q$ all have continuous partial derivatives.

Let $\mu_0$ be a bifurcating value. According to theorem 4.1 the bifurcation of system (4.7) can be divided into three types.

**1. Bifurcations pertaining to equilibrium points**

Let system (4.7) have a non-hyperbolic equilibrium point $(x_0, y_0)$ and $A$ be the matrix of linearisation at $\mu = \mu_0$. If $A$ has a zero eigenvalue, then there are higher order equilibrium points. As shown in Fig. 4.4(a), there is a saddle node point $(x_0, y_0)$ when $\mu = \mu_0$. When $\mu < \mu_0$, there is no equilibrium point, but when $\mu > \mu_0$, the system has a saddle point and a node point. Bifurcation of such a kind is called saddle-node bifurcation.

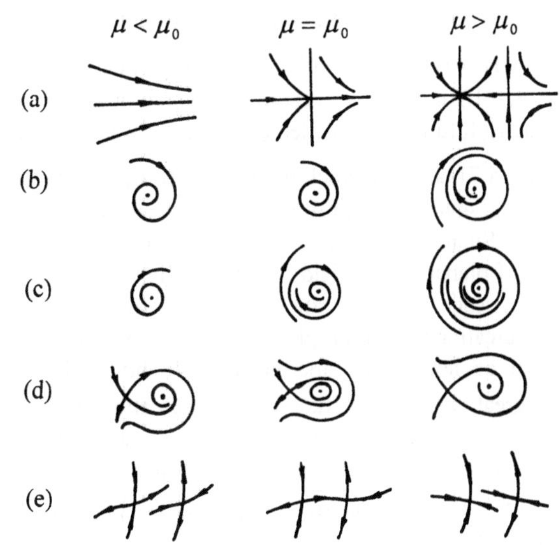


Fig. 4.4

If $A$ has a pair of purely imaginary eigenvalues and $(x_0, y_0)$ is the fine focus of system (4.7) when $\mu = \mu_0$, then a limit cycle may emerge from the equilibrium point (called a Hopf bifurcation) when $\mu$ changes. As in Fig. 4.4(b), when $\mu < \mu_0$ the system has a stable focus (it is particularly a stable focus at $\mu = \mu_0$), but no closed orbit is nearby. When $\mu > \mu_0$, this equilibrium point becomes an unstable focus and near it there is a limit cycle. When $\mu \to \mu_0 + 0$, this cycle tends to the equilibrium point.

If $A$ has a pair of purely imaginary roots and $(x_0, y_0)$ is the real centre of system (4.7) when $\mu = \mu_0$, that is, near $(x_0, y_0)$, both of them are closed orbits. A limit cycle may bifurcate to one of the closed orbits when $\mu$ changes, and the equilibrium point will no longer be the centre. Such a bifurcation is called a Poincaré bifurcation [93].

**2. Bifurcations of closed orbits**

Let system (4.7) have a non-hyperbolic closed orbit $\Gamma$ when $\mu = \mu_0$. Using the stability theorem of periodic solutions, we know the characteristic exponent of the periodic orbit $\Gamma$: $\oint_\Gamma \mathrm{div}(P, Q)\mathrm{d}t = 0$. That is, $\Gamma$ is a multiple cycle. The closed orbits may appear or disappear suddenly when $\mu$ changes. Such a change is called the bifurcation of multiple cycles. As in Fig. 4.4(c), the system has two semi-stable limit cycles when $\mu = \mu_0$. A non-closed orbit is found when $\mu < \mu_0$, but two limit cycles emerge when $\mu > \mu_0$. The two limit cycles tend to one cycle as $\mu \to \mu_0 + 0$, and such a bifurcation is called the bifurcation of a double semi-stable cycle.

**3. Bifurcations pertaining to homoclinic and heteroclinic orbits**

If system (4.7) has a homoclinic orbit at $\mu = \mu_0$, then this homoclinic orbit may disappear (Fig. 4.4(d)) and a limit cycle may bifurcate from the homoclinic orbit with changing $\mu$. Such a bifurcation is called bifurcations of homoclinic orbits.

If system (4.7) has a heteroclinic orbit as $\mu = \mu_0$, then this heteroclinic orbit may disappear (Fig. 4.4(e)) or a limit cycle may bifurcate from the heteroclinic orbit with changing $\mu$. Such a bifurcation is called a bifurcation of heteroclinic orbits. Of the bifurcations of planar vector fields discussed above, some are in the category of local bifurcation, such as bifurcation of high-order equilibrium points, Hopf bifurcations, bifurcations of multiple cycles, etc. Others belong to the global bifurcation, such as homoclinic and heteroclinic bifurcations. The results thus obtained are of great value in the study of the bifurcation of vector fields on $\mathcal{R}^2$. In addition, according to the Peixoto theorem we can obtain the corresponding classification of the bifurcation of vector fields on the two-dimensional directional compact manifolds $\mathcal{M}^2$. However, the bifurcations of high-order dimensional systems are very complicated.

It should be noted that bifurcations may appear one after the other at locations of different parameters within the range of parameter change. For the static bifurcation shown in Fig. 4.5, for $\mu = \mu_1$ an initial bifurcation arises from the trivial solution $x = 0$. $\mu$ continues to vary, and when $\mu = \mu_2$, a second-order bifurcation appears along the initial bifurcation branch.

**Andronov's Definition of the Order of Bifurcating Points in Planar Vector Fields**

If the vector field $f$ in the neighbourhoods of $y_0$ is always structurally stable, $y_0$ is a bifurcating point of zero order.

If $y_0$ is not the bifurcating point of order zero, and there exists in $\mathcal{C}^1(\mathcal{D})$ a neighbourhood $\mathcal{U}$ of $f$ (here $\mathcal{D}$ is neighbourhood $y_0$). In $\mathcal{U}$, $g$ has only bifurcating points of zero order, that is, when $g \in \mathcal{U}$ is structurally stable, then $g \sim f$, or when $g$ has bifurcating points of zero order, $y_0$ is the first-order bifurcating point of $f$.

If $y_0$ is not the bifurcating point of zero order or first order, and there is a neighbourhood $\mathcal{U}$ of $f$ in $\mathcal{C}^1(\mathcal{D})$ so that $g \in \mathcal{U}$ has a bifurcating point of zero order or first order, then $y_0$ is the second-order bifurcating point of $f$.

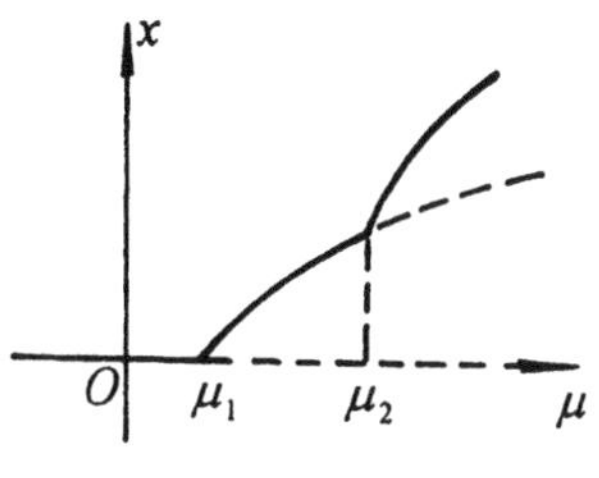


Fig. 4.5

If $y_0$ is not the ($k$–1)-order bifurcating point, and there is a neighbourhood $\mathcal{U}$ of $f$ in $\mathcal{C}^1(\mathcal{D})$ so that $g \in \mathcal{U}$ has a $k$–1 order bifurcating point, then $y_0$ is the $k$-order bifurcating point of $f$.

## 4.3 The Implicit Function Theorem

The implicit function theorem is introduced so as to find the solution $x$ with μ as a function.

**Theorem 4.2 Local implicit function theorem**

Let $\mathcal{X}$, $\mathcal{Y}$ and $\mathcal{Z}$ be Banach spaces and $\mathcal{U} \subseteq \mathcal{X} \times \mathcal{Y}$ be an open set, $(x,y) \in \mathcal{X} \times \mathcal{Y}$ and $f: \mathcal{U} \to \mathcal{Z}$ is continuously differentiable. If $f(x_0, y_0) = 0$, then $(x_0, y_0) \in \mathcal{U}$ is a trivial solution of eq. (4.6). If Jacobian matrix $A \equiv D_x f(x_0, y_0)$ is a homomorphism (or a topological isomorphism), then in the neighbourhoods $\mathcal{U}$ of $(x_0, y_0)$ eq. (4.6) has a unique solution $x = u(y)$ so that $f(u(y), y) = 0$. Moreover, from

$$D_x f(u(y), y) u_y(y) + f_y(u(y), y) = 0$$

then $$u_x = -[D_y f(u(y), y)]^{-1} f_y(u(y), y) \tag{4.8}$$

where $u(y): \mathcal{Y} \to \mathcal{X}$ is continuous and differentiable.

**Proof** First we prove the uniqueness, and then the continuity. Without loss of generality, suppose that $(x_0, y_0) = (0,0)$.

Since $Ax = Ax - f(x,y) = R(x,y)$, then $x = x - A^{-1} f(x,y) = g(x,y)$.

To prove the uniqueness of $x$, we must prove that $g(x,y)$ is contracting:

$$R(x_1,y)-R(x_2,y)=Ax_1-Ax_2+f(x_2,y)-f(x_1,y)$$
$$=A(x_1-x_2)-\int_0^1 f_x(tx_1+(1-t)x_2,y)(x_1-x_2)\mathrm{d}t$$
$$=\int_0^1[D_xf(0,0)-D_xf(tx_1+(1-t)x_2,y)]\mathrm{d}t(x_1-x_2)$$

Since $D_xf(x,y)$ is continuous at (0,0), there exist $\delta_0$ and $r_1$ related to ε so that

$$||x||<\delta,\quad ||y||<r_1$$

and so $||D_xf(0,0)-D_xf(tx_1+(1-t)x_2,y)||<\varepsilon$

then $\quad ||R(x_1,y)-R(x_2,y)||=\int_0^1\varepsilon\mathrm{d}t||x_1-x_2||=\varepsilon||x_1-x_2||$

Since $\;g(x_1,y)=A^{-1}R(x,y),\; A^{-1}||R(x_1,y)-R(x_2,y)||=A^{-1}\varepsilon||x_1-x_2||$

Taking $A^{-1}=\dfrac{1}{2\varepsilon}$ gives

$$||g(x_1,y)-g(x_2,y)||=A^{-1}\varepsilon||x_1-x_2||\le\frac{1}{2}||x_1-x_2||$$

Therefore, $g(x,y)$ is contracting.

From the fixed point theorem if there is a contracting map $g$ which maps a set to itself, i.e. $g:\mathcal{B}_\delta\to\mathcal{B}_\delta$, then there exists a unique fixed point, where

$$\mathcal{B}_\delta=\{x|\;||x||<\delta\}\subseteq\mathcal{X}$$

To verify that $g:\mathcal{B}_\delta\to\mathcal{B}_\delta$, we are only to prove that $g(x,y)=x$ can satisfy $||x||<\delta$. Write $g(x,y)=g(x,y)-g(0,y)+g(0,y)$. Since $g(0,0)=0$ and $g$ is continuous, then there exists $r_2$, so that $||g(0,y)||<\dfrac{\delta}{2}$ for $||y||<r_2$, then

$$||g(x,y)||\le||g(x,y)-g(0,y)||+||g(0,y)||\le 1/2\;||x||+\delta/2$$

which implies $||x||\le\delta/2$. This proves that there is a unique $x=u(y)$, so that $f(u(y),y)=0$.

Now we will prove that $u(y)$ is continuous.

$$u(y_1)-u(y_2)=g(u(y_1),y_1)-g(u(y_2),y_2)$$
$$=g(u(y_1),y_1)-g(u(y_2),y_1)+g(u(y_2),y_1)-g(u(y_2),y_2)$$

Taking absolute value for both sides of the equation, we have:

$$||u(y_1)-u(y_2)||\le||g(u(y_1)),y_1)-g(u(y_2),y_1)||+||g(u(y_2),y_1)-g(u(y_2),y_2)||$$

Therefore $\quad ||u(y_1)-u(y_2)||\le 2||g(u(y_1),y_1)-g(u(y_2),y_2)||$

As $g$ is a continuous function, the right-hand side of the equation is zero for $y_1\to y_2$, therefore, $u(y)$ is a continuous function.

**Theorem 4.3 Generalized implicit function theorem**

If $A=D_xf(x_0,y_0)$ is well defined and irreversible with $A(x_0,y_0)=0$, $A^{-1}$ has no definition. Since det $A=\prod_{i=1}^{n}\lambda_i=0$, there exists at least one eigenvalue of zero. It is clear that the number of zero eigenvalues is less than $n$. Usually, the number of zero eigenvalues is finite. To solve the equation $f(x_0,y_0)=0$, we must introduce the generalized implicit function theorem.

Suppose that the map $f$ is defined as the above, that is, $f:\mathcal{U}\subseteq\mathcal{X}\times\mathcal{Y}\to\mathcal{Z}$, $f\in C^p(U)$, $p\geq 1$, $f(0,0)=0$. Let $\eta(A)=\{x\in\mathcal{X}|\,Ax=0\}$ be the null space of $A$, and $\mathcal{R}(A)=\{z\in\mathcal{Z}\,|\,Ax=z$, for some $x\in\mathcal{X}\}$ be the value space (or the domain space of the value). Assuming $\eta(A)=\mathcal{X}_1\subset\mathcal{X}$ and $\mathcal{X}_1$ has a complement space $\mathcal{X}_2$ in $\mathcal{X}$, then $\mathcal{X}=\mathcal{X}_1+\mathcal{X}_2$. Any $x\in\mathcal{X}$ can be written as $x=x_1+x_2$, $x_1\in\mathcal{X}_i$. Suppose that there exist $a>0$ and $b>0$, sufficiently small so that

$$||x||_{x_1}<a,\ \text{for}\ x_1\in\mathcal{X}_1\ \text{and}\ ||y||_Y<b\ \text{for}\ y\in\mathcal{Y}$$

then there exist $u\in\mathcal{C}^1:\mathcal{B}_a(0)\times\mathcal{B}_b(0)\to\mathcal{X}_2$, where $\mathcal{B}_a(0)\subset\mathcal{X}_1$, $\mathcal{B}_b(0)\subset\mathcal{Y}$, that is $x_2=u(x_1,y)$, which satisfies

$$f(x_1+u(x_1,y),y)=0\ \text{and}\ u(0,0)=0 \tag{4.9}$$

Using the implicit function theorem from eq. (4.9), we can get $x_2=u(x_1,y)$.

**Proof** Define a new parameter space

$\widetilde{\mathcal{Y}}=\mathcal{X}_1\times\mathcal{Y}$, $\widetilde{y}$ denotes $(x_1,y)$.

Map $G(x_2,\widetilde{y})=f(x_1+x_2,y)=0$

$G:\widetilde{\mathcal{U}}\subset\mathcal{X}_1\times\widetilde{\mathcal{Y}}\to\mathcal{Z}$

then $D_{x_2}G(0,0):\mathcal{X}_2\to\mathcal{Z}$ is a homomorphism. Applying the local implicit function theorem, we obtain $x_2=u(x,y)$.

## 4.4 Liapunov—Schmidt(LS) Reduction

In the study of static bifurcation problems of local solutions near singular points, eq. (4.6) is often transformed to a lower dimension by Liapunov—Schmidt Reduction. Before LS reduction is studied, we introduce the necessary condition of bifurcation.

In the study of the fixed-point bifurcation of system (4.5), the static bifurcation is of great importance. The study will not only provide us with information about changes of the number of fixed points with parameter, but also has a close relationship with the change of the stability of the fixed point itself. Consider the static eq. (4.6), where vector field $f\,\mathcal{U}\times\mathcal{J}\subseteq\mathcal{R}^n\times\mathcal{R}^m\to\mathcal{R}^n$. Suppose that $(x_0,\mu_0)\in\mathcal{U}\times\mathcal{J}$ is a solution of eq. (4.6), then $f(x_0,\mu_0)=0$. We are interested in the change of the number of solutions (zero point of $f$) of eq. (4.6) in the neighbourhood of $(x_0,\mu_0)$. Let $\Omega\subset\mathcal{U}\times\mathcal{J}$ be a sufficiently small neighbourhood of $(x_0,\mu_0)$, $n(\mu)$ be the number of the solutions of eq. (4.6) in $\Omega$, then $(x_{0,}\mu_0)$ is called a static bifurcation point if $n(\mu)$changes suddenly as $\mu$ varies through $\mu_0$ in the neighbourhood, where $\mu_0$ is a static bifurcation value. We shall call the set $(x,\mu)$ of eq. (4.6) the bifurcation diagram or solution set neighbouring $(x_{0,}\mu_0)$. Generally speaking, the study of static bifurcation is to deal with the multiple solution of eq. (4.6). We assume that $f$ is smooth enough for $x$ and $\mu_0$ in the following statement.

Next, some methods of determining static bifurcation points will be introduced.

**Theorem 4.4** Suppose $(x_0,\mu_0) \in \mathcal{U} \times \mathcal{J}$, such that $f(x_0,\mu_0)=0$, where $f$ is differentiable with respect to $x$ neighbouring $(x_{0,}\mu_0)$, and $f(x,\mu)$ and $D_x f(x,\mu)$ are continuous for $x$ and $f$. If $(x_{0,}\mu_0)$ is a static bifurcation point, then $D_x f(x_0,\mu_0)$ is irreversible or singular.

**Proof** The method is reduction to absurdity. Suppose that $D_x f(x_0,\mu_0)$ is reversible. From the implicit function theorem we know that in a neighbourhood of point $(x_{0,}\mu_0)$, eq. (4.6) has a continuous unique solution $x=\varphi(\mu)$, $x_0=\varphi(\mu_0)$ such that $f(\varphi(\mu),\mu)=0$ neighbouring $(x_{0,}\mu_0)$. This contradicts the assumption that $(x_{0,}\mu_0)$ is a bifurcation point.

Theorem 4.4 provides a necessary condition to prove that $(x_{0,}\mu_0)$ is a static bifurcation point.

For later use we give another form of the necessary condition.

**Definition 4.3** The point $(x_{0,}\mu_0)$ is called a singular point of vector field $f(x,\mu)$ if $f(x_0,\mu_0)=0$ and $f(x_0,\mu_0)=0$ is irreversible. Note: the singular point here is different from the equilibrium point of the vector field. A singular point must be an equilibrium point, but thus not necessarily true conversely.

The following are the equivalent necessary conditions of bifurcation obtained from theorem 4.4, definition 4.3 and the properties of the inverse operator $D_x f$:

(1) $(x_{0,}\mu_0)$ is a singular point of $f$:

(2) $f(x_0,\mu_0)=0$ and $\det\left[\dfrac{\partial f_i}{\partial x_j}\right]_{(x_0,\mu_0)}=0$ (4.10)

(3) $f(x_0,\mu_0)=0$, and $D_x f(x_0,\mu_0)$ has at least one zero eigenvalue;

(4) $f(x_0,\mu_0)=0$, and $D_x f(x_0,\mu_0)$, the dimension of the null space (kernel) of $D_x f(x_0,\mu_0)$, is no less than 1, that is dim $(D_x f(x_0,\mu_0)) \geq$ .

**Note** The conditions given above are necessary but not sufficient.

**Example** Consider equation

$$f(x_0,\mu)=x^3-\mu=0,\ x\in\mathcal{R},\ \mu\in\mathcal{R} \tag{4.11}$$

(0,0) is a singular point, because $f(0,0)=0$ and $D_0 f(0,0)=3x^2|_{(0,0)}=0$. But the point (0,0) is not a bifurcation point, and eq. (4.11) has a unique solution $x=\mu^{\frac{1}{3}}$ for any $\mu\in\mathcal{R}$.

To study static bifurcation, we first find the singular points where the static bifurcation points probably appear. Then the behaviour of eq. (4.6) is studied near the singular points, in order to decide whether they are actually the static bifurcation points. Generally, we often minimize the dimension of the original equation by LS reduction or centre manifold method, and study the local solution of a lower dimensional equation.

Without loss of generality, we always assume that the singularity is at (0,0), and take $\mathcal{U}$ as $F\mathcal{R}^n$. This can be done by the transformation $\bar{x} = x - x_0$, $\bar{\mu} = \mu - \mu_0$.

Now let us come to the Liapunov—Schmidt(LS) reduction.

LS reduction is the method of reducing an equation of a higher or infinite dimension to one of lower or finite dimension.

Consider the following equation given by a $\mathcal{C}^r$ $(r \geq 1)$ map $\Phi: \mathcal{X} \times \mathcal{R}^{k+1} \rightarrow \mathcal{Y}$

$$\Phi(x,\lambda) = 0 \tag{4.12}$$

where $\mathcal{X}$ and $\mathcal{Y}$ are Banach spaces. $\mathcal{R}^{k+1}$, $k \geq 0$ is the set of parameters. Suppose that (0,0) is a bifurcation point, then

$$\Phi(0,0) = 0 \tag{4.13}$$

and the dimension of the null space of differentiable operator $D_x\Phi(0,0): \mathcal{X} \rightarrow \mathcal{Y}$ is larger than zero, that is,

$$\dim \eta(D_x\Phi(0,0)) > 0 \tag{4.14}$$

The basic idea of LS reduction is to express $\mathcal{Y}$ as the direct sum of two subspaces, on which eq. (4.6) will be projected respectively, to obtain two equations. One of them satisfies the implicit function theorem, and has a unique solution. Then substitute this into the other equation, and a problem regarding a lower dimensional one is obtained.

In many cases the differential operators are of the Fredholm type. A bounded linear operator $A \in L(\mathcal{X},\mathcal{Y})$ is called a Fredholm operator if the dimension of the null space and the codimension of the range space are finite, that is $\dim \eta(A) = n$, $\dim(\mathcal{Y} / \mathcal{R}(A)) = m, n, m < \infty$, where $n$, $m$ is called the index of $A$. It can be proved that there exist closed complementary subspaces $\mathcal{M} \subset \mathcal{X}$ and $\eta(A) \subset \mathcal{X}$ and $\mathcal{R}(A) \subset \mathcal{Y}$, such that

$$\begin{cases} \mathcal{X} = \eta(A) \oplus \mathcal{M} \\ \mathcal{Y} = \mathcal{N} \oplus \mathcal{R}(A) \end{cases} \tag{4.15}$$

Especially if $\mathcal{X}$ and $\mathcal{Y}$ are Hilbert spaces, the closed complementary space would be the orthogonal.

In the following section we shall see that in the Fredholm case, the bifurcation problem can be reduced to that of an algebraic equation in finite dimensional space.

### 4.4.1 Reduction of Bifurcation Equation

Suppose that the non-linear operator $\Phi: \mathcal{X} \times \mathcal{R}^{k+1} \rightarrow \mathcal{Y}$ satisfies conditions (4.13) and (4.14), and that the differentiable operator $D_x\Phi(x,\lambda)$ with respect to $x$ neighbouring the origin is of a Fredholm type. Write $A = D_x\Phi(0,0) \in L(\mathcal{X},\mathcal{Y})$, so that $\mathcal{X}$ and $\mathcal{Y}$ can be split into the form of eq. (4.15).

Define a projection operator

$$E: \mathcal{Y} \rightarrow \mathcal{R}(A) \tag{4.16a}$$

and a complementary projection operator

$$I - E: \mathcal{Y} \to \mathcal{N} \tag{4.16b}$$

If $y = y_1 + y_2 \in \mathcal{N} \oplus \mathcal{R}(A) = \mathcal{Y}$, then $y = 0$ if

$$Ey = 0 \text{ and } (I - E)y = 0 \tag{4.17}$$

For $x = v + w \in \eta(A) \oplus \mathcal{M},\ v \in \eta(A),\ w \in \mathcal{M}$, then eq. (4.12) is equivalent to the following equations:

$$E\Phi(v + w, \alpha) = 0 \tag{4.18}$$

$$(I - E)\Phi(v + w, \alpha) = 0 \tag{4.19}$$

Since $EA = ED_x\Phi(0,0): \mathcal{M} \to \mathcal{R}(A)$ is limited on $\mathcal{M}$ and is invertible, then there is a unique solution $w = w(v,\alpha): \eta(A) \times \mathcal{R}^{k+1} \to \mathcal{R}(A)$, such that

$$E\Phi(v + w(v,\alpha), \alpha) = 0, \text{ with } w(0,0) = 0$$

Substitute $w$ into eq. (4.19). The map $\varphi: \eta(A) \times \mathcal{R}^{k+1} \to \mathcal{N}$ is determined as follows:

$$\varphi(x,\alpha) = (I - E)\Phi(v + w(v,\alpha), \alpha)$$

then the solutions of the following equation is one to one corresponding to eq.(4.12)

$$\varphi(v,\alpha) = 0 \tag{4.20}$$

which is called the bifurcation equation of eq. (4.12).

The process of reducing the solution of eq. (4.12) near (0,0) to the solution of a one-dimensional eq. (4.20) (equations of a lower order in general) near (0,0) is called the **Liapunov—Schmidt method** (or LS reduction). Suppose $\mathcal{X}$ is a Hilbert space, then the orthogonal complementary space of $\mathcal{R}(A)$ can be written as

$$(\mathcal{R}(A))^{\perp} = \{y \in \mathcal{X} | < y, z > = 0,\ z \in \mathcal{R}(A)\}$$

Suppose operator $A$ is a Fredholm operator with a one-dimensional null space and zero index, then there exists a non-zero $v_0 \in \eta(A)$ and $v_0^{*} \in (\mathcal{R}(A))^{\perp}$. For any $v \in \eta(A)$, there exists $x \in \mathcal{R}$ such that $v = xv_0$. Define $g: \mathcal{R} \times \mathcal{R}^{k+1} \to \mathcal{R}$, where

$$g(x,\alpha) = < v*, \varphi(xv_0, \alpha) > \tag{4.21a}$$

Since $\varphi(xv_0, \alpha) \in \mathcal{N}$ then $g(x,\alpha) = 0$, if $\varphi(xv_0,\alpha) = 0$ with their solutions corresponding one to one to each other. According to eq. (4.20) and eq. (4.16b), eq. (4.21a) can be written as

$$g(x,\alpha) = < v*, \Phi(xv_0 + w(xv_0,\alpha), \alpha) > \tag{4.21b}$$

If $$g(x,\alpha) = 0 \tag{4.21c}$$

then this is the bifurcation equation, which contains only the unknown of null space. Equation (4.20) is convenient in analysis and eq. (4.21) in computation.

**Some remarks**

1. There exists such a bifurcation equation as eq. (4.21c) in theory, but what we obtain in solving such equations in many practical problems are only some low-order terms, not the whole analytical expression.

2. It is obvious that we have $g(0,0) = 0$ and $g_x(0,0) = 0$ in eq. (4.12c). This indicates

that there will be no constant and linear terms of $x$ in the expansion of the bifurcation equation.

3. The bifurcation equations are not uniquely obtained from LS reduction, but they are the equivalent [4].

### 4.4.2 LS Reduction with Symmetry

We shall prove that an equation with symmetry will keep its symmetry unchanged while undergoing the LS reduction.

Suppose that $\Gamma$ as a compact Lie group acts on the Banach spaces $\mathcal{X}$ and $\mathcal{Y}$. Map $\Phi: \mathcal{X} \times \Lambda \to \mathcal{Y}$ and $\rho_r \in \Gamma$ are commutative. That is

$$\Phi(\rho_r y, \alpha) = \rho_r \Phi(y, \alpha) \tag{4.22}$$

Now consider the bifurcation eq. (4.22) with symmetry. We assume that (0,0) is a singular point of $\Phi(y,\alpha)$. From the definition of the Frechet derivative we can see that

$$\Phi_y(\rho_r y, \alpha) = \rho_r \Phi_y(y, \alpha) \tag{4.23}$$

Using the preceding symbols, we have $A\rho_r = \rho_r A$. It is easy to see that all the subspaces $\eta(A)$ and $\mathcal{M}, \mathcal{N}$ are invariant for $\rho_r$. Under such a condition it is easy to prove that the reduced equation $\varphi$ obtained from LS reduction is commutative with $\Gamma$, where $\varphi: \eta(A) \times \mathcal{R}^{k+1} \to \mathcal{N}$.

From eq. (4.20) we can see that

$$\rho_r \varphi(v,\alpha) = \rho_r (I - E)\Phi(v + w(v,\alpha),\alpha) = (I - E)\Phi(\rho_r v + \rho_r w(v,\alpha),\alpha) \tag{4.24}$$

Next prove that $w$ is commutative with $\Gamma$. From eq. (4.18) we know that

$$0 = \rho_r E\Phi(v + w(v,\alpha),\alpha) = E\Phi(\rho_r v + \rho_r w(v,\alpha),\alpha) \tag{4.25}$$

On the other hand, replacing $v$ by $\rho_r v$ in eq. (4.18 ), we have

$$E\Phi(\rho_r v + w(\rho_r v,\alpha),\alpha) = 0 \tag{4.26}$$

Comparing (4.25) with (4.26), from the implicit function theorem we know that $w(\rho_r v,\alpha) = \rho_r w(v,\alpha)$. Substituting this into eq. (4.24) we obtain

$$\rho_r \varphi(v,\alpha) = \varphi(\rho_r v,\alpha) \tag{4.27}$$

### 4.4.3 The Computation of Derivatives of the Bifurcation Equati

Now let us study the computation of derivatives of the bifurcation function $g(x,\alpha)$ from that of the original map $\Phi(y,\alpha)$.

From (4.20) we can see that the map $\varphi$ can be written as

$$\varphi(xv_0,\alpha) = (I - E)\Phi(xv_0 + w(xv_0,\alpha),\alpha) \tag{4.28}$$

The derivatives of eq. (4.28) can be calculated directly, but the resulting formulae contain derivatives of $w$, and this must be determined by implicit differentiation of $E\Phi(v+w(v,\alpha),\alpha)=0$.

Substituting $v = xv_0$ into eq. (4.19) and differentiating it with respect to $x$ then we obtain $E\,\mathrm{d}\Phi(v_0 + w_x) = EA\,(v_0 + w_x\,(0,0)) = 0$. Since $v_0 \in \eta(A)$ and $EA = A$, hence $Aw_x(0,0) = 0$. But $w_x(0,0) \in \mathcal{M}$ and $A{:}\mathcal{M} \to \mathcal{R}(A)$ is invertible. It follows that $w_x(0,0) = 0$ (4.29)

Before actually starting the calculations we introduce an invariant notation for multivariate high-order derivatives for the functions. If $v_1, v_2, \ldots, v_k \in \mathcal{R}^n$, then define

$$(\mathrm{d}^k\Phi)_{y,\alpha}(v_1, v_2, \ldots, v_k) = \frac{\partial}{\partial t_1}\cdots\frac{\partial}{\partial t_k}\Phi(y + \sum_{i=1}^{k} t_i v_i, \alpha)\Big|_{t_1=\ldots=t_k=0} \tag{4.30}$$

Note that $(\mathrm{d}^k\Phi)_{y,\alpha}$ is the symmetrical multilinear function of $k$ arguments. If necessary, we may represent eq. (4.30) in terms of $k$th-order partial derivatives of $\Phi$.

If $k = 2$, we have $(\mathrm{d}^2\Phi)_{y,\alpha}(v,w) = \sum_{i,j=1}^{n}\frac{\partial^2\Phi}{\partial y_i \partial y_j}(y,\alpha)v_i w_j$. For eq.(4.28), we have the derivatives of $\varphi$ as follows:

$$\begin{aligned}
&\text{a) } \varphi_x = (I-E)(\mathrm{d}\Phi(v_0 + w_x))\\
&\text{b) } \varphi_{xx} = (I-E)(\mathrm{d}\Phi(w_{xx}) + \mathrm{d}^2\Phi(v_0 + w_x, v_0 + w_x))\\
&\text{c) } \varphi_{xxx} = (I-E)(\mathrm{d}\Phi(w_{xxx}) + 3\mathrm{d}^2\Phi(v_0 + w_x, w_{xx})\\
&\qquad\qquad + \mathrm{d}^3\Phi(v_0 + w_x, v_0 + w_x, v_0 + w_x))\\
&\text{d) } \varphi_{\alpha l} = (I-E)(\Phi_{\alpha l} + \mathrm{d}\Phi(w_{\alpha l}))\\
&\text{e) } \varphi_{\alpha lx} = (I-E)(\mathrm{d}\Phi_{\alpha l}(v_0 + w_x) + \mathrm{d}\Phi(w_{\alpha lx}) + \mathrm{d}^2\Phi(v_0 + w_x, w_{\alpha l}))
\end{aligned} \tag{4.31}$$

We evaluate the above formulae at $x = 0$, $\alpha = 0$ and with $(I-E)A = 0$ and $w_x(0,0) = 0$ in mind, converting the formula to:

$$\begin{aligned}
&\text{a) } \varphi_x(0,0) = 0\\
&\text{b) } \varphi_{xx}(0,0) = (I-E)(\mathrm{d}^2\Phi(v_0, v_0))\\
&\text{c) } \varphi_{xxx}(0,0) = (I-E)3\mathrm{d}^2\Phi(v_0, w_{xx}(0,0) + \mathrm{d}^3\Phi(v_0, v_0, v_0))\\
&\text{d) } \varphi_{\alpha l}(0,0) = (I-E)(\Phi_{\alpha l}(0,0))\\
&\text{e) } \varphi_{\alpha lx}(0,0) = (I-E)(\mathrm{d}\Phi_{\alpha l}(v_0) + \mathrm{d}^2\Phi(v_0, w_{\alpha l}(0,0)))
\end{aligned} \tag{4.32}$$

Before continuing the calculation, we have to make some remarks.

1. The fact that $\varphi_x(0,0) = 0$ is totally expected. This means that the reduced equation has a singular point at the origin. To show this is so, apply the implicit function theorem to the original system $\Phi(y,\alpha) = 0$; that is, rank $(\mathrm{d}\Phi)_{(0,0)}$ would be $n$ but not $n-1$ as assumed in eq. (4.13).

2. One of the major problems in evaluating (4.32) lies in the computation of $w_{xx}(0,0)$ and $w_{\alpha l}(0,0)$. We shall give specific formulae for these quantities below. The difficulty is that these formulae require inverting a linear operator. However, the

computation of $w_{xx}(0,0)$ and $w_{\alpha l}(0,0)$ is not required in (4.32) if $(\mathrm{d}^2\Phi)_{(0,0)}$ happens to be zero. This is not usual in application. For example, when $\Phi$ is an odd function, then we have

$$\Phi(-y,\alpha) = -\Phi(y,\alpha) \tag{4.33}$$

If eq.(4.33) is satisfied, then $\Phi_{\alpha,l}(0,0) = 0$ and $(\mathrm{d}^2\Phi)_{(0,0)} = 0$. So the formulae become

$$\begin{aligned}
&\varphi_x(0,0) = 0,\ \varphi_{xx}(0,0) = 0,\ \varphi_{\alpha l}(0,0) = 0 \\
&\varphi_{xxx}(0,0) = (I-E)(\mathrm{d}^2\Phi(v_0,v_0,v_0)) \\
&\varphi_{\alpha lx}(0,0) = (I-E)(\mathrm{d}\Phi_{\alpha l}(v_0,0))
\end{aligned} \tag{4.34}$$

Note that if an odd function has a singularity (i.e.$\Phi(0,0)=0$), one of its characteristics is the disappearance of these derivatives in the pitchfork bifurcation. Therefore, pitchfork bifurcation is the minimum non-degenerate singularity of the odd function. Referring to [4] for details of the symmetry of function $\Phi$. Returning to the calculation, we must solve at $x = \alpha = 0$

$$\begin{cases}
\text{a)}\ \ w_{xx}(0,0) = -A^{-1}E\,\mathrm{d}^2\Phi(v_0,v_0) & (4.35\text{a}) \\
\text{b)}\ \ w_{\alpha l}(0,0) = -A^{-1}E\Phi_{\alpha l}(0,0) & (4.35\text{b})
\end{cases}$$

where $A^{-1}:\mathcal{R}(A)\to\mathcal{M}$ denotes the inverse of the linear map $A|\mathcal{M}$. To verify eq. (4.35), we differentiate $E\Phi(xv_0 + w(xv_0,\alpha),\alpha) \equiv 0$ twice with respect to $x$ and evaluate at $x = \alpha = 0$, and obtain

$$EA(w_{xx}(0.0) + E\mathrm{d}^2\Phi(v_0 + w_x(0.0).v_0 + w_x(0.0)) = 0$$

As $w_x(0,0) = 0$ and $EA = A$ we solve this equation for $w_{xx}(0,0)$ to obtain eq. (4.35a). We can verify eq. (4.35b) in the same way.

To complete the calculation we will substitute eq. (4.35) into eq. (4.32) and use the resulting formulae $g(x,\alpha) = < v_0^{\bullet}, \varphi(xv_0,\alpha) >$ and $< v_0^{\bullet},(I-E)v > = < v_0^{\bullet}, v >$, to obtain

$$\begin{aligned}
&\text{a)}\ g_x = 0 \\
&\text{b)}\ g_{xx} = < v_0^{\bullet}, \mathrm{d}^2\Phi(v_0,v_0) > \\
&\text{c)}\ g_{xxx} = < v_0^{\bullet}, \mathrm{d}^3\Phi(v_0,v_0,v_0) - 3\mathrm{d}^3\Phi(v_0, A^{-1}E\,\mathrm{d}^2\Phi(v_0,v_0)) > \\
&\text{d)}\ g_{\alpha l} = < v_0^{\bullet}, \Phi_{\alpha l} > \\
&\text{e)}\ g_{\alpha lx} = < v_0^{\bullet}, \mathrm{d}\Phi_{\alpha l}v_0 - \mathrm{d}\Phi^2(v_0, A^{-1}E\Phi_{\alpha l}) >
\end{aligned} \tag{4.36}$$

### 4.4.4 Morse Lemma

In order to study the form of bifurcation of eq. (4.21c), we shall introduce Morse lemma. Although the bifurcation equation is usually of low order, for general study the map is taken as $F:\mathcal{R}^n \times \mathcal{R} \to \mathcal{R}$, $F \in \mathcal{C}^p(\mathcal{U})$ $p \geq 2$, $0 \in \mathcal{U} \in \mathcal{R}^n$

The bifurcation equation is

$$F(x,\lambda) = 0 \tag{4.37}$$

If bifurcation eq. (4.37) satisfies the following conditions

(a) $D_X F(0,0) = F_X(0,0) = 0$

(b) $Q = \left[\dfrac{\partial^2 F(0)}{\partial x_i \partial x_j}\right]$, $i, j = 1,2,\ldots,n$ is non-singular and symmetrical, i.e.,

$\det Q \neq 0$ and $\dfrac{\partial^2 F(0)}{\partial x_i \partial x_j} = \dfrac{\partial^2 F(0)}{\partial x_j \partial x_i}$, where $F(0)$ = constant, then without loss of generality, supposing $F(0) = 0$, and $F_X=0$ is a $\times n$ vector, where $Q = F_{xx}(0)$ is called a Hessian matrix, then there exists a coordinate transformation as follows:

$$\xi(x) = T(x)x \tag{4.38}$$

Reducing $F(x)$ to a quadric form

$$F(x) = \frac{1}{2} < Q\xi, \xi > \equiv \frac{1}{2}\xi^T Q \tag{4.39}$$

where $\xi^T$ is the transposition of $\xi$, <...> denotes the inner product in $\mathcal{R}^n$ and Map $T(x): \mathcal{R}^n \to \mathcal{R}^{n\times n}$ $T(x) \in \mathcal{C}P^{-2}$, $T(x): n\times n$, $T(0) = I$ then $T(x)$ is a near identity transformation.

Now we shall prove the existence of the unique transformation , eq. (4.38) . Substituting eq. (4.38) into eq. (4.39), we have

$$F(x) = \frac{1}{2} < QTx, Tx > = \frac{1}{2} < T^T QTx, x > \tag{4.40}$$

$$F(x) - F(0) = \int_0^1 \mathrm{d}F(tx) = \int_0^1 \frac{\mathrm{d}}{\mathrm{d}t} F(tx)\mathrm{d}t = \int_0^1 F_x(tx)\mathrm{d}t$$

$$= \int_0^1 F_x(tx)\mathrm{d}x(t-1) = (t-1)F_x x\Big|_0^1 - \int_0^1 (t-1) \bullet$$

$$\left\{F_{xx}(tx)xx + F_x(tx)\frac{\mathrm{d}x}{\mathrm{d}t}\right\}\mathrm{d}t$$

$$= -\int_0^1 (t-1) < F_{xx}(tx)x, x > \mathrm{d}t = \frac{1}{2} < Bx, x > \tag{4.41}$$

where $B(x) = 2\int_0^1 (1-t)F_{xx}(tx)\mathrm{d}t$, and $t$ is a parameter independent of $x$. Comparing eq. (4.40) with eq. (4.41) we have $T^T QT = B$. Let

$$f(T) = T^T QT - B = 0 \tag{4.42}$$

If $f_T(T) \neq 0$, the implicit function theorem can be used to solve $T$ from eq. (4.42). Give a perturbation $H$ to $T$, substitute $T=T+H$ into eq. (4.42), and expand $f(T)$ to a Taylor series:

$$f(T+H, x) = (T+H)^T Q(T+H) - B = T^T QT + T^T QH + H^T QT + H^T QH - B$$

Since $T(0) = T^T(0) = I$, $B(0) = F_{xx}(0) = Q$, then $f(T+H,0) = QH + H^T Q + H^T QH$. Taking the linear, one has $f(T+H,0) = QH + H^T Q$. Expanding the left-hand side into a Taylor series, we have

$$f_T(T+H,0)H = QH + H^T Q \tag{4.43}$$

As $Q \neq 0$ and $H \neq 0, H$ is a perturbation, then $f_T(T+H,0) \neq 0$. The implicit function theorem can be used to obtain the unique solution $T$ from eq. (4.42). If $H^T = -H$, and $H, Q$ are commutative, then $f_T(T+H,0) = 0$. If the null space is the set of $\{H | f_T H = 0\}$, then we can use the generalized implicit function theorem to solve $T$. This completes the proof of the lemma.

Now we introduce the algebraic classification of quadratic maps. Consider a given map of quadratic type $q(x) = x^T Qx$ where $Q$ is a Hessian matrix non-singular and symmetrical.

All the eigenvalues of $Q$ are real numbers, $\lambda_1, \lambda_2, \ldots, \lambda_k, \lambda_{k+1}, \ldots, \lambda_n$ with the eigenvalues from $\lambda_1$ to $\lambda_k$ being negative and positive from $\lambda_{k+1}$ to $\lambda_n$. As $Q$ is non-singular, therefore all the eigenvalues are non-zero.

Then there exists a non-singular linear transformation $\xi = Lx$ which turns $Q$ into a diagonal form $\hat{Q} = \begin{vmatrix} \lambda_1 & 0 & \ldots & 0 \\ 0 & \lambda_2 & \ldots & 0 \\ \ldots & \ldots & \ldots & \ldots \\ 0 & 0 & \ldots & \lambda_n \end{vmatrix}$, so that $x^T Qx = \xi^T \hat{Q} \xi = \sum_{j=1}^{n} \lambda_j |\xi|_j^2$.

Rescaling $\xi$, we obtain

$$< \hat{Q}\xi, \xi >= \xi^T \hat{Q} \xi = -|\xi_1|^2 - \ldots - |\xi_k|^2 + |\xi_{k+1}|^2 + \ldots + |\xi_n|^2 = 0 \tag{4.44}$$

The above formula is the quadratic form of a bifurcation equation. $k$=$n$ is called the index of the quadratic form. If $k$=0, it is called the positive definite; if $k$=$n$, it is called the negative definite and 0< $k < n$, the quadratic form, is indefinite.

For the definite (positive or negative) case, the solutions of eq. (4.44) are trival, and for the indefinite case (4.44) may have nonzero solutions which form a cone, as shown in Fig. 4.6(a). In $x$ space $\xi = T(x)$, the solution is approximately a linear curved cone, as shown in Fig. 4.6(b).

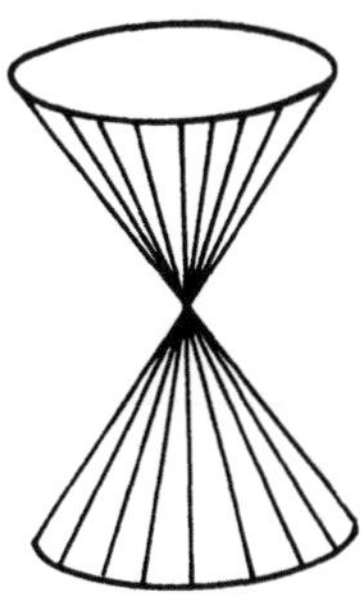

a) $\xi$ space

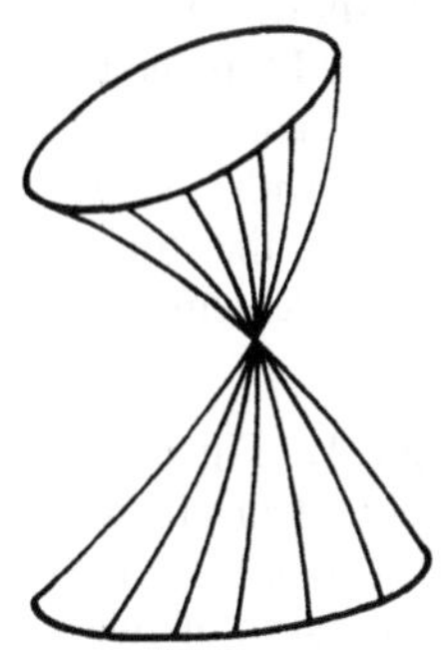

b) $x$ space

Fig. 4.6

Example. If $n = 2$, $k = 1$, from Morse lemma the quadratic form of the bifurcation equation is of the form $|\xi_1|^2 - |\xi_2|^2 = 0$. This implies that $\xi_1 = \xi_2$, and $\xi_1 = -\xi_2$ (see Fig. 4.7).

(a) $\hat{\xi}$ plane (b) $x$ plane

Fig. 4.7

## 4.5 Methods of Singularity

This section discusses local static bifurcation problems with one state variable $x \in \mathcal{R}$. For such bifurcations, we can obtain a one-dimensional bifurcation equation (or a reduced equation) produced first by LS reduction and averaging methods (see chapter 7), and then the bifurcation characteristics of the reduced equation can be studied by the methods of singularities. Developing in the middle of 1960s, the theory of singularities enables us to deal with various complex bifurcation problems. The theory comprises three parts, recognition, unfolding and classification by codimension. This section will introduce briefly the main principles of singularity theory. Readers who are interested are referred to [4].

### 4.5.1 Recognition Problem

As we know, the bifurcation equation obtained by LS reduction or the averaging method contains only finite terms. Can the equation with finite terms reveal qualitatively the bifurcation behaviour? This is the problem that recognition would solve.

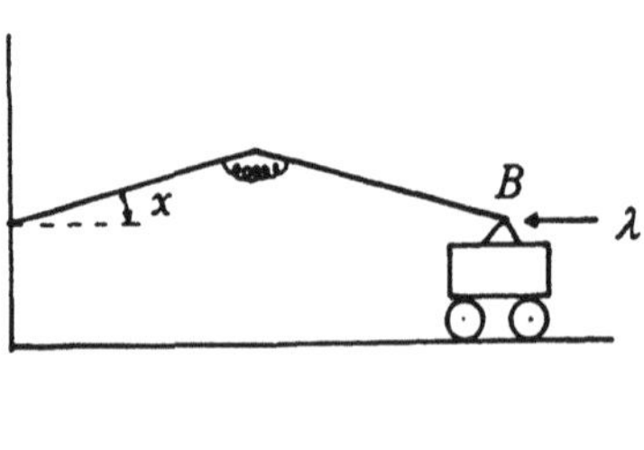


Fig. 4.8

**Example 1.** Figure 4.8 shows a simple physical system of Euler buckling, which exhibits a pitchfork bifurcation. The system consists of two rigid rods of unit length connected by hinges and a torsional spring of unit strength which permit rotation in a plane. The system is subjected to a horizontal compressive force $\lambda$. We neglect friction. Suppose the system is at rest. The state of the system is described by the angle $x$ which is taken as a state variable for measuring the deviation of the rods from the horizontal. The potential energy of this system equals

$$(x,\lambda)=\frac{1}{2}x^2+2\lambda(\cos x-1) \tag{4.45}$$

where the first term represents the stored energy in the torsional spring and the second, the work done by the external force. The steady state of the system is described by the following equation:

$$g(x,\lambda)=-\frac{\partial V}{\partial x}(x,\lambda)=-x+2\lambda\sin x \tag{4.46}$$

It is easy to check that the system has a singularity at $(x,\lambda)=\left(0,\frac{1}{2}\right)$ and an expansion near $\left(0,\frac{1}{2}\right)$:

$$g(x,\lambda)=2\left(\lambda-\frac{1}{2}\right)x-\frac{1}{6}x^3+hot \tag{4.47}$$

When we neglect the high-order terms (*hot*) of eq. (4.47), does $g(x,\lambda)$ have the same bifurcation behaviour as $\lambda x-x^3$ at $\left(0,\frac{1}{2}\right)$? Can the resulting bifurcation reflect the bifurcation behaviour of the original bifurcation $g(x,\lambda)=0$? This problem can be solved by expressing it as: under what condition do $g(x,\lambda)=0$ and its lower terms of the Taylor expansion possess the same qualitative behaviour?

**1. The Algebra on Germ**

From what was stated above, it is easy to see that the discussion of the bifurcation problem $f:\mathcal{R}\times\mathcal{R}\to\mathcal{R}$ on $\mathcal{R}^2$ is restricted in the neighbourhood of the origin. For convenience, let the bifurcation problems be $\mathcal{C}^\infty$. More strictly, a function $f(x,\lambda)$ in $\mathcal{C}^\infty$ is a germ in map space, $f:\mathcal{R}\times\mathcal{R}\to\mathcal{R}$. Two maps $f_1$ and $f_2$ (defined near the origin) are called "equal as germs", if there is some neighbourhood $V$ with the origin in it. On $V$, $f_1$ and $f_2$ are equal. Being equal as germs is an equivalent relationship in the map space. According to this relation the map space will be classified into classes and every class is a germ. The set of all germs constitutes the germ space, written as $\varepsilon(x,\lambda)$ or $\varepsilon_{x,\lambda}$. Without confusion, we denote $\varepsilon_{x,\lambda}$ as the map space.

The most important concept of the algebra on germ space $\varepsilon_{x,\lambda}$ is the ideal.

**Definition 4.4** Subspace $\mathcal{G}\subseteq\varepsilon_{x,\lambda}$ is called an ideal, if for any $f\in\varepsilon_{x,\lambda}$ and $g\in\mathcal{G}$ there exists $f\bullet g\in\mathcal{G}$. For $f\in\varepsilon_{x,\lambda}$ we use the notation $j^k(f)$ for the $k$-order Taylor polynomial of $f$ at the origin (or $k$-jet). For example $j^3(\sin x-\lambda)=x-\frac{1}{6}x^3-\lambda$.

**Corollary 1** For $k\geq$ , write $\mathcal{M}_k=\{f\in\varepsilon_{x,\lambda}|j^{k-1}f=0\}$, then $\mathcal{M}_k$ is an ideal in $\varepsilon_{x,\lambda}$. Especially, $\mathcal{M}_1=\{f\in\varepsilon_{x,\lambda}|f(0)=0\}$ is the maximum ideal in $\varepsilon_{x,\lambda}$.

Out of a group of elements $p_i\in\varepsilon_{x,\lambda},\lambda=1,2,\ldots,k$ can often generate an ideal written as

$$< p_1, p_2, \ldots, p_k >= \{\sum_{i=1}^{k} f_i p_i | f_i \in \varepsilon_{x,\lambda}\}$$

It is called the finite generated ideal with $p_1, p_2, \ldots, p_k$ as the generators. In particular, $\mathcal{M} =< x, \lambda >$.

Suppose that $\mathcal{G}$ and $\mathcal{F}$ are ideals in $\varepsilon_{x,\lambda}$. Then we can define the sum ideal and the product ideal as follows

$$\mathcal{G} + \mathcal{F} = \{g + f | g \in \mathcal{G}, f \in \mathcal{F}\}$$

$$\mathcal{G} \bullet \mathcal{F} = \{\sum_{i=1}^{m} g_i f_i | g_i \in \mathcal{G}, f_i \in \mathcal{F}\}$$

from which we can derive the definition of the $k$-th power of ideal $\mathcal{G}^k = \mathcal{G} \bullet \mathcal{G}^{k-1}$.

**Corollary 2** For any $k \geq$ we have $\mathcal{M}^k = \mathcal{M}_k$.

**Theorem 4.5** (Nakayama lemma) Suppose that $\mathcal{G}$ and $\mathcal{F}$ are ideals in $\varepsilon_{x,\lambda}$, and $\mathcal{F} =< p_1, p_2, \ldots, p_k >$ is finitely generated, then $\mathcal{F} \subset \mathcal{G}$ if $\mathcal{F} \subset \mathcal{G} + \mathcal{M}\mathcal{F}$.

**Proof** We are to prove the sufficiency only, for the necessity is clear. From $p_i = g_i + \sum_{i=1}^{k} a_{i,j} p_j$, $g_i \in \mathcal{G}$, $a_{i,j} \in \mathcal{M}$ we can see that $(I - (a_{i,j}))\mathcal{P} = \mathcal{G}$, $\mathcal{P} = (p_1, p_2, \ldots, p_k)^T$, $\mathcal{G} = (g_1, g_2, \ldots, g_k)^T$. But as $(I - (a_{i,j}))(0,0) = I$, $(I - (a_{i,j}))$ is invertible, with the inverse written as $(b_{i,j})$, then $p_i = \sum_{i=1}^{k} b_{i,j} g_j \in \mathcal{G}$.

**Corollary** If $\mathcal{F} =< p_1, p_2, \ldots, p_k >$, $q_1, q_2, \ldots, q_k \in \mathcal{M}\mathcal{F}$ then $< p_1 + q_1, \ldots, p_k + q_k > = \mathcal{F}$.

**Proof** Suppose that $\mathcal{G} =< p_1 + q_1, \ldots, p_k + q_k >$. From $p_i + q_i \in \mathcal{F} + \mathcal{M}\mathcal{F} = \mathcal{F}$ we know that $\mathcal{D} \subset \mathcal{F}$. Otherwise, from $p_i = (p_i + q_i - q_i) \in \mathcal{D} + \mathcal{M}\mathcal{F}$ and the Nakayama lemma we see that $f \subset \mathcal{G}$.

That ideal $\mathcal{F} \subset \varepsilon_{x,\lambda}$ has finite codimension means that $\mathcal{F}$ has a subspace in $\varepsilon_{x,\lambda}$, the number of codimensions of $\mathcal{F}$ is finite, codim $\mathcal{F} = \dim(\varepsilon_{x,\lambda} / \mathcal{F}) < \infty$ or equivalently there exists a subspace $\mathcal{V} \subset \varepsilon_{x,\lambda}$ such that $\dim \mathcal{V} = \text{co}\dim \mathcal{F}$ and $\varepsilon_{x,\lambda} = \mathcal{F} \oplus \mathcal{V}$. It is easy to see that codim $\mathcal{M}^k < \infty$, codim $\mathcal{M} = 1$.

**Theorem 4.6** Suppose that $\mathcal{F} \subset \varepsilon_{x,\lambda}$ is an ideal, then $F$ has finite codimension if there is $k$ such that $\mathcal{M}^k \subset \mathcal{F}$.

**Proof** Sufficiency. If $\mathcal{M}^k \subset \mathcal{F}$, then codim $\mathcal{F} \leq \text{co}\dim \mathcal{M} < \infty$.

Necessity. Suppose that codim $\mathcal{F} = l\text{-}1$, then from

$\mathcal{F} \subset \mathcal{F} + \mathcal{M}^l \subset \mathcal{F} + \mathcal{M}^{l+1} \subset \ldots \subset \mathcal{F} + \mathcal{M} \subset \mathcal{M}$ we know that

$l\text{–}1 =$ codim $\mathcal{F} \geq$ codim $(\mathcal{F} + \ell) \geq \ldots \geq 1$.

Then there exists $k \leq l$ such that $\text{codim}(\mathcal{F} + \mathcal{M}^k) = \text{condim}(\mathcal{F} + \mathcal{M}^{k+1})$. As $(\mathcal{F} + \mathcal{M}^{k+1}) = \mathcal{F} + \mathcal{M}^k$, we know $\mathcal{M}^k \subset \mathcal{F} + \mathcal{M}^{k+1}$ $\mathcal{M}^k \subset \mathcal{F} + \mathcal{M}^{k+1}$ and then $k \subset \mathcal{F}$.

## 2. The Recognition of a One State Variable

In applying singularity theory to study bifurcations with a one state variable, three important ideals are often used. They are restricted tangent space RT($f$), perturbation set P($f$) and intrinsic ideal I($f$). We shall examine the concepts one by one.

### 1) Restricted Tangent Space

Being a basic concept, the restricted tangent space is based on a strong equivalence between germs. Firstly we introduce the concept of strong equivalence.

**Definition 4.5** Two germs $f$ and $g$ are said to be strongly equivalent, written as $f \overset{s}{\sim} g$, if there is a diffeomorphism $T:(x,\lambda) \to (X(x,\lambda),\lambda)$ and $S(x,\lambda)$ such that $g(x,\lambda) = S(x,\lambda) f(X(x,\lambda),\lambda)$, where $X(0,0) = 0$, $X_x(0,0) > 0$, $S(0,0) > 0$.

**Corollary 1** Suppose that $g \overset{s}{\sim} h$, and (0,0) is a singular point of $g$ ($g(0,0)=0$, $g_x(0,0)=0$), then (0,0) is also a singular point of $h$.

**Corollary 2** Suppose that $g \overset{s}{\sim} h$, and homomorphism map $(x,\lambda) \to (X(x,\lambda),\lambda)$ is defined in definition 4.5, then the number of solutions neighbouring (0,0) of equations $g(x,\lambda) = 0$ and $h(x,\lambda) = 0$ satisfies the following relationship $n_g(\lambda) = n_h(\lambda)$ whose meaning is indicated in Fig. 4.9.

**Example 2** We can prove that

a) Bifurcation problem $f \in \varepsilon_{x,\lambda}$ is strongly equivalent to $\lambda x - x^3$ at $x = \lambda = 0$ if

$$f = f_x = f_{xx} = f_\lambda = 0 \text{ and } f_{x\lambda} > 0, f_{xxx} < 0 \tag{4.48}$$

b) Bifurcation problem $g \in \varepsilon_{x,\lambda}$ is strongly equivalent to $x^3 + \lambda^2$(Winged cusp) at $x = \lambda = 0$ if

$$g = g_x = g_\lambda = g_{xx} = g_{\lambda x} = 0 \text{ and } g_{xxx} > 0,\ g_{\lambda\lambda} > 0 \tag{4.49}$$

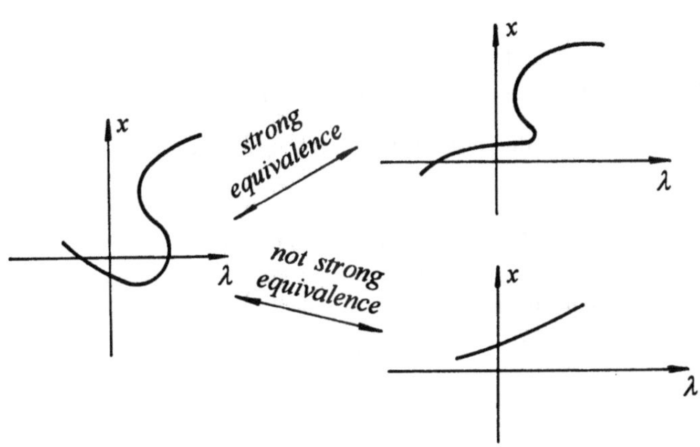


Fig. 4.9

Note that conditions (4.48) and (4.49) are called the recognition conditions for pitchfork $\lambda x - x^3$ and winged cusp $\lambda^2 + x^3$ respectively. They are Golubitsky-Schaeffer norm forms.

Under what conditions can the disturbed term $p$ make $g \sim g + tp$ for $t \in \mathcal{R}$? We introduce the concept of the restricted tangent space. If $\delta = \pm 1, k \geq 2$ is the disturbed term to satisfy the above conditions then for a sufficiently small $t$ there exist $S(x,\lambda,t)$ and $X(x,\lambda,t)$ such that

$$g(x,\lambda) + tp(x,\lambda) = S(x,\lambda,t)g(X(x,\lambda,t),\lambda) \tag{4.50}$$

where

$$X(0,0,t) \equiv 0, S(0,0,t) \equiv 1, X(x,\lambda,0) \equiv x \tag{4.51}$$

Differentiating eq. (4.50) with respect to $t$ at $t = 0$, we have

$$p(x,\lambda) = \dot{S}(x,\lambda,0)g(x,\lambda) + g_x(x,\lambda)\dot{X}(x,\lambda,0)$$

where "$\cdot$" denotes the derivative with respect to $t$. From eq. (4.51) we can see that $\dot{X}(0,0,0) = 0$.

**Definition 4.6** The restricted tangent space of germ $g \in \varepsilon_{x,\lambda}$ is the set of $p$ in the form of

$$p(x,\lambda) = a(x,\lambda)g(x,\lambda) + b(x,\lambda)g_x(x,\lambda) \tag{4.52}$$

where $a,b \in \varepsilon_{x,\lambda}$ and $b(0,0)=0$, written as

$$\mathrm{RT}(g) = \{ag + bg_x | a,b \in \varepsilon_{x,\lambda}, b(0,0) = 0\} \tag{4.53}$$

It is easy to see that $p \in \mathrm{RT}(g)$ is the necessary condition for a sufficiently small $t$ to be $g \sim g + tp$. Then follows

**Theorem 4.7** Suppose that $g$ , $p$. If for any $t \in [0,1]$ $\mathrm{RT}(g) = \mathrm{RT}(tp + g)$, then $g \overset{s}{\sim} g + tp$, $t \in [0,1]$.

As $b(0,0) = 0$, then $b(x,\lambda) = b_1 x + b_2 \lambda$ and we can rewrite the restricted tangent space as $\mathrm{RT}(g) = \{ag + (bx + c\lambda)g_x | a,b \in \varepsilon_{x,\lambda}\} = \langle g, xg_x, \lambda g_x \rangle$.

**Example 3.** Calculate the restricted tangent space $\mathrm{RT}(f)$ when $f = \varepsilon x^k + \delta\lambda\varepsilon$, $\delta = \pm 1, k \geq 2$.

From $xf_x = \varepsilon k x^k, \lambda f_x = \varepsilon\lambda k x^{k-1}$, we can see that RT($f$) is generated by $x^k$ and $\lambda$, so $\mathrm{RT}(f) = \mathrm{RT}(\varepsilon x^k + \delta\lambda) = M^k + \langle \lambda \rangle$. In the same way we can calculate the following restricted tangent spaces:

$$\mathrm{RT}(\varepsilon x^k + \delta\lambda x) = M^k + M\langle \lambda \rangle, \quad k \geq 3$$

$$\mathrm{RT}(\varepsilon x^2 + \delta\lambda^k) = M^2 + \langle k \rangle, \quad k > 2$$

$$\mathrm{RT}(\varepsilon x^3 + \delta\lambda^2) = M^3 + \langle \lambda^2 \rangle$$

**Note 1** Restricted tangent space $\mathrm{RT}(f)$ is an ideal in $\varepsilon_{x,\lambda}$.

**Note 2** Restricted tangent space $\mathrm{RT}(f)$ can be characterized by the "initial conditions" of $f$, such as

$$\mathrm{RT}(x^2+\lambda)=\{g\in\varepsilon_{x,\lambda}\,|\,g(0)=g_x(0)=0\}$$

$$\mathrm{RT}(x^3-\lambda x)=\{g\in\varepsilon_{x,\lambda}\,|\,g(0)=g_x(0)=g_\lambda(0)=g_{xx}(0)=0\}$$

**Definition 4.7** $f\in\varepsilon_{x,\lambda}$ is said to be of finite codimension, if $\mathrm{RT}(f)$ is of finite codimension in $\varepsilon_{x,\lambda}$: $\operatorname{codim}\mathrm{RT}(f)=\operatorname{codim}\{\varepsilon_{x,\lambda}/\mathrm{RT}(f)\}<\infty$.

Nakayama lemma tells us that if $\mathrm{RT}(f)$ is of finite codimension, then there exists $k$ such that $\mathcal{M}^k\subset\mathrm{RT}(f)$, then this condition is equivalent to

$$\mathcal{M}^k\subseteq\mathrm{RT}(f)+M^{k+1}$$

**2) Finite Determinacy and Perturbation Set**

As stated in the preceding sections, for the bifurcation problem $f\in\varepsilon_{x,\lambda}$, only the $k$th-order jet $j^k(f)$ need be considered. In order to keep $j^k(f)\overset{s}{\sim}f$ or let $f$ have finite determinacy, we may consider the difference $f-j^k(f)$ as a perturbation of $f$. The "perturbation set" is the description of this strong equivalence.

**Definition 4.8** $f\in\varepsilon_{x,\lambda}$ is said to be of finite determinacy if there exists $k$ such that $j^k(f)\overset{s}{\sim}f$.

The following theorem provides the condition for $f$ to be of finite determinacy.

**Theorem 4.8** Suppose that $\mathcal{M}^k\subset\mathrm{RT(g)}$, then $f\overset{s}{\sim}j^k(g)$.

**Proof.** Writint $r=j^k(g)-g\in\mathcal{M}^{k+1}$, according to theorem 4.7, we only have to prove that $\mathrm{RT}(g)=\mathrm{RT}(g+tr),t\in[0,1]$. $tr,txr_x,t\lambda\ r_x\in\mathcal{M}^{k+1}\subset\mathcal{M}$, and we know that $\mathrm{RT}(g)\supset\mathrm{RT}(g+tr)$, otherwise

$$g=(g+tr)-tr\in\mathrm{RT}(g+tr)+\mathcal{M}\mathcal{M}^k\subset\mathrm{RT}(g+tr)+\mathcal{M}\mathrm{RT}(g)$$

In a similar way, if we consider $xg_x$ and $\lambda g_x$, from the Nakayama lemma we have $\mathrm{RT}(g)\subset\mathrm{RT}(g+tr)$.

The following is the concept of a perturbation set.

**Definition 4.9** The perturbation set of the bifurcation problem $h\in\varepsilon_{x,\lambda}$ is defined as

$$P(h)=\{p\in\varepsilon_{x,\lambda}\,|\,g\overset{s}{\sim}h \text{ implies } \mathrm{RT}(g)=\mathrm{RT}(g+tr) \text{ for any } t\in\mathcal{R}\}$$

**Theorem 4.9** Suppose that $p\in P(h)$, $g\overset{s}{\sim}h$, then $g+tp\overset{s}{\sim}h$, for any $t\in\mathcal{R}$, especially $g+tp\overset{s}{\sim}h$.

Theorem 4.9 is the source of the concept of a perturbation set. The following theorems show the important properties of the perturbation set.

**Theorem 4.10** Suppose $h\in\varepsilon_{x,\lambda}$, then $P(h)$ is an ideal in $\varepsilon_{x,\lambda}$.

**Proof.** Suppose $p_1,p_2\in P(h), g\overset{s}{\sim}h$ then $g+tp\overset{s}{\sim}g\overset{s}{\sim}h$, therefore,

$$\mathrm{RT}(g+tp_1+sp_2)=\mathrm{RT}(g+tp_1)=\mathrm{RT}(g)$$

Let $s=t$, and we have $p_1+p_2 \in P(h)$. It can be seen that $P(h)$ is a linear space. Suppose $f \in \varepsilon_{x,\lambda}, p \in P(h)$. Write

$$f(x,\lambda)=s+k(x+\lambda),\ s \in \mathcal{R},\ k \in \mathcal{M}$$

From $\mathrm{RT}(g+t(sp_1))=\mathrm{RT}(g)$, we see that $sp \in P(h)$. What we have to do is to prove that $kp \in P(h)$. Since $\mathrm{RT}(g+h)=\mathrm{RT}(g)$, then $p, xp_x, \lambda p_x \in \mathcal{RT}(g)$, therefore $kp, x(kp)_x, \lambda(kp)_x \in \mathcal{M}\mathrm{RT}(g)$. Then it follows that

$$\mathrm{RT}(g+tp)=<g+tkp, xg_x+tx(kp)_x, \lambda g+t\lambda(kp)_x> = <g, xg_x, \lambda g_x> = \mathrm{RT}(g)$$

so that $kp \in P(h)$. This completes the proof.

**Theorem 4.11** Suppose $h \in \varepsilon_{x,\lambda}$ is of finite codimension, $P(h)$ is the perturbation set of $h$. If $p \in P(h), q \overset{s}{\sim} p$ then $q \in P(h)$.

**Proof** We need only prove that $\mathrm{RT}(g+tq)=\mathrm{RT}(g)$ when $g \overset{s}{\sim} h$.

Suppose $q(x,\lambda)=S(x,\lambda)p(X(x,\lambda),\lambda), S(0,0)>0, X(0,0)=0, X_x(0,0)>0$

Write $\gamma=(S,X)$ then $q=\gamma * p$ and therefore

$$\mathrm{RT}(g+tq)=\mathrm{RT}(g+t\gamma^* p)=\gamma^*\,\mathrm{RT}((\gamma^*)^{-1}g+tp)=\gamma^*\,\mathrm{RT}((\gamma^*)^{-1}g)=\mathrm{RT}(g)$$

**3) Intrinsic Ideals**

In order to solve the recognition problem, we must introduce the concept of intrinsic ideals, so as to define further the properties of perturbation set.

**Definition 4.10** Ideal $\mathcal{G} \subset \varepsilon_{x,\lambda}$ is said to be intrinsic, if for any $g,h \in \varepsilon_{x,\lambda}$, when $g \in \mathcal{G}$ and $h \overset{s}{\sim} g$, then $h \in \mathcal{G}$.

**Example 4.** $\mathcal{M}, <\lambda>$ are all intrinsic ideals and $\mathcal{M}$ has finite codimension.

If $\mathcal{F}$ and $\mathcal{G}$ are intrinsic then so are $\mathcal{G}+\mathcal{F}$ and $\mathcal{G}\bullet\mathcal{F}$, and then the ideal of the following form is intrinsic, so and is the inverse:

$$\mathcal{G}=\mathcal{M}^k+\mathcal{M}^{k_1}<l^{l_1}>+\ldots+\mathcal{M}^{k_s}<l^{l_s}> \tag{4.54}$$

Let $\mathcal{G} \subset \varepsilon_{x,\lambda}$ be an ideal of finite codimension; $\mathcal{G}^{\perp}$ is written as the finite-dimensional subspace of $\varepsilon_{x,\lambda}$ spanned out of not belonging to $\mathcal{G}$. It is clear that $\varepsilon_{x,\lambda}=\mathcal{G}\oplus\mathcal{G}^{\perp}$.

**Example 5.** $\mathcal{G}=<x+\lambda,\lambda^2>(\mathcal{G}=\mathcal{M}^2+<x+\lambda>)$ so $\mathcal{G}^{\perp}=\mathcal{R}\{1,\lambda,x\}$.

**Theorem 4.12** Suppose that $\mathcal{G} \subset \varepsilon_{x,\lambda}$ is an intrinsic of finite codimension, then

(1) if $f \in \mathcal{G}, xf_x, \lambda f_x \in \mathcal{G}$, and hence $\mathrm{RT}(f) \subset \mathcal{G}$;

(2) $p(x,\lambda)=\sum a_\alpha x^{\alpha_1}\lambda^{\alpha_2}$ so $p \in \mathcal{G}$ and $x^{\alpha_1}\lambda^{\alpha_2} \in \mathcal{G}$ if and only if for $a_\alpha \neq 0$;

(3) if $x^{\alpha_1}\lambda^{\alpha_2} \in \mathcal{G}$, then $\mathcal{M}^{\alpha_1}<\lambda^{\alpha_2}> \subset \mathcal{G}$;

(4) $\mathcal{G}$ can be written as $\mathcal{G}=\mathcal{M}^{k_0}+\mathcal{M}^{k_1}<l^{l_1}>+\ldots+\mathcal{M}^{k_s}<l^{l_s}>$ where $0<l_1<l_2<\ldots<l_s$, $k_0>k_1+l_1>k_2+l_2>\ldots>k_s+l_s>0$.

**Corollary 1.** Suppose $\mathcal{G} \subset \varepsilon_{x,\lambda}$ is an intrinsic of finite codimension, then $\varepsilon_{x,\lambda} = \mathcal{G} \oplus \mathcal{G}^{\perp}$.

Let $\mathcal{G} \subset \varepsilon_{x,\lambda}$ be an ideal. We denote by Itr$\mathcal{G}$ the greatest intrinsic ideal contained in $\mathcal{G}$, which is the union of all intrinsic ideals contained in $\mathcal{G}$.

**Corollary 2.** Suppose $\mathcal{G} \subset \varepsilon_{x,\lambda}$ is the subspace of finite codimension, then

$$\mathcal{G} = (\text{Itr}\,\mathcal{G}) \oplus V_g, \text{ where } V_g = \mathcal{G} \oplus \cap (\text{Itr}\,\mathcal{G})^{\perp}$$

We denote by $\mathcal{I}(h)$ the smallest intrinsic ideal containing $h$. It may be regarded as the intersection of all the ideals containing $h$. It can be seen that if $g \stackrel{s}{\sim} h$, then $\mathcal{I}(g) = \mathcal{I}(h)$. In fact, the definition tells us that $g \in \mathcal{I}(h)$ so $\mathcal{I}(g) \subset \mathcal{I}(h)$. Similarly, $\mathcal{I}(h) \subset \mathcal{I}(g)$, so $\mathcal{I}(g) = \mathcal{I}(h)$.

**Theorem 4.13** Suppose $h$ is of finite codimension, then

(1) $\mathcal{I}(h)$ is of finite codimension;

(2) $\mathcal{I}(h) = \sum_{\alpha} \{\mathcal{M}^{\alpha_1} < \lambda^{\alpha_2} > | D^{\alpha} h(0,0) \neq 0\}$;

(3) suppose $g \stackrel{s}{\sim} h$, then

a) $D^{\alpha} g(0,0) = 0$ when $x^{\alpha_1}\lambda^{\alpha_2} \in \mathcal{I}(h)^{\perp}$;

b) $D^{\beta} g(0,0) \neq 0$ when $x^{\beta_1}\lambda^{\beta_2}$ is the intrinsic generator of $\mathcal{I}(h)$.

**Proof** We assume that $\mathcal{M}^k \subset \text{RT}(h)$

(1) From theorem 4.8, we can see that $g \stackrel{s}{\sim} j^k h$, and hence $\mathcal{I}(h) = \mathcal{I}(j^k h)$. Then it can be assumed that $h$ is a polynomial. It can be proved that theorem 4.13 is true if $h$ is a polynomial. Then $\text{RT}(h) \subset \mathcal{I}(h)$, and hence codim $\mathcal{I}(h) < \infty$.

(2) The proof can be derived from (2) in theorem 4.12 and corollary 2.

(3) a) If $x^{\alpha_1}\lambda^{\alpha_2} \in \mathcal{I}(h)^{\perp}$, then $D^{\alpha} g(0,0) \neq 0$ from (2) and an see $x^{\alpha_1}\lambda^{\alpha_2} \in \mathcal{I}(h) = \mathcal{I}(g)$. This is a contradiction.

b) Suppose $x^{\beta_1}\lambda^{\beta_2}$ is an intrinsic generator of $\mathcal{I}(h)$, then $\mathcal{M}^{\beta_1} < \lambda^{\beta_2} > \subset \mathcal{I}(h) = \mathcal{I}(g)$, that is $D^{\beta}(g(0,0)) \neq 0$.

**Example 6.** Suppose that

$$h = \varepsilon x^k + \delta\lambda, \varepsilon, \delta = \pm 1, g \stackrel{s}{\sim} h$$

then $\quad \mathcal{I}(g) = \mathcal{I}(h) = \mathcal{M}^k + < \lambda >$,

similarly, $\mathcal{I}(\varepsilon x^k + \delta\lambda x) = \mathcal{M}^k + \mathcal{M} < \lambda >$,

$$\mathcal{I}(\varepsilon(x^2 + \delta\lambda^2)) = \mathcal{M}^2,$$

$$\mathcal{I}(\varepsilon(x^3 + \delta\lambda^2)) = \mathcal{M}^3 + < \lambda^2 >$$

The following two theorems show the relationship between an intrinsic ideal and a perturbation set.

**Theorem 4.14** Suppose $\mathcal{G}$ is an intrinsic ideal in $\varepsilon_{x,\lambda}$ and $\mathrm{RT}(h+p) = \mathrm{RT}(h)$, then $\mathcal{G} \subset \mathcal{P}(h)$ for any $p \in \mathcal{G}$.

**Theorem 4.15** Suppose $h$ is of finite codimension, then $\mathcal{P}(h)$ is an intrinsic ideal of finite condimension, and

$$\mathrm{Itr}\mathcal{M} \bullet \mathrm{RT}(h) \subset \mathcal{P}(h) \subset \mathrm{IntRT}(h)$$

**4) Recognition Problem**

We have discussed three useful ideals above. Now let us return to the solutions of recognition problems. The key to the work is to calculate the perturbation set $\mathcal{P}(h)$ of bifurcation problem $h$.

**Theorem 4.16** Suppose $h \in \varepsilon_{x,\lambda}$ is of finite codimension, then $\mathcal{P}(h) = \mathrm{Itr}\mathcal{Q}(h)$, where

$$\mathcal{Q}(h) = < xh, \lambda h, x^2 h_x, \lambda h_x > \; = \mathcal{M} \bullet \mathrm{RT}(h) + < \lambda h_x > .$$

**Proof** We need to prove that $\mathcal{P}(h) \supset \mathrm{Itr}\mathcal{Q}(h)$, for only this part will be used in our studies. Assume that $p \in \mathrm{Itr}\mathcal{Q}(h)$. From theorem 4.12, we know that

$$xp_x, \lambda p_x \in \mathrm{Itr}\mathcal{Q}(h) \subset \mathcal{Q}(h)$$

then
$$p = A_1 h + B_1 x h_x + C_1 \lambda p_x$$
$$xp_x = A_2 h + B_2 x h_x + C_2 \lambda p_x$$
$$\lambda p_x = A_3 h + B_3 x h_x + C_3 \lambda p_x$$

where $A_i(0,0) = B_i(0,0) = 0,\ i = 1,2,3,\ C_3(0,0) = 0$

Adding $(h, xh_x, \lambda h_x)^T$ on both sides and expressing the equation in the form of a matrix, we have

$$\begin{bmatrix} h+p \\ x(h+p)_x \\ \lambda(h+p)_x \end{bmatrix} = (\mathrm{Q}+\mathrm{I}) \begin{bmatrix} h \\ xh_x \\ \lambda h_x \end{bmatrix}$$

where a singular matrix **Q** is at (0,0), therefore (**Q**+I) is not singular, so that

$$\mathrm{RT}(h+p) = \mathrm{RT}(h)$$

To solve the recognition problem of the bifurcation in $\varepsilon_{x,\lambda}$ we recall what we have gone through.

Suppose $g \overset{s}{\sim} h$ and $h$ is a norm form. According to theorems 4.13 (2) and (3), when $x^{\alpha_1}\lambda^{\alpha_2}$ is the generator of

$$\mathcal{T}(h) = \sum_{\alpha} \{\mathcal{M}^{\alpha_1} < \lambda^{\alpha_2} > | D^{\alpha} h(0,0) \neq 0\}, D^{\alpha} h(0,0) \neq 0$$

but when $x^{\alpha_1}\lambda^{\alpha_2} \in \mathcal{T}(h)^{\perp}$, $D^{\alpha} h(0,0) = 0$. Thus we may obtain $g_0$ the lowest term of

$g$. $g = g_0 + p, p \in \mathcal{M}\mathcal{T}(p)$. Next, based on theorem 4.16, we have $\mathcal{P}(h) = \mathrm{Itr}\mathcal{Q}$, $\mathcal{Q} = \mathcal{M}\mathrm{RT}(h) + < \lambda h_x >$. Then when $p \in \mathcal{P}(h)$ from theorem 4.8, we know that

$$g = g_0 + p \overset{s}{\sim} g_0 \overset{s}{\sim} h$$

**Example 7**. Germ $g \in \varepsilon_{x,\lambda}$ is strongly equivalent to $h = \varepsilon x^k + \delta\lambda$ if and only if

$$g = \frac{\partial}{\partial x} g = \ \dots \ = \left(\frac{\partial}{\partial x}\right)^{k-1} g = 0,\ \varepsilon = \operatorname{sgn}\left(\frac{\partial}{\partial x}\right)^k g$$

and $\delta = \operatorname{sgn} g_\lambda$, at $x = \lambda = 0$ $(\varepsilon = \pm 1, \delta = \pm 1)$.

**Proof** Since $\mathcal{I}(h) = \mathcal{M}^k + <\lambda>$, then $g = ax^k + b\lambda + p(x,\lambda)$, where $a,b \neq 0$, $p$ $\mathcal{M}^{k+1} + \mathcal{M} <\lambda>$. Also from

$\mathcal{M} \bullet \mathrm{RT}(h) = \mathcal{M}(\mathcal{M}^k + <\lambda>) = \mathcal{M}^{k+1} + \mathcal{M} <\lambda>$ and $<\lambda h_x> = <\lambda x^{k-1}> \subset \mathcal{M} <\lambda>$

we know $\mathcal{P}(h) = \mathcal{M}^{k+1} + \mathcal{M} <\lambda>$, so $g \overset{s}{\sim} ax^k + b\lambda \overset{s}{\sim} h$.

**Example 8** Germ $g \in \varepsilon_{x,\lambda}$ is strongly equivalent to $h = \varepsilon x^k + \delta\lambda x$, if and only if

$$g = \frac{\partial}{\partial x} g = \ \dots \ = \left(\frac{\partial}{\partial x}\right)^{k-1} g = 0,\ \varepsilon = \operatorname{sgn}\left(\frac{\partial}{\partial x}\right)^k g$$

and $\quad \delta = \operatorname{sgn} \dfrac{\partial^2}{\partial x \partial \lambda} g$ at $x = \lambda = 0$ $(\varepsilon = \pm 1, \delta = \pm 1)$.

**Proof** $\mathcal{I}(h) = \mathcal{M}^k + \mathcal{M} <\lambda>$, then $g = ax^k + b\lambda x + p(x,\lambda)$, where

$$p(x,\lambda) \in \mathcal{M}^{k+1} + \mathcal{M}^2 <\lambda> + <\lambda^2>, \text{ but}$$

$$\mathcal{P}(h) = \mathcal{M}^{k+1} + \mathcal{M}^2 <\lambda> + <\lambda^2>$$

so $\quad g \overset{s}{\sim} ax^x + b\lambda x \overset{s}{\sim} h$.

If $\varepsilon = -1, \delta = 1, k = 3$, then this is the condition of the recognition of pitchfork bifurcation.

**Example 9** Germ $g \in \varepsilon_{x,\lambda}$ is equivalent to $h = \varepsilon(x^2 + \delta\lambda^2), (\varepsilon = \pm 1, \delta = \pm 1)$ if and only if $g_\chi = g_\lambda = 0, \varepsilon = \operatorname{sgn} g_{xx}$, and $\delta = \operatorname{sgn} \det D^2 g$ at $x = \lambda = 0$, where

$$D^2 g = \begin{bmatrix} g_{xx} & g_{\lambda x} \\ g_{\lambda x} & g_{\lambda\lambda} \end{bmatrix}$$ is the 2 × 2 Hessian matrix of the second-order derivatives of $g$.

**Proof** Since $\mathcal{I}(h) = \mathcal{M}^2$, then $\mathcal{I}(h) = ax^2 + p(x,\lambda)$, where $a \neq 0, p \in \mathcal{M}^3 + \mathcal{M} <\lambda>$. But $\mathcal{P}(h) = \mathcal{M}^3$, so $g$ can be written as $g = ax^2 + b\lambda x + c\lambda^2 + \tilde{p}(\lambda, x)$, for $\tilde{g} = ax^2 + b\lambda x + c\lambda^2$. Considering $\tilde{g}\left(x - \dfrac{b}{2a}\lambda, \lambda\right)$, we can attain the desired result by eliminating term $\lambda x$.

**Example 10** Germ $g \in \varepsilon_{x,\lambda}$ is equivalent to $h = \varepsilon x^3 + \delta\lambda^2$, $(\varepsilon = \pm 1, \delta = \pm 1)$, if and only if $g = g_x = g_\lambda = g_{xx} = g_{\lambda x} = 0, \varepsilon = \operatorname{sgn} g_{xxx}$ and $\delta = \operatorname{sgn} g_{\lambda\lambda}$ at $x = \lambda = 0$.

**Proof** $\mathcal{I}(h) = \mathcal{M}^3 + <\lambda^2>$, and $\mathcal{P}(h) = \mathcal{M}^4 + \mathcal{M}^2 <\lambda>$, so we can obtain the results in the same way as above.

### 4.5.2 Unfolding Theory

The object of unfolding theory is to study the behaviour of bifurcations influenced by a small disturbance. The bifurcation equation is an idealised mathematical model of the system under study. However, a small discrepancy is often found between the real practical system and the idealised model. This discrepancy is called an "imperfection" and is regarded as a perturbation of the ideal system. The study of the change in bifurcating behaviour caused by such perturbations is called imperfect bifurcation. "Imperfect bifurcation" can be described by introducing some auxiliary parameters or by unfolding theory.

**1. The Concepts of Tangent Space and Universal Unfoldings**

In the study of recognition problems, the concept of strong equivalence has been introduced and on this concept are based the basic properties of restricted tangent space. Having solved the recognition problem of bifurcation, now the concept of equivalence is introduced to develop the concept of strong equivalence so as to study the property of tangent space and unfolding problems.

**Definition 4.11** Germs $f$ and $g$ in $\varepsilon_{x,\lambda}$ are called the equivalence, written as $f \sim g$. If there is a diffeomorphism $T:(x,\lambda) \to (X(x,\lambda),\wedge(\lambda))$ and a function $S(x,\lambda)$ then

$$g(x,\lambda) = S(x,\lambda) f(X(x,\lambda),\wedge(\lambda)) \tag{4.55}$$

where $S(0,0) > 0, X(0,0) = 0, X_x'(0,0) > 0, \wedge(0) = 0, \wedge'(0) > 0$.

Note that if $f \sim g$, then the number of zero points of $f$ and $g$ at the origin will be $n_g(\wedge(\lambda)) = n_f(\lambda)$.

**Definition 4.12** Let $g$ be a germ in $\varepsilon_{x,\lambda}$. If there exists $G(x,\lambda,\alpha) \in \varepsilon_{x,\lambda,\alpha}$ such that

$$G(x,\lambda,0) = g(x,\lambda) \qquad \alpha = (\alpha_1,\alpha_2,\ldots,\alpha_k)^T \in \mathcal{R}^k$$

then $G$ is called a $k$-parameter unfolding of $g$, and $\alpha$ is called the unfolding parameter.

There may be innumerable unfoldings with $g(x,\lambda)$. What relationship exists between them?

**Definition 4.13** Let $G(x,\lambda,\alpha)$ and $H(x,\lambda,\beta)$ be the unfoldings of $g, a \in \mathcal{R}^k, b \in \mathcal{R}^m$. We say that $H$ factors through $G$ if there exists smooth mapping $S, X, \wedge$ and $A$ such that

$$H(x,\lambda,\beta) = S(x,\lambda,\beta) G(X(x,\lambda,\beta),\wedge(\lambda,\beta), A(\beta)) \tag{4.56}$$

where $S(x,\lambda,0) \equiv 1, X(x,\lambda,0) \equiv x, \wedge(\lambda,0) \equiv \lambda$ and $A(0) = 0$ as $\beta = \infty$. Then $H(x,\lambda,\beta)$ and $G(x,\lambda,A(\beta))$ are the equivalence near $\beta = 0$. For a germ $g$, that there are some unfoldings $G$ in the sense of equivalence that contain all unfoldings of $g$. This is concept of the versal and universal unfolding we are about to introduce.

**Definition 4.14** An unfolding $G$ of $g$ is called a versal, if every unfolding of $g$ factors is through $G$. A versal unfolding of $g$ with the minimum number of parameters possible is called a universal unfolding of $g$. The number of unfolding parameters in universal unfolding is called the codimension of $g$, and if $g$ does not possess a universal unfolding we say that $g$ has infinite codimension.

Now we shall discuss the concept of tangent space.

Let $g \in \varepsilon_{x,\lambda}$, $G(x,\lambda,\alpha)$ be a universal unfolding of $g$, for any $q \in \varepsilon_{x,\lambda}$, and consider a one-parameter unfolding

$$H(x,\lambda,\varepsilon) = g(x,\lambda) + \varepsilon q(x,\lambda) \tag{4.57}$$

Since $G$ is universal, $H$ factors through $G$, that is

$$H(x,\lambda,\varepsilon) = S(x,\lambda,\varepsilon)G(X(x,\lambda,\varepsilon), \wedge(\lambda,\varepsilon), A(\varepsilon)) \tag{4.58}$$

where $S(x,\lambda,0) = 1, X(x,\lambda,0) = x$, $\wedge(\lambda,0) = \lambda$ and $A(0) = 0 \in \mathcal{R}^k$..

Differentiating eq. (4.58) with respect to $\varepsilon$ at $\varepsilon = 0$ we have

$$\dot{q}(x,\lambda,0) = \dot{S}(x,\lambda,0)g(x,\lambda) + g_x(x,\lambda)\dot{X}(x,\lambda,0) + g_\lambda(x,\lambda)\dot{\wedge}(\lambda,0)$$
$$+\sum_{i=1}^{k}\dot{A}_i(0)\frac{\partial G}{\partial \alpha_i}(x,\lambda,0) \tag{4.59}$$

Then we can define tangent space.

**Definition 4.15** Supposing $g \in \varepsilon_{x,\lambda}$, we denote the tangent space of $g$ by $T(g)$, defined as follows: $T(g) = \{ag + bg_x + cg_\lambda | a,b \in \varepsilon_{x,\lambda}, c \in \varepsilon_\lambda\}$.

**Note:** the tangent space $T(g)$ does not form an ideal in $\varepsilon_{x,\lambda}$ like restricted tangent space, but we have the following conclusion.

**Theorem 4.17** RT($g$) has finite codimension if $T(g)$ has finite codimension.

The proof of the theorem refers to [4].

The following result gives the relationship of the codimension between $g$ and $T(g)$.

**Theorem 4.18** (Universal Unfolding Theorem) Let $g$ be a germ in $\varepsilon_{x,\lambda}$ and let $G$ be a $k$-parameter unfolding of $g$. Then if and only if the following is correct, $G$ is the universal unfolding of

$$\varepsilon_{x,\lambda} = T(g) + R\{\left\{\frac{\partial G}{\partial \alpha_1}(x,\lambda,0),\ldots,\frac{\partial G}{\partial \alpha_k}(x,\lambda,0)\right\}\}$$

**Corollary 1** $g \in \varepsilon_{x,\lambda}$ has finite codimension if $T(g)$ has finite codimension, and

codim ($g$)=codim $T(g)$.

**Corollary 2** If $g \in \varepsilon_{x,\lambda}$ has finite codimension $k$, and there exist $k$ germs $p_1,\ldots,p_k \in \varepsilon_{x,\lambda}$ such that

$$\varepsilon_{x,\lambda} = T(g) \oplus \mathcal{R}(p_1,\ldots,p_2) \tag{4.60}$$

then $$G(x,\lambda,\alpha) = g(x,\lambda) + \sum_{i=1}^{k}\alpha_i p_i(x,\lambda) \tag{4.61}$$

## 2. Computation of Universal Unfoldings

So far we have given the structure of a universal unfolding of a germ $g$. To find the concrete solution for the universal unfolding we have to look at the difference between tangent space and restricted tangent space. Recalling the concepts of restricted tangent space and tangent space we find $T(g) = \mathrm{RT}(g) + \varepsilon_\lambda g_\lambda$ then we have

**Theorem 4.19** Suppose $g \in \varepsilon_{x,\lambda}$ has finite codimension, then there exists $l \geq 0$, such that $T(g) = \mathrm{RT}(g) \oplus \mathcal{R}(g_x, g_\lambda, \ldots, \lambda^l g_\lambda\}$

Combining the above theorem with the structure of the universal unfolding, we can give the steps for computation of universal unfoldings as follows:

Step 1. Determine integer $l$, such that $T(g) = \mathrm{RT}(g) \oplus \mathcal{R}(g_x, g_\lambda, \ldots, \lambda^l g_\lambda\}$.

Step 2. Split $T(g)$ as follows $T(g) = [\mathrm{Itr}\, T(g)] \oplus \mathcal{V}_g$.

where $\mathcal{V}_g = T(g) \cap [\mathrm{Itr} T(g)]^\perp$.

Step 3. Find complement basis $p_1, p_2, \ldots, p_k$ in $[\mathrm{Itr}\, T(g)]^\perp$ and write the universal unfolding of $g$ $G(x,\lambda,\alpha) = g(x,\lambda) + \sum_{i=1}^{k} \alpha_i p_i(x\lambda)$.

**Example 11** Calculate the number of codimensions and universal unfoldings of $g = x^3 - \lambda x$:

$$\mathrm{RT}(g) = \mathcal{M}^3 + \mathcal{M} < \lambda >$$
$$\lambda h_\lambda = x\lambda \in \mathcal{M} < \lambda > \subset \mathrm{RT}(g)$$
$$T(x^3 - \lambda x) = (\mathcal{M}^3 + \mathcal{M} < \lambda >) \oplus \mathcal{R}\{3x^2 - \lambda, x\} \text{ and}$$
$$[\mathrm{Itr} T(x^3 + \lambda x)]^\perp = [\mathcal{M}^3 + \mathcal{M} < \lambda >]^\perp = \mathcal{R}\{1, x, \lambda, x^2\}$$

Then the complement basis of $\mathcal{V}_g = \mathcal{R}\{3x^2 - \lambda, x\}$ in $\mathcal{R}\{1, x, \lambda, x^2\}$ is $\{1, \lambda\}$ or $\{1, x^2\}$, so $G(x,\lambda,\alpha,\beta) = x^3 - \lambda x + \alpha + \beta x^2$ or $H(x,\lambda,\alpha,\beta) = x^3 - \lambda x + \alpha + \beta\lambda$ is the universal unfolding of a pitchfork.

**Example 12.** Calculate the codimension and universal unfolding of $g = x^2 + \lambda$.

If $\mathrm{RT}(x^2 + \lambda) = \mathcal{M}^2 + < \lambda >,\ T(x^2 + \lambda) = \varepsilon_{x,\lambda}$

then the codimension of $g$ is zero, and the universal unfolding is itself.

**3. Determination of Universal Unfoldings**

Suppose we want to determine the universal unfoldings of a static bifurcation of a one state variable if $g \sim h$, and $G(x,\lambda,\alpha)$ is a $k$-parameter unfolding of $g$. How do we determine that $G$ is a universal of $g$? In fact, we have given the answer above. For convenience of calculation, we will give a simple method of verification.

**Theorem 4.20** Suppose $g \in \varepsilon_{x,\lambda}$ is of codimension $k$ and $G$ is a $k$-parameter unfolding of $g$, then we denote $G_{\alpha_i} = \frac{\partial G}{\partial \alpha_i}|_{\alpha=0}$, $i = 1,2,\ldots,k$. Suppose $g \sim h$, and the definition of the projection map is $\wp : \varepsilon_{x,\lambda} \to [\mathrm{Itr}\, T(h)]^\perp$ then $G$ is a universal unfolding of $g$, if and only if $[\mathrm{Itr}\, T(h)]^\perp = \mathcal{V}_g + \mathcal{R}\{JG_{\alpha_1}, \ldots, JG_{\alpha_k}\}$ where

$$\varepsilon_{x,\lambda} = [\mathrm{Itr} T(h)] \oplus [\mathrm{Itr} T(h)]^\perp, \mathcal{V}_g = T(g) \cap [\mathrm{Itr} T(g)]^\perp$$

**Proof** From $T(g) = [\mathrm{Itr}T(h)] \oplus \mathcal{V}_g$, then $G$ is universal if $\varepsilon_{x,\lambda} = \mathrm{Itr}T(h) \oplus \mathcal{V}_g \oplus \mathcal{R}\{G_{\alpha_1},\ldots,G_{\alpha_k}\}$. Projecting both sides of the equation, then we have $[\mathrm{Itr}T(h)]^{\perp} = \mathcal{V}_g \oplus \mathcal{R}\{JG_{\alpha_1},\ldots,G_{\alpha_k}\}$.

**Note 1** $Jf = \sum_{\alpha} \frac{1}{\alpha!} D^{\alpha} f(0,0) x^{\alpha_1}\lambda^{\alpha_2}$ is the sum of all $x^{\alpha_1}\lambda^{\alpha_2}$ not belonging to $\mathrm{Itr}(T(h))$.

**Note 2** $\mathcal{V}_g = JT(g)$.

**Example 13** Suppose $g \sim h(x,\lambda) = \varepsilon x^3 + \delta\lambda, \varepsilon = \pm 1, \delta = \pm 1$. $G(x,\lambda,\alpha)$ is a one-parameter universal unfolding of $g$ at $x = \lambda = \alpha = 0$ if $\det\begin{bmatrix} g_{\lambda} & g_{\lambda x} \\ G_{\alpha} & G_{\alpha x} \end{bmatrix} \neq 0$.

**Proof** Since $T(\varepsilon x^3 + \delta\lambda) = \mathcal{M}^3 + <\lambda> \oplus \mathcal{R}\{1\}$, then $(\mathrm{Itr}\, T(h)^{\perp} = \mathcal{R}\{1,x\}$, and hence $J(g) = g(0,0) + g_x(0,0)x$. Then $G$ is a universal unfolding of $g$, if and only if $\mathcal{R}\{1,x\} = \mathcal{V}_g + \mathcal{R}\{JG_{\alpha}\}$.

Noticing that $\alpha$, and considering the conditions of recognition $g = g_x = g_{xx} = 0$ at $x = \lambda = 0$, we can see that $Jg = 0$, $Jg_x = 0$, $J(\lambda g_{\lambda}) = 0$, and hence $\mathcal{V}_g$ is generated by $g_{\lambda}(0,0) + g_{\lambda x}(0,0)x$. Writing $\mathcal{R}\{1,x\} = \mathcal{R}\{Jg_{\lambda}, JG_{\alpha}\}$ in the form of a matrix, this completes the example.

**Example 14** Suppose $g \sim h(x,\lambda) = \varepsilon x^3 + \delta\lambda x, \varepsilon = \pm 1, \delta = \pm 1$. $G(x,\lambda,\alpha,\beta)$ is a two-parameter universal unfolding of $g$ at $x = \lambda = \alpha = \beta = 0$ if and only if

$$\det\begin{bmatrix} 0 & 0 & g_{x\lambda} & g_{xxx} \\ 0 & g_{\lambda x} & g_{\lambda\lambda} & g_{\lambda xx} \\ G_{\alpha} & G_{\alpha x} & G_{\alpha\lambda} & G_{\alpha xx} \\ G_{\beta} & G_{\beta x} & G_{\beta\lambda} & G_{\beta xx} \end{bmatrix} \neq 0$$

**Example 15** Suppose $g \sim h(x,\lambda) = \varepsilon x^3 + \delta\lambda^2, \varepsilon = \pm 1, \delta = \pm 1$. $G(x,\lambda,\alpha,\beta)$ is a three-parameter universal unfolding of $g$ at $x = \lambda = \alpha = \beta = \gamma = 0$ if and only if

$$\det = \begin{vmatrix} 0 & 0 & 0 & g_{xxx} & g_{xx\lambda} \\ 0 & 0 & g_{\lambda\lambda} & g_{\lambda xx} & g_{\lambda x\lambda} \\ G_{\alpha} & G_{\alpha x} & G_{\alpha\lambda} & G_{\alpha xx} & G_{\alpha x\lambda} \\ G_{\beta} & G_{\beta x} & G_{\beta\lambda} & G_{\beta xx} & G_{\beta x\lambda} \\ G_{\gamma} & G_{\gamma x} & G_{\gamma\lambda} & G_{\gamma xx} & G_{\gamma x\lambda} \end{vmatrix} \neq 0$$

In table 4.1 later in this chapter we give all the GS normal forms and their universal unfoldings with codimensions no more than 3.

## 4. Perturbed Bifurcation Diagrams

In this section, we discuss the perturbed bifurcation diagrams of $\{(x,\lambda) | G(x,\lambda,\alpha) = 0\}$ by the method of universal unfolding.

Suppose $G(x,\lambda,\alpha)$ is a universal unfolding of $g(x,\lambda)$, and (0,0) is a singular point of $g(x,\lambda)$. Since $G$ contains all the perturbations subjected to $g$, then we can denote all the changes in bifurcation diagrams as $\alpha$ varying in $\alpha$ space.

It can be shown that, for some value of $\alpha$, the bifurcation diagram of the universal unfolding $G$ of $g$ is topologically equivalent under a small perturbation, that is the bifurcations have a generality. These bifurcation diagrams are considered persistent. But for other value of $\alpha$, the bifurcation diagram of $G$ would change its property under a small perturbation, that is, the bifurcation is degenerate. These bifurcation diagrams are considered non-persistent. We point out that non-persistent sets or transition sets in unfolding parameter space are of three cases: bifurcation point set, hysteresis set and double limit point set. Thus we put forward the following four definitions.

**Definition 4.16** Let $G:\mathcal{R}\times\mathcal{R}\times\mathcal{R}^k\rightarrow\mathcal{R}$ be a universal unfolding of $g:\mathcal{R}\times\mathcal{R}\rightarrow\mathcal{R}$. Then we can define:

a) bifurcation point set

$$B=\{\alpha\in\mathcal{R}^k\,|\,\exists(x,\lambda)\in\mathcal{R}\times\mathcal{R} \text{ such that } G=G_x=G_\lambda=0 \text{ at } (x,\lambda,\alpha)\}$$

b) hysteresis set

$$H=\{\alpha\in\mathcal{R}^k\,|\,\exists(x,\lambda)\in\mathcal{R}\times\mathcal{R}, \text{ such that } G=G_x=G_{xx}=0 \text{ at } (x,\lambda,\alpha)\}$$

c) double limit point set

$$D=\{\alpha\in\mathcal{R}^k\,|\,\exists(x_1,x_2,\lambda)\in\mathcal{R}\times\mathcal{R}\times\mathcal{R}, x_1\neq x_2 \quad \text{such that}$$

$$G=xG_x=0 \text{ at } (x_i,\lambda,\alpha), i=1,2\}$$

d) transition set $\sum=B\bigcup H\bigcup D$.

The transition set is the set of unfolding parameters corresponding to the non-persistent bifurcation diagrams of $G(x,\lambda,\alpha)$. It is a hyperplane in $\mathcal{R}^k$. The transition set includes three cases: the bifurcation, hysteresis and double limit point set. When we consider only small perturbations, or we restrict the parameter in the neighbourhood $\mathcal{W}$ of $\mathcal{R}^k$, the transition set divides parameter space into several regions, in which diagrams of $G$ are persistent. Consequently we can obtain all the perturbed bifurcation diagrams subjected to a small perturbation.

**Example 16.** Consider the perturbed bifurcation diagrams of $h(x,\lambda)=x^3-\lambda x$.

Suppose $H$ is a universal unfolding of $h$: $H(x,\lambda,\alpha,\beta)=x^3-\lambda x+\alpha+\beta x^2$.

Since $H_x=3x^2-\lambda+2\beta x,\ H_\lambda=-x,\ H_{xx}=6x+2\beta$

then $B=\{\alpha=0\}$, $H=\{\alpha=\frac{\beta^2}{27}\}$, $D=\phi$ and transition set $=\sum=B\bigcup H$.

Figure 4.10 gives all the curves of $\Sigma$ which divides: the parameter space into four regions and the diagrams on each region and $\Sigma$. All the singular points of equation $H=0$ are limit points but not double limit points. The bifurcation diagram has the bifurcation point on the bifurcation point set $B$ and the hysteresis point on the set $H$.

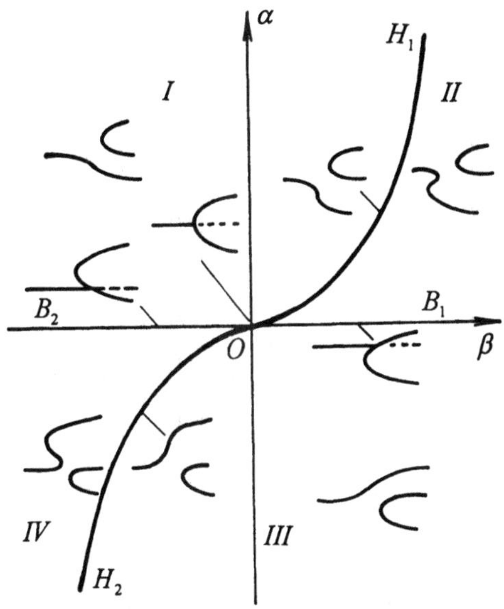


Fig. 4.10

We now return to the Euler buckling equation.

**Example 17** If the weight of the structure and the non-symmetry of the spring are taken into consideration in example 1, two parameters $\varepsilon$ and $\delta$ should be introduced. Suppose that the system is in the static state. Now consider the static bifurcation. The potential energy of the system is written as

$$\tilde{V}(x,\lambda,\varepsilon,\delta)=\frac{1}{2}(x-\delta)^2 \bullet 2\lambda(\cos x-1)+\varepsilon\sin x$$

and the static state is described by the following equation

$$G(x,\lambda,\varepsilon,\delta)=-\frac{\partial\tilde{V}}{\partial x}=-x+\delta+2\lambda\sin x-\varepsilon\cos x=0 \tag{4.62}$$

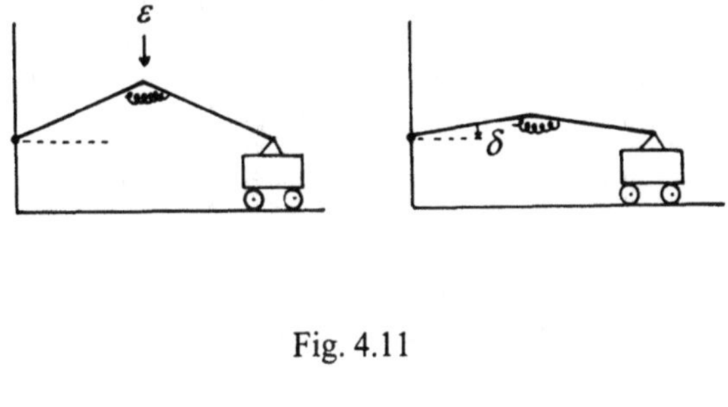


Fig. 4.11

We can identify eq. (4.62) as the perturbation of eq. (4.46). Expanding $G(x,\lambda,\varepsilon,\delta)$ at $(x,\lambda)=\left(0,\frac{1}{2}\right)$, we have

$$G(x,\lambda,\varepsilon,\delta) = -\frac{1}{6}x^3 + \frac{1}{2}\varepsilon x^2 + 2(\lambda - \frac{1}{2})x + (\delta - \varepsilon) + \text{hot} \tag{4.63}$$

Now introduce the transformation

$$y = -\frac{x}{\sqrt[3]{6}},\ \ \mu = \sqrt[3]{6}(1-2\lambda),\ \ \alpha = \delta - \varepsilon,\ \ \beta = \frac{\varepsilon}{2\sqrt[3]{36}}$$

and we have

$$H(y.\lambda.\alpha.\beta) = y^3 - \lambda y + \alpha + \beta y^2 + \text{hot} \tag{4.64}$$

Since we need to consider the diagrams only in the neighbourhood of the origin, the higher terms in $H$ can be omitted. That means we only need study the bifurcation diagrams of the equation $y^3 - \lambda y + \alpha + \beta y^2 = 0$ which is the universal unfolding of $y^3 - \lambda y$. The diagrams of this equation are shown in Fig. 4.10.

Returning to the parameters $\varepsilon$ and $\delta$, we can obtain the diagrams of the original system. To end this section we give some notes as follows:

1) If the generality of the diagrams is taken into consideration, we know that the diagrams of $G$ are generate if $\alpha \not\in \Sigma$, and degenerate if $\alpha \in \Sigma / \{0\}$ with the degenerate degree less than $\alpha = 0$.

2) Multiple parameter static bifurcation can be taken as the universal unfolding of one parameter bifurcation.

3) We have frequently placed the emphasis of our study on local static bifurcations. Nevertheless, through local results, global information on bifurcation diagrams can be derived by studying degenerate local bifurcation problems and their perturbations. For instance, the hysteresis on diagrams corresponding to $H$ in Fig. 4.10 can be taken as the coincidence of two limit points, which can also be split into two limit points subjected to some proper perturbation.

### 4.5.3 Classification by Codimension

Codimension is a very important concept in the study of static bifurcation. With the increase in the order of codimension, the bifurcation diagram will be very complicated. In consequence, it is difficult to study the systems with high-order codimension. The transition sets are complex and difficult to find if the systems are with codimensions of more than 3. However, systems are often of codimension no more than 3. So it is necessary to have an overall study of the system with codimension no more than 3. Furthermore, we can classify singularities by codimension. Thus we have the results of Table 4.1.

**Theorem 4.21** Suppose $g \in \varepsilon_{x,\lambda}$, codim $g \le 3$, then $g$ is equivalent to one of the GS normal forms shown in table 4.1.

The proof of the theorem can be found in [4]. The theorem states that there are at most eleven normal forms with codimension no more than 3. They are often the ordinary cases.

**Table 4.1** Universal Unfoldings and Diagrams of Normal Forms with Codimensions less than 3.

| | nomenclature | codim | normal form | universal unfolding | unperturbed bifurcation diagram ($\varepsilon = \pm 1$) $\delta = 1$ | $\delta = -1$ |
|---|---|---|---|---|---|---|
| (1) | limit point | 0 | $\varepsilon x^2 + \delta\lambda$ | $\varepsilon x^2 + \delta\lambda$ |  |  |
| (2) | simple bifurcation | 1 | $\varepsilon(x^2 - \lambda^2)$ | $\varepsilon(x^2 - \lambda^2 + \alpha_1)$ |  |  |
| (3) | isola centre | 1 | $\varepsilon(x^2 + \lambda^2)$ | $\varepsilon(x^2 + \lambda^2 + \alpha_1)$ |  |  |
| (4) | hysteresis | 1 | $\varepsilon x^3 + \delta\lambda$ | $\varepsilon x^3 + \delta\lambda + \alpha_1 x$ |  |  |
| (5) | asymmetric cusp | 2 | $\varepsilon x^2 + \delta\lambda^3$ | $\varepsilon x^2 + \delta\lambda^3 + \alpha_1$ $+\alpha_2\lambda$ |  |  |
| (6) | pitchfork | 2 | $\varepsilon x^3 + \delta\lambda x$ | $\varepsilon x^3 + \delta\lambda x + \alpha_1$ $+\alpha_2 x^2$ |  |  |
| (7) | quartic fold | 2 | $\varepsilon x^4 + \delta\lambda$ | $\varepsilon x^4 + \delta\lambda + \alpha_1 x$ $+\alpha_2 x^2$ |  |  |
| (8) | | 3 | $\varepsilon x^2 + \delta\lambda^4$ | $\varepsilon x^2 + \delta\lambda^4 + \alpha_1$ $+\alpha_2\lambda + \alpha_3\lambda^2$ |  |  |
| (9) | winged cusp | 3 | $\varepsilon x^3 + \delta\lambda^2$ | $\varepsilon x^3 + \delta\lambda^2 + \alpha_1$ $+\alpha_2 x + \alpha_3\lambda x$ |  |  |
| (10) | | 3 | $\varepsilon x^4 + \delta\lambda x$ | $\varepsilon x^4 + \delta\lambda x + \alpha_1$ $+\alpha_2\lambda + \alpha_3 x^2$ |  | 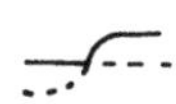 |
| (11) | | 3 | $\varepsilon x^5 + \delta\lambda$ | $\varepsilon x^5 + \delta\lambda + \alpha_1 x$ $+\alpha_2 x^2 + \alpha_3 x^3$ |  |  |

Unstable solutions ($g_x < 0$) are indicated by broken lines. Stable and unstable may be interchanged if $\varepsilon = -1$.

## 4.6 Simple Bifurcations

### 4.6.1 Simple Bifurcation

Consider a map $f(x,\lambda): \mathcal{U} \subset \mathcal{R} \times \mathcal{R} \to \mathcal{R}^n$, $f \in \mathcal{C}^p(\mathcal{U}), p \geq 2$.

Suppose $f(x,\lambda) = 0$ (4.65)

has a trivial solution $x = 0$, i.e. $f(0,\lambda) = 0$, then $(0,\lambda_0)$ is called a simple bifurcation point, if for every $\lambda \neq \lambda_0$ near the point $(0,\lambda_0)$ there exists a unique, small, non-trivial solution $x \neq 0$ See Fig. 4.12.

Note if eq. (4.65) has a trivial solution $x = x(\lambda), \lambda \in \mathcal{I}$ ($\mathcal{I}$ is an interval belonging to $\mathcal{R}$), then $f(x(\lambda),\lambda) = 0$ is satisfied. Without the loss of any generality, the trivial solution can be translated to the $\lambda$-axis through the transformation $\hat{x} = x - x(\lambda)$, i.e. $f(0,\lambda) = 0, \lambda \in \mathcal{I}$.

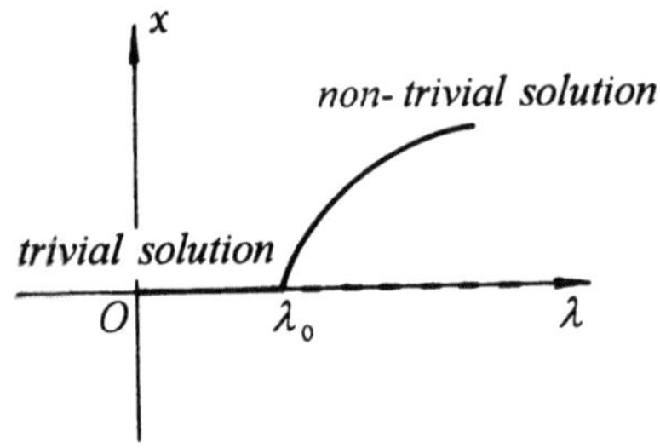


Fig. 4.12

**Theorem 4.22** Assume that the map $f(x,\lambda)$ and the equation $f(x,\lambda) = 0$ defined above satisfy:

(a) $f(0,\lambda) = 0, \lambda \in \mathcal{I} \subset \mathcal{R}$

(b) $A = f_x(0,\lambda_0)$ which has a one-dimensional nullspace $\eta(A)$ spanned by the eigenvector $\varphi$ which corresponds to the eigenvalue of single zero $\eta(A) = \{\alpha\varphi | \alpha \in \mathcal{R}\}$

(c) The range space of $A = R(A) = Y_1$, its number of codimension being one, that is $Y = Y_1 \oplus Y_{2,}$ $\dim y_2 =$ .

(d) $f_{x\lambda}(0,\lambda_0)\varphi \notin Y_1$

Then $(0,\lambda_0)$ is a simple bifurcation point of $f(x,\lambda) = 0$. The non-trivial solutions form the $\mathcal{C}^{p\text{-}2}$ curve intersecting the trivial solutions at $(0,\lambda_0)$, and the equation of the curve can be expressed as $(X(s),\lambda(s)) = (s\varphi + w(s), \lambda_0 + s\mu(s))$. Since $X = X_1 \oplus X_{2.}$ and $X_1 = \eta(A)$ then $x = x_1 + x_2 \overset{\text{def}}{=} s\varphi + w(s,\lambda)$, where $x_1 \in X_1, x_2 \in X_2$. $w(0) = 0$ and $w'(0) = 0$ which is the function of $s$ and $\lambda$ with an order of more than 2.

From the aforesaid condition (b), we have $A\varphi = 0$. $\varphi$ is called the right eigenvector. There exists a left eigenvector $\psi^T$ at the same time that satisfies $\psi^T A = 0$ (or $A^T\psi = 0$), and vector $\psi^T = [\psi_1 \ \ \psi_2 \ \ \cdots \ \ \psi_n]$. It is known from the algebra that $\psi^T\varphi \neq 0$, and furthermore, after orthonormalization we have

$$\psi^T\varphi = \psi_1\varphi + \ldots + \psi_n\varphi_n = 1.$$

In this case, we can express the null space as $\eta(A) = \{\alpha\varphi | \alpha \in \mathcal{R}\}$, and the range space as $\mathcal{R}(A) = \{y \in \mathcal{R}^n | Ax = y, \text{ for some } x \in \mathcal{R}\} = (\eta(A^\perp))^\perp = \{y \in \mathcal{R}^n | <\psi^T, y> = 0\}$.

Obviously $\eta(A)$ intersects with $\mathcal{R}(A)$ transversely, hence $\mathcal{R}^n = \eta(A) \oplus \mathcal{R}(A)$. Define a projection operator $Q = \varphi\psi^T$. Then since $x = s\varphi + w$ so $Qx = \varphi\psi^T(s\varphi + w) = \varphi\psi^T s\varphi + \varphi\psi^T w = s\varphi$ i.e. $Q: \mathcal{R}^n \to \eta(A)$ or $Q\eta(A) = \eta(A), Q\mathcal{R}(A) = 0$.

The complement projection operator $P$ of $Q$ is $P = I - Q$ where $I$ is the identity projection operator. It is obvious that

$$P\eta(A) = (I - Q)\eta(A) = 0 \text{ and } P\mathcal{R}(A) = (I - Q)\mathcal{R}(A) = \mathcal{R}(A).$$

Now we shall prove that there exists a unique solution of the simple bifurcation.

**Proof** We first apply the Liapunov—Schmidt method to reduce the vector field equations to a bifurcation equation. Equation (4.65) can equivalently be transformed to the following two scalar equations by the use of the projection operators defined above:

$$Pf(x,\lambda)=0 \tag{4.66}$$

$$Qf(x,\lambda)=0 \tag{4.67}$$

Since $x=x_1+x_2=s\varphi+w(s,\lambda), w(s,\lambda)$ can be resolved from eq. (4.66). We can also prove that $w(0)=0$ and $w'(0)=w_s(0)+w_\lambda(0)=0$. After substituting $w(s,\lambda)$ into eq. (4.67) we obtain the bifurcation equation $F=0$. Furthermore, we apply the Morse lamma to transform $F$ into a quadratic form after proving $DF=0$. Finally $w(s,\lambda)$ and $\lambda(s)$ are resolved.

Differentiating eq. (4.66) with respect to $x_2$, we have

$$D_{x_2}Pf(x_1+x_2,\lambda)|_{(0,\lambda_0)}=PD_{x_2}f(s\varphi+x_2,\lambda)|_{(0,\lambda)}=PAI_{x_2} \tag{4.68}$$

As $AI_{x_2}$ is on the complementaty set $X_2$ of $X_1=\eta(A)$, it has no zero eigenvalue, and hence it is one to one on $X_2$ and the solution $x_2=w(s,\lambda)$ is unique.

Differentiating eq. (4.66) with respect to $s$, we have

$$\frac{\partial}{\partial s}Pf(s\varphi+w(s,\lambda),\lambda)=Pf_x(0,\lambda_0)(\varphi+w_s(0))=PA(\varphi+w_S(0))=0$$

Since $A\varphi=0$, we have $Aw_s(0)=0$. The equation indicates that $w_s(0)$ belongs to $\eta(A)=X_1$. However, from the definition $w(s,\lambda)$ should belong to $X_2$. If $w_s(0)$ belongs to both $X_1$ and $X_2$, there must exist $w_s(0)=0$ because $X=X_1\oplus X_2$.

Differentiating eq. (4.66) with respect to $\lambda$, we have

$$\frac{\partial}{\partial\lambda}Pf(s\varphi+w(s,\lambda),\lambda)=Pf_x(0,\lambda_0)\frac{\partial w(0,\lambda_0)}{\partial\lambda}+Pf_\lambda(0,\lambda)=0$$

and since $f(0,\lambda_0)=0$, therefore $f_\lambda(0,\lambda)=0$. From $Aw_\lambda(0,\lambda_0)=0$, then $w_\lambda(0,\lambda_0)=0$, so we have $w'(0,\lambda_0)=w_s(0,\lambda_0)+w_\lambda(0,\lambda_0)=0$.

Substituting $x=s\varphi+w(s,\lambda)$ into eq. (4.67), we obtain the bifurcation equation:

$$Qf(s\varphi+w(s,\lambda),\lambda)=F(s\varphi+w(s,\lambda),\lambda)=0 \tag{4.69}$$

From eq. (4.69) we obtain

$$\frac{\partial F}{\partial s}=Qf_x(0,\lambda_0)(\varphi+w_s(0,\lambda_0))=QA\varphi+QAw_s(0,\lambda_0)=0$$

$$\frac{\partial F}{\partial\lambda}=Qf_x(0,\lambda_0)(w_\lambda(0,\lambda_0))+Qf_\lambda(0,\lambda_0)=0$$

and grad $F=\dfrac{\partial F}{\partial s}+\dfrac{\partial F}{\partial\lambda}=0$.

This shows that bifurcation occurs at $(0,\lambda_0)$.

Next we apply the Morse lamma to transform the bifurcation equations into quadratic form.

First we calculate the Hessian matrix $q$ of $F$

$$q=\begin{bmatrix}\dfrac{\partial^2 F}{\partial s\partial s} & \dfrac{\partial^2 F}{\partial s\partial\lambda}\\ \dfrac{\partial F}{\partial\lambda\partial s} & \dfrac{\partial^2}{\partial\lambda\partial\lambda}\end{bmatrix}=\begin{bmatrix}F_{xx}\varphi\varphi & F_{x\lambda}\varphi\\ F_{\lambda x}\varphi & F_{\lambda\lambda}\varphi\end{bmatrix}=\begin{bmatrix}a & b\\ b & c\end{bmatrix}$$

According to condition (a), $F_{\lambda\lambda}=0$, and according to (d), $f_{x\lambda}(0,\lambda)\varphi\notin Y_1=\mathcal{R}(A)$. Therefore it must belong to $\eta(A)$, so as $Qf_{x\lambda}(0,\lambda_0)\varphi\neq 0$, i.e. $b\neq 0$, hence

$$\det q=-b^2\neq 0. \tag{4.70}$$

Because $\det q=-b^2<0$, it is certain that two of the eigenvalues must be opposite signs—one is positive, the other negative such as $\rho_1>0,\rho_2<0$.

From the results of the above analysis we can transform $F$ into a quadratic form by the Morse lamma:

$$F(s,\lambda)=\frac{1}{2}(\bar{s},\bar{\lambda})Q\begin{bmatrix}\bar{s}\\ \bar{\lambda}\end{bmatrix}=\frac{1}{2}(\bar{s},\bar{\lambda})\begin{bmatrix}a & b\\ b & 0\end{bmatrix}\begin{bmatrix}\bar{s}\\ \bar{\lambda}\end{bmatrix}=\bar{s}(\frac{a\bar{s}}{2}+b\bar{\lambda})=0$$

The solutions of the bifurcation equation are

trivial solution $\bar{s}=0$ (4.71)

non-trivial solution $\bar{\lambda}=\left(\dfrac{-a}{2b}\right)\bar{s}$ (4.72)

The bifurcation diagrams are shown in Fig. 4.13.

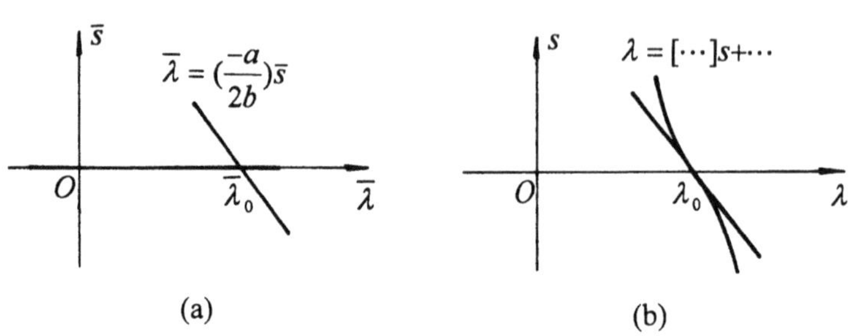


Fig. 4.13

The bifurcation solutions are as follows:

$$x(s)=s\varphi+w(s,\lambda(s)) \tag{4.73}$$

$$\lambda(s)=\lambda_0+\left(\frac{a}{2b}\right)s+o(s^2) \tag{4.74}$$

Now we shall apply the perturbation lamma about the eigenvalues to the study of the stability of simple bifurcation solutions.

Consider the linear eigenvalues in $\mathcal{R}^n$ $A\varphi = \rho\varphi$. Let $A = A(s)$, where $s \in \mathcal{R}$ is the small perturbation, when $s = 0, A(0) = A, A\varphi = 0$. Assume $\rho(0) = \rho = 0$ is a simple eigenvalue, i.e. zero is a single root. If $s$ ($\neq 0$) is arbitrarily small, then the eigenvalue problem becomes

$$A(s)\varphi(s) = \rho(s)\varphi(s) \tag{4.75}$$

Given that $\rho(s)$ and $\varphi(s)$ are existential and unique, and their smoothness is the same as $A(s)$, let $\psi^T A(0) = 0, \psi^T \varphi(0) =$ . Differentiating eq. (4.75) with respect to $s$ at $s = 0$, we have $\dot{A}(0)\varphi(0) + A(0)\dot{\varphi}(0) = \dot{\rho}(0)\varphi(0) + \rho(0)\dot{\varphi}(0) = \dot{\rho}(0)\varphi(0)$.

Multiplying the left-hand side of the equation by $\psi^T$ we have

$$\psi^T[\dot{A}(0)\varphi(0) + A(0)\dot{\varphi}(0)] = \dot{\rho}(0)\psi^T(0)\varphi(0)$$

Hence $\dot{\rho}(0) = \varphi^T \dot{A}(0)\varphi(0)$ (4.76)

Now calculate $\dot{A}(0)$

$$A(s) = D_x f(x_0 + s\varphi + w(s,\lambda), \lambda(s))$$

$$\dot{A}(0) = \frac{\mathrm{d}}{\mathrm{d}s} D_x f(x_0 + s\varphi + w, \varphi(s))|_{0,\lambda_0} = D_{xx} f(0,\lambda_0)\varphi(0)$$

so we have

$$\dot{\rho}(0) = \psi^T D_{xx} f(0,\lambda_0)\varphi(0)\varphi(0) = \mathrm{d} \tag{4.77}$$

$$\rho(s) = \mathrm{d}\, s + o(s^2) \tag{4.78}$$

where d is a constant.

$\rho(s)$ is called the critical eigenvalue if $\rho(s)$ is zero at point $(0,\lambda_0)$. The stability of the solutions are determined by the sign of the critical eigenvalue $\rho(s)$ at point $(0,\lambda_0)$.

### 4.6.2 Simple Bifurcation of the Forced Duffing Equation Without Damping

Duffing equations represent a class of important systems in non-linear vibrations. It is possible local bifurcation, globle bifurcation and chaotic motion exist in such systems. Here we shall prove that the simple local bifurcation occurs in such a system without damping. Assume that the forced Duffing equation expressed in the form of a non-linear operator is

$$f(x,\lambda) = \ddot{x} + x + \lambda(x^3 + g(t)) = 0 \tag{4.79}$$

where $\lambda$ is considered as the bifurcation parameter, and $g(t)$ is a $2\pi$-periodic function. To use the Liapunov—Schmidt method, it is necessary to introduce two periodic function spaces $\mathcal{X}$ and $\mathcal{Y}$ which are both Banach spaces. Define range $\mathcal{X}$ of the map $f(x,\lambda)$ as $\mathcal{X} = \mathcal{D}(f) = \mathcal{C}^2_{2\pi}(T,\mathcal{R})$ where $T$ is a real number set, the modulus of which is $2\pi$, $T = \{\mathcal{R}/2\pi Z\}$ and $T$ is the one-dimensional torus.

Obviously $\mathcal{X}$ is a space of $2\pi$-periodic continuous functions defined on $[0,2\pi]$, having second-order derivatives. Define the value range $\mathcal{Y}$ of the map $f(x,\lambda)$ as $\mathcal{Y} = \mathcal{R}(f) = \mathcal{C}_{2\pi}(T,\mathcal{R})$. It is obvious that $\mathcal{X} \subseteq \mathcal{Y}$.

Define the norms on the functional spaces as follows:

norm on $\mathcal{Y}$ $\|y\| = \sup_{t\in[0,2\pi]} |y(t)| = \|y\|_0$ and norm on $\mathcal{X}$ $\|x\|_2 = \|x\|_0 + \|\dot{x}\|_0 + \|\ddot{x}\|_0$. Under those norms, $\mathcal{X}$ and $\mathcal{Y}$ become Banach spaces, $f(x,\lambda)$ maps $\mathcal{X}\times\mathcal{R}$ to $\mathcal{Y}$. The Duffing equation has a trivial solution (0,0), that is $f(x,\lambda)|_{(0,0)} = 0$.

Now we shall begin to seek the non-trivial solutions (periodic solutions of simple local bifurcation) near (0,0) by the Liapunov—Schmidt reduction. Defining the Frechet derivative as $A = D_x f(x,\lambda)|_{(0,0)} = \left(\frac{\mathrm{d}^2}{\mathrm{d}t^2}+1\right)$, it is easy to prove that $A$ is a linear, bounded and invertible operator mapping $\mathcal{X}$ onto $\mathcal{Y}$. The null space $\eta(A)$ of $A$ is

$$\eta(A) = \{\varphi\in\mathcal{X} \mid A\varphi = 0\ ,\varphi = \alpha\sin t + \beta\cos t\} \tag{4.80}$$

which is a two-dimensional space. The range space $\mathcal{R}(A)$ (which is high dimensional or infinite dimensional) of $A$ is

$$\mathcal{R}(A) = \{h\in\mathcal{Y} \mid Au = h,\ \text{for some } u\in\mathcal{X}\} \tag{4.81}$$

From (4.81) we have

$$\ddot{u} + u = h(t) \qquad h\in\mathcal{R}(A) \tag{4.82}$$

The conditions that solution $u$ is periodic are $u(0) = u(2\pi) = 0$, and $\dot{u}(0) = \dot{u}(2\pi) = 0$. The general solution of eq. (4.82) is

$$u(t) = \varphi(\alpha,\beta,t) + u_p(t) \tag{4.83}$$

where $\varphi(\alpha,\beta,t)$ is the homogeneous solution and $u_p$ is the particular solution which can be obtained by the method of variation of the parameters:

$$u_p(t) = \frac{1}{2\pi}\int_0^t h(s)[\cos(s)\sin(t) - \cos(t)\sin(s)]\mathrm{d}s \tag{4.84}$$

Because $\varphi(\alpha,\beta,t)$ is $2\pi$-periodic, if $u(t)$ is to be $2\pi$-periodic, $u_p(t)$ must satisfy $u_p(0) = u_p(2\pi) = 0$ and $\dot{u}_p(0) = \dot{u}_p(2\pi) = 0$. By eq. (4.84), these conditions of periodic solutions should be

$$\begin{aligned} u_p(2\pi) &= \frac{1}{2\pi}\int_0^{2\pi} h(s)[\cos(s)\sin(2\pi) - \cos(2\pi)\sin(s)]\mathrm{d}s \\ &= \frac{1}{2\pi}\int_0^{2\pi} h(s)\sin(s)\mathrm{d}s = 0 \end{aligned} \tag{4.85}$$

Differentiating $u_p(t)$ in eq. (4.84) with respect to $t$, we have

$$\begin{aligned} \dot{u}_p(t) &= \frac{1}{2\pi}\int_0^t h(s)[\cos(s)\cos(t) + \sin(s)\sin(t)]\mathrm{d}s + \\ &\qquad \frac{1}{2\pi}h(t)[\cos(t)\sin(t) - \sin(t)\cos(t)] \\ &= \frac{1}{2\pi}\int_0^t h(s)[\cos(s)\cos(t) + \sin(s)\sin(t)]\mathrm{d}s \end{aligned}$$

Hence the periodic condition is

$$\dot{u}_p(2\pi) = \frac{1}{2\pi}\int_0^{2\pi} h(s)\cos(s)\,\mathrm{d}s = 0 \tag{4.86}$$

From eq. (4.85) and eq. (4.86), we conclude that $h(s)$ is orthogonal to both $\sin(t)$ and $\cos(t)$. Because $h(s) \in \mathcal{R}(A)$ and $\sin(t), \cos(t) \in \eta(A)$, $\mathcal{R}(A)$ is orthogonal to $\eta(A)$, i.e. $\mathcal{R}(A) = [\eta(A)]^{\perp}$. It is easy to verify that the operator $A$ is self-adjoint.

Note: $A$ is called a self-adjoint operator if for $u, v \in \mathcal{X}$, it satisfies

$$< Au, v >=< u, Av > \tag{a}$$

From the definition of inner product, we obtain

$$< Au, v >= \int vAu ds = \int v(\ddot{u}+u)\,\mathrm{d}s = \int v\ddot{u}ds + \int vu\,\mathrm{d}s \tag{b}$$

$$< u, Av >= \int Av \bullet u ds = \int (\ddot{v}+v)u ds = \int \ddot{v}u ds + \int vu ds \tag{c}$$

Therefore $A$ is self-adjoint which claims $\int \ddot{u}v ds = \int \ddot{v}u ds$. This is easy to prove:

$$\int \ddot{u}v ds = \dot{u}v|_0^{2\pi} - \int \dot{v}\dot{u} ds = -\int \dot{u}\dot{v} ds = -\left(u\dot{v}|_0^{2\pi} - \int u\ddot{v} ds\right) = \int u\ddot{v} ds$$

Thus $A$ is a self-adjoint operator.

By Fredholm's alternative theorem, Banach spaces $\mathcal{X}$ and $\mathcal{Y}$ can be resolved into the direct sums of two mutual orthogonal complementary spaces:

$$\mathcal{X} = \mathcal{X}_1 \oplus \mathcal{X}_2 \overset{def}{=} \eta(A) \oplus \mathcal{W} \text{ and } x = x_1 + x_2 = \varphi + w$$

where $\mathcal{X}_1 = \eta(A)$, $\mathcal{X}_2 = \mathcal{W}, \varphi \in \eta(A)$ and $w \in \mathcal{W}$. $\mathcal{Y} = \mathcal{Y}_1 \oplus \mathcal{Y}_2 = \mathcal{R}(A) \oplus \eta(A)$ where $\mathcal{Y}_1 = \mathcal{R}(A), \eta_2 = \eta(A)$.

Define the projection operators $P$ and $Q$ which make

$$P\mathcal{Y} = P(\mathcal{Y}_1 + \mathcal{Y}_2) = \mathcal{Y}_1,\ Q\mathcal{Y} = Q(\mathcal{Y}_1 + \mathcal{Y}_2) = \mathcal{Y}_2,\ P + Q = I.$$

The expression as for calculation of the operator $Q$ is

$$Q(x(s)) = \frac{1}{\pi}\left\{\int_0^{2\pi} x(s)\sin(s)\sin(t)ds + \int_0^{2\pi} x(s)\cos(s)\cos(t)ds\right\} \tag{4.87}$$

and
$$Q(\varphi + w) = \varphi \tag{4.88}$$

The Duffing eq. (4.79) is equivalent to the following two scalar equations

$Pf(x,\lambda) = 0$ and $Qf(x,\lambda) = 0$.

Substituting eq. (4.79) into these, we obtain

$$PAx + P\lambda(x^3+g) = (I-Q)Ax + \lambda P(x^3+g) = Ax + \lambda P(x^3+g) = 0 \tag{4.89}$$

$$QAx + Q\lambda(x^3+g) = \lambda Q(x^3+g) = 0 \tag{4.90}$$

Because $x = \varphi + w(\varphi,0)$ and $w(\varphi,0) = 0$, $D_\varphi w(\varphi,0) = 0$ when $\lambda = 0$. (4.91)

Substituting eq. (4.91) into eq. (4.89) and (4.90), we obtain

$$A(\varphi + w) + \lambda P((\varphi + w)^3 + g) = 0 \tag{4.92}$$

$$\lambda Q((\varphi+w)^3+g)=0 \tag{4.93}$$

In eq. (4.92), $A\varphi=0, w\in\mathcal{W}$, and since $A$ is invertible in space $\mathcal{W}$, a unique solution $w(\alpha,\beta,\lambda)$ can be solved from eq. (4.92) by the implicit function theorem. We then substitute $w$ into eq. (4.93) to obtain the bifurcation equation $G(\alpha,\beta,\lambda)=0$. When $\lambda\neq 0$ but near zero, the solution of eq. (4.79) is

$$x(\lambda)=\alpha(\lambda)\sin t+\beta(\lambda)\cos(t)+w(\alpha,\beta,\lambda)$$

When $\lambda=0, \alpha(0)=\alpha_0$, $\beta(0)=\beta_0$ and $w(\alpha,\beta,0)=0$. Otherwise a unique branch of solutions $(\alpha(\lambda),\beta(\lambda))$ bifurcates from $(\alpha_0,\beta_0)$ if it exists. The branch is expressed as follows $x(\lambda)=\alpha(\lambda)\sin(t)+\beta(\lambda)\cos(t)+w(\varphi(\lambda),\lambda)$ which can be expanded into a Taylor series at $\lambda=0$:

$$x(\lambda)=\varphi_0(t)+\lambda[\varphi_\lambda(0)+w_\lambda(\varphi_0,0)]+o(\lambda^2) \tag{4.94}$$

If $\varphi_\lambda(0)$ and $w_\lambda(\varphi_0,\lambda)$ can be solved, then the bifurcation solutions are determined.

Comparing eq. (4.93) with eq. (4.87) and noting that sin($t$) and cos($t$) are arbitrary about time $t$, we have

$$\begin{cases} G_1(\alpha,\beta,\lambda)=\dfrac{1}{\pi}\displaystyle\int_0^{2\pi}[(\varphi+w)^3+g]\cos(s)ds=0 \\ G_2(\alpha,\beta,\lambda)=\dfrac{1}{\pi}\displaystyle\int_0^{2\pi}[(\varphi+w)^3+g]\sin(s)ds=0 \end{cases} \tag{4.95}$$

$$G(\alpha,\beta,\lambda)=(G_1,G_2)^T$$

Integrating eq. (4.95) at $\lambda=0$, we obtain the bifurcation equations

$$\begin{cases} G_1(\alpha_0,\beta_0,0)\times\dfrac{4}{3}=\alpha_0(\alpha_0^2+\beta_0^2)+g_1=0 \\ G_2(\alpha_0,\beta_0,0)\times\dfrac{4}{3}=\beta_0(\alpha_0^2+\beta_0^2)+g_2=0 \end{cases} \tag{4.96}$$

where
$$\begin{cases} g_1=\dfrac{4}{3\pi}\displaystyle\int_0^{2\pi}g(s)\cos(s)ds \\ g_2=\dfrac{4}{3\pi}\displaystyle\int_0^{2\pi}g(s)\sin(s)ds \end{cases} \tag{4.97}$$

We can solve $\alpha_0$ and $\beta_0$ from eq. (4.96)

$$\begin{cases} \alpha_0=-\dfrac{g_1}{(g_1^2+g_2^2)^{\frac{1}{3}}} \\ \beta_0=-\dfrac{g_2}{(g_1^2+g_2^2)^{\frac{1}{3}}} \end{cases} \tag{4.98}$$

hence $\varphi_0(t)=\alpha_0\sin t+\beta_0\cos t$. When $g_1\neq 0$ and $g_2\neq 0$, $\alpha_0$ and $\beta_0$ are not zero. Therefore the trivial solution is $(\varphi_0(t),0)$. When $\lambda\neq 0$, but near zero, we seek the non-trivial solutions $\alpha(\lambda)$ and $\beta(\lambda)$ as follows:

From eq. (4.96) we have $D_{(\alpha,\beta)}G(\alpha_0,\beta_0,0)=\begin{bmatrix} 3\alpha_0^2+\beta_0^2 & 2\alpha_0\beta_0 \\ 2\alpha_0\beta_0 & \alpha_0^2+3\beta_0^2 \end{bmatrix}$

and $\det(D_{(\alpha,\beta)}G(\alpha_0,\beta_0,0) = 3(\alpha_0^2+\beta_0^2) \neq 0$ (4.99)

which shows that the Jacobian matrix is non-singular. Therefore, there exists a unique solution $(\alpha(\lambda),\beta(\lambda))$ for $\lambda$ near zero, and $\varphi_\lambda(0)$ and $w_\lambda(\varphi_0,0)$ are also unique.

From eq. (4.95) we have

$$\frac{dG}{d\lambda} = G_\alpha(\alpha_0,\beta_0,0)\alpha_\lambda(0) + G_\beta(\alpha_0,\beta_0,0)\beta_\lambda(0) + G_\lambda(\alpha_0,\beta_0,0) = 0$$

hence $\begin{bmatrix}\alpha_\lambda(0)\\ \beta_\lambda(0)\end{bmatrix} = [G_{(\alpha,\beta)}(\alpha_0,\beta_0,0)]^{-1}G_\lambda(\alpha_0,\beta_0,0)$

where $G_\lambda(\alpha_0,\beta_0,0) = \frac{1}{\pi}\int_0^{2\pi} 3(\varphi_0 + w_\lambda(0,\varphi)^2 w_\lambda(\varphi,0)\begin{bmatrix}\sin t\\ \cos t\end{bmatrix}dt$

we have $\varphi_\lambda(0) = \alpha_\lambda(0)\sin t + \beta_\lambda(0)\cos t$ (4.100)

Now we shall seek for $w_\lambda(0,\varphi_0)$. Differentiate eq. (4.92) with respect to $\lambda$:

$$\ddot{w} + w_\lambda + P[(\varphi+w)^3 + g] + P\lambda\frac{d}{d\lambda}[(\varphi+w)^3+g] = 0$$

When $\lambda = 0$, with eq. (4.93), we have $\ddot{w}_\lambda(0,\varphi_0) + w_\lambda(0,\varphi_0) = -(\varphi^3+g)$.

By the method of variation of the parameters, we obtain $w_\lambda(0,\varphi_0)$:

$$w_\lambda(0,\varphi_0) = -\frac{1}{2\pi}\int_0^{2\pi}(\varphi^3+g)(\cos(s)\sin(t) - \sin(s)\cos(t))ds + \gamma\sin(t) + \eta\cos(t) \tag{4.101}$$

Since $w \in \mathcal{Y}_1 = \mathcal{R}(A)$ we can select $\gamma$ and $\eta$ by the condition $Qw_\lambda(0,\varphi_0) = 0$. The first approximation of the bifurcation solution is obtained by substituting eq. (4.100) and (4.101) into eq. (4.94). From the above analysis, it can be seen that the bifurcation theory can explain the variation of the solutions when the parameters of the systems vary.

## 4.7 Bifurcation Solution of the 1/2 Subharmonic Resonance Case of Non-linear Parametrically Excited Vibration Systems

The study of non-linear vibration systems with parametric excitations is significant in both theory and application. The vibration of such systems can be described by Mathieu's equations or Hill's equations. For instance, the lateral vibration of the elements of structures subjected to longitudinally excited forces (Euler dynamic buckling problems) [4], the motion of a simple pendulum whose support moves up and down [6,7], the torsional vibration of diesel engines in considering the influence of the reciprocating mass on the moment of inertia [11], the vibration of systems in which the stiffness of elastic elements varies with time, and so on — all of these are parametrically vibrating systems. Theoretical analysis and application to engineering show that such systems abound with dynamic behaviours.

In this section we first discuss the symmetry of the non-linear Mathieu's equation, then seek the bifurcation equation by Liapunov—Schmidt reduction and discuss the dynamic behaviour of bifurcation to a first approximation. At the same time, we develop the corresponding method of calculating the coefficients of the bifurcation equation. Finally, the high-order approximation and universal unfolding are further discussed. Consider the non-linear Mathieu's equation in the form of

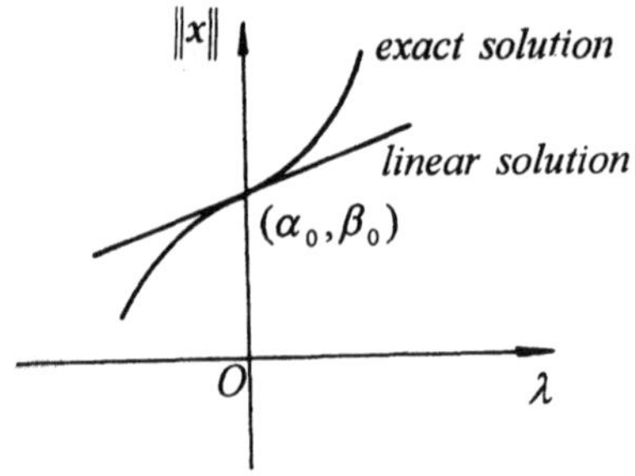


Fig. 4.14

$$u'' + \delta[u' + h(u,u',\delta)] + (1+\mu)u + f(u,u',\mu) + 2\varepsilon\cos 2t[u + g(u,u',\mu,\varepsilon)] = 0 \qquad (4.102)$$

where $\delta$ is the damping coefficient, $\varepsilon$ is the amplitude of parametric excitation, and $\mu$ the detuning parameter. $u,\delta,\varepsilon,\mu$ are sufficiently small. $f$ is a non-linear function presenting a restoring force, $h$ is a non-linear damping function and $g$ a non-linear function of parametric excitation. Here we assume that $f$ and $g$ are even functions of $u'$, and $h$ is the odd function of $u'$. In practice these assumptions are easily satisfied. We take the detuning parameter $\mu = \omega^2 -$ as the bifurcation parameter. Here is the intrinsic frequency of the linearized systems of eq. (4.102). In subsections 4.7.1 to 4.7.4, we shall seek for the $2\pi$-periodic bifurcation solutions of system (4.102). Since the first approximation corresponds to one half of the parametric excitation frequency, we call them 1/2 subharmonic bifurcation solutions in the resonant case. It is known that the form of the bifurcation equation is determined by the symmetry of the system, so we first discuss the symmetry of eq. (4.102).

### 4.7.1 Symmetry

Many non-linear vibration systems have symmetries, i.e. the properties that the systems do not change under the action of a certain subgroup. Here we only discuss some properties that are to be considered in this section. For comprehensive study of symmetry, the readers should refer to [4].

Because systems (4.102) are non-autonomous, in order to use the Liapunov—Schmidt reduction, we shall apply the same method as used in the previous section, by which we defined periodic function spaces.

The $2\pi$-periodic continuous functional space $\mathcal{C}_{2\pi}$

$$\mathcal{C}_{2\pi} = \{p \in \mathcal{C}(\mathcal{R}) | p(t+2\pi) = p(t)\} \qquad (4.103)$$

The $2\pi$-periodic continuous functional space with second-order continuous derivatives

$$\mathcal{C}^2_{2\pi} = \{p \in \mathcal{C}^2(\mathcal{R}) | p(t+2\pi) = p(t)\} \qquad (4.104)$$

Define their super-norms. On $\mathcal{C}_{2\pi}$, it is $\|p(t)\|$, and on $\mathcal{C}^2_{2\pi}$, $\|q\|_2 = \|q\| + \|q'\| + \|q''\|$.

Then both $\mathscr{C}_{2\pi}$ and $\mathscr{C}_{2\pi}^2$ become Banach spaces. Define a phase shift transformation $T_\varphi$ which acts on $p(t)$ as follows:

$$T_\varphi p(t) = p(t+\varphi) \tag{4.105}$$

Obviously it has

$$T_0 p(t) = Ip(t) \text{ and } T_\pi^2 p(t) = T_\pi \bullet T_\pi p(t) = Ip(t) \tag{4.106}$$

hence $T_0 = I$, $T_\pi^2 = I$, where $I$ is an identical operator. Define the time-reversal operator $\sigma$ by

$$\sigma p(t) = p(-t) \tag{4.107}$$

$$\sigma^2 = I \tag{4.108}$$

Generally, $T_\varphi$ and $\sigma$ are not commutative, but when $\varphi = \pi$ they are, that is

$$\sigma T_\pi = T_\pi \tag{4.109}$$

and $$\sigma T_\varphi \sigma = T_{-\varphi} \tag{4.110}$$

The operators $T_\pi$ and $\sigma$ with $I$ form groups of two elements $\{I, T_\pi\}$ and $\{I, \sigma\}$ respectively, which are isomorphic to the subgroup $O(2)$ of orthogonal linear transformations on $\mathscr{R}^2$. Define orthogonal group $O(n) = \{A \in GL(n) | AA^T = I\}$ where $GL(n)$ is a set of the reversible linear transformations from Banach space $\mathscr{X}$ to itself. Of course a closed subgroup of $O(n)$ is also a compact Lie group. On the other hand, every compact Lie group can be considered as a closed subgroup of $O(n)$.

Define the special orthogonal group $SO(n) = \{A \in O(n), \det A = 1\}$. This may be called a rotating group. $SO(2)$ especially can be composed of the rotations in the plane $SO(2) = \left\{ \begin{bmatrix} \cos\theta & -\sin\theta \\ \sin\theta & \cos\theta \end{bmatrix} \middle| 0 \le \theta \le 2\pi \right\} =$ circle group $= \mathscr{S}^1$. Let $K = \begin{bmatrix} 1 & 0 \\ 0 & -1 \end{bmatrix}$ represent the reflection, then O(2) can be generated from $SO(2)$ and $K$, i.e. $SO(2)$ and $K$ are subgroups of $O(2)$.

Define the $Z_2$ group, that is $Z_2 = \{1, -1\}$ by multiplication or $Z_2 = \{0, 1\}$ by addition. A function $f: \mathscr{R} \to \mathscr{R}$ is commutative with subgroup $Z_2$and is equivalent to which $f$ is an odd function, that is $f(-x) = -f(x)$. It is easy to verify that groups $\{I, \sigma\}$ and $\{I, T_\pi\}$ have $Z_2$-symmetry. The subgroup of $O(2)$ forms table 4.2.

**Table 4.2**

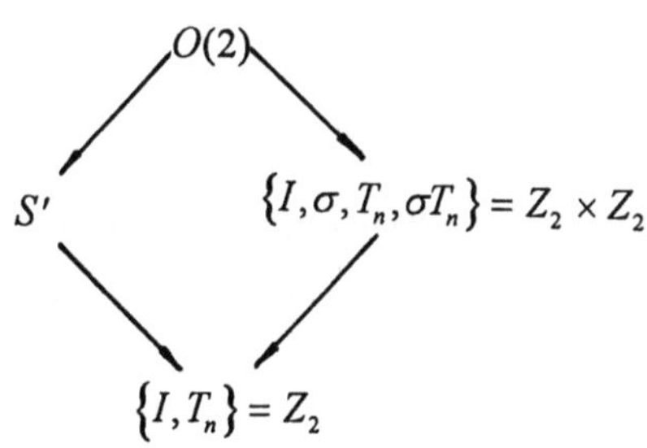


Let $M(\mu, \delta, \varepsilon)u$ stand for the left terms of eq. (4.102), which is called a non-linear Mathieu operator. It is obvious that

$$M(\mu,\delta,\varepsilon)T_\pi u = \frac{d^2u(t+\pi)}{d(t+\pi)^2} + \delta\left[\left[\frac{du(t+\pi)}{d(t+\pi)} + h(u(t+\pi),u'(t+\pi),\delta)\right]\right] + (1+\mu)u(t+\pi) + f(u(t+\pi),u'(t+\pi),\mu) + g(u(t+\pi),\frac{du(t+\pi)}{d(t+\pi)},\mu,\delta,\varepsilon) \tag{4.111}$$

$$T_\pi M(\mu,\delta,\varepsilon)u = \cdots + 2\varepsilon\cos2(t+\pi)[u(t+\pi) + g(u(t+\pi)\frac{du(t+\pi)}{d(t+\pi)},\mu,\delta,\varepsilon] + \cdots \tag{4.112}$$

In eq. (4.112), the terms not written are the same as those in eq. (4.111). Hence

$$M(\mu,\delta,\varepsilon)T_\pi u = T_\pi M(\mu,\delta,\varepsilon)u \tag{4.113}$$

Assuming a subgroup $\Gamma$ acts on spaces $\mathcal{X}$ and $\mathcal{Y}$ at the same time. The symmetry of the mapping $M:\mathcal{X} \to \mathcal{Y}$ is embodied in the exchange of the subgroup $\Gamma$ with the mapping, i.e. for any $x \in \mathcal{X}, \gamma \in \Gamma$, it holds that $M(\rho_\gamma x) = \rho_\gamma M(x)$. We can prove that the symmetric subgroups of non-linear Mathieu systems have the properties as shown in table.4.3 according to the different values of parameters $(\delta,\varepsilon)$.

**Table.4.3**

| parameters $(\delta,\varepsilon)$ | symmetric subgroups |
|---|---|
| $\delta = 0 = \varepsilon$ | $O(2)$ |
| $\delta \neq 0 = \varepsilon$ | $\mathcal{S}^1 = \{T_\varphi \mid \varphi \in [0,2\pi]\}$ |
| $\delta = 0 \neq \varepsilon$ | $\mathcal{Z}_2 \times \mathcal{Z}_2 = \{I, T_\pi, \sigma, \sigma T_\pi\}$ |
| $\delta \neq 0 \neq \varepsilon$ | $\mathcal{Z}_2 = \{I, T_\pi\}$ |

### 4.7.2 Bifurcation Equation

Denote the Frechet derivative of the non-linear Mathieu operator by

$$L = D_u M(0,0,0)u|_{u=0} : \mathcal{C}^2_{2\pi} \to \mathcal{C}_{2\pi} \tag{4.114}$$

Define the inner product on $\mathcal{C}_{2\pi}$ by

$$< q, p >= \frac{1}{2\pi}\int_0^{2\pi} \bar{p}(t)q(t)dt \tag{4.115}$$

where $\bar{p}(t)$ is the conjugation function of $p(t)$. Obviously, $L = \frac{d^2}{dt^2} + 1$ is a self-adjoint operator, its null space and range space being given by

$$N(L) = \{x\cos t + y\sin t \mid (x,y) \in \mathcal{R}^2\} \tag{4.116}$$

and $$R(L) = \{p \in \mathcal{C}_{2\pi} \mid < \cos t, p >= 0 =< \sin t, p >\} \tag{4.117}$$

For convenience, let $i = \sqrt{(-1)}$, and we complicate the spaces as follows

$$\widetilde{\mathcal{C}}_{2\pi} = \mathcal{C}_{2\pi} + i\mathcal{C}_{2\pi} \tag{4.118}$$

and $$\widetilde{\mathcal{C}}^2_{2\pi} = \mathcal{C}^2_{2\pi} + i\mathcal{C}^2_{2\pi} \tag{4.119}$$

The null space of $\mathcal{L}$ in $\widetilde{\mathcal{C}}_{2\pi}^2$ becomes $\widetilde{N}(L)=\{z_1e^{it}+z_2e^{-it}|z_1, z_2 \in \mathbf{C}\}$.

If restricting $\{z_1,z_2\}$ in a complex line $\{(z,\bar{z})|z\in\mathcal{C}\}$, we regain the null space

$$N(L)=\{ze^{it}+\bar{z}e^{-it}|z\in\mathcal{C}\} \tag{4.120}$$

and the real range space

$$\mathcal{R}(L)=\{p\in\mathcal{C}_{2\pi}|<e^{it},p>=0\} \tag{4.121}$$

Define the real projectors $Q{:}\mathcal{C}_{2\pi}\to N(L)\subset\mathcal{C}_{2\pi}$. The calculating formulae are

$$Qp(t)=<e^{it},p>e^{it}+<e^{-it},p>e^{-it}=2<\cos t,p>\cos t+2<\sin t,p>\sin t \tag{4.122}$$

and $\quad P{:}\mathcal{C}_{2\pi}\to\mathcal{R}(L)\subset\mathcal{C}_{2\pi} \qquad (4.123)$

therefore $Q=I-P$. Thus the non-linear Mathieu equation can be rewritten as the alternative pair of equations

$$PM(\mu,\delta,\varepsilon)u=0 \tag{4.124}$$

$$QM(\mu,\delta,\varepsilon)u=0 \tag{4.125}$$

The solution of eq. (4.102) can be written as

$$u(t)=ze^{it}+\bar{z}e^{-it}+w(t) \tag{4.126}$$

where $w(t)\in[N(L)]^{\perp}=\mathcal{R}(L)\cap\mathcal{C}_{2\pi}^2$. Substituting eq. (4.126) into eq. (4.124) yields

$$Lw+P\{\delta(u'+h)+\mu u+f+2\cos 2t(u+g)\}=0 \tag{4.127}$$

where $u(t)$ should be substituted by eq. (4.126). Being a linear mapping from $(N(L))^{\perp}$ to $\mathcal{R}(L)$, the operator $L$ is invertible. By the implicit function theorem, for sufficiently small $|z|,\mu,\delta$ and $\varepsilon$, we can obtain a unique, smooth, real solution

$$w=w(z,\bar{z},\mu,\delta,\varepsilon) \tag{4.128}$$

which satisfies $\qquad w(z,\bar{z},0,0,0)=0 \qquad (4.129)$

We shall calculate $w$ in the next section.

Substituting the solution w of eq. (4.128) into eq. (4,125), we obtain

$$\begin{aligned}0&=QM(\mu,\delta,\varepsilon)(ze^{it}+\bar{z}e^{-it}+w(z,\bar{z},\mu,\delta,\varepsilon))\\&=Q\{\delta(u'+h)+\mu u+f+2\varepsilon\cos 2t(u+g)\}\\&=G(z,\bar{z},\mu,\delta,\varepsilon)e^{it}+\overline{G}(z,\bar{z},\mu,\delta,\varepsilon)e^{-it}\end{aligned} \tag{4.130}$$

Comparing eq. (4.130) with eq. (4.122), we have

$$\begin{aligned}G(z,\bar{z},\mu,\delta,\varepsilon)=&<e^{it},\delta i(ze^{it}-\bar{z}e^{-it}+w')+\delta h+\\&\qquad\mu(ze^{it}+\bar{z}e^{-it}+w)+f+\varepsilon(e^{2it}+e^{-2it})\\&\qquad[(ze^{it}+\bar{z}e^{-it}+w)+g]>\\=&(\mu+i\delta)z+\varepsilon\bar{z}+\varepsilon<e^{3it},w>+<e^{it},\\&\qquad\delta h+f+2\varepsilon\cos 2tg>\end{aligned} \tag{4.131}$$

Since eq. (4.130) always holds for arbitrary time $t$, we must have

$$G(z,\bar{z},\mu,\delta,\varepsilon)=0 \tag{4.132}$$

and $\overline{G}(z,\bar{z},\mu,\delta,\varepsilon)=0$ (4.133)

Equation. (4.132) is equivalent to eq. (4.133), so we conclude that eq. (4.132) is just the scalar bifurcation equation that we are finding, and where $G:\mathcal{C}\times\mathcal{C}\times\mathcal{R}^3\to\mathcal{C}$, and $G(0)=0$. The mapping $G$ is smooth for those five variables near the origin. In section 4.4, we have proved that if a non-linear vibration system has symmetry, then the symmetry is inherited by the corresponding bifurcation equation obtained through Liapunov—Schmidt reduction, i.e. we have

$$T_\gamma G(z,\bar{z},\mu,\delta,\varepsilon)=G(T_\gamma(z,\bar{z},\mu,\delta,\varepsilon)) \tag{4.134}$$

The symmetry possessed by eq. (4.134) determines the form of bifurcation equation $G=0$. In the most general case $j\geq 1$, if taking $T_\gamma=T_\pi$, we know from eq. (4.134) that

$$G(-z,-\bar{z},\mu,\delta,\varepsilon)=-G(z,\bar{z},\mu,\delta,\varepsilon) \tag{4.135}$$

This indicates that the bifurcation equation must be the odd function of $z$ and $\bar{z}$, and that a pitchfork bifurcation will occur. If there is no damping ($\delta=0$) or parametric excitation ($\varepsilon=0$), the form of the bifurcation equations will be restricted further by larger symmetrical subgroups. Attending to the property that $h$ is an odd function of $u'$, it can take the form of the bifurcation equation in the most general case, which is shown below:

$$\begin{aligned}G(z,\bar{z},\mu,\delta,\varepsilon)=z\left[\mu+\sum a_j|z|^{2j}\right]+i\delta z\left[1+\sum b_j|z|^{2j}\right]+\\ \varepsilon\left[\bar{z}+\sum_{i+k\geq 3\ \text{odd}}C_{jk}z^j\bar{z}^k+i\varepsilon\delta\sum_{i,k\geq 3\ \text{odd}}d_{jk}z^j\bar{z}^k\right]\end{aligned} \tag{4.136}$$

Here the summations are over $j\geq 1$, and the coefficients $a_j, b_j, c_j$ and $d_j$ are all real and depend smoothly on $\mu,\delta,\varepsilon$.

We shall calculate the coefficients $a_1$, $b_1$ and $c_1$ of the first approximate bifurcation equations in the next section. The Taylor series in eq. (4.136) will be convergent in the case where $f$, $g$ and $h$ are real analytical functions. If $f$, $g$ and $h$ do not satisfy the smooth conditions, then the series in eq. (4.136) may not be convergent but can be asymptotic when $|\bar{z}|$ is sufficiently small.

### 4.7.3 Response Diagrams

Each of the small solutions $(z,\bar{z})$ of the bifurcation equation $G=0$ correspond to each of the small $2\pi$-periodic solutions of the non-linear Mathieu equation respectively:

$$u(t)=ze^{it}+\bar{z}e^{-it}+w(t,z,\bar{z},\mu,\delta,\varepsilon) \tag{4.137}$$

We shall prove that $w=O(|z|^2)$ as $z\to 0$ in the next section. Hence, $|z|$ can be used as a good measure of the amplitude of the $2\pi$-periodic solution $u(t)$ to a first approximation. Writing $z$ in polar coordinate form

$$z=re^{i\theta} \tag{4.138}$$

where $r\in\mathcal{R}, r>0, \theta\in\mathcal{S}^1$.

Now let us determine the dependence of $r$ on small $\mu$, $\delta$ and $\varepsilon$. Substituting $z = re^{i\theta}$ into the bifurcation eq. (4.136) yields

$$G(r,\mu,\delta,\varepsilon) = re^{i\theta}[\mu + A(r^2) + i\delta(1 + B(r^2))] + \varepsilon re^{-i\theta}[1 + C(r^2) + i\delta D(r^2)] = 0 \tag{4.139}$$

where we take $A = \sum a_j r^{2j}, B = \sum b_j r^{2j}, C = \sum c_j r^{2j}$ and $D = \sum d_j r^{2j}$. We can eliminate the phase $\theta$ by complex conjugation eq. (4.139), so that the bifurcation equation becomes

$$R(r^2,\mu,\delta,\varepsilon)r = r\{[(\mu + A(r^2))^2 + \delta^2(1 + B(r^2))^2] - \varepsilon^2[(1 + C(r^2))^2 + \delta^2 D^2(r^2)]\} = 0 \tag{4.140}$$

It is obvious that eq. (4.140) has a trivial solution $r = 0$ and its non-trivial solutions must satisfy the equation

$$R(r^2,\mu,\delta,\varepsilon) = (\mu + A)^2 + \delta^2(1 + B)^2 - \varepsilon^2[(1 + C)^2 + \delta^2 D^2] = 0 \tag{4.141}$$

In general, we call $R(r^2,\mu,\delta,\varepsilon)$ the response function, the figures of which with the trivial solution in the $\mu - r$ coordinate system are the bifurcation diagrams. The diagram of non-trivial solutions is called the response diagram. Because eq. (4.141) is even, the bifurcation diagram has a pitchfork form. In order to obtain the solutions of eq. (4.141), we assume that the non-degenerate bifurcation conditions at point $(\mu,\delta,\varepsilon) = (0,0,0)$

$$a_1 \neq 0, \quad \delta^2 b_1 \neq \varepsilon^2 c_1 \tag{4.142}$$

can be satisfied, so we can take a non-linear transformation

$$\overset{\wedge}{r}\ r^2 = \aleph A(r^2), \quad \aleph = \text{sgn}(a_1) = \pm 1 \tag{4.143}$$

Substituting eq. (4.143) into eq. (4.141), we obtain

$$(\mu + \aleph\hat{r}^2)^2 + \delta^2[1 + \hat{B}(r^2)] - \varepsilon^2[1 + \hat{C}(\hat{r}^2)^2 + \delta^2 D^2(\hat{r}^2)] = 0 \tag{4.144}$$

where

$$\hat{B}(\hat{r}^2) = \frac{b_1}{|a_1|}\hat{r}^2 + O(r^4),\ \hat{C}(\hat{r}^2) = \frac{c_1}{|a_1|}\hat{r}^2 + O(r^4),\ \hat{D}(\hat{r}^2) = \frac{d_1}{|a_1|}\hat{r}^2 + O(r^4) \tag{4.145}$$

For convenience we take off the cover signs, then eq. (4.144) can be written as

$$(\mu + \aleph r^2)^2 + \delta^2 - \varepsilon^2 + \frac{2}{|a_1|}(b_1\delta^2 - c_1\varepsilon^2)r^2 + O(r^4) = 0$$

or the first approximate equation is

$$R_1(r^2,\mu,\alpha,\beta) = (\mu + \aleph r^2)^2 + \alpha + \beta r^2 = 0 \tag{4.146}$$

$$\text{where}\quad \alpha = \delta^2 - \varepsilon^2,\ \beta = \frac{2}{|a_1|}[b_1\delta^2 - c_1\delta^2]R_1(r^2,\mu,\delta,\varepsilon),\ R_1(0) = 1 \tag{4.147}$$

Here $\alpha$ and $\beta$ play a double role as mechanical parameters and unfolding parameters. Now let us study the bifurcation behaviours of the first approximate eq. (4.146).

Generally we solve the eq. (4.146) for $\mu$:

$$\mu = -\aleph\, r^2 + \sqrt{-(\alpha + \beta r^2)} \tag{4.148}$$

For finding the transition sets by using definition eq. (4.16), which denotes $r^2 = s$ in eq. (4.146), we can calculate the derivatives with respect to $s$ and $u$:

$$R_{1\mu} = 2(\mu + \aleph\, s) = 0,\quad R_{1s} = 2(\mu + \aleph\, s)\aleph + \beta = 0,\quad R_{1ss} = 2.$$

Therefore we obtain the transition sets:

$$\begin{aligned} &B_0 : \alpha = 0\ \{R_1 = R_{1\mu} = 0.\ \text{at}\ (0,\mu,\alpha,\beta)\}\\ &B_1 : \beta = 0\ \text{(degenerate)}\ \{R_1 = R_{1\mu} = R_{1s} = 0.\ \text{at}\ (0,\mu,\alpha,\beta)\}\\ &H_0 : \alpha = \tfrac{-\beta^2}{4}\{R_1 = R_{1s} = 0.\ \text{at}\ (0,\mu,\alpha,\beta)\}\\ &H_1 = D = \Phi \end{aligned} \tag{4.149}$$

Drawing the curve for eq. (4.149) in the $\alpha - \beta$ plane and combining the $\alpha$ and $\beta$ axes, we see that the $\alpha - \beta$ plane is divided into six open regions in Fig. 4.15. Each of the six open regions has its own special bifurcation diagram and its form is preserved at all points in the same region. The bifurcation response diagrams have the same topological structures in any region [4].

Below we study the forms of the topological structures in each region when $= -1$.

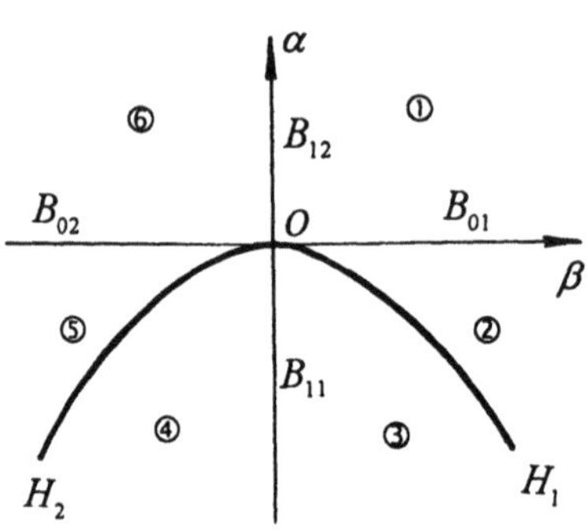


Fig. 4.15

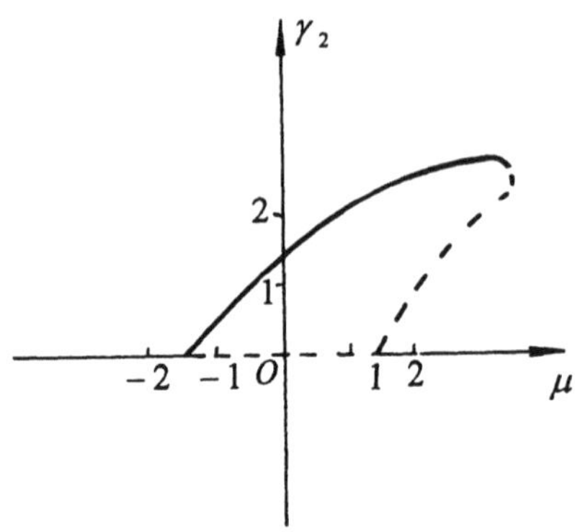


Fig. 4.16

In region (1) $\alpha > 0, \beta > 0$, as in eq. (4.148), which has no real solution, but a trivial solution $r = 0$. In region (3) $\alpha < 0, \beta > 0$, assuming $|\alpha| > |\beta$. For example, take $\alpha = -2, \beta = 1$, then $r = 0, \mu = \pm 1.414$, $\mu = 0, r^2 = (-1 \pm \sqrt{1+8})/2 = 1,-2$, hence $r = \pm 1$, $\alpha + \beta r^2 = 0$ and $\mu = 2$, so we have $r = \pm 1.414$, therefore the topological structure of the response diagram in region (3) is shown in Fig. 4.16. In the same way, we obtain the response diagrams in other regions, see Fig. 4.17.

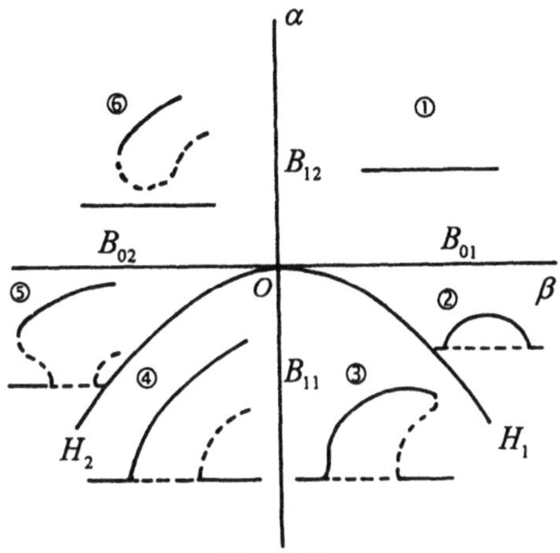


Fig. 4.17

Next we analyse the bifurcation diagrams on transition sets. The intersections of the six regions in Fig. 4.15 are called transition sets. When the parameters vary, through them the structures of the bifurcation diagrams will change. On transition curves $B_1, H_1, H_2, B_2$ and $\beta = 0$ of the sets, $\alpha$ and $\beta$ are related by a formula so that the bifurcation equation contains only one unfolding parameter apart from $\mu$. For example, $\dfrac{\partial\mu}{\partial s} = 0$ is satisfied on the transition curve $H_0$, so the response diagrams are vertical to the $\mu$-axis near $r = 0$, and we have $\mu = \pm\beta/2$, $r = 0$. In this case the response diagrams should satisfy the equation

$$\mu_{1,2} = r^2 \pm \sqrt{-\left(-\frac{1}{4}\beta^2 + \beta r^2\right)} \overset{def}{=} r^2 + R \tag{4.150}$$

where $R$ stands for the expression of the square root. When $r$ increases, $R$ decreases gradually and $\mu_1$ closes to $\mu_2$ till $\mu_1 = \mu_2$ in the case of $-\frac{1}{4}\beta^2 + \beta r^2 = 0$. The response diagrams are shown as in Fig. 4.18.

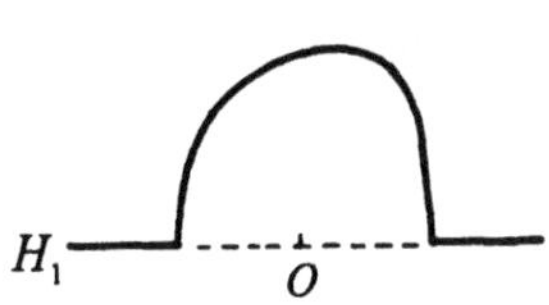


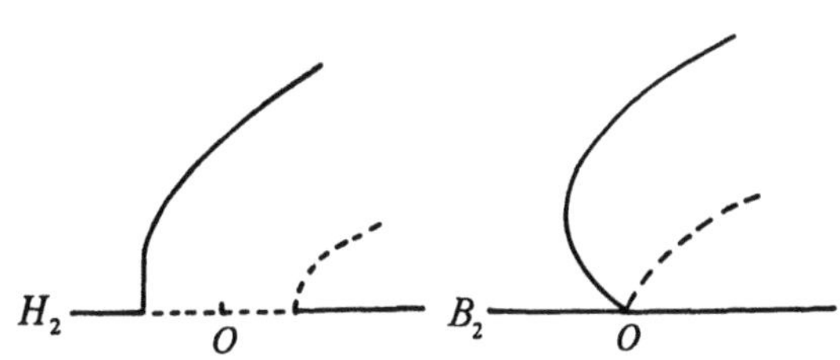


Fig. 4.18

Fig. 4.19

On $B_1, \alpha = 0, \beta > 0$, so $\mu$ has no real solutions but just a trivial solution. In the same way we can obtain the other response diagrams of codimension; they are shown in Fig. 4.19.

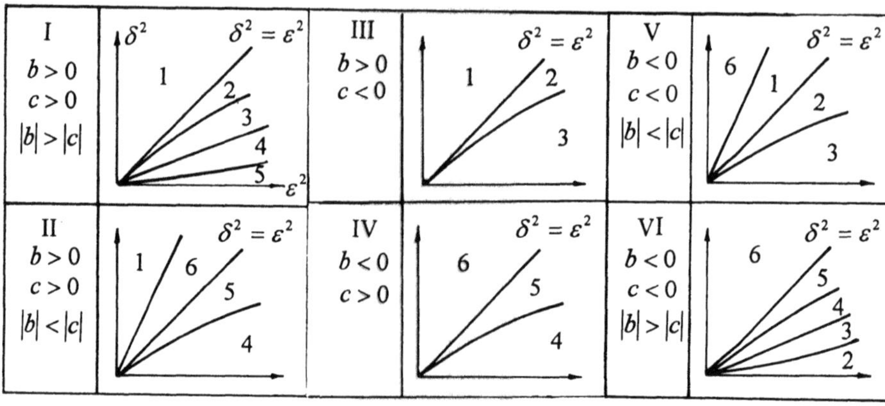


Fig. 4.20

When the non-degenerate conditions of mechanics are not satisfied, i.e., $b_1\delta^2 - c_1\varepsilon^2 = 0$, we have $\beta = 0$. In this case, we must study the high-order terms of $R(z,\bar{z},\mu,\delta,\varepsilon)$ so as to determine the forms of the response diagrams. In this case, the number of unfolding parameters of the bifurcation equation may be greater than two or even be numberless. To facilitate the experimental analysis we translate the unfolding parametric plane into the $\delta^2 - -\varepsilon^2$ physical parametric plane. We can draw out six kinds of different $\delta^2 - -\varepsilon^2$ parametric planes according to different combinations of the coefficients $b_1$ and $c_1$, in which each parametric plane corresponds to a class of special physical systems, that is, non-linear systems determined by the special functions *h*, *g* and *f*. See Fig. 4.20. The six kinds of different parametric planes in Fig. 4.20 show that the six kinds of bifurcation diagram cannot appear in special vibration systems of the same kind. Such a characteristic provides us with a means to recognise the mechanical properties of non-linear systems. From the six parametric planes in Fig. 4.20 we see that planes I, II, III and V contain parametric region (1) which corresponds to zero solution, particularly in systems V, where parametric resonance would not occur. This feature is significant in the control of vibration of non-linear parametrically excited systems. In addition, we know from the parametric planes of vibration systems IV and VI that, even though $\varepsilon < \delta$, 1/2 subharmonic resonance bifurcation is possible.

### 4.7.4 Calculation of Coefficients

If the non-degenerate conditions of mechanics are satisfied, it suffices to calculate the coefficients $a_1, b_1$ and $c_1$ for determining the bifurcation diagrams of systems (4.146). We first calculate the derivatives of $w$ with respect to $z$. Differentiating (4.127) with respect to z (subscripts denote partial derivatives) yields

$$Lw_z + \{\delta(u'_z + h_u u_z + h_{u'} u'_z) + \mu u_z + f_z u_z + f_{u'} u'_z + 2\varepsilon(u_z + g_z u_z + g_{u'} u'_z)\cos 2t\} = 0 \tag{4.151}$$

$u_z$ and $u'_z$ can be obtained from (4.126). As $f$ is the non-linear function of $u$ and even function of $u'$. Letting $(z,\bar{z},\mu,\delta,\varepsilon) = (0,0,0,0,0) = (\underline{0})$, we reduce eq. (4.151) to

$$Lw_z(\underline{0}) = 0$$

Since $w \in (\mathcal{N}(L))^{\perp}$, it must have

$$w_z(\underline{0}) = 0 \tag{4.152}$$

Similarly, we have

$$w_{\bar{z}}(\underline{0}) = w_{\mu}(\underline{0}) = w_{\delta}(\underline{0}) = w_{\varepsilon}(\underline{0}) = 0 \tag{4.153}$$

Differentiating eq. (4.151) again with respect to z and evaluating it at $(\underline{0})$, we obtain

$$Lw_{zz}(\underline{0}) = -\{f_{uu}(\underline{0}) - f_{u'u'}(\underline{0})\}e^{2it}$$

This is a second-order non-homogeneous ordinary differential equation of time $t$. Its particular solution is

$$w_{zz}(\underline{0}) = \frac{1}{3}(f_{uu} - f_{u'u'})e^{2it} \tag{4.154}$$

In the same way we get

$$\begin{aligned} w_{z\bar{z}}(\underline{0}) &= -(f_{uu} + f_{u'u'}) \\ w_{\bar{z}\bar{z}}(\underline{0}) &= \frac{1}{3}(f_{uu} - f_{u'u'})e^{-2it} \\ w_{z\varepsilon}(\underline{0}) &= \frac{1}{8}e^{3it} \\ w_{z\mu}(\underline{0}) &= 0, \ w_{z\delta}(\underline{0}) = 0 \end{aligned} \tag{4.155}$$

Integrating eq. (4.154) and (4.155) over $z$ and $\bar{z}$, we can prove

$$w = O(|z|^2) \tag{4.156}$$

Expanding eq. (4.136) to calculate $a_1, b_1$ and $c_1$

$$G(z,\bar{z},\mu,\delta,\varepsilon) = z[\mu + a_1 z\bar{z} + i\delta(1 + b_1 z\bar{z})] + \varepsilon\bar{z}(1 + c_1 z\bar{z}) + \ldots$$

From the above equation we have

$$G_{zz\bar{z}}(\underline{0}) = 2a_1, \quad G_{\delta zz\bar{z}}(\underline{0}) = 2a_{1\delta} + 2ib_1, \quad G_{\varepsilon z}\bar{z}\bar{z}(\underline{0}) = 2c_1, \quad G_{\varepsilon zzz}(\underline{0}) = 0 \tag{4.157}$$

The derivative of the left-hand side of eq. (4.157) can also be computed from eq. (4.131) $G_{zz\bar{z}}(\underline{0}) = f_{uuu} + f_{uu'u'} - \frac{5}{3}f_{uu}\left[f_{uu} + f_{u'u'} - \frac{2}{3}(f_{u'u'})^2\right]$

Comparing the above formula with the first expression of eq. (4.157) results in

$$a_1 = \frac{1}{2}(f_{uuu} + f_{uu'u'}) - \frac{5}{6}f_{uu}(f_{uu} + f_{u'u'}) - \frac{1}{3}f_{u'u'}^2 \tag{4.158}$$

Similarly

$$b_1 = -\frac{1}{2}(h_{uuu'} + h_{u'u'u'}) + \frac{1}{2}h_{uu'}(f_{uu} + f_{u'u'}) - \frac{1}{9}f_{uu}^2 - \frac{1}{9}f_{uu}f_{u'u'} + \frac{2}{9}f_{u'u'}^2 \tag{4.159}$$

$$\begin{aligned} c_1 = &\frac{1}{2}(g_{uuu'} + g_{u'u'u'}) - \frac{1}{2}g_{uu'}(2f_{uu} + f_{u'u'}) + \frac{1}{6}g_{u'u'}(7f_{uu} + 2f_{u'u'}) + \frac{1}{16}f_{uuu} \\ &+ \frac{5}{16}f_{uuu} + \frac{5}{16}f_{uu'u'} - \frac{1}{48}(21f_{uu}^2 + \frac{1}{48}(21f_{uu}^2 + 25f_{uu}f_{u'u'} + 26f_{u'u'}^2) \end{aligned} \tag{4.160}$$

All the derivatives in the equations from eq. (4.158) to eq. (4.160) are computed at $(\underline{0})$. The experiment on the mechanical model [49] corresponding to system VI in Fig. 4.20 verifies entirely the results of the theoretical analyses.

### 4.7.5 Higher Approximation and Universal Unfolding of Bifurcation Equation

Equation (4.146) is the first approximate bifurcation equation. Though the dynamic behaviours observed in the previous analysis are much richer than the results obtained by Bogolubov & Mutropoliski [6] and Nayfey & Mook [92], and may satisfy the need in engineering calculations in general, it is still necessary to recognise and unfold the equation by the use of singularity theory described in section 4.5 in order to check whether or not the equation can reflect the overall bifurcation behaviours. Besides, when parameters $\alpha$ and $\beta$ are small, a new germ equivalent to eq. (4.139) needs to be considered [99].

**1. The Choice of Germ**

The GS normal form in eq. (4.146) is

$$(\mu + a_1 r^2)^2 = 0 \tag{4.161}$$

From the theorem proved in ref. [4] we know that the germ $(\mu + a_1 r^2)^2$ has infinite codimensions, therefore we have to add infinite unfolding parameters to that germ to obtain the bifurcation equation which can reflect the entire dynamic behaviours of the systems. Obviously this is impossible and unnecessary in practice. (In application we always intentionally take the first several terms as the approximate solutions.) For this reason, we turn to seek a new germ with the aim that it can have finite codimensions.

Rewriting (4.141) in the form of

$$\begin{aligned} R = {} & (\mu + a_1 r^2 + a_2 r^4 + \cdots)^2 + \delta(1 + b_1 r^2 + \cdots)^2 - \\ & \varepsilon[(1 + c_1 r^2 + \cdots)^2 + \delta^2 (d_1 r^2 + d_2 r^4 + \cdots)^2] = 0 \end{aligned} \tag{4.162}$$

since the coefficients $a_i, b_i, c_i$ and $d_i$ are analytical functions of parameters $\mu, \delta, \varepsilon$, they can be expanded in power series, for example

$$a_j = a_j(0) + a_{j\mu}(0)\mu + a_{j\delta}(0)\delta + a_{j\varepsilon}(0)\varepsilon + \cdots$$

In this way eq. (4.162) can be rewritten as

$$\begin{aligned} R = {} & [\mu + a_1(0)\mu]^2 + 2[a_2(0) + a_1(0)a_{1\mu}(0) + b_{2\mu}(0)\delta^2 - \\ & c_{2\mu}(0)\varepsilon^2]\mu u^2 + 2[a_1(0)a_2(0) + b_1(0)b_2(0)]\delta^2 - \\ & c_1(0)c_2(0)\varepsilon^2 - d_1(0)d_2(0)\delta^2\varepsilon^2]u^3 + 2a_{1\mu}(0)u\mu^2 + \cdots = 0 \end{aligned} \tag{4.163}$$

where $u = r^2$. For convenience we denote $a_1(0) = a_1$ and

$$a = 2[a_2(0) + a_1(0)a_{1\mu}(0) + b_{2\mu}(0)\delta^2 - c_{2\mu}(0)\varepsilon^2]$$

$$b = 2[a_1(0)a_2(0) + b_1(0)b_2(0)\delta^2 - c_1(0)c_2(0)\varepsilon^2 - d_1(0)d_2(0)\delta^2\varepsilon^2]$$

$$c = 2a_1\mu(0)$$

We take a portion of the second-order approximate expression

$$g = (\mu + a_1 u)^2 + a\mu u^2 + bu^3 + cu\mu^2 \tag{4.164}$$

as the new germ. We shall prove that the germ eq. (4.164) has finite codimensions in the next section. For this germ, we have a recognition theorem to guarantee that it is $Z_2$ strongly equivalent to a normal form $h$. **Theorem 4.23** Suppose there is a normal form $h^* = rh(u,\mu) = r[\varphi u^3 + \varepsilon u^2 + 2\delta u\mu + \varepsilon\mu^2]$ where $\varphi = \mathrm{sgn}(g_{vvv}), \delta = \mathrm{sgn}(g_{u\mu}), \varepsilon = \mathrm{sgn}(g_{uu})$. Then $g^* = rg(u,\mu) \in \mathbf{E}_{u,\mu}(Z_2)$ is $Z_2$ strongly equivalent to $h^*$ if and only if

$$g = g_u = g_\mu = \det(d^2 g) = 0, g_{vvv} \neq 0 \tag{4.165}$$

at $(u,\mu) = (0,0)$. Note $\det(d^2 g) = \begin{vmatrix} g_{uu} & g_{u\mu} \\ g_{\mu u} & g_{\mu\mu} \end{vmatrix} = g_{uu}g_{\mu\mu} - g_{\mu u}^2$, $v = (v_1, v_2)$ is an eigenvector corresponding to eigenvalue zero of $\det(d^2 g) = 0$, and $v_1 > 0$ is the directional derivative along this eigenvector $\frac{\partial}{\partial v} = v_1 \frac{\partial}{\partial u} + v_2 \frac{\partial}{\partial \mu}$ which is expressed in terms of derivatives of g with respect to $u, \mu$. We have

$$g_{vvv} = v_1^3 g_{uuu} + 3v_1^2 v_2 g_{uu\mu} + 3v_1 v_2^2 g_{u\mu\mu} + v_2^3 g_{\mu\mu\mu}$$

As the proof is tedious, readers are referred to [4].

**Corollary** Suppose $g^* = rg(u,\mu)$ where $g$ is given by eq. (4.164). Then $g^*$ is $Z_2$ strongly equivalent to $h^*$ where

$$h^* = rh(u,\mu) = r[\varphi u^3 + \varepsilon u^2 + 2\delta u\mu + \varepsilon\mu^2] \tag{4.166}$$

**Proof** It is clear that at $(u,\mu) = (0,0)$ we have

$$g = 0, g_u = 0, g_\mu = 0, g_{uu} = 2a_1^2, g_{u\mu} = 2a_1, g_{\mu\mu} = 2 \text{ hence } \det(d^2 g) = 0.$$

Taking $(v_1, v_2) = (1, -a_1)$ as the eigenvectors of zero eigenvalue of $\det(d^2 g) = 0$ then $g_{uuu} = 6b, g_{u\mu\mu} = 2c, g_{uu\mu} = 2a, g_{\mu\mu\mu} = 0$, so $g_{vvv} = 6(b - aa_1 + a_1^2 c)$. Generally, because $g_{vvv} \neq 0$, so the expression

$$b - aa_1 + a_1^2 c = 2[b_1(0)b_2(0) - a_1(0)b_{2\mu}(0)]\delta^2 - 2[c_1(0)c_2(0) - a_1(0)c_{2\mu}(0)]\varepsilon^2 - 2d_1(0)d_2(0)\delta^2\varepsilon^2$$

is not zero. The conditions of theorem (4.23) are all satisfied. Since eq. (4.164) is $Z_2$ strongly equivalent to $h$ of eq. (4.166), the bifurcation diagrams are topologically equivalent, i.e.

$$g' = (\mu + ku)^2 + \varphi u^3 \tag{4.167}$$

has the same behaviours as eq. (4.164), therefore we only need to consider the properties of (4.167).

## 2. Universal Unfolding and the Transition Set

First we are going to prove a theorem which tells us that the codimension of eq. (4.167) is three.

**Theorem 4.24** The unfolding terms of germ $g' = (\mu + ku)^2 + \varphi u^3$ are $\{1, u, \mu u\}$ and its codimension is three.

**Proof** We prove the theorem in the same way described as in [4].

For the convenience of writing, taking $k = +$ for simplicity, we have

$$RT(g') = < g', ug'_u > \{r\}$$
$$= (M^4 \oplus R\{u^2 + 4u\mu + 3\mu^2, 3\varphi u^3 + 2u^2 + 2u\mathrm{u},$$
$$u^2\mu + u\mu^2, u\mu^2 + \mu^3, u^3 + u^2\mu\})\{r\}$$
$$T(g') = (RT(g') + E_\mu\{g'_\mu\})\{r\}$$
$$= (M^4 \oplus R\{u^2 + 4u\mu + 3\mu^2, 3\varphi u^3 + 2u^2 + 2u\mu, u^2\mu + u\mu^2,$$
$$u\mu^2 + \mu^3, u^3 + u^2\mu, u + \mu, \mu^2 + u\mu\})\{r\}$$

therefore $\mathrm{Itr}[T(g')] = M^4\{r\}$. To find the unfolding terms it is necessary to search for a basis $\{p_1, \cdots, p_k\}$ so that

$$\{\mathrm{Itr}[T(g')]\}^\perp = V_{g'} \oplus R\{p_1, p_2, \cdots, p_k\} \tag{4.168}$$

where $k$ is the codimension. Now we have

$$_{g'} = R\{u^2 + 4u\mu + 3\mu^2, 3\varphi u^3 + 2u^2 + 2u\mu, u^2\mu + u\mu^2,$$
$$u\mu^2 + \mu^3, u^3 + u^2\mu, u + \mu, \mu^2 + u\mu\}$$

It is easy to see that if we take $p_1 = 1$, $p_2 = u$, and $p_3 = u\mu$, then eq. (4.168) can be satisfied as follows:

$$u^3 = [(3\varphi u^3 + 2u^2 + 2u\mu) - 2(u^2 + 4u\mu + 3\mu^2) + b(\mu^2 + u\mu)] / (3\varphi)$$
$$u\mu^2 = u^2\mu + u\mu^2 - u^2\mu - u^3 + (u^3)$$
$$\mu u^2 = u^2\mu + u^3 - (u^3),\ \mu^3 = \mu^3 + u\mu^2 - (u\mu^2)$$
$$u^2 = u^2 + 4u\mu + 3\mu^2 - 3(\mu^2 + u\mu) - p_3$$
$$u\mu = p_3,\ \mu^2 = \mu^2 + u\mu - p_3$$
$$u = p_2,\ \mu = u + \mu - p_2$$
$$1 = p_1$$

So the codimension is three and the unfolding terms are $\{1, u, u\mu\}$.

The universal unfolding of eq. (4.167) is

$$R = \varphi u^3 + (\mu + ku)^2 + \alpha + \beta u + \gamma u\mu = 0 \tag{4.169}$$

Now let us calculate the transition set $\sum(Z_2)$ of eq. (4.169).

Geometrically, $\sum(Z_2)$ is a hypersurface which divides the space of unfolding parameters into several regions. Any two bifurcation diagrams are topologically equivalent in the same region and are not equivalent in any two different regions. The bifurcation diagrams are not structurally stable on $\sum(Z_2)$. Referring to [4], we have

$$\sum(Z_2) = B_1(Z_2) \cup B_0(Z_2) \cup H_1(Z_2) \cup H_0(Z_2) \cup D(Z_2) \tag{4.170}$$

The transition sets of eq. (4.169) are

$$
\begin{aligned}
&B_0(Z_2):\ \alpha = 0 \\
&B_1(Z_2):\ \begin{cases} -\gamma u^2 + 2\varphi u^3 \\ \beta = 2\gamma u - 3\varphi u^2 \end{cases} \qquad (u > 0) \\
&H_0(Z_2):\ \alpha = -\frac{\beta^2}{16(1+\gamma)^2} \\
&H_1(Z_2) = D(Z_2) = \Phi
\end{aligned}
\tag{4.171}
$$

where $\Phi$ represents an empty set.

## 3. Bifurcation Diagrams and Their Properties

We can draw the hypersurface eq. (4.171) in three-dimensional space $(\alpha,\beta,\gamma)$. We take $\gamma$ as the auxiliary parameter and draw these figures in the plane $(\alpha,\beta)$. In Fig. 4.21, (a) is the transition variety for $\varphi =$ and (b) for $\varphi = -1$. Fig. 4.22 is the corresponding bifurcation diagrams where $K = -1$.

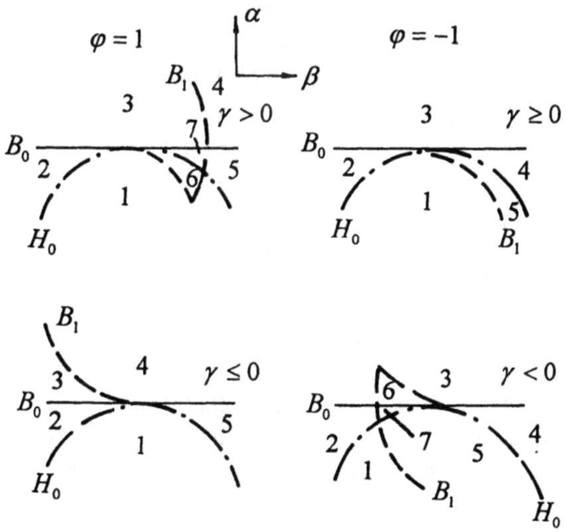


Fig. 4.21

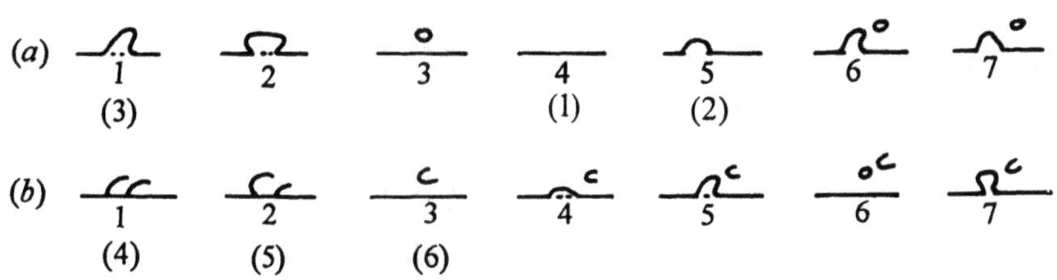


Fig. 4.22

The analysis of the mechanical behaviours of the bifurcation shows that:

1) Compared with the previous subsection and [3], this subsection has an extended result, that is, the subharmonic resonance bifurcation has fourteen persistent types. The persistent bifurcation diagrams given in [3] and the previous subsection are part of this subsection. The numbers in brackets of Fig. 4.22 are the types of bifurcation diagrams in the previous subsection.

2) The parameter $\varphi$ plays such a role as to determine whether the response amplitude $r$ is limited. $r$ is unlimited if $\varphi=-1$ and limited if $\varphi=+1$.

$$\begin{aligned}\varphi &= \operatorname{sgn}(g_{vvv})\\ &= \operatorname{sgn}\{[b_1(0)b_2(0)-a_1(0)b_{2\mu}(0)]\delta^2-[c_1(0)c_2(0)-a_1(0)a_{2\mu}(0)]\varepsilon^2-d_1(0)d_2(0)\delta^2\varepsilon^2\}\end{aligned} \tag{4.172}$$

$\varphi$ is mainly determined by the first two terms of eq. (4.172) when $\varepsilon$ and $\delta$ are small parameters. Parameters $a_1(0), b_1(0)$ and $c_1(0)$ have been given in the previous subsection and the remaining parameters can be calculated in a similar way. Here we omit these formulae since the calculations are very complex. However, we should notice that $b_1(0), b_2(0)$ and $b_{2\mu}(0)$ are determined mainly by the non-linear damping function $h$, while $c_1(0), c_2(0)$ and $c_{2\mu}(0)$ are determined mainly by the non-linear function $g$. In order to analyse the relationship between $\varphi$ and $\mu,\delta$ and $\varepsilon$, we suppose

$$b_1(0)b_2(0)-a_1(0)b_{2\mu}(0)>0 \tag{4.173a}$$

$$c_1(0)c_2(0)-a_1(0)c_{2\mu}(0)>0 \tag{4.173b}$$

When $\delta^2>\dfrac{c_1(0)c_2(0)-a_1(0)c_{2\mu}(0)}{b_1(0)b_2(0)-a_1(0)b_{2\mu}(0)}\varepsilon^2$ we have $\varphi=+1$. Mechanically, this means that the existence of a greater damping restricts the increase of the response amplitude in the system. When $\delta^2<\dfrac{c_1(0)c_2(0)-a_1(0)c_{2\mu}(0)}{b_1(0)b_2(0)-a_1(0)b_{2\mu}(0)}\varepsilon^2$, $\varphi=-1$, it means that the damping is smaller but the amplitude of parametric excitation is large. Therefore the response amplitude can increase infinitely. Of course, the actual non-linear factors of the systems always vary with the increase of the amplitude, which can restrict the increase of the response amplitude.

Moreover, we should note that when the non-linear damping is positive, the first eq. (4.173) always holds, but the second may not be so, therefore the reason why we make such an assumption is only for the convenience of discussion.

3) Algebraically, it seems that there may exist three limit cycles when we discuss the solution of eq. (4.169). However, there are at most two limit cycles in fact. Consider the function

$$y=\varphi x^3+(\mu+kx)^2+\alpha+\beta x+\gamma\mu x \tag{4.174}$$

where the number of roots of $y=0$ is the number of limit cycles $(x>0)$. It can be proved that the conditions under which eq. (4.174) has three positive real roots are

$$\left.\begin{array}{l}(1)\ \varphi=-1\\(2)\ \alpha+\mu^2>0\\(3)\ -\dfrac{1}{3}<\beta+(2+\gamma)\mu<0\\(4)\ \alpha+\mu^2<2x_1^3+x_1^2\\(5)\ \alpha+\mu^2>2x_2^3+x_2^2\end{array}\right\}\tag{4.175}$$

where $x_1=\frac{1}{3}\{1-\sqrt{1+3[\beta+(2+\gamma)\mu]}\}$, $x_2=\frac{1}{3}\{1+\sqrt{1+3[\beta+(2+\gamma)\mu]}\}$.

Obviously, conditions (4) and (5) of eq. (4.175) cannot hold simultaneously so that eq. (4.174) cannot have three real positive roots.

4) From the point of view of application, the choice of germ can be considered as the truncation of terms of eq. (4.141). The truncate terms in this subsection are more than those in the previous subsection. Clearly, the high-order solutions are different from the first approximation not only in quantity but also in quality.

The step of treating the bifurcation problems in engineering practice suggested in this subsection is first to study the dynamic behaviour of the first approximate bifurcation equation. If the equation has infinite codimensions, to study high-order approximate bifurcation equation is the second step. Proceed in this way until the universal unfolding with finite codimension is found.

5) Compared with the degenerate bifurcation with $a_1(0)=0$ [100], many results are the same. In this subsection [3,100], we find that there are two basic types of bifurcation of non-linear parametrically excited systems, as shown in Fig. 4.23. Other types of bifurcation are developed from these two types, and these two are often found in practice.

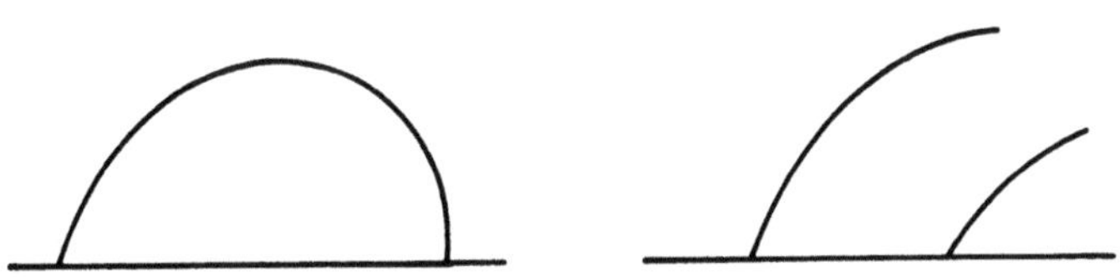

Fig. 4.23

## 4.8 Hopf Bifurcation Analyzed by Liapunov-Schmidt Reduction

The theory of Hopf bifurcation deals with the conditions in which the periodic solutions occur from an equilibrium point and with the stability of those solutions. This is one of the most important problems to be solved in non-linear vibration. In this section we describe the step and process of analysing Hopf bifurcation by Liapunov-Schmidt reduction. Though the results to be introduced are obtained mainly in the study on non-degenerate Hopf bifurcation, the method is also effective for studying the degenerate cases.

### 4.8.1 Bifurcation Equation

Consider an ordinary differential equation

$$y' = f(y,\mu),\ f(0,\mu_0) = 0 \tag{4.176}$$

where $f: \mathcal{R}^n \times \mathcal{R} \to \mathcal{R}^n$ is $\mathcal{C}^\alpha$.

The hypotheses of Hopf's theorem are

$(H\ )$ Jacobian matrix $A = D_y f(0,\mu_0)$ has simple eigenvalues $\pm i\omega_0$ and has no other eigenvalues lying on the imaginary axis and no zero eigenvalue.

When $\mu$ is near $\mu_0$, $A(\mu) = D_y f(0,\mu)$ has simple eigenvalues in the form of $\alpha(\mu) \pm i\omega(\alpha)$ which have the same smoothness as $A(\mu)$, where $\alpha(\mu_0) = 0$, $\omega(\mu_0) = \omega_0$. Without loss of generality, we take $\mu_0 = 0$ below.

$(H2)$ $\alpha'(0) = \dfrac{\mathrm{d}\alpha}{\mathrm{d}\mu}|_{\mu=0} \neq 0$

**Hopf theorem 4.25** If hypotheses $(H\ )$ and $(H2)$ are satisfied, there exists an unique branch of $2\pi/\omega$-periodic solutions to eq. (4.176) near (0,0). The solution can be parametrized by its amplitude $\varepsilon$.

In an autonomous system the period of vibration is related to the small parameter $\mu$ (or frequency of vibration $\omega$). To find the $2\pi$-periodic Hopf bifurcation solutions, we rescale time $t$ by setting

$$s = \omega_0(1+\tau)t, \qquad u(s) = y(t) \tag{4.177}$$

so that $u(s)$ becomes $2\pi$-periodic in $s$, where $\tau$ is small and related to $\mu$. Substituting eq. (4.177) into eq. (4.176), we obtain the equation in the form of a non-linear operator

$$N(u,\mu,\tau) \equiv (1+\tau)\omega_0 \frac{\mathrm{d}u}{\mathrm{d}s} - f(u,\mu) = 0 \tag{4.178}$$

Define two $2\pi$-periodic functional Banach spaces $\mathcal{C}_{2\pi}$ and $\mathcal{C}^1_{2\pi}$, with norms

$$\|u\| = \max_s |u(s)|, \quad \|u\|_1 = \|u\| + \|\frac{\mathrm{d}u}{\mathrm{d}s}\| \tag{4.179}$$

where $\|\bullet\|$ is the norm on $\mathcal{R}^n$. The mappings $N: \mathcal{C}^1_{2\pi} \times \mathcal{R} \times \mathcal{R} \to \mathcal{C}_{2\pi}$ are the functions defined in the neighbourhood of $(0,\mu_0,0)$. Define the linear operator at $(0,\mu_0,0)$

$$L = D_u(0,\mu_0,0) = -\omega_0 \frac{\mathrm{d}}{\mathrm{d}s} + A \tag{4.180}$$

where $L: \mathcal{C}^1_{2\pi} \to \mathcal{C}_{2\pi}$. The null space $\mathcal{N}(L)$ of $L$ is as follows

$$\mathcal{N}(L) = \mathrm{span}\{\varphi_1, \varphi_2\} \tag{4.181}$$

If $c$ is the right eigenvector of $A$, i.e. $Ac = i\omega_0 c$, then

$$\varphi_1(s) = \mathrm{Re}(e^{i\omega_0 s}c),\ \varphi_2(s) = \mathrm{Im}(e^{i\omega_0 s}c) \tag{4.182}$$

Define the inner product on $\mathcal{C}_{2\pi}$

$$< u, y > = \frac{1}{2\pi}\int_0^{2\pi} y^*(s)u(s)\,\mathrm{d}s \tag{4.183}$$

and the adjoint operator $L^*$ of $L$, $L^*: \mathcal{C}^1_{2\pi} \to \mathcal{C}_{2\pi}$

$$L^* = -\omega_0 \frac{\mathrm{d}}{\mathrm{d}s} - A^* \tag{4.184}$$

where $y^*$ and $A^*$ are the adjoints of $y$ and $A$ respectively. Then the null space $\mathcal{N}(L^*)$ of $L^*$ is spanned by $\psi_1$ and $\psi_2$, where

$$\psi_1(s) = \mathrm{Re}(e^{i\omega_0 s} d), \quad \psi_2(s) = \mathrm{Im}(e^{i\omega_0 s} d) \tag{4.185}$$

where $d$ is the eigenvector about $A^*$ corresponding to its eigenvalue $-i\omega_0$, and $A^* d = -i\omega_0 d$, in other words, we call $d^*$ a left eigenvector of $A$, where $d^* A = +i\omega_0 d^*$. By Fredholm's alternative theorem, the range space of $L$ is

$$\mathcal{R}(L) = \mathcal{N}(L^*)^{\perp} = \{y \in \mathcal{C}_{2\pi} | < y, \psi_i >= 0, i = 1,2\} \tag{4.186}$$

Because $i\omega_0$ is a simple eigenvalue, from linear algebra $d^* c \neq 0$. After rescaling we have $\quad d^* c = 2$ (4.187)

which can be resolved into $\mathcal{C}_{2\pi} = \mathcal{N}(L) \oplus \mathcal{R}(L)$. Defining the projection operator on

$$\mathcal{C}_{2\pi} \quad \begin{cases} Qy = < y, \psi_1 > \varphi_1 + < y, \varphi_2 > \varphi_2 & (4.188) \\ P = I - Q & (4.189) \end{cases}$$

so we have $Q\varphi = \varphi$ for $\varphi \in \mathcal{N}(L)$ as well as $P|_{\mathcal{R}(L)} = I$, $P|_{\mathcal{N}(L)} = 0$. Owing to the relationship $\mathcal{C}^1_{2\pi} \subseteq \mathcal{C}_{2\pi}$ between $\mathcal{C}^1_{2\pi}$ and $\mathcal{C}_{2\pi}$, we resolve $\mathcal{C}^1_{2\pi}$ in a similar way

$$\mathcal{C}^1_{2\pi} = \mathcal{N}(L) + \mathcal{W} \tag{4.190}$$

where $\mathcal{W} = \{w \in \mathcal{C}^1_{2\pi} | < w, \psi_i >= 0, i = 1,2\}$ (4.191)

Obviously $\mathcal{W} \subset \mathcal{R}(L) \cap \mathcal{C}^1_{2\pi}$. The solutions $u(s)$ of eq. (4.178) can be expressed as

$$u = \varepsilon_1 \varphi_1 + \varepsilon_2 \varphi_2 + w \tag{4.192}$$

where $\varepsilon_1 \varphi_1 + \varepsilon_2 \varphi_2 \in \mathcal{N}(L), w \in \mathcal{W}$. Since $u$ is symmetrical about $t$, we replace $s$ by $s + \theta$, $\varphi_1(s+\theta) = \mathrm{Re}[e^{i(\theta+s)} c] = \mathrm{Re}\, e^{i\theta}(e^{is} c) = \alpha_1 \varphi_1(s)$.

Since $\varphi \in \mathcal{N}(L)$, $\varphi(s+\theta) = \alpha_1 \varphi_1(s) + \alpha_2 \varphi_2(s)$ and $\alpha_1^2 + \alpha_1^2 = 1$. Rewrite $\varepsilon_1$ and $\varepsilon_2$ in eq. (4.192) in polar coordinate form $\varepsilon_1 = \varepsilon \cos\theta$ and $\varepsilon_2 = \varepsilon \sin\theta$. Without loss of generality, we select a proper $\theta$ which makes $\varepsilon_1 = \varepsilon$ and $\varepsilon_2 = 0$, so eq. (4.192) changes to $\quad u = \varepsilon \varphi_1 + w$ (4.193)

Defining a shift operator $T_\theta$, $T_\theta u(s) = u(s - \theta)$, it is easy to prove that $N(T_\theta u, \mu, \tau) = T_\theta N(u, \mu, \tau)$, for $\forall \theta \in \mathcal{S}^1$. By Liapunov-Schmidt reduction, eq. (4.178) is equivalent to the following two equations:

$$QN(\varepsilon \varphi_1 + w, \mu, \tau) = 0 \tag{4.194}$$

$$PN(\varepsilon \varphi_1 + w, \mu, \tau) = 0 \tag{4.195}$$

Because the Frechet derivative $D_w PN = L|_{R(L)}$ of eq. (4.195) with respect to $w$ is reversible, by the implicit function theorem we can seek a unique solution from eq. (4.195), which satisfies $w(0,0,0) = 0$. Substituting this into eq. (4.194), we obtain the bifurcation equation as follows

$$g(\varepsilon,\mu,\tau) = Q\{-\omega_0(1+\tau)\frac{\mathrm{d}u}{\mathrm{d}s} + f(u,\mu)\} = 0 \tag{4.196}$$

where $u = \varepsilon\varphi_1 + w(\varepsilon,\mu,\tau)$. Comparing $g{:}\mathcal{R}\times\mathcal{R}\times\mathcal{R}\rightarrow\mathcal{N}(L)$ with eq. (4.189) results in

$$g_i(\varepsilon,\mu,\tau) = <(1+\tau)\omega_0\frac{\mathrm{d}u}{\mathrm{d}s} - f(u,\mu),\ \psi_i> = 0, i = 1,2 \tag{4.197}$$

where $g = \begin{bmatrix} g_1 \\ g_2 \end{bmatrix}$.

### 4.8.2 Hopf Bifurcation Solutions

From eq. (4.197) we can obtain bifurcation solutions $\mu(\varepsilon)$ and $\tau(\varepsilon)$ which can be expressed in Taylor expansions

$$\begin{aligned} \mu(\varepsilon) &= \mu_0 + \mu'(0)\varepsilon + \frac{1}{2}\mu''(0)\varepsilon^2 + \cdots \\ \tau(\varepsilon) &= \tau_0 + \tau'(0)\varepsilon + \frac{1}{2}\tau''(0)\varepsilon^2 + \cdots \end{aligned} \tag{4.198}$$

We will compute the coefficients of the expansions below. For convenience, the computations are carried out in complex form. We know that

$$\begin{aligned} & e^{is}d = \psi_1(s) + i\psi_2(s) \\ & g_i = <\psi_i, N> = 0, \qquad i = 1,2 \\ & G = <e^{is}d, N> = <\psi_1, N> - i <\psi_2, N> \\ & \qquad = g_1 - ig_2 \end{aligned} \tag{4.199}$$

and

$$\begin{aligned} \psi_1 &= \frac{1}{2}[e^{is}c + e^{-is}\bar{c}] \\ \psi_2 &= \frac{1}{2i}[e^{is}c - e^{-is}\bar{c}] \end{aligned} \tag{4.200}$$

where $\bar{c}$ stands for the conjugation of $c$. From eq. (4.196), we choose

$$g(\varepsilon,\mu,\tau) = \varepsilon\hat{g}(\varepsilon,\mu,\tau) \tag{4.201}$$

When $\varepsilon \neq 0$, $\hat{g} = \frac{1}{\varepsilon}g(\varepsilon,\mu,\tau)$ (4.202)

When $\varepsilon = 0$, $\hat{g} = \lim_{\varepsilon\to 0}(\frac{g}{\varepsilon})$ =derivative of $g$ with respect to $\varepsilon$

$$= Q\left\{-\omega_0(1+\tau)\frac{\mathrm{d}(\varphi + w_\varepsilon)}{\mathrm{d}s} + A(\mu)(\varphi + w_\varepsilon)\right\} \tag{4.203}$$

Next we proceed as in the previous section. By use of the differential rules of the implicit function, from eq. (4.195) we resolve $w_\varepsilon(0) = w_\mu(0) = w_\tau(0) = 0$ at $\varepsilon = 0$.

In order to find $\mu(\varepsilon)$ and $\tau(\varepsilon)$, $\frac{\partial g}{\partial(\mu,\tau)}$ needs to be computed first. Differentiating eq. (4.199) with respect to $\varepsilon$, we have

$$G_\varepsilon|_{\mu,\tau\text{ considered as a constant}}$$
$$=<e^{is}d,\omega_0(1+\tau)\frac{\mathrm{d}(\varphi+w_\varepsilon)}{\mathrm{d}s}+D_uf(u,\mu)[\varphi_1+w_s]>$$

$$G_{\varepsilon\mu}|_0=<e^{is}d,0+A'(0)[\varphi_1+0]>$$
$$=<e^{is}d,A'(0)\left(\frac{e^{is}c+e^{-is}\overline{c}}{2}\right)>$$
$$=\frac{1}{2\pi}\bullet\frac{1}{2}\int_0^{2\pi}e^{is}d^*A'(0)(e^{is}c+e^{-is}\overline{c})\,\mathrm{d}s$$
$$=\frac{2\pi}{2\pi}\frac{d^*A'(0)c}{2}=\frac{d^*c}{2}[\alpha'(0)+i\omega'(0)]$$
$$=\alpha'(0)+i\omega'(0)$$

$$G_{\varepsilon\tau}|_0=<e^{is}d,-\omega_0\frac{\mathrm{d}\varphi_1}{\mathrm{d}s}>=<e^{is}d,\omega_0\varphi_2>$$
$$=<e^{is}d,\omega_0\frac{e^{is}c-e^{-is}\overline{c}}{2i}>=\frac{\omega_0}{2i}d^*c$$
$$=-i\omega_0$$

so we obtain

$$\frac{\partial(\widetilde{g}_1,\widetilde{g}_2)}{\partial(\mu,\tau)}|_0=\begin{bmatrix}\alpha'(0) & 0\\ -\omega'(0) & \omega_0\end{bmatrix} \tag{4.204}$$

$$\det\frac{\partial(\widetilde{g}_1,\widetilde{g}_2)}{\partial(\mu,\tau)}=\omega_0\alpha'(0)\neq 0 \tag{4.205}$$

Since the determinant is not equal to zero, a unique group of solutions $(\mu(\varepsilon),\tau(\varepsilon))$ can be solved. Now we shall begin to calculate the coefficients of the Taylor expansions in eq. (4.198) from $G(\varepsilon,\mu(\varepsilon),\tau(\varepsilon))=0$. The derivative of G with respect to $\varepsilon$ is

$$\frac{\mathrm{d}G}{\mathrm{d}\varepsilon}=G_\varepsilon+G_\mu\mu'+G_\tau\tau'$$
$$=<e^{is}d,-\omega_0(1+\tau)\frac{\mathrm{d}(\varphi_1+w_\varepsilon)}{\mathrm{d}s}-\omega_0\tau'\frac{\mathrm{d}u}{\mathrm{d}s}+D_uf(u,\mu)[\varphi_1+w_\varepsilon]+D_\mu f(u,\mu)\mu'>$$

$$\left.\frac{\mathrm{d}G}{\mathrm{d}\varepsilon}\right|_0=<e^{is}d,L[\varphi_1+w_s]>$$

$$\frac{\mathrm{d}^2G}{\mathrm{d}\varepsilon^2}|_0=<e^{is}d,-2\omega_0\tau'\frac{\mathrm{d}\varphi}{\mathrm{d}s}-\left[\omega_0\frac{\mathrm{d}w_{\varepsilon\varepsilon}}{\mathrm{d}s}+A_0w_{\varepsilon\varepsilon}\right]+$$
$$D_{uu}f|_0\varphi_1\varphi_1+2D_{\mu u}f|_0\varphi_1\mu'> \tag{4.206}$$
$$=<e^{is}d,2\omega_0\tau'\varphi_1+2A'(0)\varphi_1\mu'+D_{uu}f|_0\varphi_1\varphi_1>$$

Here it is easy to verify that

$$<e^{is}d,D_{uu}f|_0\varphi_1\varphi_1>=$$
$$<e^{is}d,D_{uu}f|_0\left[\frac{e^{is}c+e^{-is}\overline{c}}{2}\right]\left[\frac{e^{is}c+e^{-is}\overline{c}}{2}\right]>=0 \tag{4.207}$$

After computing the inner products in eq. (4.206), we separate the real and imaginary parts and obtain

$$2\begin{bmatrix} \alpha'(0) & 0 \\ -\omega'(0) & \omega_0 \end{bmatrix}\begin{bmatrix} \mu'(0) \\ \tau(0) \end{bmatrix} = \begin{bmatrix} 0 \\ 0 \end{bmatrix} \tag{4.208}$$

From eq. (4.208) we get

$$\begin{aligned} &\mu'(0) = 0 \\ &\tau'(0) = 0 \end{aligned} \tag{4.209}$$

$\mu''(0)$ and $\tau''(0)$ are computed as follows:

$$\frac{d^3G}{d\varepsilon^3}|_0 = < e^{is}d, 3\varphi_2\omega_0\tau'' + 3A'(0)\mu'' + D_{uuu}f|_0\,\varphi_1\varphi_1\varphi_1 + 3D_{uu}f|_0\,\varphi_1 w'' >$$

Separating its real and imaginary parts , we also have

$$3\begin{bmatrix} \alpha'(0) & 0 \\ -\omega'(0) & \omega_0 \end{bmatrix}\begin{bmatrix} \mu'' \\ \tau'' \end{bmatrix} = -\begin{bmatrix} \mathrm{Re} < e^{is}d, & D_{uuu}f\varphi_1\varphi_1\varphi_1 + 3D_{uu}f|_0\,\varphi_1 w'' > \\ \mathrm{Im} < e^{is}d, & D_{uuu}f|_0\,\varphi_1\varphi_1\varphi_1 + 3D_{uu}f|_0\,\varphi_1 w'' > \end{bmatrix} \tag{4.210}$$

where $w''(0) = \dfrac{d^2}{d\varepsilon^2} w(\varepsilon, \mu(\varepsilon)v(\varepsilon))|_{\varepsilon=0} = w_{\varepsilon\varepsilon}$. $w_{\varepsilon\varepsilon}$ can be solved from eq. (4.195):

$$PLw_{\varepsilon\varepsilon} + PD_{uu}f|_0\,\varphi_1\varphi_1 = 0$$

From (4.207), $QD_{uu}f|_0\,\varphi_1\varphi_1 = 0$ so that $Lw_{\varepsilon\varepsilon} + D_{uu}f|_0\,\varphi_1\varphi_1 = 0$, or

$$\begin{aligned} &\left[-\omega_0\frac{d}{ds} + A_0\right]w_{\varepsilon\varepsilon} + \frac{1}{4}\left[e^{2is}D_{uu}f|_0\,cc + \right. \\ &\qquad \left. 2D_{uu}f|_0\,c\bar{c} + e^{-2is}D_{uu}f|_0\,\bar{c}\bar{c}\right] = 0 \end{aligned} \tag{4.211}$$

Assuming the solution of the linear differential eq. (4.211) is

$$w_{\varepsilon\varepsilon}(s) = a_2 e^{2is} + a_0 + a_{-2}e^{-2is} \tag{4.212}$$

we can easily obtain $A_0 a_0 = -\dfrac{1}{2}D_{uu}f|_0\,c\bar{c}$, and

$$[A_0 - 2\omega_0 iI]a_2 = -\frac{1}{4}D_{uu}f|_0\,cc,\ a_{-2} = \bar{a}_2 \tag{4.213}$$

Substituting $w_{\varepsilon\varepsilon}$, we have

$$\begin{aligned} < e^{is}d, D_{uu}f|_0\,\varphi_1 w_{\varepsilon\varepsilon} > &= < e^{is}d, D_{uu}f|_0\left[\frac{e^{is}c + e^{-is}\bar{c}}{2}\right]\left[a_2e^{2is} + \right. \\ &\qquad \left. a_0 + a_{-2}e^{-2is}\right] > = \frac{d^*}{2}[D_{uu}f|_0\,ca_0 + D_{uu}f|_0\,\bar{c}a_2] \end{aligned} \tag{4.214}$$

and in the same way

$$< e^{is}d, D_{uuu}f|_0\left[\frac{e^{is}c + e^{-is}\bar{c}}{2}\right]\left[\frac{e^{is}c + e^{-is}\bar{c}}{2}\right]\left[\frac{e^{is}c + e^{-is}}{2}\right] >$$

$$= \frac{3}{8} d^* D_{uuu} f|_0 \, c c \bar{c} \tag{4.215}$$

Substituting eq. (4.214) and eq. (4.215) into eq. (4.210) yields

$$3\begin{bmatrix} \alpha'(0) & 0 \\ -\omega'(0) & \omega_0 \end{bmatrix}\begin{bmatrix} \mu''(0) \\ \tau''(0) \end{bmatrix} = -3\begin{bmatrix} \mathrm{Re}(H) \\ \mathrm{Im}(H) \end{bmatrix} \tag{4.216}$$

$$\overset{def}{=} d^* \left[ \frac{1}{6} D_{uu} f|_0 \, c a_0 + \frac{1}{6} D_{uu} f|_0 \, \bar{c} a_2 + \frac{1}{8} D_{uuu} f|_0 \, c c \bar{c} \right]$$

Then $\mu''(0)$ and $\tau''(0)$ are resolved from eq. (4.216) as follows:

$$\begin{cases} \mu''(0) = -\dfrac{1}{\alpha'(0)} \mathrm{Re}(H) \\ \tau''(0) = -\dfrac{1}{\omega_0} (\mathrm{Im}(H) - \omega'(0)\mu''(0)) \end{cases} \tag{4.217}$$

In general $H \neq 0$, therefore the local bifurcation solutions are

$$\begin{aligned} \mu(\varepsilon) &= \mu_0 + \frac{1}{2}\mu''(0)\varepsilon^2 + o(\varepsilon^4) \\ \tau(\varepsilon) &= \frac{1}{2}\tau''(0)\varepsilon^2 + o(\varepsilon^4) \\ u(s) &= \varepsilon\varphi_1(s) + \varepsilon^2 w_{\varepsilon\varepsilon} + o(\varepsilon^4) \end{aligned} \tag{4.218}$$

The bifurcation diagrams are shown in Fig. 4.24.

Fig. 4.24

### 4.8.3 Stability of the Hopf Bifurcation Solution

In this subsection we shall study the stability of Hopf's periodic orbits by means of Floquet's theory introduced in chapter 1.

**Theorem 4.26 (Andronov-Vitt's theorem)** For an $n$-dimensional autonomous system of ordinary differential equations, the periodic orbits are asymptotically stable if there exist $(n-1)$ Floquet exponents $\gamma$ which satisfy $\mathrm{Re}(\gamma) < 0$ and the remaining one equals zero.

According to this theorem we need to compute the Floquet exponents of eq. (4.178) in order to determine the stability of Hopf periodic orbits.

Here we further state the third hypothesis ($H3$).

($H3$) Assuming a Jacobian matrix $A_0 = D_u f(0,\mu_0)$ has a pair of purely imaginary roots $\pm i\omega_0$, all the real parts of the remaining $(n-2)$ eigenvalues are negative.

From ($H3$), $N(L)$ is two dimensional and its complementary space is $(n-2)$ dimensional. With ($H3$), we can prove that for sufficiently small $\varepsilon$'s, the perturbed periodic systems have $(n-2)$ Floquet exponents with negative real parts, and one of the Floquet exponents corresponding to null space $\mathcal{N}(L)$ is always equal to zero, which corresponds to the tangential direction of periodic orbits. The other $\gamma$ is called the critical Floquet exponent, which corresponds to the normal direction. The critical exponent is just what we have to calculate.

Before proceeding, it is necessary to review the Floquet theorem briefly. Consider a periodic system

$$y' = A(t)y \qquad A(t+T) = A(t) \tag{4.219}$$

let its general solution be

$$y(t) = p(t)e^{Bt}y_0 \tag{4.220}$$

where $p(t+T) = p(t)$, and $B$ is a $n \times n$ matrix, all the elements of which are constants. The eigenvalues of $B$ are just the Floquet exponents $\gamma$ and the corresponding eigenvectors are $v$. From $Bv = \gamma v$, $e^{Bt}v = e^{\gamma t}v$. Multiplying by $p(t)$, we obtain $p(t)e^{Bt}v = e^{\gamma t}p(t)v$. If we put $p(t) = p(t)v$, then the solution of eq. (4.219) can be written as

$$y = e^{\gamma t}p(t) \tag{4.221}$$

Substituting eq. (4.221) into eq. (4.219), we have

$$e^{\gamma t}p'(t) + \gamma e^{\gamma t}p(t) - A(t)e^{\gamma t}p(t) = 0 \tag{4.222}$$

or
$$\left[\frac{\mathrm{d}}{\mathrm{d}t} - A(t)\right]p(t) + \gamma p(t) = 0$$

$$p(0) = p(T)$$

We can study the stability of Hopf bifurcation solutions by means of eq. (4.222). First, seek the pertubation equation of eq. (4.178) about $u(s) = \varepsilon\varphi_1 + w(\varphi,s)$. Taking $t = s$, $T = 2\pi$, $\widetilde{A}(s) = D_y f(u(s))$, we assume that the solutions of the pertubated periodic system have the same form as that of eq. (4.221). Then

$$\left[-\omega_0(1+\tau)\frac{\mathrm{d}}{\mathrm{d}s} + \widetilde{A}(s)\right]p(s) - \gamma p(s) = 0 \tag{4.223}$$

$$p(0) = p(2\pi)$$

We are going to solve $\gamma$ and $p(0)$ near $\varepsilon = 0$. When $\varepsilon = 0$, eq. (4.223) changes to

$$\left[-\omega_0\frac{\mathrm{d}}{\mathrm{d}s} + A_0\right]p(s) - \gamma p(s) = 0 \tag{4.224}$$

$$p(0) = p(2\pi)$$

or $\quad Lp - \gamma p = 0, \quad p(0) = p(2\pi) \qquad (4.225)$

When $\gamma = 0$, the solution of the above equation is $\{\varphi_1, \varphi_2\}$, that is, when $\varepsilon = 0$, for $\gamma = 0$ we have a two-dimensional eigenvector space spanned by $\{\varphi_1, \varphi_2\}$, which is the null space of $L$. From (*H*3) we know that one Floquet exponent corresponding to the null space is always zero; the remaining one which is close to zero is called the critical Floquet exponent.

For the exponent that is always equal to zero, we differentiate eq. (4.178) with respect to $s$:

$$\omega_0(1+\tau)\left(\frac{du}{ds}\right)' = D_y f(u(s))\frac{du}{ds} \qquad (4.226)$$

Taking $p(t) = \dfrac{du}{ds}$, then when $\gamma = 0$, we conclude that eq. (4.226) is the same as eq. (4.223). Since $u = \varepsilon\varphi_1 + w$ when $\varepsilon \to 0$, $u \to 0$. To solve this problem, we rescale $u(s)$. Defining

$$U = \frac{1}{\varepsilon}\left(\frac{du}{ds}\right) = \frac{d\varphi_1}{ds} + \frac{1}{\varepsilon}\frac{dw}{ds} = -\varphi_2 + \frac{1}{\varepsilon}\frac{dw}{ds} \qquad (4.227)$$

since $w = o(|\varepsilon|^2)$ so $U \to -\varphi_2$ when $\varepsilon \to 0$.

Below we seek the second eigenfunction when $\varepsilon \to 0$ and $\gamma$ is close to zero. When the eigenvalues are repeated, there are two possibilities about the eigenvectors $v_1$ and $v_2$: there exists either two independent eigenvectors or one eigenvector and one generalized vector, that is

$$(M-\gamma I)v_1 = 0, \quad (M-\gamma I)v_2 = \alpha v_1 \qquad (\alpha,\ \beta \text{ allowed to be } 0) \qquad (4.228)$$

Hence the second eigenfunction can be expressed as

$$\left[-\omega_0(1+\tau)\frac{d}{ds} + \widetilde{A}(s)\right]p(s) - \gamma p(s) = \alpha U(s)$$

or $\quad Mp(s) - \gamma p(s)) = \alpha U(s) \qquad (4.229)$

where $M = \left[-\omega_0(1+\tau)\dfrac{d}{ds} + \widetilde{A}(s)\right]$.

Let the solution of (4.229) be

$$p(s) = \varphi_1(s) + v(s), \quad v(s) \in \mathcal{W} = \mathcal{R}(L) \cap \mathcal{C}^1_{2\pi} \qquad (4.230)$$

The solution does not include $\varphi_2(s)$ for $MU = 0$; eq. (4.229) guarantees that $p(s)$ and $v(s)$ are independent of each other when $\varepsilon \to 0$. Substituting eq. (4.230) into eq. (4.229), then

$$N(\varepsilon,\alpha,\gamma,s) = M(\varphi_1 + v) - \gamma(\varphi + v) - \alpha U(s) = 0 \qquad (4.231)$$

mapping $N: \mathcal{R} \times \mathcal{R} \times \mathcal{R} \times \mathcal{W} \to \mathcal{C}_{2\pi} \equiv \mathcal{N}(L) \oplus \mathcal{R}(L)$

From the implicit function theorem, when $\varepsilon = 0$, we obtain

$$\left[-\omega_0\frac{d}{ds} + A_0\right][\varphi_1 + v] - \gamma[\varphi_1 + v] - \alpha\varphi_2 = 0 \qquad (4.232)$$

There exists a unique trivial solution $(\alpha = 0, \gamma = 0, \nu = 0)$. At $\varepsilon = 0$, differentiating the above equation yields $\frac{\partial N}{\partial \alpha}|_0 = -\varphi_2$, $\frac{\partial N}{\partial \gamma}|_0 = -\varphi_1$, and $\frac{\partial N}{\partial \nu}|_0 = L$ where $\varphi_1, \varphi_2 \in \mathcal{N}(L)$, $L \in \mathcal{R}(L)$. So $\frac{\partial N}{\partial(\alpha, \gamma, \nu)}$ is invertible on $\mathcal{N}(L) \oplus \mathcal{R}(L)$, therefore there exists a unique solution $(\alpha(\varepsilon), \gamma(\varepsilon), \nu(\varepsilon))$ in the neighbourhood of (0,0,0) from the implicit function theorem. Through the above analysis, when $\varepsilon$ becomes closer to zero, the critical Floquet exponent can be expressed as

$$\gamma(\varepsilon) = 0 + \gamma'(0)\varepsilon + \frac{\gamma''(0)}{2}\varepsilon^2 + \cdots \tag{4.233}$$

The computation of $\gamma(\varepsilon)$ is the same as that of $\mu(\varepsilon)$. For Hopf bifurcation, through the computation we have $\gamma'(0) = 0$, while

$$\gamma''(0) = -2\alpha'(0)\mu''(0) \tag{4.234}$$

(4.234) is called the exchange formula of stability of Hopf bifurcation solutions. By ($H1$), $\alpha'(0) \neq 0$; in the non-degenerate case of Hopf bifurcation, $\mu''(0) \neq 0$. Therefore, we can judge the stability of the Hopf bifurcation solution by eq. (4.234), as shown in Fig. 4.24.

### 4.8.4 Degenerate Hopf Bifurcation

The degenerate Hopf bifurcation occurs when any of the three assumptions $H1$, $H2$ and $H3$ cannot be satisfied. For example, if $H1$ is not satisfied, that is, there exist more than two critical eigenvalues, its multiplicity is numerable. The dimensions of the null space of $L$ will be larger than two, and the bifurcation equations obtained by Liapunov-Schmidt reduction will be more than two as well. But the analysis is not applicable and chaos is likely to occur in this case.

If $H1$ but not $H2$ nor $H3$ is satisfied, in this degenerate case the null space of $L$ is two dimensional, although Liapunov-Schmidt reduction is still effective under such a degenerate condition. In two-dimensional space the behaviours of the dynamical systems are still periodic orbits, no chaos will appear, but there may exist behaviours of multiple periodic orbits, hysteresis, multiple Hopf bifurcation, the combination of two periodic orbits, etc.

The LS reduction can be used to study degenerate Hopf bifurcations even if $H1$ but not $H2$ nor $H3$ is satisfied. At this point in expression eq. (4.218) there may be degeneration as $\mu''(0) = 0$ and $\mu''(0) = \mu^{(4)}(0) = 0$. The $\mu^{2k}(0)_{k=1,2,3,\cdots}$ are called the numbers of focus. In 1978, the methods of centre manifold were used to calculate the second number of focus [94], The Liapunov-Schmidt reduction was also used to solve the same problem in 1981 [2]. In refs [95,96,97] published in 1987, 1985 and 1989 respectively, the LS method was further applied to calculate the third number of focus. The Hilbert 16th problem was also calculated by this method.

If we take the same symbols $N, P, Q, \varphi_j, \psi_j, c, d$ as defined in the above subsection, and let $\varepsilon = x$, then we have

$$g(x, \mu, \tau) = QN(x\varphi_1 + w(x, \mu, \tau), \mu, \tau) = 0 \tag{4.235}$$

where $N = -\omega_0(1+\tau)\dfrac{\mathrm{d}u}{\mathrm{d}s} + f(u,\mu)$, $u = x\varphi_1 + w$, $w \in \mathcal{W} \subseteq \mathcal{R}(L)$.

Because the systems have symmetries about time, according to the same theorem used in the previous subsection, $g$ is certainly the odd function of $x$, i.e. $g(-x,\mu,\tau) = -g(x,\mu,\tau)$, the components of $g$ is

$$g_i = <\psi_j, N> \quad J = 1,2 \tag{4.236}$$

According to the theory in [4], if eq. (4.178) is $\mathcal{C}^\alpha$, then

$$\begin{cases} g_1(x,\mu,\tau) = p(x^2,\mu,\tau)x \\ g_2(x,\mu,\tau) = q(x^2,\mu,\tau)x \end{cases} \tag{4.237}$$

where $p,q \in \mathcal{C}^\alpha$. In the same way as in the preceding subsection, $\hat{g}$ corresponds to $(p,q)$. We have $\dfrac{\partial q}{\partial \tau} = \omega_0 \neq 0$ from eq. (4.204), therefore we can solve $\tau = \tau(x^2,\mu)$ from $q = 0$, which are even functions of $x$. Substituting this into eq. (4.237), we obtain $h(x^2,\mu) = p(x^2,\mu,\tau(x^2,\mu)) = 0$ (4.238)

Finally we obtain the one-dimensional bifurcation equation

$$h(x^2,\mu)x = 0 \tag{4.239}$$

From eq. (4.239) a unique solution $\mu$ can be solved near $\mu = 0$. This is because $h_\mu = p_\mu + p_\tau \tau_\mu = \alpha'(0) + 0 \neq 0$, so the implicit function theorem guarantees the uniqueness of the solution $\mu = \mu(x^2)$:

$$Z = x^2,\ h(x^2,\mu) = h(Z,\mu) \tag{4.240}$$

Assuming that $h_z = \dfrac{\partial h}{\partial z}(0,0) \neq 0$, from $h = 0$, we get

$$\mu = -\left(\frac{h_z}{h_\mu}\right)x^2 + \cdots \tag{4.241}$$

and it is easy to verify that $h_z = \mu''(0)$; when *H*3 holds, $h \neq 0$ and when *H*2 holds, $h_\mu \neq 0$. Therefore, for non-degenerate Hopf bifurcations, $h_\mu \neq 0 \neq h_z$ at point (0,0), and for degenerate Hopf bifurcations, $h_\mu$ or $h_z$ is zero at (0,0). In order to solve the degenerate problem, we need to use singularity theory. The steps to be taken are first to recognize the equation so as to find a set of equivalent GS normal forms, and then to unfold the normal forms to find all possible bifurcation solutions.

As for the restriction of space, the calculations of the coefficients of the Taylor expansions in eq. (4.237) and eq. (4.238) are neglected here. Interested readers should refer to [2,97].

# Chapter 5

# Centre Manifold Theorem And Normal Form Of Vector Fields

In this chapter, section 5.1 studies another main method for the local bifurcation of dynamical systems: the Centre Manifold Theorem. In section 5.2, the centre manifold theorem is used to analyse simple bifurcation. In the section 5.3, the Normal Form theory of vector fields is introduced.

## 5.1 Centre Manifold Theorem

Suppose there exists an ordinary differential equation $\dot{x} = f(x)$, so that $f(0) = 0$. If the linearized portion of $f$ at the origin has no pure imaginary roots, then the Hartman-Grobman theorem (section 2.4) states that the number of eigenvalues with positive and negative real parts determines the topological structure of the flow near 0. If there are eigenvalues with zero real parts, then the flow near the origin can be quite complicated, including non-uniqueness and loss of smoothness in the invariant centre manifold. We shall illustrate these with examples.

### 5.1.1 Definition of Manifold

**Definition 5.1** Manifold is a type of mapping and a mapping from a one-dimensional subspace to an $n$-dimensional subspace, and the mapping is diffeomorphic. For $1 \le n$, this type of mapping is a mapping of reduced dimension.

If $k \ge 0$, $1 \le n$, there exists a $\mathcal{C}^k$ mapping $H{:}\,\mathcal{U} \to \mathcal{R}^n$, and $H$ is a one to one map. As $\mathcal{U} \subseteq \mathcal{R}^1$ is open, the set $\mathcal{W} \subseteq \mathcal{R}^n$, $\mathcal{W} = H(\mathcal{U}) \equiv \{x \in \mathcal{R}^n | x = H(u), u \in \mathcal{U}\}$ is called a one-dimensional $\mathcal{C}^k$ local manifold. If $k = 0$, then $H$ is homomorphic; if $k > 0$, then $H$ is diffeomorphic.

**Definition 5.2** If for every $x \in \mathcal{W}$, there exists neighbourhoods $\mathcal{U}(x) \subseteq \mathcal{R}^1$ and $\mathcal{V}(x) \subseteq \mathcal{R}^n$, so that the above definition can be applied to $\mathcal{U}(x)$ and $\mathcal{V}(x) \cap \mathcal{W}$, then $\mathcal{W}$ in $\mathcal{R}^n$ is a one-dimensional non-local manifold of $\mathcal{C}^k$ ( where $\mathcal{U}, \mathcal{V}$ and $\mathcal{H}$ vary with $x$).

### 5.1.2 Centre Manifold and Centre Subspace

Suppose that there exists a linear differential equation $y' = Ay = Df(p)y$. If $A$ has $n_S$ eigenvalues with negative real parts whose corresponding eigenvectors are $V_1, V_2, \cdots, V_{N_S}$ and their corresponding eigenvalues with positive real parts whose eigenvectors are $U_1, U_2, \cdots, U_{N_U}$ ; and also $N_c$ eigenvalues with zero real parts whose corresponding eigenvectors are $W_1, W_2, \cdots, W_{N_C}$

then the stable subspace $\mathcal{E}^s = \text{span}\{V_1, V_2, \cdots, V_{N_S}\}$

the unstable subspace $\mathcal{E}^u = \text{span}\{U_1, U_2, \cdots, U_{N_S}\}$

the centre subspace $\mathcal{E}^c = \text{span}\{W_1, W_2, \cdots, W_{N_S}\}$

where $N_s + N_u + N_c = \mathcal{N}$; $\mathcal{E}^s \oplus \mathcal{E}^u \oplus \mathcal{E}^c = \mathcal{R}^n$.

**Theorem 5.1 Centre Manifold Theorem**

If $f \in \mathcal{C}^k(\mathcal{U})$ is a vector field in $\mathcal{R}^n$ $k \geq 1, \mathcal{U} \in \mathcal{R}^n$ is an open set, $A = Df(p)$, the spectrum of $A$ can be divided into three parts $\sigma_s, \sigma_u$ and $\sigma_c$, and

$$\text{Re}\,\lambda \begin{cases} < 0 & \text{if } \lambda \in \sigma_s \\ = 0 & \text{if } \lambda \in \sigma_c \\ > 0 & \text{if } \lambda \in \sigma_u \end{cases}$$

Let the generalized eigenspaces corresponding to $\sigma_s, \sigma_u$ and $\sigma_c$ be $\mathcal{E}^s, \mathcal{E}^u$ and $\mathcal{E}^c$ respectively. Then there exist $\mathcal{C}^k$ stable and unstable invariant manifolds $\mathcal{W}^u$ and $\mathcal{W}^s$ tangential to $\mathcal{E}^u$ and $\mathcal{E}^s$ at $p$, and a $\mathcal{C}^{k-1}$ centre manifold $\mathcal{W}^c$ tangential to $\mathcal{E}^c$ at $p$. The manifolds $\mathcal{W}^u, \mathcal{W}^s$ and $\mathcal{W}^c$ are all invariant for the flow of $f$. The stable and unstable manifolds are unique, but $\mathcal{W}^c$ is not, i.e. the direction of $\lim_{t \to \pm\infty} \varphi' y_0$ is not determined. For example, consider the system

$$\begin{cases} x' = -x \\ y' = y^2 \end{cases} \tag{5.1}$$

and its linear system

$$\begin{bmatrix} x' \\ y' \end{bmatrix} = \begin{bmatrix} -1 & 0 \\ 0 & 0 \end{bmatrix} \begin{bmatrix} x \\ y \end{bmatrix} = A \begin{bmatrix} x \\ y \end{bmatrix} \tag{5.2}$$

The eigenvalues of $A$ are $\lambda = 0$, and $\lambda = -1$, and their corresponding eigenvectors are $\begin{bmatrix} 0 \\ 1 \end{bmatrix}$ and $\begin{bmatrix} 1 \\ 0 \end{bmatrix}$. The subspaces spanned by the eigenvectors are

$\mathcal{E}^s = \text{span}\left\{\begin{bmatrix} 1 \\ 0 \end{bmatrix}\right\}$, $\mathcal{E}^c = \text{span}\left\{\begin{bmatrix} 1 \\ 0 \end{bmatrix}\right\}$, $\mathcal{E}^u = \Phi$, and $n_s = n_c = 1$, $n_u = 0$.

The exact solution of the linearized system is

$$\begin{cases} x(t) = x(0)e^{-t} \\ y = \text{const} \end{cases} \tag{5.3}$$

The exact solution of the original system is

$$y(t) = \frac{y(0)}{1 - y(0)t} \tag{5.4}$$

Eliminating $t$ from eqs. (5.3) and (5.4) we have

$$x = Ke^{1/y} \tag{5.5}$$

where $K = x_0 e^{-1/y_0}$, and $\chi_0$ and $y_0$ are the mean values of $x$ and $y$ when $t = 0$.

The manifold of the linearized system is
$$\mathcal{W}^s = \{(x,y)|x,y \text{ satisfy } (5.3)\}$$
$$\mathcal{W}^c = \{(0,y)|x,y \in \mathcal{R}\}$$

The phase portrait is shown in Fig. 5.1.

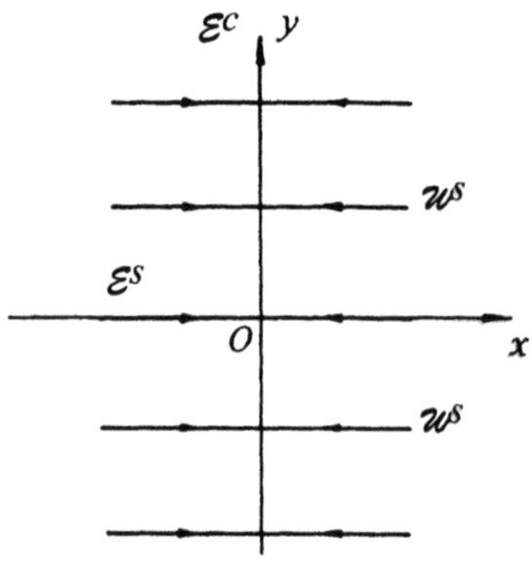


Fig. 5.1

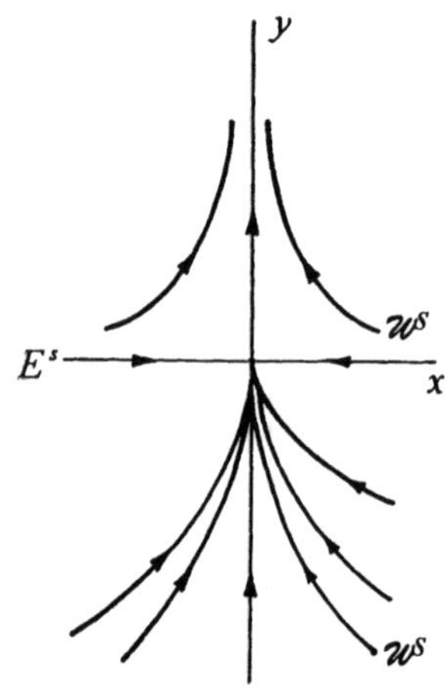


Fig. 5.2

We shall now discuss the phase portrait from eq. (5.5). When $y$ tends to 0 in the negative direction, that is $y \approx -0$, $\frac{1}{y} = -\frac{1}{0} = -\infty$, so that $x = Ke^{1/y} = Ke^{-\infty} \approx 0$. When $y$ tends to 0 in the positive direction, $x \approx \infty$, so that the derivative of the phase portrait is not continuous. It is easy to see that of $\mathcal{W}^c$ is not unique, see Fig. 5.2 .

In order to ensure that the derivative of the phase portrait at (0,0) is continuous, take the expression for the centre manifold $\mathcal{W}^c$ as

$$\mathcal{W}^c = \left( (x,y)\Big|_{x=Ke^{1/y},\ \text{when } y\le 0}^{x=0,\ \text{when } y>0} \right) \tag{5.6}$$

For $\mathcal{W}^c$ when $y > 0, x = 0$ $\frac{\mathrm{d}x}{\mathrm{d}y} = -\frac{0}{y^2} = 0$, i.e. $\mathcal{W}^c$ is a tangent to $\mathcal{E}^c$; for $y \le 0$, $x = Ke^{1/y}$, when $y$ tends to zero in the negative direction, we obtain $\frac{\mathrm{d}x}{\mathrm{d}y} = 0$, so that $\mathcal{W}^c$ is a tangent to $\mathcal{E}^c$ too. The derivative of $\mathcal{W}^c$ at (0,0) is unique, see Fig. 5.3.

There are different centre manifolds for different $K$ at (0,0), i.e. $\mathcal{W}^c$ is not unique, but its derivative at (0,0) is unique. In this manner, we are sure that the Taylor series of $\mathcal{W}^c$ is unique, but may not be convergent. This series is asymptotic if the amplitude is small.

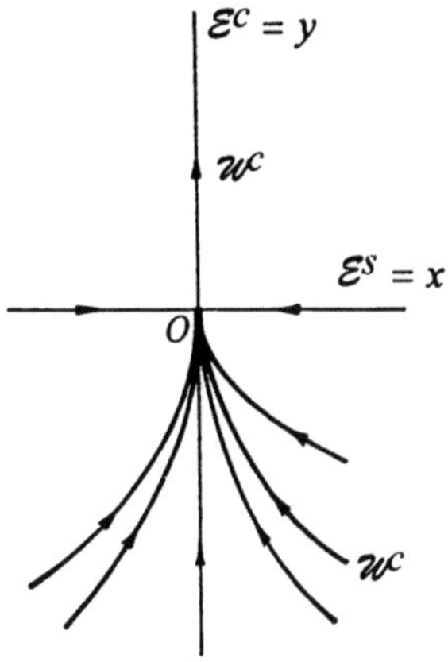


Fig. 5.3

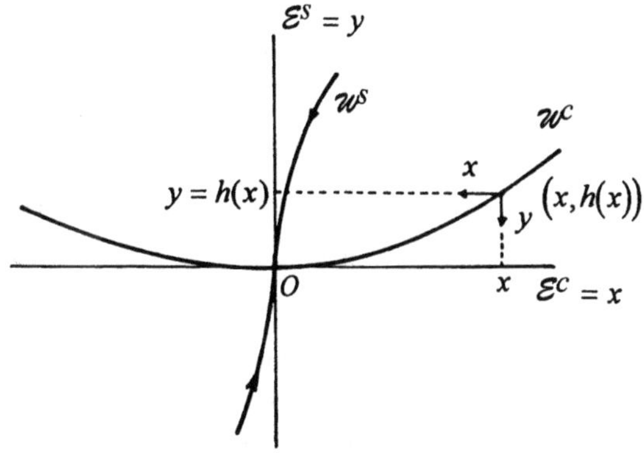


Fig. 5.4

### 5.1.3 Calculation of the Centre Manifold

For convenience, suppose that $\mathcal{E}^u = \{0\}$, $n_u = 0$, $n_s + n_c = n$. Consider the system

$$y' = f(y),\ f \in \mathcal{C}^k(\mathcal{U}),\ k \geq 1,\ \mathcal{U} \in \mathcal{R}^n \text{ is an open set,}$$
$$f(0) = 0,\ A = D_y f(0) \tag{5.7}$$

Linearizing eq. (5.7), we obtain

$$y' = Ay \tag{5.8}$$

The spectrum of $A$ can be divided into two parts, $\sigma_s$ and $\sigma_c$ in this case.

Step 1: take a linear transformation to make $A$ a block diagonal

$$y = T\tilde{y},\quad T^{-1}AT = \begin{bmatrix} B & 0 \\ 0 & C \end{bmatrix} \tag{5.9}$$

Where $B$ is a $n_c \times n_c$ matrix, all $\text{Re}(\sigma) = 0$; $C$ is a $n_s \times n_s$ matrix, all $\text{Re}(\sigma) < 0$, written as $\tilde{y} = \begin{bmatrix} x \\ y \end{bmatrix}$, $x \in \mathcal{E}^c \equiv \mathcal{R}^{n_c}$, $y \in \mathcal{E}^s \equiv \mathcal{R}^{n_s}$. Substituting eq. (5.9) into eq. (5.7), we obtain

$$\begin{bmatrix} x \\ y \end{bmatrix}' = \begin{bmatrix} B & 0 \\ 0 & C \end{bmatrix}\begin{bmatrix} x \\ y \end{bmatrix} + \begin{bmatrix} f(x,y) \\ g(x,y) \end{bmatrix} \tag{5.10}$$

where $f$ and $g$ are non-linear functions after the transformation, and $f(0,0) = g(0,0) = 0$. $Df(0,0) = Dg(0,0) = 0$, $\|\begin{bmatrix} f \\ g \end{bmatrix}\| = o(\|\begin{bmatrix} x \\ y \end{bmatrix}\|^2)$.

Step 2: Calculating $\mathcal{W}^c$

**Theorem 5.2** Suppose that there exists $\mathcal{W}^c$ and it is a tangent to the space $\mathcal{E}^c \equiv x$, then there exists a diffeomorphism $h: \mathcal{R}^n \to \mathcal{W}^s$ or $h: \mathcal{E}^c \to \mathcal{W}^s$, so the centre manifold is $\mathcal{W}^c = \{(x,y) | y = h(x), x \in \mathcal{R}^n\}$, i.e. $\mathcal{W}^c$ is the "graph" of $h$, so that $\mathcal{W}^c$ is $y = h(x)$, which is the phase portrait on $xy$ space, see Fig. 5.4 . Here $h(x)$ satisfies $h(0) = 0$

and $D_x h(0) = 0$, in other words, the centre manifold is a mapping which maps the centre subspace $\mathcal{E}^c$ to the stable subspace $\mathcal{E}^s$.

Calculate $y = h(x)$ near (0,0). Because $h(0) = 0$, therefore $D_x h(0) = 0$. Now calculate $D^2 h(0)$. Substituting $y = h(x)$ into eq. (5.10), then we obtain

$$x' = Bx + f(x, h(x)) \qquad (5.11a)$$

$$y' = D_x h(x) x' = ch(x) + g(x, h(x)) \qquad (5.11b)$$

When $h(x)$ is known, then eq. (5.11a) is the equation of the flow on the centre manifold, i.e. may be obtained from the bifurcation equation. Generally speaking $n_c << n_s$, so eq. (5.11a) transforms the $n$-dimensional system into a $n_c$-dimensional system. This method is much the same as Liapunov—Schmidt (LS) reduction. First we calculate $h(x)$.

Substituting eq. (5.11a) into eq. (5.11b) and taking the $M(h)$ operator, then

$$M(h) = D_x h(x)(Bx + f(x, h(x) - ch(x) - g(x, h(x)) = 0 \qquad (5.12)$$

We then expand $h(x)$ into the formal series

$$h(x) = \sum_{n_i + n_j = 2} a_{n_i n_j} x_i^{n_i} x_j^{n_j} \qquad (5.13)$$

Substituting eq. (5.13) into eq. (5.12) and comparing the coefficients, we can solve $a_{n_i n_j}$ etc., up to order $k-1$.

Step 3: Calculating the equation of flow

Substitute eq. (5.13) into eq. (5.11a), and the solution can be calculated. This solution is the equation of flow on the centre manifolds.

**Theorem 5.3** If a dynamical system defined by eq. (5.11a) at the origin $x = 0$ is locally asymptotically stable (or unstable) then system (5.7) is also locally asymptotically stable (or unstable).

**Theorem 5.4** If a function $\Phi(x): \mathcal{U} \subseteq \mathcal{R}^n \to \mathcal{R}^n$, with $\Phi(0) = 0$, and such that $M(\Phi(x)) = o(|x|^p)$, for some $p > 1$ and $p \le k-1$, as $|x| \to 0$, then $\Phi(x)$ is an asymptotically expression of $h(x)$, that is

$$h(x) = \Phi(x) + o(|x|^p) \text{ as } |x| \to 0 \qquad (5.14)$$

We can calculate $\Phi(x)$ by the formal Taylor series according to the above theorem:

$$\Phi(x) = Ax + Bx^2 + Bx^3 + \cdots \qquad (5.15)$$

where $A$ is a linear operator and $B$ is a bi-linear operator, etc. Substituting eq. (5.15) into eq. (5.12), then $M(\Phi(x)) = 0$. If we expand this and compare the coefficients of $x$, $A, B, C$ can be calculated.

**Corollary of theorem 5.4** If $f \in \mathcal{C}^k(\mathcal{U})$ and $h_1, h_2$ are any two arbitrary centre manifolds, then $h_1(x) - h_2(x) = o(|x|^{k-1})$, and if $f \in \mathcal{C}^\infty(\mathcal{U})$ then all the derivatives of $[h_1(x) - h_2(x)]$ at origin (0) are zero, therefore the Taylor expansion of the centre manifold is unique.

### 5.1.4 Bifurcation of the Lorenz System

Now we will explain the analytical method of the centre manifold theorem through study of the Lorenz system. Consider the Lorenz system

$$\left.\begin{aligned} x' &= \sigma(y-x) \\ y' &= \rho x - y - xz \\ x' &= -\beta z + xy \end{aligned}\right\} x' \overset{def}{=} f(x) = A(x) + g(x) \tag{5.16}$$

where $\sigma, \rho, \beta$ are the parameters greater than zero. Evidently the equilibrium point $\begin{bmatrix} x \\ y \\ z \end{bmatrix} = \begin{bmatrix} 0 \\ 0 \\ 0 \end{bmatrix}$ is the trivial solution. The linearized system of eq. (5.16) is

$$\begin{bmatrix} x \\ y \\ z \end{bmatrix} = \begin{bmatrix} -\sigma & \sigma & 0 \\ \rho & -1 & 0 \\ 0 & 0 & -\beta \end{bmatrix} \begin{bmatrix} x \\ y \\ z \end{bmatrix} \tag{5.17}$$

and its eigenvalues $\lambda_1 = -\beta$ and $\lambda_2, \lambda_3$ satisfy $\lambda^2 + (1+\sigma)\lambda + \sigma(1-\rho) = 0$

Case 1. Take $\rho = 1$, then the three eigenvalues are: $\lambda_1 = -\beta$, $\lambda_2 = 0$, $\lambda_3 = -(1+\sigma)$. In this case the bifurcation will occur. The eigenvectors can be calculated by means of the following equations: $\begin{bmatrix} \sigma+\lambda & -\sigma & 0 \\ -1 & 1+\lambda & 0 \\ 0 & 0 & \beta+\lambda \end{bmatrix} \begin{bmatrix} x \\ y \\ z \end{bmatrix} = \begin{bmatrix} 0 \\ 0 \\ 0 \end{bmatrix}$.

$\lambda_1$'s corresponding eigenvector is $(0,0,1)^T$, $\lambda_2$'s corresponding eigenvector is $(1,1,0)^T$, and $\lambda_3$'s corresponding eigenvector is $(\sigma,-1,0)^T$. The stable subspace $\mathcal{E}^s$ is spanned by the eigenvectors of $\lambda_1$ and $\lambda_3$, the centre subspace $\mathcal{E}^s$ is spanned by the eigenvector of $\lambda_2$.

Step one. Take linear transformation $T(u,v,w)^T$ to eigenbasis $(x,y,z)^T$, i.e.

$$\begin{bmatrix} x \\ y \\ z \end{bmatrix} = u \begin{bmatrix} 1 \\ 1 \\ 0 \end{bmatrix} + v \begin{bmatrix} \sigma \\ -1 \\ 0 \end{bmatrix} + w \begin{bmatrix} 0 \\ 0 \\ 1 \end{bmatrix} \equiv T \begin{bmatrix} u \\ v \\ w \end{bmatrix} \equiv TU \tag{5.18a}$$

$$T = \begin{bmatrix} 1 & \sigma & 0 \\ 1 & -1 & 0 \\ 0 & 0 & 1 \end{bmatrix}, \quad T^{-1} = \begin{bmatrix} \dfrac{1}{1+\sigma} & \dfrac{1}{1+\sigma} & 0 \\ \dfrac{1}{1+\sigma} & \dfrac{1}{1+\sigma} & 0 \\ 0 & 0 & 1 \end{bmatrix} \tag{5.18b}$$

Substituting eq. (5.18a) into eq. (5.16), we obtain $u' = T^{-1}ATu + T^{-1}g(Tu)$.

Substituting eq. (5.18b) into the above equation, we obtain

$$\begin{bmatrix} u \\ v \\ w \end{bmatrix}' = \begin{bmatrix} 0 & 0 & 0 \\ 0 & -(1+\sigma) & 0 \\ 0 & 0 & -\beta \end{bmatrix} \begin{bmatrix} u \\ v \\ w \end{bmatrix} + \begin{bmatrix} -\dfrac{\sigma}{1+\sigma}(u+\sigma v)w \\ \dfrac{\sigma}{1+\sigma}(u+\sigma v)w \\ (u+\sigma v)(u-v) \end{bmatrix} \tag{5.19}$$

The eigenvalues do not change for the variable $u$, and its eigenvectors are: $(0,0,1)^T$ to $\lambda_1 = -\beta$; $(1,0,0)^T$ to $\lambda_2 = 0$; $(0,1,0)^T$ to $\lambda_3 = -(1+\sigma)$, so that

$$\mathcal{E}^c = u\text{–axis}, \quad \mathcal{E}^s = \text{span}\left\{\begin{bmatrix} 0 \\ v \\ w \end{bmatrix}\right\}, \quad \mathcal{E}^u = \phi$$

The linear parts of eq. (5.19) have been decoupled:

$$u' = -\frac{\sigma}{1+\sigma}(u+\sigma v)w \tag{5.19a}$$

$$\begin{bmatrix} v \\ w \end{bmatrix}' = \begin{bmatrix} -(1+\sigma) & 0 \\ 0 & -\beta \end{bmatrix} \begin{bmatrix} v \\ w \end{bmatrix} + \begin{bmatrix} \dfrac{1}{1+\sigma}(u+\sigma v)w \\ (u+\sigma v)(u-v) \end{bmatrix} \tag{5.19b}$$

Centre manifold $\mathcal{W}^c$ is the curve tangent to $\mathcal{W}^c$.

Step two. Calculation of the equation of $\mathcal{W}^c$

When $\mathcal{W}^c$ is tangent to $\mathcal{W}^c$, the functions $h_1(u)$ and $h_2(u)$ deviating from the axis $u$ to the axes $v$ and $w$ should be calculated at the neighbourhood of 0.

Centre manifold

$$\mathcal{W}^c = \left\{\begin{bmatrix} u \\ v \\ w \end{bmatrix} \middle| \begin{bmatrix} v \\ w \end{bmatrix} = h(u) = \begin{bmatrix} h_1(u) \\ h_2(u) \end{bmatrix}\right\} \tag{5.20}$$

where $\underline{h}(0) = 0$, $D\underline{h}(0) = 0$, $h{:}\,\mathcal{R} \to \mathcal{R}^2$ is a diffeomorphism at $\underline{\mathbf{0}}$. Now compute $\underline{h(u)}$ by the formal power series method.

Suppose that
$$\begin{bmatrix} v \\ w \end{bmatrix} = h(u) = \begin{bmatrix} a_1u^2 + b_1u^3 + c_1u^4 + \cdots \\ a_2u^2 + b_2u^3 + c_3u^4 + \cdots \end{bmatrix} \tag{5.21}$$

Substituting eq. (5.21) into eq. (5.19b), we obtain

$$\begin{bmatrix} h'_1(u) \\ h'_2(u) \end{bmatrix}\left(\frac{-\sigma}{1+\sigma}[u+\sigma h_1(u))h_2(u)]\right)$$
$$= \begin{bmatrix} -(1+\sigma)h_1(u) + \dfrac{1}{1+\sigma}(u+\sigma h_1(u))h_2(u) \\ -\beta h_2(u) + (u+\sigma h_1(u))(u-h_1(u)) \end{bmatrix} \tag{5.22}$$

where $h_i'(u) = \dfrac{\partial h_i(u)}{\partial(u)} = \dfrac{\mathrm{d}h_i}{\mathrm{d}u} = 2a_i u + 3b_i u^2 + 4c_i u^3 + \cdots,\ i = 1,2$. Substituting eq. (5.21) into eq. (5.22) and comparing the coefficients, we obtain $V$-equations

$$o(u^2):\ 0 = -(1+\sigma)a_1,\ \ \therefore a_1 = 0$$

$$o(u^3):\ 0 = -(1+\sigma)b_1 + \left(\frac{1}{1+\sigma}\right)a_2$$

$$o(u^4):\ 2\left(\frac{-\sigma}{1+\sigma}\right)a_1 a_2 = -(1+\sigma)c_1 + \left(\frac{\sigma}{1+\sigma}\right)a_1 a_2 + \left(\frac{1}{1+\sigma}\right)b_2$$

......

and $W$-equations

$$o(u^2):\ 0 = -\beta a_2 + 1,\ \therefore a_2 = 1/\beta$$

$$o(u^3):\ 0 = -\beta b_2 + \sigma a_1 - a_1,\ \therefore b_2 = 0$$

$$o(u^4):\ 2\left(\frac{-\sigma}{1+\sigma}\right)a_2^2 = -\beta c_2 + \sigma c_2 + \sigma c_1 - c_1$$

$$\therefore a_1 = 0,\ a_2 = 1/\beta,\ b_1 = \frac{1}{(1+\sigma^2)\beta},\ b_2 = 0,$$

$$c_1 = 0,\ c_2 = \frac{2\sigma}{(1+\sigma)\beta^3}$$

The centre manifold in order 4 is

$$\mathcal{W}^c = \left\{ \begin{bmatrix} u \\ v \\ w \end{bmatrix} \middle| \begin{bmatrix} v \\ w \end{bmatrix} = \begin{bmatrix} h_1(u) = \dfrac{u^3}{(1+\sigma)^2\beta} + o(u^5) \\ h_2(u) = \dfrac{1}{\beta}u^2 + \dfrac{2\sigma}{(1+\sigma)\beta^3} + o(u^5) \end{bmatrix} \right\} \tag{5.23}$$

Step three. Substituting $h_1$ and $h_2$ into eq. (5.19a), we can determine the equation of the flow on the centre manifold:

$$u' = -\frac{\sigma}{1+\sigma}[u + \sigma h_1(u)]h_2(u) = \frac{1}{\beta}\left(-\frac{\sigma}{1+\sigma}\right)\left[u^3 + \frac{\sigma(\beta + 2(1+\sigma)}{2\beta(1+\sigma)^2})u^5\right] + o(u^6) \tag{5.24}$$

The coefficient of $u^3$ is negative, so the flow on the centre manifold is asymptotically stable, and point (0,0,0) is also asymptotically stable by theorem 5.2.

Case 2. Consider $\rho$ near 1. Now, the eigenvalues are: $-\beta$ and

$-\dfrac{1+\sigma}{2} \pm \sqrt{\left(\dfrac{1+\sigma}{2}\right)^2 - \sigma(1-\rho)}$. Let $\rho = 1+\varepsilon, \varepsilon$ be a small parameter, and substituting $\rho$ into the expression for the eigenvalues, we obtain

$$\lambda = -\frac{(1+\sigma)}{2} \pm \frac{1}{2}\sqrt{(1+\sigma)^2 + 4\sigma\varepsilon} = \frac{(1+\sigma)}{2}\left\{-1 \pm \sqrt{1 + \frac{4\sigma\varepsilon}{(1+\sigma)^2}}\right\}$$

$$= \frac{(1+\sigma)}{2}\left\{-1 \pm \left(1 + \frac{2\sigma\varepsilon}{(1+\sigma)^2} + \cdots\right)\right\}$$

$$\therefore \begin{cases} \lambda_2 = \dfrac{\lambda\varepsilon}{(1+\sigma)} + o(\varepsilon^2) \\ \lambda_3 = -(1+\sigma) + \dfrac{\sigma\varepsilon}{(1+\sigma)} + o(\varepsilon^2) \end{cases} \tag{5.25}$$

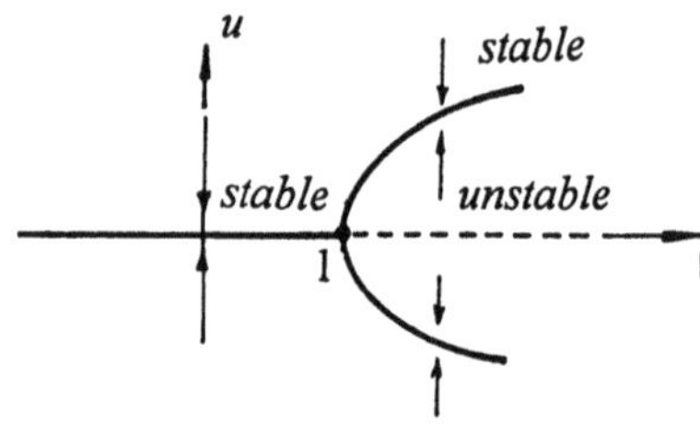

Fig. 5.5

According to the above analysis we come to the following conclusion. If origin (0,0,0) is: $\rho < 1$, i.e. $\varepsilon < 0$ then $\lambda_1$, $\lambda_2$ and $\lambda_3$ all have negative real parts, and the origin is hyperbolic and stable. There exists a three-dimensional stable flow; for $\rho > 1$, i.e. $\varepsilon > 0$, the origin is hyperbolic and unstable saddle, with a two-dimensional stable flow and a one-dimensional unstable flow; for $\rho = 1$, i.e. $\varepsilon = 0$ then $\lambda_2 = 0$, the origin is non-hyperbolic, a vague attractor and with a stable flow. When $\rho$ varies from less than 1 to 1, then to greater than 1, the flow near the origin varies from hyperbolic to non-hyperbolic, then to hyperbolic, and bifurcation occurs. See Fig. 5.5.

### 5.1.5 Centre Manifold and Bifurcation of Parameterised Systems

Consider differential equations with parameters

$$u' = F(u,\mu), \ \mu \in \mathcal{R}^p \tag{5.26}$$

$F: \mathcal{R}^n \times \mathcal{R}^p \to \mathcal{R}^n$, $F \in \mathcal{C}^K$, $K > 1$. Assume $F(0,0) = 0$. Let $A = D_u F(0,0)$. Suppose $A$ has a centre subspace $\mathcal{E}^c$ with $n_c$ dimension, and a stable subspace $\mathcal{E}^s$ with $n_s$ dimension, and $n_c + n_s = n$.

Take the linear transformation to convert $A$ into the form of a diagonal block $\begin{bmatrix} B & 0 \\ 0 & C \end{bmatrix}$ where $B$ has $n_c$ eigenvalues with zero real parts, and $C$ has $n_s$ eigenvalues with negative real parts at $\mu = 0$. Then eq. (5.26) becomes

$$\begin{cases} x' = B(\mu)x + f(x,y,\mu) \\ y' = C(\mu)y + g(x,y,\mu) \end{cases} \tag{5.27}$$

For $\mu = 0$, eq. (5.27) is the same as that stated above. For $\mu \neq 0$ there are two approaches: (1) prove the parameterised centre manifold theorem; (2) redefine $\mu$ as a "state variable" and consider the system in $\mathcal{R}^{n+p}$, but here $\mu' = 0$. Now we use the second method, and eq. (5.26) becomes

$$\begin{cases} u' = F(u,\mu) \\ \mu' = 0 \end{cases} \tag{5.28}$$

Equation. (5.28) has $n + p$ equations with $n + p$ state variables. We apply the centre manifold theorem described in the previous discussion, then eq. (5.28) becomes

$$\left.\begin{aligned} x' &= (B(0) + B_1(\mu) + \cdots)x + f(x, y, \mu) \\ y' &= (C(0) + C_1(\mu) + \cdots)y + g(x, y, \mu) \end{aligned}\right\} \qquad \mu' = 0 \tag{5.29}$$

where $B_1(0) = 0, C_1(0) = 0$. The linearized system of eq. (5.29) is

$$\begin{bmatrix} x \\ \mu \\ y \end{bmatrix}' = \begin{bmatrix} B & 0 & 0 \\ 0 & 0 & 0 \\ 0 & 0 & C \end{bmatrix} \begin{bmatrix} x \\ \mu \\ y \end{bmatrix} \tag{5.30}$$

where $\begin{bmatrix} B & 0 \\ 0 & 0 \end{bmatrix}$ is an $(n_c + p) \times (n_c + p)$ matrix, all the eigenvalues having zero real parts. $C$ is an $n_s \times n_s$ matrix, all the eigenvalues having negative real parts. In this way we can apply the centre manifold theorem as before. For example, consider a parameterised system:

$$x' = \mu x - x^2, \; y' = -y + x^2, \; \mu' = 0 \tag{5.31}$$

where the third equation is an additional equation after $\mu$ has become a state variable. Now $\mu x$ is a "quadratic". So the $x$ equation has zero eigenvalues for all $\mu$. Equation (5.31) will have a two-dimensional centre subspace and a one-dimensional stable subspace for all $\mu$ that are nearly close to 0 (not just $\mu$=0):

$$\mathcal{E}^c = \left\{ \begin{bmatrix} x \\ \mu \\ 0 \end{bmatrix} \right\}, \quad \mathcal{E}^s = \left\{ \begin{bmatrix} 0 \\ 0 \\ y \end{bmatrix} \right\}$$

Then eq. (5.31) can be written in the form of a matrix:

$$\begin{bmatrix} x \\ \mu \\ y \end{bmatrix}' = \begin{bmatrix} 0 & 0 & 0 \\ 0 & 0 & 0 \\ 0 & 0 & -1 \end{bmatrix} \begin{bmatrix} x \\ \mu \\ y \end{bmatrix} + \begin{bmatrix} \mu x - x^2 \\ 0 \\ x^2 \end{bmatrix} \tag{5.32}$$

Assume a two-dimensional centre manifold

$$\mathcal{W}^c = \{[x \quad \mu \quad y]^T \,|\, y = h(x.\mu), \; h(0.0) = 0. Dh(0.0) = 0$$

Now compute $h(x, \mu)$ by power series:

$$y = h(x, \mu) = ax^2 + b\mu x + c\mu^2 + dx^3 + \cdots \tag{5.33a}$$

Substituting eq. (5.33a) into the $y$ equation of eq. (5.31), then obtain

$$y' = h_x x' + h_\mu \mu' = -h(x, \mu) + x^2$$

Substituting $x'$ and $\mu'$ into the above equation, and comparing the coefficients, we get

$$o(x^2) \quad 0 = -a + 1, \quad \therefore a = 1; \; o(\mu x) \quad 0 = -b, \quad \therefore b = 0;$$
$$o(\mu^2) \quad 0 = c, \quad \therefore c = 0; \quad o(x^3) \quad -2a = d, \quad d = 2$$

So the equation of centre manifold is

$$y = x^2 + 2x^3 + \cdots \tag{5.33b}$$

If we substitute eq. (5.33b) into the $x',\mu'$ equations of eq. (5.31), we obtain the differential equations of flow restricted to the centre manifold:

$$x' = \mu x - x^2,\ \mu' = 0 \tag{5.34}$$

From the above equations we know that the trivial solutions of the flow are

$$\begin{cases} x = 0 \\ \mu = \text{const} \end{cases} \qquad \begin{cases} \mu = x \\ \mu = \text{const} \end{cases} \tag{5.35}$$

The phase portrait on the two-dimensional centre manifold is shown in Fig. 5.6. If $\mu$ is taken as a control variable and $x$ remains a state variable, then Fig. 5.6 is a bifurcation diagram. The solution is stable on three-dimensional subspace, because $\mathcal{E}^u$ is a null subspace and $y$ is asymptotically stable. If $\mu = 0$ in eq. (5.31), there is a two-dimensional system

$$\begin{cases} x' = -x^2 \\ y' = -y + x^2 \end{cases} \tag{5.36}$$

The eigenvalues at the origin are 0 and $-1$ whose corresponding eigenvectors are $(1,0)^T$ and $(0,1)^T$, so their centre subspace $\mathcal{E}^c$ and stable subspace $\mathcal{E}^s$ are

$$\mathcal{E}^c = \left\{ \begin{bmatrix} \alpha \\ 0 \end{bmatrix} \middle| \alpha \in \mathcal{R} \right\}, \qquad \mathcal{E}^s = \left\{ \begin{bmatrix} 0 \\ \beta \end{bmatrix} \middle| \beta \in \mathcal{R} \right\} \tag{5.37}$$

Seek a one-dimensional centre manifold. From eq. (5.36) we obtain

$$x(t) = \frac{x_0}{1 + x_0 t},\ y(t) = y_0 e^{-t} + x_0^2 \int_0^t \frac{e^{-(t-s)}}{1 + x_0 s} \mathrm{d}s \tag{5.38}$$

where $(x_0, y_0) = (x(0), y(0))$. See Fig. 5.7 for the phase portrait on the $x-y$ plane.

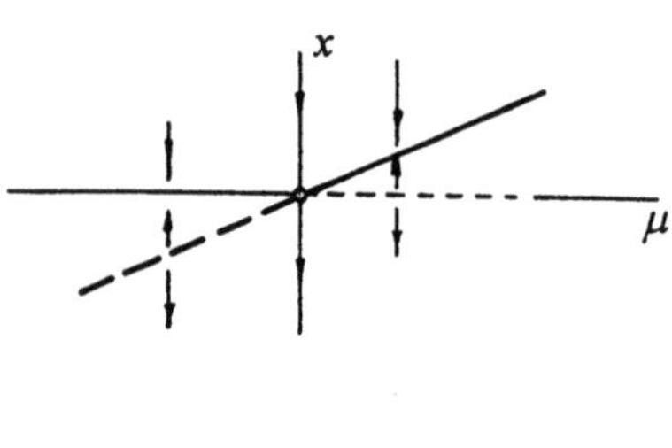


Fig. 5.6

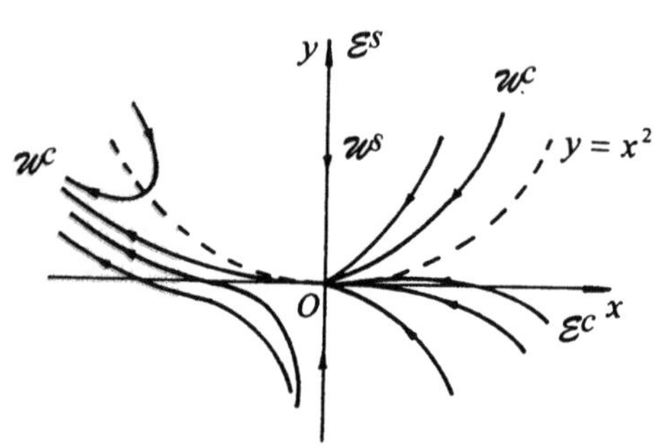


Fig. 5.7

Suppose a one-dimensional centre manifold is

$$\mathcal{W}^c = \left\{ \begin{bmatrix} x \\ y \end{bmatrix} \middle| y = h(x),\ h{:}\,\mathcal{R} \to \mathcal{R},\ h(0) = 0,\ Dh(0) = 0 \right\} \tag{5.39}$$

Calculate $h(x)$ by the power series method, and suppose that

$$y = h(x) = ax^2 + bx^3 + cx^4 + \cdots \tag{5.40}$$

If we substitute eq. (5.40) into the second equation of eq. (5.36), and $x'$ is substituted by the first equation of eq. (5.36) we obtain

$$x^2(2ax + 3bx^2 + 4cx^3 + \cdots) = (ax^2 + bx^3 + cx^4 + \cdots) - x^2$$

To compare the same order terms on both side of the above equation we get

$o(x^2)$ $\quad 0 = a - 1, \quad \therefore a = 1$

$o(x^3)$ $\quad 2a = b, \quad \therefore b = 2!$

$o(x^4)$ $\quad 3b = c, \quad \therefore c = 3!$

...... ......

It's easy to see that the coefficients of $x^n$ are $(n-1)!$. Substituting the above coefficients into eq. (5.40), we get

$$y = h(x) = x^2 + 2x^3 + 3!x^4 + \cdots + K!x^{K+1} + \cdots \tag{5.41}$$

Thus is divergent for any $x \neq 0$. $\sum_{K=1}^{\infty} K!x^{K+1}$ is meaningless. However, we can prove that eq. (5.41) is asymptotic and valid if $x$ is small enough. If $f \in \mathcal{C}^1$, there exists $h$ and $h \in \mathcal{C}^{-1}$; and $h \in \mathcal{C}^l$ for any l if $f \in \mathcal{C}^\infty$. Applying Taylor's theorem to $h$ for $1 = K + 1$, we get $h(x) = x^2 + 2x^3 + 3!x^4 + \cdots + (K-1)!x^K + R_K(x)$. Where $R_K(x) = h^{(K+1)}(\varsigma)x^{K-1}/(K+1)!, (\varsigma \in [0, x])$ is called the Taylor remainder. So in a given interval $|x| < \delta$, it follows that $|\varsigma| \leq \delta$. Because $h^{(K+1)}$ is continuous on $[-\delta, \delta]$, therefore $|h^{(K+1)}(\varsigma)| \leq B$ (constant); for $x \in [-\delta, \delta]$, we get

$$|R_K(x)| \leq B|x|^{K+1}/(K+1)! = B|x|^{K+1},$$

Therefore $\quad h(x) = x^2 + 2x^3 + 3!x^4 + \cdots + (K-1)!x^K + o(x^{K+1}) \quad (5.42)$

The above expression is an asymptotic expression rather than a series when $x \to 0$. This example illustrates that $h$ may not be analytic even though $f$ is.

Another approach is used to calculate $h(x)$. Substituting eq. (5.40) into the second equation of eq. (5.36) where $x'$ is substituted by the first equation of eq. (5.36), then $x^2 \dfrac{dh}{dx} = h - x^2$, or $\dfrac{dh}{dx} - \dfrac{1}{x^2}h = -1$. Multiplying both sides of the above expression by $e^{1/x}$, we get $\dfrac{d}{dx}\left[he^{\frac{1}{x}}\right] = -e^{\frac{1}{x}}$. The integral of the above expression is $\quad h(x) = e^{-\frac{1}{x}}\left[C - \int e^{\frac{1}{x}}dx\right]$. When $x \to 0^+$, then $e^{-\frac{1}{x}} \to 0$; and when $x \to 0^-$, then $e^{-\frac{1}{x}} \to \infty$. So $h(x)$ is not continuous at $x = 0$.

## 5.2 Saddle-Node Bifurcation

Saddle-node bifurcation is the simplest case of the equilibrium state. Consider the simplest one-dimensional example:

$$x' = f(x,\mu), \quad f: \mathcal{R} \times \mathcal{R} \to \mathcal{R} \tag{5.43}$$

Without loss of generality, suppose that (0,0) is an equilibrium point, i.e. $f(0,0) = 0$. There is a necessary condition for bifurcation, $D_x(0,0) = 0$. Its eigenvalue is zero because the system is one-dimensional. The Taylor series of vector field $f$ is

$$f(x,\mu) = a\mu + bx^2 + c\mu x + \cdots \tag{5.44}$$

The simplest case: suppose $a, b \neq 0$ [$a = 1,\ b = -1$ after scaling]. Because $D_\mu(0,0) = a \neq 0$, $\mu = \mu(x)$ can be solved from $f(x,\mu) = 0$ by the implicit function theorem, i.e. let eq. (5.44) equal zero, and we get

$$\mu = -\frac{b}{a}x^2 - \frac{c}{a}\mu x + \cdots = -\frac{b}{a}x^2 + o(x^3) = x^2 + o(x^3) \tag{5.45}$$

The bifurcation diagram is depicted in Fig. 5.8.

Now we analyse the stability of $x$. Because $x' = a\mu + bx^2 + c\mu x + \cdots$, when $\mu = 0$, if $b = -1$, $x' = -x^2 + \cdots$, so $x$ is a decreasing function whether $x$ is positive or negative; if $\mu < 0, a = 1, b = -1$, then $x$ is a decreasing function; if $\mu > 0, \mu < x^2$, then $x$ is a decreasing function; if $\mu > 0, \mu > x^2$, then $x$ is an increasing function, see Fig. 5.8.

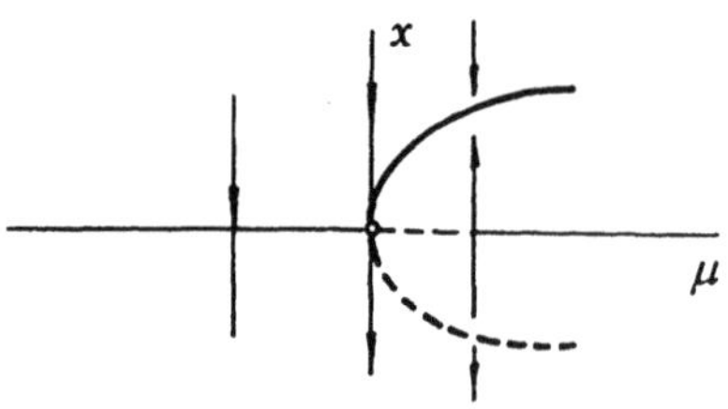

Fig. 5.8

Now let us come to the $n$-dimensional system saddle—node bifurcation

$$y' = f(y,\mu), \quad f: \mathcal{R}^n \times \mathcal{R} \to \mathcal{R}^n, f \in \mathcal{C}^p, p \geq 2 \tag{5.46}$$

Suppose that the equilibrium point is (0,0), i.e. $f(0,0) = 0, A = D_y(0,0)$ is an $n \times n$ matrix. $A$ has a simple zero eigenvalue and the other eigenvalues have negative real parts. So $y' = Ay$ has a one-dimensional $\mathcal{E}^c$, an $(n-1)$-dimensional $\mathcal{E}^s$ and no $\mathcal{E}^u$, so the linear part of eq. (5.46) can be decoupled.

Let $\varphi_1$ and $\psi_1^T$ be the right and left eigenvectors of $A$ corresponding to zero eigenvalue:

$$\psi_1^T A = 0 = A\varphi_1 \tag{5.47}$$

Let $\varphi_2, \cdots, \varphi_n$ be the bases of $\mathcal{E}^s$, and $\varphi_1$ the basis of $\mathcal{E}^c$, $\mathcal{E}^c \oplus \mathcal{E}^s = \mathcal{R}^n$. Let

$$M = (\varphi_1, \varphi_2, \cdots, \varphi_n) \tag{5.48}$$

where every $\varphi_i$ is a column vector $(\varphi_{i1}, \varphi_{i2}, \cdots, \varphi_{in})^T$, then

$$M^{-1}AM = \begin{bmatrix} 0 & 0 & \cdots & 0 \\ 0 & & & \\ \cdots & & B & \\ 0 & & & \end{bmatrix}$$

where $B$ is an $(n-1)\times(n-1)$ matrix, and all its eigenvalues have real parts. Take the transformation $y = M\begin{bmatrix} x \\ v \end{bmatrix}$, $x \in \mathcal{R}$, $v \in \mathcal{R}^{n-1}$ (5.49)

It is not difficult to prove the following orthogonal relationship between $\psi^T$ and $\varphi$: $\psi_1^T\varphi_j = 0$, $j = 2,3,\cdots,n$, $\psi_1^T\varphi_1 \neq 0$ after normalisation so that $\psi_1^T\varphi_1 = 1$ (5.50)

Let $\psi_1^T$ be the first row of $M^{-1}, M^{-1}M = I$. Then the first row of the product $\begin{bmatrix} \psi_1^T & 0 & \cdots & 0 \\ \cdots & \cdots & \cdots & \cdots \\ \cdots & \cdots & \cdots & \cdots \end{bmatrix}(\varphi_1,\varphi_2,\cdots,\varphi_n)$ is $(1,0,\cdots,0)$. Substituting eq. (5.49) into eq. (5.46), we obtain $\begin{bmatrix} x \\ y \end{bmatrix}' = \begin{bmatrix} 0 & 0 & \cdots & 0 \\ 0 & & & \\ \cdots & & B & \\ 0 & & & \end{bmatrix}\begin{bmatrix} x \\ v \end{bmatrix} + M^{-1}f\left(M\begin{bmatrix} x \\ v \end{bmatrix},\mu\right)$

According to the treatment method of the parameterised system, then

$$x' = F(x,v,\mu),\ v' = Bv + G(x,v,\mu),\ \mu' = 0 \tag{5.51}$$

where $M^{-1}f\left(M\begin{bmatrix} x \\ v \end{bmatrix},\mu\right) = \begin{bmatrix} F \\ G \end{bmatrix}$, $F = \psi_1^T f_1\left(\begin{bmatrix} x \\ v \end{bmatrix},\mu\right)$. The centre Manifold theorem is used to eliminate $v$ using $v = h(c,\mu)$. We substitute this into the first equation of eq. (5.51) and obtain

$$x' = F(x,h(x,\mu),\mu)) = \tilde{F}(x,\mu),\ \mu' = 0 \tag{5.52}$$

Now expand $F(x,h(x,\mu),\mu)$ and $\tilde{F}(x,\mu)$ into the Taylor series at $(x,\mu) = (0,0)$:

$$F(x,v,\mu) = F_\mu(0)\mu + \frac{1}{2}F_{xx}(0)x^2 + F_{x\mu}(0)x\mu + F_{xv}(0)xv + F_{\mu v}(0)\mu v + \cdots \tag{5.53}$$

If we take

$$v = h(x,\mu) = ax^2 + bx\mu + c\mu^2 \tag{5.54}$$

and substitute eq. (5.54) into eq. (5.53), then $xv$ and $\mu v$ become the third-degree term in $x$ and $\mu$. The Taylor series of $\tilde{F}(x,\mu)$ is

$$\tilde{F}(x,\mu) = F_\mu(0)\mu + F_{x\mu}(0)x\mu + \frac{1}{2}F_{xx}(0)x^2 + \frac{1}{2}F_{\mu\mu}(0)\mu^2 + hot \tag{5.55}$$

$v = h$ does not affect the second-degree term of the Taylor series. If $F_\mu(0) \neq 0 \neq F_{xx}\mu(x)$ this can be solved from $\widetilde{F}(x,\mu)$ using the implicit function theorem, and the bifurcation diagram, which is a pitchfork, is shown in Fig. 5.9.

$$\widetilde{F}(x,\mu) = \psi_1^T f_1\left(\begin{bmatrix} x \\ v \end{bmatrix}, \mu\right)$$

$$F_\mu(0) = \psi_1^T f_\mu(0,0) h_\mu(0) + \psi_1^T f_\mu(0,0)$$
$$= \psi_1^T f_\mu(0)$$

$$F_{xx}(0) = \psi_1^T f_{yy}(0,0) M\begin{bmatrix} 1 \\ 0 \end{bmatrix} M\begin{bmatrix} 1 \\ 0 \end{bmatrix}$$
$$= \psi_1^T f_{yy}(0,0)\varphi_1\varphi_1$$

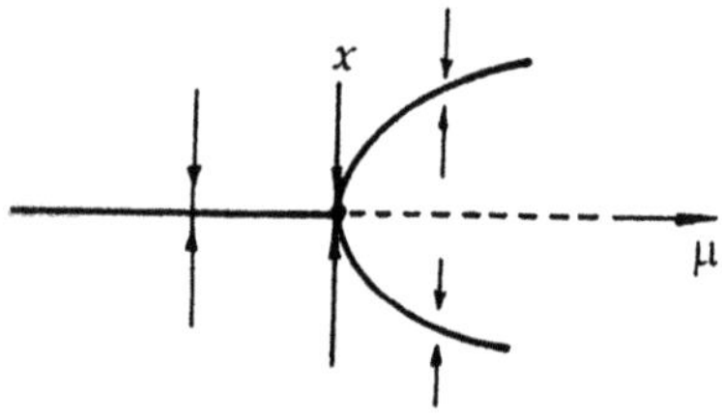


Fig. 5.9

Without any loss of generality, supposing

$F_\mu(0) = 1$, $F_{xx}(0) = -1$ exists as a saddle node bifurcation (see Fig. 5.9).

Phase portraits in $x$—$v$ space are shown in Fig. 5.10.

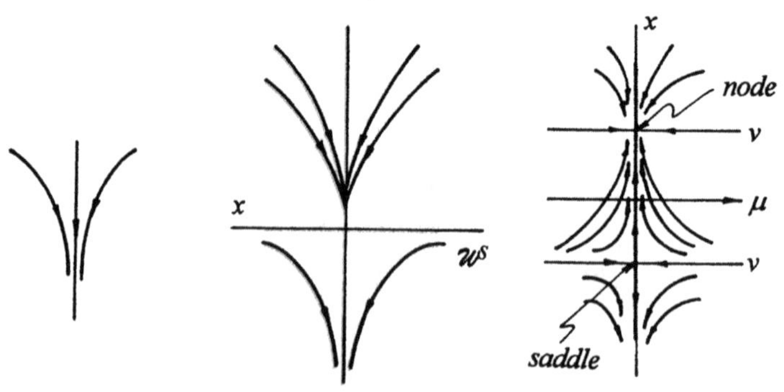


(a) $\mu < 0$ (b) $\mu = 0$ (c) $\mu > 0$

Fig. 5.10

Now we use the centre manifold method to prove other bifurcation theorems.

1. Transcritical Bifurcation (or "Simple Bifurcation")

Consider system (5.46) and use the same notation as employed in the preceding section. Assume $A$ has only a simple zero eigenvalue. If $a = \psi^T D_{y\mu} f(0,0)\varphi \neq 0$ and $2b = \psi^T D_{yy} f(0,0)\varphi\varphi \neq 0$ then the centre manifold equation is

$$x' = ax\mu + bx^2 + hot$$

Let $a = 1$, $b = -1$, then the above equation becomes

$$x' = x\mu - x^2 + hot$$

The bifurcation diagram is depicted in Fig. 5.11.

Other signs of $a$, $b$ are treated in the same way.

2. Pitchfork Bifurcation

Still considering system (5.46), assume that there exists a linear symmetrical operator $S:\mathcal{R}^n \to \mathcal{R}$, $S\varphi = \varphi, S\mathcal{E}^s = \mathcal{E}^s$. Since there is a simple zero eigenvalue, $A\varphi = 0$. The dimension of $\mathcal{E}^s$ is $n-1$. Also assume that $S$ and $f$ are changeable, i.e. $f(Sy,\mu) = S(fy,\mu)$. If $S = -$ , then $f$ is an odd function in $y$. The centre manifold equation can be proved as $x' = F(x+v(x,\mu),\mu)$ and it satisfies

$$F(-x+v(-x,\mu),\mu) = -F(x+v(x,\mu),\mu)$$

As the Taylor expression of $F$ has no even terms, it is required that $a = \psi^T D_{y\mu} f(0,0)\varphi \neq 0$ and $c = \psi^T D_{yyy} f(0,0)\varphi\varphi\varphi \neq 0$. Now the centre manifold equation has the form $x' = ax\mu + cx^3 + hot = x(a\mu + cx^2 + \cdots +$ even terms). And the bifurcation diagram is composed of a trivial solution $x = 0$ and a non-trivial solution $\mu = -\frac{c}{a}x^2 + hot$. They are called either supercritical for $ac < 0$ or subcritical for $ac > 0$ (see Fig. 5.12).

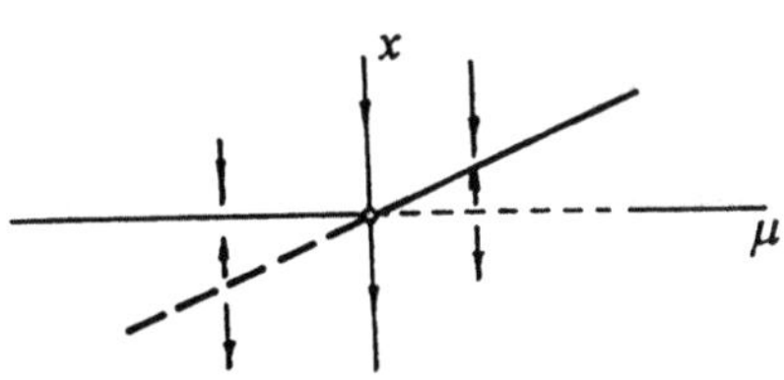


Fig. 5.11

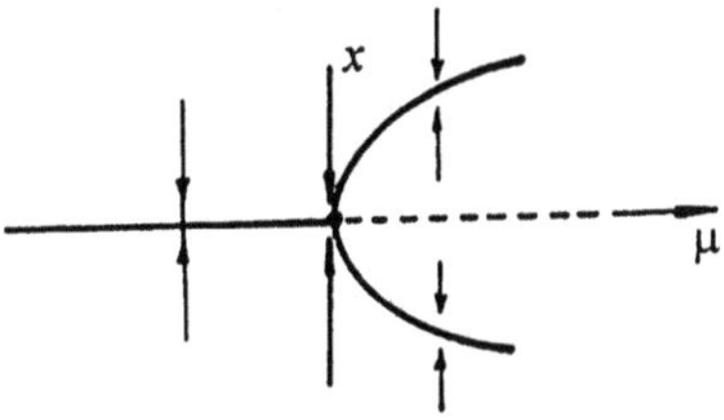


Fig. 5.12

## 5.3 Normal Form of Vector Fields

The normal form of vector fields (or Poincaré normal form, or standard form) is a simplified analytic expression. Such a simplified expression is obtained by transforming the nearly identical non-linear transformation of variables. The qualitative geometric structure can be acquired for every bifurcation of flows by means of analysis of the dynamical character of normal forms. Successful introduction of non-linear transformation of variables to simplify the analytic expression is a very efficient academic idea. Being the basis for the KAM (Kolmogorov—Arnold—Moser) study of quasi-period phenomena, this academic idea is applied to the averaging method for calculating normal forms. The local limit truncated normal form of differential equations is also very beneficial for global bifurcation theory. These normal forms considerably simplify the process of seeking and studying bifurcation problems, and also simplify the proof and analysis of the results obtained. On the other hand, normal form theory is part of global bifurcation theory. This theory is very important in practical application. A lot of work has been involved in determining the relationship between normal form theory and global bifurcation theory. After the initial work of Poincaré, Birkhoff and others, Arnold, Takens, Chow and Iooss have made important developments in normal form theory in recent years. Methods of solving the normal form of vector fields are the matrix representation method, a method based on the representation theory of $sl(2,\mathcal{R})$ and the conjugate adjoint operator method. In this section we will only introduce the matrix representation method.

### 5.3.1 Homological Operator

Consider the following non-linear differential equations:

$$y' = f(y),\ f \in \mathcal{C}^K(\mathcal{R}^n),\ K \text{ large integer} \tag{5..56}$$

We expand the vector field $f(y)$ into a Taylor series

$$f(y) = Ay + B(y,y) + C(y,y,y) + \cdots \tag{5.57}$$

where $A = D_y f(0)$ can be changed into a diagonal block form; $B$ is bi-linear, $B{:}\mathcal{R}^n \times \mathcal{R}^n \to \mathcal{R}^n$; $C$ is tri-linear, $C{:}\mathcal{R}^n \times \mathcal{R}^n \times \mathcal{R}^n \to \mathcal{R}^n$. Now we shall illustrate how to obtain the simplest form of $B(y,y)$ by a nearly identical non-linear transformation of coordinates. Take a near identity change of coordinates

$$y = x + P_2(x), \qquad x \in \mathcal{R}^n \tag{5.58}$$

$$\text{where } P_2(x) = \begin{bmatrix} a_1 x_1^2 + b_1 x_1 x_2 + \cdots \\ a_2 x_1^2 + b_2 x_1 x_2 + \cdots \\ \cdots\cdots\cdots\cdots\cdots\cdots \\ a_n x_1^2 + b_n x_1 x_2 + \cdots \end{bmatrix} \quad x = \begin{bmatrix} x_1 \\ x_2 \\ \cdots \\ x_n \end{bmatrix} \tag{5.59}$$

$P_2(x)$ is a homogeneous second-degree polynomial. If we substitute eq. (5.59) into eq. (5.57), the left side becomes

$$y' = x' + [D_x P_2(x)]x' = [I + D_x P_2(x)]x' \tag{5.60}$$

and the right side

$$f(y) = A(x + P_2(x)) + B(x,x) + hot \tag{5.61}$$

$[I + D_x P_2(x)]^{-1}$ is existent for sufficiently small $x$. According to the binomial theorem of the matrix we get $[I + D_x P_2(x)]^{-1} = I - D_x P_2(x) + [D_x P_2(x)]^2 + \cdots$. Let eq. (5.60) be equal to eq. (5.61),then

$$x' = Ax + AP_2(x) - D_x P_2(x) Ax + B(x,x) + h.o.t. \tag{5.62}$$

Its linear part is the same as that before transformation. The ideal simplified case of the second-order terms is to let them be zero. For this reason, the function $P_2(x)$ must be that quadratic terms can vanish, i.e.

$$D_x P_2(x) Ax - AP_2(x) = B(x,x) \tag{5.63}$$

Equation (5.63) is called a homological equation. Define a homological operator

$$L_A(P) = [D_x P(x)]Ax - AP(x) \tag{5.64}$$

then eq. (5.63) can be written as

$$L_A(P_2) = B \tag{5.65}$$

The properties of the homological operator are: (a) $L_A$ is linear; (b) if $P$ is homogeneous of degree $K$, so is $L_A(P(x))$. All homogeneous terms with a degree less than $K-$ (and including $K-$ ) are the same as those before the transformation of coordinates.

**Definition 5.3** $H_k(\mathcal{R}^n)$ =Space of all homogeneous degree $K$ bases = spanned by every monomial of all $K$-order homogeneous polynomials in $n$ variables:

$$\dim H_k(\mathcal{R}^n) = n \times (n+k-1) \times (n-1) \tag{5.66}$$

Example for $n=2,\ k=2$: $P_2(\mathcal{R}^n) = \begin{bmatrix} a_1 x_1^2 + b_1 x_1 x_2 + c_1 x_2^2 \\ a_2 x_1^2 + b_2 x_1 x_2 + c_2 x_2^2 \end{bmatrix}$

$\dim H_2(\mathcal{R}^2) = 2(2+2-1) \bullet (2-1) = 6$, so the dimension of $H_2(\mathcal{R}^2)$ is 6. Every term of the above two homogeneous degrees-2 polynomial is a basis of $H_2(\mathcal{R}^2)$. The set $\left\{ \begin{bmatrix} 0 \\ x_1^2 \end{bmatrix}, \begin{bmatrix} 0 \\ x_1 x_2 \end{bmatrix}, \begin{bmatrix} 0 \\ x_2^2 \end{bmatrix}, \begin{bmatrix} x_1^2 \\ 0 \end{bmatrix}, \begin{bmatrix} x_1 x_2 \\ 0 \end{bmatrix}, \begin{bmatrix} x_2^2 \\ 0 \end{bmatrix} \right\}$ is a basis in reverse lexicographical order. We note that two-dimensional $\mathcal{R}^2$ becomes six-dimensional $H_2(\mathcal{R}^2)$, so the dimension of $H_2(\mathcal{R}^2)$ is higher than $\mathcal{R}^2$. $H_k(\mathcal{R}^n)$ is always finite-dimensional if $x, k$ is finite: $L_A : H_k(\mathcal{R}^n) \to H_k(\mathcal{R}^n)$. Because mapping $L_A$ is linear, $L_A$ is a "matrix", $(\dim H_k) \times (\dim H_k)$. From eq. (5.65) we know that if $L_A$ is invertible, or $B$ is in the range of $L_A$, then solution $P_2$ is unique.

**The Poincaré lemma about the spectrum of $L_A$**

Suppose that $A$ has eigenvalues $\lambda_1, \lambda_2, \cdots, \lambda_n$ on $\mathcal{R}^n$(containing multiple and complex), then eigenvalues of $L_A$ on $H_k(\mathcal{R}^n)$ are

$$m_1\lambda_1 + m_2\lambda_2 + \cdots + m_n\lambda_n - \lambda_s = \sum_{i=1}^{n} m_i\lambda_i - \lambda_s = < m, \lambda > - \lambda_s \tag{5.67}$$

where $m_1, m_2, \cdots, m_n$ are integers $\geq 0$; $S \in \{1, 2, \cdots, n\}$. (The Theorem will be proved later.)

### 5.3.2 Non-resonant Case

**Theorem 5.5 Poincaré's linearization theorem**

If all $S \in \{1, 2, \cdots, n\}$ and all $m = (m_1, \cdots, m_n)$ satisfy $|m| = \sum_{i=1}^{n} m_i \geq 2$, then there exists a non-resonant condition

$$< m, \lambda > - \lambda_s \neq 0 \tag{5.68}$$

Then eq. (5.65), $L_A P = B$, can be solved for any right side of the equation, where $B \in H_k(\mathcal{R}^n)$. In this case, take a successive transformation

$$y = x + P_k(x),\ k = 2, \cdots,\ l-1 \tag{5.69}$$

and the ordinary differential equation, $y' = f(y)$, $f \in \mathcal{C}^1$, can be transformed into

$$x' = Ax + o(|x|^l) \tag{5.70a}$$

where $P_k(x) \in H_k(\mathcal{R}^n)$. If we ignore the higher term in eq. (5.70a), then

$$x' = Ax \tag{5.70b}$$

If the non-resonant condition (5.68) is satisfied, then the linear differential eq. (5.70b) can replace the non-linear eq. (5.56), i.e. in the non-resonant case, the non-linear equation can be linearized.

**Definition 5.4** (Definition of Resonance for Poincaré Normal Form)

If there exist integers $m_1, m_2, \cdots, m_n \geq 0$ with $|m| = \sum m_i \geq 2$, and there exists

$S \in \{1, 2, \cdots, n\}$, so that $\qquad < m, \lambda > - \lambda_s = 0 \qquad$ (5.71)

then $\lambda_1, \lambda_2, \cdots, \lambda_n$ are called resonant, where $\lambda_i \in \mathcal{C}$.

Note 1. $< m, \lambda >= \lambda_S$ in eq. (5.71) shows that if the combinative frequency $< m, \lambda >$ of the non-linear terms in eq. (5.56) is equal to the $S$th natural frequency, then resonance happens. This kind of resonance is call internal resonance.

Note 2. The Lie bracket (Guckenheimer and Holmes) or Poincaré bracket (Arnold) is: $[P, Q] = (DQ)P - (DP)Q$, where $P$ and $Q$ are vector fields. If we take $Q(x) = Ax$ as in the previous discussion, then $DAx = A$, so

$$[P(x), Ax] = AP(x) - (DP(x))Ax = -L_A(P)$$

If we add a negative sign in front of operator $L_A$, then it is called a Lie bracket. If $P \in H_m(\mathcal{R}^n)$, then $L_A(P) \in H_m(\mathcal{R}^n)$.

Now we prove the Poincaré lemma. Assume that $A$ is diagonalizable. Let $e_1, e_2, \cdots, e_n$ be the eigenvectors corresponding to the eigenvalues $\lambda_1, \lambda_2, \cdots, \lambda_n$; let $x_1, x_2, \cdots, x_n$ be the coordinates corresponding to the eigenbasis; so $x^m e_s$ is also an eigenvector. For example, $x = (x_1, x_2, \cdots, x_n)$, $x^m = x_1^{m_1} \cdot x_2^{m_2} \cdot \cdots \cdot x_n^{m_n}$. This product is a monomial of degree $|m| = \sum_{i=1}^{n} m_i$ and is a scale.

Suppose $P_{m,s}(x) = (0, \cdots, 0, x^m, 0, \cdots, 0)^T = x^m e_s \qquad$ (5.72)

and that monomial $x^m$ is at the $S$ th position.

Note: $\{P_{m,S}(x) \| m| = K, 1 \leq S \leq n\}$ is a basis of $H_k$. So $H_k(\mathcal{R}^n)$ is a space spanned by every monomial of homogeneous polynomial of degree $K$.

$$L_A[P_{m,S}(x)] = \left[\frac{\partial P_{m,S}(x)}{\partial x}\right] Ax - AP_{m,S}(x) \tag{5.73}$$

From eq. (5.72) we know that all elements are zero except the element $x^m$ in the $S$th position in the column $P_{m,S}(x)$, so

$$\left[\frac{\partial P_{m.S}(x)}{\partial x}\right]=\frac{\partial}{\partial x}\begin{bmatrix}0\\ \cdots\\ 0\\ x^m\\ 0\\ \cdots\\ 0\end{bmatrix}=\begin{bmatrix}0 & \cdots & \cdots & 0\\ \cdots & \cdots & \cdots & \cdots\\ 0 & \cdots & \cdots & 0\\ \frac{\partial x^m}{\partial x_1} & \frac{\partial x^m}{\partial x_2} & \cdots & \frac{\partial x^m}{\partial x_n}\\ 0 & \cdots & \cdots & 0\\ \cdots & \cdots & \cdots & \cdots\\ 0 & \cdots & \cdots & 0\end{bmatrix}\leftarrow S\text{th row} \qquad (5.74)$$

For $x^m = x_1^{m_1}x_2^{m_2}\cdots x_n^{m_n}$

and $$\frac{\partial x^m}{\partial x_i}=m_i x_1^{m_1}x_2^{m_2}\cdots x_i^{m_i-1}\cdots x_n^{m} \qquad (5.75)$$

substituting eqs. (5.72), (5.74) and (5.75) into eq. (5.73), we obtain

$$L_A[P_{m.S}(x)]=\left[\frac{\partial P_{m.S}(x)}{\partial x}\right]Ax-AP_{m.S}(x)$$

$$=\begin{bmatrix}0 & \cdots & \cdots & \cdots & 0\\ \cdots & \cdots & \cdots & \cdots & \cdots\\ 0 & \cdots & \cdots & \cdots & 0\\ \cdots & \cdots & m_i(x_1^{m_1}\cdots x_i^{m_i-1}\cdots x_n^{m_n}) & \cdots & \cdots\\ 0 & \cdots & \cdots & \cdots & 0\\ \cdots & \cdots & \cdots & \cdots & \cdots\\ 0 & \cdots & \cdots & \cdots & 0\end{bmatrix}\begin{bmatrix}\lambda_1x_1\\ \lambda_2x_2\\ \cdots\\ \lambda_Sx_S\\ \cdots\\ \lambda_nx_n\end{bmatrix}=\begin{bmatrix}\lambda_1 & & & & & 0\\ & \lambda_2 & & & & \\ \cdots & \cdots & \cdots & \cdots & \cdots & \cdots\\ & & & \lambda_S & & \\ & & & & \cdots & \\ & & & & & \lambda_n\end{bmatrix}\begin{bmatrix}0\\ \cdots\\ 0\\ x^m\\ 0\\ \cdots\\ 0\end{bmatrix}$$

$$=(m_1x^m\lambda_1+m_2x^m\lambda_2+\cdots+m_nx^m\lambda_n)e_s-x^m\lambda_se_s$$

$$=\left(\sum_{i=1}^{n}m_i\lambda_i-\lambda_S\right)x^me_S=[<m.\lambda>-\lambda_S]x^me_s \qquad (5.76)$$

So $<m,\lambda>-\lambda_S$ are the eigenvalues of $L_A$, and $x^me_S$ are the eigenvectors of $L_A(P_{m,S}(x))$. The Poincaré lemma has been proved. The above method can be extended to non-diagonal (also complex) $A$ by means of the Jordan Normal Form. The Poincaré theorem can be proved by the mathematical induction method. We have proved that it is true for $k=2$. If it is also true for $k=2,3,\cdots,r-1$, then we verify that it must also be true for $k=r$ i.e.

$$y'=Ay+V_r(y)+hot \qquad (5.77)$$

where $V_r\in H_r(\mathcal{R}^n)$. To eliminate $V_r$, let $y=x+P_r(x)$, and substitute this into eq. (5.77), so that $x'+\frac{\partial P_r(x)}{\partial x}x'=Ax+AP_r(x)+V_r(x+P_r(x))+hot$, or

$$x'=\left[I+\frac{\partial P_r(x)}{\partial x}\right]^{-1}[Ax+AP_r(x)+V_r(x+P_r(x))+hot]$$

$$=Ax-\frac{\partial P_r(x)}{\partial x}Ax+AP_r(x)+V_r(x+P_r(x))+hot$$

To eliminate $_r(y)$ in eq. (5.77), let

$$L_A[P_r(x)] = V_r(x) \tag{5.78}$$

The above equation is solvable for all $V_r$, if $L_A$ is invertible. Original non-linear differential eq. (5.56) can be made linear after a series of nearly identical transformations of coordinates, if all the eigenvalues are not equal to zero and the non-resonant condition (5.68) is satisfied, i.e. $L_A$ is invertible. If $f(y)$ in eq. (5.56) is analytic or $\mathscr{C}^\infty$ then the limit transformation for $K \to \infty, y = x + P^K(x)$ is analytic and eq. (5.56) becomes $x' = Ax$ which is just linear.

**The Poincaré resonant relationship and bifurcation problems**

Case 1. If the system has only one simple zero eigenvalue, for example $\lambda_1 = 0$, then eq. (5.71) $m\lambda_1 = \lambda_1 = (m-1)\lambda_1 = 0$ is satisfied. Because $\lambda_1 = 0, m-1 \neq 0$. According to the definition $\sum m = m \geq 2$, so the monomial power of bifurcation equation is $m \geq 2$. The form of the bifurcation equation is probably $z' = c_1 z^2 + c_2 z^3$, so simple bifurcation may occur from the above equation.

Case 2. The system has a pair of simple purely imaginary roots $\lambda_1 = i\omega = -\lambda_2$, the and resonant relationship $m_1\lambda_1 + m_2\lambda_2 - \lambda_1 = m_1 i\omega - m_2 i\omega - i\omega = 0$ is satisfied. Eliminating $i\omega$, we obtain

$$m_1 - m_2 - 1 = 0 \tag{a}$$

and $$\sum m = m_1 + m_2 \geq 0 \tag{b}$$

From (a) and (b) we know that $m_1$ and $m_2$ need to satisfy $\begin{bmatrix} m_1 \\ m_2 \end{bmatrix} = \begin{bmatrix} 2,3,4,\cdots \\ 1,2,3,\cdots \end{bmatrix}$. The possible form of the bifurcation equation is $z' = i\omega z + c_1 z^2 \bar{z} + c_2 z^3 \bar{z}^2 + \cdots$. We have found that bifurcation may occur if the Poincaré resonant condition is satisfied only.

### 5.3.3 Resonant Case—Poincaré Normal Form (or Poincaré—Birkhoff Normal Form)

If $A$ has eigenvalues $\lambda_1, \lambda_2, \cdots, \lambda_n$ which satisfy the Poincaré resonant relationship (5.71)

$$< m, \lambda > - \lambda_S = \left( \sum_{i=1}^{n} m_i \lambda_i \right) - \lambda_S = 0$$

then $P_{m,S}(x) \in H_m(\mathscr{R}^n)$ is the Poincaré resonant case for $A$, where the definition of $m, S$ is the same as before. In this case, homological eq. (5.78) becomes

$$L_A(P_k(x)) = V_k(x)$$

Because $(L_A)^{-1}$ does not exist, we cannot obtain $P_k(x)$ to eliminate the $V_k(x)$ term of the vector field in the resonant case. This means that $V(x) \notin \mathscr{R}(L_A)$, and we can use the same method as the generalized implicit function theorem to treat:

$$H_k(\mathcal{R}^n) = \mathcal{R}(L_A) \oplus \mathcal{W}_k$$

$$v_k = r_k + w_k, \quad r_k \in \mathcal{R}(L_A), \; w_k \in \mathcal{W}_k$$

so $$L_A P_k(x) = r_k + w_k. \tag{5.79}$$

As $L_A$ is a linear operator and is invertible for $r_k$, the $r_k$ term in a non-linear differential can be eliminated because $L_A P_k(x) = r_k$ and there only the $w_k \in \mathcal{W}_k(x)$ term is left. So in the Poincaré resonant case, the vector normal form is

$$x' = Ax + w_2(x) + \cdots + w_{k-1}(x) + o(\|x\|^k) \tag{5.80}$$

Note 1. When $k$ is a fixed number, if $\|x\| \to 0$, then $o(\|x\|^k) = \mathcal{R}_k(x) \to 0$.

Note 2. For fixed $x$, if $k \to \infty$, $\mathcal{R}_k(x) \to \infty$. So what we have calculated is a formal normal form (i.e. it is an approximate normal form, but not a convergent normal form):

$$x' = Ax + w_2(x) + w_3(x) + \cdots \tag{5.81}$$

Normal form eq. (5.80) is called the Poincaré—Birkhoff normal form (PB normal form for short).

Now we shall discuss the choice of $\mathcal{W}_k(x)$, so that $H_k = \mathcal{R}(L_A) \oplus \mathcal{W}_k$. According to the Fredholm alternative theorem, let $L_A: \mathcal{H} \to \mathcal{G}$ be a linear and bounded operator in a finite-dimensional inner product space $\mathcal{H}$ and $\mathcal{G}$ (i.e. Hilbert space). Suppose $L_A^{\bullet} = L_A{}^*$ is the conjugate operator of $L_A$, i.e. $(L_A x, y) = (x, L_A{}^* y)$ is satisfied, then

$$(\mathcal{R}(L_A))^{\perp} = \eta(L_A{}^*), \text{ where } L_A{}^*: \mathcal{G} \to \mathcal{H}.$$

Proof: As $(L_A x, y) = (x, L_A{}^*)$, $x \in \mathcal{H}$, $y \in \mathcal{G}$, and when $y \in \eta(L_A{}^*)$, then $L_A{}^* y = 0$, so $(L_A x, y) = 0$, and for any $x \in \mathcal{H}$ is satisfied, hence $y \in (\mathcal{R}(L_A))^{\perp}$, i.e.

$$\eta(L_A{}^*) \subset (\mathcal{R}(L_A))^{\perp}$$

Additionally for any $y \in (\mathcal{R}(L_A))^{\perp}$, then for any $x \in \mathcal{H}$, we obtain

$$(L_A x, y) = 0 = (x, L_A{}^* y)$$

and also $L_A{}^* y = 0$, so $y \in \eta(L_A{}^*)$, i.e. $(\mathcal{R}(L_A))^{\perp} \subset \eta(L_A{}^*)$, so that

$$(\mathcal{R}(L_A))^{\perp} = \eta(L_A{}^*)$$

$$\mathcal{H}_k = \mathcal{R}(L_A) \oplus \eta(L_A{}^*)$$

therefore we can choose $\mathcal{W}_k = \eta(L_A{}^*)$.

It should be pointed out that the choice of $\mathcal{W}_k$ is not unique. The expression of $\mathcal{W}_k$ may be different if different methods are used, but they are equivalent in a certain sense.

# Chapter 6

# Hopf Bifurcation

Periodic vibration phenomena can be found in many non-conservative systems in nature. Hopf bifurcation theory is a theory that studies the modern development of period vibration phenomena. In this chapter we introduce the method of studying the Hopf bifurcation of autonomous systems by the normal form theory.

## 6.1 Hopf Bifurcation Theorem

### 6.1.1 Example of Hopf Bifurcation

First of all, consider a single degree of freedom vibration system with non-linear damping:

$$x'' + (x^2 + x'^2 - \mu)x' + x = 0 \tag{6.1}$$

If the bifurcation parameter $\mu > 0$, then non-linear damping $(x^2 + x'^2 - \mu)x'$ may change its sign when vibration increases; i.e. the non-linear damping is negative for a small value of vibration amplitude, and causes the vibration to increase; the non-linear damping is positive for a large value, and causes the vibration to decrease. So the system may keep the vibration in a certain steady-state.

Transforming eq. (6.1) into standard equations. Suppose $y_1 = x$, $y_2 = x'$ then

$$\begin{bmatrix} y_1 \\ y_2 \end{bmatrix}' = \begin{bmatrix} 0 & 1 \\ -1 & \mu \end{bmatrix} \begin{bmatrix} y_1 \\ y_2 \end{bmatrix} - \begin{bmatrix} 0 \\ y_2(y_1^2 + y_2^2) \end{bmatrix} \tag{6.2}$$

and the linearized equation of the above equation is

$$\begin{bmatrix} y_1 \\ y_2 \end{bmatrix}' = \begin{bmatrix} 0 & 1 \\ -1 & \mu \end{bmatrix} \begin{bmatrix} y_1 \\ y_2 \end{bmatrix} \tag{6.3}$$

Its eigenvalues are $$\lambda_{1,2} = \frac{\mu}{2} \pm i\sqrt{1 - \left(\frac{\mu}{2}\right)^2} \tag{6.4}$$

We shall only study the case where $|\mu| << 1$ (see Fig. 6.1 where the phase portrait for $\mu$ near zero is shown). When $\mu > 0$ and $\mu < 0$ the phase portrait is structurally stable; when $\mu = 0$ it is structurally unstable, and there is a family of periodic orbits of single parameter. Now we shall study non-linear differential eq. (6.2) in polar coordinate. Suppose

$$y_1 = r\cos\theta, \quad y_2 = r\sin\theta \tag{6.5}$$

Substituting eq. (6.5) into eq. (6.2), we obtain

$$r'\sin\theta + r\cos\theta' = -r\cos\theta + \mu r\sin\theta - r^3\sin\theta$$

From the above two equations we get

$$r' = r(\mu - r^2)\sin\theta^2, \quad \theta' = -1 + (\mu - r^2)\sin\theta\cos\theta \tag{6.6}$$

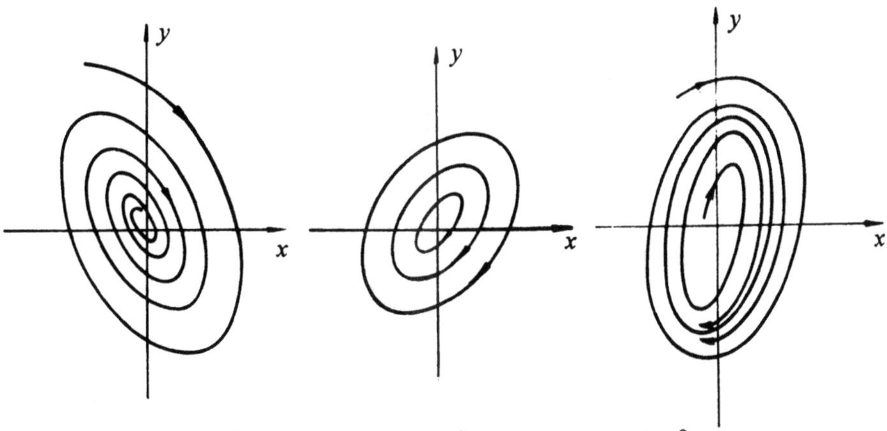


$\mu < 0$ $\mu = 0$ $\mu > 0$

Fig. 6.1

From the second equation of eq. (6.6) we know that if $r$ and $\mu$ are small values near zero, then $\|\mu - r^2\| << 1$, so $\theta'$ is negative. Therefore the flow turns clockwise. Now we shall study the first equation of eq. (6.6).

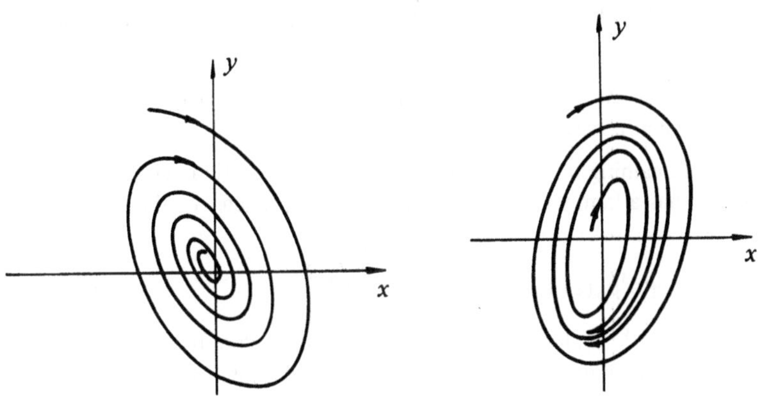


$\mu < 0$ $\mu > 0$

Fig. 6.2

If $\mu > r^2$, when $\theta \neq 0$ and $\theta \neq \pi$, then $r' > 0$ and the amplitude increases continuously, i.e. the phase portrait is an outward spiral; if $\mu < r^2$, then $r' < 0$, and the phase portrait is an inward spiral; if $\mu = r^2$, then $r = \sqrt{\mu}(\mu > 0), r' = 0$ and $\theta' = -$ , and the phase portrait is a periodic orbit which turns clockwise, see Fig. 6.2. The bifurcation diagram in Fig. 6.3 shows the change of $\mu$ in system (6.6). The linearized system has a family of periodic orbits of single parameters if $\mu = 0$, but the non-linear system has a family of periodic orbits of single parameters only when $\mu > 0$, i.e. $\mu = r^2$. All orbits are asymptotically stable.

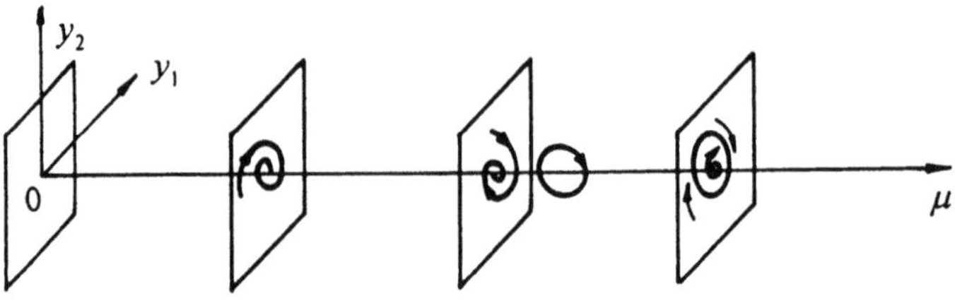


$\mu < 0$ $\mu = 0$ $\mu > 0$

Fig. 6.3

### 6.1.2 Hopf Bifurcation Theorem

To meet the needs of this chapter, we need to describe the Hopf bifurcation theorem again. Suppose the non-linear differential equations exist:

$$y' = f(y,\mu),\ f:\mathcal{R}^n \times \mathcal{R} \to \mathcal{R}^n,\ n \geq 2,\ f \in \mathcal{C}^\infty \tag{6.7}$$

1. If there is an equilibrium point $y_0 \in \mathcal{R}^n$ at $\mu_0$, then $f(y_0,\mu_0) = 0$ and $D_y f(y_0,\mu_0) \equiv A_{n\times n}$ are invertible. According to the implicit function theorem, a curve $y = y(\mu)$ passes through the equilibrium point $y_0$ in the neighbourhood of $\mu_0$, and $y(\mu_0) = y_0$ can move the $\mu$—axis without any loss of generality so that $f(0,\mu) = 0$.

2. $A(\mu) \equiv D_y f(0,\mu)$ has a pair of conjugate complex eigenvalues $\alpha(\mu) \pm i\beta(\mu)$, $\alpha(0) = 0$, $\beta(0) = \beta_0 > 0$ when $\mu = \mu_0$.

3. The transversal condition $\alpha'(\mu_0) \neq 0$, means that $\alpha(\mu) \pm i\beta(\mu)$ crosses the imaginary axis when $\mu$ passes through $\mu = \mu_0$;

4. The remaining $n-2$ eigenvalues of $A(\mu_0)$ have strictly negative real parts, so system (6.7) has a family of periodic solutions of parameter $\varepsilon$: there exists a $\mu(\varepsilon)$ function of $\mathcal{C}^4$

$$\mu(\varepsilon) = \mu_0 + \varepsilon^3 \mu_2(\varepsilon) + o(\varepsilon^4),\quad 0 < \varepsilon < \varepsilon_p > 0 \tag{6.8a}$$

so that for each $\varepsilon \in [0,\varepsilon_p]$, there exists a periodic solution $y(s,\varepsilon)$ occurring from $\mu(\varepsilon)$, that is, there is a neighbourhood $\eta$ of $y_0 = 0$ and an open interval $\theta$ containing $\mu_0$ so that for any $\mu \in \theta$, there is a non-constant periodic solution $y(s,\varepsilon)$ of eq. (6.7) in $\eta$ which is a member of the family

$$y(s,\varepsilon) = \varepsilon\varphi(s) + \varepsilon^2\omega(s,\varepsilon) + o(\varepsilon^4) \tag{6.8b}$$

The period of the solution is

$$T(\varepsilon) = \frac{2\pi}{\beta_0} + \varepsilon^2 T_2(\varepsilon) + o(\varepsilon^4) \tag{6.8c}$$

where $S = \frac{2\pi}{T}t$ is the rescaled time, and $T(\varepsilon)$ is the period of the orbit. $\varphi(S) = \text{Re}[ve^{is}]$, where $v$ is the eigenvector corresponding to $i\beta$. $\mu_2, T_2$ and $\omega$ are continuous functions of $\varepsilon, \omega$ is in the complementary subspace of $i\beta$.

If the vector fields $f(y,\mu)$ in eq. (6.7) are analytic to $y$ and $\mu$ in the neighbourhood $(y_0, \mu_0)$, then the analytic expression of the solution is

$$\mu(\varepsilon) = \sum_{i=2}^{\infty} \mu_i \varepsilon^i \tag{6.9a}$$

$$T(\varepsilon) = \frac{2\pi}{\beta_0}[1 + \sum_{i=2}^{\infty} \tau_i \varepsilon^i] \tag{6.9b}$$

$$y(S,\varepsilon) = \sum_{i=1}^{\infty} y_i(S)\varepsilon^i \tag{6.9c}$$

The method of calculation and the stability of the solution will be discussed later.

## 6.2 Complex Normal Form of the Hopf Bifurcation

Suppose that a two-dimensional system is obtained by simplifying eq. (6.7) by means of the centre manifold theorem:

$$y' = f(y) = Ay + hot \qquad y \in \mathcal{R}^2 \tag{6.10}$$

where $A = D_y f(y)|_{y=0} = D_y f(0)$.

Suppose also that $A$ has a pair of pure imaginary eigenvalues $\pm i\omega$. After the change of coordinate we obtain the real number expression:

$$y' = \begin{bmatrix} 0 & -\omega \\ \omega & 0 \end{bmatrix} y + hot \tag{6.11}$$

In order to change from real space $\mathcal{R}$ to complex space $\mathcal{C}$, we take

$$z = y_1 + iy_2$$
$$\bar{z} = y_1 - iy_2$$

If we derive the above equations and substitute eq. (6.10), then we obtain

$$z' = y_1' + iy_2' = -\omega y_2 + i\omega y_1 + hot = i\omega(y_1 + iy_2) + hot = i\omega z + hot \tag{6.12a}$$

and in the same way, we obtain

$$\bar{z}' = -i\omega\bar{z} + hot \tag{6.12b}$$

So the bifurcation equation in the complex space is

$$z' = i\omega z + g(z,\bar{z}) \tag{6.13a}$$

$$\bar{z}' = -i\omega\bar{z} + \bar{g}(z,\bar{z}) \tag{6.13b}$$

where $g(z,\bar{z}) = o(|z|^2)$ is the non-linear function.

As eq. (6.13a) is the conjugate of eq. (6.13b), we need only study one of them. Consider eq. (6.13a). To change to a Poincaré—Birkhoff (PB) normal form, let

$$g(z,\bar{z}) = \sum_{2\le l+m\le n} g_{lm} z^l \bar{z}^m + o(|z|^{n+1}) \tag{6.14}$$

Suppose $g(z,\bar{z})$ of eq. (6.13a) is reduced to the normal form of degree $k-1$, now we transform the terms of degree $k$ into normal form, and take the nearly identical non-linear change

$$z = \xi + P_k(\xi,\bar{\xi}) = \xi + \sum_{l+m=k} P_{em} \xi^l \bar{\xi}^m \tag{6.15}$$

Substituting (6.15) into the left and right side of eq. (6.13a), we have

left-hand side:

$$z' = \xi' + \sum_{l+m=k} P_{lm} (l\xi^{l-1}\bar{\xi}^m \xi' + m\xi^l \bar{\xi}^{m-1} \bar{\xi}') \tag{6.16a}$$

right-hand side:

$$i\omega\xi + i\omega \sum_{l+m=k} P_{lm}\xi^l \bar{\xi}^m + \sum_{l+m=k} g_{lm}\xi^l\bar{\xi}^m + \text{terms higher than } k \tag{6.16b}$$

where the coefficient $g_{lm}$ is still the same as in eq. (6.14).

Comparing eq. (6.16a) and eq. (6.16b), we see that

$$\begin{aligned} \xi' &= i\omega\xi + hot \\ \bar{\xi}' &= -i\omega\bar{\xi} + hot \end{aligned} \tag{6.17}$$

Substituting eq. (6.17) into eq. (6.16a), we obtain

$$z' = i\omega\xi + \sum_{l+m} \{l\xi^{l-1}\bar{\xi}^m i\omega\xi + m\xi^l \bar{\xi}^{m-1}(-i\omega\bar{\xi})\} P_{lm} + \cdots = i\omega\xi + \sum_{l+m=k} i\omega P_{lm}(l-m)\xi^l\bar{\xi}^m + \cdots \tag{6.18}$$

Let the coefficient of $l+m=k$ in eq. (6.18) be equal to the same order of the coefficient of eq. (6.16a), then

$$i\omega P_{lm}(l-m) = i\omega P_{lm} + g_{lm} \quad \text{or} \quad i\omega P_{lm}(l-m-1) = g_{lm} \tag{6.19}$$

If $l-m-1 \neq 0$, then $P_{lm}$ is unique.

In the similar way as in a homological equation, we define

$$l-m-1=0 \tag{6.20a}$$

as a resonant condition, and the non-resonant condition is

$$l-m \neq 1 \tag{6.20b}$$

Because we study the $k$-order term, we have

$$l+m=k \tag{6.21}$$

In a resonant case, from eq. (6.20a) and eq. (6.21) we obtain

$$2m+1=k \tag{6.22}$$

From eq. (6.22) we know that if $k$ is an even number, $2m+1\neq k$, and hence $-m\neq$ . This is the non-resonant case, so we can get $P_{lm}$ from eq. (6.19), i.e. the $k$-order even term in $g(z,\bar{z})$ can be eliminated.

If $k$ is an odd number, eq. (6.22) is satisfied, i.e. the resonant condition eq. (6.20a) is satisfied. Consequently, we cannot get $P_{lm}$ from eq. (6.19), since the $k$-order term is

$$z^l\bar{z}^m=z^{m+1}\bar{z}^m=z|z|^{2m} \tag{6.23}$$

which cannot be eliminated. Though $k=1+m$ is odd, the term which does not satisfy $l=m+1$ can also be eliminated. For example, if $k=3$, the terms which satisfy $l+m=k$ are

$$\begin{bmatrix} l \\ m \end{bmatrix}=\begin{bmatrix} 0,1,2,3 \\ 3,2,1,0 \end{bmatrix}$$

Only when $l=2$ and $m=1$, then $l-m=1$ is satisfied. Therefore in the four types of combination of $k=3$, all terms but $z^2\bar{z}$ can be eliminated. If $k$ is odd and the resonant condition is satisfied, the general form that cannot be eliminated is $z^{m+1}\bar{z}^m=z|z|^{2m}$; if $k$ is even, all the corresponding terms can be eliminated. So the PB normal form in complex form is

$$z'=i\omega z+c_1z|z|^2+c_2z|z|^4+\cdots+c_mz|z|^{2m}+o(|z|^{k+1}) \tag{6.24}$$

where $k=2m+1$.

It is easy to translate the above equation into polar coordinate form. Let

$$z=re^{i\theta} \tag{6.25a}$$

then $$z'=r'e^{i\theta}+i\theta' re^{i\theta} \tag{6.25b}$$

Substituting eq. (6.25) into eq. (6.24), we obtain

$$r'e^{i\theta}+ir\theta'e^{i\theta}=i\omega re^{i\theta}+c_1re^{i\theta}r^2+c_2re^{i\theta}r^4+\cdots+c_mre^{i\theta}r^{2m}+o(|z|^{k+1}) \tag{6.26}$$

where $o(|z|^{k+1})$ is the higher order term. In the resonant case, as $k=2m+1$, $k+1=2m+2=l+m$, i.e. $l=m+2$ is the higher order term, whereas

in the non-resonant case, $l\neq m+2$. Let $l=m$, then $z^m\bar{z}^m=r^{2m}$.

Eq. (6.26) can be written as

$$r'+i\theta'r=i\omega r+c_1r^3+c_2r^5+\cdots+c_mr^{2m+1}+r^{2m+2}e^{i\theta}\ (\text{or } r^{2m}e^{-i\theta})$$

We divide the above equation into two parts, the real part and the imaginary part, then the following two real number equations are obtained:

$$\begin{aligned} r' &= \mathrm{Re}[c_1]r^3+\mathrm{Re}[c_2]r^5+\cdots+hot \\ &= a_1r^3+a_2r^5+\cdots+hot \end{aligned} \tag{6.26a}$$

$$\theta'=\omega+\mathrm{Im}[c_1]r^2+\mathrm{Im}[c_2]r^4+\cdots+hot$$

$$= \omega + b_1 r^2 + b_2 r^4 + \cdots + hot \tag{6.26b}$$

where $hot$ have something to do with $r$ and $\theta, a_1 = \mathrm{Re}[c_1], a_2 = \mathrm{Re}[c_2], \cdots$, $b_1 = \mathrm{Im}[c_1]$, $b_2 = \mathrm{Im}[c_2], \cdots$.

For sufficiently small $r$, because $\theta' > 0, \theta$ is increasing monotonously. From eq. (6.20a) we know that, supposing $a_1 = \mathrm{Re}[c_1] \neq 0$, when $r$ is sufficiently small, the sign of $r'$ is determined by the sign of $a_1$. The stability of the solution of eq. (6.24) is determined by the sign of $a_1$: if $a_1$ is negative, then $r$ is asymptotically stable; if $a_1$ is positive, then $r$ is unstable.

## 6.3 Normal Form of the Hopf Bifurcation in Real Numbers

In this section, we shall study the real normal form which is convenient for practical applications.

Suppose that we have obtained the two-dimensional system by simplifying the $n$-dimensional non-linear differential equations using the centre manifold theorem:

$$y' = Ay + G(y), \quad y \in \mathcal{R}^2 \tag{6.27}$$

where $A = \begin{bmatrix} 0 & \omega \\ -\omega & 0 \end{bmatrix}$, $\omega > 0$. As the square term in the expansion of non-linear function $G(y)$ does not correspond to the Poincaré resonant case, it has been eliminated. Equation (6.27) can be written as

$$y' = Ay + C(y, y, y) \tag{6.28}$$

where $C(y, y, y)$ is a homogeneous polynomial of order-3, which includes the "contribution" to the third-order term when two order terms have been eliminated.

To simplify $C(y, y, y)$, we take

$$y = x + P_3(x) \qquad x \in \mathcal{R}^2 \tag{6.29}$$

where $P_3(x) \in H_3(\mathcal{R}^2)$

$$P_3(x) = \begin{bmatrix} P_{31}(x) \\ P_{32}(x) \end{bmatrix} = \begin{bmatrix} p_1 x_1^3 + p_2 x_1^2 x_2 + p_3 x_1 x_2^2 + p_4 x_2^3 \\ q_1 x_1^3 + q_2 x_1^2 x_2 + q_3 x_1 x_2^2 + q_4 x_2^3 \end{bmatrix} \tag{6.30}$$

The eight coefficients $p_1, p_2, \cdots, q_4$ in the above equations are determined by homological equations

$$L_A P_3 = C \text{ or } \left(\frac{\partial P_3}{\partial x}\right) Ax - AP_3(x) = C(x) \tag{6.31}$$

where $\frac{\partial P_3}{\partial x} = \begin{bmatrix} \frac{\partial P_{31}}{\partial x_1} & \frac{\partial P_{31}}{\partial x_2} \\ \frac{\partial P_{32}}{\partial x_1} & \frac{\partial P_{32}}{\partial x_2} \end{bmatrix} = \begin{bmatrix} 3p_1 x_1^2 + 2p_2 x_1 x_2 + p_3 x_2^2 & p_2 x_1^2 + 2p_3 x_1 x_2 + 3p_4 x_2^2 \\ 3q_1 x_1^2 + 2q_2 x_1 x_2 + q_3 x_2^2 & q_2 x_1^2 + 2q_3 x_1 x_2 + 3q_4 x_2^2 \end{bmatrix}$

$$Ax = \begin{bmatrix} 0 & \omega \\ -\omega & 0 \end{bmatrix} \begin{bmatrix} x_1 \\ x_2 \end{bmatrix} = \begin{bmatrix} -\omega x_2 \\ \omega x_1 \end{bmatrix}$$

$$AP_3(x) = \begin{bmatrix} 0 & -\omega \\ \omega & 0 \end{bmatrix} \begin{bmatrix} p_1x_1^3 + p_2x_1^2x_2 + p_3x_1x_2^2 + p_4x_2^3 \\ q_1x_1^3 + q_2x_1^2x_2 + q_3x_1x_2^2 + q_4x_2^3 \end{bmatrix}$$

$$= \omega \begin{bmatrix} -(q_1x_1^3 + q_2x_1^2x_2 + q_3x_1x_2^2 + q_4x_2^4) \\ p_1x_1^3 + p_2x_1^2x_2 + p_3x_1x_2^2 + p_4x_2^4 \end{bmatrix}$$

$$C(x) = \begin{bmatrix} c_1x_1^3 + c_2x_1^2x_2 + c_3x_1x_2^2 + c_4x_2^3 \\ d_1x_1^3 + d_2x_1^2x_2 + d_3x_1x_2^2 + d_4x_2^3 \end{bmatrix}$$

Substituting the above equations into eq. (6.31), we obtain

$$\omega \begin{bmatrix} x_1^3(p_2+q_1) + x_1^2x_2(-3p_1+2p_3+q_2) + x_1x_2^2(-2p_2+q_3+3p_4) + x_2^3(p_4-p_3) \\ x_1^3(q_2-q_1) + x_1^2x_2(-3q_1+2q_3-p_2) + x_1x_2^2(-2q_2-p_3+3q_4) - x_2^3(q_4+q_3) \end{bmatrix}$$
$$= \begin{bmatrix} c_1x_1^3 + c_2x_1^3x_2 + c_3x_1x_2^2 + c_4x_2^3 \\ d_1x_1^3 + d_2x_1^3x_2 + d_3x_1x_2^2 + d_4x_2^3 \end{bmatrix} \tag{6.32}$$

Let the coefficients of the same order terms on both side be equal, then

| | First Row | Second Row |
|---|---|---|
| $x_1^3$: | $p_2 + q_1 = \dfrac{c_1}{\omega}$ | $q_2 - q_1 = \dfrac{d_1}{\omega}$ |
| $x_1^2x_2$: | $-3p_1 + 2p_3 + q_2 = \dfrac{c_2}{\omega}$ | $-3q_1 + 2q_3 - p_2 = \dfrac{d_2}{\omega}$ |
| $x_1x_2^2$: | $-2p_2 + q_3 + 3p_4 = \dfrac{c_3}{\omega}$ | $-2q_2 + p_3 + 3q_4 = \dfrac{d_3}{\omega}$ |
| $x_2^3$: | $-p_3 + p_4 = \dfrac{c_4}{\omega}$ | $-(q_3 + q_4) = \dfrac{d_4}{\omega}$ |

We can calculate $p_1, p_2, \cdots, q_4$ from the above eight algebraic equations. For convenience we write the equations in matrix form.

$$\begin{bmatrix} 0 & 1 & 0 & 0 & 1 & 0 & 0 & 0 \\ -3 & 0 & 2 & 0 & 0 & 1 & 0 & 0 \\ 0 & -2 & 0 & 3 & 0 & 0 & 1 & 0 \\ 0 & 0 & -1 & 0 & 0 & 0 & 0 & 1 \\ -1 & 0 & 0 & 0 & 0 & 1 & 0 & 0 \\ 0 & -1 & 0 & 0 & -3 & 0 & 2 & 0 \\ 0 & 0 & -1 & 0 & 0 & -2 & 0 & 3 \\ 0 & 0 & 0 & -1 & 0 & 0 & -1 & 0 \end{bmatrix} \begin{bmatrix} p_1 \\ p_2 \\ p_3 \\ p_4 \\ q_1 \\ q_2 \\ q_3 \\ q_4 \end{bmatrix} = \frac{1}{\omega} \begin{bmatrix} c_1 \\ c_2 \\ c_3 \\ c_4 \\ d_1 \\ d_2 \\ d_3 \\ d_4 \end{bmatrix}$$

or

$$\begin{bmatrix} B & I \\ -I & B \end{bmatrix} \begin{bmatrix} p \\ q \end{bmatrix} = \frac{1}{\omega} \begin{bmatrix} c \\ d \end{bmatrix} \tag{6.33}$$

where $B=\begin{bmatrix}0&1&0&0\\-3&0&2&0\\0&-2&0&3\\0&0&-1&0\end{bmatrix}$; $I$ is the identity matrix. From eq. (6.33) we can solve

$$Bp+q=\frac{c}{\omega}\qquad\qquad \therefore q=\frac{c}{\omega}-Bp \tag{6.34}$$

$$-p+Bq=\frac{d}{\omega},\ -p+B(\frac{c}{\omega}-Bp)=\frac{d}{\omega},\ \therefore p=\frac{8c-d}{(I+B^2)\omega} \tag{6.35}$$

Now we calculate eq. (6.34) and eq. (6.35)

$$B^2=\begin{bmatrix}0&1&0&0\\-3&0&2&0\\0&-2&0&3\\0&0&-1&0\end{bmatrix}\begin{bmatrix}0&1&0&0\\-3&0&2&0\\0&-2&0&3\\0&0&-1&0\end{bmatrix}=\begin{bmatrix}-3&0&2&0\\0&-7&0&6\\6&0&-7&0\\0&2&0&-3\end{bmatrix}$$

$$B^2+I=\begin{bmatrix}-2&0&2&0\\0&-6&0&6\\6&0&-6&0\\0&2&0&-2\end{bmatrix} \tag{6.36}$$

The first and the third rows, and the second and the fourth rows are linearly dependent on the matrix, so their ranking is two

$$Bc-d=\begin{bmatrix}0&1&0&0\\-3&0&2&0\\0&-2&0&3\\0&0&-1&0\end{bmatrix}\begin{bmatrix}c_1\\c_2\\c_3\\c_4\end{bmatrix}-\begin{bmatrix}d_1\\d_2\\d_3\\d_4\end{bmatrix}=\begin{bmatrix}c_2-d_1\\-3c_1+2c_3-d_2\\-2c_2+3c_4-d_3\\-c_3-d_4\end{bmatrix} \tag{6.37}$$

Substituting (6.36) and (6.37) into eq. (6.35), we obtain

$$\begin{bmatrix}-2&0&2&0\\0&-6&0&6\\6&0&-6&0\\0&2&0&-2\end{bmatrix}\begin{bmatrix}p_1\\p_2\\p_3\\p_4\end{bmatrix}=\frac{1}{\omega}\begin{bmatrix}c_2-d_1\\-3c_1+2c_3-d_2\\-2c_2+3c_4-d_3\\-c_3-d_4\end{bmatrix}$$

Using the above method we have avoided algebraic equations. The first line in the matrix is divided by 2, the second line by 6; the first line is multiplied by 3 and then added to the third line, the second line is divided by 3 and then added to the fourth line. Thus, we have the following:

$$\begin{bmatrix}-1&0&1&0\\0&-1&0&1\\0&0&0&0\\0&0&0&0\end{bmatrix}\begin{bmatrix}p_1\\p_2\\p_3\\p_4\end{bmatrix}=\frac{1}{\omega}\begin{bmatrix}\frac{1}{2}(c_2-d_1)\\-\frac{1}{2}c_1+\frac{1}{3}c_3-\frac{1}{6}d_2\\c_2+3c_4-3d_1-d_3\\-\left(c_1+\frac{1}{3}c_3+\frac{1}{3}d_2+d_4\right)\end{bmatrix} \tag{6.38}$$

First line: $$-p_1 + p_3 = \frac{1}{2\omega}(c_2 - d_1) \tag{6.39a}$$

Second line: $$-p_2 + p_4 = \frac{1}{\omega}\left(-\frac{c_1}{2} + \frac{c_3}{3} - \frac{d_2}{6}\right) \tag{6.39b}$$

Third line: $$c_2 + 3c_4 - 3d_1 - d_3 = 0 \tag{6.39c}$$

Forth line: $$c_1 + \frac{c_3}{3} + \frac{d_2}{3} + d_4 = 0 \tag{6.39d}$$

From eq. (6.39c) and eq. (6.39d) we know that the 8 numbered expressions in $c$ and $d$ should be satisfied for $c_1, c_2, \cdots, d_3, d_4$, so only six quantities are independent. The range space of $L_A$ is six-dimensional, i.e.

$$\dim(\mathrm{P}(L_A)) \tag{6.40a}$$

$$\operatorname{codim}(\mathrm{P}(L_A)) = 2 \tag{6.40b}$$

$p,q$ can be derived from eq. (6.31) according eq. (6.40a). They are then substituted into $P_3(x)$, and a part of $c(x)$ can be eliminated.

A convenient choice of a two-dimensional basis and a third-order basis complementary space of $(\mathcal{R}(L_A))$ is

$$\begin{bmatrix} c \\ d \end{bmatrix} = \begin{bmatrix} 0 & -1 & 0 & -1 & 1 & 0 & 1 & 0 \end{bmatrix}^T; \text{ and } \begin{bmatrix} c \\ d \end{bmatrix} = \begin{bmatrix} 1 & 0 & 1 & 0 & 0 & 1 & 0 & 1 \end{bmatrix}^T$$

Their corresponding resonant monomials in $c(x)$ are

$$\begin{bmatrix} -x_1^2 x_2 - x_2^3 \\ x_1^3 + x_1 x_2^2 \end{bmatrix} = (x_1^2 + x_2^2)\begin{bmatrix} -x_1 \\ x_2 \end{bmatrix} \tag{6.42a}$$

and $$\begin{bmatrix} x_1^3 + x_1 x_2^3 \\ x_1^2 x_2 + x_2^3 \end{bmatrix} = (x_1^3 + x_2^2)\begin{bmatrix} x_1 \\ x_2 \end{bmatrix} \tag{6.42b}$$

All terms but the two resonant polynomials of eq. (6.42) in $c(x)$ can be eliminated. The real Poincaré normal form of a Hopf bifurcation is, therefore

$$x' = Ax + a(x_1^3 + x_2^2)\begin{bmatrix} x_1 \\ x_2 \end{bmatrix} + b(x_1^2 + x_2^2)\begin{bmatrix} -x_2 \\ x_1 \end{bmatrix} + hot$$

$$= \begin{bmatrix} 0 & \omega \\ \omega & 0 \end{bmatrix}\begin{bmatrix} x_1 \\ x_2 \end{bmatrix} + (x_1^2 + x_2^2)\begin{bmatrix} a & -b \\ b & a \end{bmatrix}\begin{bmatrix} x_1 \\ x_2 \end{bmatrix} + hot \tag{6.43}$$

In the above equation, if $z = x_1 + ix_2$; $x_1 = r\cos\theta$, $x_2 = r\sin\theta$, it is not difficult to prove that eq. (6.43) is exactly the same as eq. (6.24), eq. (6.26) (here $a = a_1$, $b = b_1$).

## 6.4 Hopf Bifurcation with Parameters

Suppose we have simplified the $n$-dimensional non-linear differential equations to a two-dimensional system by means of the centre manifold theorem:

$$y' = f(y,\mu),\ f:\mathcal{R}^2\times\mathcal{R}\to\mathcal{R}^2,\ f\in\mathbf{C}^k,\ k\geq 2 \tag{6.44}$$

The assumptions are as follows:

1. $f(0,\mu)=0$, $\mu\in\mathcal{R}$ is the bifurcation parameter, $y=0$ is a trivial solution;
2. $A(\mu)\equiv D_y f(0,\mu)$;
3. $A(0)$ has a pair of purely imaginary eigenvalues $\pm i\omega_0$, $\omega_0\neq 0$, so $A(\mu)$ has a pair of eigenvalues $\alpha(\mu)\pm i\omega(\mu)$ of $\mathcal{C}^{k-1}$. When $\mu=0$, $\alpha'(0)\neq 0$, $\omega(0)=\omega_0$, and $\alpha(0)=0$, there also exists the complex eigenvector of $\mathcal{C}^{k-1}$.

First step. Translate $A$ into a standard form, so for this reason take the linear transformation $y=Px$, where $P$ is a $2\times 2$ matrix, the columns of which are composed of the real and imaginary parts of the eigenvectors, so

$$P^{-1}AP=\begin{bmatrix}\alpha(\mu) & -\omega(\mu)\\ \omega(\mu) & \alpha(\mu)\end{bmatrix}$$

Substituting the linear transformation into eq. (6.44), we have

$$x' = P^{-1}f(Px,\mu)=\begin{bmatrix}\alpha(\mu) & \omega(\mu)\\ \omega(\mu) & \alpha(\mu)\end{bmatrix}x+F(x,\mu) \tag{6.45}$$

where $D_xF(0,\mu)=0$, i.e. $F(x,\mu)=o(\|x\|^2)$. When $\|x\|\to 0$, $F(0,\mu)=0$, so it is uniform for $\mu$.

Second step. Rescaling $x$, we take $x=\varepsilon v$, $\varepsilon<<1$, $v\in\mathcal{R}^2$ (not small, but bounded). Substituting $x=\varepsilon v$ into (6.45), and dividing it by $\varepsilon$, then

$$v' = \begin{bmatrix}\alpha(\mu) & -\omega(\mu)\\ \omega(\mu) & \alpha(\mu)\end{bmatrix}v+G(v,\mu,\varepsilon) \tag{6.46}$$

where $G(v,\mu,\varepsilon)=\frac{1}{\varepsilon}F(v,\mu,\varepsilon)$, and $G\in\mathcal{C}^k$ (except for $\varepsilon=0$).

Define $G(v,\mu,0)=\lim\limits_{\varepsilon\to 0}\frac{1}{\varepsilon}F(\varepsilon v,\mu)=D_xF(0,\mu)v=0$ (6.47)

and choose the initial condition $v(0)=\begin{bmatrix}1\\0\end{bmatrix}$, corresponding to $x(0)=\begin{bmatrix}\varepsilon\\0\end{bmatrix}$.

Third step. The calculation of the fundamental solution matrix of the linear part of eq. (6.46)

$$v' = \begin{bmatrix}\alpha(\mu) & -\omega(\mu)\\ \omega(\mu) & \alpha(\mu)\end{bmatrix}v \tag{6.48}$$

Its fundamental solution matrix is

$$\Phi_\mu(t) = \exp\begin{bmatrix} \alpha & -\omega \\ \omega & \mu \end{bmatrix} t = e^{\left[\begin{bmatrix} \alpha & 0 \\ 0 & \alpha \end{bmatrix} t + \begin{bmatrix} 0 & -\omega \\ \omega & 0 \end{bmatrix} t\right]}$$

$$= e^{\alpha t}\left[ I + \omega \begin{bmatrix} 0 & -1 \\ 1 & 0 \end{bmatrix} t + \frac{\omega^2 t^2}{2!}\begin{bmatrix} 0 & -1 \\ 1 & 0 \end{bmatrix}^2 + \frac{\omega^3 t^3}{3!}\begin{bmatrix} 0 & -1 \\ 1 & 0 \end{bmatrix}^3 + \cdots \right]$$

$$= e^{\alpha t}\left\{ I + \begin{bmatrix} -\dfrac{\omega^2 t^2}{2!} + \cdots & -\omega t + \dfrac{(\omega t)^3}{3!} + \cdots \\ \omega t - \dfrac{(\omega t)^3}{3!} + \cdots & -\dfrac{(\omega t)^2}{2!} + \cdots \end{bmatrix} \right\}$$

$$= e^{\alpha t}\begin{bmatrix} \cos\omega t & -\sin\omega t \\ \sin\omega t & \cos\omega t \end{bmatrix} \tag{6.49}$$

Applying the variation of parameters to eq. (6.46), we have

$$v(t) = \Phi_\mu(t)v(0) + \int_0^t \Phi_\mu(t-S)G(v(S),\mu,\varepsilon)dS \tag{6.50}$$

In this way the differential eq. (6.46) is converted to the integral eq. (6.50). We seek periodic solutions in $T(\mu)$ period, $T(\mu)$ is the undetermined function in the neighbourhood of $\dfrac{2\pi}{\omega_0}$. According to the periodic condition $v(t) = v(0)$ we get

$$[I - \Phi_\mu(T)]v(0) = \int_0^T \Phi_\mu(T-S)G(v(S),\mu,\varepsilon)\,\mathrm{d}S \tag{6.51}$$

For sufficiently small $\varepsilon$ and $\mu$, $v$ is defined on $[0,T]$ so the right-hand side of eq. (6.51) is also defined on $[0,T]$.

Fourth step. Suppose that eq. (6.51) is the equation of real functions $T,\mu$ and $\varepsilon$. We shall solve for $(T,\mu) = (T(\varepsilon),\mu(\varepsilon))$ from eq. (6.51).

First calculate $T$ and $\mu$ corresponding to the zero solution $(\varepsilon = 0)$. Because $G(v,\mu,0) = 0$, the right-hand side of eq. (6.51) is equal to zero. Substituting $\Phi_\mu(T)$ and $v(0)$ into the left-hand side of eq. (6.51), we obtain

$$\left\{\begin{bmatrix} 1 & 0 \\ 0 & 1 \end{bmatrix} - e^{\alpha\omega T}\begin{bmatrix} \cos\omega(\mu)T & -\sin\omega(\mu)T \\ \sin\omega(\mu)T & \cos\omega(\mu)T \end{bmatrix}\right\}\begin{bmatrix} 1 \\ 0 \end{bmatrix} = 0 \tag{6.52}$$

From the equation we get

$$1 - e^{\alpha(\mu)T}\cos\omega(\mu)T = 0 \tag{6.53a}$$

$$-e^{\alpha(\mu)T}\sin\omega(\mu)T = 0 \tag{6.53b}$$

We can get the smallest $T$ from eq. (6.53). From eq. (6.53b) we obtain

$$\omega(\mu)T = 2\pi, \quad \therefore T = 2\pi/\omega(\mu)$$

from eq. (6.53a) $\quad 1 - e^{\alpha(\mu)T}\cos 2\pi = 1 - e^{\alpha(\mu)T} = 0$

$$\therefore \alpha(0) = 0, \qquad \mu = 0$$

So when $\varepsilon = 0$, $(T,\mu) = \left(\frac{2\pi}{\omega_0}, 0\right)$. Calculate $(T,\mu)$ when $\varepsilon \neq 0$. For this purpose calculate the Jacobian determinant of eq. (6.53) with respect to $(T,\mu)$:

$$J \equiv \left|\frac{\partial(6.53)}{\partial(T,\mu)}\right|_{\varepsilon=0,\ \mu=0}$$

$$= \begin{vmatrix} -\alpha e^{\alpha T_0}\cos\omega_0 T_0 + e^{\alpha T_0}\omega_0 \sin\omega_0 T_0 & -T_0\alpha' e^{\alpha T_0}\cos\omega_0 T_0 + e^{\alpha T_0}\omega' T_0 \sin\omega_0 T_0 \\ -\alpha e^{\alpha T_0}\sin\omega_0 T_0 + e^{\alpha T_0}\omega_0\cos\omega_0 T_0 & -\alpha' e^{\alpha T_0}\sin\omega_0 T_0 - e^{\alpha T_0}\omega' T_0 \sin\omega_0 T_0 \end{vmatrix}$$

$$= \begin{vmatrix} 0 & -T_0\alpha' \\ \omega_0 & \omega' T_0 \end{vmatrix} = \alpha' T_0 \omega_0 \qquad (6.54)$$

where $\omega_0 T_0 = 2\pi$. If $\alpha'(0) \neq 0$ then eq. (6.54) does not equal zero. Then we can obtain $T = T(\varepsilon)$ and $\mu = \mu(\varepsilon)$ from eq. (6.51) by the implicit function theorem, and $v(t)$ from eq. (6.50), and hence $x(t)$ and $y(t)$.

Now we shall discuss the relationship between the coefficients of the normal form and the coefficients of the original equations [50]. Suppose that we have reduced the $n$-dimensional non-linear differential equations to a two-dimensional system by means of the centre manifold theorem:

$$y' = A(\mu)y + \mu + F^2(y) + F^3(y) + hot \qquad (6.55)$$

where $y \in \mathcal{R}^2$, $F^{(i)}(y) \in \mathcal{H}^{k/2}$: homogeneous polynomials in order $i$ of real variables.

$$y = \begin{bmatrix} y_1 \\ y_2 \end{bmatrix},\quad A(\mu) = \begin{bmatrix} d\mu & -(c\mu + \omega_0) \\ c\mu + \omega_0 & d\mu \end{bmatrix},\quad A_0 = A(0) = \begin{bmatrix} 0 & -\omega_0 \\ \omega_0 & 0 \end{bmatrix}$$

$$F^k(X) = \begin{bmatrix} \sum_{i=0}^{k} a_{i,j} x_1^i x_2^j \\ \sum_{i=0}^{k} b_{i,j} x_1^i x_2^j \end{bmatrix};\ k = 2,3;\ i = k - j$$

From the above discussion we know that if $a$ and $b$ have been calculated, the character of the bifurcation solution can be recognized for the non-degenerate system. When we eliminate the homogeneous polynomials in order-2 in eq. (6.55) by the nearly identical non-linear change, this kind of change has "contribution" to higher order terms. So we should seek out this change, and make successive changes to obtain $a$ and $b$. Choose the change of the variables for eq. (6.55)

$$y = x + P_2(x),\quad P_2(x) \in H_2^2(\mathbf{R}^2) \qquad (6.56)$$

where $x = \begin{bmatrix} x_1 \\ x_2 \end{bmatrix}$; $P_2(x) = \begin{bmatrix} c_{20}x_1^2 + c_{11}x_1x_2 + c_{02}x_2^2 \\ d_{20}x_1^2 + d_{11}x_1x_2 + d_{02}x_2^2 \end{bmatrix}$. Substituting (6.56) into eq. (6.55), we obtain the following after rearrangement:

$$x' = Ax + [F^2(x) + AP_2(x) - DP_2(x) - DP_2(x)Ax] + \{F^3(x) + DF^2(\mathrm{x})P_2(\mathrm{x}) + [DP_2(\mathrm{x})]^2 Ax - DP_2(\mathrm{x})[AP_2(x) + F^2(\mathrm{x})]\} + hot \qquad (6.57)$$

The square bracket in the equation represents the homogeneous polynomial of degree-2 which is equal to zero. The first brace in the equation represents the new homogeneous polynomial of degree-3, which contains the effect of the transformation. We begin by setting up the relationship between the coefficients of $P^2(x)$ and those of $F^2(x)$. Let the first bracket in eq. (6.57) equal zero, then

$$L_A P_2(x) = F^2(x) \tag{6.58}$$

We define a new basis to simplify (6.58): $\{y^\alpha e_j = y_1^{\alpha_1} y_2^{\alpha_2} e_j | \alpha = \alpha_1 + \alpha_2\}$, and its defined order is the reversed lexicographic ordering. Let $y^\alpha e_i$ and $y^\beta e_j$ be any two monomials in $H_n^k$, where $e_i$ and $e_j$ are the standard unit column vectors in $\mathcal{R}^n$. We say that $y^\alpha e_i < y^\beta e_j$, if $i > j$ or $i = j$ and the first different components $\alpha_1$ of $\alpha$ and $\beta_1$ of $\beta$ satisfy $\alpha_1 > \beta_1$. According to the above definition the set

$$\{e'_s = y^i y^j e_m | i + j = k;\ m = 1,2;\ s = (3 - m)(k + 1) - i\}$$

is a basis in $H_2^k$ in the reversed lexicographic order, its dimension being $2(k+1)$. Linear operator $L_A^k$ can be expressed by a matrix $[L_A^k]$ under this basis. The column vector of $L_A^k$ is expressed by $L_S^k$:

$$L_S^k = L_A^k e'_S = [-ie'_{S+1} + je'_{S-1} + (-1)^m e'_{m(k+1)-i}]\omega_0 \tag{6.59}$$

On $H_2^2$, the set $\left\{\begin{bmatrix}0\\y_1^2\end{bmatrix}, \begin{bmatrix}0\\y_1y_2\end{bmatrix}, \begin{bmatrix}0\\y_2^2\end{bmatrix}, \begin{bmatrix}y_1^2\\0\end{bmatrix}, \begin{bmatrix}y_1y_2\\0\end{bmatrix}, \begin{bmatrix}y_2^2\\0\end{bmatrix}\right\}$ is a basis in reversed lexicographic ordering. Equation (6.58) can be rewritten in the following matrix form according to eq. (6.59):

$$\omega_0 \begin{bmatrix} 0 & 1 & 0 & -1 & 0 & 0 \\ -2 & 0 & 2 & 0 & -1 & 0 \\ 0 & -1 & 0 & 0 & 0 & -1 \\ 1 & 0 & 0 & 0 & 1 & 0 \\ 0 & 1 & 0 & -2 & 0 & 2 \\ 0 & 0 & 1 & 0 & -1 & 0 \end{bmatrix} \begin{bmatrix} d_{20} \\ d_{11} \\ d_{02} \\ c_{20} \\ c_{11} \\ c_{02} \end{bmatrix} = \begin{bmatrix} b_{20} \\ b_{11} \\ b_{02} \\ a_{20} \\ a_{11} \\ a_{02} \end{bmatrix} \tag{6.60}$$

Solving the above equation we get the algebraic relationship between the coefficients of $P_2(Y)$ and (those of) $F^2(Y)$:

$$\begin{bmatrix} d_{20} \\ d_{11} \\ d_{02} \\ c_{20} \\ c_{11} \\ c_{02} \end{bmatrix} = \frac{1}{3\omega_0} \begin{bmatrix} a_{20} - b_{11} + 2a_{02} \\ 2b_{20} - a_{11} - 2b_{02} \\ 2a_{20} + b_{11} + a_{02} \\ -b_{20} - a_{11} - 2b_{02} \\ 2a_{20} + b_{11} - 2a_{02} \\ -2b_{20} + a_{11} - b_{02} \end{bmatrix} \tag{6.61}$$

Secondly calculate the relationship between the coefficients of $F^2(x)$ and (those of) $F^3(x)$. In eq. (6.57), let

$$\overline{F}_1^3(x) = F^3(x) + DF^2(x)P_2(x) + [DP_2(x)]^2 Ax - DP_2(x)[AP_2(x) + F^2(x)] \quad (6.62)$$

where: $\overline{F}_1^3 = \begin{bmatrix} \overline{a}_{30}x_1^3 + \overline{a}_{21}x_1^2x_2 + \overline{a}_{12}x_1x_2^2 + \overline{a}_{03}x_2^3 \\ \overline{b}_{30}x_1^3 + \overline{b}_{21}x_1^2x_2 + \overline{b}_{12}x_1x_2^2 + \overline{b}_{03}x_2^3 \end{bmatrix}$

$$\overline{a}_{30} = a_{30} + a_{11}d_{20} + \omega_0[2c_{20}c_{11} + c_{11}d_{11}] - [-2c_{20} - d_{20}\omega_0 + c_{11}(b_{20} + \omega_0 c_{20})]$$

$$\begin{aligned}\overline{a}_{21} = {} & a_{21} + a_{11}c_{20} + 2a_{20}c_{11} + a_{11}d_{11} + 2a_{02}d_{20} + \omega_0[4c_{20}(c_{02} - c_{20}) + c_{11}^2 \\ & + 2c_{11}(d_{02} - d_{20}) + 2c_{02}d_{11}] - [2c_{20}(a_{11} - \omega_0 d_{11}) + c_{11}(a_{20} - \omega_0 d_{20}) \\ & + c_{11}(b_{11} + \omega_0 c_{11}) + 2c_{02}(b_{20} + \omega_0 c_{20})]\end{aligned}$$

$$\begin{aligned}\overline{a}_{12} = {} & a_{12} + 2a_{20}c_{02} + a_{11}c_{11} + a_{11}d_{02} + 2a_{02}d_{11} + \omega_0[-2c_{20}c_{11} + 2c_{11}(c_{02} - c_{20}) \\ & - c_{11}d_{11} + 4c_{02}(d_{02} - d_{20})] - [2c_{20}(a_{02} - \omega_0 d_{02}) + c_{11}(a_{11} - \omega_0 d_{11}) \\ & + c_{11}(b_{02} + \omega_0 c_{02}) + 2c_{02}(b_{11} + \omega_0 c_{11})]\end{aligned}$$

$$\begin{aligned}\overline{a}_{03} = {} & a_{03} + a_{11}c_{02} + 2a_{02}d_{02} - \omega_0(c_{11}^2 + 2c_{02}d_{11}) - [c_{11}(a_{02} - d_{02}\omega_0) + 2c_{02}(b_{02} \\ & + \omega_0 c_{02})]\end{aligned}$$

$$\begin{aligned}\overline{b}_{30} = {} & b_{30} + 2b_{20}c_{20} + b_{11}d_{20} - [2d_{20}(a_{20} - \omega_0 d_{20}) + d_{11}(b_{20} + \omega_0 c_{20})] \\ & + \omega_0[4d_{20}(c_{02} - c_{20}) + d_{11}c_{11} + d_{11}(2(d_{02} - 2d_{20}) + 2d_{02}d_{11}]\end{aligned}$$

$$\begin{aligned}\overline{b}_{21} = {} & b_{21} + 2b_{20}c_{11} + b_{11}c_{20} + b_{11}d_{11} + 2b_{02}d_{20} - [2d_{20}(a_{11} - \omega_0 d_{11}) + d_{11}(a_{20} \\ & - \omega_0 d_{20}) + 2d_{02}(b_{20} + \omega_0 c_{20}) + d_{11}(b_{11} + \omega_0 c_{11})] + \omega_0[4d_{20}(c_{02} - c_{20}) \\ & + d_{11}c_{11} + 2d_{11}(d_{02} - d_{20}) + 2d_{02}d_{11}]\end{aligned}$$

$$\begin{aligned}\overline{b}_{12} = {} & b_{12} + 2b_{20}c_{02} + b_{11}c_{11} + b_{11}d_{02} + 2b_{02}d_{11} + \omega_0[-2d_{20}c_{11} + 2d_{11}(c_{02} - c_{20}) \\ & - d_{11}^2 + 4d_{02}(d_{02} - d_{20})] - [2d_{20}(a_{02} - \omega_0 d_{02}) + d_{11}(a_{11} - \omega_0 d_{11}) \\ & + d_{11}(b_{02} + \omega_0 c_{02}) + 2d_{02}(b_{11} + \omega_0 c_{11})]\end{aligned}$$

$$\overline{b}_{03} = b_{03} + b_{11}c_{02} + \omega_0 d_{11}(c_{11} + 2d_{02}) - [d_{11}(a_{02} - \omega_0 d_{02}) + 2d_{02}c_{02}\omega_0]$$

Substituting (6.62) into eq. (6.57), we have

$$x' = Ax + \overline{F}_1^3(x) + hot \quad (6.63)$$

To simplify the order-3 terms, we take the nearly identical transformation

$$x = u + P_3(u) \quad (6.64)$$

where $u = \begin{bmatrix} u_1 \\ u_2 \end{bmatrix}$; $P_3(u) = \begin{bmatrix} c_{30}u_1^3 + c_{21}u_1^2u_2 + c_{12}u_1u_1^2 + c_{03}u_2^3 \\ d_{30}u_1^3 + d_{21}u_1^2u_2 + d_{12}u_1u_1^2 + d_{03}u_2^3 \end{bmatrix}$. Substituting (6.64) into eq. (6.63), we have

$$u' = Au + F_1^3(u) - L_A P_3(u) + hot \quad (6.65)$$

Comparing eq. (6.65) with eq. (6.43), we obtain

$$L_A P_3(u) = \overline{F}_1^3(u) - W(u) \quad (6.66)$$

where $(u)$ represents the homogeneous polynomials of order-3 in eq. (6.43). On $H_2^3$, the set $\left\{\begin{bmatrix}0\\u_1^3\end{bmatrix},\begin{bmatrix}0\\u_1^2u_2\end{bmatrix},\begin{bmatrix}0\\u_1u_1^2\end{bmatrix},\begin{bmatrix}0\\u_2^3\end{bmatrix},\begin{bmatrix}u_1^3\\0\end{bmatrix},\begin{bmatrix}u_1^2u_2\\0\end{bmatrix},\begin{bmatrix}u_1u_2^2\\0\end{bmatrix},\begin{bmatrix}u_2^3\\0\end{bmatrix}\right\}$ is a basis in reversed lexicographic ordering, and eq. (6.66) can be written as the following matrix form according to eq. (6.59):

$$\begin{bmatrix} 0 & 1 & 0 & 0 & -1 & 0 & 0 & 0 \\ -3 & 0 & 2 & 0 & 0 & -1 & 0 & 0 \\ 0 & -2 & 0 & 3 & 0 & 0 & -1 & 0 \\ 0 & 0 & -1 & 0 & 0 & 0 & 0 & -1 \\ 1 & 0 & 0 & 0 & 0 & 1 & 0 & 0 \\ 0 & 1 & 0 & 0 & -3 & 0 & 2 & 0 \\ 0 & 0 & 1 & 0 & 0 & -2 & 0 & 3 \\ 0 & 0 & 0 & 1 & 0 & 0 & -1 & 0 \end{bmatrix} \begin{bmatrix} d_{30} \\ d_{21} \\ d_{12} \\ d_{03} \\ c_{30} \\ c_{21} \\ c_{12} \\ c_{03} \end{bmatrix} = \frac{1}{\omega_0} \begin{bmatrix} \bar{b}_{30} - b \\ \bar{b}_{21} - a \\ \bar{b}_{12} - b \\ \bar{b}_{03} - a \\ \bar{a}_{30} - a \\ \bar{a}_{21} + b \\ \bar{a}_{12} - a \\ \bar{a}_{03} + b \end{bmatrix} \tag{6.67}$$

From the above equation we get

$$\begin{cases} a = (\bar{b}_{21} + 3\bar{a}_{30} + 3\bar{b}_{03} + \bar{a}_{12})/8 \\ b = (-\bar{a}_{21} + 3\bar{b}_{30} - 3\bar{a}_{03} + \bar{b}_{12})/8 \end{cases} \tag{6.68a}$$

Substituting the coefficients $\bar{a}_{30},\cdots,\bar{b}_{30}$ into eq. (6.68a), we can simplify the result which is derived by the "REDUCE" language:

$$\begin{cases} a = (b_{21}\omega_0 + 3a_{30}\omega_0 + 3b_{03}\omega_0 + a_{12}\omega_0 - b_{02}b_{11} + 2a_{02}b_{02} + a_{11}a_{02} \\ \qquad + a_{11}a_{20} - b_{20}b_{11} - 2b_{20}a_{20})/(8\omega_0) \\ b = (3b_{12}\omega_0 - 9a_{03}\omega_0 - 3a_{21}\omega_0 + 9b_{30}\omega_0 + 4b_{02}^2 + 5b_{02}a_{11} \\ \qquad - 10b_{02}b_{20} - a_{11}^2 + a_{11}b_{20} - 10b_{20}^2 - 10a_{02}^2 + a_{02}b_{11} \\ \qquad - 10a_{02}a_{20} - b_{11}^2 + 5b_{11}a_{20} - 4a_{20}^2)/(24\omega_0) \end{cases} \tag{6.68b}$$

Equation (6.68b) sets up the algebraic relationship between the coefficients in normal form eq. (6.43) and eq. (6.55). These expressions can also be used to calculate the steady-state solution of the autonomous non-linear vibration system.

**Example solving a van der Pol equation**

Consider a van der Pol equation: $x'' + \varepsilon(x^2 - \alpha)x' + x = 0$.

Let $x' = \mu x - y$; when $\mu = \frac{\alpha}{2}$, the above equation can be written as the first-order system of differential equations of standard form

$$\begin{bmatrix} \dot{x} \\ \dot{y} \end{bmatrix} = \begin{bmatrix} \mu & -1 \\ 1 & \mu \end{bmatrix} y + \begin{bmatrix} 0 \\ -\varepsilon x^2 y \end{bmatrix}$$

Comparing the equation with eq. (6.55), we get $c = 0, d = 1, b_{21} = -1$, and the other coefficients are equal to zero. Substituting these into eq. (6.68), we obtain

$a = -1/8, \qquad b = 0$

Because $a < 0$, the steady-state solution is stable. Substituting $a$ and $b$ into eq. (6.26), we obtain $x = r\cos\theta,\ r' = (\frac{\alpha}{2} - \frac{1}{8}r^2)r,\ \theta' = 1$. The steady-state solutions are: $r' = 0$, $r = 2\sqrt{\alpha}$, when $\alpha = 1$, $\gamma = 2$, period $T = 2\pi$. The results obtained in this manner are the same as those in [1] which are calculated by the $K - B$ averaging method.

## 6.5 Calculating Formula for the Hopf Bifurcation Solution

Now we shall discuss the practical method of calculating the Hopf bifurcation. We retain study eq. (6.44), and add the parametric equation $\mu' = 0$. Suppose that we have the PB normal form in complex form

$$z' = \lambda(\mu)z + \sum_{j=1}^{m} c_j(\mu)z|z\bar{z}|^j + o(|z|^{2m+2}) \tag{6.69}$$

where $z, \lambda$ and $c$ are all complex numbers, $\lambda(\mu) = \alpha(\mu) + i\omega(\mu)$, $\lambda(0) = i\omega_0$ that is $\alpha(0) = 0$, $\omega_0 = \omega_0$, $\alpha'(0) \neq 0$, $z = x_1 + ix_2$.

If $z = re^{i\theta}, r, \theta \in \mathcal{R}$, then a polar form of the normal form can be obtained:

$$\begin{cases} r' = \alpha(\mu)r + \sum_{i=1}^{m} \mathrm{Re}[C_j]r^{2j+1} + o(r^{2m+2}) \\ \theta' = \omega(\mu)r + \sum_{i=1}^{m} \mathrm{Im}[C_j]r^{2j+1} + o(r^{2m+1}) \end{cases} \tag{6.70}$$

where $o(r^{2m+2})$ and $o(r^{2m+1})$ are all functions of $r$ and $\theta, \mu$.

The trivial solution of the equation is as follows: $r = 0$, $\theta = \omega t$.

The non-trivial solution: let the first equation of eq. (6.70) be equal to zero, that is, $r$ is a constant, and let $r = \varepsilon > 0$ be small. If we consider only the first $m$ terms, then

$$\alpha(\mu) + \mathrm{Re}[C_1]\varepsilon^2 + \cdots + \mathrm{Re}[C_m]\varepsilon^{2m} = 0 \tag{6.71}$$

Derive the equation with respect to $\mu$, then

$$\frac{\mathrm{d}}{\mathrm{d}\mu}[\alpha(\mu) + \mathrm{Re}[C_1]\varepsilon^2 + \cdots] = \alpha'(0) + 0 \neq 0$$

So $\mu = \mu(\varepsilon)$ can be solved from eq. (6.71) by the implicit function theorem.

Expanding $\alpha(\mu)$ into a Taylor series of $\mu$ at zero, from eq. (6.71) we get

$$\mu(\varepsilon) = -\{\mathrm{Re}[C_1(0)]/\alpha'(0)\}\varepsilon^2 + hot \tag{6.72}$$

We know $\mu(\varepsilon)$ is the even function of $\varepsilon$ from the type of normal form, and the highest order is $2m$. Substituting this into equation $\theta'$, and truncating to order $2m$, we get the solution

$$\theta(t) = \theta(0) + \{\omega_0(\mu(\varepsilon)) + \text{Im}[C_1(\mu(\varepsilon))]\varepsilon^2] + \cdots\}t \tag{6.73}$$

Removing higher order terms from eq. (6.72) and eq. (6.73), we obtain two coupled equations of $\mu$ and $\varepsilon$, which are unique solutions of bifurcation corresponding to the normal form:

$$\begin{aligned} r &= \varepsilon \\ \mu &= -\left\{\frac{\text{Re}[C_1(0)]}{\alpha'(0)}\varepsilon^2 + \cdots\right\} \\ \theta &= \{\omega_0(0) + \text{Im}[C_1(0)]\varepsilon^2 + \cdots\}t \end{aligned} \tag{6.74}$$

When solving eq. (6.74) we can calculate to any precision, which corresponds to the unique periodic solution of the original non-linear equation. In order to obtain the period $T$, we need to integrate the second equation of eq. (6.74) from 0 to $2\pi$:

$$2\pi - 0 = T[\omega(0) + \text{Im}(C_1)\varepsilon^2 + \cdots]$$

where $\omega = \omega(\mu(\varepsilon)) = \omega_0 + \omega'(0)\mu + \cdots$. Substituting $\mu$ of eq. (6.74) into this, we obtain $\omega = \omega_0 + \omega'(0)\mu_2\varepsilon^2 + \cdots$, where $\mu_2 = -\left[\frac{\text{Re}(C_1(0))}{\alpha'(0)}\right]$. $T$ is given by

$$\begin{aligned} T &= 2\pi / \{\omega_0 + [\omega_0'(0)\mu_2 + \text{Im}\{C_1\}\varepsilon^2\} \\ &= \frac{2\pi}{\omega_0}\left\{1 - \left[\frac{\omega_0'(0)\mu_2 + \text{Im}(C_1)}{\omega_0}\right]\varepsilon^2 + \cdots\right\} \\ &= \frac{2\pi}{\omega_0}[1 + \tau_2\varepsilon^2 + \tau_4\varepsilon^4 + \cdots] \end{aligned} \tag{6.75}$$

where $\tau_2 = -(\omega_0'(0)\mu_2 + \text{Im}\,C_1(0))/\omega_0$, $T$ is the even function of $\varepsilon$. From the above analysis we know that the formulae for calculating the bifurcation solutions are

$$\mu(\varepsilon) = \mu_2\varepsilon^2 + \mu_4\varepsilon^4 + \cdots,\quad T(\varepsilon) = \frac{2\pi}{\omega_0}[1 + \tau_2\varepsilon^2 + \tau_4\varepsilon^4 + \cdots] \tag{6.76}$$

where $\mu_2 = -\text{Re}[C_1(0)]/\alpha'(0)$, $\tau_2 = -\frac{1}{\omega_0}(\omega_0'\mu_2 + \text{Im}\,C_1(0))$. Formulae (6.76) are sometimes called HKW (B.D. Hassard, N.D. Kazarinoff and Y-H. Wan) Hopf bifurcation formulae. $\mu_2$ determines the direction of bifurcation ,$\tau_2$ determines the change in period.

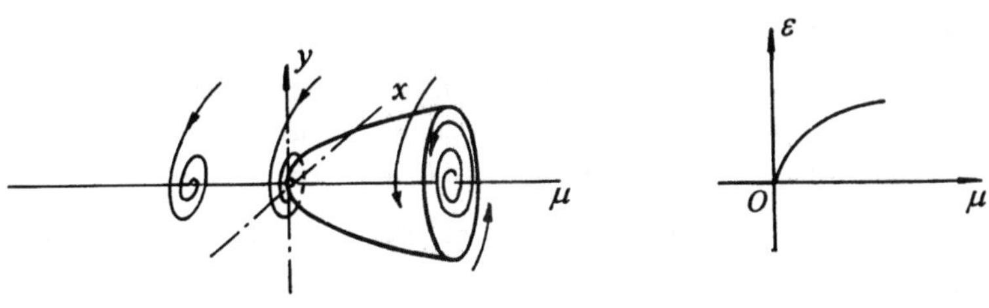


Fig. 6.4

When $\mu_2 \neq 0 \neq \tau_2$, it is the general or non-degenerative case. If $\mu_2 = 0$ is degenerate, $\mu_4$, ... can be computed, and the numbers $\mu_2, \mu_4$ are called focal numbers. The Hopf bifurcation figure for the non-degenerate case is shown in Fig. 6.4.

In the degenerate case, we may study the property of bifurcation according to universal unfolding theory. Refer to [4].

## 6.6 Stability of the Hopf Bifurcation Solution

So far, we have studied the existence and the computing method for the bifurcation solution in non-linear ordinary differential systems under the condition of the Hopf bifurcation theorem. Now when we study the stability of the bifurcation solution, we need to use Floquet theory which has been introduced in chapter 1. In this section we shall discuss another example of the application of this theory.

Suppose that we have obtained the periodic solution $p(t)$ of the Hopf bifurcation of a non-linear autonomous equation, its period being $T$, i.e.

$$p(t) = p(t+T), \quad \text{for any } t,\ T > 0$$

In order to study its stability, give $p(t)$ a perturbation $hq(t)$, then $y = p(t) + hq(t)$. To derive the above equation, we have to expand the vector function of eq. (6.10) into a Taylor series at $p(t)$, so

$$y' = p'(t) + hq'(t)$$

$$f(y) = f(p(t) + hq(t)) = f(p(t)) + f_y(p(t))hq(t) + \cdots$$

Substituting the two equations into eq. (6.10) we obtain the perturbation equation, its linear parts being

$$q'(t) = A(t)q(t) \tag{6.77}$$

where $A(t) = f_y(p(t))$, and $A(t) = A(t+T)$.

Now let us study the stability of the solution of linear differential equations with periodic coefficients. Floquet proved that the fundamental solution matrix of eq. (6.77) is

$$\Phi(t) = p(t)e^{Bt} \tag{6.78}$$

where $p(t)$ is a periodic matrix with $T$ period, and $B$ is a $n \times n$ constant matrix.

Note 1. If $B$ has a zero eigenvalue, and an eigenvector corresponding to it, $v$, then $Bv = 0$, and $e^{BT}v = Iv = v$. As $|x| \to 0$, so $\Phi(t)v$ is also a periodic solution of eq. (6.77), that is $\Phi(T)v = p(T)e^{BT}v = p(0)v = \Phi(0)v$. This equation shows that if $p(t)$ is a periodic function, then $\Phi(t)$ is also a periodic function.

Note 2. If $B$ has an eigenvalue whose real part is greater than zero, then there exists a solution $|x| \to 0$ when $t \to \infty$ ; if $B$ has an eigenvalue whose real part is smaller than zero, when $t \to \infty$, there exists $|x| \to 0$, in which $x(t) = p(t)e^{Bt}v$.

Note 3. If $p(t)$ is a periodic solution of $y' = f(y)$, then $p' = f(p(t))$. Let $q(t) = p'(t)$, then $q'(t) = P''(t) = \dfrac{\mathrm{d}}{\mathrm{d}t}[f(p(t)] = D_y[p(t)]q(t) = A(t)q(t)$. If $p(t)$ is the solution of eq. (6.10), then $q(t) = p'(t)$ is the solution of eq. (6.77), and $T$ is also the period of $q(t)$. The derivative of eq. (6.78) is

$$\Phi'(t) = p'(t)e^{Bt} + p(t)e^{Bt}B \tag{6.79}$$

When $B$ has at least one zero eigenvalue, then $B = 0$, so $\Phi'(t) = p'(t)$, that is, we have proved that $p'(t)$ is the solution of eq. (6.77). Otherwise, if the derivative $p'(t)$ of solution $p(t)$ of eq. (6.10) is the solution of eq. (6.77), then $B$ has at least one zero eigenvalue.

Substituting $t+T$ into $\Phi(t)$, then $\Phi(t+T)$ is still the fundamental solution matrix of eq. (6.77), and there is a non-singular constant matrix $R$, so that

$$\Phi(t+T) = \Phi(t)R \tag{6.80}$$

Suppose that $\Phi_{ij}(t)$ satisfies the initial condition

$$\Phi_{ij}(0) = \delta_{ij} \tag{6.81}$$

then the element of $R$ is $r_{ij} = \Phi_{ij}(T)$.

Suppose the characteristic equation

$$|R - \rho I| = 0 \tag{6.82}$$

has $n$ different eigenvalues $\rho_1, \rho_2, \cdots, \rho_n$, then $R$ can be transformed into a diagonal

type $R = \begin{bmatrix} \rho_1 & & & 0 \\ & \rho_2 & & \\ & & \rho_3 & \\ 0 & & & \rho_4 \end{bmatrix}$.

So the solutions of eq. (6.77) satisfy

$$\Phi_j(t+T) = \rho_j \Phi_j(t) \tag{6.83}$$

and the solution of eq. (6.77) can also be written as eq. (6.78), i.e.

$$\Phi_j(t) = p_{j(t)} e^{\frac{t}{T}\ln\rho_j} = e^{\lambda_j t} p_j(t)$$

Where $\lambda_j$ is the eigenvalue of $B$, the equation can be written as

$$\Phi_j(t) = e^{\frac{t}{T}\ln|\rho_j|} \varphi_j(t) \tag{6.84}$$

If $\rho_k$ is an $r$-repeat root, then

$$\Phi_k^{(1)} = e^{\lambda_k t}\Phi_k^{(1)}(t), \quad \Phi_k^{(2)} = e^{\lambda_k t}\left[t\Phi_k^{(1)}(t) + p_k^{(2)}(t)\right]$$

......

$$\Phi_k^{(r)} = e^{\lambda_k t}\left[t^{(r-1)}\Phi_k^{(1)}(t) + \cdots + tp_k^{(r-1)}(t) + p_k^{(r)}(t)\right]$$

where $p_k^{(s)}(t+T) = p_k^{(s)}(t)$; $\Phi_k^{(s)}$ is in permanent terms. The eigenvalue of B is a Floquet exponent; the eigenvalue of $e^{BT}$ is the Floquet characteristic multiplier and the eigenvalue of $e^{BT} = e^{(\text{eigenvalue of } B)T}$.

If the module of all the character multipliers is smaller than 1, then the zero solution of eq. (6.77) is asymptotically stable. If the elementary factors to which the characteristic multiplier with a module of 1 are all of one order, and the module of other characteristic multipliers is smaller than 1, then the zero solution is stable. If the characteristic multiplier with at least one model of 1 has one elementary factor of an order of two or more, then the zero solution is unstable.

**Floquet Identity**

Let $\lambda_1, \lambda_2, \cdots, \lambda_n$ be eigenvalues of $B$, called Floquet exponents, so that

$$\sum_{i=1}^{n} \lambda_i = \frac{1}{T} \int_0^T \operatorname{tr}(A(s))\mathrm{d}s \tag{6.85}$$

If $\rho_1, \rho_2 \cdots, \rho_n$ are Floquet multipliers, then

$$\prod_{i=1}^{n} \rho_i = \exp \int_0^T \operatorname{tr}(A(s))\mathrm{d}s \tag{6.86}$$

For Hopf bifurcation, because $n=2$ and the system has a periodic solution, one of $\lambda_1$ and $\lambda_2$ should be equal to zero. Let $\lambda_1 = 0$, then

$$\lambda_2 = \frac{1}{T} \int_0^T \operatorname{tr}(A(s))\mathrm{d}s \tag{6.87}$$

The sign of $\lambda_2$ determines the stability of the periodic orbit, i.e. if $\mathrm{Re}\lambda_2 < 0$, the orbit is asymptotically stable; if $\mathrm{Re}\lambda_2 > 0$, the orbit is unstable. So the key to deciding the stability is to calculate $\lambda_2$.

Consider the ordinary differential equation in PB normal form in $\mathcal{R}^2$:

$$\begin{aligned} \begin{bmatrix} x_1 \\ x_2 \end{bmatrix} &= F\left\{\begin{bmatrix} x_1 \\ x_2 \end{bmatrix}, \mu\right\} \\ &= \begin{bmatrix} \alpha & -\omega_0 \\ \omega_0 & \alpha \end{bmatrix}\begin{bmatrix} x_1 \\ x_2 \end{bmatrix} + r^2 \begin{bmatrix} \mathrm{Re}[c_1] & -\mathrm{Im}[c_1] \\ \mathrm{Im}[c_1] & \mathrm{Re}[c_1] \end{bmatrix}\begin{bmatrix} x_1 \\ x_2 \end{bmatrix} + o(r^5) \\ &= \begin{bmatrix} x_1[\alpha + r^2 \mathrm{Re}[c_1] + o(r^4)] - x_2[\omega + r^2 \mathrm{Im}[c_1] + o(r^4)] \\ x_1[\omega_0 + \mathrm{Im}[c_1]r^2 + o(r^4)] + x^2[\alpha + r^2 \mathrm{Re}[c_1] + o(r^4)] \end{bmatrix} = \begin{bmatrix} F_1 \\ F_2 \end{bmatrix} \end{aligned} \tag{6.88}$$

Let $p(t)$ be the periodic solution with $T$ period, so

$$A(t) = D_x F(p(t), \mu) = \begin{bmatrix} F_{11} & F_{12} \\ F_{21} & F_{22} \end{bmatrix} \tag{6.89}$$

is a $T$-periodic matrix made up of periodic functions as elements.

Because $\|p(t)\| = \varepsilon + o(\varepsilon^2)$, let $r' = 0$ in eq. (6.70), so

$$\alpha(\mu) + \varepsilon^2 \mathrm{Re}[c(\mu)] + o(\varepsilon^4) = 0 \tag{6.90}$$

Calculate the elements in eq. (6.89):

$$F_{11} = \frac{\partial F_1}{\partial x_1} = [\alpha + r^2 \operatorname{Re}[c_1] + o(r^4)] + x_1[2x_1 \operatorname{Re}[c_1] + o(r^3)] - 2x_1x_2 \operatorname{Im}[c_1] + \cdots$$
$$= 2x_1^2 \operatorname{Re}[c_1] - 2x_1x_2 \operatorname{Im}[c_1] + \cdots,$$
$$F_{22} = x_1[2x_2 \operatorname{Im}[c_1] + o(r^3)] + x_2[2x_2 \operatorname{Re}[c_1] + o(r^3)]$$
$$= 2x_1x_2 \operatorname{Im}[c_1] + 2x_2^2 \operatorname{Re}[c_1] + \cdots \tag{6.91}$$

In the same way we can calculate $F_{12}$ and $F_{21}$.

Substituting (6.91) into eq. (6.89), we have

$$\operatorname{tr} D_x F(x,\mu)\big|_{\|x\|=\|p(t)\|=\varepsilon} = 2\varepsilon^2 \operatorname{Re}[c_1] + o(\varepsilon^4) \tag{6.92}$$

Substituting the equation into eq. (6.87) and integrating it, we obtain:

$$\lambda_2 = \frac{1}{T}\int_0^T \operatorname{tr}(A(s))\mathrm{d}s) = 2\varepsilon^2 \operatorname{Re}[c_1] + o(\varepsilon^4) \tag{6.93}$$

The sign of $\lambda_2$ determines the stability of $p(t)$. Generally we would write the expression in the form of the $\varepsilon$ power series (in eq. (6.93), $n$=2)

$$\lambda_2(\varepsilon) = \beta_2\varepsilon^2 + \beta_4\varepsilon^4 \tag{6.94}$$

Comparing eq. (6.93) with eq. (6.94), we have

$$\beta_2 = 2\operatorname{Re}[c_1(0)] \tag{6.95}$$

From eq. (6.76) we know that $\mu_2 = -\operatorname{Re}[c_1(0)]/\alpha'(0)$. Substituting this into eq. (6.95), we obtain the Hopf exchange formula of stability:

$$\beta_2 = -2\mu_2\alpha'(0) \tag{6.96}$$

From the equation we know that $\beta_2$ and $\alpha'(0)$ are not independent. The three coefficients are all key coefficients in studying Hopf bifurcation: the sign of $\alpha'(0)$ determines the stability of the zero solution; the sign of $\mu_2$ determines the direction of the bifurcation curve; sign of $\beta_2$ determines the stability of the non-zero solution.

If $\mu_2\alpha'(0) \neq 0$, the Hopf bifurcation diagrams have four different cases, see Fig. 6.5. In Fig. 6.5(a), when $\alpha'(0) > 0$, $\alpha$ increases; if $\alpha(\mu_0) = 0$, when $\mu < \mu_0$, $\alpha$ is negative, and the zero solution is stable; when $\mu > \mu_0$, $\alpha$ is positive, and the zero solution is unstable. When $\mu > 0$ and $\alpha'(0) > 0$, $\beta_2$ is negative from eq. (6.96), and the non-zero solution is stable. In the same way we can analyze the stability of the zero solution and non-zero solutions in Fig. 6.5(b), (c), (d).

It is easy to generalize the above discussion to $\mathcal{R}^n$. Suppose $A_0 = D_y(0,0)$ has $(n-2)$ eigenvalues with negative real parts. The stability of $y$=0 is determined by the sign of $\alpha(\mu)$. It is easy to show that for the periodic solution $p(t)(n-2)$ Floquet exponents will have negative real parts, a zero characteristic exponent and another Floquet exponent given by $\lambda_2 = \beta_2\varepsilon^2 + o(\varepsilon^4)$, where $\beta_2 = -2\alpha'(0)\mu_2$. In such a case, the "exchange formula of stability" prevails. For proof of the Hopf bifurcation theorem of dimension $n$, refer to [51].

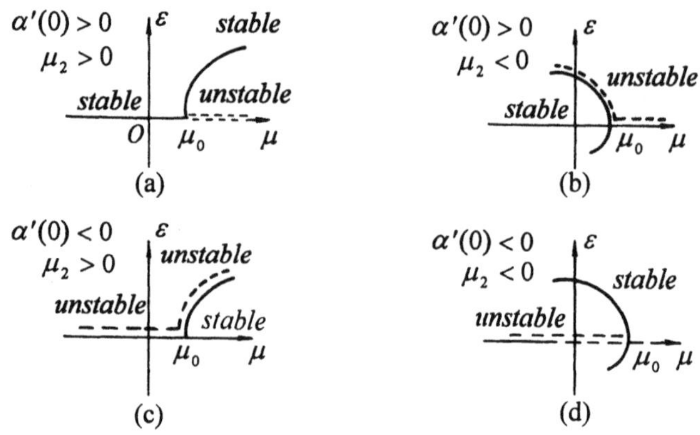


Fig. 6.5

## 6.7 Method for calculating Hopf Bifurcation Solution Coefficients

The method described in this section was proposed by Iooss and Joseph [52]. It does not need require translation of the original equation into the normal form reduced by a nearly identical non-linear transformation, but calculation of the solution of the non-linear system can be done using the power series and the Fredholm alternative method directly. Suppose

$$\mu = \sum_{k \ge 1} \mu_k \varepsilon^k \tag{6.97a}$$

$$\tau = \sum_{k \ge 1} \tau_k \varepsilon^k \tag{6.97b}$$

$$y = \sum_{k \ge 1} y(s)_k \varepsilon^k \tag{6.97c}$$

where $\varepsilon$ is amplitude.

We have proved earlier that $\mu, \tau$ and $y$ are all even functions of $\varepsilon$, and for the purpose of generalization, we suppose again that $\mu$ and $\tau$ are power series starting from the first order of $\varepsilon$, to ensure that the coefficients of the $\varepsilon$ odd terms are zero.

Consider a two-dimensional ordinary differential equation

$$y' = f(y, \mu) \tag{6.98}$$

Let $f(0,0) = 0$ and $D_y(0,0) = A_0$, the eigenvalues of $A_0$ being $\pm i\omega_0$. Defining symbol

$f^{ij} = \dfrac{D_y^i D_\mu^j f(y,\mu)}{i!j!}\Big|_{(0,0)}$ then the Taylor expansion of $f(y,\mu)$ is

$$f(y,\mu) = f(0,0) + f^{1,0}y + \mu f^{1,1}(y) + f^{20}(y,y) + f^{3,0}(y,y,y) + \cdots + \mu f^{2,1}(y,y) + \cdots$$

Let $\quad f^{1,0} = A_0, \quad f^{1,1} = \dfrac{\partial A_0}{\partial \mu}$

then $y' = A_0 y + \mu A_0^1 y + f^{2,0}(y,y) + f^{3,0}(y,y,y)$ (6.99)

Equation (6.99) is an autonomous system, whose solution is periodic of period $T$ in $t$:

$$T = T(\varepsilon) = \frac{2\pi}{\omega_0}[1 + \tau(\varepsilon)] \tag{6.100}$$

where $\tau(\varepsilon)$ is determined by eq. (6.97b). In order to get the periodic solution of eq. (6.99) in $2\pi$ period, we need to rescale the time of the variable. If we take the new variable $S = \omega_0(1 + \tau(\varepsilon)t)$ then eq. (6.98) becomes $\omega_0(1+\tau)\frac{dy}{ds} = f(y,\mu)$. Rescale the new variable once more without any loss of generality. Let $\omega_0 =$ , then

$$(1 + \tau(\varepsilon))\frac{dy}{ds} = f(y,\mu) \tag{6.101}$$

Substituting the Taylor series of $f(y,\mu)$ into the above equation, and rearranging terms, gives

$$\left[\frac{d}{ds} - A_0\right]y = -\tau\frac{dy}{ds} + \mu A_0' y + f^{2,0}(y,y) + f^{3,0}(y,y,y) + \cdots \tag{6.102}$$

Let operator $\frac{d}{ds} - A_0 = L$ and substitute $\mu, \tau, y$ of eq. (6.97) into eq. (6.102). Let the coefficients of the expression on both sides of the same $\varepsilon$ order be equal, so

$$o(\varepsilon^1) \qquad Ly_1 = 0$$

$$o(\varepsilon^2) \qquad Ly_2 = -\tau\frac{dy_1}{ds} + \mu_1 A_0^1(y_1) + f^{2,0}(y_1, y_1)$$

$$o(\varepsilon^3) \qquad Ly_3 = \cdots$$

$$\cdots$$

$$o(\varepsilon^k) Ly_k = -\tau_{k-1}\frac{dy_1}{ds} + \mu_{k-1} A_0^1 y_1$$

$$+ \text{ (inclusion terms } of\ y_1, \cdots, y_{k-2}, \tau_1, \cdots, \tau_{k-2}, \mu_1, \cdots, \mu_{k-2}) \tag{6.103}$$

Solve the equations of recurrence to the above equations. Applying the Fredholm alternative method for every $(\varepsilon^k)$, the equation of $o(\varepsilon^1)$ becomes $y_1 \in \eta(L) \subset \mathcal{C}_{2\pi}^1$ where $\eta(L)$ is the null space of $L$. Let $C$ be the complex eigenvector corresponding to the eigenvalue $i$ of $A_0$, so $e^{is}C$ satisfies equation $Le^{is}C = 0$, i.e.

$$\left(\frac{d}{ds} - A_0\right)e^{is}C = ie^{is}C - A_0 e^{is}C = 0 \tag{6.104}$$

Let $\varphi_1 = \text{Re}[e^{is}C] = \frac{1}{2}[e^{is}C + e^{-is}\overline{C}]$ and $c = a + ib$, $\overline{c} = a - ib$, then

$$\varphi_1 = a\cos(s) - b\sin(s) \tag{6.105a}$$

$$\varphi_2 = \text{Im}[e^{is}C] = \text{Im}\left\{\frac{1}{2i}[e^{is}c - e^{-is}\overline{c}]\right\} = a\sin(s) + b\cos(s) \tag{6.105b}$$

Then $\{\varphi_1,\varphi_2\}$ is a basis of $\eta(L)$. And let

$$y_1(s)=\varphi_1(s) \tag{6.106}$$

Now let us discuss the $o(\varepsilon^2)$ equation. Because it is a non-homogeneous linear ordinary differential equation, if $L$ is not invertible to the terms on the right-hand side (RHS) of the equation $o(\varepsilon^2)$, then the $o(\varepsilon^2)$ equation has no solution to these terms, so the Fredholm alternative theorem is used. The solutions exist, if RHS $\in \mathcal{R}(L)$ in $o(\varepsilon^2)$. We have proved $\mathcal{R}(L)=(\eta(L^*))^{\perp}$ where $L^*=-\frac{\mathrm{d}}{\mathrm{d}s}-aA_0^*$ is the formal adjoint of $L$, or the formal conjugate operator, and $A_0^*$ is the transposition of $A_0$. The complex inner product in $\mathcal{C}_{2\pi}$ is as follows

$$<u,v>=\frac{1}{2\pi}\int_0^{2\pi} v^*(s)u(s)\mathrm{d}s \tag{6.107}$$

where $v^*(s)$ is the conjugate transposition of $v(s)$. If $<u,v>=0$, this means that $u$ is vertical to $v$. $\{\psi_1,\psi_2\}$ is the basis of $\eta(L^*)$. Let $d$ be an eigenvector of $A^*$, so $A^*d=-id$, then $L^*(e^{is}d)=\left[-\frac{\mathrm{d}}{\mathrm{d}s}-A^*\right]e^{is}d=-ie^{is}d+ide^{is}=0$ Let $\psi_1=\mathrm{Re}[e^{is}d]$, $\psi_2=\mathrm{Im}[e^{is}d]$, so $\{\psi_1,\psi_2\}$is a basis of $\eta(L^*)$. The range of $L$ is $\mathcal{R}(L)$:

$$\mathcal{R}(L)=\{v\in\mathcal{C}_{2\pi}^0 \,|< v,\varphi_i>=0, i=1,2\}$$

i.e. the range of $L$ is spanned by $v$, and $v$ is orthogonal to $\psi_i$, or the frequency of function $v$ is greater than 1. This conclusion can also be obtained from $\mathcal{R}(L)=(\eta(L^*))^{\perp}$ directly. Applying the above to the $o(\varepsilon^2)$ equation, we find the necessary condition for $o(\varepsilon^2)$ to have solutions

$$<\mathrm{RHS}(y_2),\psi_i>=0,\ \ i=1,2$$

where RHS$(y_2)$ shows that the right hand side of the $o(\varepsilon^2)$ equation, or the above equation an be written as

$$<-\tau\frac{\mathrm{d}\varphi_1}{\mathrm{d}s}+\mu_1 f^{1,1}\varphi_1+f^{2,0}(\varphi_1,\varphi_1),\psi_i>=0, i=1,2 \tag{6.108}$$

Equation (6.108) contains two scalar equations, from which we can determine two real unknown qualities $\tau_1$ and $\mu_1$. As $\psi_1+i\psi_2=e^{is}d$, eq. (6.108) can be written as one complex equation:

$$<-\tau_1\frac{\mathrm{d}\varphi_1}{\mathrm{d}s}+\mu_1 f^{1,1}\varphi_1+f^{2,0}(\varphi_1,\varphi_1),e^{is}d>=0 \tag{6.109}$$

Now calculate the three inner products of the above equation respectively:

$$\begin{aligned}<\frac{\mathrm{d}\varphi_1}{\mathrm{d}s},e^{is}d> &=\frac{1}{2\pi}\int_0^{2\pi} e^{-is}d\frac{1}{2}(ie^{is}c-ie^{-is}\bar{c})\mathrm{d}s\\ &=\frac{i}{4\pi}\int_0^{2\pi}(d^*c-d^*\bar{c}e^{-2is})\mathrm{d}s \qquad (6.110a)\\ &=\frac{1}{2}d^*ci\end{aligned}$$

We have proved that $d^* c \neq 0$. Let $d^* c = 2$ after rescaling, so

$$< \frac{d\varphi_1}{ds}, e^{is} d >= i \tag{6.110b}$$

The second inner product is

$$< f^{1,1}\varphi_1, e^{is} d >= \frac{1}{2\pi}\int_0^{2\pi} e^{-is} d^* \frac{1}{2}(f^{1,1}e^{is}c + f^{1,1}e^{-is}\bar{c})ds$$
$$= \frac{d^* f^{1,1}c}{4\pi}\int_0^{2\pi} ds + \frac{d^* f^{1,1}\bar{c}}{4\pi}\int_0^{2\pi} e^{-2is}ds$$
$$= \frac{1}{2} d^* f^{1,1} c$$

The perturbation lemma: $d^* f^{1,1} c = \lambda'(0)[d^* c]$ (6.111)

Because $A(\mu)c(\mu) = \lambda(\mu)c(\mu)$, the derivative of the left equation to $\mu$ at $\mu = 0$ is as follows $A'(0)c(0) + A_0(0)c'(0) = \lambda'(0)c(0) + \lambda(0)c'(0)$ As $A(0)c'(0) = \lambda(0)c'(0)$, we have $A'(0)c(0) = \lambda'(0)c(0)$ and multiplying the left-hand side of the above equation by $d^*$, we obtain $d^* A'(0)c(0) = d^*\lambda'(0)c(0) = \lambda'(0)c(0)d^* = \lambda'(0)d^* c(0)$. As $f^{11} = A'(0)$, the perturbation lemma has been proved.

Substituting (6.111) into the second inner product, yields

$$< f^{11}\varphi_1, e^{is} d >= \frac{1}{2\pi}\lambda' d^* c = \lambda'(0) = \alpha'(0) + i\omega'(0) \tag{6.112}$$

The third inner product

$$< f^{2,0}(\varphi_1,\varphi_1), e^{is} d >= \frac{1}{2\pi}\int_0^{2\pi}\frac{1}{4} e^{-is} d^* f^{2,0}(e^{is}c + e^{-is}\bar{c}, e^{is}c + e^{-is}\bar{c})ds$$
$$= \frac{1}{8\pi}\int_0^{2\pi} e^{-is} d^*[f^{2,0}(c,c)e^{is} + f^{2,0}(c,\bar{c}) + f^{2,0}(\bar{c},c) + e^{-2is} f^{2,0}(\bar{c},\bar{c})]ds = 0 \tag{6.113}$$

If we substitute (6.110), (6.112) and (6.113) into eq. (6.109), then the $o(\varepsilon^2)$ equation is $-i\tau_1 + [\alpha'(0) + i\omega'(0)]\mu_1 = 0$. Dividing the above equation into a real part and an imaginary part, we get real form equations:

$$\begin{bmatrix} \alpha'(0) & 0 \\ \omega'(0) & -1 \end{bmatrix}\begin{bmatrix} \mu_1 \\ \tau_1 \end{bmatrix} = \begin{bmatrix} 0 \\ 0 \end{bmatrix} \tag{6.114}$$

Because $\begin{bmatrix} \alpha'(0) & 0 \\ \omega'(0) & -1 \end{bmatrix} = -\alpha'(0) \neq 0$ there exist unique solutions

$$\mu_1 = 0, \ \tau_1 = 0 \tag{6.115}$$

i.e. we have proved the previous conclusion that the coefficients of order-1 in the expansion of $\mu(\varepsilon)$ and $\tau(\varepsilon)$ are equal to zero. In order to calculate $\mu_2$ and $\tau_2$ in the expansion of $\mu(\varepsilon)$ and $\tau(\varepsilon)$, we now study the $o(\varepsilon^2)$ equation. Firstly calculate $y_2$. From eq. (6.103) we know that

$$\left[\frac{\mathrm{d}}{\mathrm{d}s}-A_0\right]y_2=f^{2,0}(\varphi_1,\varphi_1)=\frac{1}{4}[e^{2is}f^{2,0}(c,c)+e^{-2is}(\bar{c},\bar{c})+2f^{2,0}(c,\bar{c})]$$

The equation is a linear ordinary differential equation, and its $2\pi$-period $y_2$ is as follows

$$y_2=a_0+a_2e^{2is}+\bar{a}_2e^{-2is} \tag{6.116}$$

where $a_0=-\frac{1}{2}A_0f^{2,0}(c,\bar{c})$, $a_2=-\frac{1}{4(A_0-2iI)}f^{2,0}(c,c)$ (6.117)

Considering eq. (6.115), then the $o(\varepsilon^3)$ equation is

$$Ly_3=-\tau_2\frac{\mathrm{d}\varphi_1}{\mathrm{d}s}+\mu_2f^{1,1}\varphi_1+2f^{2,0}(\varphi_1,y_2)+f^{3,0}(\varphi_1,\varphi_1,\varphi_1) \tag{6.118}$$

the solvable condition of the equation is $<\mathrm{RHS}(y_3),e^{is}d>=0$ where $\mathrm{RHS}(y_3)$ shows the right-hand side of eq. (6.118), i.e.

$$-\tau_2<\frac{\mathrm{d}\varphi_1}{\mathrm{d}s},e^{is}d>+\mu_2<f^{1,1}\varphi_1,e^{is}d>+2<f^{2,0}(\varphi_1,y_2),e^{is}d>+$$
$$<f^{3,0}(\varphi_1,\varphi_1,\varphi_1),e^{is}d>=0$$

Substituting eq. (6.110b) and eq. (6.112) into the above equation, and calculating the third and fourth inner products, then we get

$$-i\tau_2+[\alpha'(0)+i\omega'(0)]\mu_2+\frac{1}{2\pi}\int_0^{2\pi}e^{-is}d^*[e^{is}f^{2,0}(c,a_0)+e^{-is}f^{2,0}(\bar{c},a_0)+e^{3is}f^{2,0}(c,a_2)$$
$$+e^{is}f^{2,0}(\bar{c},a_2)+e^{-is}f^{2,0}(c,\bar{a}_2)+e^{-3is}f^{2,0}(\bar{c},\bar{a}_2)]\mathrm{d}s+\frac{1}{2\pi}\frac{d^*}{8}\int_0^{2\pi}e^{-is}[e^{3is}f^{3,0}(c,c,c)$$
$$+3e^{is}f^{3,0}(c,c,\bar{c})+\cdots]\mathrm{d}s=-i\tau_2+[\alpha'(0)+i\omega'(0)]\mu_2+d^*[f^{2,0}(c,a_0)+f^{2,0}(\bar{c},a_2)$$
$$+\frac{3}{8}f^{3,0}(c,c,\bar{c})]=0$$

Dividing the above equation into a real and an imaginary part, we obtain

$$\begin{bmatrix}\alpha'(0) & 0\\ \omega'(0) & -1\end{bmatrix}\begin{bmatrix}\mu_2\\ \tau_2\end{bmatrix}=\begin{bmatrix}-\mathrm{Re}\,d^*\left[f^{2,0}(c,a_0)+f^{2,0}(\bar{c},a_2)+\frac{3}{8}f^{3,0}(c,c,\bar{c})\right]\\ -\mathrm{Im}\,d^*\left[f^{2,0}(c,a_0)+f^{2,0}(\bar{c},a_2)+\frac{3}{8}f^{3,0}(c,c,\bar{c})\right]\end{bmatrix}$$

After solving it, we get

$$\mu_2=-\frac{1}{\alpha'(0)}\mathrm{Re}\,d^*\left[f^{2,0}(c,a_0)+f^{2,0}(\bar{c},a_2)+\frac{3}{8}f^{3,0}(c,c,\bar{c})\right]$$

$$\tau_2=\mathrm{Im}\,d^*\left[f^{2,0}(c,a_0)+f^{2,0}(\bar{c},a_2)+\frac{3}{8}f^{3,0}(c,c,\bar{c})\right] \tag{6.119}$$

We can continue the calculation if necessary.

## 6.8 Bifurcation Problem Introducing Double Zero Eigenvalues

Consider a non-linear ordinary differential equation

$$y' = f(y,\mu), \quad f: \mathcal{R}^n \times \mathcal{R}^2 \to \mathcal{R}^n \tag{6.120}$$

where $\mu \in \mathcal{R}^2$. Since it has double zero eigenvalues, two unfolding parameters must be added for a universal unfolding bifurcation equation.

Suppose $f(0,0) = 0$, i.e. the equilibrium point is $(0,0)$, and $A = D_y f(0,0)$ has a zero eigenvalue with an algebraic multiplicity of 2. And let the real parts of other eigenvalues be smaller than zero. Firstly according to the centre manifold theorem, take the linear change of coordinates, to convert the variable to $\begin{bmatrix} x \\ z \end{bmatrix}$, $x \in \mathcal{R}^2, z \in \mathcal{R}^{n-2}$. $A$ can be changed to a diagonal form $\begin{bmatrix} B & 0 \\ 0 & C \end{bmatrix}$, where $B$ is a $2 \times 2$ matrix, $C$ is an $(n-2) \times (n-2)$ matrix, so eq. (6.120) becomes

$$x' = BX + g_1(x,z,\mu), \quad x \in \mathcal{R}^2 \tag{6.121a}$$

$$z' = CX + g_2(x,z,\mu), \quad z \in \mathcal{R}^{n-2} \tag{6.121b}$$

$$\mu' = 0 \quad \mu \in \mathcal{R} \tag{6.121c}$$

where $B$ has two zero eigenvalues, and all the negative real parts of the eigenvalues of $C$ are less than zero. According to the centre manifold theorem, it is not difficult to calculate the centre manifold equation

$$z = h(x,\mu) = o(\|(x,\mu)\|^2) \tag{6.122}$$

Substituting $z$ into eq. (6.121a), we obtain the two-dimensional manifold equations $\quad x' = B(\mu)x + g(x,\mu), \ \mu' = 0 \quad$ (6.123)

where $g(x,\mu) = g_1(x, h(x,\mu), \mu)$.

### 6.8.1 The Standard Form of *B*

Now let us discuss the Jordan's normal form of *B*. When eq. (6.120) has two zero eigenvalues, matrix $B$ has two possibilities:

$$B = \begin{bmatrix} 0 & 0 \\ 0 & 0 \end{bmatrix} \tag{6.124a}$$

and

$$B = \begin{bmatrix} 0 & 1 \\ 0 & 0 \end{bmatrix} \tag{6.124b}$$

At this point, the eigenvalue $\lambda$ is zero, i.e.

$$\begin{bmatrix} \lambda & 0 \\ 0 & \lambda \end{bmatrix} = 0 \tag{6.125a}$$

and

$$\begin{bmatrix} \lambda & 1 \\ 0 & \lambda \end{bmatrix} = 0 \tag{6.125b}$$

Equation (6.125a) is semi-simple and eq. (6.125b) is non-semi-simple. The two different cases of eq. (6.125) will correspond to the dynamical behaviours with great discrepancy. Which one is more generalised? To resolve this, let us study an arbitrary $2 \times 2$ matrix $A$

$$A = \begin{bmatrix} a & b \\ c & d \end{bmatrix} \tag{6.126}$$

Its characteristic equation is $\lambda^2 - (a+d)\lambda + (ad - bc) = 0$ or

$$\lambda^2 - (\text{tr } A)\lambda + (\det A) = 0 \tag{6.127}$$

For the double zero eigenvalue, we get $\text{tr}A = \lambda_1 + \lambda_2 = 0$, $\det A = \lambda_1 \cdot \lambda_2 = 0$.

To change $A$ to a semi-simple case as $B$, take the change of variable $(S)_{2\times 2}$,

i.e. $S^{-1}AS = B = \begin{bmatrix} 0 & 0 \\ 0 & 0 \end{bmatrix}$

Since it is impossible that all elements in $S$ are zero, in order to satisfy the above equation, there must have $a = b = c = d = 0$, being semi-simple is, therefore a strictly restricted case, while being non-semi-simple is a more ordinary case. We can clarify this as follows:

$$\lambda_1 = \frac{1}{2}(\text{tr } A) + \sqrt{\Delta} \tag{6.128a}$$

$$\lambda_2 = \frac{1}{2}(\text{tr } A) - \sqrt{\Delta} \tag{6.128b}$$

where $\Delta = \frac{1}{4}(\text{tr } A)^2 - \det A$. For the non-semi-simple case, we have the following:

If $b \neq 0$, from $(\text{tr } A) = 0$, $\det A = 0$, so $c$ and $d$ can be expressed by $a$ and $b$, i.e.

$$A = \begin{bmatrix} a & b \\ -\dfrac{a^2}{b} & -a \end{bmatrix}$$

If $c \neq 0$, from $(\text{tr } A) = 0$, $\det A = 0$, so $b$ and $d$ can be expressed by $a$ and $c$, i.e.

$$A = \begin{bmatrix} a & -\dfrac{a^2}{c} \\ c & -a \end{bmatrix}$$

Thus when there are double zero eigenvalues, there exist two possible parameter sets, and there are two arbitrary parameters in every set. In a diagonal form they may be $\begin{bmatrix} 0 & b \\ 0 & 0 \end{bmatrix}$ and $\begin{bmatrix} 0 & 0 \\ c & 0 \end{bmatrix}$. After rescaling, we can take $b = 1$ and $c =$ , and the corresponding Jordan's normal form is $\begin{bmatrix} 0 & 1 \\ 0 & 0 \end{bmatrix}$ or $\begin{bmatrix} 0 & 0 \\ 1 & 0 \end{bmatrix}$, i.e. eq. (6.124b) is a more general case. In contrast, if $A \neq 0$, but $A$ has double zero eigenvalues, $A$ must be a non-semi-simple case.

**Theorem 6.1 Cayley-Hamilton theorem**

If $A$ has double zero eigenvalues, i.e. $(\operatorname{tr} A)=0$ and $\det A=0$, and $A$ is non-semi-simple, then $A$ is the root of its characteristic polynomial. It is not difficult to prove this theorem. Substituting $(\operatorname{tr} A)=0=(\det A)$ into eq. (6.127), then $\lambda^2=0$. If $A=\lambda$ and is a non-semi-simple case

$$A=\begin{bmatrix}0 & 1\\ 0 & 0\end{bmatrix},\ \text{so } A^2=\begin{bmatrix}0 & 1\\ 0 & 0\end{bmatrix}\begin{bmatrix}0 & 1\\ 0 & 0\end{bmatrix}=\begin{bmatrix}0 & 0\\ 0 & 0\end{bmatrix}=0$$

### 6.8.2 Null Space and Generalised Eigenspace of $A$

There must be a non-zero eigenvector $\varphi_1 \neq 0$ corresponding to the zero eigenvalue of $A$, so that $A\varphi_1=0\ \varphi_1=0$, where we let $\varphi_1=\begin{bmatrix}1\\ 0\end{bmatrix}$. There is another eigenvector $\varphi_2$.

Let $\varphi_2=\begin{bmatrix}0\\ 1\end{bmatrix}$, so that $A=D_y f(0,0)$, so $A\varphi_2$ is another eigenvector corresponding to the zero eigenvalue of $A$, i.e. (6.129)

where $\varphi_1$ and $\varphi_2$ are linearly independent. But only when

$$A\varphi_2=\alpha\varphi_1,\ \alpha\in\mathcal{R}, \alpha\neq 0,\ \text{or } A\varphi_2=\varphi_1 \tag{6.130}$$

is eq. (6.129) correct. We can prove eq. (6.129), by multiplying eq. (6.130) by the left-hand side of $A$. Because the two vectors $\varphi_1$ and $A\varphi_2$ that span null space $\eta(A)$ are equal, so $\eta(A)$ is only spanned by one independent eigenvector, so $\eta(A)$ is one-dimensional

$$y'=0,\ A\varphi_2=\varphi_1 \tag{6.131}$$

By eq. (6.131) we get $\varphi_1\in\mathcal{R}(A)\cap\eta(A)\neq\{0\}$ where $\varphi_1$ is called an eigenvector, and $\varphi_2$ is called a generalised eigenvector. The generalised eigenspace is spanned by eigenvectors $\{\varphi_1,\varphi_2\}$. As $\varphi_1$ and $\varphi_2$ are linearly independent, the dimension of the generalised eigenspace is 2, and the centre manifold is a tangent to the generalised eigenspace.

To sum up, when a two-dimensional system has double zero eigenvalues, there is a flow equation on the two-dimensional centre manifold with a non-semi-simple manifold, and the centre manifold is a tangent to the two-dimensional generalised eigenspace.

### 6.8.3 Bifurcation of the Steady-State Solution

Suppose that a two-dimensional system has a trivial solution (0,0), i.e. $f(0,0)=0$, and $A=D_y f(0,0)$ has double zero eigenvalues. In this case, the null space of $A$, $\eta(A)$, is one dimensional as proved earlier, and $A^T$ (transposition of $A$) also has a one-dimensional null space $\eta(A^\perp)$, i.e. $\eta(A^\perp)=\operatorname{span}\{\psi | A^\perp\psi=0\}$. As

$$A=\begin{bmatrix}0 & 1\\ 0 & 0\end{bmatrix},\ \psi=\begin{bmatrix}0\\ 1\end{bmatrix},\text{so } A^\perp\psi=\begin{bmatrix}0 & 0\\ 1 & 0\end{bmatrix}\begin{bmatrix}0\\ 1\end{bmatrix}=0$$

According to the Fredholm alternative theorem $\mathcal{R}(A) = (\eta(A^T))^{\perp} = \{y | <y, \psi> = 0\}$.

**Theorem 6.2** If $<D_{y\mu}f(0,0)\varphi, \psi> \neq 0$, and $<D_{yy}f(0,0)\varphi\varphi, \psi> \neq 0$, then there is a simple steady bifurcation solution in the system.

Comparing this with eq. (6.99), we know that when it satisfies the conditions of this theorem, the terms on its right-hand side, $\mu f^{1,1}y$ and $f^{2,0}(y,y)$, are all non-resonant terms, so they cannot be eliminated. We can conclude that $\mu$ is the linear function of $y$ from the stationary solution equation $y' = 0$, i.e. there is a stationary simple bifurcation. Similarly it can be proved that, there is saddle node bifurcation when $<D_{\mu}f(0,0), \psi> \neq 0$; there is a pitchfork bifurcation when the vector field function of the system is an odd function of $y$. Note: The basis of the centre manifold theorem is that for the linearized part of the non-linear ordinary differential equation, $\mathcal{R}^n$ can be decomposed into invariant subspaces, but for generalised eigenvector $\psi$, because $A\psi = \varphi$, and $\varphi$ is the eigenvector corresponding to the zero eigenvalue, so $\mathcal{R}^n = \eta(A) \oplus (\eta(A))^{\perp}$ where $\eta(A)$=span$\{\varphi\}$, $(\eta(A))^{\perp}$ is an $(n-1)$-dimensional state space. As $\psi \in (\eta(A))^{\perp}$, and $A\psi = \varphi \notin (\eta(A))^{\perp}$, $(\eta(A))^{\perp}$ is not an invariant subspace, and we need to seek another invariant subspace. If the generalised eigenspace is span$\{\varphi, \psi\}$, then $\mathcal{R}^n$ can be expressed as

$$\mathcal{R}^n = \text{span}\{\varphi, \psi\} \oplus \text{span}\{\text{other } (n-2) \text{ generalised eigenvectors}\} \overset{def}{=} ⓐ \oplus ⓑ$$

where ⓐ and ⓑ are invariant under the action of mapping $A$ and flow $e^{At}$.

If $A\psi = \varphi$ and $A\varphi = 0$ are all in ⓐ, then ⓐ is invariant. There may be very interesting dynamical phenomena in the two-dimensional generalised eigensubspace span$\{\varphi, \psi\}$. Now let us study the two-dimensional centre manifold equation by the PB normal form.

### 6.8.4 Two-dimensional Centre Manifold Equation and its Dynamical Behaviour

We have obtained a two-dimensional manifold eq. (6.123) in the preceding section: $x' = B(\mu)x + g(x,\mu)$, $\mu' = 0$, where $g(x,\mu) = g_1(x, h(x,\mu), \mu)$.

In bifurcation problems, system (6.123) with double zero eigenvalues can completely retain the dynamical behaviour studied by PB normal form theorem.

**1. Firstly, calculate the PB normal form when** $\mu = 0$.

Take the nearly identical non-linear change of coordinate

$$x = u + P_k(u) \tag{6.132}$$

where $P_k(u)$ is the $k$-order homogeneous polynomial of $u$. Using the same method as in chapter 5, we obtain a homological equation when $\mu = 0$:

$$D_u P_k(u,0)Au - AP_k(u) = g_k(u,0) \tag{6.133}$$

where $g_k(u,0)$ is the $k$-order homogeneous polynomial of $u$. To get $P_k(u)$ from eq. (6.133), let

$$P_k\begin{bmatrix}u_1\\u_2\end{bmatrix}=\begin{bmatrix}\sum_{j=0}^{k}p_ju_1^ju_2^{k-j}\\ \sum_{j=0}^{k}q_ju_1^ju_2^{k-j}\end{bmatrix} \tag{6.134a}$$

Similarly the expression of $g_k(u,0)$ can be written as

$$g_k(u,0)=\begin{bmatrix}\sum_{j=0}^{k}a_ju_1^ju_2^{k-j}\\ \sum_{j=0}^{k}b_ju_1^ju_2^{k-j}\end{bmatrix} \tag{6.134b}$$

The Jacobian matrix $D_uP_k(u)$ of eq. (6.133) is

$$D_uP_k(u)=\begin{bmatrix}\sum_{j=0}^{k}jp_ju_1^{j-1}u_2^{k-j} & \sum_{j=0}^{k}(k-j)p_ju_1^ju_2^{k-j-1}\\ \sum_{j=0}^{k}jq_ju_1^{j-1}u_2^{k-j} & \sum_{j=0}^{k}(k-j)q_ju_1^ju_2^{k-j-1}\end{bmatrix} \tag{6.134c}$$

and $$Au=\begin{bmatrix}0&1\\0&0\end{bmatrix}\begin{bmatrix}u_1\\u_2\end{bmatrix}=\begin{bmatrix}u_2\\0\end{bmatrix} \tag{6.134d}$$

Substituting (6.134) into eq. (6.133), and writing it as two equations

$$\sum_{j=0}^{k}jp_ju_1^{j-1}u_2^{k-j+1}-\sum_{j=0}^{k}q_ju_1^ju_2^{k-j}=\sum_{j=0}^{k}a_ju_1^ju_2^{k-j} \tag{6.135a}$$

and $$\sum_{j=0}^{k}jq_ju_1^{j-1}u_2^{k-j+1}=\sum_{j=0}^{k}b_ju_1^ju_2^{k-j} \tag{6.135b}$$

eq. (6.135b) can be rewritten as

$$\sum_{j=0}^{k}jq_ju_1^{j-1}u_2^{k-j+1}+(k-2)q_{k-2}u_1^{k-3}u_2^3+(k-1)q_{k-1}u_1^{k-2}u_2^2+kq_ku_1^{k-1}u_2$$

$$=\sum_{j=0}^{k-3}b_ju_1^ju_2^{k-j}+b_{k-2}u_1^{k-2}u_2^2+b_{k-1}u_1^{k-1}u_2+b_ku_1^k$$

Comparing the coefficients of $u_1^ju_2^{k-j}$ on both sides, we have $kq_ku_1^{k-1}u_2=b_{k-1}u_1^{k-1}u_2$

so $$kq_k=b_{k-1} \tag{6.136}$$

and $b_ku_1^k$ cannot be eliminated. From eq. (6.135a) we get

$$a_k=q_k \tag{6.137}$$

From eq. (6.136) and eq. (6.137) we know that $q_k$ has two possible choices: one is to eliminate $a_k$, and the other is to eliminate $b_{k-1}$. Assuming that $a_k$ has been eliminated, then $b_{k-1}u_1^{k-1}u_2$ of eq. (6.135b) cannot be eliminated, i.e. $b_{k-1}u_1^{k-1}u_2$ is the Poincaré resonant term. If $b_{k-1}$ has been eliminated, then $a_ku_1^k$ in eq. (6.135a) is the Poincaré resonant term. So for every determined $k$, the resonant term has two possible choices, i.e.

$$\begin{bmatrix} a_k u_1^k \\ b_k u_1^k \end{bmatrix} \quad (6.138a)$$

or
$$\begin{bmatrix} 0 \\ b_k u_1^k + b_{k-1} u_1^{k-1} u_2 \end{bmatrix} \quad (6.138b)$$

In this section we shall use eq. (6.138b) as an example for discussion. For convenience, use $\begin{bmatrix} x \\ y \end{bmatrix}$ to express $\begin{bmatrix} u_1 \\ u_2 \end{bmatrix}$. When $\mu = 0$, we get the PB normal form

$$\begin{bmatrix} x \\ y \end{bmatrix}' = \begin{bmatrix} 0 & 1 \\ 0 & 0 \end{bmatrix} \begin{bmatrix} x \\ y \end{bmatrix} + \{\text{resonant terms}\}$$

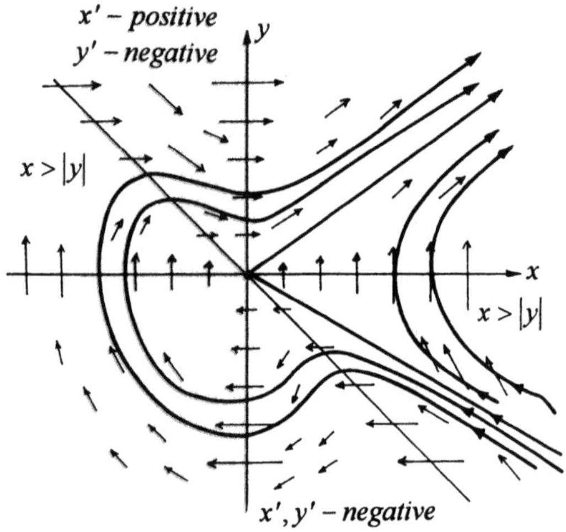


Fig. 6.6

or $\quad x' = y + hot\,,\ \ y' = b_{20}x^2 + b_{11}xy + b_{30}x^3 + b_{21}x^2y + h.o.t. \quad (6.139)$

where $i, j$ express the indices of $x$ and $y$ respectively in $b_{ij}$, and we take $j + i \geq 2$, because we are interested in the non-linear terms higher than order-2. If the Poincaré resonant term chooses eq. (6.138a), then the PB normal form is

$$x' = y + a_{20}x^2 + a_{30}x^3 + \cdots,\ \ y' = b_{20}x^2 + b_{30}x^3 + \quad (6.140)$$

**Theorem 6.3** The terms of order- and higher than 3 of eq. (6.139) do not affect the topological structure of the phase portrait of dynamical systems [53]. According to this theorem, the continuous change of coordinate can eliminate the terms of higher than 3 in eq. (6.139). To discuss the behaviour of the phase portrait, we only study the order-2 term. Rescale $x, y$ in eq. (6.139), then let $b_{20} = 1, b_{11} = \pm 1$, so

$$x' = y,\ \ y' = x^2 \pm xy$$

If we permit $t \to -t$, then in the above equation we can choose only $+xy$, i.e.

$$x' = y,\ \ y' = x^2 + xy \quad (6.141)$$

The unique equilibrium point of the phase portrait is $(x, y) = (0,0)$. Secondly, we seek the isoclinal lines $x' = 0$ and $y' = 0$, and locate the direction of the phase portrait on the isoclinal lines. Phase portraits are parallel to the $y$-axis at $x' = 0$, and parallel to the $x$-axis at $y' = 0$. From eq. (6.141), if $x' = 0$, then $y = 0$, and the $x$-axis is the isoclinal line, i.e. the direction along the $x$-axis of phase portraits does not change with $t$. At the same time for $y' = x^2$, at different values of $x$ the velocity change of phase portraits in the $y$ direction is directly proportional to $x^2$. When $y' = 0$, then $x = 0$ (the $y$ axis) and $x = -y$ are both isoclinal lines. The phase portraits along the isoclinal line are parallel to the $x$-axis. According to $x' = y$ we know that the velocity change of phase portraits along the $x$-axis is directly proportional to $y$. In different regions of the phase plane $xy$, the phase direction can be determined by the positive or negative signs of $x'$ and $y'$ of eq. (6.141). Then we can draw out phase portraits qualitatively, as shown in Fig. 6.6.

**2. Universal unfolding of a two-dimensional and order-2 system**

Now let us discuss the dynamical behaviour of the system when $\mu \neq 0$. When $\mu \neq 0$, we still assume that $f(0,\mu)=0$, and so eq. (6.139) becomes

$$\begin{bmatrix} x \\ y \end{bmatrix}' = A(\mu)\begin{bmatrix} x \\ y \end{bmatrix} + \begin{bmatrix} hot \\ b_{20}(\mu)x^2 + b_{11}(\mu)xy + hot \end{bmatrix} \tag{6.142}$$

$$\text{where}\quad A(\mu) = \begin{bmatrix} 0 & 1 \\ 0 & 0 \end{bmatrix} + \begin{bmatrix} a_1(\mu) & a_2(\mu) \\ b_1(\mu) & b_2(\mu) \end{bmatrix}. \tag{6.143}$$

Here we introduce four new parameters $a_1(\mu), a_2(\mu), b_1(\mu)$ and $b_2(\mu)$. To seek the universal unfolding of eq. (6.142), we need to reduce the number of parameter to two. Because when $\mu = 0$, $A(\mu) = \begin{bmatrix} 0 & 1 \\ 0 & 0 \end{bmatrix}$, i.e. $a_i(0) = b_i(0) = 0$, from the non-semi-simple case condition we know that: $\operatorname{tr} A(\mu) = 0$ and $\det A(0) = 0$. When $f(0,\mu) = 0$, but $\mu \ll 1$, then $\operatorname{tr} A(\mu) \neq 0$ and $\det A(\mu) \neq 0$, and both are small quantities. We can take $a_i$ and $b_i$ as the functions of $\det A(\mu)$ and $\operatorname{tr} A(\mu)$. If

$$J = \left| \frac{\partial(\det A(\mu), \operatorname{tr} A(\mu))}{\partial(\mu_1, \mu_2)} \right|_{\mu_1=\mu_2=0} \neq 0 \tag{6.144}$$

is satisfied, we can calculate two parameters $\mu_1$ and $\mu_2$ as functions of $\det A$ and $\operatorname{tr} A$.

Transversality (crossing) condition. If $J \neq 0$ is satisfied, then curve $\det A(\mu) = 0$ crosses $\operatorname{tr} A(\mu) = 0$, but is not tangential, as shown in Fig. 6.7.

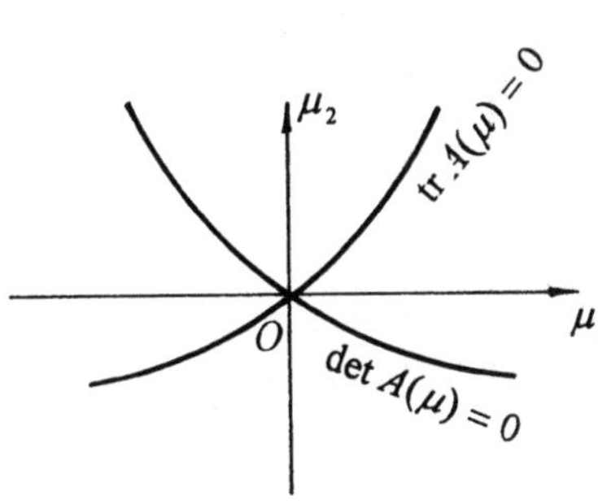


Fig. 6.7

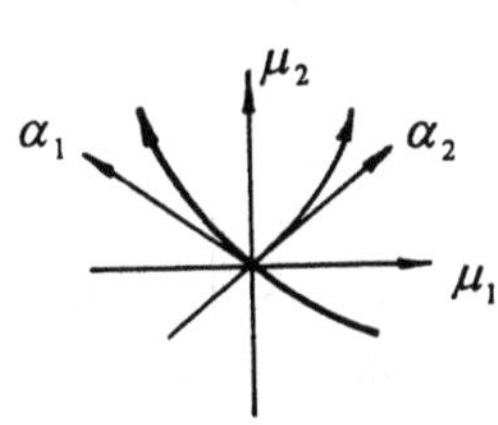


Fig. 6.8

Suppose $x = \pm\sqrt{-\alpha_1}$ and $\alpha_2 = \operatorname{tr} A(\mu)$ (6.145) then if $\dfrac{\partial(\alpha_1,\alpha_2)}{\partial(\mu_1,\mu_2)} \neq 0$ this is the same as making a one to one change of variables, and transforming parameters $\mu_1$ and $\mu_2$ into $\alpha_1$ and $\alpha_2$ (see Fig. 6.8). So the universal unfolding of eq. (6.142) can be taken as

$$\begin{bmatrix} x \\ y \end{bmatrix}' = \begin{bmatrix} 0 & 1 \\ \alpha_1 & \alpha_2 \end{bmatrix} \begin{bmatrix} x \\ y \end{bmatrix} + \begin{bmatrix} hot \\ x^2 + xy + hot \end{bmatrix} \tag{6.146}$$

where $b_{20}(0) = 1$, $b_{11}(0) = 1$, $A(\mu) = \begin{bmatrix} 0 & 1 \\ \alpha_1(\mu) & \alpha_2(\mu) \end{bmatrix}$ (6.147)

By eq. (6.145) we know that $\alpha_1 = -\det\begin{bmatrix} 0 & 1 \\ \alpha_1 & \alpha_2 \end{bmatrix}$, and $\alpha_2 = \operatorname{tr}\begin{bmatrix} 0 & 1 \\ \alpha_1 & \alpha_2 \end{bmatrix}$, when $\mu = 0$, $\det A(0) = 0 = \operatorname{tr} A(0)$, so $\alpha_1 = 0, \alpha_2 = 0$, and by eq. (6.147) we have $A(0) = \begin{bmatrix} 0 & 1 \\ 0 & 0 \end{bmatrix}$, so universal unfolding eq. (6.146) is correct.

It is not difficult to prove that $\alpha_1$ and $\alpha_2$ can be removed from the higher-order terms of the normal form by the non-linear change of coordinate, and hence higher-order terms in universal unfolding equations may not contain $\alpha_1$ and $\alpha_2$.

**3. Universal unfolding of saddle node bifurcation**

Firstly we shall discuss the saddle node bifurcation with double zero eigenvalues. In such a case, we cannot assume that $f(0,\mu) = 0$, but can assume that $f(0,0) = 0$, and $D_\mu(0,0) \neq 0$. When $\mu \neq 0$, there must be a constant independent of time in the universal unfolding equation, so we can choose one of the following two forms:

Takens's form [53] $x' = y,\ y' = \alpha_1(\mu) + \alpha_2(\mu)y + ax^2 + bxy$ (6.148)

Bogdanov's form [55] $x' = y,\ y' = \alpha_1(\mu) + \alpha_2(\mu)x + ax^2 + bxy$ (6.149)

In the two cases, $\dfrac{\partial(\alpha_1,\alpha_2)}{\partial(\mu_1,\mu_2)} \neq 0$, so the universal unfolding of the normal form is not unique. Now let us discuss the saddle node bifurcation in detail. For this purpose we use Takens's universal unfolding eq. (6.148), and write it as

$$\begin{bmatrix} x \\ y \end{bmatrix}' = f(x,y) \tag{6.150}$$

We now seek the equilibrium point. Let the right-hand side of eq. (6.148) be equal to zero, then $y = 0$, $\alpha_1 + x^2 = 0$. From the second equation we get: $x = \pm\sqrt{-\alpha_1}$, so when $\alpha_1 < 0$, we have two equilibrium points:

$$(x_+,0) = (\sqrt{-\alpha_1},0) \tag{6.151a}$$

$$(x_-,0) = (-\sqrt{-\alpha_1},0) \tag{6.151b}$$

In order to study the stability of the solution, we linearized eq. (6.148) at the equilibrium point $(x_\pm,0)$:

$$D_{(x,y)}f(x_\pm,0)=\begin{bmatrix}0 & 1\\ 2x_\pm & \alpha_2+x_\pm\end{bmatrix} \tag{6.152}$$

$$\det D_{(x,y)}f(x_\pm,0)=-2x_\pm$$

So for $(x_-,0)$, $\det D_{(x,y)}f(x_-,0)>0$

for $(x_+,0)$, $\det D_{(x,y)}f(x_+,0)<0$

From eq. (6.152) we can obtain an eigen-problem $\lambda^2-(\alpha_2+x_\pm)\lambda-2x_\pm=0$. So the eigenvalues are as follows:

$$\lambda_1=\frac{1}{2}(\alpha_2+x_\pm)+\frac{1}{2}[(\alpha_2+x_\pm)^2+8x_\pm]^{1/2}$$

$$\lambda_2=\frac{1}{2}(\alpha_2+x_\pm)-\frac{1}{2}[(\alpha_2+x_\pm)^2+8x_\pm]^{1/2} \tag{6.153}$$

**(1) Discussion of the stability in the neighbourhood of the equilibrium point $(x_-,0)=(-\sqrt{-\alpha_1},0)$.**

(a) If $\alpha_2+x_-=0$, from eq. (6.151b), $\alpha_2=\sqrt{-\alpha_1}$, $\alpha_1<0$. Substituting this into eq. (6.153), yields

$$\lambda_{1,2}=\pm i[2\sqrt{-\alpha_1}]^{1/2} \tag{6.154}$$

If $\alpha_1<0$, there are two purely imaginary roots. At this point the transition set

$$\alpha_1=-\alpha_2^2 \tag{6.155}$$

The Hopf bifurcation is derived from the equilibrium point $(-\sqrt{-\alpha_1},0)$.

Take $\alpha_2$ as the parameter and verify the transversal condition by eq. (6.153)

$$\frac{d}{d\alpha_2}(\mathrm{Re}\,\lambda)|_{x=x_-,\ y=0}=\frac{1}{2}>0 \tag{6.156}$$

so the transversal condition of Hopf bifurcation is satisfied. From the Hopf exchange formula $\beta_2=-2\mu_2\left(\frac{d\,\mathrm{Re}\,\lambda}{d\alpha_2}\right)$, after calculating $\mu_2$ at the equilibrium point $(-\sqrt{-\alpha_1},0)$, we know that $\mu_2$ is negative, and the bifurcation solution is a subcritical bifurcation. Considering eq. (6.156), $\beta_2$ is positive, so the non-zero solution is unstable. We know from eq. (6.156) that $\frac{d\,\mathrm{Re}\,\lambda}{d\alpha_2}$ is positive, so when $\alpha_2=\sqrt{-\alpha_1}$, $\mathrm{Re}\,\lambda=0$; also when $\alpha_2<\sqrt{-\alpha_1}$. $\mathrm{Re}\,\lambda$ is negative, therefore the zero solution is stable; when $\alpha_2>\sqrt{-\alpha_1}$, $\mathrm{Re}\,\lambda$ is positive, so the zero solution is unstable. Figure 6.9 is the bifurcation diagram.

(b) If $\alpha_2+x_-<0$, $\alpha_2<\sqrt{-\alpha_1}$ $(\alpha_1<0)$, then $\mathrm{Re}\,\lambda<0$, so the zero solution is stable and it forms a sink.

(c) If $\alpha_2 + x_- > 0, \alpha_2 < \sqrt{-\alpha_1}$, then $\mathrm{Re}\lambda > 0$, so the zero solution is unstable and it forms a source.

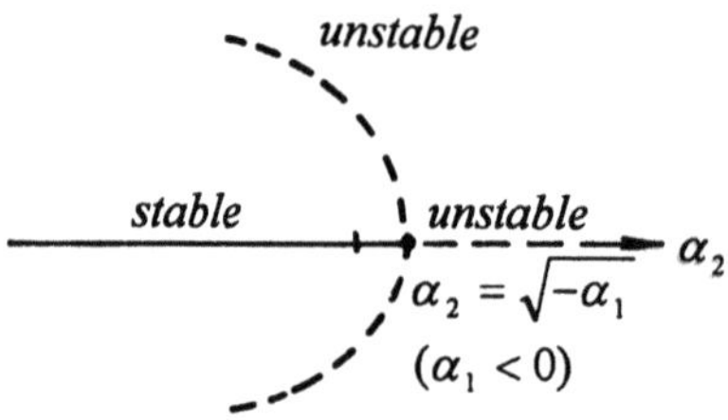


Fig. 6.9

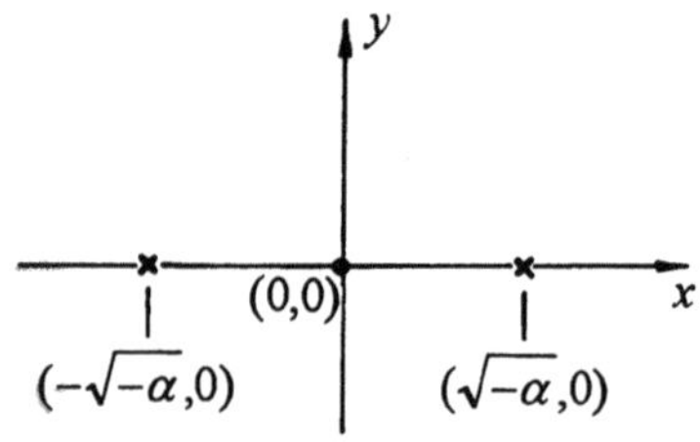


Fig. 6.10

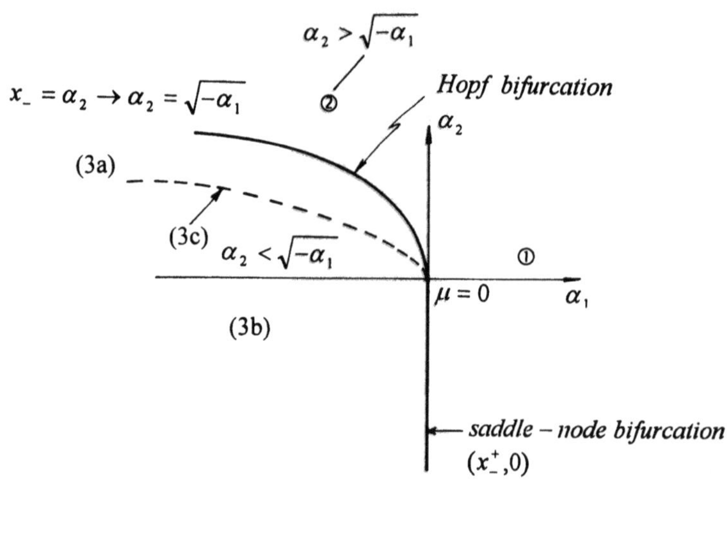


Fig. 6.11

**(2) Equilibrium point** $(x_+,0) = (\sqrt{-\alpha_1},0)$

From eq. (6.153) we know that there are two real roots of eigenvalues with opposite signs in this case. Therefore for eq. (6.153), $(x_+,0)$ is a saddle. To sum up, for $(x_-,0)$ we have a source $(\alpha_2 > \sqrt{-\alpha_1}, \alpha_1 < 0)$ and a sink $(\alpha_2 < \sqrt{-\alpha_1}, \alpha_1 < 0)$ which are separated by the Hopf bifurcation; and for $(x_+,0)$ a saddle, shown in Fig. 6.10. If the transition set $\alpha_1 = -\alpha_2^2$ curve and $\alpha_1 = -\frac{49}{25}\alpha_2^2 + o(\alpha_2^3)$, as well as $\alpha_1 = 0, \alpha_2 \neq 0$, are drawn on the universal unfolding parameter plane $(\alpha_1,\alpha_2)$, then the $\alpha_1 - \alpha_2$ plane is separated into several open sets (or domains). The bifurcation diagram in every open set has the same topological structure (see Fig. 6.11).

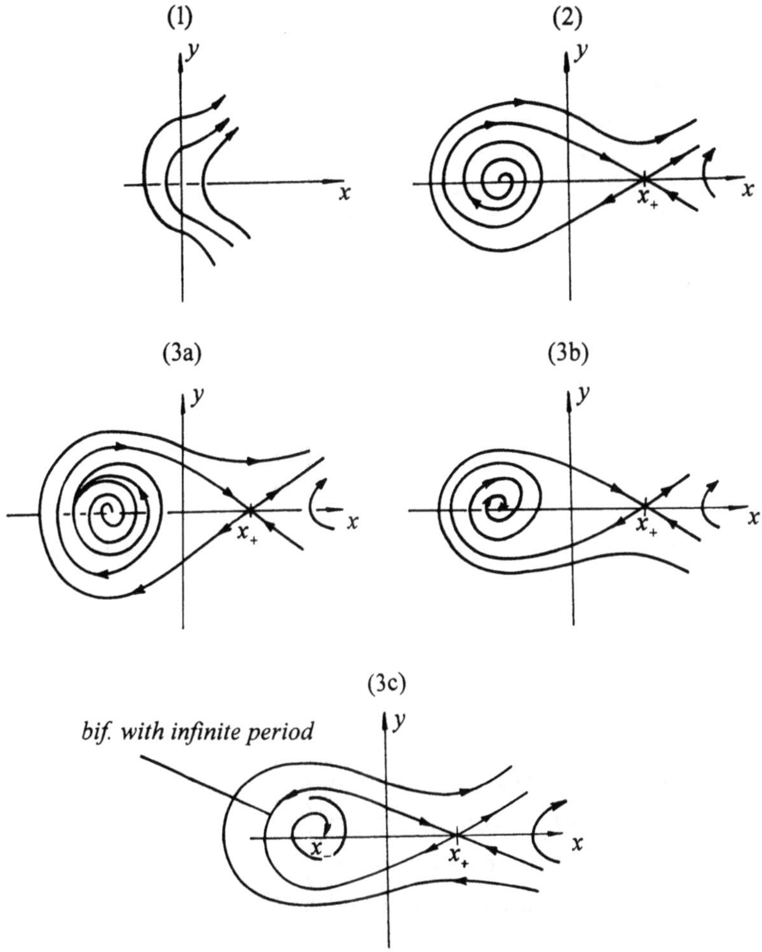


Fig. 6.12

A Hopf bifurcation will occur on transition set $\alpha_1 = -\alpha_2^2$ and a saddle loop will occur on transition set $\alpha_1 = -\frac{49}{25}\alpha_2^2 + o(\alpha_2^3)$. When $\alpha_1 = 0$ and $\alpha_2 \neq 0$, if $x > 0$, there is a saddle; if $x < 0$, there is a node. In domain ①, $\alpha_1 > 0$, so there is no equilibrium point and limit circle. In domain ②, $\alpha_1 < 0$ and $\alpha_2 > \sqrt{-\alpha_1}$, the equilibrium point $(x_+,0)$ is a saddle, and $(x_-,0)$ is an unstable node or focus. In domain (3a), for $\alpha_2 < \sqrt{-\alpha_1}$ and $\alpha_1 < 0$, point $(x_+,0)$ is a saddle, point $(x_-,0)$ is a focus. From the unstable focus in domain ②, a Hopf bifurcation occurs when the parameter varies and crosses transition set $\alpha_1 = -\alpha_2^2$, i.e. a limit circle happens. As $\beta_2$ is positive, this limit circle is unstable, and as $\alpha_2 < \sqrt{-\alpha_1}$, the zero solution $(x_-,0)$ is stable. In domain (3b),

$\alpha_2 < \sqrt{-\alpha_1}$, $\alpha_1 < 0$, the equilibrium point $(x_+,0)$ is a saddle, $(x_-,0)$ is a stable focus. Therefore there is a limit circle in (3a), and there is no limit circle in (3b), so there must be another transition set between (3a) and (3b). The topological structure of both the saddle $(x_+,0)$ and sink $(x_-,0)$ in the whole domain of ③ does not vary; as the phase portraits change in nature outside the local domain of those two changing points, so a global bifurcation must occur. The disappearance of the limit circle from the loop may be due to the appearance of the saddle node loop. The stable manifold crosses the unstable manifold at the saddle point. In other words, this bifurcation is a bifurcation of an infinite period (see Fig. 6.12).

To prove our guess that transition set (3c) is a saddle loop (or bifurcation with infinite period), we shall discuss eq. (6.148) again, using singular transformation and rescaling eq. (6.148): $x = \varepsilon^2 u$, $y = \varepsilon^3 v$, $\alpha_1 = \varepsilon^4 \gamma_1$, $\alpha_2 = \varepsilon^2 \gamma_2$ and $t = \frac{1}{\varepsilon}\tilde{t}$, where $\varepsilon > 0$ is small. Substituting every transformation into eq. (6.148) yields

$$u' = v, \text{ and } v' = \gamma_1 + u^2 + \varepsilon(\gamma_2 v + uv) \tag{6.157}$$

System (6.157) becomes a Hamilton system when $\varepsilon = 0$, so we have Hamilton function:

$$H(u,v) = \frac{v^2}{2} - \gamma_1 u - \frac{u^3}{3} \tag{6.158}$$

and

$$\frac{\partial H}{\partial v} = v = u', \quad -\frac{\partial H}{\partial u} = \gamma_1 + u^2 = v' \tag{6.159}$$

Thus we have proved that the system is a Hamilton system when $\varepsilon = 0$.

From eq. (6.158) and eq. (6.159) we have

$$\frac{\mathrm{d}H}{\mathrm{d}\tilde{t}} = \frac{\partial H}{\partial u}u' + \frac{\partial H}{\partial v}v' = \frac{\partial H}{\partial u}\frac{\partial H}{\partial v} - \frac{\partial H}{\partial v}\frac{\partial H}{\partial u} = 0$$

We know from the equation that the Hamilton function is independent of time, so

$$H(u,v) = \frac{v^2}{2} - \gamma_1 u - \frac{u^3}{3} = \text{const} \tag{6.160}$$

The solution of system (6.159) is the contour line of $H(u,v)$ =constant. According to the preceding discussion, when $\alpha_1 \le 0$, the system has equilibrium points. When $\alpha_1 = \varepsilon^4 \gamma_1$, we take $\gamma_1 = -1$, thus eq. (6.158) becomes

$$H(u,v) = \frac{v^2}{2} + u - \frac{u^3}{3} \tag{6.161}$$

and eq. (6.159) becomes

$$u' = v \text{ and } v' = u^2 - 1 \tag{6.162}$$

The equilibrium points of eq. (6.162) are: $(u_+,0) = (+1,0)$, $(u_-,0) = (-1,0)$. The linear part of the variational equation of system (6.162) is $u_0' = v_0$, $v_0' = 2uu_0$. The eigen-problem of equilibrium point (+1,0) is $\begin{vmatrix} \lambda & 1 \\ 2 & \lambda \end{vmatrix} = \lambda^2 - 2 = 0$, $\therefore \lambda_{1,2} = \pm\sqrt{2}$.

Thus equilibrium point (+1,0) is a saddle. For equilibrium point (−1,0), the eigen-problem is $\begin{vmatrix} \lambda & 1 \\ -2 & \lambda \end{vmatrix} = \lambda^2 + 2 = 0$, $\therefore \lambda_{1,2} = \pm i\sqrt{2}$. Thus equilibrium point (−1,0) is a centre.

For saddle point (+1,0), we obtain a Hamilton constant from eq. (6.161): $H(1,0) = \dfrac{2}{3}$. The contour line equation of the saddle point is

$$\frac{v^2}{2} + u - \frac{u^3}{3} = \frac{2}{3} \tag{6.163}$$

Now we determine the crossing points of $v = 0$ and contour line of eq. (6.163). Substituting $v = 0$ into the above equation, yields

$$u^3 - 3u + 2 = 0$$

or $$(u-1)(u^2 + u - 2) = 0$$

i.e. $$u^2 + u - 2 = (u-1)(u+2), \quad \therefore u = 1, -2.$$

Therefore, $u =$  and $u = -2$ are the crossing points of $v = 0$ and the contour line.

According to the above discussion, we can draw the phase portrait of eq. (6.159), as shown in Fig. 6.13.

Give a perturbation ε to the global solution curve of eq. (6.159), and this can show the dynamical behaviour of system (6.148) when $\alpha_1$ and $\alpha_2$ are close to zero. $\Gamma_0$ in Fig. 6.13 corresponds to contour line $H(u,v) = 2/3$, allowing $r_1 = -1$ to correspond to $\alpha_1 \le 0$. In this case the Hamilton vector field of eq. (6.159) contains all behaviours of universal unfolding.

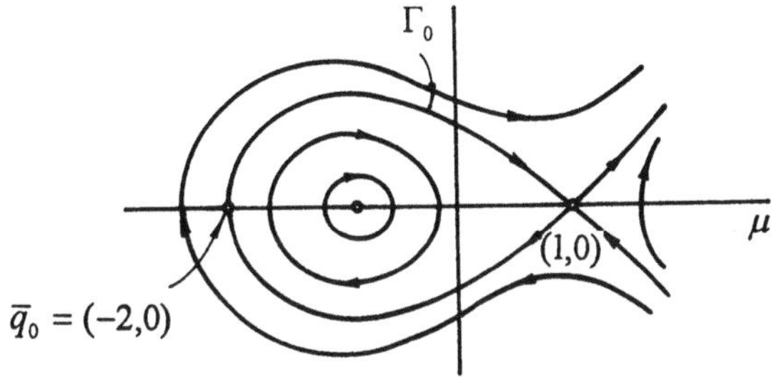


Fig. 6.13

To obtain the equation of saddle loop $\Gamma_0$ which passes through $q_0 = (-2,0)$, we expand the right-hand side of eq. (6.162) at $q_0$ to a Taylor series:

$$u_0' = v_0 \tag{6.164a}$$

$$v_0' = 3 - 4u_0 + u_0^2 \tag{6.164b}$$

Deriving eq. (6.164a) and substituting (6.164b) into it, we obtain

$$u_0'' = 3 - 4u_0 + u_0^2$$

As $$\frac{\mathrm{d}^2 u_0}{\mathrm{d}t^2} = \frac{\mathrm{d}}{\mathrm{d}t}\left(\frac{\mathrm{d}u_0}{\mathrm{d}t}\right) = \frac{\mathrm{d}u_0 \mathrm{d}}{\mathrm{d}t\mathrm{d}u_0}\left(\frac{\mathrm{d}u_0}{\mathrm{d}t}\right)$$

If we suppose $\frac{\mathrm{d}u_0}{\mathrm{d}t} = \hat{u}$,

then $$\frac{\mathrm{d}^2 u_0}{\mathrm{d}t^2} = \hat{u}\frac{\mathrm{d}\hat{u}}{\mathrm{d}t} = 3 - 4u_0 + u_0^2 \tag{6.165}$$

and integrating the above equation yields

$$(\hat{u})^2 = \left(\frac{\mathrm{d}u_0}{\mathrm{d}t}\right)^2 = 6u_0 - 4u_0^2 + \frac{2}{3}u_0^3 + c \tag{6.166}$$

i.e. eq. (6.166) is the first integration, and at $q_0$ we have

$$\left.\frac{\mathrm{d}u_0}{\mathrm{d}t}\right|_{t=0} = u_0'|_{t=0} = v_0|_{t=0} = 0$$

and $$\left(\frac{\mathrm{d}u_0}{\mathrm{d}t}\right)^2 = 0 = -12 - 16 - \frac{16}{3} + c$$

We obtain $c = 100/3$ from the above equation, so

$$\left(\frac{\mathrm{d}u_0}{\mathrm{d}t}\right)^2 = \frac{2}{3}(u_0^3 - 6u_0^2 + 9u_0 + 50) \tag{6.167}$$

We can make the following factoring of the right hand side of the equation:

$$(u_0 + 2)(u_0 - (4 + 3i))(u_0 - (4 - 3i))$$

Let $\beta = -2,\ \alpha = 4 + 3i,\ \gamma = 4 - 3i$

Then eq. (6.167) can be written as

$$\left(\frac{\mathrm{d}u_0}{\mathrm{d}t}\right)^2 = h^2(u_0 - \beta)(u_0 - \alpha)(u_0 - \beta)(u_0 - \gamma) \tag{6.168}$$

where $h^2 = \frac{2}{3}$, then

$$K^2 = \frac{\beta - \gamma}{\alpha - \gamma} = \frac{1}{2} + i$$

$$M^2 = \frac{\alpha - \gamma}{\frac{1}{2}(\alpha + \gamma)} = \frac{3i}{2} \qquad \therefore M = \sqrt{\frac{3i}{2}}$$

$$u_0 = \gamma + (\alpha - \gamma)Z^{-2}$$

$$Z^2 = \delta_n^2\left\{\left(\sqrt{\frac{2}{3}} \times \sqrt{\frac{3i}{2}}\right)t, \sqrt{\frac{1}{2} + i}\right\},\ \therefore n = (4 - 3i) + (6i)S_n^{-2}\left(\sqrt{i}t, \sqrt{\frac{1}{2} + i}\right)$$

or $$u_0 = 1 - 3\operatorname{sec}h^2(t/\sqrt{2}) \tag{6.169a}$$

$$v_0 = u_0' = 3\sqrt{2}\sec h^2(t/\sqrt{2})\tanh(t/\sqrt{2}) \tag{6.169b}$$

and eq. (6.169) is the very solution that we need. In this case, as perturbation value $(0\gamma_2 v + uv)^T$ is a constant vector field, Melnikov function $M_0(t_0)$ is independent

$$\begin{aligned} M(\gamma_2) &= \int_{-\infty}^{\infty}(f_1 h_2 - f_2 h_1)\mathrm{d}t \\ &= \int_{-\infty}^{\infty} v_0(t)(\gamma_2 v_0(t + u_0(t)v_0 ct))\mathrm{d}t \end{aligned} \tag{6.170}$$

$$= \frac{1}{\sqrt{2}}\left[\gamma_2 \int_{-\infty}^{\infty}(18\sec h^4 18\sec h^4 \tanh^2\tau)\mathrm{d}\tau + \int_{-\infty}^{\infty}(1 - 3\sec h^2\tau \sec h^4\tau \tanh h^2\tau)\mathrm{d}\tau\right]$$

where $\tau = t/\sqrt{2}$.

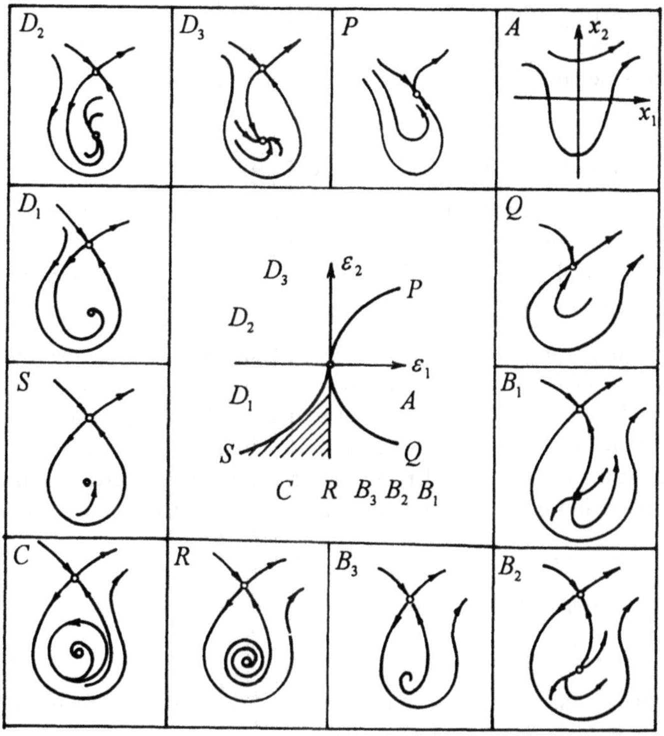


Fig. 6.14

If there exists a saddle link, when $M(\gamma_2) \equiv 0$, bifurcation will occur.

$$\gamma_2 = \frac{\int_{-\infty}^{\infty}(1-3\sec h^2\tau)\sec h^4\tau\tan h^2\tau d\tau}{\int_{-\infty}^{\infty}\sec h^4\tau\tan h^2\tau d\tau}$$

As $\sec h^2\tau = 1-\tan h^2\tau$ and $\int_{-\infty}^{\infty}\sec h^2\tau\tan h^2 d\tau = \left.\frac{\tan h^{k+1}(\tau)}{k+1}\right|_{-\infty}^{\infty} = \frac{2}{k+1}$.

So we obtain $\gamma_2 = \frac{5}{7}$ (6.171)

According to the singular transformation $\alpha_1 = \varepsilon^4\gamma_1, \alpha_2 = \varepsilon^2\gamma_2$, if $\gamma_1 = -$ , then $\alpha_1 = -\varepsilon^4$. If we substitute (6.166) into $\alpha_2$, and $\alpha_2 = \varepsilon^2\frac{5}{7}$, taking the square yields

$$\alpha_2^2 = \varepsilon^4\frac{25}{49}, \text{ so } \varepsilon^4 = \frac{49}{25}\alpha_2^2, \text{ and } \alpha_1 = -\frac{49}{25}\alpha_2^2, \alpha_2 \geq 0 \qquad (6.172)$$

Equation (6.172) is the equation of the transition set of (3c). The actual bifurcation curve will be tangential to the semi-hyperbola eq. (6.172) at $\alpha_1 = \alpha_2 = 0$. If the universal unfolding of the saddle node is in the form of the Bogdanov type eq. (6.149), then the bifurcation diagrams are as shown in Fig. 6.14.

### 6.8.5 The Universal Unfolding of the Pitchfork

In 1974, Takens studied the universal unfolding of a vibration system with rotation symmetry as follows: $g(-x,\mu) = -g(x,\mu)$ that is, $g(x,\mu)$ is the odd non-linear function of $x$, with $g(0,\mu) = 0$. The universal unfolding of the normal form is as follows $x' = y$ and $y' = \mu_1 x + \mu_2 y + a_3 x^3 + xy^3$ (6.173)

where $\mu_1$ and $\mu_2$ are universal unfolding parameters. Recalling the variable time, without loss of generality, we assume that $b_3 = -1$, but $a_3$ belongs to one of the two cases $+1$ and $-1$. We begin with the case of $a_3 = +1$. After local analysis we can show the bifurcation set and the phase diagrams in Fig. 6.15, with a degenerate saddle singularity at the origin. In order to check for the global bifurcation in the second quadrant shown in Fig. 6.15, we introduce a rescaling transformation as follows:

$$x = \varepsilon u,\ y = \varepsilon^2 v,\ \mu_1 = \varepsilon^2\gamma_1,\ \mu_2 = \varepsilon^2\gamma_2,\ t \to \varepsilon t \qquad (6.174)$$

Substituting (6.174) into eq. (6.173), we get

$$\mu' = v, \text{ and } v' = \gamma_1 u + \gamma_2 \varepsilon v + u^3 - \varepsilon\mu^2 v \qquad (6.175)$$

Suppose that $\gamma_1 = -1\ (\mu_1 \leq 0)$. If $\varepsilon = 0$ then eq. (6.175) is the Hamilton system, with its Hamilton function as follows:

$$H(\mu,v) = \frac{v^2}{2} + \frac{u^2}{2} - \frac{u^4}{4} \qquad (6.176)$$

There exists a symmetric heteroclinic orbit connecting a pair of saddle singularities $(u,v) = (\pm 1,0)$ on $H(u,v) = 1/4$. Using the same procedure as above, we obtain the transition set shown in Fig. 6.16.

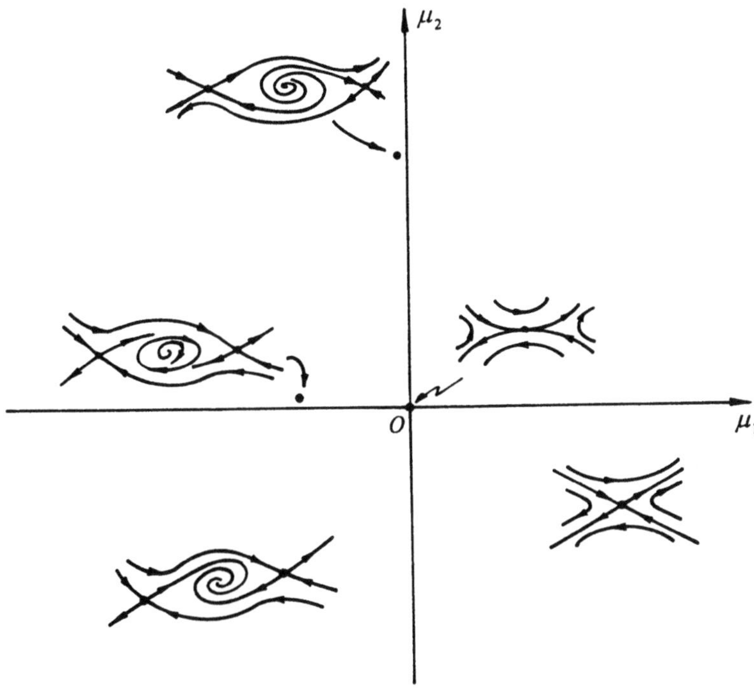


Fig. 6.15

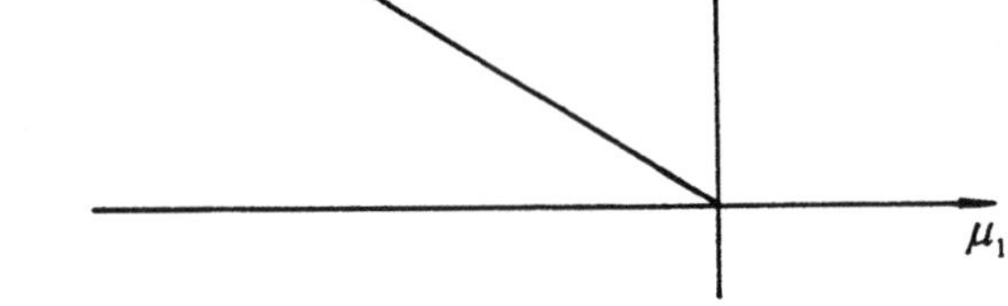


Fig. 6.16

If $a = -1$, there will be more complex forms of global bifurcation, such as the packing between closed orbits and the connection between saddle singularities. We also use the singular transformation of eq. (6.174) and let $\gamma_1 = +1$, then the non-trivial fixed point $(x, y) = (\pm\sqrt{\mu_1}, 0)$ or $(\mu, v) = (\pm 1, 0)$, and eq. (6.175) can be reduced to

$$\mu' = v \text{ and } v' = u + \varepsilon\gamma_2 v - u^3 - \varepsilon u^2 v \quad (6.177)$$

The Hamilton function at $\varepsilon = 0$ is

$$H(\mu, v) = \frac{v^2}{2} - \frac{u^2}{2} + \frac{u^4}{4} \tag{6.178}$$

and the phase portrait is shown in Fig. 6.17.

The solution of the double homoclinic orbit $\Gamma_0$ through $(\pm\sqrt{2}, 0)$ is

$$(u_0(t), v_0(t)) = (\pm\sqrt{2}\operatorname{sec}ht, \mp\sqrt{2}\operatorname{sec}ht \tan ht) \tag{6.179}$$

and the connecting curve between the double saddle singularities is on the curve

$$\gamma_2 = 4/5 \tag{6.180}$$

or on the curve at $(\mu, v) = (0,0)$ tangential to

$$\mu_2 = 4\mu_1/5 \tag{6.181}$$

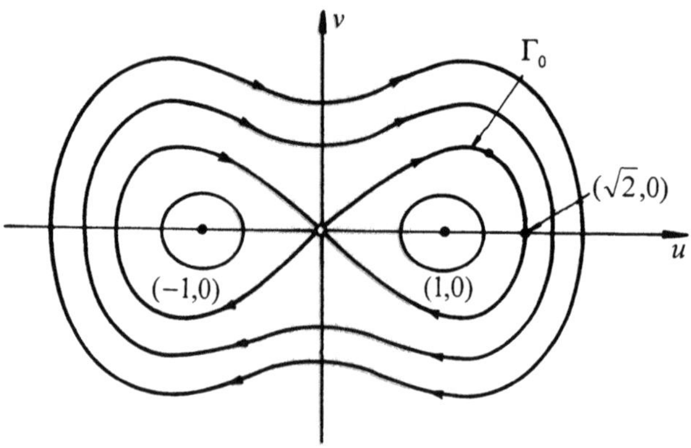


Fig. 6.17

As a result of the analysis of linearization and the computation of Hopf bifurcation, we can obtain part of the transition sets and phase portrait diagrams as shown as in Fig. 6.18.

In order to continue our analysis, we notice that $\operatorname{tr}(Df) = \mu_2 - x^2 < 0$, when $\mu_2 < 0$, then from the Bendixson criterion, there should exist no closed orbit. That means the closed orbit containing three fixed points will disappear in the first quadrant above the $\mu_1$ axis and below $\mu_2 = 4\mu_1/5$. In order to explain this proposition, we shall consider the Hamilton system

$$u = v \text{ and } v' = u - u^3 \tag{6.182}$$

Equation (6.177) is the perturbation of eq. (6.182).

In addition to the discussion of the perturbation of the homoclinic orbit $\Gamma_0$, we must consider the perturbations of the contour lines near the homoclinic orbit $\Gamma_0$. Some complex computations are omitted in this book, and interested readers can refer to references [54,57].

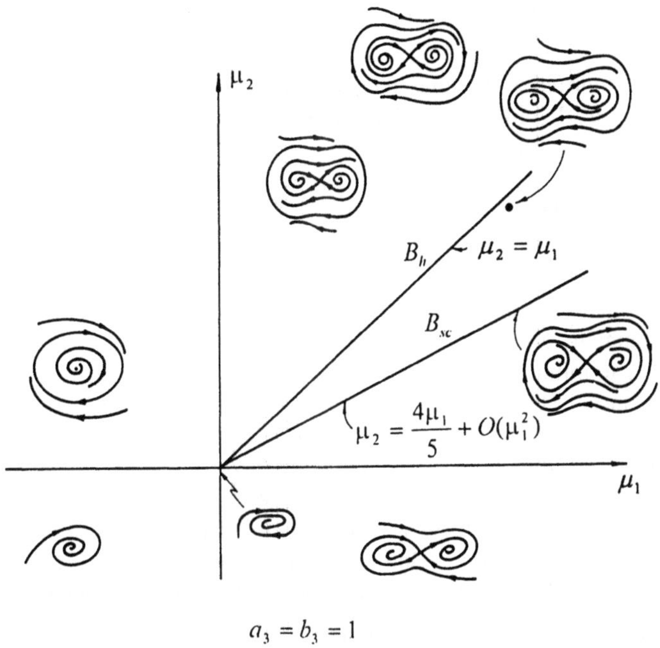


Fig. 6.18

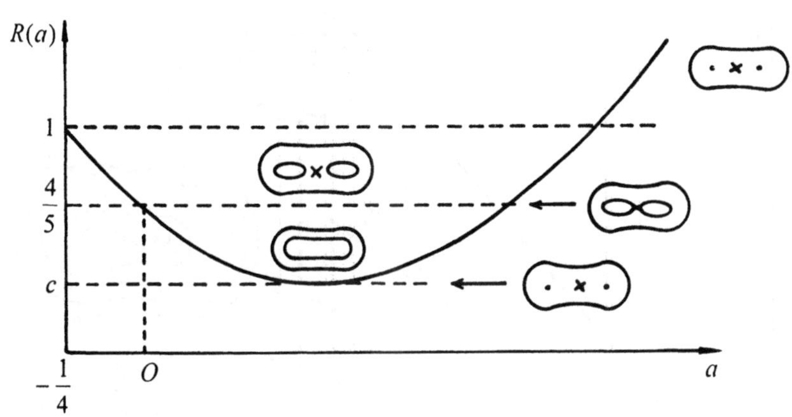


Fig. 6.19

If we take a contour line $\Gamma^\alpha = H_0^{-1}(\alpha)$, its solution can be represented as $(\mu_\alpha(t), v_\alpha(t))$, and the Melnikov function can be written as

$$M^\alpha(\gamma_2) = \int_0^{T_\alpha} v_\alpha(t)\left[\gamma_2 v_\alpha(t) - u_\alpha^2(t) v_\alpha(t)\right]\mathrm{d}t$$

$$= \gamma_2 \int_{\Gamma^\alpha} v\mathrm{d}u - \int_{\Gamma^\alpha} u^2 v\mathrm{d}u \tag{6.183}$$

Here we have transferred the time integration into the orbit integration along $\Gamma^\alpha$ by letting $v = \mathrm{d}u/\mathrm{d}t$. In order to calculate eq. (6.183), we represent $v$ as a function of $u$ and $\alpha$, and if $M^\alpha = 0$, then

$$\gamma_2 = \frac{\int_{\Gamma^\alpha} u^2 v\mathrm{d}u}{\int_{\Gamma^\alpha} v\mathrm{d}u} \overset{def}{=} R(\alpha) \tag{6.184}$$

By means of the properties of integration, J. Carr proved in [58] that $R(\alpha)$ has a unique minimum $c \approx 0.752$ at some finite value of $\alpha > 0$, and $R(\alpha)$ may tend monotonously to infinity as $\alpha$ increases. Figure 6.19 shows the diagrams when $R(\alpha)$ has contour lines. Then there exists a unique closed orbit near a contour line for $\gamma_2 \in (1,\infty)$, which is preserved for $\alpha > 0$. For $\gamma_2 \in (4/5,1)$, there exist three orbits, two of them are for $\alpha \in (-1/4,0)$, and the other is for $\alpha > 0$. If $\gamma_2 = 4/5$, there is a homoclinic saddle-shaped connecting orbit, as stated above, and a closed orbit at the same time. If $\gamma_2 \in (c,4/5)$, there will be two closed orbits, one being an attractor enclosing another unstable orbit. Both of the orbits enclose three fixed points. When $\gamma_2$ varies through $c$, these orbits are packed and disappear as illustrated in Fig. 6.20.

As an unfolding of Fig. 6.18, a transition set $\beta_{po}$ is added to Fig. 6.20, and on $\beta_{po}$ the closed orbits tend to packing.

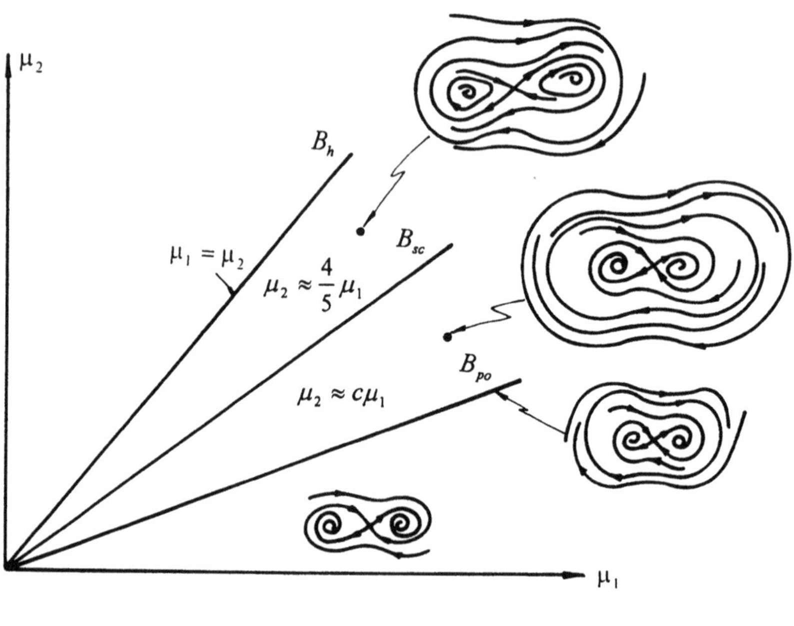


Fig. 6.20

### 6.8.6 The Bifurcation with Semi-simple Double Zero Eigenvalues

Given a non-linear system $y' = f(y,\mu)$, which has a trivial solution, that is $f(0,\mu) = 0$. Suppose that the linear operator $A(\mu) = D_y f(0,\mu)$ has $n$ eigenvalues $\{\lambda_1(\mu), \lambda_2(\mu), \ldots, \lambda_n(\mu)\}$. Bifurcation will occur if the real part of some eigenvalue $\mathrm{Re}(\lambda_i(\mu)) = 0$. Now we shall consider the case $\mu \in \mathcal{R}^2$, and let $\mu_1 = \mu$, $\mu_2 =$ , namely $\mu \in (\mu,\sigma)$. If $\lambda_i(\mu,\sigma) = 0$ or $\mathrm{Re}(\lambda_i(\mu,\sigma)) = 0$, bifurcations will occur as shown in Fig. 6.21. Generally, these curves intersect at isolated simple points, and when the parameter $\mu$ varies and intersects with $\lambda_i = 0$ or $\mathrm{Re}(\lambda_i) = 0$, the sign of $\lambda_i$ will change, so there may be simple or Hopf bifurcation. We can see that the trivial solution $y = 0$ is stable for all the parameters $(\mu,\sigma)$ below the curve or equivalently $\lambda_i < 0$ and $\mathrm{Re}(\lambda_i) < 0$.

**Definition 6.1** The curve $\lambda_i = 0$ or $\mathrm{Re}(\lambda_i) = 0$ is said to be neutrally stable if the boundary ABCD of the shadowed region enclosed by the curves $\lambda_i = 0$, $\mathrm{Re}(\lambda_i) = 0$ is stable.

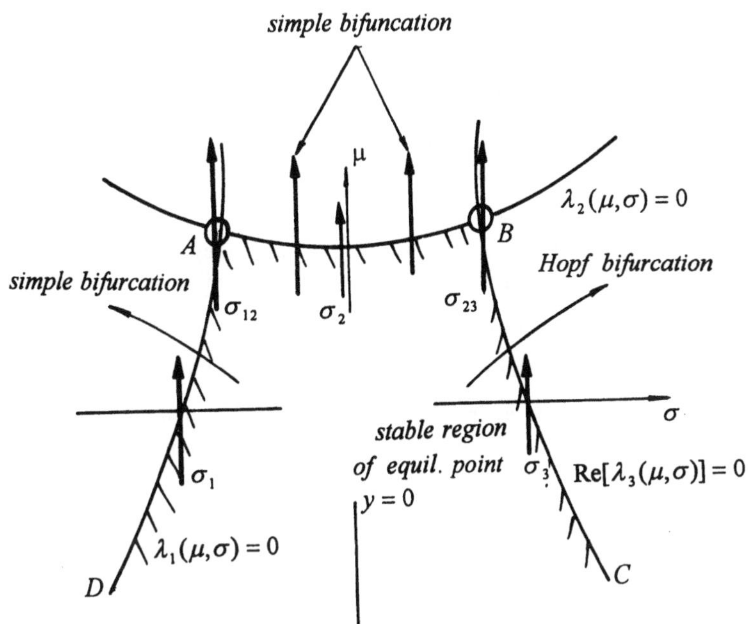


Fig. 6.21

For a fixed , let $\mu$ increase through $\lambda_j = 0$ or $\mathrm{Re}(\lambda_j) = 0\,(j = 1,2,3)$ and simple or Hopf bifurcations will occur as shown in Fig. 6.21. But if $\sigma = \sigma_{ij}$ (as $\sigma_{12}$ or $\sigma_{23}$ in Fig. 6.21), the theory described above will not hold. There will be two zero eigenvalues for $\sigma_{12}$, and for $\sigma_{23}$ there will be a pair of conjugate complexes with real zero part and a zero eigenvalue. Fixing $\sigma = \sigma_j$, let us consider the initial value problem: $y' = f(y,\mu)$, $y \in \mathcal{R}^n$, $\mu \in \mathcal{R}$, with $f(0,\mu) = 0$, $A(\mu) = D_y(0,\mu)$, $A(0)$ having a two-dimensional null space. Suppose that the system has been reduced to the following two-dimensional system by using LS reduction or the centre manifold theorem: $x' = g(x,\mu),\ \ g{:}\mathcal{R}^2 \times \mathcal{R} \to \mathcal{R}^2,\ g(0,\mu) = 0,$

and $$g_\mu(0,0) = 0,\ B = D_x g(0,0) = \begin{bmatrix} 0 & 0 \\ 0 & 0 \end{bmatrix} \tag{6.185}$$

Expanding $g$ into the Taylor series, we can write

$$x = \begin{bmatrix} x \\ y \end{bmatrix},\ g = \begin{bmatrix} g_1 \\ g_2 \end{bmatrix},\ g_k(x,y,\mu) = \mu(\alpha_k x + \beta_k y) + a_k x^2 + 2b_k xy + c_k y^2 \tag{6.186}$$

and $$B' = D_{x\mu} g(0,0) = \begin{bmatrix} \alpha_1 & \beta_1 \\ \alpha_2 & \beta_2 \end{bmatrix}.$$

In order to solve the steady-state solution, suppose that

$$g_k = 0,\ \ k = 1,2 \tag{6.187}$$

Rescaling variables and taking $\mu = \varepsilon << 1,\ x = \varepsilon\hat{x},\ y = \varepsilon\hat{y}$, we have

$$\begin{aligned} g_k(\varepsilon\hat{x},\varepsilon\hat{y},\varepsilon) &= \varepsilon^2 \hat{g}_k(\hat{x},\hat{y},\varepsilon) + o(\varepsilon^3) \\ &= \varepsilon^2 \left\{ (\alpha_k\hat{x} + \beta_k\hat{y}) + a_k\hat{x}^2 + 2b_k\hat{x}\hat{y} + c_k\hat{y}^2 \right\} + o(\varepsilon^3) \end{aligned} \tag{6.188}$$

and we can obtain the solution $\hat{g} = 0$ when $\varepsilon = 0$ by using the implicit function theorem, namely $\hat{g}_k(\hat{x},\hat{y},0) = 0,\ k = 1,2$ (6.189)

Equation (6.189) is an equation of two variables $(\hat{x},\hat{y})$. They are second curves passing through the trivial solution (0,0). In general, they are ellipses, parabolas or two straight lines. Using higher algebra we can solve the intersecting points.

Example 1. Figure 6.22 shows the cases of ellipses and a group of hyperbolas.

Example 2. Figure 6.23 shows the cases of two groups of hyperbolas.

Generally, we can see from Fig. 6.22 and Fig. 6.23 that the two quadric curves of $(\hat{x},\hat{y})\hat{g}_m = 0$ and $\hat{g}_l = 0$ intersect at zero, two or four points. In the case we are discussing, as the zero solution exists forever, the nontrivial solution is one or three.

Figure 6.24(a), (b) shows the degenerate cases which two groups of curves are tangential.

In this case, there will be one or three solutions, namely, there will be zero or two non-trivial solutions. Since the degenerate case is structurally unstable, these cases are very easy to be disturbed.

From these considerations we see that the quadric eq. (6.189) can be totally solved.

**Theorem 6.4.** Suppose that $(x^0, y^0)$ is the solution of the following equation

$$\hat{g}(\hat{x}, \hat{y}, 0) = 0$$

so that $J = \frac{\partial(g_1, g_2)}{\partial(x, y)}\Big|_{(x^0, y^0)}$ (6.190)

is non-singular. Then there exists a unique differential solution for $g = 0$ written as

$$\mu = \varepsilon,\ x = \varepsilon\hat{x}(\varepsilon),\ y = \varepsilon\hat{y}(\varepsilon)$$

with the initial conditions

$$\hat{x}(0) = x^0 \text{ and } \hat{y}(0) = y^0$$

This theorem can be proved by the implicit function theorem [52].

**Conclusion** The number of the bifurcation may be one or three as shown in Fig. 6.25(a), (b).

**Stability** $J$ has two senses, as shown in eq. (6.190), one is the transversality of curve $g = 0$, and the other is the stability of the solution

$$\frac{\partial(g_1, g_2)}{g(x, y)} = \varepsilon J + o(\varepsilon^3) \quad (6.191)$$

where the eigenvalues of $\varepsilon J$ are the first approximations of the eigenvalues of the system; $\det J \neq 0$ means that the eigenvalues may be positive or negative. When the sign of $J$ changes as $\varepsilon$ or $\mu$ varies, the stability will change accordingly, as shown in Fig. 6.26.

If is fixed, and only $\mu$ varies, we may have one or three bifurcation solutions. However, there

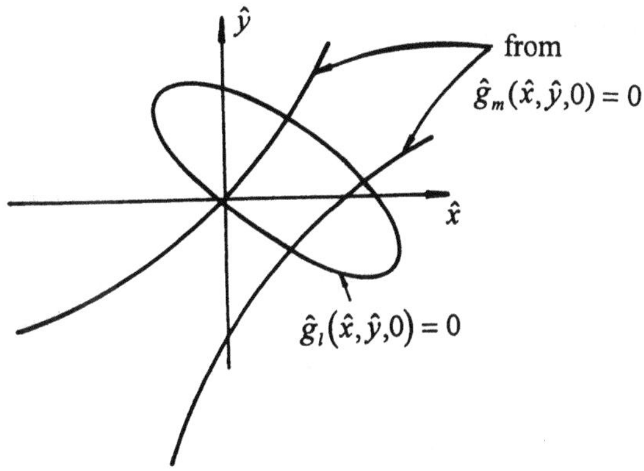


Fig. 6.22

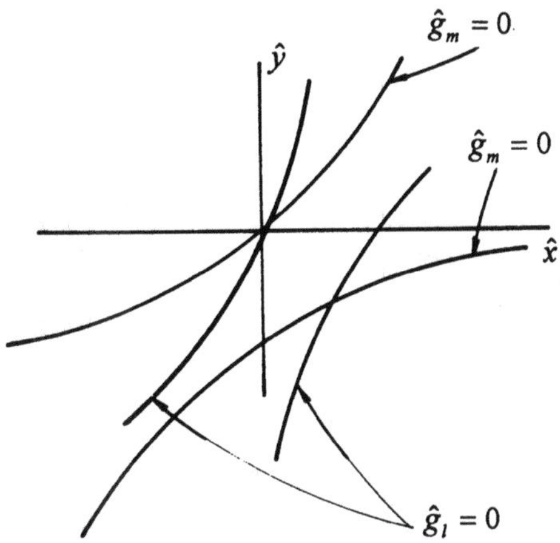


Fig. 6.23

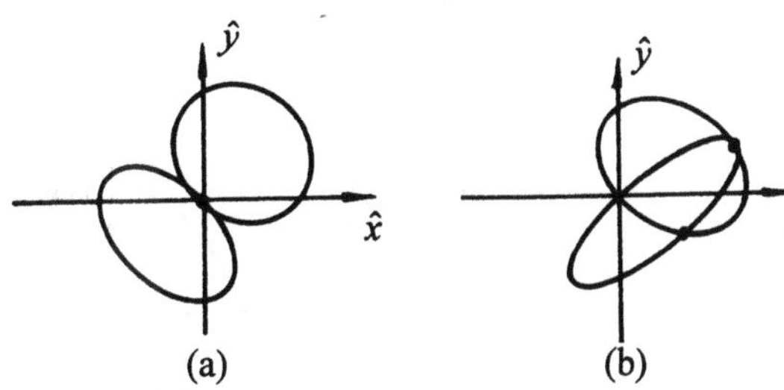


Fig. 6.24

may be exceptions, as shown in Fig. 6.27. If increases from ① to ② or decreases from ③ to ②, we can see that there are two paths of the bifurcation as shown in Fig. 6.27. The bifurcation behaviour at point ② is shown in Fig. 6.28.

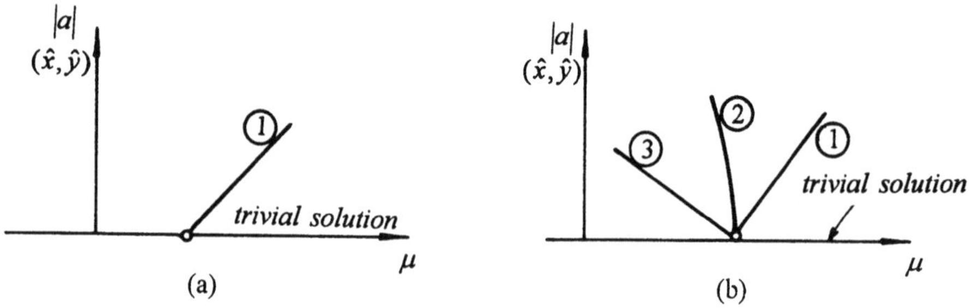


Fig. 6.25

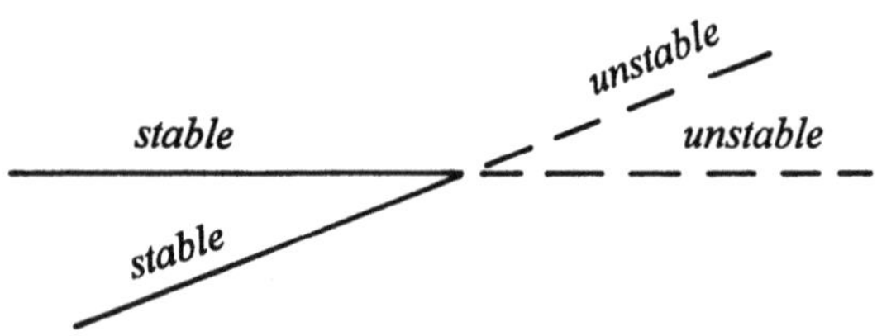


Fig. 6.26

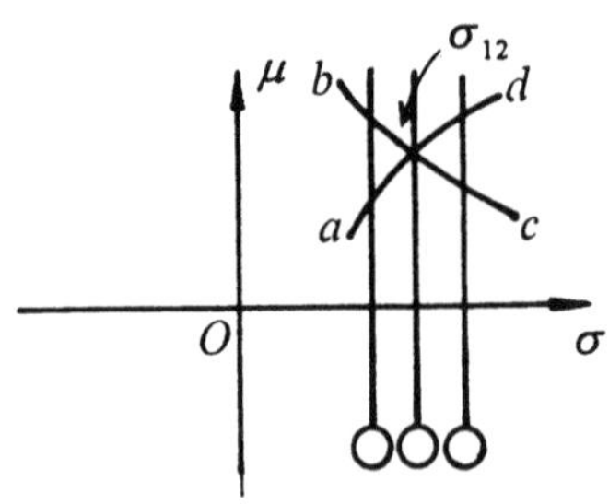


Fig. 6.27

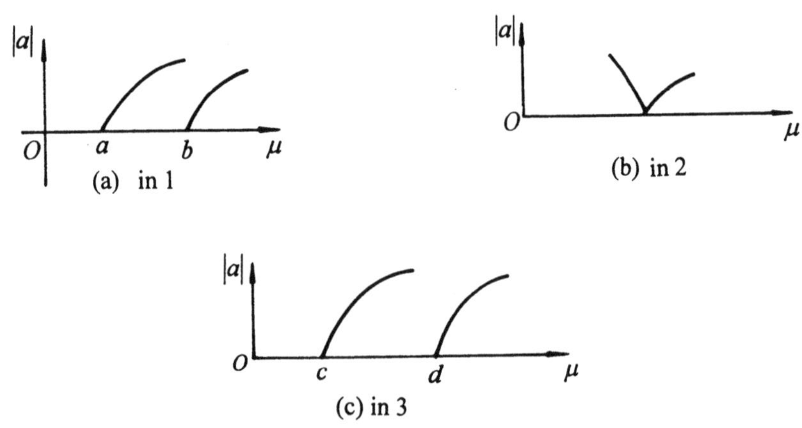


Fig. 6.28

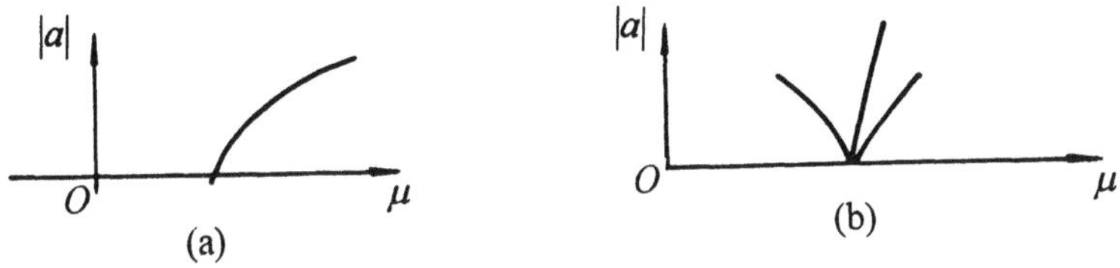


Fig. 6.29

But according to the discussion, we know that there are one or three non-trivial solutions at point ② (or $\sigma=\sigma_{12}$ as shown in Fig. 6.29). That is probably a contradiction. Further studies tell us that there is a second bifurcation here.

### 6.8.7 The Universal Unfolding of Semi-simple Zero eigenvalues

Consider the following system with two families of parameters:

$$y' = f(y,\mu,\sigma)$$

where is the split parameter. In the same way, we can reduce the system into a two-dimensional system:

$$\frac{dx}{dt} = g(\hat{x},\mu,\sigma), \; g: \mathcal{R}^2 \times \mathcal{R}^2 \to \mathcal{R}^2 \tag{6.192}$$

and $g(0,\mu,\sigma) \equiv 0$, $\hat{x} = \begin{bmatrix} x \\ y \end{bmatrix}$, $g = \begin{bmatrix} g_1 \\ g_2 \end{bmatrix}$.

$$A_0 = D_x g(0,0,0) = \begin{bmatrix} 0 & 0 \\ 0 & 0 \end{bmatrix} \tag{6.193}$$

The eigenvalues $\lambda_i(\mu,\sigma), i=1,2$ of a linear operator $A(\mu,\sigma) = D_x g(0,\mu,\sigma)$ satisfy $\lambda_i(0,0)=0$ at $\mu=\sigma=0$. The transversality condition is

$$J = \left| \frac{\partial(\lambda_1,\lambda_2)}{\partial(\mu,\sigma)} \right|_{(0,0)} \neq 0 \tag{6.194}$$

Now, consider the two-parameter families $(\mu,\sigma)$ neighbouring (0,0). In order to solve the stationary solution we let $x' = y' = 0$, then

$$g_k(x,y,\mu,\sigma) = 0 \tag{6.195}$$

Let $\sigma = \varepsilon$, $\mu = \varepsilon\hat{\mu}$, $x = \varepsilon\hat{x}$, $y = \varepsilon\hat{y}$ then

$$\hat{g}_k(\hat{x},\hat{y},\hat{\mu},\varepsilon) = \hat{\mu}(\alpha_k\hat{x} + \beta_k\hat{y}) + r_k\hat{x} + \delta_k\hat{y} + a_k\hat{x}^2 + 2b_k\hat{x}\hat{y} + c_k\hat{y} + o(\varepsilon) \tag{6.196}$$

where $D_{\mu x}g(0,0) = \begin{bmatrix} \alpha_1 & \beta_1 \\ \alpha_2 & \beta_2 \end{bmatrix}$, and $D_{\sigma x}g(0,0) = \begin{bmatrix} \gamma_1 & \delta_1 \\ \gamma_2 & \delta_2 \end{bmatrix}$. The solution of the system $\hat{g}_k(\hat{x},\hat{y},\hat{\mu},\varepsilon) = 0$ (6.197) can be obtained in the same way when $\varepsilon = 0$. The eq. (6.197) is a quadric equation with linear terms. (0,0) is still its trivial solution, and its non-trivial solution may still be one or three. But when $\mu$ varies, the number of the solution may change.

**Theorem 6.5** Given $\mu^0$, suppose that $(x^0, y^0)$ is the solution of eq. (6.197) and $\left|\frac{\partial(g_1, g_2)}{\partial(x^0, y^0)}\right| \neq 0$. If $|\varepsilon|$ is sufficiently small, and $\hat{\mu}$ sufficiently close to $\mu^0$, then there exists a solution branch for $g_k = 0$:

$$\sigma = \varepsilon,\ \mu = \varepsilon\hat{\mu}, x = \varepsilon\hat{x}(\varepsilon, \hat{\mu}), y = \varepsilon\hat{y}(\varepsilon, \hat{\mu}) \tag{6.198}$$

where $\hat{x}(0, \mu_0) = x^0$, $\hat{y}(0, \mu_0) = y^0$.

**Example 1:** $\dfrac{dx}{dt} = g_1(x, y, \mu, \sigma) = \mu x - x^2 + y^2$, $\dfrac{dy}{dt} = g_2(x, y, \mu, \sigma) = (\mu - \sigma)y - 2xy$.

Its linearized system is $\begin{bmatrix} \dfrac{dx}{dt} \\ \dfrac{dy}{dt} \end{bmatrix} = \begin{bmatrix} \mu & 0 \\ 0 & \mu - \sigma \end{bmatrix}\begin{bmatrix} x \\ y \end{bmatrix}$ with the eigenvalues $\lambda_1 = \mu$, $\lambda_2 = \mu - \sigma$. At (0,0) $\left|\dfrac{\partial(\lambda_1, \lambda_2)}{\partial(\mu, \sigma)}\right| = \begin{vmatrix} 1 & 0 \\ 1 & -1 \end{vmatrix} = -1 \neq 0$. Therefore it is transversal. The stationary solution is $(x' = y' = 0)$

$$\mu x - x^2 + y^2 = 0 \tag{a}$$

$$(\mu - \sigma)y - 2xy = 0 \tag{b}$$

and from (b) we obtain $y = 0$ or $x = (\mu - \sigma)/2$. Substituting $y = 0$ to (a), we have $x = 0$ or $x = \mu$, so the solution is

$$\begin{bmatrix} x \\ y \end{bmatrix} = \begin{bmatrix} 0 \\ 0 \end{bmatrix}, \begin{bmatrix} \mu \\ 0 \end{bmatrix} \tag{c}$$

Substituting $x = (\mu - \sigma)/2$ into (a) then we have $\dfrac{\mu^2 - \mu\sigma}{2} - \left(\dfrac{\mu^2 - 2\mu\sigma + \sigma^2}{4}\right) + y^2 = 0$, so $y = \pm\frac{1}{2}\sqrt{\sigma^2 - \mu^2}$ and the solution is

$$(x, y) = \left(\frac{\mu - \sigma}{2}, \frac{1}{2}\sqrt{\sigma^2 - \mu^2}\right) \tag{d}$$

Therefore, when $\sigma^2 - \mu^2 > 0$, we know from (c) and (d) that there are three non-zero solutions. From (d), an elliptic equation can be made. The ellipse meets $\mu = \sigma$ at solution $(x, y) = (0,0)$ and $(x, y) = (\mu, 0)$ with $\mu = -\sigma$. Since there is no real solution for $y$, the ellipse tends to the origin as $\sigma \to 0$, as shown in Fig. 6.30.

**Example 2:** $x' = \mu x - x^2 - y^2$, $y' = (\mu - \sigma)y - 2xy$.

The stationary solution is $(x, y) = (0,0)$, $(x, y) = (\mu, 0)$ and $\left(\dfrac{\mu - \sigma}{2}, \pm\sqrt{\sigma^2 - \mu^2}\right)$, $\sigma > 0$.
In this system, there exist three non-zero solutions for $\sigma = 0$ as shown in Fig. 6.31.

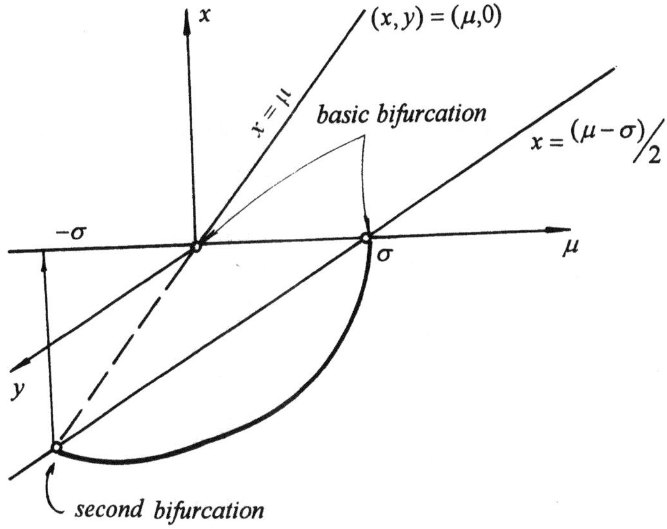


Fig. 6.30

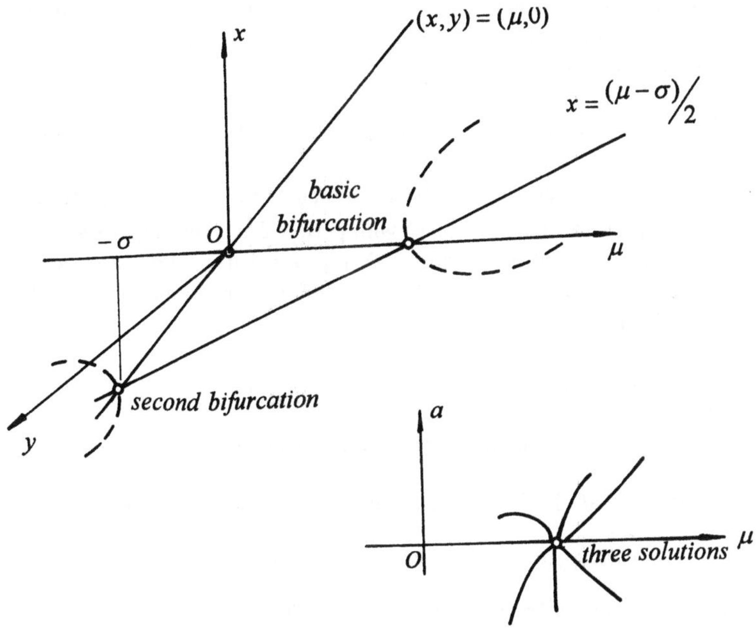


Fig. 6.31

# Chapter 7

# Application of the Averaging Method in Bifurcation Theory

The averaging method is very famous in non-linear vibration. This chapter describes it from the viewpoint of geometry and Poincaré mapping. We will introduce it according to the theory of KBM (Krylov, Bogoliubov, Mytropolisky). This method is very effective for the weak non-linear system and the linear disturbed system [1]. We shall prove that in certain condition the use of the averaging method enables us to obtain information on the global dynamical behaviour of the semi-infinite time domain. The theory of the averaging method has various forms. When we describe the averaging method in the theory of bifurcation, we base our statements on the work of KBM and Hale (J.K. Hale) [30].

## 7.1 Standard Equation

If a vector field and a small parameter $\varepsilon$ are in proportion, then the equation describing the dynamical system is called a standard equation. Suppose the non-linear equations in standard form are

$$\frac{dx}{dt} = Ax + f(\gamma t) + \varepsilon F(\gamma t, x, \varepsilon), \quad x \in R^n, 0 \le \varepsilon << 1 \tag{7.1}$$

where, $A = (a)_{n\times n}$ is a matrix with constant elements; $f, F$ are $n$-dimensional column vectors and are periodic functions of $\gamma t$ with period $2\pi$; $F$ belongs to $\mathcal{C}^r$, $r \ge 2$, and is bounded on the bounded set. Next we shall prove that in some cases eq. (7.1) can be transformed to a standard equation by an averaging transformation. The linearized system of eq. (7.1) is

$$\frac{dx}{dt} = Ax + f(\gamma t) \tag{7.2}$$

Supposing $A$ has only simple purely imaginary eigenvalues $\pm i\omega_1, \pm i\omega_2, \cdots, \pm i\omega_q$, $\left(q = \frac{\pi}{2}\right)$, where $i = \sqrt{-1}$. The special solution of the homogeneous part of eq. (7.2) with $\omega_i$ is as follows: $\varphi_{si}(\omega_i t) = P_{si}\cos\omega_i t - Q_{si}\sin\omega_i t$

$$\varphi^{\bullet}_{si}(\omega_i t) = P_{si}\sin\omega_i t + Q_{si}\cos\omega_i t, \quad s = 1, 2, \ldots, n \tag{7.3}$$

where $P_{si}$ and $Q_{si}$ are real constants.

The conjugate equation system of the homogeneous part of eq. (7.2) is

$$\frac{dy}{dt} = -A^T y \tag{7.4}$$

where $A^T$ is the transposition of $A$.

The special solution of eq. (7.4) corresponding to $\omega_j$ is

$$\psi_{si}(\omega_j t) = C_{sj}\cos\omega_j t - D_{sj}\sin\omega_j t$$

$$\psi^*_{si}(\omega_j t) = C_{sj}\sin\omega_j t + D_{sj}\cos\omega_j t \tag{7.5}$$

Express the basic solution system of the homogeneous parts of eq. (7.4) by $x_{sj}(t)$ and taking the initial condition

$$x_{sj}(0) = \delta_{sj} = \begin{cases} 1 & s = j \\ 0 & s \neq j \end{cases} \tag{7.6}$$

It is easy to prove that if the general solution of function $f_s(\gamma t)$ satisfies (7.2) it is the periodic condition with period $T = \dfrac{2\pi}{\gamma}$:

$$\int_0^T \sum_{s=1}^n f_s(\gamma t)\varphi_{sj}(\omega_i t)\mathrm{d}t = 0 \tag{7.7}$$

Then eq. (7.2) will have a periodic special solution:

$$\int_0^T \sum_{\alpha=1}^n x_{s\alpha}(t-\tau)f_\alpha(\gamma t)\mathrm{d}\tau \tag{7.8}$$

**Proof** Assume the general solution of eq. (7.2) is

$$x_s(t) = c_1 x_{s1}(t) + \cdots + c_n x_{sn}(t) + \int_0^t \sum_{\alpha=1}^n x_{s\alpha}(t-\tau)f_\alpha(\gamma\tau)\mathrm{d}\tau \tag{7.9}$$

where $c_1,\cdots,c_n$ are arbitrary constants defined by the initial conditions. We may choose $c_1,\cdots,c_n$ in order that eq. (7.9) can be a periodic function with a period $T = \dfrac{2\pi}{\gamma}$. Use the periodic condition

$$x_s(T) - x_s(0) = 0 \qquad s = 1,2,\cdots,n$$

The linear non-homogeneous equation defining $c_i$ is obtained:

$$c_1 x_{s1}(t) + \cdots + c_n x_{sn}(t) + \int_0^T \sum_{\alpha=1}^n x_{s\alpha}(T-\tau)f_\alpha(\gamma\tau)\mathrm{d}\tau = 0$$

Denote the basic solution system of eq. (7.4) with $y_{si}(t)$. Because there is a relationship between the basic solution systems of conjugate equations

$$\sum_s \frac{\mathrm{d}x_{sj}}{\mathrm{d}t}y_{si} + \sum_s x_{sj}\frac{\mathrm{d}y_{si}}{\mathrm{d}t} = \sum_{s=1}^n\sum_{\beta=1}^n a_{s\beta}x_\beta y_{si} - \sum_{s=1}^n x_{si}\sum_{\beta=1}^n a_{\beta s}y_\beta = 0$$

this gives $\sum\limits_{s=1}^n x_{sj}(t)y_{si}(t) = \text{const} \qquad i,j = 1,2,\ldots \qquad (7.10)$

If we assume $y_{si}(t) = \psi_{si}(t)$, and take an arbitrary constant $h$, we have

$$\sum_{\alpha=1}^n x_{\alpha j}(t-h)\psi_{\alpha i}(t) = \text{const} \tag{7.11}$$

Assume $t=h$, then $\sum_{\alpha} x_{\alpha j}(h-k)\psi_{\alpha i}(h)=$ constant. From eq. (7.6), we know $\varphi_{ji}(h)=$constant. Substituting the above equation into eq. (7.11) gives

$$\sum_{\alpha=1}^{n} x_{\alpha j}(t-h)\psi_{\alpha i}(t)=\psi_{ji}(h) \tag{7.12}$$

Assuming $t=T$, $h=0$, we have

$$\sum_{\alpha=1}^{n} x_{\alpha j}(T)\psi_{\alpha i}(T)=\psi_{ji}(0) \tag{7.13}$$

Assuming $t=T$, $h=\tau$, we have

$$\sum_{s=1}^{n} x_{s\alpha}(T-\tau)\psi_{si}(T)=\psi_{\alpha i}(\tau) \tag{7.14}$$

Multiplying eq. (7.9) by $\varphi_{si}(T)$, and summing $s$ from 1 to $n$, we have

$$c_1\sum_{s=1}^{n} x_{s1}(T)\varphi_{si}(T)+\ldots+c_n\sum_{s=1}^{n} x_{sn}(T)\varphi_{si}(T)-$$
$$\sum_{s=1}^{n}\varphi_{si}(T)c_s+\int_0^T\sum_{\alpha=1}^{n}\sum_{\alpha=1}^{n} x_{x\alpha}(T-\tau)\varphi_{si}(T)f_\alpha(\gamma\tau)\mathrm{d}\tau=0$$

Substituting eqs. (7.13) and (7.14) into the above equation gives

$$c_1[\varphi_{1i}(0)-\varphi_{1i}(T)]+\cdots+c_n[\varphi_{ni}(0)-\varphi_{ni}(T)]$$
$$+\int_0^T\sum_{\alpha=1}^{n} f_\alpha(\gamma\tau)\varphi_{\alpha i}(\tau)\mathrm{d}\tau=0$$

Because $\psi_{si}(T)=\psi_{si}(0)$, we have $\int_0^T\sum_{\alpha=1}^{n} f_\alpha(\gamma\tau)\varphi_{\alpha i}(\tau)\mathrm{d}\tau=0$ i.e. eq. (7.7) is proved.

We now use variable transformation

$$x_s=\sum_{i=1}^{q} a_i(t)\varphi_{si}(\theta_i)+x_s^*(t) \tag{7.15}$$

Next we introduce some new variables $a_1(t),\ldots,a_q(t),\theta_1(t),\ldots,\theta_q(t)$, and substitute $\theta_i$ for $\omega_i t$ in eq. (7.3).

If $a_i=$constant, $\frac{\mathrm{d}\theta_i}{\mathrm{d}t}=\omega_i$, eq. (7.15) is the general solution of eq. (7.2), and therefore, $\frac{\mathrm{d}\varphi_{si}(\theta_i)}{\mathrm{d}\theta_i}=-\varphi_{si}^{\bullet}(\theta_i)$; $\sum_{i=1}^{q} a_i\varphi_{si}^{\bullet}(\theta_i)=-\sum_{\beta=1}^{n} a_{s\beta}\sum_{i=1}^{q} a_i\varphi_{\beta i}(\theta_i)$. Substituting eq. (7.15) into eq. (7.1), we obtain

$$\sum_{i=1}^{q}\frac{\mathrm{d}a_i}{\mathrm{d}t}\varphi_{si}(\theta_i)-\sum_{i=1}^{q} a_i\varphi_{si}^{\bullet}(\theta_i)\left(\frac{\mathrm{d}\theta_i}{\mathrm{d}t}-\omega_i\right)=\varepsilon F_s(\gamma t,a,\theta,\varepsilon) \tag{7.16}$$

To solve $\frac{\mathrm{d}a_i}{\mathrm{d}t}$ and $\frac{\mathrm{d}\theta_i}{\mathrm{d}t}$ in the equation, we must use the special solution (7.5) of eq. (7.4).

Next, we first prove that there is orthogonality between solution (7.3) and rescaling solution (7.5):

$$\sum_{s=1}^{n}\varphi_{si}(\theta_i)\varphi_{sj}(\theta_j)=\sum_{s=1}^{n}\varphi^*_{si}(\theta_i)\varphi^*_{sj}(\theta_j)=0$$

$$\sum_{s=1}^{n}\varphi_{si}(\theta_i)\varphi^*_{sj}(\theta_j)=\delta_{ij},\ \sum_{s=1}^{n}\varphi^*_{si}(\theta_i)\varphi_{sj}(\theta_j)=-\delta_{ij} \tag{7.17}$$

where
$$\delta_{ij}=\begin{cases}1 & i=j\\ 0 & i\neq j\end{cases}$$

$$\psi_{sj}(\theta_j)=C_{sj}\cos\theta_j-D_{sj}\sin\theta_j$$

$$\psi^*_{sj}(\theta_j)=C_{sj}\sin\theta_j+D_{sj}\cos\theta_j \tag{7.18}$$

It is known that the special solution of the homogeneous part of eq. (7.2) in complex form corresponding to frequency $\pm\omega_i$ is

$$x_{si}(t)=(P_{si}+iQ_{si})e^{i\omega_i t}$$

$$x^*_{si}(t)=(P_{si}-iQ_{si})e^{-i\omega_i t} \tag{7.19}$$

The special solution of eq. (7.4) corresponding to frequency $\pm\omega_j$ is

$$y_{sj}(t)=(C_{sj}+iD_{sj})e^{i\omega_j t}$$

$$y^*_{sj}(t)=(P_{sj}-iQ_{sj})e^{-i\omega_j t} \tag{7.20}$$

According to the character (7.10) of solutions of conjugate equations, if $i\neq j$, then $\sum\limits_{s=1}^{n}(P_{si}+iQ_{si})(C_{sj}+iD_{sj})=\sum\limits_{s=1}^{n}(P_{si}+iQ_{si})(C_{sj}-iD_{sj})\equiv 0$.

Expanding the above equation, we get

$$\sum_{s=1}^{n}P_{si}C_{sj}=\sum_{s=1}^{n}P_{si}D_{sj}=\sum_{s=1}^{n}Q_{si}C_{sj}=\sum_{s=1}^{n}Q_{si}D_{sj}=0 \tag{7.21}$$

Next we study one of the sums in (7.17) in detail, such as the first sum. Substituting $\varphi_{sj}(\theta_j)$ and $\psi_{sj}(\theta_j)$ gives

$$\sum_{s=1}^{n}\varphi_{si}(\theta_i)\psi_{sj}(\theta_j)=\sum_{s=1}^{n}(P_{si}\cos\theta_i-Q_{si}\sin\theta_i)(C_{sj}\cos\theta_j-D_{sj}\sin\theta_j)$$
$$=\cos\theta_i-\cos\theta_j\sum_s P_{si}C_{sj}-\cos\theta_i\sin\theta_j\sum_s P_{si}D_{sj}$$
$$-\sin\theta_i\cos\theta_j\sum_s Q_{si}C_{sj}+\sin\theta_i\sin\theta_j\sum_s Q_{si}D_{sj}$$

It is known from relations(7.21) that the above equation is identically equal to zero. Similarly, it can be proved that the other sums in eq. (7.17) are identically equal to zero if $i\neq j$. Now we calculate the value of the sums in eq. (7.17) when $i=j$. It is known from condition (7.10) that when $i=j$, we have

$$\sum_{s=1}^{n}(P_{si}+iQ_{si})(C_{sj}+iD_{sj})=0$$

or $$\sum_{s=1}^{n}(P_{si}C_{sj}-Q_{si}D_{sj})=\sum_{s=1}^{n}(P_{si}D_{sj}+Q_{si}C_{sj})=0 \tag{7.22}$$

and $$\sum_{s=1}^{n}(P_{si}+iQ_{si})(C_{sj}-iD_{sj})\neq 0$$

or $$\sum_{s=1}^{n}(P_{si}C_{sj}+Q_{si}D_{sj})+i\sum_{s=1}^{n}(Q_{si}C_{sj}-P_{si}D_{sj})\neq 0.$$

Next we calculate the first two sums of eq. (7.17) when $i=j$:

$$\begin{aligned}\sum_{s=1}^{n}\varphi_{si}(\theta_i)\psi_{si}(\theta_i)&=\sum_{s=1}^{n}(P_{si}\cos\theta_i-Q_{si}\sin\theta_i)(C_{si}\cos\theta_i-D_{si}\sin\theta_i)\\&=\cos^2\theta_i\sum_{s=1}^{n}P_{si}C_{si}-\cos\theta_i\sin\theta_i\sum_{s=1}^{n}(P_{si}D_{si}+Q_{si}C_{si})+\sin^2\theta_i\sum_{s=1}^{n}Q_{si}D_{si}\end{aligned}$$

$$\begin{aligned}\sum_{s=1}^{n}\varphi_{si}(\theta_i)\psi^{\bullet}_{si}(\theta_i)&=\sum_{s=1}^{n}(P_{si}\sin\theta_i+Q_{si}\cos\theta_i)(C_{si}\sin\theta_i+D_{si}\cos\theta_i)\\&=\sin^2\theta_i\sum_{s=1}^{n}P_{si}C_{si}+\cos^2\theta_i\sum_{s=1}^{n}Q_{si}D_{si}+\sin\theta_i\cos\theta_i\sum_{s=1}^{n}(P_{si}D_{si}+Q_{si}C_{si})\end{aligned}$$

Consider condition (7.22), then

$$\sum_{s=1}^{n}\varphi^{\bullet}_{si}(\theta_i)\varphi^{\bullet}_{si}(\theta_i)=\sum_{s=1}^{n}\varphi_{si}(\theta_i)\psi_{si}(\theta_i)=\sum_{s=1}^{n}Q_{si}D_{si}$$

Choosing the coefficient of $y_{si}$ suitably can also make $\sum_{s}Q_{si}P_{si}=0$. If we multiply eq. (7.20) by complex number $\alpha+i$, we obtain

$$\begin{aligned}y_{si}(t)&=(C_{si}+iD_{si})(\alpha+i)e^{i\omega_i t}\\&=(\alpha C_{si}-D_{si})\cos\omega_i t-(\alpha D_{si}+C_{si})\sin\omega_i t\\&\qquad+i[(\alpha C_{si}-D_{si})\sin\omega_i t+(\alpha D_{si}+C_{si})\cos\omega_i t]\end{aligned}$$

Assuming

$$\psi_{si}(\theta_i)=(\alpha C_{si}-D_{si})\cos\theta_i-(\alpha D_{si}+C_{si})\sin\theta_i$$

and choosing

$$\alpha=-\frac{\sum_{s=1}^{n}Q_{si}C_{si}}{\sum_{s=1}^{n}Q_{si}D_{si}}$$

then $$\sum_{s=1}^{n}\varphi_{si}(\theta_i)\psi_{si}(\theta_i)=\sum_{s=1}^{n}\varphi^{*}_{si}(\theta_i)\psi^{*}_{si}(\theta_i)=0 \tag{7.23}$$

We can calculate the last two sums of eq. (7.17) when $i=j$:

$$\begin{aligned}\sum_{s=1}^{n}\varphi_{si}(\theta_i)\psi^{*}_{si}(\theta_i) &= \sum_{s=1}^{n}(P_{si}C_{si}-Q_{si}D_{si})\sin\theta_i\cos\theta_i \\ &+\cos^2\theta_i\sum_{s=1}^{n}P_{si}D_{si}-\sin^2\theta_i\sum_{s=1}^{n}Q_{si}C_{si}=-\sum_{s=1}^{n}Q_{si}C_{si}\end{aligned}$$

$$\begin{aligned}\sum_{s=1}^{n}\varphi^{*}_{si}(\theta_i)\psi_{si}(\theta_i) &= \sum_{s=1}^{n}(P_{si}C_{si}-Q_{si}D_{si})\sin\theta_i\cos\theta_i \\ &+\cos^2\theta_i\sum_{s=1}^{n}Q_{si}C_{si}-\sin^2\theta_i\sum_{s=1}^{n}P_{si}D_{si}=\sum_{s=1}^{n}Q_{si}C_{si}\end{aligned}$$

Multiplying $\varphi_{si}(\theta_i)$ and $\varphi^{*}_{si}(\theta_i)$ by real number $-\dfrac{1}{\sum_{s}Q_{si}C_{si}}$, we obtain

$$\sum_{s=1}^{n}\varphi_{si}(\theta_i)\psi^{*}_{si}(\theta_i)=1,\ \ \sum_{s=1}^{n}\varphi^{*}_{si}(\theta_i)\psi_{si}(\theta_i)=-1,\ i=1,2,\ldots,q \tag{7.24}$$

In eq. (7.17), $n$ sums are not equal to zero as shown in eq. (7.24), although the rest are equal to zero. Thus, the character of eq. (7.24) can be expressed as

$$\sum_{s=1}^{n}\varphi_{si}(\theta_i)\psi^{*}_{sj}(\theta_j)=\delta_{ij}\ \text{, and, }\ \sum_{s=1}^{n}\varphi^{*}_{si}(\theta_i)\psi_{sj}(\theta_j)=-\delta_{ij} \tag{7.25}$$

Equation (7.17) is proved.

After having dealt with eqs. (7.17), (7.23) and (7.25), and multiplying eq. (7.16) by $\psi^{*}_{sj}(\theta_j)$ and $\psi_{sj}(\theta_{sj})$ respectively, after summing $s$ from 1 to $n$, we obtain the standard form of eq. (7.1) as follows:

$$\begin{aligned}\frac{da_i}{dt}&=\varepsilon R_i(\gamma t,a_1,\ldots,a_q;\theta_1,\ldots,\theta_q)\\ \frac{d\theta_i}{dt}&=\omega_i+\varepsilon S_i(\gamma t,a_1,\ldots,a_q;\theta_1,\ldots,\theta_q,\varepsilon)\end{aligned} \tag{7.26}$$

where $$\begin{aligned}R_i&=\sum_{s=1}^{n}F_s(\gamma t,a,\theta,\varepsilon)\psi^{*}_{si}(\theta_i)\\ S_i&=\frac{1}{a_i}\sum_{s=1}^{n}F_s(\gamma t,a,\theta,\varepsilon)\psi_{si}(\theta_i)\end{aligned} \tag{7.27}$$

It is known from the above equation, that $R_i$ and $S_i$ are periodic functions of $\gamma t,\theta$ with period $2\pi$. The generalized Fourier serieses of $R_i$ and $S_i$ are

$$R_i(\gamma t,a,\theta,0)=R_{i0}(a)+\sum_{s=1}^{n}[R_{i\delta}(a)\cos\alpha_\delta+R'_{i\delta}\sin\alpha_\delta]$$

$$S_i(\gamma t,a,\theta,0)=S_{i0}(a)+\sum_{s=1}^{n}[S_{i\delta}(a)\cos\alpha_\delta+S'_{i\delta}\sin\alpha_\delta] \tag{7.28}$$

where $\delta$ is a finite integer.

$$\alpha_\delta=m_1^\delta\theta_1+m_2^\delta\theta_2+\ldots+m_q^\delta\theta_q+n^\delta\gamma t \tag{7.29}$$

$m_i^\delta=m_{Fi}\pm1$, $m_{Fi}$ is the combinative coefficient of the $i$th harmonic frequency in the Fourier series of $F_s$. and $n$ are positive or negative integers (similar to the integers of the condition for Poincaré resonance). $R_{i\delta}$ and $R'_{i\delta}$ are coefficients:

$$R_{i\delta}=\frac{2\gamma}{(2\pi)^{q+1}}\int_0^{2\pi}\ldots\int_0^{2\pi}\int_0^{\frac{2\pi}{\gamma}}R(\gamma t,a,\theta,0)\cos(m_1^\delta\theta_1+$$

$$\ldots+m_q^\delta\theta_q+n^\delta\gamma t)\mathrm{d}\theta_1\ldots\mathrm{d}\theta_q\mathrm{d}t$$

$$R'_{i\delta}=\frac{2\gamma}{(2\pi)^{q+1}}\int_0^{2\pi}\ldots\int_0^{2\pi}\int_0^{\frac{2\pi}{\gamma}}R(\gamma t,a,\theta,0)\sin(m_1^\delta\theta_1+ \tag{7.30}$$

$$\ldots+m_q^\delta\theta_q+n^\delta\gamma t)\mathrm{d}\theta_1\ldots d\theta_q\mathrm{d}t$$

Similarly, the coefficient of the general Fourier series of $S_i(\gamma\tau,a,\theta,0)$ can be written. In eq. (7.2), if $A$ has $q_1$ pairs of simple purely imaginary eigenvalues $\pm i\omega_1,\ldots,\pm i\omega_{q_1}$ and $n-n_1$ complex eigenvalues which have strictly non-positive real parts, assume they are conjugate eigenvalues $\lambda_{2q_i+1}=\overline{\lambda}_{2q_i+2}$. The special solution of the homogeneous part of eq. (7.2) corresponding to $\omega_k(k=1,\ldots,q_1)$ and $-u_\beta\pm i\omega_\beta$ $(\beta=q_1+1,q_1+2,\ldots,q)$ are

$$\overline{\varphi}_{s\beta}=(P_{s\beta}\cos\omega_\beta t-Q_{s\beta}\sin\omega_\beta t)e^{-u_\beta t}=\varphi_{s\beta}(\theta_\beta)e^{-u_\beta t}$$

$$\overline{\varphi}^*_{s\beta}=(P_{s\beta}\sin\omega_\beta t+Q_{s\beta}\cos\omega_\beta t)e^{u_\beta t}=\varphi^*_{s\beta}(\theta_\beta)e^{u_\beta t} \tag{7.31}$$

where $P_{s\beta}$ and $Q_{s\beta}$ are conatants, $u_k=0$ $(k=1,\ldots,q)$ but the special solutions of eq. (7.4) are

$$\psi_{s\beta}=(C_{s\beta}\cos\omega_\beta t-D_{s\beta}\sin\omega_\beta t)e^{u_\beta t}=\psi_{s\beta}(\theta_\beta)e^{-u_\beta t}$$

$$\overline{\psi}^*_{s\beta}=(C_{s\beta}\sin\omega_\beta t+D_{s\beta}\cos\omega_\beta t)e^{u_\beta t}=\psi^*_{s\beta}(\theta_\beta)e^{u_\beta t}$$

In such a case, the variable transformation can be taken as the following form:

$$x_s=\sum_{s=1}^{n}a_i(t)\varphi_{si}(\theta_i)e^{-u_i t}+x_s^*(t) \tag{7.32}$$

By the same method used above, the standard equation can be obtained:

$$\frac{\mathrm{d}a_i}{\mathrm{d}t}=\varepsilon\sum_{s=1}^{n}F_i(-ut,\gamma t,a,\theta,\varepsilon)\psi^*_{i\beta}(\theta_i)e^{u_\beta t}\overset{def}{=}\varepsilon R_\beta(\gamma t,a,\theta,\varepsilon) \tag{7.33}$$

$$\frac{d\theta_i}{dt} = \omega_\beta + \frac{\varepsilon}{u_\beta}\sum_{s=1}^{n} F_i(-ut,\gamma t,a,\theta,\varepsilon)\psi_{i\beta}(\theta_i)e^{u_{\beta}t} \overset{def}{=} \omega_\beta + \varepsilon S_\beta(\gamma t,a,\theta,\varepsilon)$$

It is easy to transform the second equation of eq. (7.33) to a form similar to the first equation by a nearly identical transformation, then eq. (7.33) can be considerd as the standard form.

## 7.2 Averaging Method and Poincaré Maps

Now, we shall discuss the equations in the standard form

$$x' = \varepsilon f(x,t,\varepsilon) \quad x \in \mathcal{U} \subseteq \mathcal{R}^n, 0 \le \varepsilon << 1 \tag{7.34}$$

where $f{:}\mathcal{R}^n \times \mathcal{R} \times \mathcal{R}^+ \to \mathcal{R}^n$ is $\mathcal{C}^r$, $r \ge 2$ and is bounded on the bounded set, and a $T$-periodic function in $t$. The first approximation order approximating an averaged system of eq. (7.34) is

$$y' = \varepsilon \frac{1}{T}\int_0^T f(y,y,0)dt \overset{def}{=} \varepsilon F(y) \tag{7.35}$$

The above equation is autonomous or known as an autonomous averaged system. Using order-2 approximating KBM transformation

$$a_i = y_i + \varepsilon U_i^{(1)}(t,y,\phi) + \varepsilon^2 U_i^{(2)}(t,y,\phi)$$

$$\theta_i = \omega_i + \phi_i + \varepsilon V_i^{(1)}(t,y,\phi) + \varepsilon^2 V_i^{(1)}(t,y,\phi) \tag{7.36}$$

eq. (7.26) can be transformed to

$$\frac{dy_i}{dt} = \varepsilon Y_i^{(1)}(y,\phi) + \varepsilon^2 Y_i^{(2)}(y,\phi) + \varepsilon^3 Y_i^{\bullet}(t,y,\phi,\varepsilon)$$

$$\frac{d\phi_i}{dt} = \varepsilon Z_i^{(1)}(y,\phi) + \varepsilon^2 Z_i^{(2)}(y,\phi) + \varepsilon^3 Z_i^{\bullet}(t,y,\phi,\varepsilon) \tag{7.37}$$

where $Y_i^{(1)}, Y_i^{(2)}, Z_i^{(1)}, Z_i^{(2)}$ do not contain time $t$ apparently, but $U_i, V_i$ and $Y_i^{\bullet}, Z_i^{\bullet}$ are $2\pi$-periodic functions in $\phi$ and periodic functions in $t$. For the method of calculation of these functions and the high-order approximating solutions, see [1].

The first-order approximating system of eq. (7.37) is

$$\frac{dy_i}{dt} = \varepsilon Y_i^{(1)}(y,\phi)$$

$$\frac{d\phi_i}{dt} = \varepsilon Z_i^{(1)}(y,\phi) \tag{7.38}$$

With the precision of the first-order approximation, we transform the non-autonomous system (7.1) into the autonomous system (7.38). For simplicity, in the following discussion, we use eqs. (7.35) and (3.34) instead of eqs. (7.38) and (7.33) respectively. According to the discussion, the following theorem can be given.

**Theorem 7.1** (The Averaging Theorem): There exists a change of coordinates $x = y + \varepsilon W(y,t,\varepsilon)$. After the reduction, eq. (7.34) becomes

$$y' = \varepsilon \bar{f}(y) + \varepsilon^2 f_1(y,t,\varepsilon) \tag{7.39}$$

where $f_1$ is a $2\pi$-periodic function in $t$. Moreover:

(1) $x(t)$ and $y(t)$ are the solutions of eqs. (7.34) and (7.35). $x_0$ and $y_0$ are the initial values at $t=0$; if $|x_0 - y_0| = o(\varepsilon)$, then $|x(t)-y(t)| = o(\varepsilon)$ on a time scale $t \sim \frac{1}{\varepsilon}$.

(2) If $p_0$ is a hyperbolic fixed point of eq. (7.35), then there exists $\varepsilon_0 > 0$ so that, for all $0 < \varepsilon \le \varepsilon_0$, eq. (7.34) possesses a unique hyperbolic periodic orbit $r_\varepsilon = p_0 + o(\varepsilon)$ of the same stability type as $p_0$.

(3) If $x^s(t) \in \mathcal{W}^s(r_\varepsilon)$ is a solution of (7.34) lying in the stable mainfold of the hyperbolic periodic orbit $r_\varepsilon = p_0 + o(\varepsilon)$, $y^s(t) \in \mathcal{W}^s(p_0)$ is a solution of eq. (7.35) lying in the stable mainfold of the hyperbolic fixed point $p_0$, and if $|x^s(0) - y^s(0)| = o(\varepsilon)$, then $|x^S(t) - y^S(t)| = o(\varepsilon)$ for $t \in [0,\infty)$. Similar results can be obtained from the solution lying in the unstable mainfold at the time interval $t \in (-\infty, 0]$.

**Note** Conclusions (2) and (3) can be extended to more complicated hyperbolic sets. In particular, Hale [30] showed that if eq. (7.35) has a hyperbolic closed orbit $\Gamma$, then eq. (7.34) has a hyperbolic invariant torus $T_r$. Generalizations to almost periodic functions $f$ are also available (Hale [30]). Conclusion (3) implies that the averaging theorem can be used to study asymptotically stable and unstable mainfolds in bounded sets as well as to study the global structure of the Poincaré map of eq. (7.34); this will be demonstrated in our examples.

**Proof** We will prove the first two parts using standard results from differential equations; for the last part, it is more convenient to use the ideas of Poincaré maps and invariant manifolds. The generalized Fourier series of $f(x,t,0)$ is on the right-hand side of eq. (7.34):

$$f(x,t,0) \overset{def}{=} f(\gamma t, x, \theta, 0) = R_0(x) + \sum_\delta [R_\delta(x)\cos\alpha_\delta + R'_\delta(x)\sin\alpha_\delta]$$

$$\overset{def}{=} R_0(x) + \bar{f}(x,t,0) \tag{7.40}$$

where $\delta$ is a definite integer.

$$\alpha_\delta = m_1^\delta\theta_1 + m_2^\delta\theta_2 + \ldots + m_q^\delta\theta_q + n^\delta\gamma t \tag{7.41}$$

$m$ and $n$ are positive or negative integers, and the coefficients of $R_\delta$ and $R'_\delta$ are

$$R_\delta = \frac{2\gamma}{(2\pi)^{q+1}} \int_0^{2\pi}\ldots\int_0^{2\pi}\int_0^{\frac{2\pi}{\gamma}} f(\gamma t, x, \theta, 0)\cos(m_1^\delta\theta_1 + \ldots + m_q^\delta\theta_q + n^\delta\gamma)\mathrm{d}\theta_1\ldots\mathrm{d}\theta_q\mathrm{d}t \tag{7.42a}$$

$$R'_\delta = \frac{2\gamma}{(2\pi)^{q+1}} \int_0^{2\pi}\ldots\int_0^{2\pi}\int_0^{\frac{2\pi}{\gamma}} f(\gamma t, x, \theta, 0)\sin(m_1^\delta\theta_1 + \ldots + m_q^\delta\theta_q + n^\delta\gamma)\mathrm{d}\theta_1\ldots\mathrm{d}\theta_q\mathrm{d}t \tag{7.42b}$$

Using the first-order approximation KBM transformation:

$$x = y + \varepsilon W(y,t,\varepsilon) \tag{7.43}$$

Substituting eq. (7.43) into eq. (7.34) and considering eq. (7.37) and eq. (7.40), then we have

$$\varepsilon\bar{f}+\varepsilon^2 f_1+\varepsilon\frac{\partial W}{\partial t}(\varepsilon\bar{f}+\varepsilon^2 f_1)=\varepsilon(R_0+\bar{f})+\varepsilon^2()+\ldots \tag{7.44}$$

Let the coefficients of $\varepsilon$ on both sides of the above equation be equal to each other, then

$$\bar{f}+\frac{\partial W}{\partial t}=R_0+\bar{f} \tag{7.45}$$

In order to make $\bar{f}$ not contain time $t$, suppose

$$\bar{f}=R_0 \tag{7.46}$$

and

$$\frac{\partial W}{\partial t}=\bar{f} \tag{7.47}$$

or

$$W=\sum_{\delta=1}^{n}\frac{R_\delta\sin\alpha_\delta-R'_\delta\cos\alpha_\delta}{m_1^\delta\omega_1+\ldots+m_q^\delta\omega_q+n^\delta\gamma} \tag{7.48}$$

To obtain conclusion (1), we use Gronwill's lemma.

**Lemma** If $u,v$ and $c\geq 0$ on $[0,t]$, $c$ is differentiable, and $v(t)\leq c(t)+\int_0^t u(s)v(s)\mathrm{d}s$

then

$$v(t)\leq c(0)\exp\int_0^t u(s)\mathrm{d}s+\int_0^t c'(s)[\exp\int_s^t u(\tau)\mathrm{d}\tau]\mathrm{d}s \tag{7.49}$$

To prove the lemma, let $R(t)=\int_0^t u(s)v(s)\mathrm{d}s$ and $R'-uR\leq uc$. After integrating this differential inequality and some integrations by parts, we obtain the result (refer to [31]). Now consider eqs. (7.35) and (7.39). Integrating and subtracting, we obtain

$$y_\varepsilon(t)-y(t)=y_{\varepsilon 0}-y_0+\varepsilon\int_0^t[\bar{f}(y_s(s))-\bar{f}(y(s))]\mathrm{d}s+\varepsilon^2\int_0^t f_1(y_2(s),s,\varepsilon)\mathrm{d}s$$

where $y_\varepsilon(t)$ is the solution of eq. (7.39) based on $y_\varepsilon(0)=y_{\varepsilon 0}$. Let $y_\varepsilon-y=\varsigma$, then if $L$ is in the Lipschitz constant of $\bar{f}$ and $C$ is the maximum value of $\bar{f}$, the equation becomes

$$|\varsigma(t)|\leq|\varsigma(0)|+\varepsilon L\int_0^t|\varsigma(s)|\mathrm{d}s+\varepsilon^2 ct \tag{7.50}$$

Applying Gronwall's lemma, with $c(t)=|\varsigma(0)|+\varepsilon^2 ct$ and $u(s)=\varepsilon L$, we have

$$|\varsigma(t)|\leq|\varsigma(0)|e^{\varepsilon Lt}+\varepsilon^2 c\int_0^t e^{\varepsilon L(t-s)}\mathrm{d}s\leq\left[|\varsigma(0)|+\frac{\varepsilon C}{L}\right]e^{\varepsilon Lt} \tag{7.51}$$

Thus, if $|y_{\varepsilon 0}-y_0|=o(\varepsilon)$, we conclude that $|y_\varepsilon(t)-y(t)|=o(\varepsilon)$ for $t\in\left[0,\frac{1}{\varepsilon}\right]$ via the transformation eq. (7.43), and we have $|x(t)-y(t)|\leq\varepsilon W(y_\varepsilon,t,\varepsilon)=o(\varepsilon)$. Then using the triangle inequality

$$|x(t)-y(t)|\leq|x(t)-y_\varepsilon(t)|+|y_\varepsilon(t)-y(t)| \tag{7.52}$$

we obtain the desired result.

To prove (2) we consider the Poincaré maps $P_0$ and $P_\varepsilon$ associated with eqs. (7.35) and (7.39). Rewrite these systems as

$$y' = \varepsilon f(y) \quad \varphi' = 1 \tag{7.53}$$

$$y' = \varepsilon f(y) + \varepsilon^2 f(y,\varphi,\varepsilon) \quad \varphi' = 1 \tag{7.54}$$

where $(y,\varphi) \in \mathcal{R}^n \times \mathcal{S}^1$ and $\mathcal{S}^1 = \frac{\mathcal{R}}{T}$ is a circle with a length $T$. We define a global cross-section $\sum = \{(y,\varphi)|\varphi = 0\}$, and the first return or time $T$. Poincaré maps $\mathcal{P}_0 : \mathcal{U} \to \sum$, $\mathcal{P}_e : \mathcal{U} \to \sum$ are defined by eqs. (7.53) and eq. (7.54) in the usual way, where $\mathcal{U} \subseteq \sum$ is an open set. Since the solution of eq. (7.54) remains $o(\varepsilon)$ closed to that of eq. (7.53) for times of $\frac{1}{\varepsilon}$, and $P_0$ is the time $T$ map, $P_e$ is ε-close to $P_0$. If $P_0$ is a hyperbolic fixed point of eq. (7.53) then $P_0$ is also a hyperbolic fixed point for $P_0$, because $DP_0(p_0) = e^{TDf(p_0)}$ and thus no eigenvalue of $DP_0(p_0)$ lies on the unit circle. Hence $(Id - DP_0(p_0))$ is invertible. This means that the equation $x = P_\varepsilon(x)$ has a unique solution $p_\varepsilon$ which is ε-close to $P_0$. Moreover, since the eigenvalue of $DP_\varepsilon(p_0)$ depends continuously on $\varepsilon$, the two fixed points have the same forms of stability. Thus eq. (7.54) has a periodic orbit $r_\varepsilon$ ε-closed to $P_0$ and, via the change of coordinates eq. (7.43), eq. (7.1) has a similar orbit.

The condition for the existence of a periodic orbit in eq. (7.1) is the absence of any eigenvalues equal to one in the spectrum of $DP_0(p_0)$. However, the stability of $P_0$ and $r_\varepsilon$ may not correspond if any eigenvalue of $DP_0(p_0)$ lies on the unit circle.

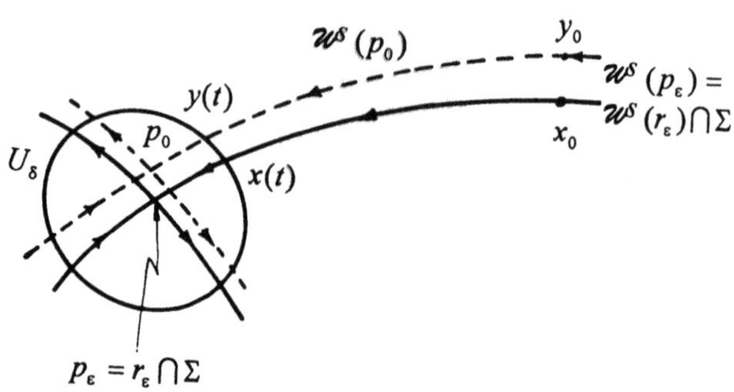


Fig. 7.1

To prove (3), we suppose that eq. (7.35) has a hyperbolic saddle point $P_0$ and consider solutions $y(t) \in \mathcal{W}^s(p_0)$ and the corresponding precise solutions $y_\varepsilon \in \mathcal{W}^s(r_\varepsilon)$ of system (7.39). The cases in which $P_0$ is a sink or a source and $\mathcal{W}^s$ is replaced by $\mathcal{W}^u$ can be dealt with similarly. The proof is divided into two parts: an outer region in which the averaging vector field $\varepsilon\bar{f}(y)$ is large in comparison with the remainder term

$\varepsilon^2 f_1(y,t,\varepsilon)$, and an inner region in which the "perturbations" $\varepsilon^2 f_1$ and $\varepsilon\bar{f}$ are of comparable order. For more details see Sanders and Verhulst [32]. We fix a $\mathcal{U}_\delta$ neighbourhood of $P_0$, and, outside $\mathcal{U}_\delta$, we have $|\bar{f}(y)| >> \varepsilon|f_1(y,t,\varepsilon)$. As with the above discussion, Gronwall's estimates show that $|y_\varepsilon - y_0| = o(\varepsilon)$ for times of order $\frac{1}{\varepsilon}$ outside $\mathcal{U}_\delta$. On the other hand, in $\mathcal{U}_\delta$ the (local) stable mainfold theorem guarantees that the stable mainfold $\mathcal{W}^s_{eoc}(r_\varepsilon)$ is $\mathcal{C}^r, \varepsilon$ is close to $\mathcal{W}^s_{eoc}(p_0) \times [0,T]$. Moreover, within $\mathcal{W}^s_{eoc}(r_\varepsilon)$ and $\mathcal{W}^s_{eoc}(p_0)$ solutions are contracting towards $r_\varepsilon$ and $p_0$ respectively. This contraction is dominated by an exponential term in the form $e^{-\lambda t}$. Based on this fact, we can prove that, if $y_\varepsilon$ and $y_0$ enter $\mathcal{U}_\delta o(\varepsilon)$, they remain within $o(\varepsilon)$ at all times, see Fig. 7.1. Piecing the two estimates together and using the transformation eq. (7.43), we obtain the desired result.

## 7.3 The Geometric Description of The Averaging Method

### 7.3.1 Autonomous Systems

Consider a two-dimensional system

$$x'' + \omega_0^2 x = \varepsilon g(x,x') \qquad \varepsilon << 1 \tag{7.55}$$

Using the method of the first part of this chapter, and taking the transformation

$$x = a\cos\varphi$$
$$y = -a\omega_0 \sin\varphi \tag{7.56}$$

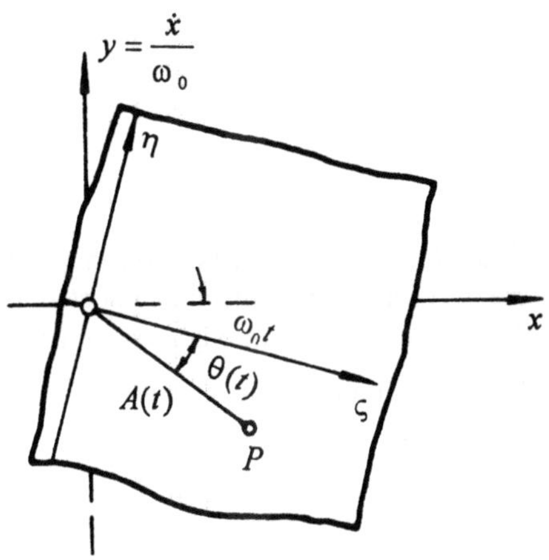


Fig 7.2

where $a$ and $\varphi$ are functions in $t$, the averaging equation of the first approximation is

$$\frac{da}{dt} = -\frac{\varepsilon}{2\pi\omega_0}\int_0^{2\pi} g(a\cos\varphi, -a\omega_0\sin\varphi)\sin\varphi d\varphi$$

$$\frac{d\varphi}{dt} = \omega_0 - \frac{\varepsilon}{2\pi\omega_0 a}\int_0^{2\pi} g(a\cos\varphi, -a\omega_0\sin\varphi)\cos\varphi d\varphi \tag{7.57}$$

Next, we will give the geometric explanation. See Fig. 7.2. The coordinate of phase point $P$ on the phase plane $\left(x, y = \dfrac{x'}{\omega_0}\right)$ is $(x,y)$. The coordinate on the rotating phase plane $\varsigma$ with an angular speed $\omega_0$ is $(\varsigma,\eta)$, so

$$\varsigma = a\cos\theta,\ \eta = -a\sin\theta,\ \theta = \varphi - \omega_0 t$$

Point $P$ will also draw the relevant track. The differential equation in polar coordinate $(a,\theta)$ is

$$\frac{da}{dt} = -\frac{\varepsilon}{\omega_0} g(a\cos(\omega_0 t+\theta), -a\omega_0\sin(\omega_0 t+\theta))\sin(\omega_0 t+\theta)$$

$$\frac{d\theta}{dt} = -\frac{\varepsilon}{a\omega_0} g(a\cos(\omega_0 t+\theta), -a\omega_0\sin(\omega_0 t+\theta))\cos(\omega_0 t+\theta)$$

These are non-autonomous equations on the plane $\varsigma$ . Generally speaking, the track can be self-transversal on the plane $\varsigma$ . After averaging the above equation in a period, we can obtain eq. (7.57). This is an autonomous equation. Therefore, its track cannot be self-transversal on the plane $\varsigma$ .

### 7.3.2 Non-autonomous System

Consider a two-dimensional nonautonomous system

$$\frac{dx}{dt} = X(x, y, \omega t, \varepsilon)$$

$$\frac{dy}{dt} = Y(x, y, \omega t, \varepsilon) \tag{7.58}$$

where $X,Y$ are $2\pi$-periodic functions in $\omega t$; because they apparently contain $t$, the vector fields on the plane $(x,y)$ defined by them are not stationary. At this point, the characters that have only one track passing through a point $(x,y)$ are not correct.

We illustrate this by a simple example below. Consider a forced vibrating system of one degree of freedom:

$$x'' + 0.1x' + x = \cos 1.6t \tag{7.59}$$

i.e.
$$\frac{dx}{dt} = 1.6y$$

$$\frac{dy}{dt} = -\frac{x}{1.6} - 0.1y + \frac{1}{1.6}\cos 1.6t \tag{7.60}$$

Supposing the initial values are $o\varsigma$ . The solutions are

$x = 0.634\cos 1.6t + 0.065\sin 1.6t + e^{-0.05t}(0.634\cos 0.999t - 0.072\sin 0.999t)$ and

$$y = \frac{x'}{1.6} = 0.634\sin 1.6t + 0.065\cos 1.6t - e^{-0.05t}(0.394\sin 0.999t + 0.065\cos 0.999t)$$

When $t \to +\infty$, the solution tends to the non-homogeneous solution or limiting periodic solution:

$$x \to -0.634\cos 1.6t + 0.065\sin 1.6t \text{ and } y \to 0.634\sin 1.6t + 0.065\cos 1.6t .$$

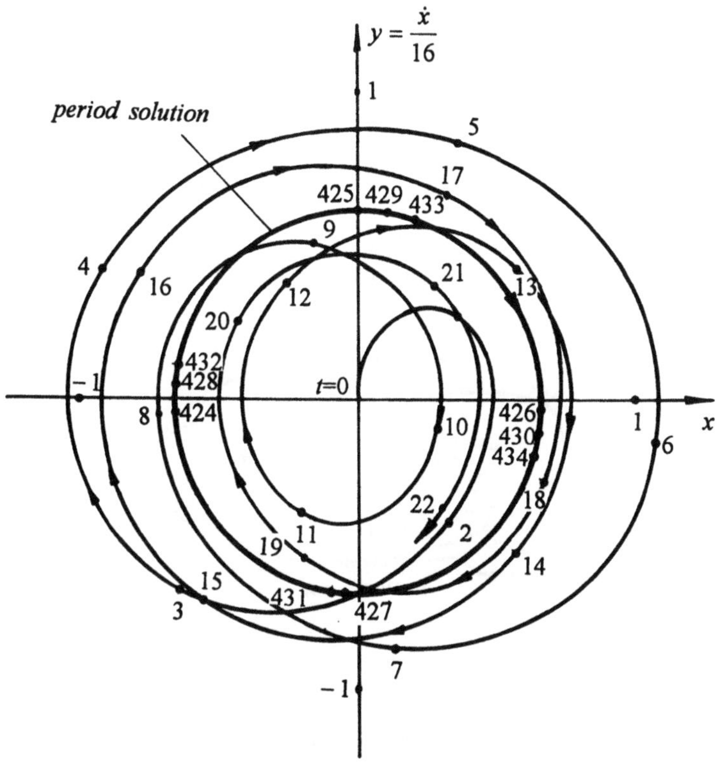


Fig 7.3

These tracks are shown in Fig. 7.3. It can be seen that the track of limit periodic solution can be self-transversal. The tracks from different initial values can be self-transversal too. If we study eq. (7.59) or (7.60) in three-dimensional space, we can avoid the above self-transversality. Rewite eq. (7.60) as

$$\frac{dx}{dt} = 1.6y$$

$$\frac{dy}{dt} = -\frac{x}{1.6} - 0.1y + \frac{1}{1.6}\cos\theta$$

$$\frac{d\theta}{dt} = 1.6, \quad (\theta(0) = 0) \tag{7.61}$$

Thus, system (7.61) becomes an autonomous system. No track can be self-transversal in the space $xy\theta$.

Because $\dfrac{d\theta}{dt} = 1.6$, the system has no singular point. The limit periodic solution in space $xy\theta$ is a helical line. In fact, the right-hand side of eq. (7.61) is $2\pi$-periodic in $\theta$ (in the general case, as is eq. (7.58)). So we can only take $\theta \in [0,2\pi)$, in space $xy\theta$ and connect $\theta = 2\pi$ with $\theta = 0$. Equation (7.61) can be taken as defined in the torus space $\mathcal{R}^2 \times \mathcal{S}^1$, and the limit periodic solution is a closed track in this space. See Fig. 7.4.

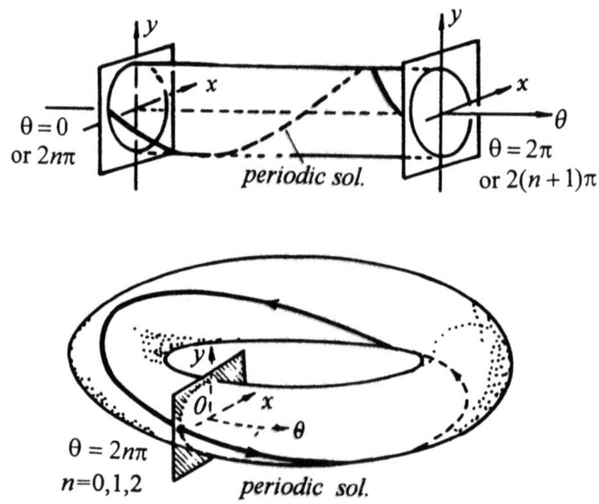


Fig 7.4

We can also study the closed orbit in Fig. 7.4 from another viewpoint. Take a point $(x_0, y_0)$ on the cross-section $\theta = 0$ as an initial point. The orbit will return to the point $(x_1, y_1)$ on the section at $t = \dfrac{2\pi}{1.6}$, or $\theta = 2\pi$. Generally, $(x_1, y_1)$ is different from $(x_0, y_0)$. Only when $(x_0, y_0) = (-0.634, +0.065)$ can $(x_1, y_1)$ be equal to $(x_0, y_0)$. Therefore, to find a periodic solution is to find whether this return mapping $(x_0, y_0) \rightarrow (x_1, y_1)$ has any fixed point. This is the method of Poincaré mapping. We shall not use this method in this section, but the averaging method.

Suppose the periodic solution of system (7.58) $x' = X(x, y, \omega t)$ and $y' = Y(x, y, \omega t)$ is an approximately sinusoidal vibration, with a frequency of , that is, the periodic solution on plane $xy$ is a circle approximately. A rotating plane $o\varsigma$ is introduced (Fig. 7.5) and it rotates on the plane $xy$ with a constant angular speed $\omega$. The equation on a rotating plane is

$$\varsigma' = \Xi(\varsigma, \eta, t) \text{ and } \eta' = H(\varsigma, \eta, t) \tag{7.62}$$

Generally speaking, system (7.62) is still non-autonomous. Its terms on the right-hand side contain time $t$ in appearance. If this equation has a stable stationary solution, it will be corresponding to a stable periodic solution of the initial equation. Notice the differences between the following three (refer to Fig. 7.4 and 7.6):

(a) The periodic solution on the plane $xy$ is denoted by a circle.

(b) The periodic solution at cross-section $t = 0, \frac{2\pi}{\omega}, \ldots$ (that is, the shadowed plane in Fig. 7.4) is denoted by the fixed point of the mapping.

© The periodic solution on the rotating plane $0\varsigma\eta$ is a fixed point (Fig. 7.6).

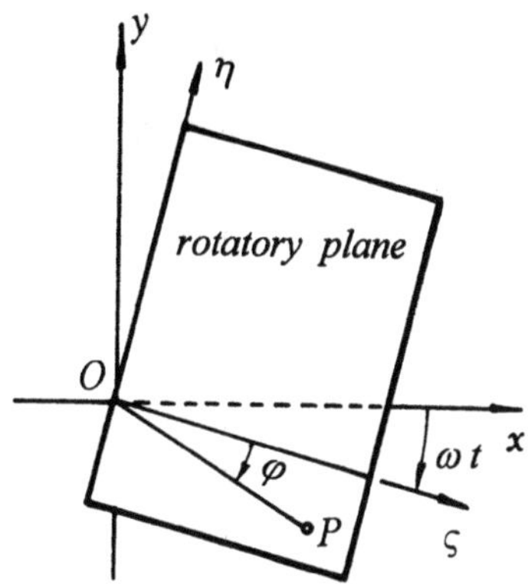

Fig. 7.5

Only when $t = 0, \frac{2\pi n}{\omega}, \ldots,$ are the point coordinates of these three cases are the same.

Take the solution ($x(0) = 0, y(0) = 0$) of the forced linear vibration eq. (7.59) as an example (see Fig. 7.3). The equation on the revelent rotating plane is

$$\varsigma' = -(0.1\varsigma + 0.975\eta)\sin^2 1.6t + (0.975\varsigma + 0.1\eta + 0.625)\sin 1.6t \cos 1.6t$$

$$\eta' = (0.975\varsigma + 0.1\eta + 0.625)\cos^2 1.6t + (0.1\varsigma + 0.975\eta)\sin 1.6t \cos 1.6t \quad (7.63)$$

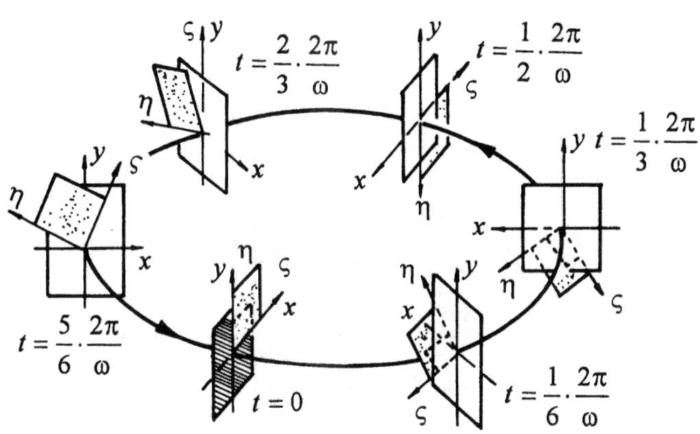

Fig. 7.6

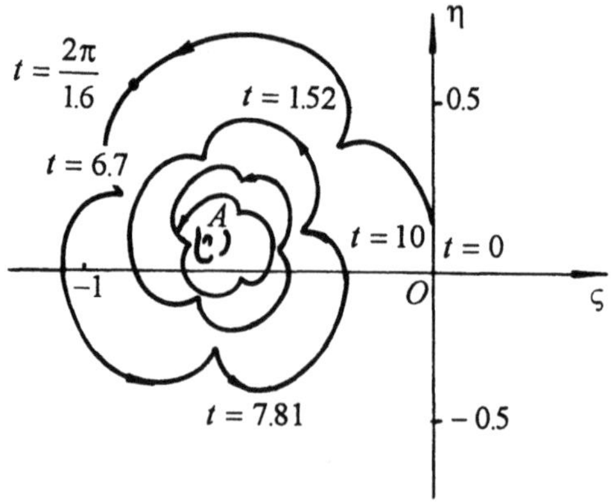


Fig. 7.7

The solution corresponding to $\varsigma(0) = 0, \eta(0) = 0$ is shown in Fig. 7.7. Its fixed points are $\varsigma = \varsigma_A = -0.634, \eta = \eta_A = 0.06$. This satisfies eq. (7.63). When $t \to \infty$, every solution (as the curve in the figure) tends to this fixed point. Now, we average eq. (7.63), and obtain

$$\begin{aligned} \varsigma' &= -0.065\varsigma - 0.4875\eta \\ \eta' &= 0.4875\varsigma + 0.05\eta + 0.3125 \end{aligned} \tag{7.64}$$

The solution with the initial condition $\varsigma(0) = 0$ and $\eta(0) = 0$ is shown in Fig7.8. Comparing it with Fig. 7.7, we can see the averaging results. The fixed points are the same in both figures.

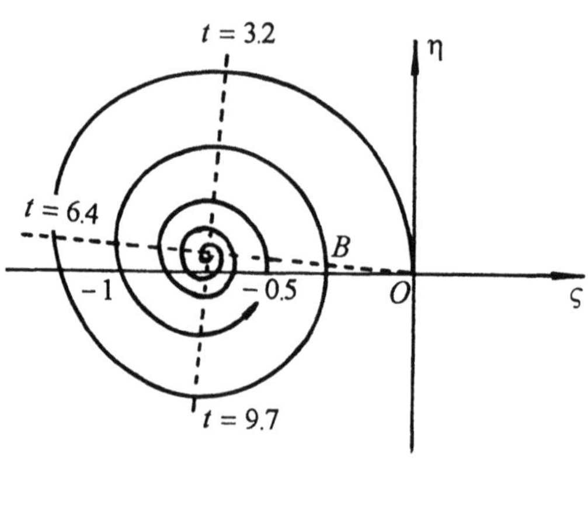


Fig 7.8

The general form of the averaging equation of the non-linear forced vibrating system is

$$x'' + \omega_0^2 x = \varepsilon f(x, x') + \varepsilon F \cos\omega t \tag{7.65}$$

Using the coordinate transformation

$$x = \varsigma \cos\omega t + \eta \sin\omega t$$

$$\frac{x'}{\omega} = y = \varsigma \sin\omega t + \eta \cos\omega t \tag{7.66}$$

This means adding a condition

$$\varsigma' \cos\omega t + \eta' \sin\omega t = 0 \tag{7.67}$$

Substituting (7.66) into (7.65), we have

$$(\omega_0^2 - \omega^2)\varsigma\cos\omega t + (\omega_0^2 - \omega^2)\eta\sin\omega t - \omega\varsigma'\sin\omega t + \omega\eta'\cos\omega t$$

$$= \varepsilon f(\varsigma\cos\omega t + \eta\sin\omega t, -\varsigma\omega\sin\omega t + \eta\omega\cos\omega t) + \varepsilon F\cos\omega t \tag{7.68}$$

Solving for $\varsigma'$ and $\eta'$ from eqs. (7.67) and (7.68), we have

$$\varsigma' = \frac{\omega_0^2 - \omega^2}{\omega}\varsigma\cos^2\omega t + \frac{\omega_0^2 - \omega^2}{\omega}\eta\sin^2\omega t - \frac{\varepsilon}{\omega} f(\varsigma\cos\omega t$$

$$+ \eta\sin\omega t, -\omega\varsigma\sin\omega t + \omega\eta\cos\omega t) - \frac{\varepsilon F}{\omega}\cos\omega t\sin\omega t$$

$$\eta' = \frac{\omega_0^2 - \omega^2}{\omega}\varsigma\cos^2\omega t - \frac{\omega_0^2 - \omega^2}{\omega}\eta\cos\omega t\sin\omega t + \frac{\varepsilon}{\omega} f(\varsigma\cos\omega t$$

$$+ \eta\sin\omega t, -\omega\varsigma\sin\omega t + \omega\eta\cos\omega t) + \frac{\varepsilon F}{\omega}\cos^2\omega t \tag{7.69}$$

So far, we have not introduced approximation. Equation (7.67) is accurate, and is equivalent to eq. (7.65). The right-hand side of eq. (7.69) is a $2\pi$-periodic function in $\alpha = \omega t$. If we take the average of eq. (7.69) in one circle, we then get the first approximate averaging equation:

$$\frac{d\varsigma}{dt} = \frac{\omega_0^2 - \omega^2}{\omega}\eta - \frac{\varepsilon}{2\pi\omega}\int_0^{2\pi} f(\varsigma\cos\alpha + \eta\sin\alpha, -\omega\varsigma\sin\alpha$$

$$+ \omega\eta\cos\alpha)\sin\alpha d\alpha$$

$$\frac{d\eta}{dt} = \frac{\varepsilon F}{2\omega} - \frac{\omega_0^2 - \omega^2}{\omega}\varsigma + \frac{\varepsilon}{2\pi\varsigma}\int_0^{2\pi} f(\cos\alpha + \eta\sin\alpha, -\omega\varsigma\sin\alpha$$

$$+ \omega\eta\cos\alpha)\cos\alpha d\alpha \tag{7.70}$$

(7.70) is the averaging equation of eq. (7.65) on a rotating plane. It is an autonomous equation. We change the transformation of eq. (7.66) to

$$x = A\cos(\omega t + \varphi)$$

$$y = -A\sin(\omega t + \varphi) \tag{7.71}$$

where $A(t)$ and $\varphi(t)$ are the polar coordinates on the rotating plane. We can easily write the averaging equation of $\frac{dA}{dt}$ and $\frac{d\varphi}{dt}$.

## 7.4 An Example of the Averaging Method — the Duffing Equation [33]

The Duffing equation is a forced vibration system with a non-linear restoring force. It is typical in most fields of engineering. Now let us analyze it by the method introduced above.

### 7.4.1 The Periodic Solution of the Duffing Equation

Consider a Duffing equation

$$x'' + 2nx' + \omega_0^2 x + \varepsilon x^3 = F\cos\omega t \tag{7.72}$$

where $n, F, \varepsilon$ are small values, the parameter $\varepsilon > 0$ represents the hard property of a spring, and $\varepsilon < 0$ represents the soft property. Substituting eq. (7.72) into eq. (7.70), we obtain a first approximate averaging equation of plane $\varsigma$ :

$$\frac{d\varsigma}{dt} = -\eta\varsigma - \frac{\omega_0^2 - \omega^2}{2\omega}\eta + \frac{3\varepsilon}{8\omega}(\varsigma^2 + \eta^2)$$

$$\frac{d\eta}{dt} = \frac{F}{2\omega} + \frac{\omega_0^2 - \omega^2}{2\omega}\varsigma - n\eta - \frac{3\varepsilon}{8\omega}\varsigma(\varsigma^2 + \eta^2) \tag{7.73}$$

Let the right-hand side of the above equations be equal to zero. The roots are therefore the first approximate stationary solution (or periodic solution) of eq. (7.72)

$$x = \varsigma^{\bullet}\cos\omega t + \eta^{\bullet}\sin\omega t \tag{7.74}$$

Let $A = \sqrt{\eta^{\bullet 2} + \varsigma^{\bullet 2}}$ be the amplitude of a stationary vibration, then $A$ must satisfy the equation

$$\Phi(A^2, \omega^2, F) \equiv A^2\left[\frac{3\varepsilon}{4}A^2 + (\omega^2 - \omega_0^2)\right]^2 + 4n^2\omega^2 A^2 - F^2 = 0 \tag{7.75}$$

where $F, n, \omega$ are known. This is a cubic equation of $A^2$ and usually there may be one or three periodic solutions. We can illustrate eq. (7.75) from a geometrical point of view.

If $n, \varepsilon$ are known, the frequency $\omega$ of the disturbed force and the amplitude $F$ are controlling parameters. $A$ is the state variable. $A, F,$ are shown as curved surfaces in Fig. 7.9.

In Fig. 7.9, the values of $F$,0,1,2,3,4 represent the relative sizes. Several solutions may be obtained only when $F \geq 2$. Of the three surfaces, the middle one is unstable. The dividing line between stability and instability is the curve $ABCDE$. If we make a projection of $ABCDE$ onto the plane $F\omega$, we get $abcde$. The bifurcation set looks sharp in shape. In the projection on plane $c$ is the sharp point (projection of all points in any curved surface can be classified into three kinds: non-singular point, folding point and sharp point). Generally, we call the figure of $A$ and $\omega$ the response curve when $F \geq 2$ (Fig. 7.10). We can easily find that "jumping" and "hysteresis" exist.

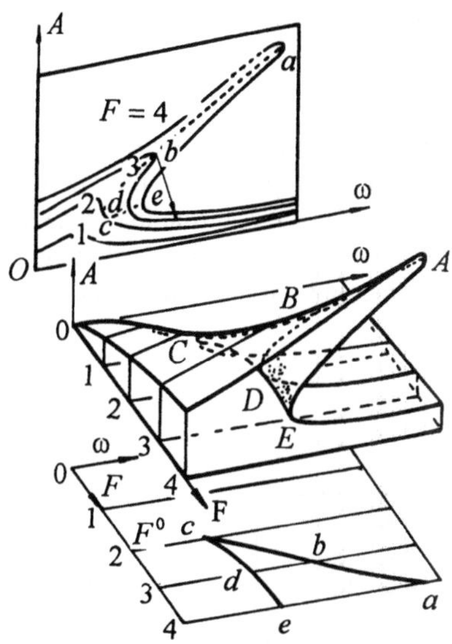


Fig. 7.9

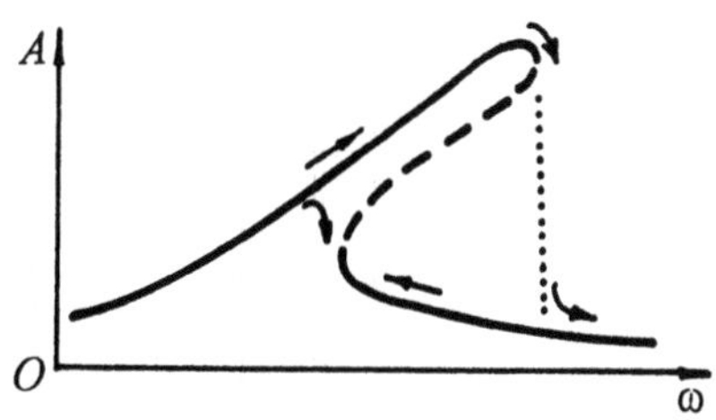


Fig. 7.10

### 7.4.2 Transition Process of the Duffing Equation

The stationary solution on the rotating plane represents the periodic solution on plane $xy$. If the track on plane $\varsigma$ approaches a fixed point, the motion on plane $xy$ approaches the periodic solution. The track line on plane $\varsigma$ can be used to study the process of transition. Under the condition that there is only one stable periodic solution, the structure of the phase diagram on plane $\varsigma$ is the same as that of the linear system. That is to say, there is only one stable fixed (singular) point. When damping exists, the whole plane becomes the attractive domain. A complicated phenomenon appears under the condition of three periodic solutions.

Firstly, let us discuss the following Duffing equation without damping:

$$x'' + x + \frac{2}{15}x^3 = \cos 1.6t \tag{7.76}$$

The averaging equations on the rotating plane are as follows:

$$\frac{d\varsigma}{dt} = -0.4875\eta - 0.05\varsigma + 0.03125\eta(\varsigma^2 + \eta^2)$$

$$\frac{d\eta}{dt} = 0.3125 + 0.4875\varsigma - 0.03125\varsigma(\varsigma^2 + \eta^2) \tag{7.77}$$

Let the right-hand side of the equations be equal to zero. We obtain singular points: centres $S(-0.66,0)$, $S'(4.24,0)$ and saddle (-3.58,0). And we get an equation for the integral curve:

$$40\varsigma + 31.25(\varsigma^2 + \eta^2) - (\varsigma^2 + \eta^2)^2 = \text{const} \tag{7.78}$$

This is a generalized conservative system. The two homoclinic orbits passing through $H_1$ and $H_2$ (Fig. 7.11) classify all the track lines into three kinds: the periodic track $c_1$, around $S$; $c_2$ around $S'$; and $c_3$ around both $S$ and $S'$. These three kinds of tracks are all closed or periodic solution on plane $\varsigma$. But in general, their periods are not just the multiple of rationals of time expended by one circle rotating plane in $\varsigma$. They are almost periodic motions in coordinates $x-t$. This kind of homoclinic motion $H_1$ or $H_2$ is just a double approaching homoclinic solution in Poincaré's sense, as shown in Fig. 7.12.

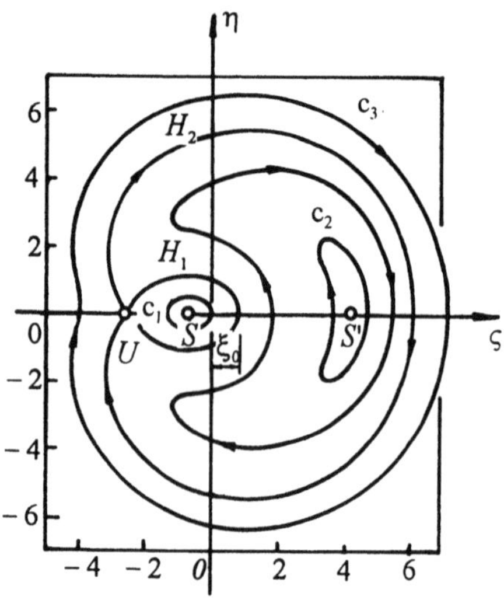


Fig. 7.11

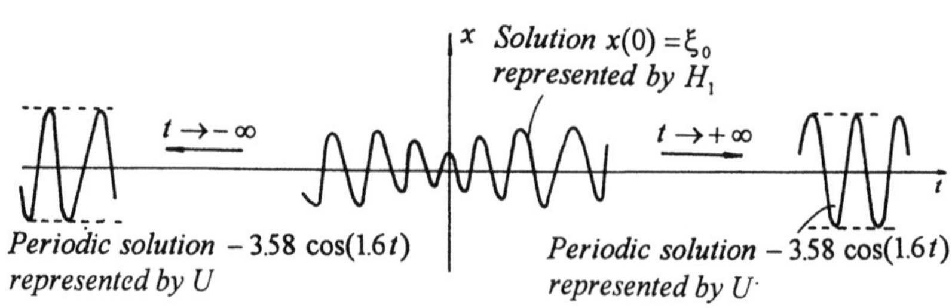


Fig. 7.12

We cannot reach this solution by physical experiment because it is unstable. Furthermore, the structure of the conservative system is also unstable, but theoretically for the occurrence of chaos it is more important.

Let us discuss the following Duffing equation when damping exists:

$$x'' + 0.1x' + x + \frac{2}{15}x^3 = \cos 1.6t \tag{7.79}$$

The averaging equations on a rotating plane are:

$$\frac{\mathrm{d}\varsigma}{\mathrm{d}t} = -0.05\varsigma - 0.4875\eta + 0.03125\eta(\varsigma^2 + \eta^2)$$

$$\frac{\mathrm{d}\eta}{\mathrm{d}t} = 0.4875\varsigma - 0.05\eta + 0.3125 - 0.03125\varsigma(\varsigma^2 + \eta^2) \tag{7.80}$$

The singular points of the system are $S(-0.65, 0.07)$, $S'(3.10, 2.78)$ and $U(-2.97, 2.14)$. The corresponding amplitudes $A = \sqrt{\varsigma^2 + \eta^2}$ are 0.65, 4.16 and 3.66. However, there is no first intergral as in eq. (7.78). $\varsigma$ systems are not conservative. By comparing the corresponding phase diagrams in Fig. 7.12 and Fig. 7.13, we find that the original homoclinic orbits $H$ and $H'$ break into two curves respectively. They divide plane $\varsigma$ into two regions (Fig. 7.14). They are two stable singular points, and are also attractor regions. For the original vibration system, the motion finally tends to $S$ or $S'$, depending upon the initial conditions. These two regions are twinned to each other. The proper change of the initial condition will cause great variation in the final stationary states.

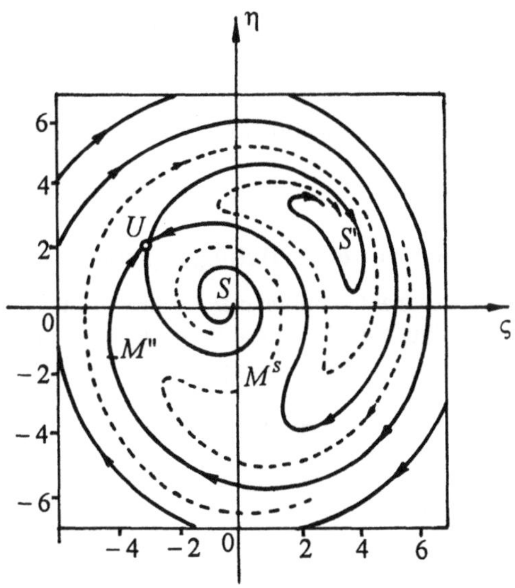


Fig. 7.13

The twinning of the two-attractor regions, however, is not the whole of the problem. Fig. 7.13 is the phase diagram of the averaging equation on the rotating plane. Let us discuss again the phase diagram of the averaging equation and the real equation as shown in Fig. 7.8 and Fig. 7.7. Can we guess the type of Fig. 7.7 from Fig. 7.8 ? It is possible in linear systems, because there is only one equilibrium point $A$. However, it is rather complicated if we guess the real $\varsigma$ plane curves from Fig. 7.13. Roughly speaking, the periodic wave added to the track line in Fig. 7.13 will not be large if the amplitude is not large. So even if the shape of the dividing line is not as smooth as that in Fig. 7.14, we can still divide the plane into two parallel bands, that is to say, there exist two attractor regions. But, if the amplitude of the disturbing force is large enough, then the amplitude of the wave to be added is large. Furthermore, as the $\varsigma$ system is not autonomous, a track line may intersect itself or track each other. Besides, there is a certain kind of motion that approaches neither $S$ nor $S'$, but wanders between $S$ and $S'$. This is what we call "non-periodic motion", and even damping exists. It is different in property from the curve $c_3$ where damping does not exist (Fig. 7.11), which is the process of chaos that we are going to discuss. Chaos depends upon the initial conditions. The dependence is extremely sensitive.

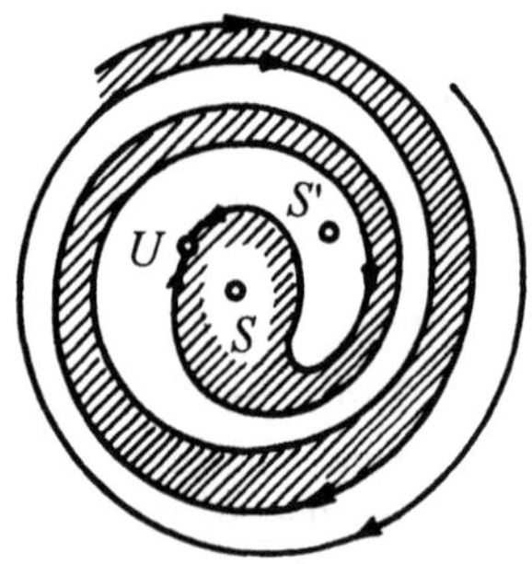

Fig. 7.14

The above statement is only a guess or an estimation. It indicates that the averaging method (together with other methods of approximation, such as the method of perturbation, the asymptotic method and so on) may fail, because its results are not qualitatively correct in comparison with practice. This was not recognized until the middle of the 1970s. It is necessary to take this into acount in future so as to determine the range of utility.

### 7.4.3 The Subharmonic Solution of the Duffing Equation

One third subharmonic resonant vibration may occur when the disturbing frequency satisfies $\omega = 3\omega_0$. The amplitude of the subharmonic solution can be obtained by the use of the averaging method on the rotating plane, except that now the rotating angular velocity is $\frac{\omega}{3}$, instead of $\omega$. So, the fixed point on the rotating plane equals the subharmonic solution, and the disturbing force turns one circle when the plane rotates three turns. Assuming

$$x = \varsigma \cos\frac{1}{3}\omega t + \eta \sin\frac{1}{3}\omega t \tag{7.81}$$

$$y = \frac{x'}{\omega/3} = -\varsigma \cos\frac{1}{3}\omega t + \eta \sin\frac{1}{3}\omega t \tag{7.82}$$

the amplitude can be solved by the equation

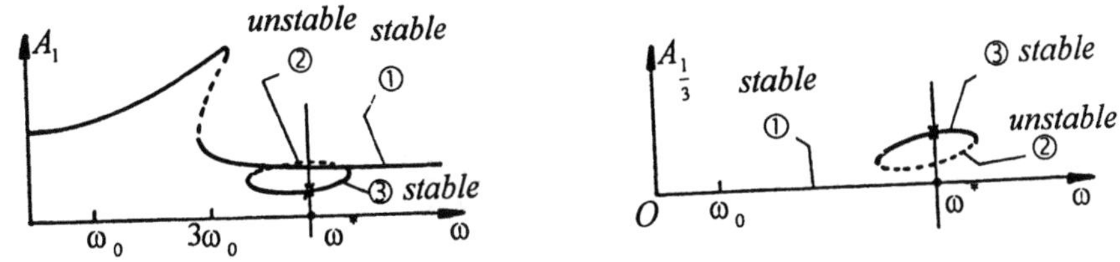


Fig. 7.15

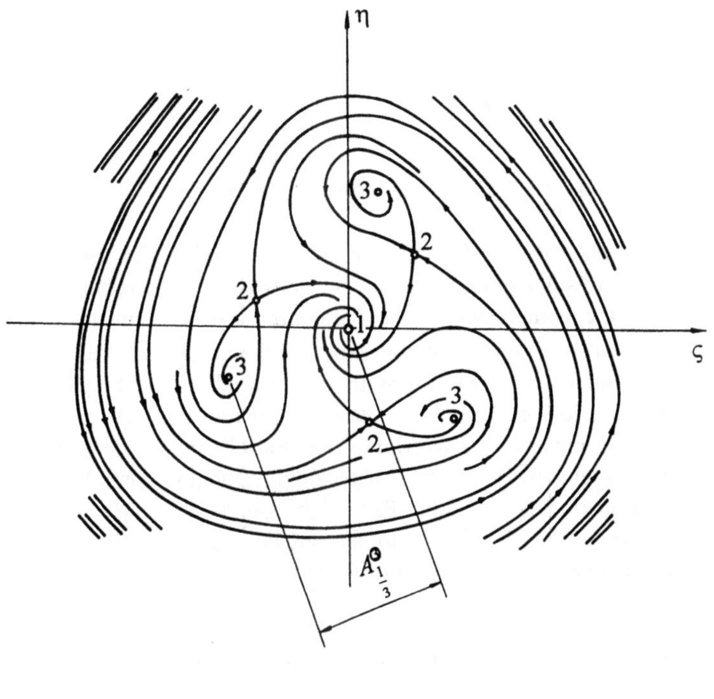


Fig. 7.16

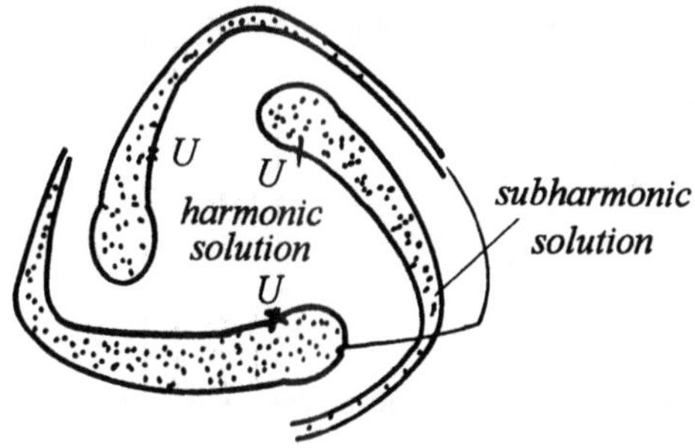


Fig. 7.17

$$A_{\frac{1}{3}} = \sqrt{\varsigma^{*2} + \eta^{*2}} \tag{7.83}$$

so, the solution is

$$x = A_{\frac{1}{3}} \cos\left(\frac{1}{3}\omega t + \beta_{\frac{1}{3}}\right) + A_1 \cos(\omega t + \beta_1) + \ldots \tag{7.84}$$

The subharmonic solution can be obtained only when the value of damping is small; $\omega$ is a little bit larger than $3\omega_0$, when $\varepsilon > 0$. As shown in Fig. 7.15, there are possibly three solutions when $\omega = \omega^*$.

① $$x = A_{\frac{1}{3}}^{(1)} \cos\left(\frac{1}{3}\omega^* t + \beta_{\frac{1}{3}}^{(1)}\right) + A_1^{(1)} \cos(\omega^* t + \beta_1^{(1)}) + \ldots$$

② $$x = A_{\frac{1}{3}}^{(2)} \cos\left(\frac{1}{3}\omega^* t + \beta_{\frac{1}{3}}^{(2)}\right) + A_1^{(2)} \cos(\omega^* t + \beta_1^{(2)}) + \ldots$$

③ $$x = A_{\frac{1}{3}}^{(3)} \cos\left(\frac{1}{3}\omega^* t + \beta_{\frac{1}{3}}^{(3)}\right) + A_1^{(3)} \cos(\omega^* t + \beta_1^{(3)}) + \ldots$$

The first harmonic solution has already been obtained. The second solution is unstable. The third subharmonic solution is what we want to determine. We can see, therefore, that the component of the harmonic wave is somewhat weaker, i.e. $A_1^{(3)} < A_1^{(1)}$, when harmonic solution exists. When $\omega = \omega^*$, the occurrence of subharmonic solutions is determined by the initial conditions. The phase diagram of the averaging equation on the rotating plane is shown in Fig. 7.16. There are seven singular points on plane $\varsigma$. The origin gives out a harmonic solution; of the three saddle points two give out an unstable subharmonic solution; three focus points give out a stable subharmonic solution. Part of the curves pass through two boundary lines, which divide the $\varsigma$ plane into the attractive regions of a harmonic wave solution and the attractive regions of a subharmonic solution respectively.

### 7.4.4 The Present State of the Art of the Duffing Equation Study

It is now more than 60 years [34] since the Duffing equation was first studied. The nature of its solution is still not quite yet clear.

The Duffing equation can be of the following types:

$$x'' + \delta x' + x + x^3 = F\cos\omega t \tag{7.85}$$

$$x'' + \delta x' + x - x^3 = F\cos\omega t \tag{7.86}$$

$$x'' + \delta x' + x^3 = F\cos\omega t \tag{7.87}$$

$$x'' + \delta x' - x + x^3 = F\cos\omega t \tag{7.88}$$

Equation (7.87) is of the Japanese type, which was studied in great detail by C. Hayashi and others [35], and eq. (7.88) was studied by P. Holmes [36]. There are only three parameters in the above listed equations. $\delta > 0$ is the damping coefficient; $F$ and  are the amplitude and frequency of a disturbing force respectively. The parameter space $(\delta, F, \omega)$ is three-dimensional. We need to analyze and check the types of solution corresponding to each point in the parameter space. This can be a harmonic periodic solution, a subharmonic solution, chaos, or exist in all types. We can, therefore, divide the parameter space into several regions. The type of solution in each region is definite. After making sure of these fundamental differences, it is possible to choose the method for determining their quantity (e.g. the resonant case or the non-resonant case of the averaging method). The pity is that we do not yet completely understand how to divide the parameter space.

Even if we keep a certain parameter constant and consider two-dimensional (or one-dimensional) parametric space only, the problem is still not well solved, for example, when does a subharmonic solution appear? Is there a new bifurcation to chaos? What is the mechanism leading to chaos through continuous bifurcation of the subharmonic solution? As we know that it is the key factor during the process of bifurcation, is there any great difference between eq. (7.85) and eq. (7.86)? (There is homoclinic motion in free vibration in eq. (7.86), but this does not appear in eq. (7.85).

## 7.5 The Averaging Method and Local Bifurcation

Suppose that we have an equation:

$$x' = \varepsilon f(x,t,\mu,\varepsilon) \qquad \mu \in \mathcal{R} \tag{7.89}$$

The averaged system is

$$y' = \varepsilon f'(y,\mu) \tag{7.90}$$

When the parameter varies, bifurcation occurs in eq. (7.90). The following theorem will prove that similar bifurcation will occur in eq. (7.89) too.

**Theorem 7.2** If $\mu = \mu_0$ and there is a saddle node or Hopf bifurcation in eq. (7.90), a Poincaré map of system (7.89) will occur as a saddle node or a Hopf bifurcation, when $\varepsilon$ is sufficiently small and $\mu$ is in the neighbourhood of $\mu_0$.

**Proof** Assuming that $\varepsilon$ is a fixed small value, $P_\varepsilon^\mu$ and $P_0^\mu$ indicate the Poincaré maps of eq. (7.89) and (7.90), the difference between them being $\varepsilon$.

consider the bifurcation equation for saddle-node bifurcation

$$\left(Id - P_0^\mu\right)(y) = 0 \tag{7.91a}$$

and

$$\left(Id - P_\varepsilon^\mu\right)(x) = 0 \tag{7.91b}$$

According to the given conditions, there is a pair of fixed points $y_+, y_-$ in space $(y,\mu)$. When $\mu = \mu_0$, they coalesce into the smooth bifurcation curve shown in Fig. 7.18. There is a local coordinate transformation near the points $(y_\pm(\mu_0),\mu_0) \in \mathcal{R}^n \times \mathcal{R}$. These points can be expressed in the form:

$$y_\pm(\mu) = (\pm C\sqrt{\mu_0 - \mu}, 0) \in \mathcal{R} \times \mathcal{R}^n = \mathcal{R}^n$$

As $P_\varepsilon^\mu$ and $P_0^\mu$ approach $\varepsilon$, in the case of the same coordinate, $P_\varepsilon^\mu$ has a pair of bifurcations of fixed points. $x_\pm(\mu) = (\pm\sqrt{\mu_0 - \mu} + \theta(\varepsilon), o(\varepsilon))$. As the matrixes

$DP_{\varepsilon}^{\mu}(x\pm(\mu))$ and $DP_{0}^{\mu}(y\pm(\mu))$ are also close to each other, the linearized stability of bifurcations is the same for these two systems.

Because there is one unique steady-state curve of $\bar{y}(\mu)$ in $P_0^{\mu}$ for Hopf bifurcation, the spectrum of $DP_0^{\mu}(\bar{y}(\mu))$ does not contain 1. Near the stationary curve, the curve $\bar{x}(\mu)=\bar{y}(\mu)+o(\varepsilon)$ is still correct for $P_{\varepsilon}^{\mu}$. Because the spectra of $DP_0^{\mu}(\bar{y}(\mu))$ and $DP_{\varepsilon}^{\mu}(\bar{x}(\mu))$ are very close, a pair of simple complex roots must pass through the unit circle. If the corresponding eigenvalue of $D\bar{f}(\bar{y}(\mu_0),\mu_0)$ is $\pm i\omega$, the eigenvalue of $DP_0^{\mu}(\bar{y}(\mu))$ is $e^{\pm i\omega T}$ and the eigenvalue of perturbation linearized mapping $DP_{\varepsilon}^{\mu}(\bar{x}(\mu))$ near the bifurcation point is $e^{\pm(\varepsilon\omega T+o(\varepsilon^2))}$. Therefore, as $T$ is a fixed value, when $\varepsilon \ll 1$, $e^{im2\varepsilon T}\neq 1$ can be satisfied for $m=1,2,3,4$.

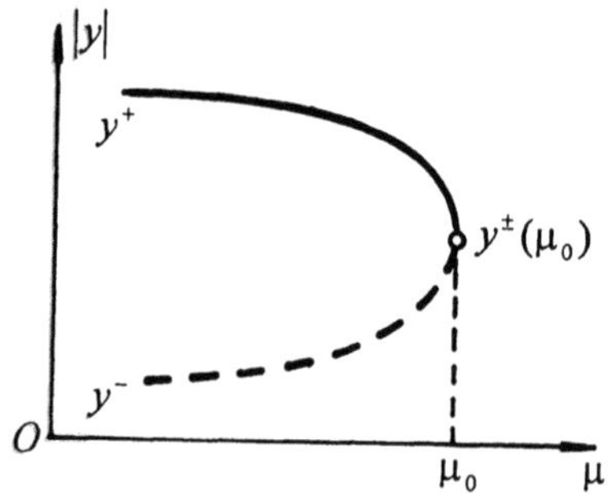


Fig.7.18

This result can be used in conjunction with the Duffing equation already discussed. In this case, the saddle-node bifurcation occurs on the folding point or jumping resonant vibration point. In averaging the Duffing equation the saddle-node point responds to the saddle-node bifurcation of the periodic orbit of precise systems.

The above theorem can be expanded to multi-parameter systems and other more complicated bifurcations. For example, there is a certain point on the plane $\Omega\gamma$. For this point the Duffing system has a fixed degenerate point which can be split into one, two or three points. On this point, two points of the frequency response curve fold into one point with an orthogonal tangent. Thus, near this point there is a bifurcation in the precise system in which three periodic orbits have coalesced. However, the averaging method cannot obtain all the information of codimension 2, because of the sensitive global influence on the homoclinic orbit. This influence will be discussed in the next paragraph.

As an example, we will study a bifurcation in self-vibrating system with the hysteresis non-linearity as shown in Fig.7.19(a). The equation of this system is

$$x''-(\beta-x^2)x'+x+F(x)=0 \tag{7.92}$$

where $\beta$ is the linear damping coefficient which is chosen as the bifurcation parameter. $F$ is a bi-linear hysteresis function, as shown in Fig.7.19(b). When $\beta<0$, the trivial solution of the system is stable. When $\beta=0$, bifurcation will occur [36]. Add a small parameter $\varepsilon$ in front of relatively small terms, then

$$x''+x=\varepsilon\left[(\beta-x^2)x'-F(x)\right] \tag{7.93}$$

Because of the discontinuity of the derivative of hysteresis, it is not convenient to use the LS method or the centre manifold method to study this directly. So, we study it by the averaging method.

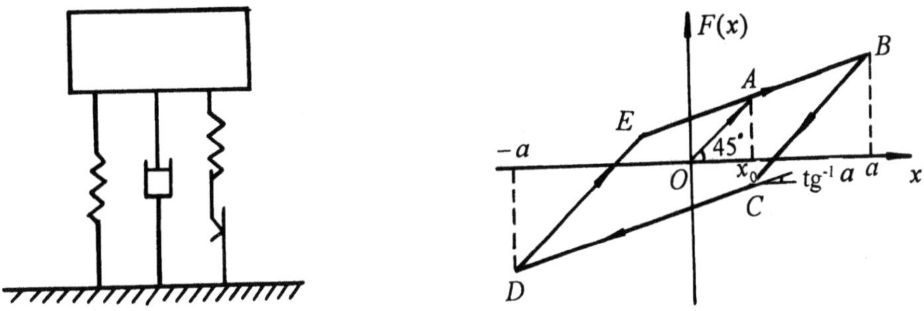


(a) (b)

Fig.7.19

Suppose the bifurcation solution of the above equation is

$$x = a\cos\varphi \qquad x' = -a\sin\varphi \tag{7.94}$$

where $\varphi = t+\theta$. Taking $a$ and $\theta$ as new variables, on the basis of the averaging method introduced in section 7.1, we obtain the first-order approximate averaging equation

$$x = z\cos(t+\gamma),\ Z' = \varepsilon Y_1(A),\ \text{and}\ \gamma' = \varepsilon Z_1(A) \tag{7.95}$$

where $Y_1 = -\frac{1}{2}z\left(\frac{1}{4}z^2 - \beta\right) + A(z)$ (7.96a)

$$Z_1 = B(z) \tag{7.96b}$$

$$A(z) = \frac{1}{2\pi}\int_0^{2\pi} f(z\cos\varphi)\sin\varphi d\varphi \tag{7.96c}$$

$$B(z) = \frac{1}{2\pi}\int_0^{2\pi} f(z\cos\varphi)\cos\varphi d\varphi \tag{7.96d}$$

Now, let us calculate $A(z)$ and $B(z)$. Suppose $\theta_1 = \arccos\left(1-\frac{2x_0}{z}\right)$ and consider the symmetry of bi-linear hysteresis character, so

$$A(z) = \frac{1}{z}\{\int_0^{\theta_1}[(1-\alpha)(x_0-z)+z\cos\varphi]\sin\varphi d d\varphi$$
$$+\int_0^{\pi}[(\alpha-1)x_0+\alpha z\cos\varphi]\sin\varphi d\varphi\} \tag{7.97}$$
$$=\begin{cases} -\dfrac{1-\alpha}{2\pi}z\sin^2\alpha & z > x_0 \\ 0 & z < x_0 \end{cases}$$

$$B(z) = \frac{1}{\pi z}\{\int_0^{\theta_1}[(1-\alpha)(x_0-z)+z\cos\varphi]\cos\varphi d d\varphi$$
$$+\int_0^{\pi}[(\alpha-1)x_0+\alpha z\cos\varphi]\cos\varphi d\varphi\}$$

$$= \begin{cases} \dfrac{1}{\pi z}[\alpha\pi + (1-\alpha)\theta_1 - \dfrac{1}{2}(1-\alpha)\sin 2\theta_1 & z > x_0 \\ -\dfrac{1}{z} & z < x_0 \end{cases} \tag{7.98}$$

The stationary solution of the averaging equation is as follows:

$$\begin{aligned} &-\frac{1}{2}z\left(\frac{1}{4}z^2 - \beta\right) - \frac{(1-\alpha)}{2\pi} z\sin^2\theta = 0 \qquad z > x_0 \\ &\frac{1}{2}z\left(\frac{1}{4}z^2 - \beta\right) = 0 \qquad z < x_0 \end{aligned} \tag{7.99}$$

where

$$\begin{aligned} \theta_1 &= \arccos\left(1 - \frac{2x_0}{z}\right) \\ \sin^2\theta_1 &= \frac{4x_0}{z}\left(1 - \frac{x_0}{z}\right) \end{aligned} \tag{7.100}$$

Substituting (7.100) into eq. (7.99), we have a bifurcation equation

$$z^4 - 2\beta z^2 + \frac{16(1-\alpha)x_0}{\pi}(z - x_0) = 0 \qquad z > x_0 \tag{7.101a}$$

$$z(z^2 - 4\beta) = 0 \qquad z < x_0 \tag{7.101b}$$

If we express eq. (7.101) as $G = 0$, the stability of the stationary solution $z = z_0$ can be determined as follows. If $G_z(z_0, \lambda, \alpha) < 0$, $z = z_0$ is asymptotically stable; if $G_z(z_0, \lambda, \alpha) > 0$, $z = z_0$ is unstable. Assuming $\lambda = 4\beta$, $\varepsilon_1 = \dfrac{16(1-\alpha)x_0}{\pi}$, and $\varepsilon_2 = -\varepsilon_1 x_0$ then the bifurcation equation is

$$G(z, \lambda, \alpha) = z^4 - \lambda z^2 + \varepsilon_1 z + \varepsilon_2 \tag{7.102}$$

Next, we shall discuss the bifurcation diagram. According to the method of singularity introduced above, we can calculate the bifurcation set $B$, hysteresis set $H$ and double limit point set $D$.

We know from the definition of the bifurcation set that

$$G_\lambda = -z^2 = 0 \tag{7.103}$$

$$G_z = 4z^3 - 2\lambda z + \varepsilon_1 = 0 \tag{7.104}$$

From eqs. (7.102), (7.103) and (7.104), we obtain $z = 0$, $\varepsilon_1 = 0$, $\varepsilon_2 = 0$, that is

$$\frac{16(1-\alpha)x_0}{\pi} = 0 \quad \text{so } \alpha = 1 \text{ or } x_0 = 0$$

Therefore, the bifurcation set B is

$$\begin{aligned} \mathcal{B} &= \{\alpha \in \mathcal{R}^k : \exists (x, \lambda) \in \mathcal{R} \times \mathcal{R}, \text{ on } (x, \lambda, \alpha),\ G = G_x = G_\alpha = 0\} \\ &= \{\alpha = 1,\ x_0 \in \mathcal{R} \text{ or } \alpha \in \mathcal{R},\ x_0 = 0\} \end{aligned} \tag{7.105}$$

Similarly, the hysteresis set $\mathcal{H}$ is

$$\begin{aligned} \mathcal{H} &= \{\alpha \in \mathcal{R}^k : \exists (x,\lambda) \in \mathcal{R}\times\mathcal{R},\ \text{on } (x,\lambda,\alpha),\ G = G_x = G_\alpha = 0\} \\ &= \{\alpha = 1,\ or\ x_0 = 0 \text{ or } x_0^2 = \frac{27}{256\pi}(1-\alpha)\} \end{aligned} \tag{7.106}$$

and the double limit point set is

$$\begin{aligned} D &= \{\alpha \in \mathcal{R}^k : \exists (x_1,x_2,\lambda) \in \mathcal{R}\times\mathcal{R}\times\mathcal{R},\ x_1 \neq x_2,\ \text{on } (x_i,\lambda,\alpha),\ G = G_x = 0,\ i = 1,\ 2\} \\ &= \{\varepsilon_1 = 0,\ \varepsilon > 0\} = \Phi \end{aligned} \tag{7.107}$$

Because of the discontinuity of the connection point of eq. (7.101a) and (7.101b), we define an additional persistent set TL.

If $\quad G_1,\ G_2 : \mathcal{R}\times\mathcal{R}\times\mathcal{R}^k \to \mathcal{R}$

$$G_1(x,\lambda,\alpha) = 0 \qquad \lambda < \lambda_1$$
$$G_2(x_1,\lambda,\alpha) = 0 \qquad \lambda > \lambda_1$$

Define: $TL = \{\alpha \in R^k : \exists (x,\lambda) \in \mathcal{R}\times\mathcal{R};\ \text{at } (x,\lambda,a),\ G_1 = G_2 = G_{2x} = 0\}$

$$= \left\{x_0^2 = \frac{8}{\pi}(\alpha - 1)\right\} \tag{7.108}$$

So the transition set $\sum$ is

$$\sum = \mathcal{B}\cup\mathcal{H}\cup\mathcal{D}\cup TL \tag{7.109}$$

**Table 7.1** Results of numerical integration

| Bifurcation parameter $\beta$ | Numerical integral solution $x_n$ | Theoretical solution $Y_n$ of this process | Error $(X_n - Y_n)\times 100\% / X$ |
|---|---|---|---|
| 0.200 | 0.8944 | 0.8944 | 0.00 |
| 0.400 | 1.1487 | 1.1210 | 2.41 |
| 0.600 | 1.3767 | 1.3208 | 4.05 |
| 0.800 | 1.6081 | 1.5447 | 3.94 |
| 1.000 | 1.8263 | 1.7675 | 3.22 |
| 1.200 | 2.0312 | 1.9771 | 2.67 |
| 1.400 | 2.2199 | 2.1712 | 2.16 |
| 1.600 | 2.3964 | 2.3513 | 1.88 |
| 1.800 | 2.5620 | 2.5192 | 1.67 |
| 2.000 | 2.7196 | 2.6769 | 1.57 |
| 2.200 | 2.8698 | 2.8257 | 1.54 |
| 2.400 | 3.0124 | 2.9671 | 1.51 |
| 2.600 | 3.1494 | 3.1018 | 1.51 |
| 2.800 | 3.2781 | 3.2308 | 1.44 |
| 3.000 | 3.4053 | 3.3547 | 1.49 |
| 3.200 | 3.5268 | 3.4741 | 1.49 |
| 3.400 | 3.6435 | 3.5894 | 1.54 |

Transition set $\sum$ divides the parameter space into eight regions, as shown in Fig. 7.20. We have different bifurcation figures for different regions, and the topological structures of the bifurcation figures are the same in the same region, as shown in Fig. 7.21. We are interested in the case of $x_0 > 0$ from the viewpoint of physics. That is, the regions I—IV. In this case, the physical transition set (7.109) is as shown in Fig. 7.22 for $z > 0$. For the cases of $z < x_0$ and $z > x_0$, the bifurcation figures are as shown in Fig. 7.23. In Fig. 7.21 and Fig. 7.23, the solid line denotes the stable solution, and the broken line denotes the unstable solution. We know from Fig. 7.23 that when $\alpha >$ , there is a second bifurcation at the point $z = x_0$. The jumping and the hysteresis exist only in regions I and IV. The results of numeric integration are listed in table 7.1.

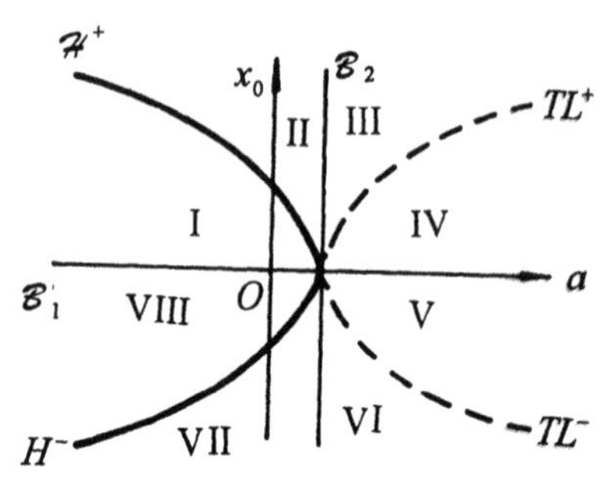


Fig. 7.20

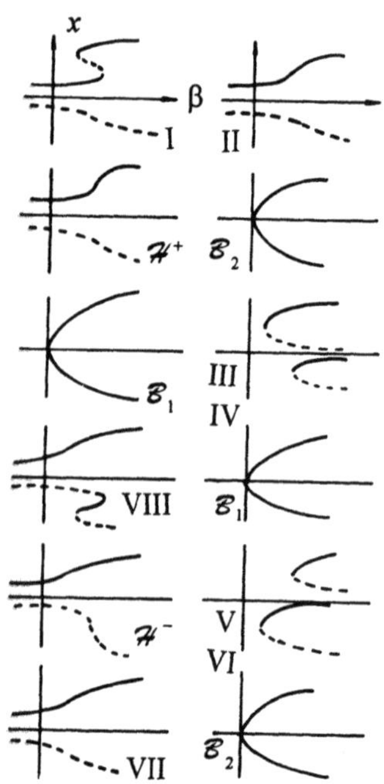


Fig. 7.21

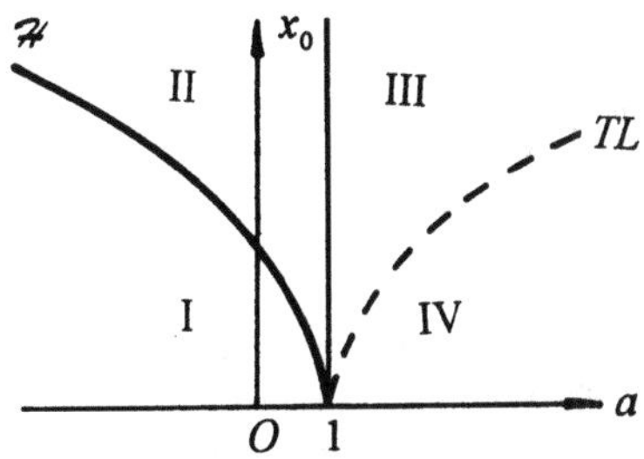


Fig. 7.22

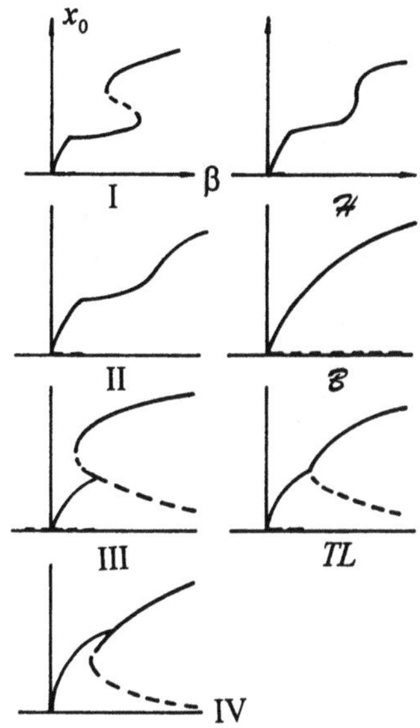


Fig. 7.23

## 7.6 The Averaging Method, Hamiltonian Systems and Global Behaviour

From the above discussion, we know that averaging theory provides a method that approximate actual Poincaré mapping, or the method of approximation in which the $n$-dimensional mapping is replaced by the $n$-dimensional vector field. According to theorem 7.1, the averaging method preserves the local behaviour in the system. For example, the fixed point of the averaging vector field $\bar{f}$ corresponds to the periodic orbit of eq. (7.34) and the fixed point of $P_\varepsilon$. But, because there is great difference of generality between the $n$-dimensional value vector field and the $n$-dimensional mapping, the global behaviour cannot be preserved directly. Therefore, the property of the global behaviour in $\bar{f}$ may not reflect that in $P_\varepsilon$, nevertheless, under certain conditions, the global behaviour can still be preserved.

**Theorem 7.3** If the first approximate averaging system (7.35) is limited to the internal domain $\mathcal{D} \subset \mathcal{R}^n$, if the time $T$ flow mapping $P_0$ of this system has a limit set that is composed only of hyperbolic fixed points, and if the intersections of the stable and unstable manifolds are all transversal, then, for sufficiently small $\varepsilon > 0$, the Poincaré mapping $P_\varepsilon|_D$ of precise system (7.34) is topologically equivalent to $P_0|_D$.

**Proof** This theorem is a corollary deduced directly from the averaging theorem (2) and (3). When explained in terms of dynamical systems, the result is derived from the structural stability of the flow of eq. (7.35) when $\varepsilon > 0$. Yet, the perturbation method cannot be applied directly to the standard equation, because, in eq. (7.39), when $\varepsilon \to 0$, undisturbed system $x' = 0$ is fully degenerate. On the other hand, let us discuss the equation after scaling $\tau = \varepsilon t$ again.

$$y' = \bar{f}(y) + \varepsilon f_1\left(y, \frac{\tau}{\varepsilon}, \varepsilon\right) \tag{7.110}$$

where $y' = \frac{dy}{dt}$; in eq. (7.110) there is a disturbing term $\varepsilon T$. However, this does not affect the use of Growall's estimation about the proof of theorem 7.1. So, we can obtain the same result as in theorem 7.1.

Now, let us analyze the example of the Duffing equation again. Checking the phase portrait of eq. (7.73) by the Bendixson criterion, we know that it does not contain any closed orbit. That is to say, there is no closed loop or homoclinic loop. For all $(\varsigma, \eta) \in \mathcal{R}^2$, we have

$$\text{trace } Df = \frac{\partial f_1}{\partial \varsigma} + \frac{\partial f_2}{\partial \eta} = -2n < 0 \tag{7.111}$$

So the phase portrait contains fixed points only; as what we analyze here is a two-dimensional system, there is no homoclinic loop, but a saddle point only, and moreover, there is no intersection of the stable and unstable manifolds. The assumption of theorem 7.3 is satisfied, and $P_\varepsilon$ and $P_0$ are topologically equivalent when $\varepsilon$ is very small.

In many averaging equations, there is also a periodic orbit in addition to the fixed point. For example, the averaging van der Pol equation produces a Hopf bifurcation, and the periodic orbits are disappeared one after the other at the connection of the saddle points. This is more difficult to treat, but under certain conditions the periodic orbit of the averaging equation is preserved like that of the invariant torus.

**Theorem 7.4** If there is a hyperbolic periodic orbit $\gamma_0$ in eq. (7.35), then the flow of the suspended system (7.54) has a hyperbolic invariant torus $T_\varepsilon$ near $\gamma_0 \times \mathcal{S}^1$. That is to say, the Poincaré mapping $P_\varepsilon$ of precise system (7.34) has an invariant closed curve $\gamma_\varepsilon$ near $\gamma_0$.

For a detailed proof of the above theorem refer to [30]. The normal hyperbolicity of the invariant set $\gamma_0 \times \mathcal{S}^1$ guarantees its persistence (M.W. Hirsh). But while smooth invariant closed curves are existent for both $P_0$ and $P_\varepsilon$, the dynamics of $P_\varepsilon|_{\gamma_\varepsilon}$ are generally quite complex owing to the effects of the resonant vibration.

Now, let us analyze eq. (7.110). When $\varepsilon = 0$, there is an orbit of period $\bar{\tau}$ in eq. (7.110). When $\varepsilon \to 0$, we have the resonance relationship

$$\bar{\tau} = \frac{m\varepsilon T}{n} \qquad m, n \in Z \tag{7.112}$$

Hence, according to the general principle of mapping on the circle, a smooth closed curve $\gamma_\varepsilon$ is obtained. This is the set containing the periodic points, whose periods $m - \frac{1}{\varepsilon}$ depend on $\varepsilon$. When $\varepsilon \to 0$ they appear or disappear in countable bifurcating sequence. The analysis of such resonance motion depends on the small denominators; for further discussion, please see [36].

If a precise system (7.34) is a Hamilton system with energy $\varepsilon H(\varsigma, \eta, t, \varepsilon)$, then another Hamilton system can be obtained by means of transformation eq. (7.36). When $x \in R^2$, it is a two-dimensional autonomous system. Its solution curves are level curves of the averaged Hamilton function

$$\overline{H}(\varsigma, \eta) = \frac{1}{T}\int_0^T H(\varsigma, \eta, t, 0)\mathrm{d}t \tag{7.113}$$

If the averaging system has a compact level curve containing a saddle point, then necessarily this curve is the non-transversal intersection of the stable and unstable manifolds for the unperturbed Poincaré mapping and we con not expect it to be preserved upon restoration of the time-dependent term $\varepsilon^2 f_1$. Now let us discuss the Duffing eq. (7.72) again. Let $n = 0$, then the precise system becomes a Hamiltonian system. We can obtain the averaged Hamiltonian equation directly from eq. (7.73):

$$\varepsilon\overline{H}(\varsigma, \eta) = -\frac{1}{4\omega}\{(\omega_0^2 - \omega^2)(\varsigma^2 + \eta^2) - 2F\varsigma + \frac{3\varepsilon}{8}\left(\varsigma^2 + \eta^2\right)^2\} \tag{7.114}$$

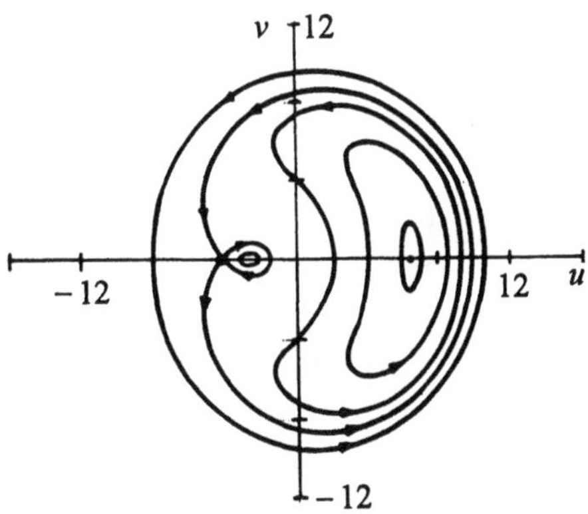


Fig 7.24

The corresponding phase portrait is shown in Fig. 7.24. We can see that besides the double homoclinic loop, there are continuous families of three periodic orbits. These level curves are invariant curves for the (unperturbed) Poincaré mapping $P_0$ of the averaging system. In the generic case we expect the homoclinic loops to break, yielding transverse intersection of the perturbed stable and unstable manifolds. We also expect the resonant closed curves to break up in some complicated way. None of the closed curves is hyperbolic, and we have to resort to the Kolmogorov-Arnold-Moser theory to prove that sufficiently irrational ones close enough to the elliptic centres are preserved for $P_\varepsilon$, with $\varepsilon$ being small. We shall return to the general topic of perturbation of integrable Hamiltonian systems.

Even when the original system is not the Hamiltonian system, the homoclinic bifurcations may occur also in an averaged equation. For example, the averaged van der Pol system with weak damping and forcing has a bifurcation, with phase portraits as shown in Fig. 7.25.

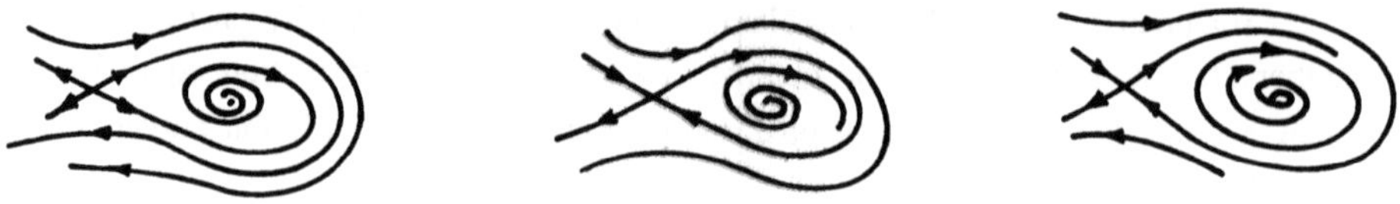

(a) (b) (c)

Fig. 7.25

According to the principle discussed above, when ε is sufficiently small, and depending upon other parameters, the phase portraits in Fig. 7.25(a), (c) are preserved for the precise system because the structure of time $T$ flow mapping for the averaged system is structurally stable. However, as we approach the parameter value for which the homoclinic bifurcation occurs, the ε range for which theorem 7.4 holds, shrinks. Clearly, at the bifurcation point, we cannot expect the stable and unstable manifolds of the saddle for the perturbed mapping $P_\varepsilon$ to be identified. As in Fig. 7.25(b), and in the generic case one expects to find transversal intersections occurring near the bifurcation point for the unperturbed system, with (generic) quadratic tangency of the manifolds at discrete points occurring at the boundaries of the region, see Fig. 7.26.

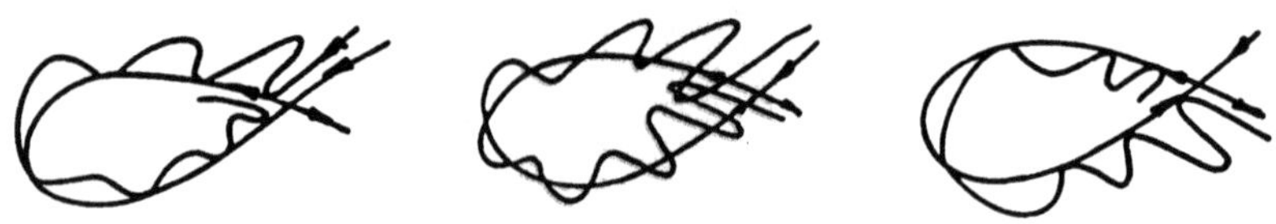

(a) (b) (c)

Fig. 7.26

# Chapter 8

# Brief Introduction to Chaos

## 8.1 What Is Chaos?

As an important feature of the solutions of a deterministic non-linear dynamic system (such as non-linear differential equations, difference equations, algebraically iterative equations, etc.) chaos has attracted wide attention. So far no strict general mathematical definition of chaos has been available, but it is depicted in many different ways. It is found in a wide variety of fields, such as mathematics, physics, mechanics, astronomy, chemical engineering, optics, bio-mechanics, ecology and sociology. Even philosophers are interested in it.

What are the basic features of chaos now that it is rearranged as a common phenomenon? At present, the acceptable three conclusions are (1) macroscopically irregular, (2) regularity in microcosm and (3) mutual compensation of regularity and irregularity. First,viewed from a macroscopic point of view, the chaotic movement appears disorderly, seemingly random, and sensitive to the initial condition (known as the butterfly effect). Although the process is strictly deterministic, its long-term behaviour cannot be predicted, and it is as random as casting coins. Such a disorder can be approached in three aspects:

**(i) Stochasticity.** The internal stochasticity is used mainly to distinguish it from the external stochasticity. (The external stochasticity is the random motion caused by external random excitation.) Usually, if some state of a system can appear or disappear under a certain condition, the system is considered to be stochastic. The internal stochasticity is the stochastic character produced spontaneously by a completely deterministic system under a certain system parameter condition. In the 1960s, Aleksiev and others [59] found the following fact. When they were calculating the motion of a celestial body *m* with a small mass moving on the vertical line of a plane having two large celestial bodies $M_1$ and $M_2$ with equal mass. The behaviour of the small body *m*, after it had swung to and fro a number of times, because its location was so stochastic that its speed and return time could no longer be predicted. That is to say, the internal stochasticity appeared. This is the important property of the disorder of chaos.

**(ii) Non-periodicity.** The periodic movement transforms to an non-periodic movement. Periodic movements are common in nature, and so are non-periodic ones. But the emphasis here is placed on the transformation from the periodic to the non-periodic. Take the feature of the double periodic bifurcation of ordinary differential equations as an example. The eventual development of periodic movement will result in non-periodic ones. The often-cited example, the insect model [59], says that the change in the number of insects can be described by the following one-dimensional mapping $x_{n+1} = 1 - \mu x_n^2$ (8.1)

After determining a certain initial value $x_0$, one can calculate the change in the number of insects by means of eq. (8.1). To a certain extent, the change in number of insects is periodic. With increase of $\mu$, the period of the number of insects changes from 1 to 2, 4, 8, . . . . When $\mu$ changes to 1.4015, the period turns to $2^k$. When $k \to \infty$, it changes to an non-periodic movement. This is a breakthrough in the traditional thoughts about periodic movement.

**(iii) Local instability (sensitivity to initial conditions)**. The relative stability is a prerequisite for the existence of a system process. A very obvious feature in chaotic movement is sensibility to the initial value in some aspects. Lorenz once vividly referred to the butterfly effect [60]. He stated that, owing to the sensibility of the system to the initial value of a certain parameter, "Only a little flap of the butterfly's wings can make the meteorologist unable to predict the weather in a month". This kind of local instability or so-called sensibility to the initial value is sometimes used to describe or judge the occurrence or existence of chaos.

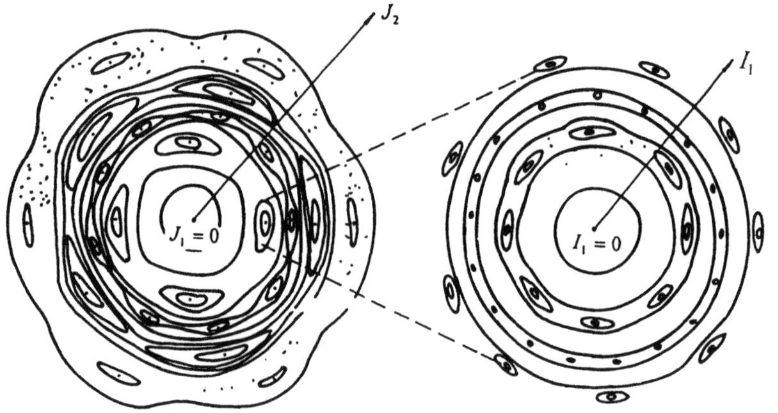


Fig. 8.1

The above is the feature of disorder of chaotic movement. What should be pointed out is that this disordered movement comes via the completely deterministic non-linear equation. This contradicts the Laplace deterministic idea which dominated people's mind for hundreds of years: a deterministic system can give a deterministic result for a definite initial value. This model of thinking has been rooted deeply in people's minds. Especially since Newton, deterministic thinking has been fixed in men's thinking. In mathematics, the $\varepsilon-\delta$ language is the mathematical description of this deterministic theory. The fully-developed analytical system based on it provides deterministic theory with a very convenient analytical method. The discovery and research of chaotic movement, not only challenges deterministic theory but also breaks through the bondage of deterministic theory. And this is a big leap in people's understanding of nature. The second feature of chaotic movement is regularity in microcosm. Continuous research shows that chaos is not a completely stochastic movement. It has a regularity with a long background. This regularity manifests itself in three aspects:

**(i) Geometric structure with self-resemblance**. The geometric structure referred to here is a characteristic of chaotic behaviour. This characteristic is somewhat similar to holograms. For example, in the non-linear Hamiltonian system, the gap between the two KAM tori is full of chaos, see Fig. 8.1. We can shift the origin to the periodic point with a period $s$ by a coordinate transformation, and study the motional orbit near it. Generally, we take $s$ iterations continuously as one map. We find that between the periodic orbits there are high-order periodic fixed points, i.e. there are periodic solutions with higher period. They have global features similar to those described above, but on a small scale. This is a typical self-resemble embedded structure.

**(ii) Universality**. Chaotic movement is not completely irregular, but has some universal regularity. The regularity includes two main aspects: structural universality and scaling universality. Structural universality means that the bifurcation of portraits and the quantitative character the process tending to chaos are independent of the concrete content of this process, but related to its mathematical structure only. For one-dimensional mapping

$$x_{n+1} = f(x_n) \tag{8.2}$$

where $f:I \to I$ is a self to self map, $I$ is the defining region. If only $f$ is single convex and continuous, its result is universal. Scaling universality means the same embedded structure between different scaling levels for the same mapping; the characteristics of the structure are dependent only on the order of the non-linear function. This adds many convenient conditions to the study of chaos. We can study the most simple model and then obtain results applicable to other complicated forms.

**(iii) New natural constants**. For example, for the mapping described by eq. (8.2), if the bifurcation point from period $2^{n-1}$ to period $2^n$ in the major sequence is $\mu_n$, and its expanding order is 2, we can define the convergence speed of the bifurcation sequence $\delta_n$:

$$\delta_n = \frac{\mu_n - \mu_{n+1}}{\mu_{n+1} - \mu_{n+2}} \tag{8.3}$$

The limit is a universal constant:

$$\delta = \lim_{n\to\infty} \delta_n = 4.669201660910299097 \tag{8.4}$$

It is also suitable for the second, third-level bifurcation sequences in the chaotic region. It was first described by Feigenbaum [61]. Another well-known universal constant is the scaling transforming factor $\alpha$. This describes invariant scaling, i.e. when it tends to chaos, the size of the "islands" in the chaotic region diminishes in every bifurcation, with its reducing factor also tending to a constant:

$$\alpha = 2.502907875095892848 5 \tag{8.5}$$

This reveals the deeper regularity in chaos. The discovery of universal constants in the research of chaotic movement marks the relative maturity of chaotic theory. The third feature of chaotic motion is the mutual compensation of regularity and irregularity. Formerly, a clear separation was made between order and disorder of a system. Through the study of chaos, it has been found that the two are combined in chaotic movement, and are complementary. Chaos is the connecting point of the two.

At present, the study of chaos, especially the study of theory, is not very profound. Therefore, it is still difficult to make a meaningful generalization about the basic features of chaotic movement. Further progress on the theory is needed.

It has been no more than 30 years since the discovery and study of modern chaos. The first man to discover chaos was Lorenz—the meteorologist. In 1963, he studied a simplified convection model of flow which was a completely deterministic order-3 non-linear differential equation:

$$x' = -10x + 10y$$

$$y' = 28x - x - xz$$

$$z' = -\frac{8}{3}z + xy \tag{8.6}$$

He discovered by calculation [60] that the solution of the equations within a certain parameter extent was not periodic and seemed very disorderly. Meanwhile, in the field of analytical mechanics, Arnold [62] and Moser [63] proved separately the so-called KAM theorem suggested by Kolmogorov in 1954 [64]. This theorem points out some characteristics of the near integral Hamiltonian system to the effect that the addition of a perturbed term to an integrable Hamiltonian system leads to non-integrability. If the perturbed term is very small, and the frequency corresponding to the unperturbed Hamiltonian function does not satisfy the resonant condition, the qualitative picture of the perturbed system is the same as that of the unperturbed system. Chaos will appear when the non-resonant conditions are not satisfied. In 1964 Henon and others found that, for a non-integrable Hamiltonian system with two degrees of freedom, the distribution of its orbits in the phase space becomes more and more stochastic if the energy is increasing.

Because of the complexity of chaotic movement, an enormous amount of time has been devoted to numerical analysis to find and study these movements. At present, the numerical analytical method is still an important techniques, especially in the field of application. The theoretical analytical method is still limited. However, there are some theoretical methods which are accepted universally, such as the KAM theorem for the Hamilton system, the Melnikov method, Smale horseshoe mapping, the Fegenbaum theory, the geometrical method, etc.

In this chapter we shall only introduce some basic concepts of chaos, and not consider the application of chaos to different branches of physics and mechanics nor the mathematical theory. Interested readers should refer to the following works: Guckenheimer & Holmes discuss the chaotic theory from the viewpoint of dissipative theory [36], Lichtonberg & Liberman mainly consider chaos in the Hamiltonian system [89], and Sparrow analyses the change of dynamic behaviour when a parameter changes in a family of equations [87].

## 8.2 Some Examples of Chaos

In this section we introduce some typical examples of dynamical systems that produce chaotic motions, and also some basic concepts of chaos.

(1) Non-linear forced vibration

In a two-dimensional autonomous system

$$x' = f_1(x,y)$$
$$y' = f_2(x,y) \qquad (8.7)$$

there may be singular points and limit cycles, i.e. there may be equilibrium states and periodic solutions, but no strange attractor because the portrait described by eq. (8.7) on the *xy* plane cannot be self-transversal.

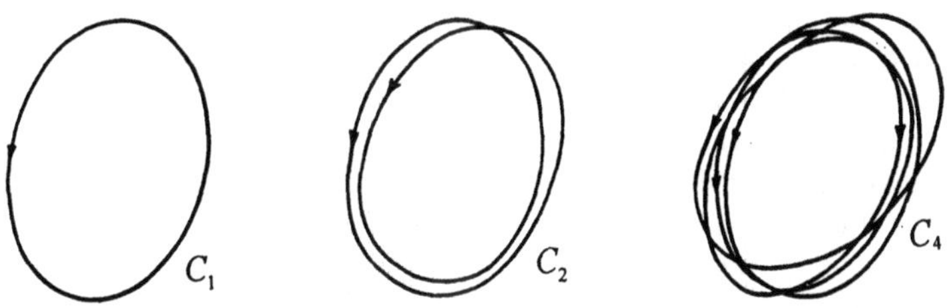


Fig. 8.2

The closed orbit $C_1$ in Fig. 8.2 is divided from 1 into 2 as $C_2$ and from 2 into 4 as $C_4$. And it must be self-transversal, which violates the uniqueness theorem of solution. Therefore, at least a three-dimensional model must be used as a chaotic model in differential equations. If we add the time $t$ as the third dimension, then we have a non-autonomous system

$$x' = F_1(x,y,\omega t)$$
$$y' = F_2(x,y,\omega t) \qquad (8.8)$$

where $F_1$ and $F_2$ are $2\pi$-periodic functions in $\omega t$. Equation (8.8) is a non-linear forced vibration equation with one degree of freedom. Its subharmonic resonance solution can be solved by the KBM method (averaging method). But in the van der Pol equation

$$x'' + \varepsilon(x^2 - 1)x' + x = f\cos\omega t \qquad (8.9)$$

Sometimes there appears a frequency of very low order (for example, 1/200). This is a phenomenon that cannot be explained by the KBM method. It was to explain this type of phenomenon that Smale put forward his horseshoe theory [67, 68].

The relationship between the forced vibrating system (8.8) and the two-dimensional map is as follows. Suppose the point $(x,y)$ of time $n\left(\frac{2\pi}{\omega}\right)$ is $(x_n, y_n)$, the value of the solution determined by this initial value and eq. (8.8), at time $(n+1)\left(\frac{2\pi}{\omega}\right)$, is $(x_{n+1}, y_{n+1})$. Then eq. (8.8) entirely corresponds to a two-dimensional map $T:(x_n, y_n) \to (x_{n+1}, y_{n+1}) \qquad (8.10)$

We can obtain the places of solution after period 1, period 2 , etc., starting from an arbitrary initial value $(x_0, y_0), (x_1, y_1), (x_2, y_2)$. In fact, this is equivalent to taking sections at $t=0$, $\frac{2\pi}{\omega}$, $\frac{4\pi}{\omega}$ on *xyt* space (called a Poincaré map), and then projecting onto the *xy* plane together (two-dimensional space). The fixed point (or periodic point) in the two-dimensional map is equivalent to the solution with a frequency of $\omega$ in a forced vibration. The set of two-dimensional fixed points (the invariant set consists of $m$ isolated points) is equivalent to the solution with a period of $m\left(\frac{2\pi}{\omega}\right)$ in a forced vibration, i.e. the $\frac{1}{m}$-order subharmonic wave. The chaos induced from periodic double bifurcation in a two-dimensional map is equivalent to the continuous splitting of $\omega$ in the forced vibration, resulting in $\frac{\omega}{2},\frac{\omega}{4},\ldots$, and finally tending to the continuous spectrum and forming chaos. If there appears a certain window (i.e. periodic solution) in a chaotic region, the continuous spectrum changes to a discrete spectrum. This proves the existence of subharmonic periodic waves.

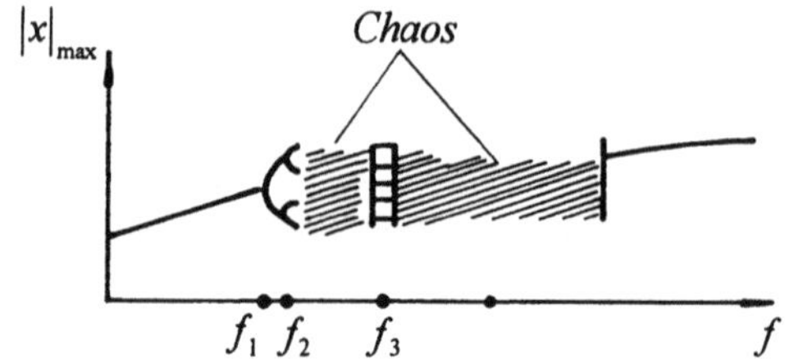


Fig. 8.3

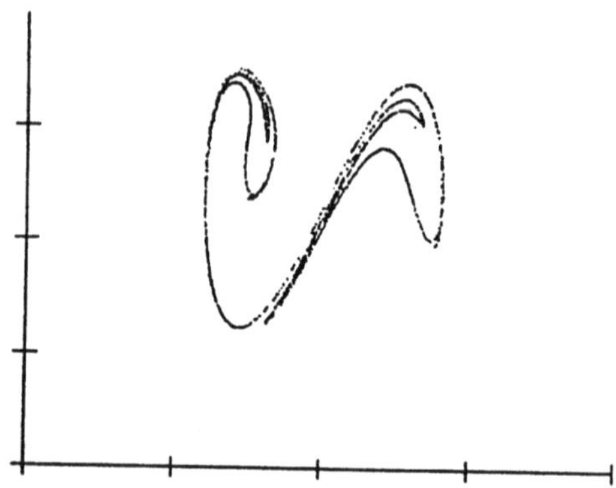

Fig. 8.4

Holmes studied the Duffing equation with negative elastic restoring force in 1979 [36]

$$x''+x'-10x+100x^3=f\cos 3.76t \tag{8.11}$$

The physical model of this equation can be a forced vibration of an elastic buckling column (the axial force is larger than the Euler critical value), or a forced vibration of an inverted pendulum. Equation (8.11) can be written as a three-dimensional autonomous system

$$\begin{aligned} x' &= y \\ y' &= 10x-100x^3-y+f\cos 3.76\theta \\ \theta' &= \end{aligned} \tag{8.12}$$

where $(x,y,\theta) \in \mathcal{R}^2 \times \mathcal{S}^1$. If we take the amplitude $f$ of a forced force as a parameter, then the relationship between the value $|x$ and $f$ is something like that of Fig. 8.3.

When the force $f$ is small, we obtain a solution with a frequency of $\omega$. The result obtained by the KBM method is correct. When $f$ is very large, we also obtain a solution with a frequency of $\omega$. The result obtained by the KBM method is also correct! But for $f$ in the middle section of values, the KBM method cannot give the correct result. A bifurcation appears at $f_1 \cong 0.8$, and then followed by the $\frac{1}{2}$ subharmonic solution. The bifurcation splitting from 2 to 4 appears at $f_2 \cong 0.9,\ldots$. We have found a window in the region of chaos. When $f = f_* \cong 1.2$, the stationary solution of eq. (8.12) (it can converge to this solution starting from an arbitrary initial value) is the solution with a period of $5\left(\frac{2\pi}{\omega}\right)$. Its Poincaré section has five points: $(-0.208, 0.735)$, $(0.335, 0.118)$, $(0.214, 0.583)$, $(0.272, -0.436)$, $(-0.247, 0.394)$, and $(-0.208, 0.735)$ again.

(2) The Japanese attractor

The Japanese scientist Y. Ueda *et al.* studied the changed Duffing equation in the 1960's [69, 70]

$$x'' + \delta x' + x^3 = P\cos t \tag{8.13}$$

When $\delta > 0$, this is a dissipative system. Rewriting it in the form of a state equation, we can easily calculate that $\mathrm{div} = -\delta < 0$. After the calculation, the points in the Poincaré section are $\{x(0), x'(0)\}, \{x(2\pi), x'(2\pi)\}, \ldots \{x(2k\pi), x'(2k\pi)\}, \ldots$ . The strange attractor is shown in Fig. 8.4 ($\delta = 0.1, P = 12$). These points are distributed on an infinitely long curve in a finite region non-repeatedly. This is termed a strange attractor (or an Japanese attractor). This is a typical case of chaotic motion.

(3) The Lorenz attractor

The first example of chaos was given by Lorenz in 1963 when he studied air turbulent flow [60]. For the study of Benard's convection flow, by non-dimensionalizing it, making a Fourier expansion from the Baussinesq equation and taking the first two terms, we can obtain a group of differential equations—the Lorenz equations.

$$\begin{aligned} x' &= \sigma(y - x) \\ y' &= \rho x - y - xz \\ z' &= -\beta z + xy \end{aligned} \tag{8.14}$$

where $\sigma, \rho, \beta$ are three parameters. The physical background of the variables and parameters are: $x$ denotes the convection intensity, $y$ the temperature difference between the rising flow and the descending flow, and $z$ stands the non-linear scale of temperature distributed in the vertical direction; parameter $\rho$ denotes the proportion of the Rayleigh number and its critical value $\rho = 1$ is the simple bifurcation point when the convection flow loses stability; $\delta$ is the Prandtl number related to the medium, while $\beta$ stands for the torus form of the field of flow. Equation (8.14) is a three-dimensional autonomous system, a dissipative system, because $\mathrm{div} = -(\sigma + \beta + 1) < 0$, and there are two non-linear terms. Its singular point $O(0,0)$ corresponds to having no convection; other singular points correspond to simple convection. Singular point $O(0,0)$ is a fixed point. When $\rho = 0$, this fixed point is not hyperbolic, but it is stable (a vague attractor); when $0 < \rho < 1$, it is a stable hyperbolic sink; when $\rho > 0$, it is an unstable saddle. It has a Hopf bifurcation when $\rho \cong 24.74$. The integral curve passing through every point (non-singular) is unique in three-dimensional space $xyz$, but its projection can be self-transversal on a certain coordinate plane, so chaos may exist. Figure 8.5 shows the space figure of the Lorenz attractor. This figure gives a typical phase portrait calculated by numerical integral. It starts from the initial point in the neighbourhood of $O(0,0,0)$. The parameters are $\sigma = 10, \beta = 8/3, \rho = 28$. It can be seen from Fig. 8.5 that the phase portrait circles between two discs, and will never repeat, but also will never go far away. Figure 8.6 is the projection of the phase portrait onto the $xy$ plane. This phase portrait is self-transversal.

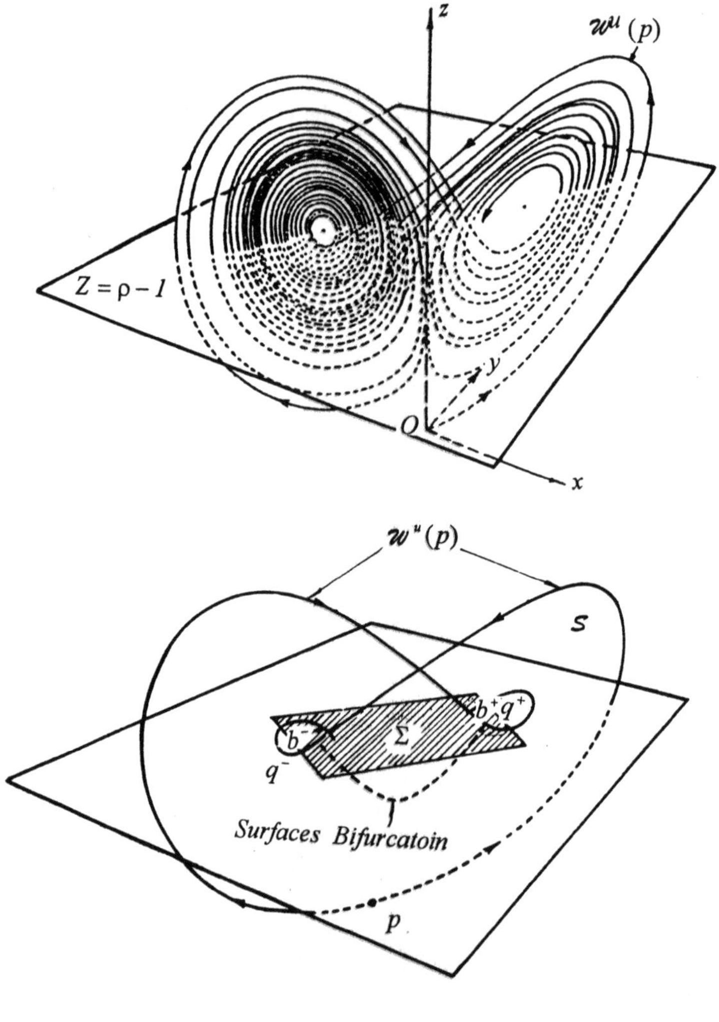


Fig. 8.5

With the development of the study on chaos and strange attractors, much work has been done on the properties of the Lorenz equation at different parameter values. Most work is based on the assumption that $\sigma = 10$ and $\beta = 8/3$, with $\rho$ as an adjustable parameter. Whether chaos or stable convection takes place at $\rho > 24.74$ depends upon the initial value, as pointed out by Guckenheimer in 1976. In 1980, Tomitak: described the various types of attractor when the value of $\rho$ is greater and confirmed the presence of the windows of different periodic solutions in the chaotic region of the Lorenz system.

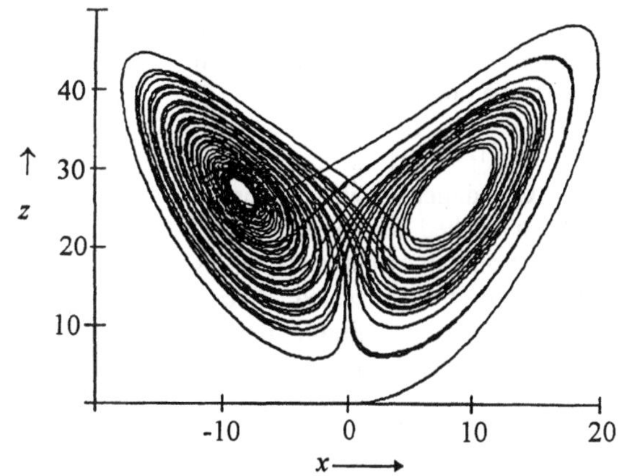


Fig. 8.6

There are windows of arbitrary period in the chaotic region of the Lorenz system. Pomeau pointed out in 1980 [73] that when $\rho = 166.6$, a saddle-node bifurcation appears and hence so does an intermittent turbulent flow. In 1982, Ueza *et al.* studied the state of the Lorenz system under the action of a periodic forced force.

## 8.3 A Brief Introduction to the Analytical Method of Chaotic Study

In this section we study the strange attractor and the chaos from three viewpoints: discrete, continuous and statistical. In the discussion of discrete methods, we introduce the semi-dynamical system with one-dimensional mapping and two-dimensional mapping. They are visual models in the study of chaotic motion; in the study of continuous or analytical methods, we introduce the Melnikov method. It is a powerful tool in the study of transversal homoclinic points connected by saddles and produced by non-integrable small perturbations. The resulting horseshoe structure is the omen for the occurrence of chaotic motion. Finally, we discuss the measure characteristics of strange attractors and chaos from a statistical viewpoint.

### 8.3.1 Discrete Method

(1). One-dimensional mapping

At present, one-dimensional mapping is the field studied most deeply and fully:

$$x_{n+1} = f(\lambda, x_n), \quad n = 0,1,2,\ldots \tag{8.15}$$

This can be taken as a dynamical system, where $\lambda$ is a real parameter, and $f$ is a non-linear function. Without any loss of generality, the given $x_n$ and $x_{n+1}$ are in line segment $\mathcal{I}$, i.e. the mapping is the mapping of the line segment onto itself: $f: \mathcal{I} \to \mathcal{I}$.

For a defined function $f$, we can obtain a defined process or time sequence from an arbitrary initial value ($n$ at a time)

$$x_0, x_1, x_2, x_3, \ldots \tag{8.16}$$

One-dimensional mapping has been studied for more than 60 years, but only in recent years has its rich contents been revealed. Below, we introduce some simple results. For the details, see [75].

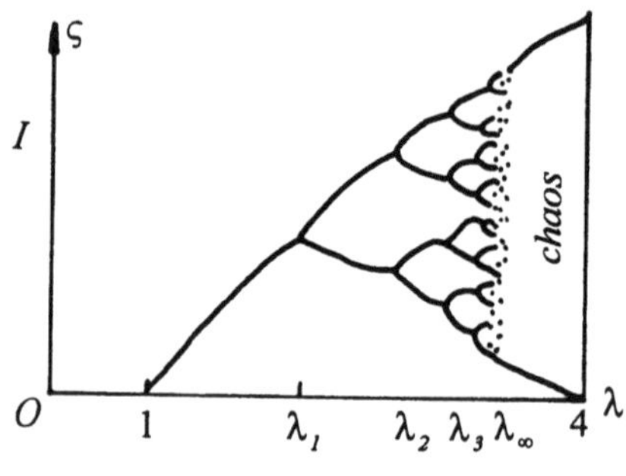


Fig. 8.7

Consider a simple one-dimensional mapping

$$f(x) = \lambda x(1-x) \tag{8.17}$$

where $f: \mathcal{I} \to \mathcal{I}$, $\lambda \in (0,1]$, $x \in \mathcal{I}$. It may produce a semi-dynamical system:

$$x_{n+1} = \lambda x_n (1 - x_n) \tag{8.18}$$

$\varsigma \in \mathcal{I}$ is called period $k$ point, the point satisfying $f \cdot f \cdots f(x) = f^k(\varsigma) = \varsigma$; when $|f'(\varsigma)| < 1$, this point is stable, when $|f'(\varsigma)| > 1$, it is unstable; when $|f'(\varsigma)| = 1$, bifurcation may occur. Next, we will discuss the behaviour of the time sequence of system (8.16) when $n \to \infty$, for different values of parameter $\lambda$ which will be constant in eq. (8.18) once given. When $0 \le \lambda < 1$, eq. (8.18) has a fixed point: $\varsigma = 0$. It satisfies $\varsigma = f(\varsigma)$, i.e. eq. (8.18) maps $\varsigma$ onto itself. Because

$$\left|\frac{\mathrm{d}}{\mathrm{d}x} f(x)\right|_{x=0} = |\lambda - 2\lambda x|_{x=0} < 1 \tag{8.19}$$

this fixed point is stable, i.e. whatever the initial value $x_0$ (in $\lambda \in [0,1]$) may be, $x_n \to \varsigma = 0$. For $1 < \lambda \le 3$, equilibrium state $\varsigma = 0$ is unstable, but another equilibrium state

$$\varsigma = 1 - \frac{1}{\lambda} \tag{8.20}$$

is stable, because $|f'(\varsigma)| < 1$. When $\lambda_1 = 3$, this point becomes unstable, and it bifurcates to a pair of period-2 points. When $3 < \lambda \le \lambda_2 = 1 + \sqrt{6}$, the equilibrium states $\varsigma = 0$ and $1 - \frac{1}{\lambda}$ are unstable. For any initial value (except the equilibrium state), when $n \to \infty$, the sequence unlimitedly tends to

$$\varsigma_1, \varsigma_2, \varsigma_1, \varsigma_2, \varsigma_1, \varsigma_2, \quad \text{or } \varsigma_1 \rightarrow\leftarrow \varsigma_2 \tag{8.21}$$

where
$$\varsigma_1, \varsigma_2 = \frac{2\lambda(1+\lambda) \pm \sqrt{(\lambda+1)(\lambda-3)}}{2\lambda} \tag{8.22}$$

Equation (8.21) is a solution with period 2 of eq. (8.18), i.e. the twice iteration map of $\varsigma_1, \varsigma_2$ returns to itself. $\varsigma = f(f(\varsigma))$, but $\varsigma \ne f(\varsigma)$, or $\varsigma_1, \varsigma_2$ satisfy $x_{n+2} = x_n$. We can take the equilibrium state as the solution with period 2, because it satisfies $x_{n+2} = x_n$. When the change of parameter $\lambda$ of system (8.18) is over $\lambda_1 = 3$, the stable solution of the system is the solution with period 2. This is a process of bifurcation from 1 to 2. When $\lambda_2 = 1+\sqrt{6} \cong 3.45$, the two points bifurcate to a pair of period-4 points separately. The period double bifurcates in the sequence of $\lambda_1 < \lambda_2 < \lambda_3 <\ldots$ . The width of $\lambda$ in which a periodic solution exists is called the periodic window. These windows will quickly tend to zero. When $\lambda_m \to \lambda_\infty = 3.56994$, chaos exists, see Fig. 8.7. In the chaotic region $(\lambda_\infty, 4)$, when $\lambda = 4$, there exists an uncountable set $S$ for the system. The points in the system wander "randomly" over the interval $I$. Their orbits can join up to form a large slice. When the value of $\lambda$ decreases from $\lambda = 4$, this slice decreases too; when $\lambda_{(1)} = 3.6786$, it divides into two slices; when $\lambda_{(2)} = 3.5926$, it divides into four slices; then, 8 slices, 16 slices etc. — this process of inverse bifurcation does not stop till $\lambda_{(k)} \to \lambda_\infty$, see Fig. 8.8(a). Moreover if we amplify this process of inverse bifurcation, new periodic windows will be found. They are very narrow. The widest are period-3 windows: $1+\sqrt{8} = 3.828 < \lambda < 3.8415$, see Fig. 8.8(b). This chaos state with windows can occur alternatively in the chaotic region, in the proportion of 1:9.

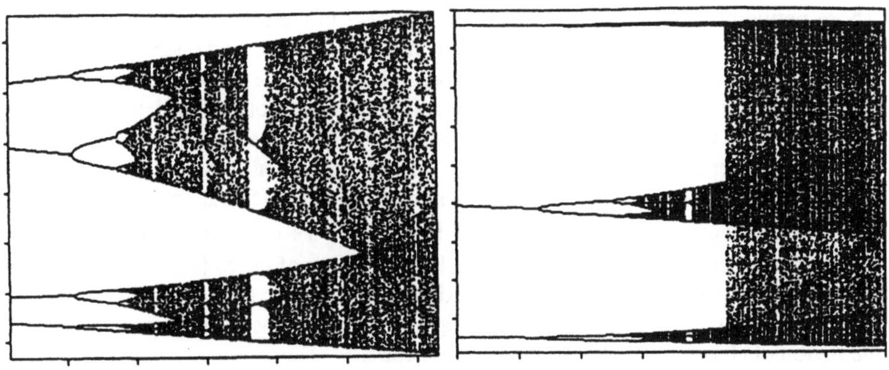

(a) *Inverse Bifurcation* $1+\sqrt{8}$ (b) *Period-3 window*

Fig. 8.8

According to the analyses above, five points are worthy of mention:

1. In the region of the double period bifurcation $(0, \lambda_\infty)$, the period is $2^k$, $k = 0,1,2, \ldots$, and the bifurcation point $\{\lambda_m\}$ has the following universal constant:

$$\delta = \lim_{m\to\infty} \frac{\lambda_m - \lambda_{m-1}}{\lambda_{m+1} - \lambda_m} = 4.669201661\ldots \tag{8.23}$$

This is called the Feigenbaum number [61]. It is similar to the process of inverse bifurcation. The above results are obtained from studying the single convex function $y = f(x) = \lambda x(1-x)$. For other single convex functions, the Feigenbaum number is the same. Feigenbaum number has a universal feature. For example, in a certain experiment, if bifurcation occurs from 1 to 2 when the parameter is $\lambda_1$, and then from 2 to 4 when the parameter is $\lambda_2$, and if a Feigenbaum number 4.6692 applies, we can estimate that chaos will occur when the parameters change continuously by 13% of $(\lambda_2 - \lambda_1)$.

Moreover, the height of the bifurcation point also shrinks, and the proportion of shrinkage tends to a limit too:

$$\alpha = 2.502907897 \tag{8.24}$$

This is also a universal constant. Giesel and Nierwetberg obtained the third universal constant in 1981 [76]: $r = 2.94805$. They discussed the other universal characteristic in 1982.

2. The Shacovsky sequence [77]

Shacovsky proved that for a general type of one-dimensional iteration $x_{n+1} = \lambda f(x_n)$, when its parameters change continuously, there is a rule for the appearance order of the period of solution 1, 2, 4, 8, ...

$$1,\ 2,\ 4,\ \ldots,\ 7\times 2^2,\ 5\times 2^2,\ 3\times 2^2,\ldots$$

$$7\times 2,\ 5\times 2,\ 3\times 2,\ \ldots,\ 7,\ 5,\ 3 \tag{8.25}$$

If the system has a solution with some period, it must contain those solutions whose period numbers are before this period number in the sequence. We have seen above that there is a solution with period 4 in system (8.18), at the same time there exist solutions with periods 2 and 1, and periods 1 and 2 are before period 4 in the order of sequence (8.25). However, if a solution is unstable, it will not be seen physically. In the order of sequence (8.25), period 3 is the last, i.e. if there is a solution with period 3 in the system, then at the same time (for the same $\lambda$) there exist solutions with all the periods. Li and Yorke stated in 1975 that "period 3 means chaos" [78]. For parabola iteration (8.18), when $\lambda = 1+\sqrt{8} = 3.838$, there exists a solution with period 3. This value is indeed in the region of chaos $(\lambda_\infty, 4)$. Of course, even though the solution with period 3 exists, the attractor of this system can be a periodic solution, or ever a strange attractor. The motions when the parameters are in the region of chaos have very complicated structure.

3. When $\lambda$ is in the region of chaos, solutions with arbitrary periods exist. (this can be expressed by a Fourier series), and all these solutions are unstable. So, on $\mathcal{I}=[0,1]$, there are countable infinite unstable periodic points $p$. Besides these points, there are uncountable infinite points $S$ in $\mathcal{I}$, i.e. $S, p \in \mathcal{I}$, where the orbit is dependent on the initial value sensitively: for the two different points $x$ and $y$ in $S$, this takes the form

$$\overline{\lim_{n\to\infty}} |f^n(x) - f^n(y)| > 0$$

$$\underline{\lim_{n\to\infty}} |f^n(x) - f^n(y)| = 0$$

These points form a strange attractor. If the initial value is an arbitrary point in this point set, all the points in this sequence are kept in this point set, and will never repeat. In this case, the stable value $\varsigma$ is a region of chaos, and this region is connected as a slice.

4. The arrangement of the periodic windows accords to various rules.

Iterate from the critical point $x_0 = \frac{1}{2}$, $x_n = f^n(x_0)$, $n = 1,2,\ldots$, and stop when the point returns to the initial point $x_0$. For the point with period $k$, when $\lambda = 3.236$, $k = 2, P = R$; when $\lambda = 3.499$, $k = 4, P = RLR,\ldots$. This model constitutes a regular

MSS sequence, as proved by MSS (N. Metropolis, M.L. Stein & P.R. Stein) in 1973 [79].

5. The complicated accurate structures in the region of chaos

We have mentioned above that when $\lambda = 1+\sqrt{8} = 3.828,\ldots$, there is a solution with period 3. In fact, if $\lambda$ increases to $\lambda = 1+\sqrt{8} = 3.828$ and over the value $1+\sqrt{8}$, there will be a saddle-node bifurcation; in a certain region of $\lambda \geq 1+\sqrt{8}$, and this system is attracted to the stable solution with period 3. Thus, there is a periodic window with period 3 in the region of chaos. The windows of period 3 are wider, i.e. when $1+\sqrt{8} \leq \lambda \leq 3.841499008$, the solution of period 3 is stable. In addition, there are two windows of period 4, more windows of period 5, windows of period 6 and so on, i.e. there are countable infinite windows. These windows are arranged in a certain order [79, 80]. Some windows are so narrow that it is hard to find them even with high precision computers. Processes of bifurcation exist in every window. For example, in the window of period 3 (see Fig. 8.8(b)), the solution of period 3 divides into the solution of period 6, at $\lambda = 3.8415$, 6 into 12 etc., and there is a sequence of inverse bifurcations, finally it returns to the initial chaotic structure. It can be seen that for the three parts of the solution of period 3, every part repeats the structure of the whole, and in its region of inverse bifurcation there will occur some new hierarchy. If it is hard to confirm this using a computer, it is even more difficult to confirm it by experiment, but it does exist in theory.

(2) Two-dimensional mapping and the Smale horseshoe

1. The general form of a two-dimensional mapping or a two-dimensional homeomorphism is

$$x_{n+1} = X(x_n, y_n)$$

$$y_{n+1} = Y(x_n, y_n),\ n = 0,1,2,\ldots \tag{8.26}$$

This can be written in the form of a point transformation

$$T\colon (x,y) \to (X(x,y), Y(x,y)) \text{ or } T(x_n, y_n) = (x_{n-1}, y_{n-1}).$$

We are mostly interested in whether there is an invariant set for this transformation, especially a stable invariant set, i.e. an attractor. The stable fixed point is the simplest form of this set. There is one point $(\varsigma, \eta)$ in such a set. It is transformed to itself by transformation $T$: $T(\varsigma,\eta) = (\varsigma,\eta)$. The state near the fixed point $(\varsigma,\eta)$ can be explained by the linearized equation

$$x' - \varsigma' = \frac{\partial X}{\partial x}(\varsigma,\eta)(x-\varsigma) + \frac{\partial X}{\partial y}(\varsigma,\eta)(y-\eta)$$

$$y' - \varsigma' = \frac{\partial Y}{\partial x}(\varsigma,\eta)(x-\varsigma) + \frac{\partial Y}{\partial y}(\varsigma,\eta)(y-\eta) \tag{8.27}$$

At a position far from the fixed point of the non-linear system, there is a great difference between it and the corresponding linearized system. At present, there is no general method of studying this case. The stable periodic solution is another form of a stable invariant set. For example, it can be composed of three different points

$$\to (\varsigma_1,\eta_1) \xrightarrow{T} (\varsigma_2,\eta_2) \xrightarrow{T} (\varsigma_3,\eta_3) \xrightarrow{T} (\varsigma_1,\eta_1) \longrightarrow$$

and the points near the fixed point will tend to this fixed point by repeated mapping. Besides the above two kinds of common attractors, there is also a type of strange attractor. What we term the strange attractor is really an infinite point set. No mapping can return to itself for an arbitrary point in it, but it is in the set. The strange attractor corresponds to the chaos in the physical process. A typical example of a strange attractor in two-dimensional mapping is the Henon map:

$$x_{n+1} = 1 + by_n - ax_n^2$$

$$y_{n+1} = x_n \tag{8.28}$$

where $a$, $b$ are constants. Equation (8.28) can be expressed as a two-order iteration

$$x_{n+2} = 1 + bx_n - ax_{n+1}^2 \tag{8.29}$$

When $X$ and $Y$ in eq. (8.26) are order-2 polynomials, this can be transformed into a Henon map (8.28) in general. When $a = 1.4$, $b = 0.3$, there is a strange attractor in the mapping (8.28), the so-called Henon attractor. There are two unstable fixed points in this case: $A$ (0.631, 0.631) and $B$ $(-1.131, -1.131)$.

There are two invariant sets $M$ and $M'$ passing through point $A$, but they do not contain point $A$ itself, because point $A$ itself is an independent invariant set. The points in $M$ will be far from point $A$ as a result of the repetitive mapping or iterating, and $M$ is called an unstable manifold. The points in $M'$ will tend to point A as a result of repetitive mapping and $M'$ is called the stable manifold. Almost all the orbits in the positive direction on the plane tend to $M$. The image of the unstable manifold itself is shown in Fig. 8.9.

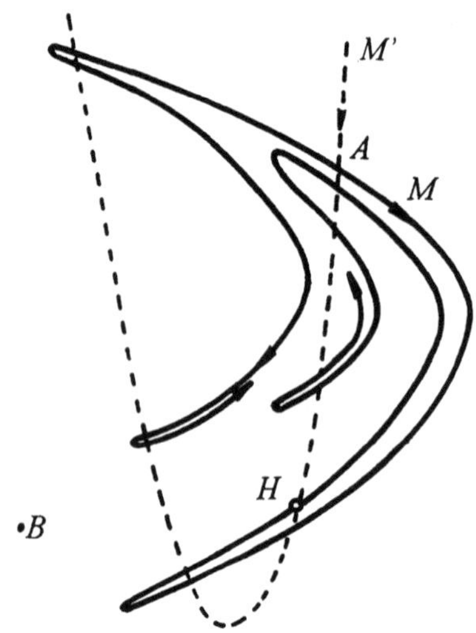

Fig. 8.9

The formation of a Henon attractor is as follows: suppose $b = 0.3$. When $a$ rises from 0, a double period bifurcation occurs; when $a_1 = 0.3675$, the solution with period 1 bifurcates into period 2, when $a_2 = 1$, 2 into 4, when $a_3$, 4 to 8, etc., and finally to $a_\infty = 1.058$ and then into the region of chaos. When $a = 1.153$, two slices of chaos (the strange attractor is two discontinuous line segments) form one big slice (deleting a point). When $a = 1.4$, there is a strange attractor $M$ in this system, and when $a > 1.427$, the invariant set is no longer an attractor. The well-studied window is the window of period 7 in the region of chaos. It starts from $a = 1.2991$. When it reaches 1.3038, period 7 bifurcates into period 14. Bifurcation stops when it reaches 1.3064. After that begins the process of the second inverse bifurcation, and finally it returns to the Henon attractor at 1.30865.

This example shows that the sequence of bifurcation, chaos, inverse bifurcation, window, etc. is a universal attribute to the exhibition of dimensional mapping. But there are many different characteristics between two-dimensional mapping and one-dimensional mapping. For example, for a Henon map, the following features exist when $b = 0.3$.

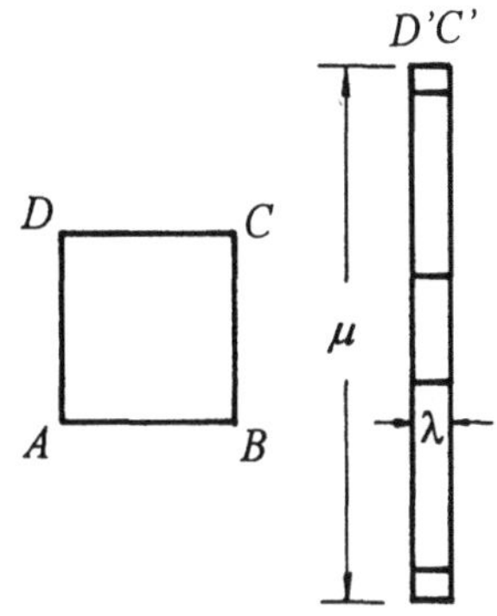

Fig. 8.10

(a) The results of a two-dimensional mapping may be dependent on the initial value. If we divide the plane $(x, y)$ into basins, the initial values will fall into different basins and the iterative process will converge to different periodic orbits or non-periodic orbits. While in a one-dimensional mapping, there is one unique stable period at most, no matter where the initial value is chosen, so there is only one stable attractor.

(b) For some parameters and initial values, the mapping results converge rapidly to the attractor, which is almost a one-dimensional attractor on the plane $(x, y)$. This attractor is like a smooth curve, but with a certain width, and the interior of the attractor ( in the case of a strange attractor) is of the self-resemble structure.

Conservative two-dimensional mappings have different universal constants. For example, the Henon map at $b = 1$, they are $\delta = 8.7210$ and $\alpha = 4.018$ [81].

Another example for a mapping

$$x_{n+1} = 2\mu(x_n + x_n^2) - y_n$$
$$y_{n+1} = bx_n \tag{8.30}$$

when $b = 1$, then $\delta = 8.7210$, and when $b < 1$, then $\delta = 4.66920$ [82].

If we introduce an effective parameter $b_{eff} = b^{2^n}$, then when $n \to \infty$, $\delta$ as a function of $b_{eff}$ changes continuously from 4.66920 to 8.7210. This is a new scaling property in two-dimensional mapping. So far there has been no effective method for the study of two-dimensional mapping although it has much in common with high-dimensional mapping. The flow of high dimension corresponds to many two-dimensional mappings naturally, via a Poincaré mapping. For the occurrence of chaos and some basic concepts of two-dimensional mapping, refer to [5]

In what follows, we will introduce the most important mathematical theory in studying the two-dimensional mapping—Smale horseshoe theory.

2. Smale horseshoe theory

Define a horseshoe mapping

$$f: \mathcal{R} \to \mathcal{R} \quad (\mathcal{R} = \mathcal{R}^2)$$

A square plane is transformed into a horseshoe by stretching and folding, $f$ being a diffeomorphism.

The mapping $f$ contains two steps:

(a) to stretch $\mathcal{R}$ to a long strip; see Fig. 8.10;

(b) to fold the long strip in the middle so that a horseshoe $f(\mathcal{R})$ is formed; see Fig. 8.11.

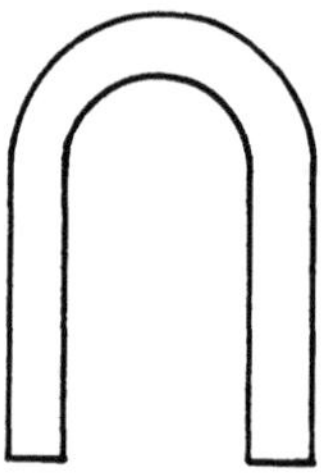

Fig. 8.11

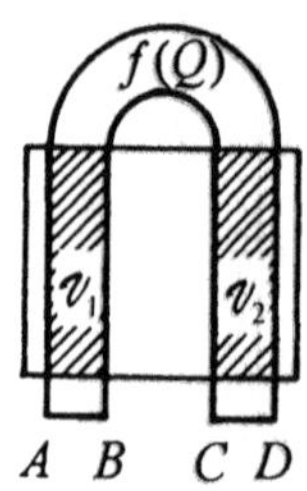


Fig. 8.12

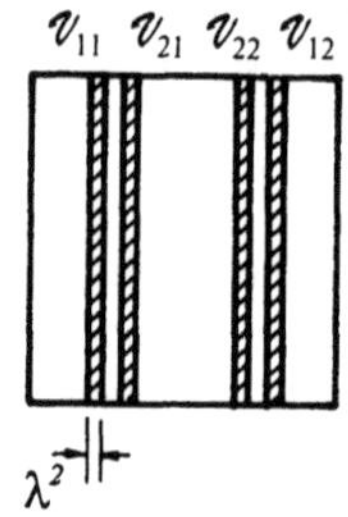


Fig. 8.13

$\mathcal{R}\cap f(\mathcal{R})$ is composed of two lines, that is, $\mathcal{V}_1 \cup \mathcal{V}_2$. See Fig. 8.12.

Apply mapping $f$ again to $\mathcal{R}\cap f(\mathcal{R})$, that is, $f(\mathcal{R})\cap\mathcal{R}$. See Fig. 8.13. $\mathcal{V}_1$ splits into two strips $\mathcal{V}_{11}, \mathcal{V}_{12}$ and $\mathcal{V}_2$ into $\mathcal{V}_{21}, \mathcal{V}_{22}$. There are 4! kinds of arrangement model for these four strips, and $k$th horseshoe mapping. $f^k(\mathcal{R})\cap\mathcal{R} = 2^k$ vertical strip set.

When $k$ tends to $\infty$, the width of the strip tends to zero. It forms a (Cantor's set) $\times$ (vertical line). Defining an inverse mapping $f^{-1}: \mathcal{R}\to\mathcal{R}$ then, $f^{-1}: \mathcal{R}\to\mathcal{R}$ as shown in Fig. 8.14.

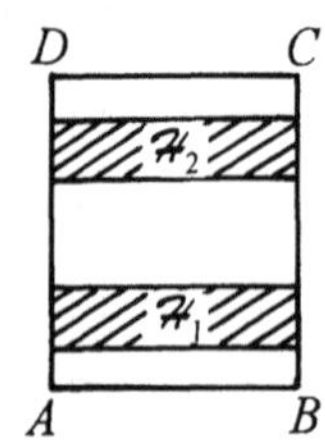


Fig. 8.14

The set of $f^{-k}(\mathcal{R})\cap\mathcal{R} = 2^k$ horizontal strip set. When $k$ tends to $\infty$, it forms a (Cantor's set)$\times$(horizontal line) $f^{-1}(\mathcal{R})\cap\mathcal{R}\cap f(\mathcal{R}) = 4$ squares, see Fig. 8.15.

Similarly, there are

$$f^{-2}(\mathcal{R})\cap f^{-1}(\mathcal{R})\cap\mathcal{R}\cap f^{1}\cap f^{2}(\mathcal{R}) = 16 \text{ squares}$$

So, the set of $f^{-k}(\mathcal{R})\cap\ldots\cap f^{k}(\mathcal{R}) = 4^k$ squares.

**Definition 8.1** $\wedge = \{x | f^k(x) \in \mathcal{R}, \vee k \in \mathcal{Z}\}$ where, $-\infty < k < \infty$.

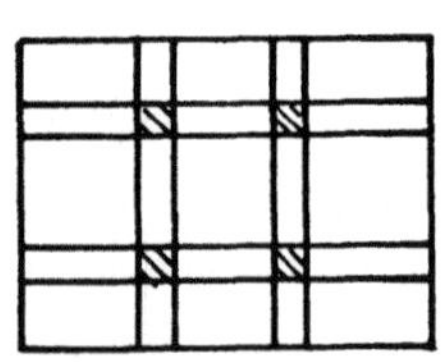

Fig. 8.15

The structure of $\wedge = \ldots\cap f^{-1}(\mathcal{R})\cap\mathcal{R}\cap f(\mathcal{R})\cap$ is given below. Because point set $\wedge$ is always in $\mathcal{R}$, so $\wedge$ is an invariant set. This corresponds to the chaotic state that we have studied previously. In order to study its characteristics, we will introduce the symbolic dynamical method.

3. Symbolic dynamical method

To begin with, we will state a rule for numbering the remainder parts of the horseshoe mapping. Let the two vertical strips of $\mathcal{R} \cap f(\mathcal{R})$ be 1 (left) and 2 (right). Similarly, we denote the two horizontal strips as .1 (below) and .2 (upper). After the horseshoe mapping, the positive mapping is equivalent to one place of the right shift of the point; the inverse mapping is equivalent to one place of the left shift of the point. See Fig. 8.16.

If $A = \{1,2\}$, $a_i \in A$, $i = 1,2$, and supposing there is a bi-infinite sequence

$$a = \{\ldots a_{-2}a_{-1}a_0a_1a_2\ldots\}_{i=-\infty}^{i=+\infty}$$

and its definition is

$$a_i = \begin{cases} 1 & \text{where } f^i(x) \in \mathcal{V}_1 \\ 2 & \text{where } f^i(x) \in \mathcal{V}_2 \end{cases}$$

Another bi-infinite sequence is $b = \{b_i\}_{i=-\infty}^{i=+\infty}$.

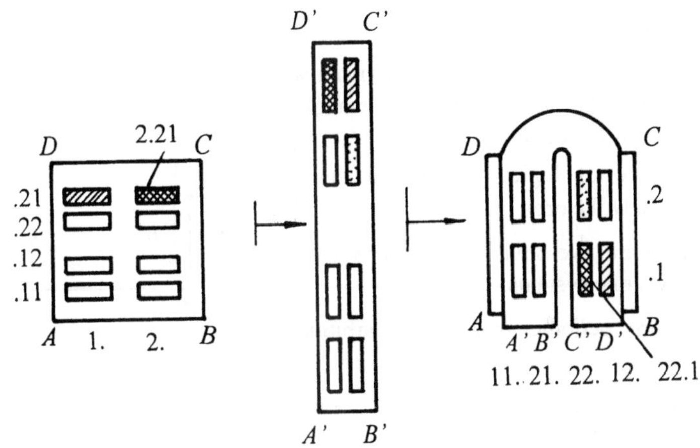


Fig. 8.16

**Definition 8.2** $\Sigma$ stands for the set of the bi-infinite sequence of two symbols.

**Theorem 8.1** Suppose $k$ is a mapping from $x \in \wedge$ to $a$ $\Sigma$, then

(a) $h$ is onto and one to one set.

(b) if the distance on $\Sigma$ is given by $d(a,b) = \sum_{-\infty}^{\infty} \frac{|a_i - b_i|}{2^{|i|}} = \sum_{-\infty}^{\infty} \delta_i 2^{-|i|}$, where

$$\delta_i = \begin{cases} 0 & \text{when } a_i = b_i \\ 1 & \text{when } a_i \neq b_i \end{cases}$$

then $h$ is a homeomorphism;

(c) if $a = h(x)$ and $b = h(f(x))$, for all $i \in \mathfrak{z}$, then $b_i = a_{i+1}$ is true, that is the sequence $b$ can be obtained from the sequence $a$ by shifting the subscript one place;

**Definition 8.3** The shift map$\sigma: \sum \to \sum$ if and only if $b_i = a_{i+1}, \vee i$ (such as $i = 0$, $b_0 = a_1$), and it has

$$\begin{aligned} b = \sigma(a) &= \sigma\{\ldots a_{-2}a_{-1}\overset{*}{a_0}a_1a_2\ldots\} \\ &= \{\ldots a_{-2}a_{-1}\overset{*}{a_0}\overset{*}{a_1}a_2\ldots\} = b \\ &= \{\ldots b_{-2}b_{-1}b_0b_1b_2\ldots\} \end{aligned}$$

where $b_0 = a_1$, that is, the element is shifted one place to the left. is a homeomorphism map, i.e. the map is single-valued on both sides and a continuous map.

**Theorem 8.2** $\forall x \in \wedge, h(x) \in \sum, h \cdot f(x) = \sigma h(x)$ or $f = h^{-1}\sigma h, f^n = h^{-1}\sigma^n h$

Therefore $f$ and are topologically conjugate or topologically equivalent. That is, $f$ and satisfy the exchanged figure. See Fig. 8.17. $h$ maps the orbit of $f$ to the orbit of . That is, there is a one to one relationship between the orbit of $f$ and , and the period remains invariant.

Let $\{\mathcal{R}, f\}$ defines the dynamical systems, and $\{\sum, \sigma\}$ define the symbolic dynamic systems. As $h: \wedge \to \sum$, then $\{\wedge, f\}$ and $\{\sum, \sigma\}$ are topologically equivalent, but it is easier to study the orbit of than $f$.

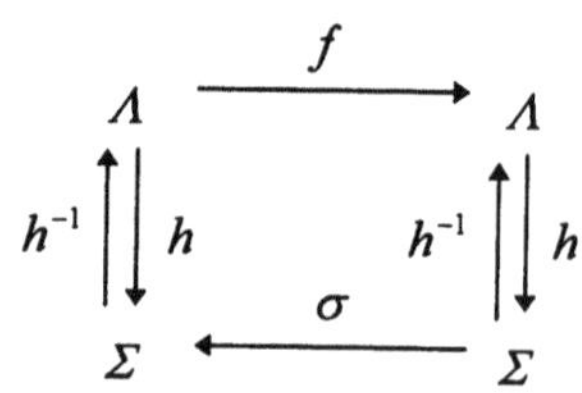


Fig. 8.17

There are five obvious characteristics of symbolic dynamic systems.

(a) There are periodic points with arbitrary periods:

There are two fixed points 111, 111, ..., ... 2 2 2, 2 2 2, ...,

There is one period-2 point. ...,12 12 12, 12 12 12, ...

There are two period-3 points. 221, 221, ... and 211, 211, ...

There are three period-4 points; six period 5 points, etc. That is, the circulative sequence is a periodic point on $\sum$. As there are arbitrarily long periodic circulative sequences, so there are periodic points with arbitrary periods.

(b) The periodic points are dense everywhere in $\sum$. Because there is a periodic circulative sequences in the arbitrary column subset in $\sum$, the periodic points are dense everywhere.

(c) There is a dense orbit in $\sum$.

This orbit can be written in the following manner. Write all the combinations of the two numbers 1 and 2 to places, (11, 12, 21, 22), all the combinations to three places (111, 112, 121, 122, 211, 212, 221, 222), etc. Continuously write them to infinity, then any orbit can approach to any extent. The extent to which two orbits approach to each other is measured by the length of their common parts. Suppose this dense orbit is required to have $n$ elements ($n$ may be arbitrarily large) in common with another orbit. This requirement can be satisfied because all possible combinations on $n$ places, including 1 and 2, are in the dense orbits.

(d) Uncountable infinite periodic points. The sum of the first three terms contains only countable infinite elements, but there are uncountable infinite elements in $\wedge_\infty$ obviously. Therefore, excepting the countable infinite elements in $\wedge_\infty$, the rest are not periodic points.

(e) The periodic orbit is saddle-type and dense in $\wedge_\infty$.

The horseshoe mapping is topologically conjugate with the shifting mapping of two symbols, so the Smale horseshoe contains typical complicated characteristics.

### 8.3.2 Continuous Method

The methods of discrete dynamical systems have revealed the basic phenomena of chaotic motion. Dynamical systems in mathematical physics are often described by differential equations. Continuous or analytical methods are often used to reveal their chaotic phenomena and to transform the continuous system into discrete systems. For example, the Duffing equation and the van der Pol equation describe a typical vibrating process when its forced-vibration core is replaced by the Melnikov method. We determine whether a transversal-homoclinic point exists. According to the Smale-Birkhoff theorem, the transversal-homoclinic point contains the Smale horseshoe structure. This is the reason for creating chaos.

#### 1. Homoclinic Orbit

**Definition 8.4** Suppose $p_1$ and $p_2$ are two hyperbolic fixed points on $\mathcal{R}_n$, and $\mathcal{W}^s(p_2)\bigcap\mathcal{W}^u(p_2)$ non-empty or $\mathcal{W}^s(p_2)\bigcap\mathcal{W}^u(p_2)$ non-empty; when $p_1 = p_2$, the point of intersection set is called a homoclinic point, and the orbit passing through these points is called a homoclinic orbit; when $p_1 \neq p_2$, the points in the intersection set are called heteroclinic points, and their orbits are called heteroclinic orbits.

Next, we study an example of a homoclinic orbit. Assume a two-dimensional system $x_1' = x_2$ and $\quad x_2' = x_1 + \mu x_2 - x_1^2$ (8.31)

We know from the form of the system that this is a dissipative system, and when $\mu = 0$ it is transformed into a Hamiltonian system

$$x_1' = x_2 \text{ and } x_2' = x_1 - x_1^2 \tag{8.32}$$

It is easy to find its Hamiltonian function

$$H(x_1,x_2) = \frac{1}{3}x_1^3 + \frac{1}{2}x_2^2 - \frac{1}{2}x_1^2 \tag{8.33}$$

Suppose $x_1(t)$ and $x_2(t)$ are the solutions of this Hamiltonian system, then

$$H' = \frac{\partial H}{\partial t} = \frac{\partial H}{\partial x_1}\frac{\partial x_1}{\partial t} + \frac{\partial H}{\partial x_2}\frac{\partial x_2}{\partial t} = 0$$

So, $H(x_1,x_2)=$constant is the orbit of the solutions, that is

$$H(x_1,x_2)=\frac{1}{3}x_1^3+\frac{1}{2}x_2^2-\frac{1}{2}x_1^2=\text{const} \tag{8.34}$$

We know from eq. (8.31), when $\mu=0$, (0,0) is a hyperbolic saddle point, and $(x_1^0,0)$ is a centre point. Suppose that there is a homoclinic orbit at point $O$. It is the unstable manifold of this saddle, and also the stable manifold of this point, that is to say, the starting phase point from any point of this orbit, when $t\to\pm\infty$, tends to the same point $O$ (0, 0). See Fig. 8.18(a), where $(x_1^0,0)$ is centre.

When $\mu<0$, the Hamiltonian system is transformed into a positive dissipative system, and the center changes to a sink. See Fig. 8.18(b).

When $\mu>0$, the Hamiltonian system is transformed into a negative dissipative system, and the centre changes to a source. See Fig. 8.18(c).

We know from Fig. 8.18, when $\mu=0$, the phase structure of this orbit has undergone a change in nature, and hence a bifurcation occurs.

We shall study an example of a heteroclinic orbit once again. Suppose

$$x_1'=\mu+x_1^2-x_1x_2 \text{ and } x_2'=x_2^2-x_1^2-1 \tag{8.35}$$

When $\mu=0$, let $x_1'=x_2'=0$, so we can obtain two saddles. For the heteroclinic orbit, see Fig. 8.19.

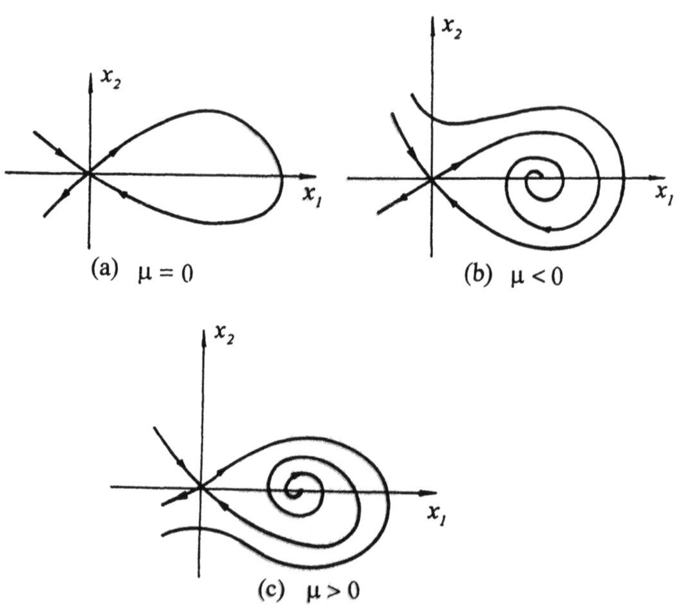


Fig. 8.18

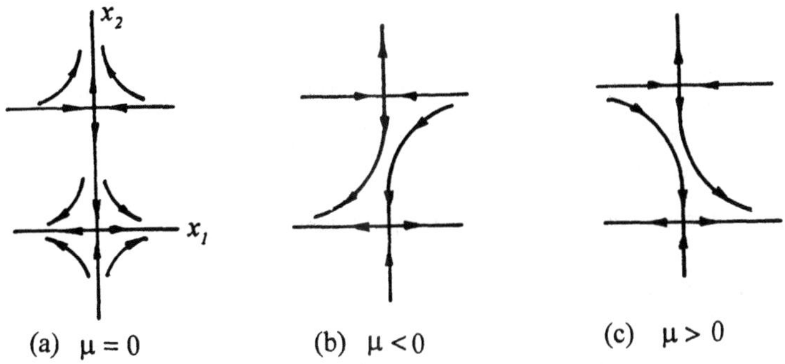


Fig. 8.19

The heteroclinic portrait is a stable manifold of a saddle point, but an unstable manifold of another saddle point. Here, what we called the manifold, generally speaking, means the curve or curved surface in Euclid space.

To generalize, consider a two-dimensional system

$$x' = f(x),\ f \in \mathcal{C}^r(\mathcal{R}^2, \mathcal{R}^2),\ r \geq 1 \tag{8.36}$$

Suppose that eq. (8.36) has a saddle point $p_0 \in \mathcal{R}^2$, that is $f(p_0) = 0$, and the two eigenvalues of $Df(p_0)$ are one positive and the other negative, and further suppose that there exists a homoclinic orbit $q^0(t)$, passing through $p_0$ and that when $t \to \pm\infty, q^0(t) \to p^0$. Now assume eq. (8.36) is a Hamiltonian system, i.e. there exists $H(x)$ so that $f_1 = \dfrac{\partial H}{\partial x_2}, f_2 = -\dfrac{\partial H}{\partial x_1}$, and suppose there exists a critical point in the closed curve $\mathbf{P}^0 = \{q^0(t) | t \in \mathcal{R}\} \cup \{p^0\}$of $H$. Thus, there is a continuous family $q^\alpha(t), \alpha \in (-1,0)$ of periodic orbits which fill the whole $\mathcal{P}^0$ when $\alpha \to 0$, $q^\alpha(t)$ tends to $\mathcal{P}^0$, that is

$$\lim_{\alpha \to 0} \sup_{t \in \mathcal{R}} \mathrm{d}(q^\alpha(t), p^0) = 0 \text{ where } \mathrm{d}(x, p^0) = \operatorname*{Int}_{q \in \mathcal{P}^0} |x - q|$$

As we know, there is no chaos in a two-dimensional space, such as an insect flying on an aircraft, and it is required that the flying orbit does not repeat. Thus, the orbit must either fly away or reduce to a point. See Fig. 8.18(c), (b). Now, let us make a small perturbation with a period $T$ to system (8.36). Suppose $g \in \mathcal{C}^r(\mathcal{R} \times \mathcal{R}, \mathcal{R}^2)$: $\quad g(x, t+T) = g(x,t), \quad \vee (x,t) \in \mathcal{R}^2 \times \mathcal{R}$.

Then eq. (8.36) is transformed into a non-autonomous system after the small disturbance

$$x' = f(x) + \varepsilon g(x,t) \tag{8.37}$$

Now, let us consider the change in the structure of the orbit of eq. (8.37) near the orbit $\mathcal{P}^0$, especially the change of the unstable manifold $\mathcal{W}_\varepsilon^u$ and the stable manifold $\mathcal{W}_\varepsilon^s$ at $\mathcal{P}^0$. To do this, we must transform the non-autonomous system (8.37) into the equivalent suspensory autonomous system:

$$x' = f(x) + \varepsilon g(x,\theta) \text{ and } \theta' = 1,\ (x,\theta) \in \mathcal{R}^2 \times \mathcal{S}^1 \tag{8.38}$$

where $\mathcal{S}^1 = \dfrac{\mathcal{R}}{T}$ is a circle of circumference $T$. Note that the flow $(x(t),\theta(t))$ generated by system (8.38) is on a three-dimensional manifold, and there may be chaos in this three-dimensional system.

Transform a two-dimensional mapping into a one-dimensional mapping $\mathcal{S}_1$ by using the Poincaré mapping, see Fig. 8.20. The plane orbit is a circle, but a fixed point on plane $\mathcal{S}$. We can transform the continuous systems into discrete systems and decrease the dimension of the systems by using Poincaré mapping. We take a section of this three-dimensional manifold at $t_0 \in [0,T]$: $\sum^{t_0} = \{(x,\theta\} \in \mathcal{R}^2 \times \mathcal{S}^1 | \theta = t_0\}$.

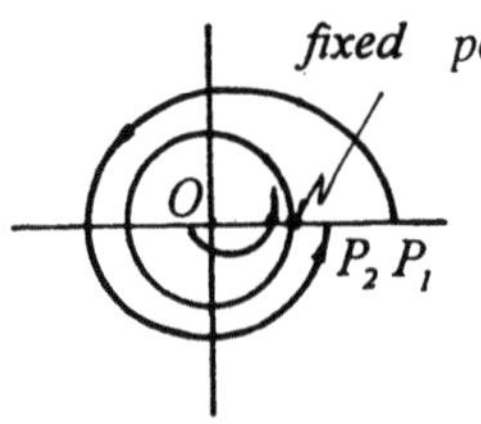


Fig. 8.20

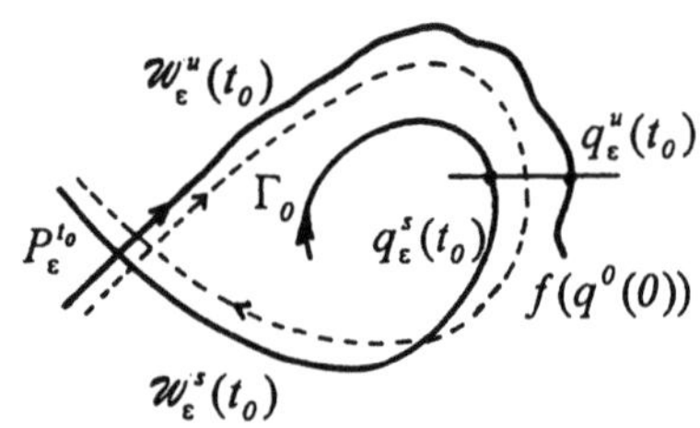


Fig. 8.21

Define the Poincaré mapping $\mathcal{P}_\varepsilon^{t_0}: \sum^{t_0} \to \sum^{t_0}$, where mapping $\mathcal{P}_\varepsilon^{t_0}$ transforms the point $(x(t_0),t_0) \in \sum^{t_0}$ into the point returning to $\sum^{t_0}$ that starts from the point $(x(t_0),t_0)$ along $(x(t+t_0),\theta(t+t_0))$, so that we simplify the discussion of system (8.37) into the discussion of mapping $\mathcal{P}_\varepsilon^{t_0}$. Especially, because $p_0$ is a saddle, we know that id$-D\mathcal{P}_0^{t_0}(\mathcal{P}_0^{t_0}(p_0))$ is reversible. This is obtained by means of the implicit function theorem.

## 2. Melnikov Function

**Proposition 8.1** For the saddle point $\mathcal{P}_0$ of system (8.36), if $\varepsilon$ is sufficiently small, there exists a unique hyperbolic closed orbit $\gamma_\varepsilon^0(t) = p^0 + o(\varepsilon)$ in system (8.37). Correspondingly, there exists a unique saddle $\mathcal{P}_\varepsilon^{t_0} = p_0 + o(\varepsilon)$ in mapping $\mathcal{P}_\varepsilon^{t_0}$.

Suppose the stable and unstable manifolds are $\mathcal{W}_\varepsilon^s(t_0)$ and $\mathcal{W}_\varepsilon^u(t_0)$ at saddle point $\mathcal{P}_\varepsilon^{t_0}$ on the section $\sum^{t_0}$. When $\varepsilon = 0$, they coincide. Consider $\varepsilon \neq 0$. The crack at the $q^0$ point is $d(t_0) = q_\varepsilon^u(t_0) - q_\varepsilon^s(t_0)$, where $q_\varepsilon^u(t_0)$ and $q_\varepsilon^s(t_0)$ are the unique points on the normal $\mathcal{P}^0$of $\mathcal{W}_\varepsilon^u(t_0)$ and $\mathcal{W}_\varepsilon^s(t_0)$ at $q^0(0)$. See Fig. 8.21. Note that the wedge product is defined by $f \wedge g = f_1 g_2 - f_2 g_1$. We can prove [36] that

$$d(t_0) = \frac{\varepsilon M(t_0)}{|f(q^0(0))|} + o(\varepsilon^2) \tag{8.39}$$

where $M(t_0)$ is called the Melnikov function, its expression being

$$M(t_0) = \int_{-\infty}^{\infty} f(q^0(t)) \wedge g(q^0(t), t + t_0)\mathrm{d}t \tag{8.40}$$

We can see from eq. (8.39) that the crossing of $\mathcal{W}_\varepsilon^u(t_0)$ and $\mathcal{W}_\varepsilon^s(t_0)$ can be defined by the zero point of the Melnikov function $M(t_0)$. See below.

**Proposition 8.2** If $M(t_0)$ has a simple zero independent of ε, then for sufficiently small $\varepsilon > 0$, $\mathcal{W}_\varepsilon^u(t_0)$ and $\mathcal{W}_\varepsilon^s(t_0)$ intersect transversely non-tangentically. If $M(t_0)$ remains away from zero, $t_0 \in [0, T]$, then $\mathcal{W}^u(p_\varepsilon^{t_0}) \cap \mathcal{W}^s(p_\varepsilon^{t_0}) = \Phi$, or

$$M(t_0) = \begin{cases} M'(t_0) \neq 0, \text{ transveral} \\ M'(t_0) = 0, \text{ tangential} \end{cases}$$

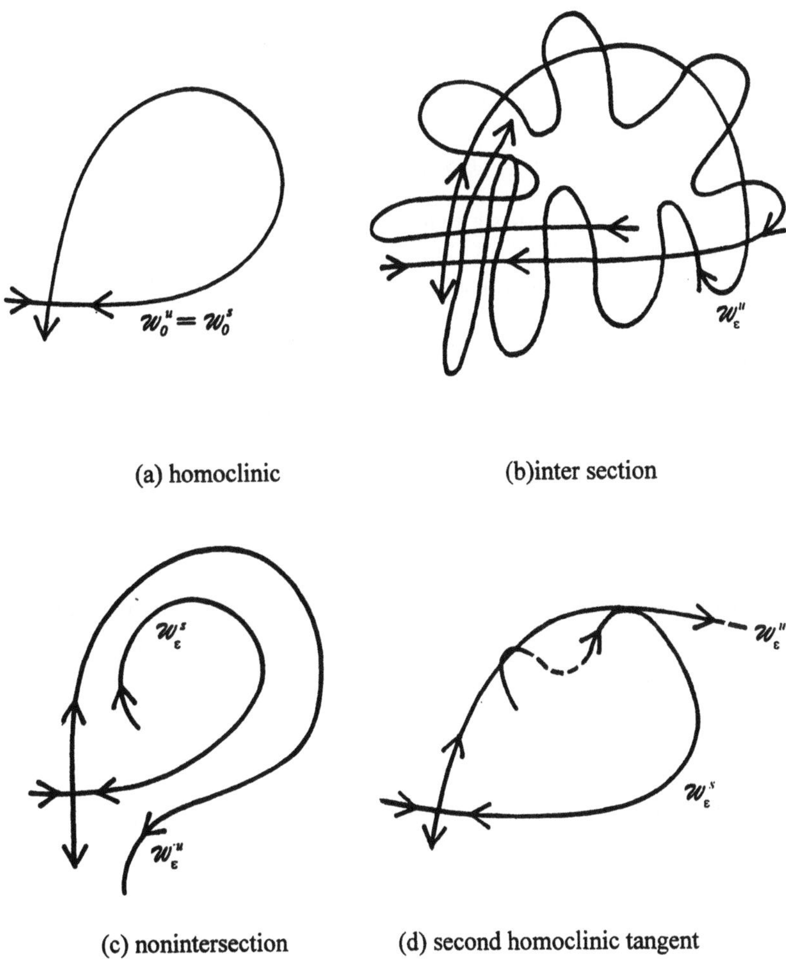


(a) homoclinic (b)inter section

(c) nonintersection (d) second homoclinic tangent

Fig. 8.22

If the disturbance term $\varepsilon g(x,t)$ of system (8.37) is dependent on parameter $\mu \in \mathcal{R}$, that is for the parameter family

$$x' = f(x) + \varepsilon g(x,t,\mu), \text{ then } \mu \in \mathcal{R} \tag{8.41}$$

If $g$ satisfies the conditions above, see below.

**Proposition 8.3** Suppose that the Melnikov function $M(t_0,\mu)$ has a quadratic zero point, i.e. $M(t,\mu_0) = \left(\frac{\partial H}{\partial t_0}\right)(\tau,\mu_b) = 0$, but $\left(\frac{\partial^2 M}{\partial t_0^2}\right)(\tau,\mu_0) \neq 0$ and $\left(\frac{\partial M}{\partial \mu}\right)(\tau,\mu_b) \neq 0$. Then system (8.41) has a bifurcation value $\mu_b + o(\varepsilon)$ which is a quadratic homoclinic tangent. The conclusion of propositions 8.2 and 8.3 can be seen in Fig. 8.22.

Suppose the periodic orbit $q^\alpha(t-t_0)$ of system (8.36) on the section $\sum^{t_0}$ has a period of $T_\alpha$, and for $h_\alpha = H(q^\alpha(t))$, $\frac{dT_\alpha}{dh_\alpha} \neq 0, \alpha \in (-1,0)$. Suppose also that $T_\alpha = \frac{m}{n}T$, where $m$ and $n$ are prime numbers and $T$ is also the period of disturbance term $\varepsilon g$. Then we can define the subharmonic Melnikov function

$$M^{m/n}(t_0) = \int_0^{mT} f(q^\alpha(t)) \wedge g(q^\alpha(t), t+t_0)\mathrm{d}t \tag{8.42}$$

**Proposition 8.4** If $M^{m/n}(t_0)$ has a simple zero point which is independent of $\varepsilon$ and when $\varepsilon$ is sufficiently small, eq. (8.37) has a subharmonic orbit with a period of $mT$. If $n=1$ then the result is uniformly valid in $0<\varepsilon\le o(1)$. Let $M^{m/1}(t_0) = M^m(t_0)$, then

$$\lim_{m\to\infty} M^m(t_0) = M(t_0) \tag{8.43}$$

See [36] for details.

Example of the Duffing equation

$$x_1' = x_2$$
$$x_2' = x_1 - x_1^3 + \varepsilon(\gamma\cos\omega t - \delta x_2)$$

If $\varepsilon = 0$, there exist a centre $(\pm 1,0)$ and a saddle $(0,0)$, as well as the Hamiltonian function $H(x) = \frac{1}{2}x_2^2 - \frac{1}{2}x_1^2 + \frac{1}{4}x_1^4$. The horizontal set of $H(x)=0$ is composed of two homoclinic orbits $\mathcal{P}_+^0$ and $\mathcal{P}_-^0$ and the point (0,0). $\mathcal{P}_\pm^0$ can be defined by

$$q_+^0(t) = (x_1^0, x_2^0) = (\sqrt{2}\,\mathrm{sech}\,t, -\sqrt{2}\,\mathrm{sech}\,t\tanh t)$$
$$q_-^0(t) = -q_+^0(t)$$

The Melnikov function described by eq. (8.40) is

$$M(t_0,\gamma,\delta,\omega) = \int_{-\infty}^{\infty} x_2^0(t)[\gamma \cos\omega(t+t_0) - \delta x_2^0(t)]\mathrm{d}t$$
$$= -\sqrt{2}\gamma \int_{-\infty}^{\infty} \operatorname{sech} t \tanh t \cos\omega(t+t_0)\mathrm{d}t$$
$$-2\delta \int_{-\infty}^{\infty} \operatorname{sech}^2 t \tanh^2 t\mathrm{d}t$$
$$= -\frac{4\delta}{3} - \sqrt{2}\gamma\pi\omega \operatorname{sech}\left(\frac{\pi\omega}{2}\right) \sin\omega t_0$$

Writing $\mathcal{R}^0(\omega) = \dfrac{4\cosh(\frac{\omega\pi}{2})}{3\sqrt{2}\pi\omega}$, we know from proposition 8.2 that when $\frac{\gamma}{\delta} < \mathcal{R}^0(\omega)$, $\mathcal{W}_\varepsilon^s \cap \mathcal{W}_\varepsilon^s =$ ; when $\frac{\gamma}{\delta} = \mathcal{R}^0(\omega)$, $M(t_0,\gamma,\delta,\omega)$ has twice the zero points. We know from proposition 8.3 that for a given $\omega$, a bifurcation curve on the plane $(\gamma,\delta)$ crosses with $\gamma = \mathcal{R}^0(\omega)\delta$ at $\gamma = \delta = 0$. By calculating the subharmonic Melnikov function, we can obtain the corresponding case of proposition 8.4 (refer to [36])

**Note**: When eq. (8.36) is not a Hamiltonian system, $\mathrm{tr} Df(q^0(t)) \neq 0$. The conclusion of proposition 8.2 remains valid only if we change the Melnikov function. Moreover, for a heteroclinic orbit, i.e. the orbit connecting two different saddles in system (8.36), system (8.37) obtained by a small disturbance has a conclusion similar to propositions 8.2 and 8.3.

The Melnikov method discloses the relationship between the stable and unstable manifolds of the homoclinic orbit after a disturbance. As described above, the most important here is the transversality between the two manifolds (see Fig. 8.22(b)). Their intersection point (besides point $p_0$) is the transversal homoclinic point. So long as there is one transversal homoclinic point, there will be infinite transversal homoclinic points. The concept of transversal homoclinic points was first proposed in the discussion of the double approximate solution by Poincaré, where he estimated the complexity of this problem. Poincaré stated: "he even does not want to draw it". Afterwards Birkhoff made some advances. Smale pointed out that the existence of a transversal homoclinic point makes possible the appearance of the hyperbolic invariant set with horseshoe structure beside the saddle.

**Theorem 8.3 Smale—Birkhoff homoclinic theorem**

Let $\mathcal{F}:\mathcal{R}^2 \to \mathcal{R}^2$

(1) Assume a hyperbolic equilibrium point $p$.

(2) Suppose $q \neq p$ makes $q \in \mathcal{W}_\varepsilon^s(p) \cap \mathcal{W}_\varepsilon^s(p)$, that is, there is a transversal homoclinic point.

Then, for some $k$, there exists an invariant set $\wedge$ of $\mathcal{F}^k$; it has all the characteristic of a horseshoe.

## 8.4 Hamiltonian System

A non-linear dynamical system, whether it be a continuous dynamical system expressed by an ordinary differential equation, or a discrete dynamical system, can be divided into a conservative system and a non-conservative system, or into a Hamiltonian system and a dissipative system. The discrete system derived from the differential equation is described as a point map, the Hamiltonian system is a conservation-area map, and the dissipative system is a contracting map. There are chaotic phenomena under certain values of parameters for these two systems. The above examples belong to the dissipative systems. In this section, we deal with some important concepts of the Hamiltonian system. More details can be found in [83].

### 8.4.1 The Elliptic Point and the Hyperbolic Point

To introduce some concepts, let us consider a simple linear iteration system

$$\varphi:\begin{cases} x_{n+1} = y_n \\ y_{n+1} = -x_n + \lambda y_n \end{cases} \tag{8.44}$$

and write it as a form of mapping

$$(x,y)\overset{\varphi}{\rightarrow}(y,-x+\lambda y) \tag{8.45}$$

Its Jacobian determinant is

$$\det\begin{bmatrix} 0 & 1 \\ -1 & \lambda \end{bmatrix} = 1 \tag{8.46}$$

and it is a conservation area. The product of two eigenvalues is always 1. When $|\lambda| < 2$, the eigenvalues are a conjugate complex pair, of modulus 1. When $|\lambda| > 2$, there are two real eigenvalues, the modulus of one of the eigenvalues being greater than 1. The former shows that the fixed point (0, 0) is stable, and the latter shows that it is unstable. For a conservation-area map, the stable fixed point is called an elliptic point, and the unstable fixed point is called a hyperbolic point. For the hyperbolic point refer to [83]. Below we discuss the elliptic point that we are interested in. When $-2 < \lambda < 2$, the fixed point (0,0) is an elliptic point, its eigenvalues being

$$k_{1,2} = -\frac{\lambda}{2} \pm (\sqrt{(1 - \frac{\lambda^2}{4}})i$$

Make a variable transformation for eq. (8.44)

$$(\sqrt{1-m^2})x = X + mY$$

$$(\sqrt{1-m^2})y = mX + Y \tag{8.47}$$

where $m = \frac{2}{\lambda} \mp \sqrt{\frac{4}{\lambda^2} - 1}$ $(\lambda > 0 \text{ or } \lambda < 0)$ (8.48)

$$m = 0 \quad (\lambda = 0)$$

The geometric sense of this transformation is to transform the rectangular coordinate into an oblique Descartes coordinate (when $\lambda = 0$ it does not change.)

In $(X,Y)$ coordinate system, eq. (8.44) can be transformated into

$$\varphi : \begin{cases} X_{n+1} = X_n \cos\varphi + Y_n \sin\varphi \\ Y_{n+1} = -X_n \sin\varphi + Y_n \cos\varphi \end{cases} \tag{8.49}$$

where $\cos\varphi = \dfrac{\lambda}{2}$ (8.50)

Then $X_{n+1}^2 + Y_{n+1}^2 = X_n^2 + Y_n^2$ (8.51)

That is $X^2 + Y^2$ is an invariant variable of transformation $\varphi$. Returning to the $(x, y)$ coordinate system, this invariant variable is

$$x^2 - \lambda xy + y^2 = x_0^2 - \lambda x_0 y_0 + y_0^2 \tag{8.52}$$

This shows that the motion is confined to an ellipse (8.52) starting from the initial value $(x_0, y_0)$. This is the reason why we call it the elliptic point. Equation (8.49) shows that every iteration is equivalent to rotating an angle around the centre on the plane $XY$ (see Fig. 8.23).

$$\varphi = \arccos\frac{\lambda}{2} \tag{8.53}$$

This rotating angle or the corresponding rotating number

$$N = \frac{\varphi}{2\pi} = \frac{1}{2\pi}\arccos\frac{\lambda}{2} \tag{8.54}$$

plays an important role in the theory of chaos. When $N$ is a rational number $\frac{p}{q}$ ($p$ and $q$ are prime numbers) and any point returns to the initial point after $q$ iterations, i.e. all solutions of system (8.44) have a period of $q$. An invariant set comprises a certain point and its $(q-1)$ points of iteration. When $N$ is an irrational number, its point of multiple iterations, starting from any initial point except the $o$ point, can return to any small neighbourhood of the initial point an infinite number of times, but not to the initial point. In such a case the invariant set is the whole ellipse (or the circle $X^2 + Y^2 = X_0^2 + Y_0^2$), but not the set composed of the finite point. The solution is no longer periodic, but spread all over.

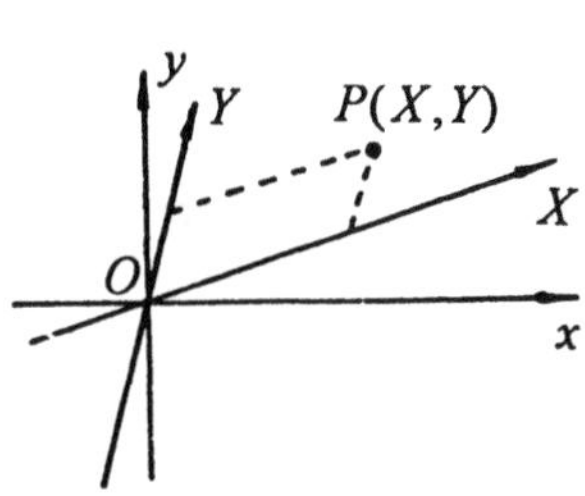


Fig. 8.23

**Table 8.1**

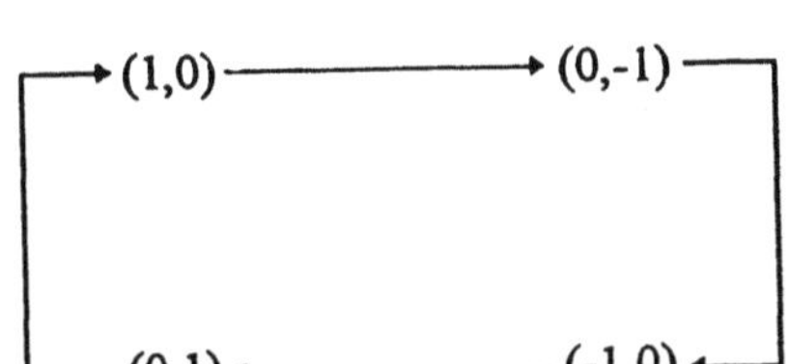

The simplest example is $\lambda = 0$, i.e. $(x,y) \to (y,-x)$. See Fig. 8.24. The four points having the same sign in the Figure constitute an invariant set, as shown in Table 8.1. The rotating angle is $90°$, the rotating number is $\frac{1}{4}$.

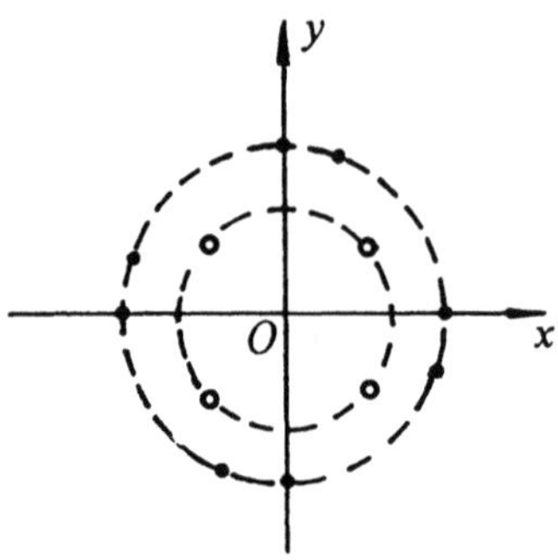


Fig. 8.24

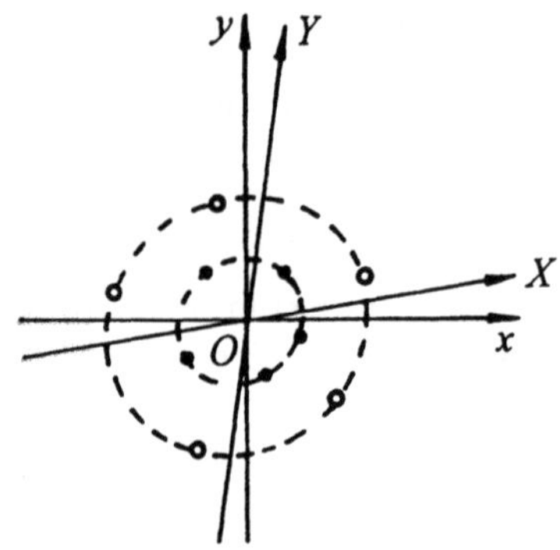


Fig. 8.25

If $\lambda = 2\cos 72° = \frac{1}{2}(\sqrt{5}-1) \cong 0.618$, then the invariant set is composed of five points on the same ellipse, as shown in Fig. 8.25, where the slopes of $X, Y$ are $m$ and $\frac{1}{m}$, but $m$ is defined by eq. (8.48) as 0.1584. If $X, Y$ are the rectangular coordinates, then every iteration is equal to $\frac{1}{5}$ or $72°$ of a rotating circle.

If $\lambda = 0.6$, then we have

$$\varphi: (x,y) \to (y, -x + 0.6y) \tag{8.55}$$

At this time, the rotating angle is $72.54°$, the rotating number $N = 0.20150...$ is an irrational number and the invariant set is an ellipse, as shown in Fig. 8.26. We will denote the rational case by broken line, as shown in Fig. 8.24 and Fig. 8.25, and the irrational case by solid line as shown in Fig. 8.26. Sometimes we refer to them as a rational ellipse and an irrational ellipse respectively.

In the linear system (8.44), all ellipses are either rational or irrational. But in non-linear systems, even in the small domain of the fixed point, rational ellipses and irrational ellipses are embedded in each other. Here "the ellipse" means invariant closed orbit, its shape approximates an ellipse, and rational and irrational also mean its number of rotation. For example, consider the non-linear system:

$$\varphi: \begin{cases} x_{n+1} = y_n \\ y_{n+1} = -x_n + 0.6y_n + 2y_n^2 \end{cases} \tag{8.56}$$

At the fixed point (0,0), the characteristic equation of the linearized part is equal to the characteristic equation of eq. (8.44) at $\lambda = 0.6$. Its number of rotation is 0.20150... . In the neighbourhood of (0,0), the influence of the non-linear term $2y_n^2$

will make the rotation number decrease with increase of the distance from the origin. The process of diminution is continuous. When the rotation number decreases to $403/2000 = 0.20150$, the invariant set on the rational ellipse is made up of 2000 points. When the ellipse increases to the point when its rotation number is $141/700 = 0.20142857...$, the invariant set on the rational ellipse is made up of 700 points, and between the two rational ellipses there are infinite rational or irrational ellipses. This structure cannot be drawn to scale, so we have to use Fig. 8.27 as a sketch map. Of course, if the rotation number decreases to 1/5 = 0.2, there will be a periodic solution. When it is further from the fixed point, the structure of the wide scope will be more complicated. Next we will analyze system (8.56) in detail.

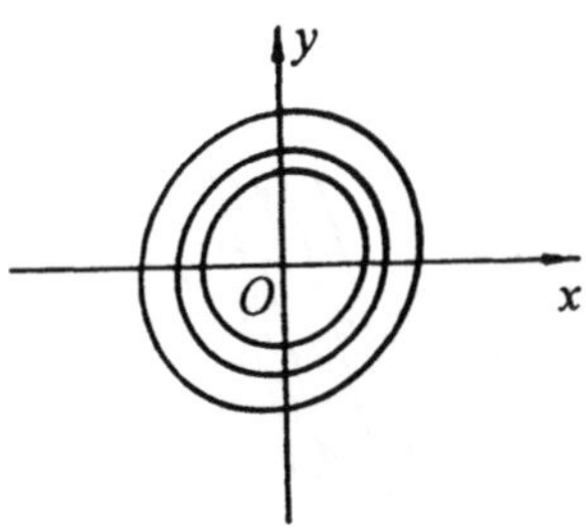


Fig. 8.26

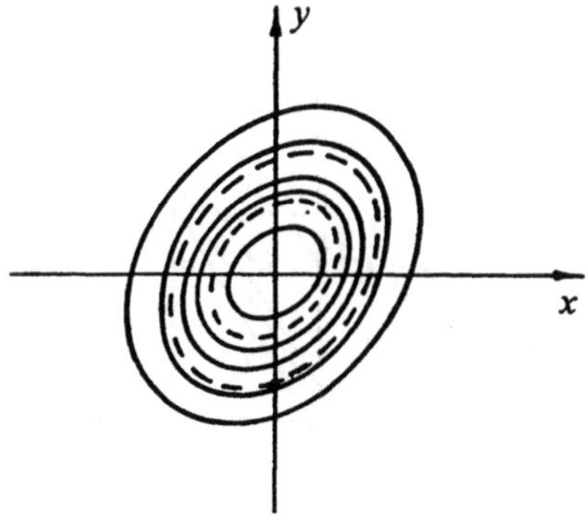


Fig. 8.27

### 8.4.2 Typical Structure of the Conservation—Area System

We use a non-linear system to illustrate:

$$\varphi: \begin{cases} x_{n+1} = y_n \\ y_{n+1} = -x_n + \lambda y_n + 2y_n^2, \ |\lambda| < 2 \end{cases} \tag{8.57}$$

This system is a conservation area system, and the Jacobian matrix is

$$\begin{bmatrix} 0 & +1 \\ -1 & \lambda + 4\lambda \end{bmatrix} \tag{8.58}$$

its determinant value being 1. It has two fixed points $o\,(0,0)$ and $A\,(1-(1/2)\lambda, 1-(1/2)\lambda)$. Changing $x$ for $y$, we obtain the inverse retransformation of eq. (8.57) $\varphi^{-1}$ and the motion behaviours on the $(O)$-plane are symmetrical to the diagonal line $x = y$. We change eq. (8.57), $X = -(\lambda + 4y)/2\mu$, $Y = -(\lambda + 4x)/2\mu$, where $\mu = \lambda(\lambda - 4)/4$ and obtain $(X, Y) \to (1 - y - \mu x^2, y)$. This is a Henon equation of the conservation area. When $|\lambda| < 2$, from the characteristic equation we know that the fixed point (0,0) is an elliptic point, and the rotation angle $\varphi = \arccos(\lambda/2)$; the fixed point $A\,(1-(1/2)\lambda, 1-(1/2)\lambda)$ is a hyperbolic point. Therefore, eq. (8.57) is a system containing an elliptic point and a hyperbolic point. When $\lambda = 0.6$, the eigenvalues are 3.075 and 0.325 at the hyperbolic fixed point

$A$ (0.7, 0.7) and the gradients of the characteristic direction are 1:3075 and 1:0.325; the characteristic line is the stable and unstable manifolds of the linearized system at point $A$. In order to obtain the unstable manifold of the non-linear system we take a certain point on the small segment $AB$ of the characteristic line of 1:3.075. After many iterations, this will expand into an unstable manifold as shown in Fig. 8.28, namely the solid line $ABCDEFG...$ . At the beginning this curve approximates to a straight line, but then it gradually begins to bend owing to the influence of the non-linear term. When it returns to $A$, the bending amplitude and oscillating amplitude become bigger and bigger. We can draw the stable manifold by the inverse transformation $\varphi^{-1}$ in the same way. But for the symmetry, we make the solid line $ABCDEFG...$ a mirror reflection in $OA$ and then obtain the stable manifold $AIHGFED$, represented by the broken line.

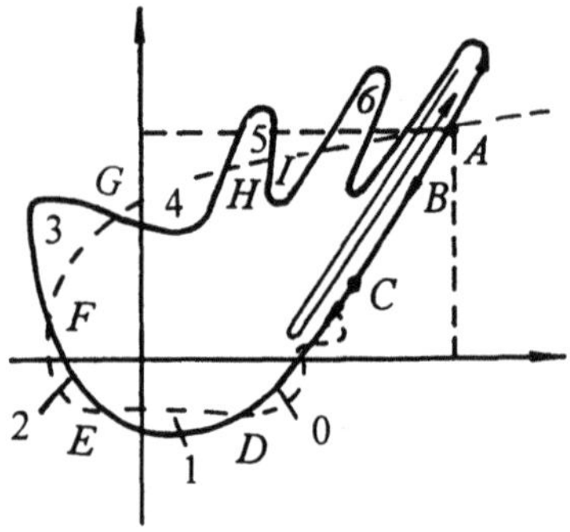


Fig. 8.28

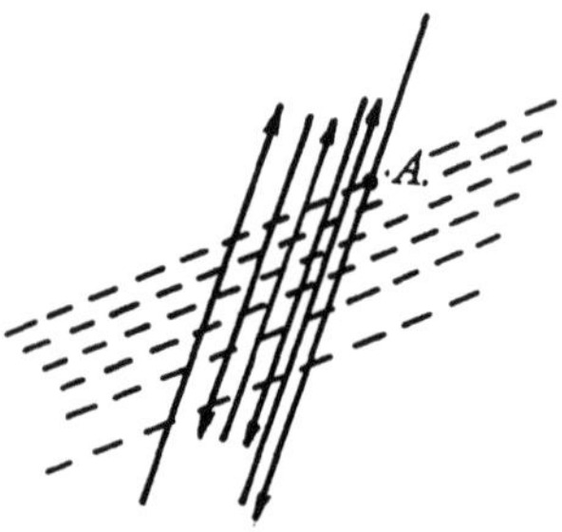


Fig. 8.29

The two invariant manifolds are invariant to the transformation $\varphi$. They have many intersections. They are homoclinic points $D, E, F, ...$ and so on. In these homoclinic points, the accumulated iteration result of $D$ is $F, H, ...,$ namely $\varphi D = F$, $\varphi^2 D = H$ and so on, while $\varphi E = G$, $\varphi^0 E = I$ and so on. $D$ is the area enclosed by the two invariant manifolds, ①, ②, ③, ..., which have the relation equations:

$$\varphi ① = ③,\ \varphi^2 ① = ⑤$$

$$\varphi ② = ④,\ \varphi^2 ③ = ⑥$$

$$\vdots \qquad \vdots$$

From its symmetry, the area of ① is equal to that of ②. From the characteristic of the conservation area of the transformation, we know that the area of ① is equal to that of ③, so all the areas ①,②,③,④,⑤ are the same. This is an important characteristic of the conservation—area system. Along the broken line $HI...$, approaching the hyperbolic point $A$, "the bases" of these areas (we regard them as triangles) become smaller and smaller. Every iteration will decrease by 1/3.075 or so. In order to agree with the characteristic of the conservation area, the relevant height must increase by 3.075 times. Finally, nearer to point $A$, this area becomes narrower and longer. Then a situation like that of Fig. 8.29 will develop at point $A$.

The figure shows Poincaré bars. The Poincaré bars at the hyperbolic point are the reason why there are chaos and random phenomena in the conservation—area system.

The dynamical behaviour of system (8.56) at the elliptic fixed point $O(0, 0)$ has been represented in Fig. 8.27. Combining Fig. 8.27 with 8.28 we obtain a portion of the result of the global dynamical behaviour of the system. When the initial point is outside the region enclosed by the two invariant manifolds of point $A$, the iteration result will diffuse rapidly. When the initial point is in these regions, which are netted by the two invariant manifolds after many iterations, it will go to infinity finally. The points in the domain of $O(0, 0)$ will be in these regions forever in the course of iteration. We generalize the above results in Fig. 8.30. We term those original value regions that will dissipate eventually (it may take a long time) chaos seas (here it is actually a divergent region). The points that cannot dissipate are on the island of the regular solution.

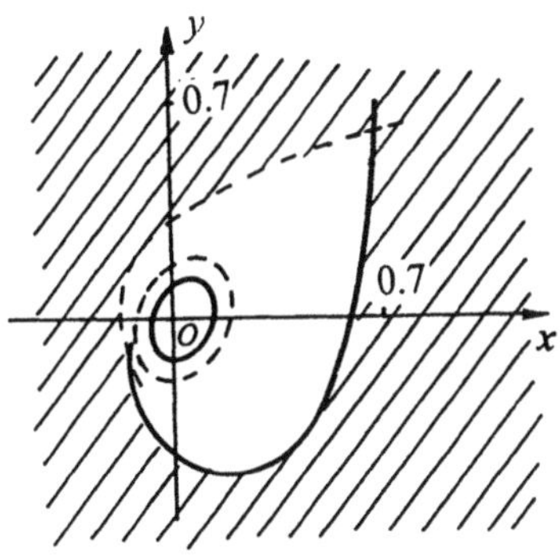

Fig. 8.30

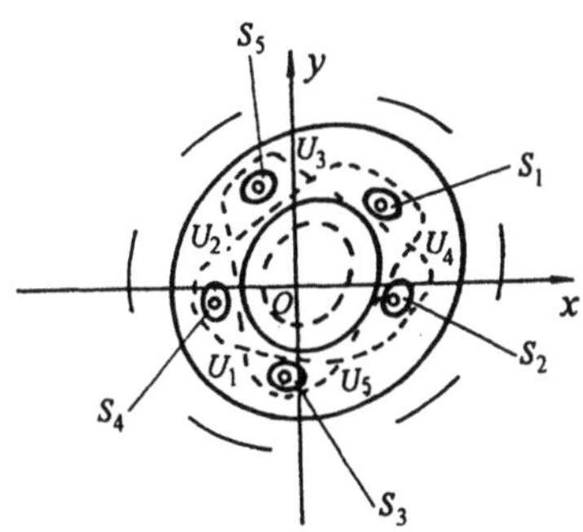

Fig. 8.31

We have seen that the rotation angle at the fixed point $O$ is $\varphi = \arccos(0.6/2) = 72.54°$; away from $O$, the rotation angle will decrease, and when it decreases to $72°$, we obtain the period-5 solutions. In linear systems, there are infinite groups of period-5 solutions; they form many ellipses. But in non-linear systems, there are only two groups of period-5 solutions. Their computation result is as follows:

$$S_1\ (0.0844,\ 0.0844) \to S_2\ (0.0844,\ -0.0195) \to S_3\ (-0.0195,\ -0.0953) \to$$
$$S_4\ (-0.0953,\ -0.195) \to S_5\ (-0.0195,\ 0.0844) \to S_1;$$
$$_1\ (-0.0702,\ -0.0702) \to V_2\ (-0.0702,\ 0.0379) \to V_3\ (-0.0379,\ -0.0958) \to$$
$$_4\ (0.0958,\ 0.0379) \to V_5\ (0.0379,\ -0.0702) \to V_1;$$

These ten period-5 points are fixed points to transformation $\varphi^5$. If we linearize $\varphi^5$ around the fixed point and obtain its eigenvalues, we can determine whether these period-5 points are elliptic points (stable) or hyperbolic points (unstable). Pay attention to the fact that the transformation $\varphi^5$ is a conservation area, so these characteristic equations still take the form of $k^2 + \mu k_{+1} = 0$. If $|\mu| > 2$, they are hyperbolic points, and if $|\mu|<2$ they are elliptic points. From the numerical computation, for the period-5 solutions, $S_1, S_2, S_3, S_4, S_5$, the value of $\mu$ is 1.9999<2. They are elliptic periodic points. On the other hand, the five period-5 points $U_1$, $U_2$, $U_3$, $U_4$, $U_5$ are hyperbolic periodic points. The value of $\mu$ is 2.00172. Near every hyperbolic point there is also a Poincaré bar. The difference is that it is formed by heteroclinic structure instead of homoclinic structure; for example, $U_1$, according to Fig. 8.30(a). These four lines, the unstable and stable manifolds which go from $U^5$ forward or backward to the vicinity of $U_1$, and the two invariant manifolds from $U_2$, build the bar at $U_1$. If the initial value is in these bars and the four invariant manifolds of $U_1$ (it has not been drawn), after accumulative iterations of $\varphi^5$, these points will drift in these regions. Such regions constitute the chaotic river network. Just as in Fig. 8.32(b), the width of the river is amplified by a few orders of magnitude. Actually, it is very difficult to locate the point exactly in the chaotic river because the value of $\mu$ is close to 2. An important fact is that these river networks form five small islands around the stable periodic points.

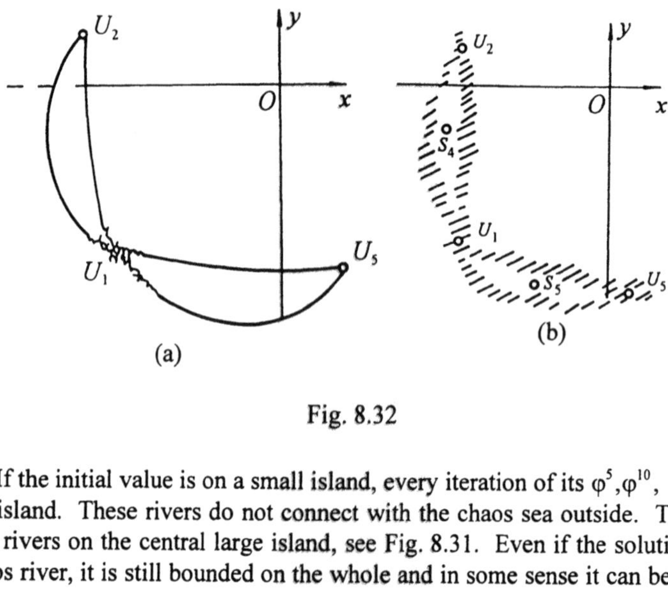


Fig. 8.32

If the initial value is on a small island, every iteration of its $\varphi^5, \varphi^{10}$, is always on this island. These rivers do not connect with the chaos sea outside. They are the internal rivers on the central large island, see Fig. 8.31. Even if the solution drifts in the chaos river, it is still bounded on the whole and in some sense it can be said to be stable.

Similarly we can analyze the period-6 solution, the period-7 solution and so on. These solutions are still further from origin $O(0, 0)$. The size of the island around the stable period point is too small for the small island of the period-7 solution to exist.

The period 5, 6 corresponds to the rotation numbers 1/5, 1/6. In fact, between 1/5 and 1/6, there are still many rational numbers, such as 2/11. There will be a period-11 solution and a series of islands of the period-11 solution. These rational numbers are infinite, and hence there are infinite numbers of various periodic small islands. Some are inside the big islands and are islands in the inland river; others are beside the big island. The boundaries of the big islands are like real geographical coastlines. The structure of every small island is the same as that of the big island: rivers in the islands and island in the rivers. These structures will go on forever. The above structures are typical of the constitution formed by elliptic points and hyperbolic points in conservation—area systems. The change of parameter $\lambda$ has an obvious influence on the structure. For example, when $\lambda = a_4$, there exist a group of stable period-5 solutions. Around these periodic points there are five small islands; while the unstable period-5 solutions sink into the chaos sea, see Fig. 8.33.

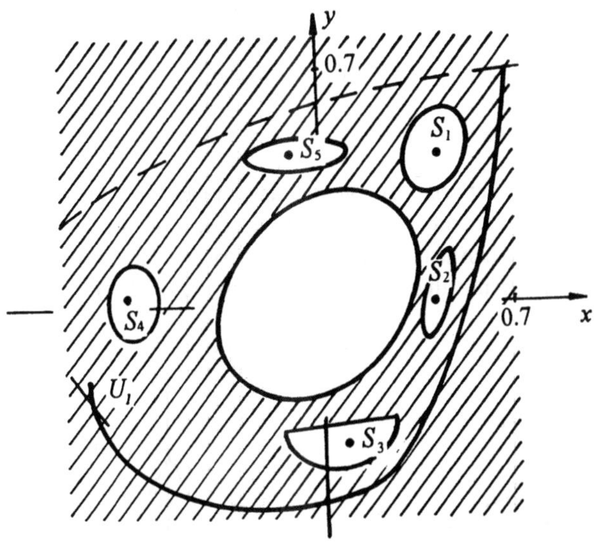


Fig. 8.33

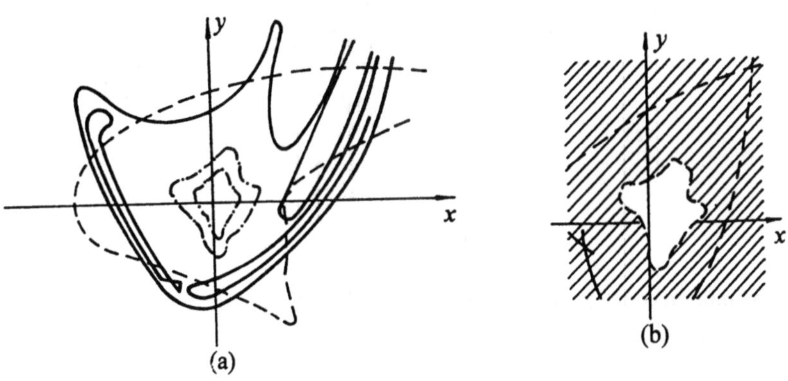


Fig. 8.34

If $\lambda = 0$, the fixed point $A$ (1,1) is a hyperbolic point and $O(0,0)$ is an elliptic point. The rotative angle is 90°, and the shape of the invariant closed curve around $O$ deviates greatly from an ellipse but obviously retains the trace of a period-4 solution (the rotative number is 1/4). When it is very close to $O$, the closed curve looks like a cruciform. There are two groups of period-5 solutions, but they are all inside the bar and network. As they are unstable, there is no island of period-5 solution. Fig. 8.34(a) shows the structure of the manifolds. Figure 8.34(b) shows the approximate shape of the "land".

### 8.4.3 Conservation Systems with Two Degrees of Freedom

Examples of Henon and Heiles are used here to explain some typical characteristics of conservative non-linear systems with two degrees of freedom. Suppose a particle of unit mass moves on the $x$—$y$ plane with a potential of

$$=\frac{1}{2}(x^2+y^2)+x^2y-\frac{1}{3}y^2 \tag{8.59}$$

The relationship between the potential and $xy$ can be represented by the curved surface in Fig. 8.35(a). It approximates a rotative parabolic surface when it is near the lowest point (0,0,0). When it is cut by $v=1/6$, we get three lines $AB, BC$ and $CA$. The contours with different values are represented in Fig. 8.35(b).

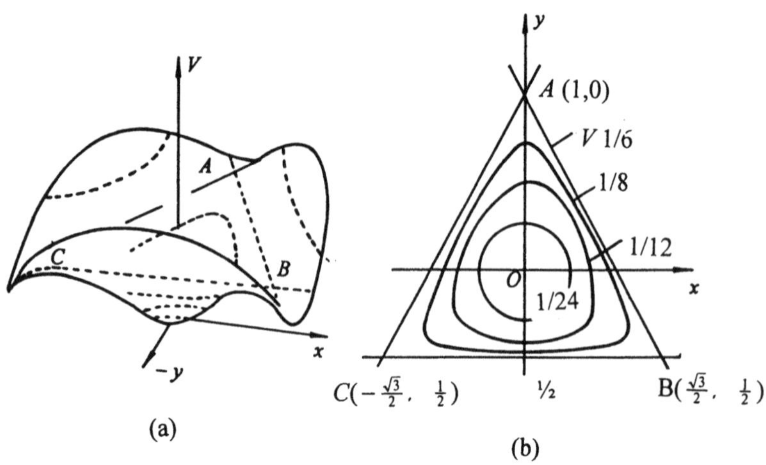


Fig. 8.35

The motion equation of the particle under the action of the potential is:

$$x''=-x-2xy$$

$$y''=-y+y^2-x^2 \tag{8.60}$$

Suppose the motion of the particle is limited to the triangular path $ABC$ shown in Fig. 8.35. The total amount of motion energy is $E \le 1/6$. The degree of freedom of the system is two. The relevant phase space is four dimensional, but because the Hamiltonian function

$$\frac{1}{2}(x'^2+y'^2)+\frac{1}{2}(x^2+y^2)+x^2y-\frac{1}{3}y^3=E \tag{8.61}$$

exists, the motion can be represented by the integral curve of a three-dimensional space, such as $(x,y,y')$. In order to explain the motional characteristic, we take the section $(x,z)$ of the integral curve on a plane such as $x=0$ where $z=y'$. The motion portrait consists of many discrete points. The following example belongs to a two-dimensional iteration:

$$y_{n+1}=f(y_n,z_n) \text{ and } z_{n+1}=g(y_n,z_n) \tag{8.62}$$

Figure 8.36 represents the distribution of these sections in cases where $E=1/12,\ 1/8,\ 1/6$. The outermost solid lines in the figure represent the relationship between $y$ and $z$, when motion is limited to the $y$—axis $(x(l)\equiv 0)$. The broken lines are made up of sections of several portraits for different initial values. The periodic solution is represented by the periodic points in the figure.

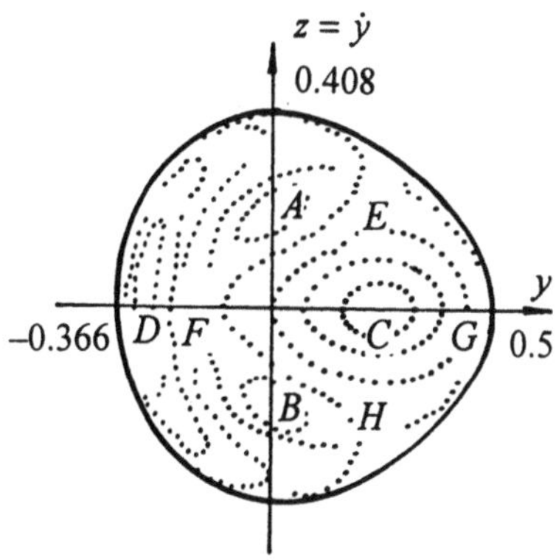


(a) $E=1/12$

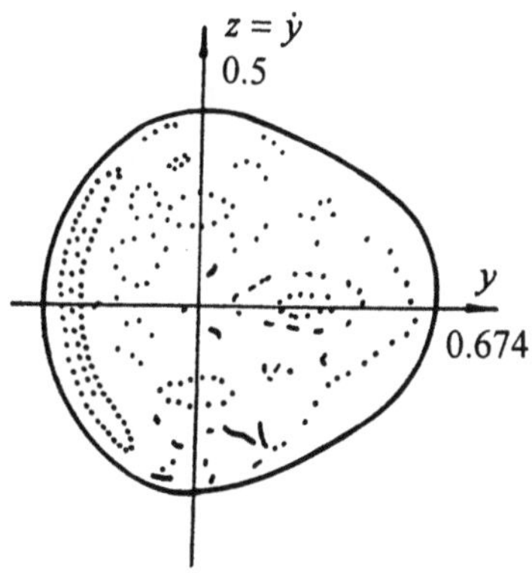


(b) $E=1/8$

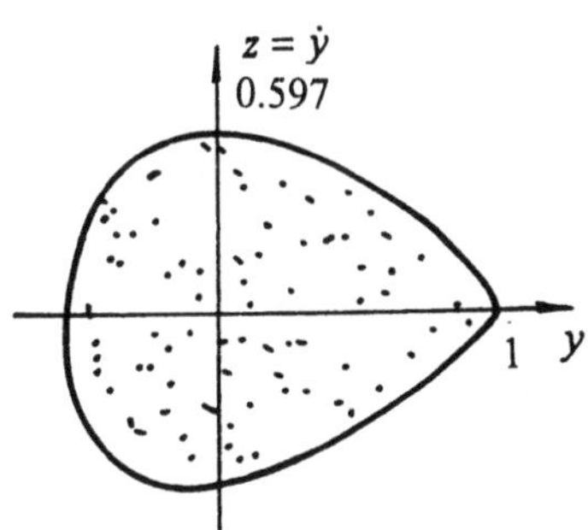


(c) $E=1/6$

Fig. 8.36

For instance, when $E = 1/12$, the motion of plane *ozy* must be limited to the isopotential line $V = 1/12$: $\frac{1}{2}z^2 + \frac{1}{2}y^2 - \frac{1}{3}y^3 = \frac{1}{12}$, namely inside the outermost solid line in Fig. 8.36.(a). A possible periodic solution is the particle vibration between $M(0, 0.5)$ and $N(0, -0.366)$ on the *xy* plane, shown as ① in Fig. 8.37. This motion is represented by the outermost solid line in Fig. 8.36(a).

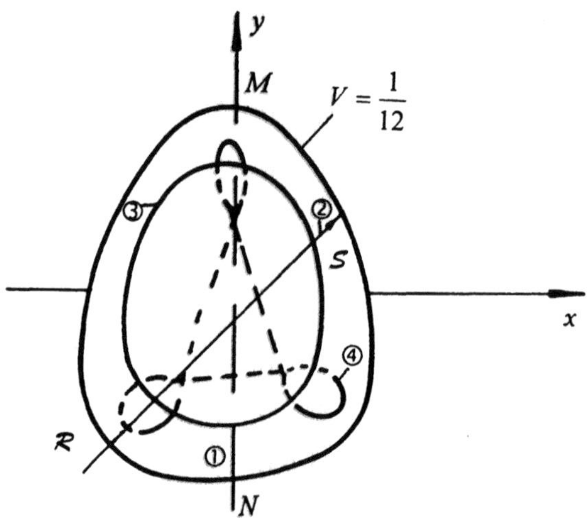


Fig. 8.37

Equivalent to this periodic motion is ② in Fig. 8.37. The particle vibrates between $R$ and $S$. The motion corresponds to the period-2 points $A$ and $B$ in Fig. 8.36(a). The trace ③, which corresponds to the period-2 points $C$ and $D$, is a periodic solution. In addition, there is an unstable periodic solution. Its trace on the *xy* plane is shown by ④ in Fig. 8.37. Its section on the $y$—axis and its corresponding $z\ (= \dot{y})$ give four hyperbolic points, namely the four points $F, E, G, H$ in Fig. 8.36(a). The solutions around the elliptic periodic points $A, B$ or $C, D$ are almost periodic solutions. There are chaos regions around the hyperbolic periodic points $E, F, G, H$, which are similar in structure to the chaos river networks presented in Fig. 8.31, but the chaos river is very narrow. The initial value $(y_0, z_0)$ has little chance of dropping into the chaos river. On the whole, the motion of the system is basically regular, an almost periodic solution.

When $E$ is increasing, the width of the chaos river also increases, as shown in Fig. 8.36(b). The chaos river actually becomes a chaos sea. The regular solution exists only when the initial point $(y_0, z_0)$ falls on the island in the sea. Besides the original big island, some small islands can be seen. In Fig. 8.36(c), nearly all of them are chaos solutions. The possibility of the appearance of regular solutions is very small. The non-linear Hamiltonian system with two degrees of freedom has obvious internal random characteristics. This example illustrates that the unintegrable Hamiltonian system will induce chaos solutions and has internal random characteristics generally.

### 8.4.4 KAM Theorem

The KAM theorem was first presented by A.N. Kolmogorov in 1954 [64] and then proved by V.I. Arnold [62] and J. Moser [63] in the early 1960s. It is concerned with some important results regarding the nature of the solution of approximately integrable Hamiltonian systems. Together with the Lorenz equation it marks the beginning of chaos theory. The former made contributions to the area of mathematical theory of the Hamiltonian system while the latter contributed to the other in the area of numerical calculation of dissipative systems.

**Definition 8.5** If $J_i$ (action variable of Hamilton ) and $Q_i$ (relevant angle variable) are Hamiltonian canonical variables, when the Hamiltonian function $H$ is composed only of $J_1,\ldots,J_N$

$$\overline{H} = \overline{H}(J) \tag{8.63}$$

then the Hamiltonian function is said to be integrable.

In the above equation:

$$J_i' = \frac{\partial H}{\partial \theta_i} = 0,\ J_i = \alpha_i = \text{const}\,,\ \theta_0' = \frac{\partial \overline{H}}{\partial J_i} = \omega_i = \text{const}\,,\ \theta_i = \omega_i t + \beta_i \tag{8.64}$$

**Definition 8.6** An approximately integrable system is a very small perturbation of an Hamiltonian integrable system

$$H(\theta, J) = H_0(J) + \varepsilon H_1(\theta, J) \tag{8.65}$$

where $\varepsilon$ is a small parameter.

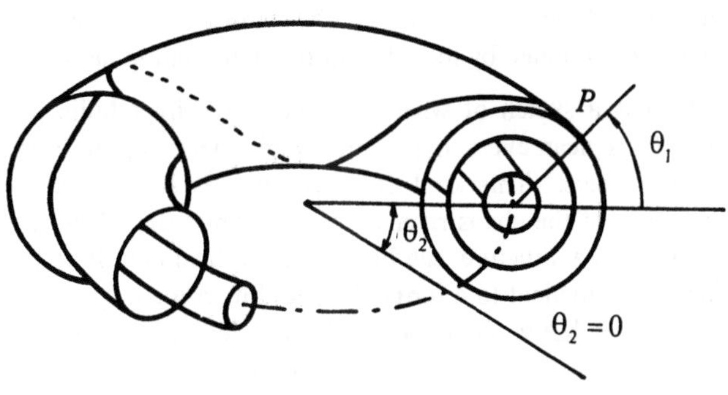


Fig. 8.38

The following brief statement applies only to a system with two degrees of freedom. In a system with two degrees of freedom, the Hamilton equation can be represented by two action variables $J_1$ and $J_2$ and their relevant angle variables $\theta_1$ and $\theta_2$. Because there exists a Hamiltonian function as an initial integration, we can consider variables $J_1$ and $\theta_1,\theta_2$ only (similar to the $x,y,y'$ in Henon and Heiles problems). As to the original undisturbed integrable case, we can select a proper change, so that $J_1$ =constant, and the solution is $\theta_1=\theta_1(t),\theta_2=\theta_2(t)$. Its portrait is always on a torus as shown in Fig. 8.38. This torus is called an invariant torus. It cuts $\theta_1$ =constant and yields a group of concentric circles. For a particular portrait, its sections with $\theta_2$ =constant are on a circle. These sections will compose a set of finite points (when the rotation number is rational ) or fill the line with circles (when the rotation number is irrational).

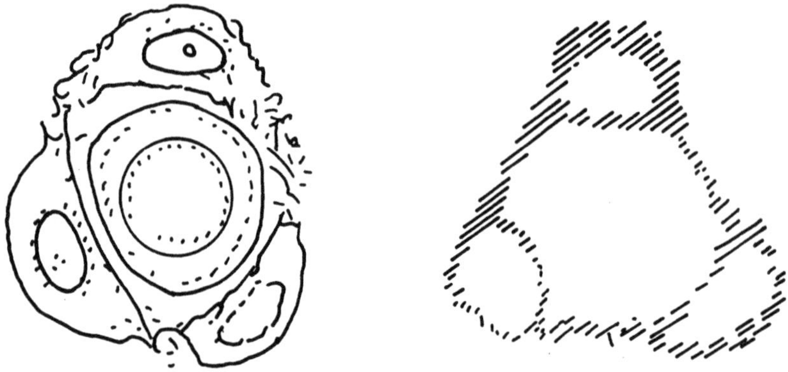

Fig. 8.39

When the system is becoming approximately integrable from non-integrable as a result of a small disturbance, these concentric circles will deform. The KAM theorem points out that because of the conservation of the area, if hyperbolic points appear their number will be the same as the elliptic points. The number is appropriate to the denominator represented by the relevant rational rotation number.

Figure 8.39 is a sketch of when there are three hyperbolic points and three elliptic points. Thus near every hyperbolic point there appear Poincaré bars and homologous chaos rivers. This kind of chaos river is surrounded by land on the outside. The network of the chaos river is infinite, while the island in every river, if amplified, repeats such a structure, see Fig. 8.1, such land is infinite too. The closed orbit on the land is equivalent to a torus of three-dimensional space $(J_1,\theta_1,\theta_2)$ called the KAM torus. The KAM theorem can be applied to approximately integrable systems, if its cases are not far from the integrable system. If the deviation is too big, the theorem is untenable and the KAM torus in the figure breaks down. In the outer layer there are chaos seas, and some of these and the chaos rivers are connected. When a KAM torus exists, from the viewpoint of the whole motion behaviour of the mechanical system, although the local motion of the approximately integrable system has random characteristics, the global motion is bounded and regular. Only when the system deviates far from the integrable system does the chaos sea occupy most of the parts.

On the whole, the motion is mostly random. Moreover, the conclusion that the chaos region is surrounded by KAM is applicable only to mechanical systems with two degrees of freedom. When the degree of freedom is three or more, such a conclusion cannot be made. All the chaos regions (no matter how narrow they are) form a network (homologous to the chaos river mentioned above). If the initial point falls into such a region, it moves regularly in a short time, but in fact in a certain sense it will drift along the network. The speed of this drift is extremely slow and finally it induces a qualitative change of motion, such as unbounded, unstable, or obviously irregular. This extra slow drift is called Arnold diffusion in the literature.

The KAM theorem is qualitative in nature. It does not indicate to what extent the case is "approximately integrable". What is the qualitative estimation of the Arnold diffusion speed? All such questions can be determined by numerical experiment now, the theorem merely points out a possible result that helps us to avoid losing ourselves in a large quantity of numeral results.

## 8.5 Some Statistical Characteristics

Chaos appears irregular, but it is not really so. Although it cannot be predicted over a long period, it has some definite statistical characteristics. The attractor can be divided into two classes, one is regular, such as stable fixed point, periodic orbit and limit loop plane; the other is strange. Although the strange attractor has no recognized definition, we can determine its their common characteristics from a statistical viewpoint. For attractor $\wedge$, we determine these from aspects of dimension, Liapunov characteristic exponents and topological entropy [36].

### 8.5.1 Fractal Dimension

Suppose the space is $m$-dimensional, $\varepsilon > 0$. If at least the $N(\varepsilon)$th $m$ dimensions for square bodies with sides of length $\varepsilon$ are needed to cover the attractor $\wedge$, the definition of the Hausdorff dimension of $\wedge$ [85] is

$$d = \lim_{\varepsilon \to 0} \left( \frac{\ln N(\varepsilon)}{\ln\left(\frac{1}{\varepsilon}\right)} \right)$$

For a normal attractor, such as a stable fixed point, a periodic orbit and a limit torus, the Hausdorff dimension is 0, 1 and an integer greater than 1. Because the strange attractor always has a structure similar to that of a Cantor set, the dimension of the strange attractor generally is not an integer. For example, the dimension of the Henon map is 1.2 (when $a = 1.2$, $b = 0.3$) or 1.26 (when $a = 1.4$, $b = 0.3$). Another example: the dimension of the Lorenz equation is 2.06 (when $\beta = 4, \sigma = 10, \rho = 40$).

### 8.5.2 Liapunov Characteristic Index

Suppose $x(t)$ is a bounded portrait of the system

$$x' = f(x), \quad x \in \mathcal{R}^n \tag{8.66}$$

We give a disturbance to $x(t)$, $x(t) + \delta x(t)$. From

$$x' + \delta x' = f(x + \delta x) = f(x) + Df(x)\delta x + o(|\delta x|^2)$$

we hav $\delta x' = Df(x)\delta x + o(|\delta x|^2) \overset{def}{=} A(t)\delta x + o(|\delta x|^2)$

The linearized system is

$$\omega' = A(t)\omega \tag{8.67}$$

Suppose $\omega(t)$ is the solution of eq. (8.67), when $\omega(0) = \omega_0$, $\omega(t) = v(t)\omega_0$, where $v(t)$ is the basic solution matrix, and the definition of Liapunov characteristic index of the system (8.66) is $x(\omega_0) = \overline{\lim_{t \to \infty}} t^{-1} \ln|\omega(t)|$.

It can obviously be seen that $x(\omega_0) < \infty$, and it has nothing to do with the selection of the initial value $x(0)$ of trace $x(t)$. For $\omega_1$, or $\omega_2 \in \mathcal{R}^n$, $c \in \mathcal{R}$, we have

$$x(c\omega_1) = x(\omega_1), \; x(\omega_1 + \omega_2) \le \max\{x(\omega_1), x(\omega_1)\}$$

Proposition: $\mathcal{E}_r = \{x \in \mathcal{R}^n | x(t) = r\}$

It is clear that $\mathcal{E}_r$ is a linear subspace. If $r \le r'$, this follows implicitly from $\mathcal{E}_r \subset \mathcal{E}_{r'}$. When $r$ increases from $-\infty$ to $+\infty$, $\dim \mathcal{E}_r$ increases from 0 to $n$, so that $\dim \mathcal{E}_r$ jumps at $r = r_1, r_2, \ldots$ . Writing $\mathcal{E}_j = \mathcal{E}_{rj}$, then $0 = \mathcal{E}_0 \subset \mathcal{E}_1 \subset \ldots \subset \mathcal{E}_n = \mathcal{R}^n$. Therefore, for a one to one (bounded) portrait of eq. (8.66) we have $n$ Liapunov characteristic indices written as $L\mathcal{E}_j$, $j = 1, 2, \ldots, n$ and arranged from big to small habitually: $L\mathcal{E}_1 \ge L\mathcal{E}_2 \ge \ldots \ge L\mathcal{E}_n$.

As to attractors, $\{L\mathcal{E}_j\}$ has characteristics:

| Stable Condition | Attractor Type |
|---|---|
| $L\mathcal{E}_1 < 0$ | Fixed point (equilibrium state) |
| $L\mathcal{E}_1 = 0, L\mathcal{E}_2 < 0$, | Periodic orbit |
| $L\mathcal{E}_1 = \ldots = L\mathcal{E}_k = 0, L\mathcal{E}_{k+1} < 0$, | *k*-dimensional limit torus (quasi-periodic motion) |
| $L\mathcal{E}_m > 0, L\mathcal{E}_{m+1} = 0, L\mathcal{E}_{m+2} < 0$, | attractor (chaos) |

### 8.5.3 Topological Entropy

Suppose $f$ is a continuous map from compact metric space $(\mathcal{X}, d)$ to itself. For every $n \ge$ , we can define a new metric on $\mathcal{X}$:

$$d_n(x, y) = \max_{0 \le i \le n-1} \{d(f^i(x), f^j(y))\}$$

Given $\varepsilon > 0$, the subset $\mathcal{S}$ of $\mathcal{X}$ is called a $(n,\varepsilon)$ separated set. If for any $x, y \in \mathcal{S}$, $x \neq y$, we have $d_n(x,y) > \varepsilon$. Obviously, the cardinal number of separated sets is limited. We write the biggest cardinal number of $(n,\varepsilon)$—separated sets of $\mathcal{X}$ as $\mathcal{S}(n,\varepsilon)$ expressed as $h(f,\varepsilon) = \overline{\lim_{n\to\infty}} \frac{1}{n} S(n,\varepsilon)$ and $h(f) = \lim_{\varepsilon\to\infty} h(f,\varepsilon)$.

Then $h(f)$ is called the topological entropy of $f$. For instance, the topological entropy shift of the symbol dynamical systems is ln2. We can prove that, for regional maps: $f: \mathcal{I} \to \mathcal{I}$, the existence of the not two power period is equivalent to topological entropy $h(f) > 0$ of $f$. This means that chaotic motion exists.

So far we have discussed how to use the Liapunov index, the fractal dimension and topological entropy to appraise a chaotic motion. The kernel of the discussion is the computation of the Liapunov index, because the calculation of the separate dimension and topological entropy usually lead to the calculation of the Liapunov index. It is worth noticing that this computation is done in the tangent space of a phase space, while computations of periodic motion and quasi-periodic motion can be done in the phase space itself. We often modify them by means of the Newton method or quasi-Newton method. Although this involves the tangent space, it is only auxiliary. The judgment as to whether we have gained periodic solutions or quasi-periodic solutions is made in phase space. In contrast, we cannot appraise chaotic motion quantitatively in phase space. The main characteristics of the four steady motions are shown in table 8.2.

### 8.5.4 Frequency Spectrum Analysis

Spectrum analysis is the most common statistical characteristic. We analyze the sequence that has been obtained by Fourier analysis, FFT or spectrum analysis under amplitude or power. For instance, if the frequency spectrum is constant and continuous and reappears again, the solution can be determined to be chaos instead of an almost periodic solution. In actual calculation, the data taken are always limited. The spectrum is described by finite resolution. From the viewpoint of numerical calculation and physical experiment, it is very difficult to distinguish the solution with very long period from the chaotic solution. The sequence can be obtained by Fourier analysis, FFT or spectrum analysis (in amplitude or power).

## 8.6 Conclusions

In this section we will discuss some questions, such as the route to chaos, the distribution of stationary motion in parametrical space and, looking ahead, the conclusions of this chapter. The discussion will be far from complete. Those who are interested should refer to [5, 27, 83, 86, 87]

### 8.6.1 The Route to Chaos

For such a complicated movement as chaos, we hope to establish some connections between the simple movements (equilibrium, periodic, quasi-periodic motion) and chaotic movement. That is to say, to explain how the system transits from the state of

**Table 8.2** Four kinds of stationary movement states and their numerical characteristics

| Name | Trajectory | Time History | Power spectrum | Liapunov exponent $\lambda_1 \geq \lambda_2 \geq \lambda_3$ | Fractional number | Poincaré point set | Poincaré multiplier $\sigma_1 \geq \sigma_2$ | Fractional number of Poincaré point set |
|---|---|---|---|---|---|---|---|---|
| Equilibrium | 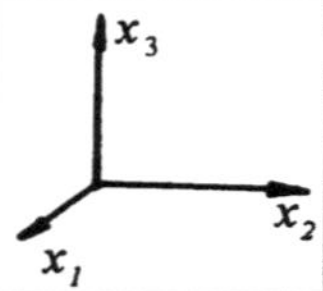  | 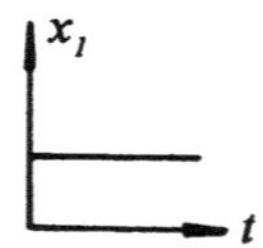  | 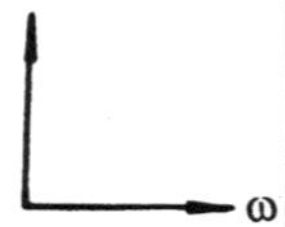  | - - - | 0 | | | |
| Periodic |  | 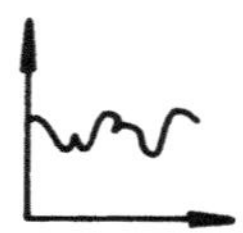 |  | 0 - - | 1 | | $1_-1_-$ | 0 |
| Quasi-periodic | 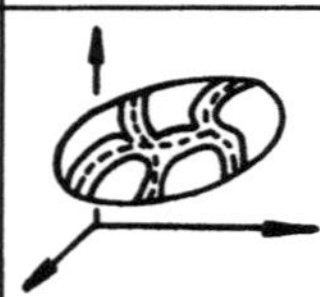 | 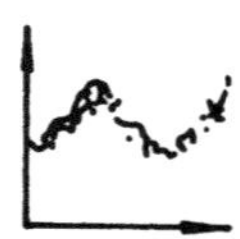 |  | 0 0 - | 2 |  | $1\ 1_-$ | 1 |
| Chaotic | 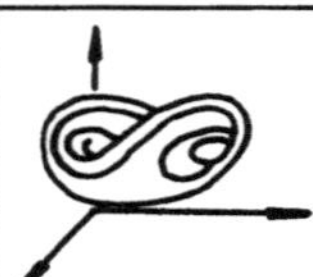 | 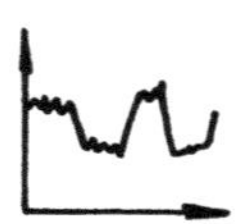 |  | + 0 - | 2<d<3 |  | $1_+1_-$ | 1<dp<2 |

simple movement to the chaos movement when the parameter changes? This approach is commonly called the route to chaos. At present, much work is focused on the period double bifurcation, the second Hopf bifurcation, intermittence and the broken KAM torus.

As we have discussed above, the period double bifurcation may lead to chaos. In this connection, universal constants have been obtained, such as the Feigenbaum constant. When the value of a parameter is fixed, period-3 implies chaos. This is a kind of saddle-node bifurcation. As can be seen from Fig. 8.8(b). From the frequency spectrum, intermittence means period-3 or chaos, see Fig. 8.40.

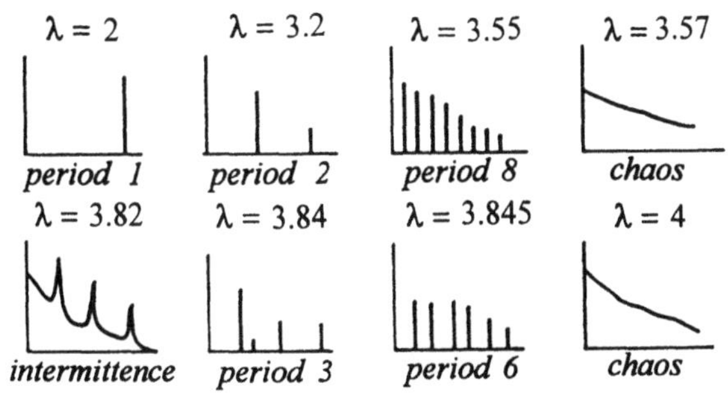


Fig. 8.40

As we know, when the parameter of the system changes, and Hopf bifurcation takes place in the system, then the limit cycle will appear. We can consider further the motion of bifurcation from a limit cycle to a two-dimensional torus $(T_2)$(quasi-periodic motion with prime frequencies). Of course, it will appear in three-dimensional dynamical systems, often called second Hopf bifurcation. The motion in $T_2$ can bifurcate into the motion in $T_2$, and so on. S.E. Newhouse and others proved [88] that quasi-periodic solutions in $T_3$ are generally unstable. Therefore, a periodic solution can be in a chaotic state if it passes the Hopf bifurcation or second Hopf bifurcation. However, stable periodic solutions in $T_3$ are found on the computing model and in experiments of hydrodynamics. The process from quasi-periodic solutions in $T_3$ to chaos requires further study.

Problems of the broken KAM torus remain. For the quasi-Hamiltonian system, the KAM theorem tells us that the portraits of the quasi-Hamiltonian system are distributed on two tori (one embedded in the other), but between the two tori is the chaotic region. The cross-line in the normal direction plane is called a KAM curve. For an integrable Hamilton, such as a pendulum, the phase portrait consists of elliptic points and hyperbolic points alternatively. The phase plane is divided continuously by saddle points. The motions of all parts in the phase space do not cross one another. In the case of an non-integrable small disturbance, a change appears near the saddle point. The connecting line for saddle breaks and violent oscillations takes place near the saddles. (Fig. 8.41)

This kind of oscillation leads to structure similar to the Smale horseshoe, and hence to chaos. The relevant region is called the chaos layer. Here, this marks an important difference between systems with two degrees of freedom and multi degrees of freedom. See Fig. 8.1. The KAM torus in the system with two degrees of freedom is the two-dimensional torus in a three-dimensional phase space. Therefore, the motion in a random layer is confined to a certain region, and cannot depart much from it. But for the system with three degrees of freedom, the phase space is five dimensional and the KAM torus is three dimensional, so its track cannot be confined to it. The track in the random layer can drift freely, and can go far from it at a slow speed. Thus is called Arnold diffusion.

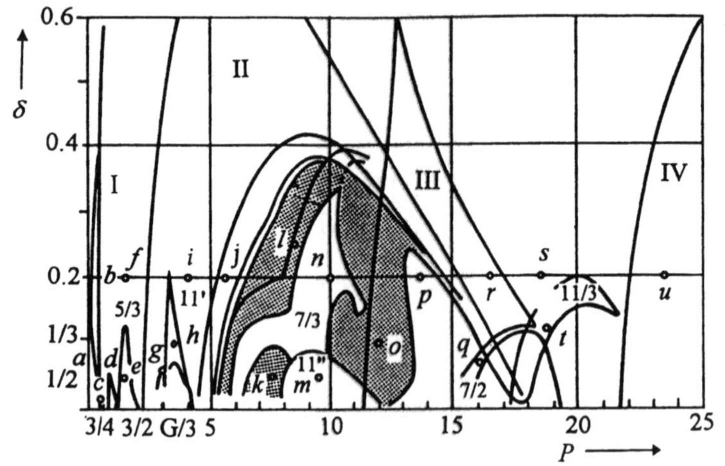


Fig. 8.41

### 8.6.2 The Distribution of Stationary Movement in Parameter Space

We hope to know what will be the final state of stationary movement (including chaos) when the parameter in a non-linear system changes. It is clear that this problem is very difficult to solve. One reason is that it is difficult to distinguish a chaotic movement, or at very least it takes much time to compute this. Another reason is that several attractors can exist at the same time, that is to say, starting from different initial conditions, the same system may have many different endings, or each attractor may have its own attractor region. There are two ways to study the distribution of stationary movements under all types of conditions. The third way is found in Chapter 9.

(1) Experimental Research

Y. Ueda has done a lot of experimental research using analogue computers, especially for the equation $x'' + \delta x' + x^3 = P\cos t$.

He has obtained some very detailed movement distribution sketches (see Fig. 8.41). This is often cited, even at the present date. However, it was determined by experiment, and is not very precise.

(2) Numerical Simulation

The standard method is used to perform a power spectrum analysis of the results of numerical simulation. We can also calculate the Liapunov index according to the results of the simulation. In this way, we can define chaotic motion well, but cannot give the period number of the periodic motions.

The steps of analysis are:

(1) To define the dynamical behaviour of some points in the parameter plane or curves by analyzing the numerical simulation, the power spectrum or the Liapunov index.

(2) To plot the bifurcation curves of periodic and quasi-periodic solutions in the parameter plane by the direct numerical method. In this case, not just the results in (1) are used as the starting point of the bifurcation curve, but also the knowledge of the behaviour of the solution, such as the coexistence of some solutions.

(3) To find the region of chaos by using the Melnikov method, when possible.

(4) To analyze the shape of basins by the cell mapping method when there are multi-attractors coexisting.

(5) To analyze the features of the chaos region by the spectrum analysis method, such as the existence of multi-slice chaos.

### 8.6.3 The Prospect of Chaos Research

Through rapid development over many years, the research on chaos has now entered the stage of steady development. At present, the achievements already made are being summarized and future work being assessed. The value of the development of theory lies in its application, now that there are many fields in which applications of chaos are being studied. The applications of chaos are called chaotic control. The key point of this application research is to arm it with new concepts and new methods of chaos theory to make more profound studies of the non-linear model and the evolution process in various fields, and thence to reach the aim of quantitative control and application of chaos. Opinions on the development of chaos research can be summed up as follows:

1. Application to the problem of turbulence. To understand the true mechanism of turbulence and to improve the developed turbulence, we must study its behaviour in time and space, that is, we cannot confine our study only to the change of time of a system with several degrees of freedom.

2. Research into the transition process. Most of the studies on chaos in the past belonged to the steady dynamic state. But the transition process may be more varied. The practical problem that we are studying is a kind of relative transition process. Limitation to infinite time is less important in these problems.

3. Further classification of the chaos process itself. There should be further classification of chaos movement. Strange attractors can have different levels of strangeness. The method of describing chaos attractors is still in the initial stage of research, particularly on how to this from experimental data.

4. Pattern recognition. The numerical study of some concrete models should be changed to bifurcation and the spectrum of chaos, that is, the whole structure in parameter space, combined with pattern recognition in quantity to all kinds of attractors, including the transition process. Various kinds of methods should be used in the numerical experiment.

5. Research into chaos must be done in such a way as to shift theory to practice and practice to theory constantly. On the one hand we should popularize the new concepts and the new methods that have been accumulated to all walks of life. Let chaotic control and application be extended gradually. On the other hand we should prove and improve various practical methods that the theory researchers have created mathematically.

Recently, S. Wiggins attempted to popularize and apply the Melnikov method and the Smale horse shoe theory to high-dimensional systems. This is also a new tendency.

# Chapter 9

# Construction of Chaotic Regions

A new method of numerical simulation is given in this chapter to describe the characteristics of the solutions of dynamical systems. The characteristics of the solutions, such as the number of solutions, the type and periodicity of the solutions and, more importantly, the existence of chaotic solutions, are of great interest in a physical parametric space. The boundaries separating qualitatively different solutions in a physical parametric space, constituting a characteristic diagram, are constructed below. The critical points in the initial bifurcation graph are used as initial approximations. Depending on the nature of the critical points, additional constraints are used to trace the neighbouring critical points by the method of parametric unfolding. As a result, the characteristics of the steady-state solutions can be read directly from the characteristic diagram. Non-linear governing differential equations arise from various types of physical and engineering problems. Owing to the presence of non-linearities, analytical solutions are rarely obtained. Apart from their deterministic nature, random-like solutions are observed for certain parameters and initial conditions. The occurrence of these chaotic motions has been studied by many authors.

Ueda [69] visually surveyed the chaotic solutions for a wide range of parameters with zero linear stiffness in the Duffing equation. Holmes and Moon [151] reviewed several examples of non-linear systems and the related mathematical models exhibiting chaotic motions to illustrate the sensitivity of the initial conditions. Tongue [152] showed that the existence of multiple static equilibria was not a necessary requirement for chaos to occur. Dowell and Pezeshki [153] established a relationship between chaotic oscillations and forced excitation by studying the dynamics of a buckled beam, and used physical experimental results to compare with the theoretical model results. Without exception, all the results were generated by using a time integration method. Tongue [154] showed that an improper choice of time step in the time integration can give spurious results with regard to the existence of chaotic motion. Since the non-existence of periodic solutions is of interest in the study of chaotic motion, the solution of the non-linear differential equation is reported completely by means of the incremental harmonic balance method in this chapter. Higher order subharmonic periodic solutions can be determined by including sufficient number of harmonic terms. Steady-state periodic solutions, both stable and unstable, are assumed initially to eliminate the time variable and the associated initial conditions. The response $u(t)$ is first replaced by its Fourier components and a set of non-linear algebraic equations expressed in terms of the Fourier components derived. The unknown Fourier coefficients are solved by the Newtonian algorithm, resulting in a set of linear algebraic equations in terms of the increments of the unknowns. The right-hand side to be eliminated is called the residual vector. The residual vector is in fact the unbalanced error of the previous approximation. The use of the residual vector keeps the solutions from error accumulation by truncation as well as round-offs. In the method both the total non-linear stiffness and the tangential stiffness are included so that large non-linear can be considered.

To start with, only one parameter is allowed to be changed to determine the initial boundary points. When the residual vector vanishes, an equilibrium steady-state results. Taking the increments, one obtains a linear scalar incremental differential equation with periodic varying coefficients which can be used to determine the stability properties of the solution by Floquet theory. By removing all the unstable branches, one obtains the bifurcation graph. When there is no stable equilibrium steady-state solution which corresponds to a gap in the bifurcation diagram, chaotic solution is the only possibility if the solution remains bounded [155]. The boundary points are used as an initial approximation for constructing the transition boundaries in the physical parametric space. The boundaries are extended to cover the parametric space of interest by simultaneous variation of parameters. As a result, the characteristics of the steady-state solutions can be read directly from the physical parametric diagram — the characteristic diagram.

## 9.1 Incremental Harmonic Balance Method (IHB Method)

In non-linear dynamic analyses, the IHB method is one of the best choices. This method does not make the assumption of small parameters and is capable of treating strongly non-linear systems. The IHB method is simple and convenient for computer implementation. In this method non-linear differential equations are transformed into a set of linearized incremental algebraic equations in terms of the Fourier coefficients, and so only linear equations have to be formed and solved iteratively in each incremental step. The IHB method with a variable parameter is ideally suited to parametric studies for obtaining solution diagrams. After obtaining the solution of a particular value of the parameter, this parameter can then be incremented. The new solution may be sought again by iteration using the previous solution as an approximation. By this mean, solution diagrams of a dynamic system can easily be traced. The IHB formulation of a generalised autonomous system with $n$ degrees of freedom is described in this chapter [149]. Generally a system of non-dimensional instantaneous equations can be expressed as follows:

$$\omega^2 \ddot{x}_i + f_i(x_1,\cdots,x_n,\dot{x}_1,\cdots,\dot{x}_n,\omega,\gamma) = 0 \qquad i = 1,\ldots,n \tag{9.1}$$

where a dot denotes differentiation with respect to the dimensionless time $\tau$ ( $=\omega t$) and $f_i(x_1,\ldots,x_n,\dot{x}_1,\cdots,\dot{x}_n,\omega,\gamma)$ is a non-linear function of the dimensionless oscillation frequency $\omega$, the time $t$, the parameter $\gamma$, the dependent variable $x_i$, and its velocity $\dot{x}_i$.

After the time transformation $\tau = \omega t$, the periodic functions, $f_1,\ldots, f_w$, will be of period $2\pi$. The procedure of the IHB method for seeking periodic solutions is divided into two main steps.

The first step is the Newton—Raphson procedure, where small increments (symbolised by $\Delta$) are added to the current solution (or a solution guessed at the start of the procedure) of eq. (9.1) i.e. $\omega_0$ , $\gamma_0$ and $x_{i0}$ ($i$ = 1,2,...,$n$) to obviously a neighbouring solution:

$$\begin{aligned} \omega &= \omega_0 + \Delta\omega \\ \gamma &= \gamma_0 + \Delta\gamma \qquad\qquad\qquad\qquad\qquad\qquad (9.2) \\ x_i &= x_{i0} + \Delta x_i \qquad\qquad i = 1,\ldots,n \end{aligned}$$

Expanding eq. (9.1) by the Taylor's series about the initial state and neglecting all the non-linear terms of small increments will yield the linearized incremental equation:

$$\omega^2 \Delta\ddot{x}_i + \sum_{j=1}^{n}\left(\frac{\partial f_i}{\partial \dot{x}_j}\right)_0 \Delta\dot{x}_j + \sum_{j=1}^{n}\left(\frac{\partial f_i}{\partial x_j}\right)_0 \Delta x_j$$

$$= -\left(\omega_0^2\ddot{x}_{i0} + f_{i0}\right) - \left(\left(\frac{\partial f_i}{\partial \omega}\right)_0 + 2\omega_0\ddot{x}_{i0}\right)\Delta\omega - \left(\frac{\partial f_i}{\partial \gamma}\right)_0 \Delta\gamma \qquad i = 1,2,\ldots,n \quad (9.3)$$

This variational equation is equivalent to the linear matrix equation for the increments

$$\omega^2 \Delta\ddot{X} + D\Delta\dot{X} + L\Delta X = R - \left(2\omega\ddot{X}_0 + Q\right)\Delta\omega - G\Delta\gamma \qquad (9.4)$$

where $\Delta X = \left[\Delta x_1, \Delta x_2, \cdots, \Delta x_n\right]^T$ (9.5)

$$X_0 = \left[x_{10}, x_{20}, \cdots, x_{n0}\right]^T \qquad (9.6)$$

The elements of the matrices L and D are given by

$$L_{ij} = \left(\frac{\partial f_i}{\partial x_j}\right)_0 \qquad (9.7)$$

$$D_{ij} = \left(\frac{\partial f_i}{\partial \dot{x}_j}\right)_0 \qquad (9.8)$$

and the elements of the vectors $Q$, $G$ and $R$ are

$$Q_i = \left(\frac{\partial f_i}{\partial \omega}\right)_0 \qquad (9.9)$$

$$G_i = \left(\frac{\partial f_i}{\partial \gamma}\right)_0 \qquad (9.10)$$

$$R_i = -\left[\omega_0^2\ddot{x}_{i0} + f_{i0}\right] \qquad (9.11)$$

In eq. (9.4), $R$ is a corrective term to prevent the incrementation process drifting away from the actual solution. Note that the derivatives $D$, $L$, $Q$ and $G$ are in general varying with time, and so eq. (9.4) represents a linear ordinary differential equation with time-varying coefficients.

The second step of the IHB method is the Galerkin procedure. Assuming that the unknown $x_0$ can be expanded into a truncated Fourier series

$$x_{i0} = \sum_{n=n_1,n_2,\cdots} \Delta a_{in}\cos n\tau + \sum_{m=m_1,m_2,\cdots} \Delta b_{im}\sin m\tau \qquad i = 1,\ldots,n \qquad (9.12)$$

where the progressions $n_1, n_2,$ and $m_1, m_2,$ are positive integers. They may be 0, 1, 2, ... or 1, 3, 5, ... etc., according to circumstances.

Similarly, $\Delta x$ is expanded as:

$$\Delta x_i = \sum_{n=n_1,n_2,\cdots} \Delta a_{in} \cos n\tau + \sum_{m=m_1,m_2,\cdots} \Delta b_{im} \sin m\tau \qquad i=1,\ldots,n \tag{9.13}$$

For conciseness, the functions $x_0$ and $\Delta x$ can be rewritten as:

$$x_{i0} = T\, a_{i0} \tag{9.14}$$

$$\Delta x_i = T\, \Delta a_i \qquad i=1,\ldots,n \tag{9.15}$$

where $T = \left[\cos n_1\tau, \cos n_2\tau, \ldots, \sin m_1\tau, \sin m_2\tau, \ldots\right]$ (9.16)

$$a_{i0} = \left[a^0_{in_1}, a^0_{in_2}, \ldots, b^0_{im_1}, b^0_{im_2}, \ldots\right]^T \qquad i=1,\ldots,n \tag{9.17}$$

$$\Delta a_i = \left[\Delta a_{in_1}, \Delta a_{in_2}, \ldots, \Delta b_{im_1}, \Delta b_{im_2}, \ldots\right]^T \qquad i=1,\ldots,n \tag{9.18}$$

Therefore,

$$X_0 = Y A_0 \tag{9.19}$$

$$\Delta X = Y \Delta A \tag{9.20}$$

where
$$Y = \begin{bmatrix} T & & & 0 \\ & T & & \\ & & \ddots & \\ 0 & & & T \end{bmatrix} \tag{9.21}$$

$$A_0 = \left[a_{10}, a_{20}, \ldots, a_{n0}\right]^T \tag{9.22}$$

$$\Delta A = \left[\Delta a_1, \Delta a_2, \ldots, \Delta a_n\right]^T \tag{9.23}$$

With these notations, we have

$$\dot{X}_0 = \dot{Y} A_0 \tag{9.24}$$

$$\ddot{X}_0 = \ddot{Y} A_0 \tag{9.25}$$

By substituting eqs. (9.19), (9.20), (9.24) and (9.25) into eq. (9.4), and applying the Galerkin procedure for one period, we get

$$\delta \Delta A^T \left\{ \int_0^{2\pi} Y^T \left[\omega_0^2 \ddot{Y} + D\dot{Y} + LY\right] \mathrm{d}\tau \right\} \Delta A$$

$$= \delta \Delta A^T \left\{ \int_0^{2\pi} Y^T R\, d\tau - \int_0^{2\pi} Y^T \left[Q + 2\omega_0 \ddot{Y} A\right] \mathrm{d}\tau\, \Delta\omega - \int_0^{2\pi} Y^T G\, \mathrm{d}\tau\, \Delta\gamma \right\} \tag{9.26}$$

Or simply,

$$k\, \Delta A = r + q\, \Delta\omega + g\, \Delta\gamma \tag{9.27}$$

where the matrix
$$k = \int_0^{2\pi} Y^T \left[\omega_0^2 \ddot{Y} + D\dot{Y} + LY\right] \mathrm{d}\tau \tag{9.28}$$

and the vectors

$$r = \int_0^{2\pi} \left[\omega_0^2 \dot{Y}^T \dot{Y} A_0 - Y^T F_0\right] d\tau \tag{9.29}$$

$$q = \int_0^{2\pi} \left[2\omega_0 \dot{Y}^T \dot{Y} A_0 - Y^T Q\right] d\tau \tag{9.30}$$

$$g = -\int_0^{2\pi} Y^T g d \, d\tau \tag{9.31}$$

Referring to eq. (9.27), the number of unknowns exceeds the number of equations by two; these two are the oscillation frequency and the parameter $\gamma$. Depending on the situation, the value of an unknown may be fixed. If the limit cycle is concerned, for example, one of the Fourier coefficient has to be fixed (e.g. $a_{i1}$=0 or $b_{i1}$=0). This will not have any consequences as it merely represents a shift of $x_i$ on the time axis. To make the number of equations equal to that of the unknowns, $\gamma$ has to be taken as the control parameter (i.e. $\Delta\gamma = 0$) while $\Delta\omega$ should be regarded as an unknown.

With the column corresponding to coefficient $b_{11}$ say, replaced by $q$ in $k$ and $\Delta b_1$ replaced by $\Delta\omega$ in $\Delta c$, we obtain

$$k' \Delta A' = r \tag{9.32}$$

where $k'$ and $\Delta A'$ are the corresponding matrix $k$ and vector $\Delta A$ respectively after the replacement. Equation (9.32) is now ready to be solved by any method. A good approximation can be achieved by iteration. The iterative process is carried out with updated $k'$ and $r$ each time until $\delta$ is less than a given tolerance $T$, where $\delta$ is defined by

$$\delta = \frac{\|\Delta A'\|}{\max(A_i')} \tag{9.33}$$

in which $\|\Delta A'\|$ denotes the norm of the amplitudes $\Delta A'$. After obtaining the solution for a particular $\gamma$, it is then added by increment, $\Delta\gamma$ (known as the control increment). This is known as an argumentation process. The iterative process is repeated again using the previous solution as an approximation until the solution for the neighbouring $\lambda$ is sought. By this mean, the neighbouring solution can be sought. By successive use of argumentation and iteration processes, a solution diagram may be traced.

Alternatively, the frequency $\omega$, or one of the Fourier coefficients $a_{ij}$ and $b_{ij}$ may be used as the control increment in the above argumentation process. The choice of the control increment depends on the rate of the variation of the increments; usually the one with the greatest increment (absolute values) will be chosen as the control increment. This is extremely useful when a solution diagram is very sensitive to the variation of parameter $\gamma$ or similar. The number of steps required to trace the solution diagram will be reduced. The harmonic balance method is employed to transform the ordinary differential equation in the time domain into a set of non-linear algebraic equations in terms of the Fourier coefficients of the periodic response. The set of non-linear algebraic equations is solved by the Newtonian algorithm. The gradient matrix is obtained by taking increments of the original equation and the harmonic balance method is then applied to the incremental equations again. This is known as the incremental harmonic balance method [155]. The resulting linear algebraic equations in terms of the increments of the Fourier coefficients can be solved by any method and the coefficients are updated until convergence to a periodic solution is reached.

Consider an ordinary differential equation in the form

$$\phi(\ddot{u},\dot{u},u,\lambda) - f\cos t = 0 \tag{9.34}$$

where $\phi$ is analytic, $u(t)$ is the unknown response, $\{\lambda\}$ is a vector of system parameters and $f$ is the amplitude of harmonic excitation. A dot denotes a derivative with respect to the non-dimensional time $t$. Equation (9.34) can easily be generalised to periodic excitation by replacing $f\cos t$ by $[1 \ \cos t \ \cos 2t \quad ]\{f\}$. Since periodic solutions are of interest, the following expansion is admissible

$$u(t) = \sum_{i=0}^{\infty} a_i \cos\frac{it}{m} + \sum_{i=0}^{\infty} b_i \sin\frac{it}{m} \tag{9.35}$$

where $a_i$ and $b_i$ are the unknown Fourier coefficients and $m$ is an integer representing the order of subharmonic resonance so that $u(t)$ is of period $2m\pi$. For primary and superharmonic resonance $m=1$. The period is longer for higher subharmonics. When $m$ is large enough, the corresponding periodic solutions are numerically unstable as well as physically unstable. Non-periodic chaotic motion results. It will be shown later in dealing with the Duffing equation that chaotic motion occurs when the periodic solutions bifurcate from $m=4$ to $m=8$ and to higher order subharmonics at an accelerated rate. The motion regains its periodic stability when $m=3$ has solutions. Therefore, to include $m$ in eq. (9.35) is of primary importance in chaotic region construction, as well as the construction of transition boundaries.

When substituting eq. (9.35) into eq. (9.34) and collecting similar harmonic terms, one has

$$\sum_{i=0}^{\infty} R_i^c \cos\frac{it}{m} + \sum_{i=1}^{\infty} R_i^s \sin\frac{it}{m} = R(t) = 0 \tag{9.36}$$

where $R(t)$ is the residue which is zero for an actual solution and non-zero for an approximation. $R_i^c$ and $R_i^s$ are its Fourier components. The harmonic balance method requires one to multiply eq. (9.36) by $\cos\frac{jt}{m}$ and $\sin\frac{jt}{m}$ respectively and integrate over $t$ from zero to $2m\pi$ so that:

$$\int_0^{2m\pi} R(t)\cos\frac{jt}{m}\,dt = 0 \quad \text{and} \quad \int_0^{2m\pi} R(t)\sin\frac{jt}{m}\,dt = 0 \tag{9.37}$$

After carrying out the integration, one obtains

$$[K(q,\lambda)]\{q\} - f\{F\} = \{R\} = \{0\} \tag{9.38}$$

where $\{R\} = [R_0^c, R_1^c, R_2^c, \ldots, R_1^s, R_2^s, \ldots]^T$, $\qquad \{q\} = [2a_0, a_1, a_2, \ldots, b_1, b_2, \ldots]^T$,

$\{F\}$ is the unit column vector with unity at the $m$th position, and $[K(q,\lambda)]$ is the total stiffness matrix, and can be evaluated explicitly when the form of $\phi(\ddot{u},\dot{u},u,\lambda)$ is given. It is non-symmetrical in general but symmetrical for non-gyroscopic conservative systems.

## 9.2 The Newtonian Algorithm

To solve eq. (9.38) by the Newtonian algorithm, one takes the following multi-parametric increments on eq. (9.34), $\ddot{u}$ to $\ddot{u}+\Delta\ \ddot{u}$, $\dot{u}$ to $\dot{u}+\Delta\ \dot{u}$, $u$ to $u+\Delta\ u$, $\lambda$ to $\lambda+\Delta\ \lambda$, and $f$ to $f+\Delta f$, and hence

$$\phi(\ddot{u}+\Delta\ddot{u},\dot{u}+\Delta\dot{u},u+\Delta u,\lambda+\Delta\lambda)-(f+\Delta f)\cos t=0$$

After neglecting higher order terms of smallness, one has

$$\frac{\partial\phi}{\partial\ddot{u}}\Delta\ddot{u}+\frac{\partial\phi}{\partial\dot{u}}\Delta\dot{u}+\frac{\partial\phi}{\partial u}\Delta u+\{\Delta\lambda\}^T\frac{\partial\phi}{\partial\lambda}-\Delta f\cos t+R(t)=0 \tag{9.39}$$

where $R(t)$ is given by eq. (9.36). It is essential to keep the residue as $u(t)$ may just be an approximation to eq. (9.36). Since the increment of $u(t)$ is also periodic, the Fourier expansion on $\Delta u(t)$ is admissible,

$$\Delta u(t)=\sum_{i=0}^{\infty}\Delta a_i\cos\frac{it}{m}+\sum_{i=1}^{\infty}\Delta b_i\sin\frac{it}{m} \tag{9.40}$$

Substituting eq. (9.40) into eq. (9.39) and applying the harmonic balance method, one obtains, after integration:

$$[k(q,\Delta q,\lambda)]\{\Delta q\}-[A(q,\Delta q,\lambda)]\{\Delta\lambda\}-\Delta f\{F\}+\{R\}=\{0\} \tag{9.41}$$

Here $\{\Delta q\}=[2\Delta a_0, \Delta a_1, \Delta a_2,..., \Delta b_1, \Delta b_2,...]^T$, and $\{F\}$ and $\{R\}$ are the same as in eq. (9.38). $[k(q, \Delta q, \lambda)]$ is the tangential matrix and $[A(q, \Delta q, \lambda)]$ is the parametric gradient matrix which can be evaluated explicitly when the form $\phi(\ddot{u},\dot{u},u,\lambda)$ is given. Equation (9.41) is the product of the incremental harmonic balance method.

Suppose both $\{\Delta F\}$ and $\{\Delta\lambda\}$ are zero. If $\{q\}$ is an exact solution of eq. (9.38), the residual vector $\{R\}$ vanishes and eq. (9.41) ensures that the increment $\{\Delta q\}$ vanishes. If $\{q\}$ is an approximation, $\{\Delta R\}$ is evaluated from eq. (9.38) and $\{\Delta q\}$ is solved from eq. (9.41). A better approximation is given by $\{q+\Delta q\}$ instead of $\{q\}$, as guaranteed by the nature of the Newtonian algorithm. The process is repeated until $\{R\}$ is sufficiently small and a solution is obtained. $\{R\}$ corresponds to the function evaluation in the usual Newtonian algorithm and $[k]$ corresponds to the Jacobian matrix. The above process is called a Newtonian iteration, or simply an iteration.

Suppose the system parameters vary slightly so that $\{\Delta F\}$ and $\{\Delta\lambda\}$ are zero. By using the previous solution as an approximation, the solution increment is obtained by one iteration. The process is called an argumentation. By applying the corresponding increments to the system parameters and the approximate solution, the solution for the new system parameters is obtained by iterations. An alternative application of argumentations and iterations can be used for parametric studies.

In fact, the prescribed increments are not restricted to the system parameters. One of the Fourier coefficients can be specified. It is called a control parameter. One of the system parameters is dependent on the other and is solved from eq. (9.41).

## 9.3 Number of Harmonic Terms

In eqs. (9.35) and (9.36), only a finite number of harmonic terms can be used. In general, the harmonic terms used in these equations are the same. This approximation is equivalent to the Galerkin method with trigonometric functions as the complete basis functions. The mathematical theories of the existence, error bound and stability analysis can be found in [157, 158]. Mickens [159] showed that the upper bounds of the absolute magnitudes of the harmonic coefficients decrease exponentially. Van Dooren [160] studied the transition from regular periodic motion to chaotic motion in Duffing oscillator with soft characteristics and found the corresponding constants for the existence of unique solutions, error bounds and the stability properties. In the present discussion, the non-vanishing residual terms that are included in eq. (9.36) are examined to prevent large unbalanced residual errors. When large unbalanced residual terms are found, the corresponding harmonic terms are included in eq. (9.35) for annihilation.

## 9.4 Stability Characteristics

When a solution is obtained, its stability characteristics can be checked by means of the Floquet theory. Unstable solutions are not physically observable. When the solution is perturbed by $\Delta u$, the equation is

$$\frac{\partial \phi}{\partial \ddot{u}} \Delta \ddot{u} + \frac{\partial \phi}{\partial \dot{u}} \Delta \dot{u} + \frac{\partial \phi}{\partial u} \Delta u = 0 \tag{9.42}$$

which is linear in $\Delta u$ but with periodic varying coefficients in general. Rewrite eq. (9.42) as

$$\{\dot{x}\} = [B(t)]\{x\} \tag{9.43}$$

where $\{x\} = [\Delta \dot{u} \Delta u]^T$, and $[B] = \begin{bmatrix} -\dfrac{\partial \phi / \partial \dot{u}}{\partial \phi / \partial \ddot{u}} & -\dfrac{\partial \phi / \partial u}{\partial \phi / \partial \ddot{u}} \\ 1 & 0 \end{bmatrix}$ is periodic with period $T$.

The stability of eq. (9.43) is checked by evaluating the eigenvalues of the transformation matrix $[Q]$ which transforms the state vector $\{x_n\}$ at $t=nT$ to $\{x_{n+1}\}$ at $t=(n+1)T$. If the absolute magnitudes of the two eigenvalues are less than unity, the solution is stable. For any small perturbation $\Delta u$, it would be diminishing. The explicit form of $[Q]$ can be written as [161]

$$[Q] = \begin{bmatrix} Q_{11} & Q_{12} \\ Q_{21} & Q_{22} \end{bmatrix} = \prod_{i=1}^{N} \exp\left(\Delta t[B(i\Delta t)]\right), \qquad \Delta t = T/N,$$

where $N$ is the number of divisions used. The two eigenvalues are given by $\lambda_{1,2}=[-\beta \pm \sqrt{(\beta^2 - 4\gamma)}\,]/2$, where $\beta=-(Q_{11}+Q_{22})$, $\gamma=Q_{11}Q_{22}-Q_{12}Q_{21}$.

It can be shown that the moduli of the complex eigenvalues are equal to

$$\exp\left(-\int_0^T \left\{\frac{\partial \phi / \partial \dot{u}}{\partial \phi / \partial \ddot{u}}\right\} dt\right)$$

For dissipative systems, the moduli of the complex eigenvalues are less than unity and the solutions with complex eigenvalues are stable. Instability may be due to a positive real eigenvalue exceeding unity; then eq. (9.43) predicts an additional small solution with a period $T$ and a growth of amplitude monotonically without bound. The presence of the non-linearity results merely in a shift of the equilibrium position. When the instability is due to a negative real eigenvalue exceeding unity, eq. (9.43) predicts an additional small solution with period $2T$ and a growth of amplitude without bound. The presence of the non-linearity results in a change of the equilibrium position with the period being doubled. Therefore, at the transition boundary points, the eigenvalues are $+1$ or $-1$ for dissipative systems.

The eigenvalues also relate to the time required in the time integration method for the transient to die down. When the magnitude of an eigenvalue is much smaller than unity, it attracts the neighbouring trajectories quickly. When the magnitude of an eigenvalue is much larger than unity, it pushes the neighbouring trajectories away quickly. when the magnitude of an eigenvalue is around unity, it shows little attractive or repulsive power and the transition period can be very long. Near the bifurcation points, the moduli of the eigenvalues are around unity and the time integration method converges very slowly.

## 9.5 Transition Sets in Physical Parametric Space

The characteristics of the solutions, such as the number of solutions, the type and periodicity of the solutions and, more importantly, the existence of chaotic solutions, in a physical parametric space are of interest. Instead of searching the whole parametric space of interest point by point, the transition boundaries are constructed here. A parametric diagram is the end product. The system bifurcation is investigated and can be read directly from the parametric diagram. For a two-parameter system, the parametric diagram is a plane with transition curves dividing regions with qualitatively different characteristics.

The boundaries separating qualitatively different solutions in fact represent additional constraints among the system parameters. By writing the constraint in an incremental form with the residual term defining the constraint relation and combining with eq. (9.41), one of the system parameters is dependent on the others. By varying other system parameters, the boundaries in the parametric space are constructed. When a solution changes from stable to unstable, with an eigenvalue $-1$, a new stable solution would branch out having a period twice the original. Near the branching point, the newly bifurcated subharmonic components are very small as compared with the others. At the bifurcation point, the tangential matrix $[k]$ is singular when the additional bifurcated subharmonic components are taken into account. At least two sets of increments can be found for a small parametric increment, one leads to the originally stable but now unstable branch and the other leads to the new stable bifurcated branch.

In constructing these stable subharmonic solutions, the original solution is rewritten in an extended form with the additional subharmonic coefficients being zero. This solution is then used as an initial approximation with one of the subharmonic coefficients as an active parameter. In constructing the bifurcation boundaries, one of the subharmonic coefficients is kept at a very small value instead of being zero during the iteration to avoid singular gradient matrices.

Alternatively, a solution can become unstable by passing a fold, that is, some parameters reach an extremity. In such a case, a large change of amplitude would result in an infinitesimal change in parameters. In this case, the tangential matrix $[k]$ is singular with zero determinant. This can be used in addition to eq. (9.41) to solve for the unknown Fourier increments and to determine one of the system parameter increments. The additional equation is

$$\det[k(q,\lambda)] = \delta = 0 \tag{9.44}$$

The incremental form can be written as

$$\{\Delta q\}^T \{\partial\delta / \partial q\} + \{\Delta \lambda\}^T \{\partial\delta / \partial\lambda\} + \delta = 0 \tag{9.45}$$

and the incremental coefficients $\{\partial\delta / \partial q\}, \{\partial\delta / \partial\lambda\}$ can be determined as follows. Without loss of generality, consider a matrix $[D(\lambda)]$ that depends on a parameter $\lambda$ only. It is decomposed into a lower triangular matrix $[L]$ with unit diagonal and an upper triangular matrix $[U]$ so that

$$[D(\lambda)] = [L(\lambda)][U(\lambda)] \tag{9.46}$$

where $U_{ij} = \left( D_{ij} - \sum_{k=1}^{i-1} L_{ik} U_{kj} \right), \quad j \geq i$ and $L_{ij} = \left( D_{ij} - \sum_{k=1}^{j-1} L_{ik} U_{kj} \right) \Big/ U_{jj}, \quad j < i$

The determinant is given by

$$\delta = \prod U_{ii} \tag{9.47}$$

Differentiating eq. (9.46) with respect to $\lambda$,

$$[D'(\lambda)] = [L'(\lambda)][U(\lambda)] + [L(\lambda)][U'(\lambda)] \tag{9.48}$$

where the elements $L'_{ij}$ and $U'_{ij}$ are given by

$$U'_{ij} = \left( D'_{ij} - \sum_{k=1}^{i-1} L'_{ik} U_{kj} - \sum_{k=1}^{i-1} L_{ik} U'_{kj} \right), j \geq i$$

and $$L'_{ij} = \left( D'_{ij} - \sum_{k=1}^{j-1} L'_{ik} U_{kj} - \sum_{k=1}^{j} L_{ik} U'_{kj} \right) \Big/ U_{jj}, j < i$$

The derivative of the determinant is $\delta' = \partial\delta/\partial\lambda = \sum_i U'_{ii} \prod_{j \neq i} U_{jj}$. For a multi-parameter system, $\{\partial\delta/\partial q\}$ and $\{\partial\delta/\partial\lambda\}$ can be determined similarly because the corresponding derivative of $[k]$ can be evaluated accordingly.

## 9.6 Example Duffing Equation with Multi-Harmonic Excitation

Consider, as an illustrative example, the Duffing equation subjected to two harmonic excitation forces in the form

$$\ddot{u} + 2\mu\dot{u} + ku + \alpha u^3 = f_1 \cos p\omega t + f_2 \cos q\omega t$$

where $\mu, k, \alpha, \omega, f_1$ and $f_2$ are the damping coefficient, linear stiffness coefficient, non-linear stiffness coefficient, excitation frequency, and the two forcing amplitudes respectively. Assuming that $p$ is even and $q$ is odd, and they are relatively prime. The period of the excitation is $T=2\pi/\omega$. It is noted that if $u(t)$ is a solution for $(f_1, f_2)$, then $u(t+T/2)$ corresponds to $(f_1, -f_2)$. As a result, for a given $f_1$, only the positive values of $f_2$ have to be investigated. If $f_2=0$, one could assume $p=1$ and if $u(t)$ is a solution for $f_1$, $u(t=T/2)$ is a solution for $-f_1$ so that only the positive values of $f_1$ need investigation.

The contributions to eq. (9.38) are:

$$\{\lambda\} = [\mu, k, \alpha, \omega]^T,$$
$$\{F\} = f_1\{e_{mp+1}\} + f_2\{e_{mq+1}\},$$

$\{e_r\}$= unit column vector with a unity at the r-th position,

$$[K] = \frac{1}{2}\begin{bmatrix} \frac{1}{2}K_0^c & K_c & K_s \\ K_c^T & K^{cc} & K^{cs} \\ K_s^T & K^{sc} & K^{ss} \end{bmatrix} + \begin{bmatrix} \frac{1}{2}k & 0 & 0 \\ 0 & -\omega^2[J^2]+k[I] & 2\mu\omega[J] \\ 0 & -2\mu\omega[J] & -\omega^2[J^2]+k[I] \end{bmatrix}$$

$$[K_c] = \text{row}[K_1^c \; K_2^c \ldots],$$
$$[K_s] = \text{row}[K_1^s \; K_2^s \ldots],$$
$$[J^2] = \text{diag}[1^2, 2^2, 3^2, \ldots],$$
$$[J] = \text{diag}[1, 2, 3, \ldots],$$

and the elements of $[K^{cc}]$, $[K^{cs}]$, $[K^{sc}]$ and $[K^{ss}]$ are given respectively by

$$K_{ij}^{cc} = K_{i+j}^c + K_{|i-j|}^c,$$
$$K_{ij}^{cs} = K_{i+j}^s + \text{sgn}(j-i)K_{|i-j|}^s,$$
$$K_{ij}^{sc} = K_{i+j}^s + \text{sgn}(i-j)K_{|i-j|}^s,$$
$$K_{ij}^{ss} = -K_{i+j}^c + K_{|i-j|}^c,$$

where sgn($x$)=–1 if $x$<0, =0 if $x$=0, and =1 if $x$>0, and $\delta_{ij}$=0 if $i \neq j$, and =1 if $i$=$j$.

Also $K_0^c = \alpha\left(\sum_{i=0} a_i^2 + \sum_{i=0} b_i^2\right),$

$$K_i^c = \frac{\alpha}{2}\left(\sum a_k a_{k+i} + \sum a_k a_{|i-k|} + \sum b_k b_{k+i} - \sum \text{sgn}(i-k) b_k b_{|i-k|}\right),$$
$$K_i^s = \alpha\left(\sum a_k b_{k+i} + \sum \text{sgn}(i-k) a_k b_{|i-k|}\right)$$

The incremental equation is:

$$\Delta\ddot{u} + 2\mu\Delta\dot{u} + k\Delta u + 3\alpha u^2 \Delta u + R(t) = 0$$

and the contributions to eq. (9.41) are:

$$[k]=\frac{3}{2}\begin{bmatrix}\frac{1}{2}K_0^c & K_c & K_s\\ K_c^T & K^{cc} & K^{cs}\\ K_s^T & K^{sc} & K^{ss}\end{bmatrix}+\begin{bmatrix}\frac{1}{2}k & 0 & 0\\ 0 & -\omega^2[J^2]+k[I] & 2\mu\omega[J]\\ 0 & -2\mu\omega[J] & -\omega^2[J^2]+k[I]\end{bmatrix},$$

$$[A]=[A_\mu\ A_k\ A_\alpha\ A_\omega],\ [A_\mu]=[A_{\mu0}^c\ A_{\mu1}^c\cdots A_{\mu1}^s\cdots]^T,\ [A_k]=[A_{k0}^c\ A_{k1}^c\cdots A_{k1}^s\cdots]^T,$$

$$[A_\alpha]=[A_{\alpha0}^c\ A_{\alpha1}^c\cdots A_{\alpha1}^s\cdots]^T,\ [A_\omega]=[A_{\omega0}^c\ A_{\omega1}^c\cdots A_{\omega1}^s\cdots]^T,$$

$$A_{\mu0}^c=0,\ A_{\mu i}^c=2i\omega b_i,\ A_{\mu i}^s=-2i\omega a_i,\ \ A_{ki}^c=a_i,\ A_{ki}^s=b_i,$$

$$u^3(t)=\sum A_{\alpha i}^c\cos i\omega t+\sum A_{\alpha i}^s\sin i\omega t,$$

$$A_{\omega0}^c=0,\ A_{\omega i}^c=-2i^2\omega a_i+2i\mu b_i,\ A_{\omega i}^s=-2i^2\omega b_i-2i\mu a_i$$

The explicit derivatives of $[k]$ with respect to the system parameters are:

$$\frac{\partial[k]}{\partial\mu}=\begin{bmatrix}0 & 0 & 0\\ 0 & 0 & 2\omega[J]\\ 0 & -2\omega[J] & 0\end{bmatrix},\quad \frac{\partial[k]}{\partial\omega}=\begin{bmatrix}0 & 0 & 0\\ 0 & -2\omega[J^2] & 2\omega[J]\\ 0 & -2\omega[J] & -2\omega[J^2]\end{bmatrix},$$

$$\partial[k]/\partial k=\mathrm{diag}\left[\frac{1}{2}[I]\ [I]\right],$$

$\partial[k]/\partial\{q\}$ and $\partial[k]/\partial\alpha$ can be obtained by differentiating $K_0^c$, $K_i^c$ and $K_i^s$ appropriately:

$$\frac{\partial K_0^c}{\partial\alpha}=\left(\sum_{i=0}a_i^2+\sum_{i=0}b_i^2\right),$$

$$\frac{\partial K_i^c}{\partial\alpha}=\frac{1}{2}\left(\sum a_k a_{k+i}+\sum a_k a_{|i-k|}+\sum b_k b_{k+i}-\sum\mathrm{sgn}(i-k)b_k b_{|i-k|}\right),$$

$$\frac{\partial K_i^s}{\partial\alpha}=\sum a_k b_{k+i}+\sum\mathrm{sgn}(i-k)a_k b_{|i-k|},$$

$$\frac{\partial K_0^c}{\partial a_k}=2\alpha a_k,\quad \frac{\partial K_i^c}{\partial a_k}=\alpha\left(a_{i+k}+a_{|i-k|}\right),\quad \frac{\partial K_i^s}{\partial a_k}=\alpha\left(b_{k+i}+\mathrm{sgn}(i-k)b_{|i-k|}\right),$$

$$\frac{\partial K_0^c}{\partial b_k}=2\alpha b_k,\quad \frac{\partial K_i^c}{\partial b_k}=\alpha\left(b_{k+i}-\mathrm{sgn}(i-k)b_{|i-k|}\right),\quad \frac{\partial K_i^s}{\partial b_k}=\alpha\left(-a_{i+k}+a_{|i-k|}\right),$$

The stability characteristics can be studied by the Floquet theory, with

$$[B(t)]=\begin{bmatrix}-2\mu & -k-3\alpha u^2\\ 1 & 0\end{bmatrix}$$

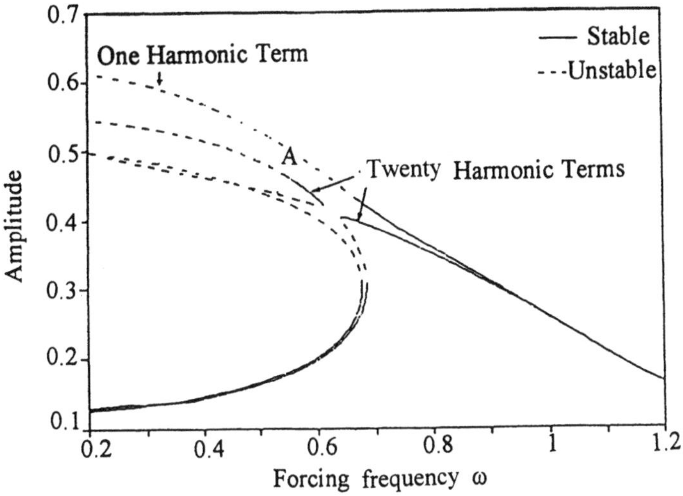


amplitude frequency curves of Duffing equation
($\mu$=0.2, $k$=1.0, $\alpha$=-4.0, $f_1$=0.115)

Fig. 9.1(a)

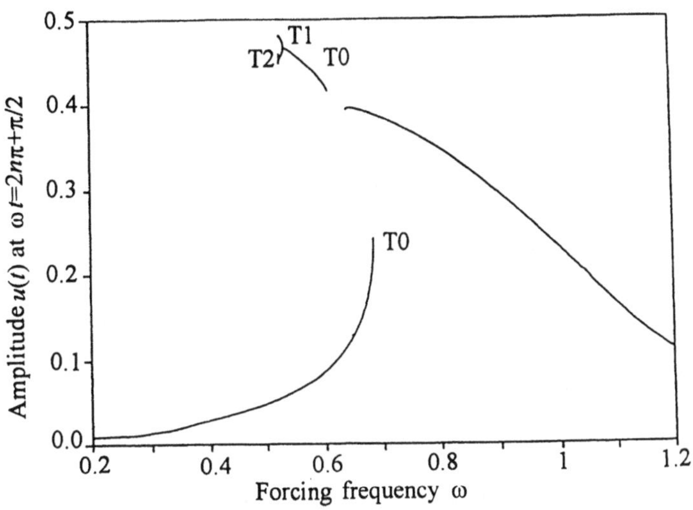


Bifurcation graph of Duffing equation
with stable branches only
($\mu$=0.2, $k$=1.0, $\alpha$=-4.0, $f_1$=0.115)

Fig. 9.1(b)

### 9.6.1 Simple Harmonic Excitation

For simple harmonic excitation, let $f_2$=0 and $p$=1. The number of harmonic terms used depends on the order of the subharmonic solutions under consideration. The number of harmonic terms is 20$m$+1 initially. The number of terms increases when large unbalanced residual terms are detected. The harmonic terms with large residual terms are included for annihilation. For symmetrical solutions, only odd harmonic terms are required.

#### 9.6.1.1 Bifurcation Graph for $\omega$

Consider the non-linear dynamics of particles in an external field [163]. The equation of motion is the softening type Duffing equation with coefficients $\mu$ = *0.2*, $k$ = 1.0, $\alpha = -4.0$ and $f_1$ = 0.115. The frequency is regarded as a variable. Van Dooren [160] used the Garlerkin method to evaluate the symmetric periodic solutions with high accuracy using the one-harmonic solutions as initial approximations. By constructing the amplitude frequency curves parametrically, these two sets of solutions are qualitatively different as shown in Fig. 9.1a. The difference was not noted by Van Doonren as the method used was point searching in contrast with the present method of continuation. The turning points at = 0.610 and 0.642 are indicated by the singular tangential matrix evaluated at these points. The stable and unstable solutions are marked with solid and broken lines respectively. For frequency between 0.610 to 0.642, only one stable solution is possible.

At point $A$, when = 0.537, the original stable solutions lose their stability with an eigenvalue exceeding unity. Asymmetric periodic solutions with all harmonic terms are obtained. As the frequency decreases, the asymmetrical solutions lose their stability at = 0.530 with an eigenvalue exceeding minus one. Stable solutions with period twice the original are obtained. The frequency range for the stable subharmonic solutions is narrow. Figure 9.1b shows the bifurcation graph with the amplitude at $t = 2n\pi+\pi/2$ being used as the reference amplitude. Only the stable branches are shown. T0, T1 and T2 denote symmetrical solutions, asymmetrical solutions and second-order subharmonic solutions respectively. The unstable solutions do not regain their stability as the frequency decreases. Chaotic solutions are found to be co-existing with the small amplitude symmetrical solutions when the frequency is just below the bifurcating frequency. The steady-state solution will depend on the initial conditions.

#### 9.6.1.2 Bifurcation Graph for $f_1$, with $f_2$=0.0

Consider the Duffing equation with $\mu$ =0.1, $k$=0.0, $\alpha$=1.0, $\omega$=1.0, $p$=1 and $f_2$=0. The period of the excitation is $2\pi$. The bifurcation graph is shown in Fig. 9.2a and the critical points are shown in Table 9.1. The amplitude of $u(t)$ at $t=2n\pi$ is used as the reference amplitude. The sign of $u(t)$ is not removed so that a negative amplitude is possible. Figures 9.2b to 9.2k show the trajectories on the phase plane for various excitation forces with the points at $t=2n\pi$ marked by small circles. The stable and unstable trajectories are shown by solid lines and dotted lines respectively.

Initially, when the forcing amplitude is small, $u(t)$ resembles the linear solution with zero stiffness. As the forcing amplitude increases, the response amplitude increases. The solution becomes unstable at $f_1$=0.46 due to the presence of non-linearity. The unstable branch continues with a decrease of forcing amplitude and a drop of response amplitude due to the shift of the phase angle difference between the response and the excitation. The solution regains its stability at $f_1$=0.23. Two stable solutions and one unstable solution are possible when $0.23 \le f_1 \le 0.46$ as shown in Fig. 9.2c. Alternatively, this can be regarded as an outbirth of a saddle node at $f_1$=0.23 giving stable and unstable branches. While the stable branch varies with the control parameter, the unstable branch collides with the other stable branches to form a saddle node. This is called a tangential bifurcation or a saddle node bifurcation. This kind of folding structure gives a jump phenomenon as the forcing amplitude is monotonically increasing or decreasing. Within this range, either of the two stable solutions is observable depending on the initial conditions.

All coefficients associated with the even harmonic terms are zero so far. As the forcing amplitude increases further, the odd harmonic (symmetrical) solutions become unstable at $f_1$=2.40. The instability results from a positive eigenvalue so that no period doubling occurs. This is checked by the vanishing of the determinant of $[k]$ in eq. (9.41) at $f_1$=2.40 by including the constant and even harmonic terms. The bifurcation is due to the birth of stable asymmetrical solutions having both odd and even harmonic terms. Two branches of solutions $u_1(t)$ and $u_2(t)$ are emitted. These two solutions have the relationship of $u_2(t) = -u_1(t+\pi)$. As the forcing amplitude increases, the solutions have unstable branches when $3.26 \le f_1 \le 3.29$. The solutions become unstable at $f_1$=5.39 with a negative eigenvalue exceeding minus one. This results in a period doubling and the solutions have a period $4\pi$. The tangential matrix $[k]$ with zero second order subharmonic Fourier coefficients ($m$=2) is singular at this point. The subharmonic coefficients associated with $m$=2 are non-trivial. The bifurcation continues as the forcing amplitude increases. Higher order subharmonic solutions emerge and the response becomes much more difficult to trace since the higher order subharmonic coefficients are several orders smaller than the lower order ones. When the forcing amplitude is further increased, no stable periodic solution is found and chaotic motions are observed.

In the range of $6.5 \le f_1 \le 7.85$, no higher order stable periodic solution can be found. In fact, the numerical time integration method shows that the solutions are chaotic in this range. As the excitation force increases beyond 7.85, a stable third-order subharmonic solution is born. Competitive solutions are co-existing and the final settlement depends on the initial conditions. Near the bifurcation points, the eigenvalue is near unity and the convergence rate is extremely slow if the time integration method is used. Over 5000 cycles are required to obtain the steady-state solution. The difficulty is overcome by the harmonic balance method here since the Fourier coefficients are determined directly and high accuracy can be obtained by including more harmonic terms. Also, for higher order subharmonic solutions, the stable and unstable solutions are closely placed in the phase space. It is difficult to construct the trajectories efficiently by the time integration method. The singular points (turning points and bifurcation points) are indicated by the singularity of the tangential matrix $[k]$ and can be determined with sufficient accuracy.

For larger damping coefficients, the bifurcation graph with $\mu$=0.2 is shown in Fig. 9.3a, with the solution amplitude at $t=2n\pi$ used as a reference amplitude. Only the stable solutions are shown by the solid line. Fig. 9.3b and 9.3c show the trajectories on the phase plane for various excitation forces with the points at $t=2n\pi$ marked by small circles. The stable and unstable trajectories are shown by solid lines and broken lines respectively. In Fig. 9.3a, the solutions have a maximum period of $4\pi$ and merge to the period $2\pi$ solutions after the bifurcation. It is also noted that no period $6\pi$ solution is obtained. The notation $xTy$ is used to name a solution in which $x$ is a uniquely assigned solution number and $y$ is the periodicity of the solution.

**Table 9.1** Critical points for Fig. 9.2a

$\omega$=0.5, μ=0.1, $k$=0.0, $\alpha$=1.0, $f_2$=0.0

| $f_1$ | |
|---|---|
| 0.0000 | stable 1T0 |
| 0.2284 | stable 2T0 and unstable 3T0 born |
| 0.4614 | stable 1T0 and unstable 3T0 collide |
| 2.3982 | 2T0 to 2T1 |
| 3.2574 | stable 4T1 and unstable 5T1 born |
| 3.2917 | stable 2T1 and unstable 5T1 collide |
| 5.3941 | 4T1 to 4T2 |
| 6.2603 | 4T2 to 4T4 |
| 7.8542 | stable 6T3 and unstable 7T3 born |
| 10.0000 | stable 6T3, unstable 2T0, 4T1, 4T2, 4T4, 7T3 |

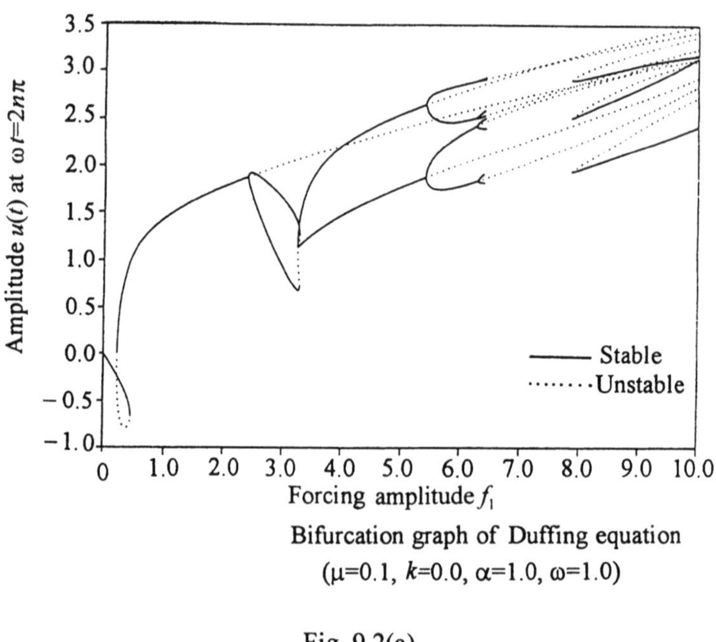


Bifurcation graph of Duffing equation
(μ=0.1, $k$=0.0, α=1.0, ω=1.0)

Fig. 9.2(a)

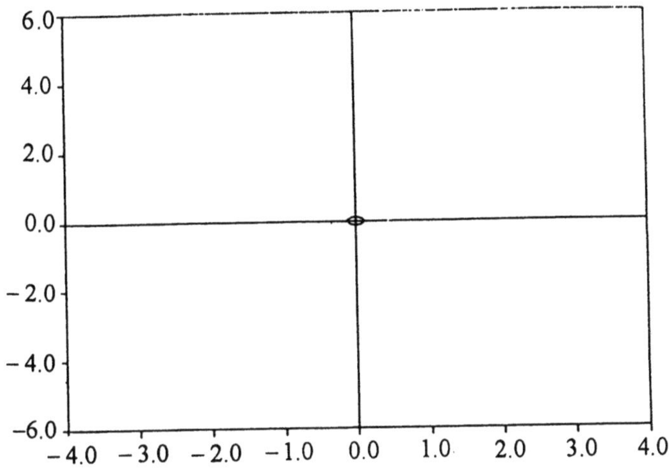


Phase trajectory of Duffing equation
($\mu$=0.1, $k$=0.0, $\alpha$=1.0, $\omega$=1.0, $f_1$=0.1)

Fig. 9.2(b)

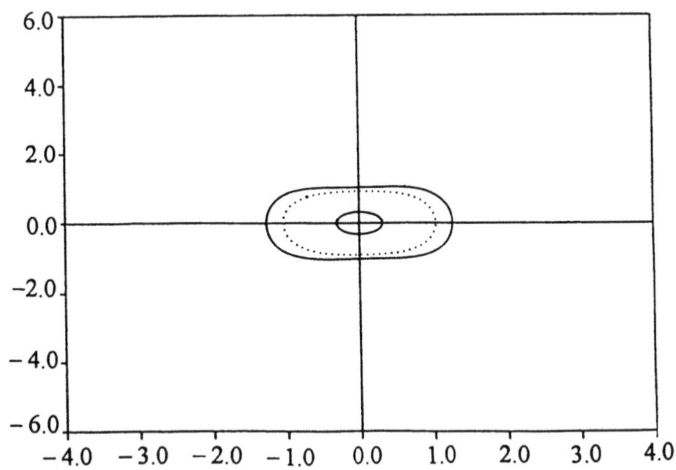


Phase trajectories of Duffing equation
($\mu$=0.1, $k$=0.0, $\alpha$=1.0, $\omega$=1.0, $f_1$=0.3)

Fig. 9.2(c)

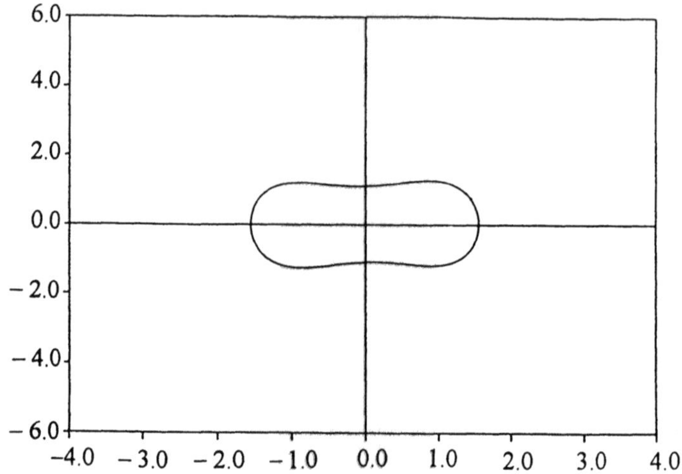


Phase trajectory of Duffing equation
($\mu$=0.1, $k$=0.0, $\alpha$=1.0, $\omega$=1.0, $f_1$=1.0)

Fig. 9.2(d)

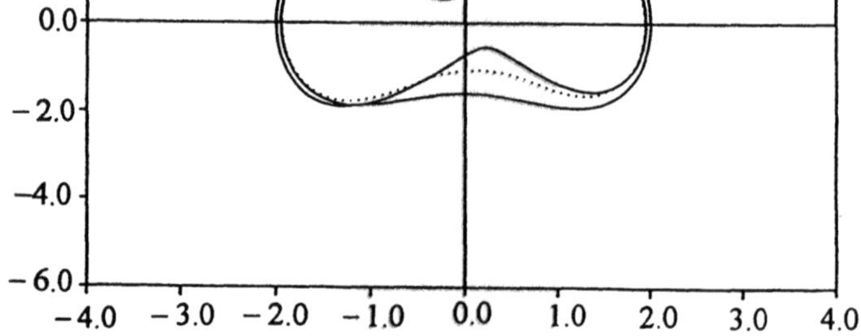


Phase trajectories of Duffing equation
($\mu$=0.1, $k$=0.0, $\alpha$=1.0, $\omega$=1.0, $f_1$=2.5)

Fig. 9.2(e)

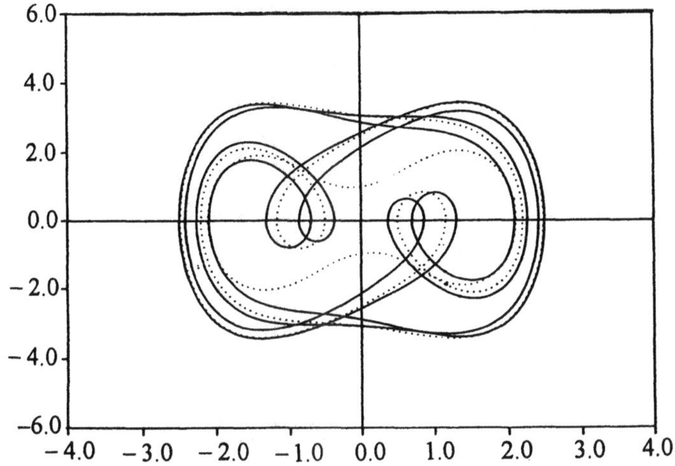


Phase trajectories of Duffing equation
($\mu$=0.1, $k$=0.0, $\alpha$=1.0, $\omega$=1.0, $f_1$=3.27)

Fig. 9.2(f)

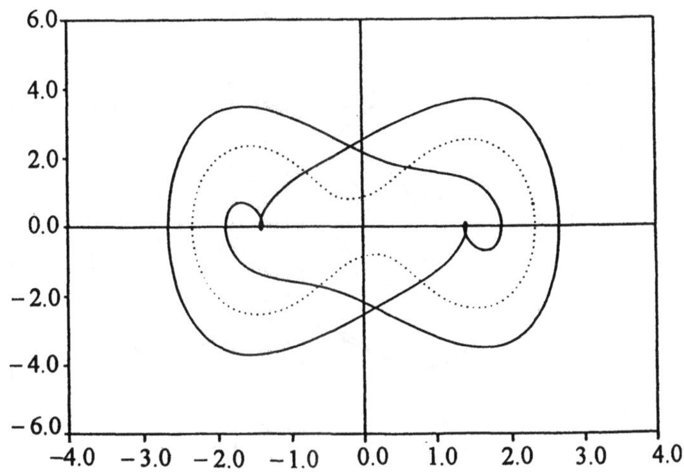


Phase trajectories of Duffing equation
($\mu$=0.1, $k$=0.0, $\alpha$=1.0, $\omega$=1.0, $f_1$=4.5)

Fig. 9.2 (g)

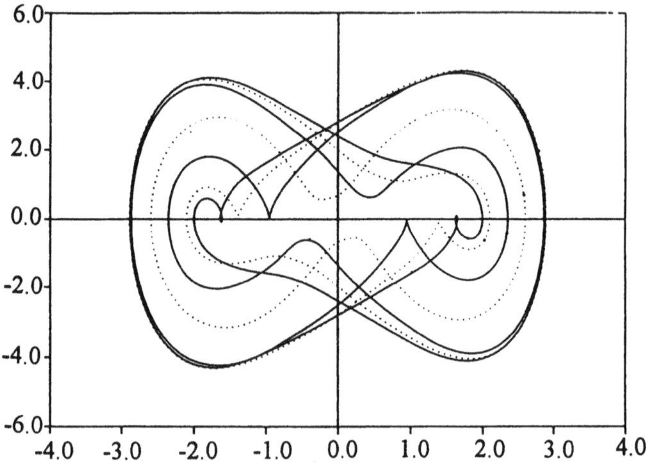


Phase trajectories of Duffing equation
($\mu$=0.1, $k$=0.0, $\alpha$=1.0, $\omega$=1.0, $f_1$=6.0)

Fig. 9.2(h)

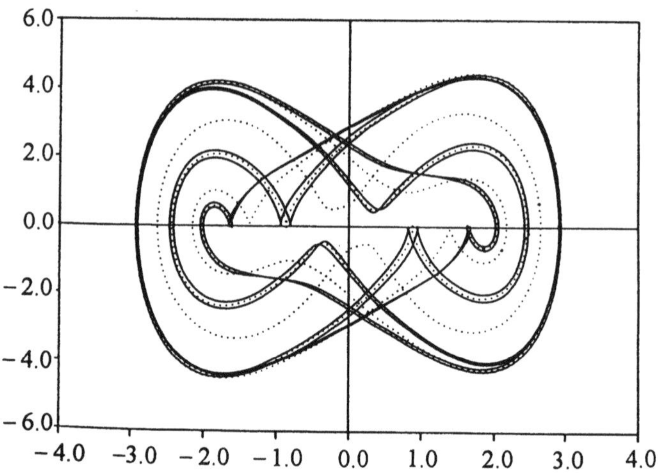


Phase trajectories of Duffing equation
($\mu$=0.1, $k$=0.0, $\alpha$=1.0, $\omega$=1.0, $f_1$=6.3)

Fig. 9.2(i)

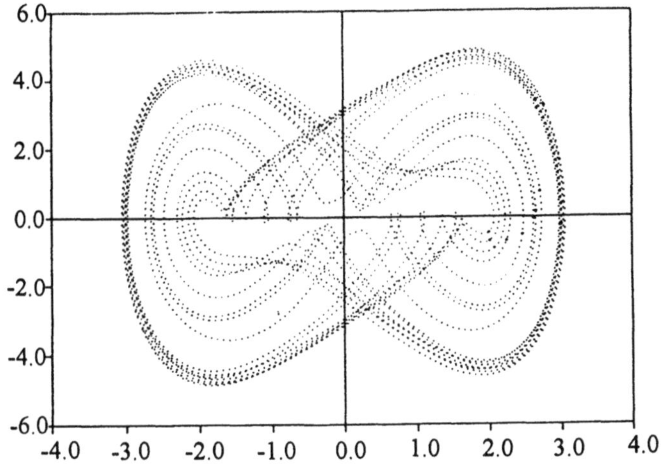


Phase trajectories of Duffing equation
($\mu$=0.1, $k$=0.0, $\alpha$=1.0, $\omega$=1.0, $f_1$ = 7.0)

Fig. 9.2(j)

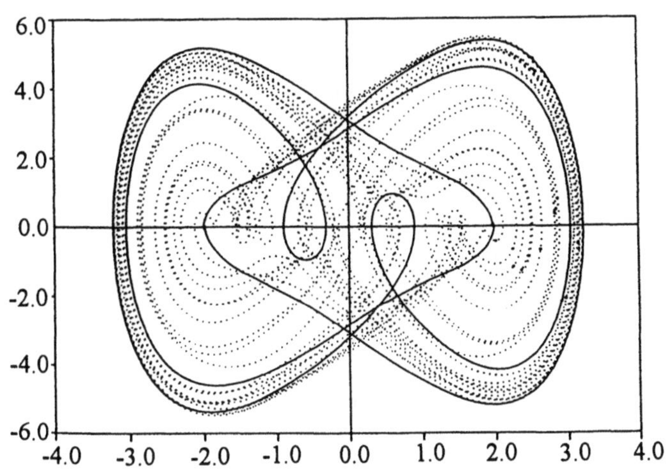


Phase trajectories of Duffing equation
($\mu$=0.1, $k$=0.0, $\alpha$=1.0, $\omega$=1.0, $f_1$ = 8.0)

Fig. 9.2(k)

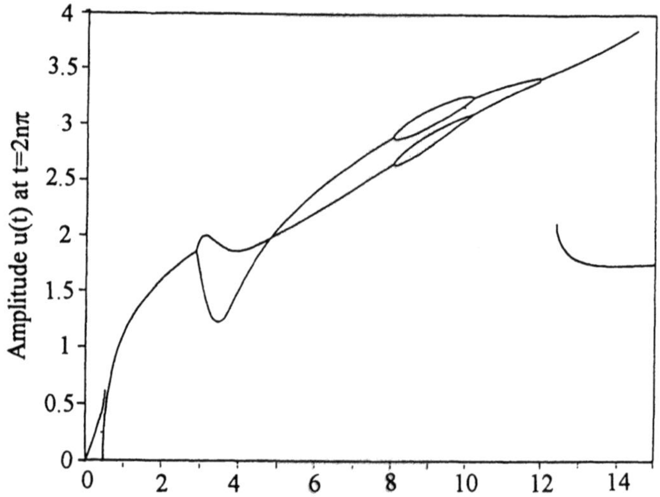


Bifurcation graph of Duffing equation
with stable branches only
(μ=0.2, $k$=0.0, α=1.0, ω=1.0)

Fig. 9.3(a)

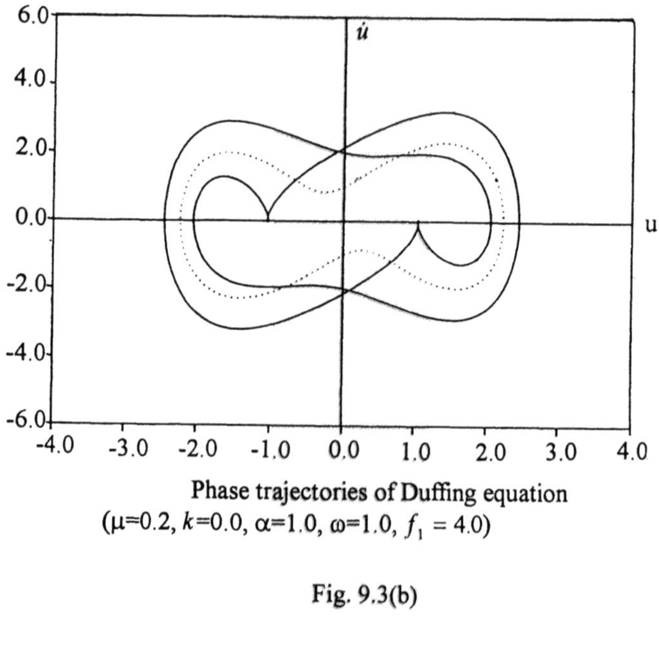


Phase trajectories of Duffing equation
(μ=0.2, $k$=0.0, α=1.0, ω=1.0, $f_1$ = 4.0)

Fig. 9.3(b)

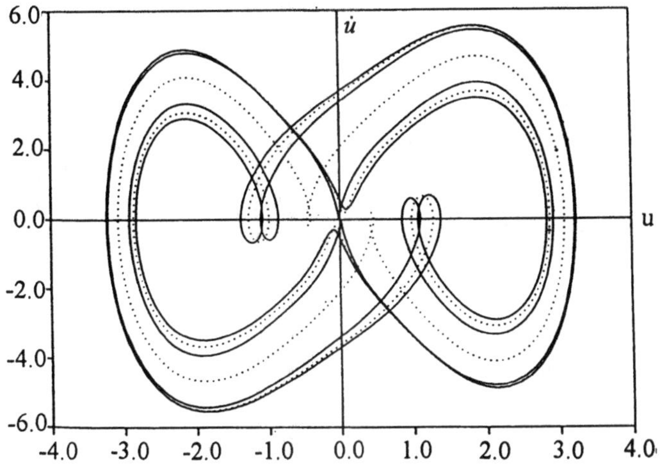


Phase trajectories of Duffing equation
($\mu$=0.2, $k$=0.0, $\alpha$=1.0, $\omega$=1.0, $f_1$ = 8.0)

Fig. 9.3(c)

#### 9.6.1.3 Characteristic Diagram for $f_1$ and $\mu$

A parametric study with the damping coefficient $\mu$ and the forcing amplitude $f_1$ is conducted while $k$=0.0, $\alpha$=1.0 and =1.0. The characteristic diagram is shown in Fig. 9.4. The diagram is comparable to the one obtained by Ueda [69], but has been obtained by using the incremental harmonic balance method instead of a point-by-point time integration search. Solutions for small damping and forcing amplitude are not shown for clarity.

Curves F1 and F2 correspond to the fold boundaries of the period $2\pi$ solutions with odd harmonic terms only and curve F3 corresponds to the fold boundary of period $2\pi$ asymmetrical solutions. Within these fold boundaries, three solutions are co-existing. Two of them are stable and either of them is observable depending on the imposed initial conditions. Curves B1, B2, B4 and B8 are the bifurcation boundaries to period $2\pi$ (with all harmonic components), period $4\pi$, period $8\pi$ and period $16\pi$ solutions respectively. On the boundaries, the originally stable solutions become unstable and new stable subharmonic solutions are formed in passing the boundaries. Curves T1 and T2 are the fold and bifurcation boundaries for period $6\pi$ solutions respectively. After the bifurcation boundaries, no stable solution is found. Just below the B8 curve, the solutions continue to bifurcate rapidly to higher order subharmonic responses. The solutions are in fact chaotic. On the other hand, below the curve B1, period $2\pi$ solutions become stable again and no other subharmonic solution is co-existing. In the regions under curve B8 when no stable solution exists, as hatched in the figure, chaotic solutions are obtained. For other regions under curve B8, because of the co-existence of other stable solutions, the final settlement may be chaotic depending on the initial conditions. However, the boundaries for the periodic solutions alone are under investigation.

Figure 9.5 shows a characteristic diagram of positive linear stiffness, with $k$=0.5 and $\alpha$=1.0. The boundary curves have shrunk and the chaotic region decreases with increasing linear stiffness. Time integration using parameters in this region confirms the existence of the chaotic solutions. On the other hand, the characteristic diagram is complicated as the linear stiffness decreases. Figure 9.6 shows a characteristic diagram of negative stiffness, with $k = -0.5$ and $\alpha$=1.0. The boundary curves are expanded and the chaotic regions cover a larger area. The arrangements are the same with additional fold boundaries T3 for period $6\pi$ solutions.

### 9.6.2 Multi-Harmonic Excitation

Consider the Duffing equation with $\mu$=0.1, $k$=0.0, $\alpha$=1.0, $\omega$=0.5, $p$=2 and $q$=3. The period of the excitation is $4\pi$. When $f_2$ =0, the bifurcation graph for $f_1$ is shown in section 9.6.1.2 and the characteristic diagram for $f_1$ and $\mu$ is shown in Section 9.6.1.3.

The number of harmonic terms used depends on the order of the subharmonic solutions under consideration. The number of harmonic terms is $40m$+1 initially. The number of terms increases when large unbalanced residual terms are detected. The harmonic terms with large unbalanced residual are included for annihilation. For symmetrical solutions, only odd harmonic terms are required. The name of a solution is written in the form $xTy$, where $x$ and $y$ are the uniquely assigned solution number and periodicity respectively.

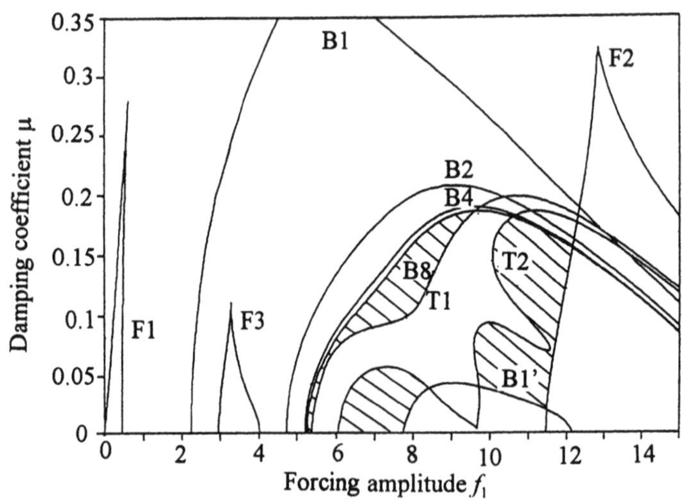


Characteristic diagram of Duffing equation
($k$=0.0, $\alpha$=1.0, $\omega$=1.0)

Fig. 9.4

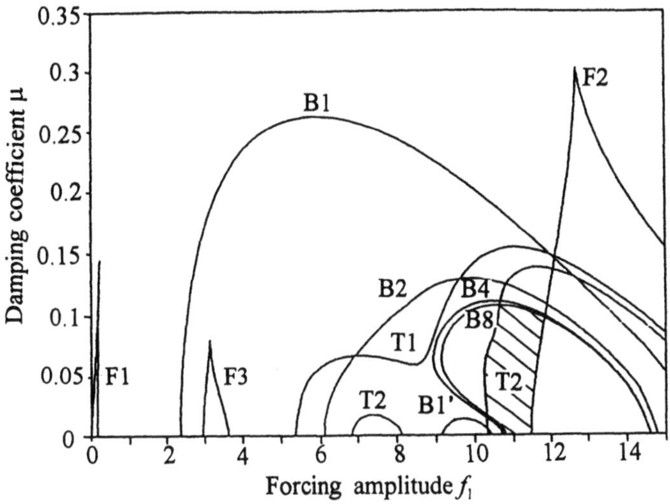


Characteristic diagram of Duffing equation
($k$=0.5, α=1.0, ω=1.0)

Fig. 9.5

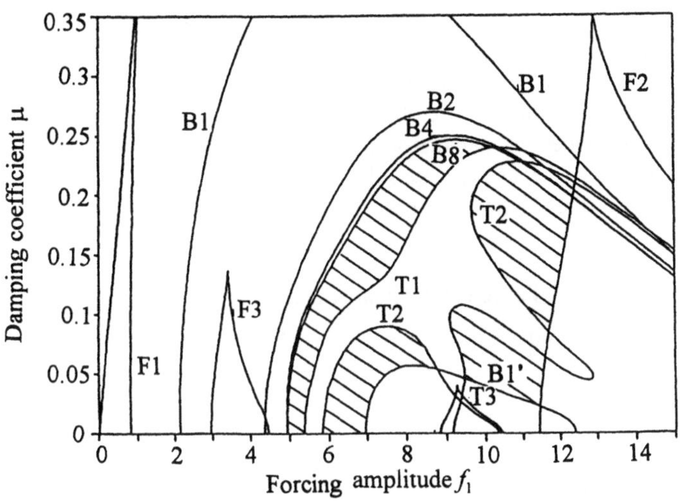


Characteristic diagram of Duffing equation
($k$=-0.5, α=1.0, ω=1.0)

Fig. 9.6

### 9.6.2.1 Bifurcation Graph for $f_2$, with $f_1$=2.0

When $f_1$ is small and $f_2$=0, only one solution with odd harmonic terms exists initially. The existence of $f_2$ destroys the symmetry. Figure 9.7 shows a bifurcation graph for $f_1$=2.0 with $f_2$ varying from 0.0 to 4.0. The solid lines and dotted lines denote the stable and unstable solutions respectively. At the six bifurcation points, the stable second-order ($m$=2) subharmonic solutions are emitted. The originally stable solution becomes unstable. Besides, two saddle nodes of periodicity $m$=1 are found. Competitive solutions are found within this range. Table.9.2 shows the critical points with the evolution indicated. Simple bifurcations to second-order subharmonic solutions only are observed. The folds give rise to competitive steady-state solutions and jump phenomena.

### 9.6.2.2 Bifurcation Graph for $f_2$, with $f_1$=5.0

For large $f_1$ amplitude, a bifurcation graph with $f_1$=5.0 is shown in Fig. 9.8. Complicated solution branches are observed. The stable solutions are drawn in solid lines. The dashed and dotted lines denote the unstable subharmonic solutions with $m$=1 and $m$=2 respectively. Unstable branches for the subharmonic solutions with $m$=4 are not shown for clarity. Thirteen bifurcation points and eleven saddle nodes are found. Table 9.3 shows the critical points and the solution types in accordance with the naming scheme discussed above.

It can be found that higher order subharmonic solutions are possible for a wide range of parameters and chaotic solutions are observed. In the ranges where stable periodic solutions are co-existing, the resultant steady-state solution depends on the initial conditions. Since many turning points exist, monotonically increasing $f_2$ will frequently give jumping phenomena. In the range of 3.5 to 4.0, simple periodic steady-state solutions with $m$=1 are obtained. In fact, this section crosses a transition boundary which divides the parametric plane into simple and complicated portions. For small and large $f_2$, the steady-state solutions are simple with small periodicity. For moderately large $f_2$, the solutions are chaotic.

**Table 9.2.** Critical points for Fig. 9.7

$\omega$=0.5, $\mu$=0.1, $k$=0.0, $\alpha$=1.0, $f_1$=2.0

| $f_2$ | |
|---|---|
| 0.0000 | stable 1T1 |
| 0.6872 | 1T1 to 1T2 |
| 1.1823 | 1T2 to 1T1 |
| 1.9209 | stable 2T1 and unstable 3T1 born |
| 2.3217 | 1T1 to 1T2 |
| 2.4884 | 2T1 to 2T1 |
| 2.6536 | 1T2 to 1T1 |
| 2.7639 | stable 1T1 and unstable 3T1 collide |
| 3.1775 | 2T2 to 2T1 |
| 4.0000 | stable 2T1 |

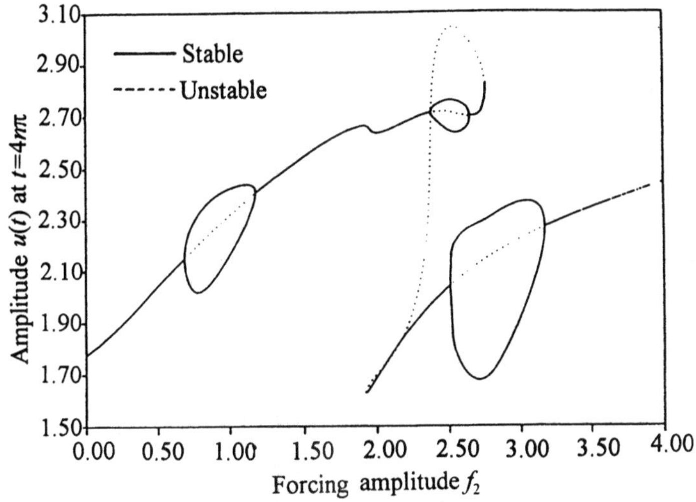


Bifurcation graph of Duffing equation
($\mu$=0.1, $k$=0.0, $\alpha$=1.0, $\omega$=0.5, $f_1$=2.0)

Fig. 9.7

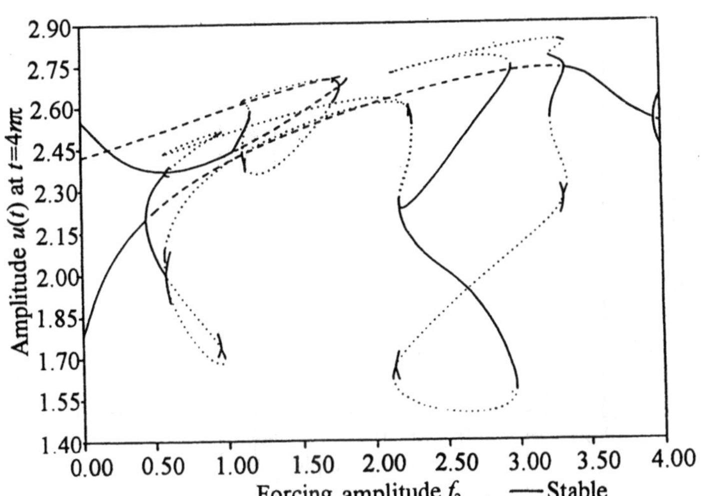


Bifurcation graph of Duffing equation
($\mu$=0.1, $k$=0.0, $\alpha$=1.0, $\omega$=0.5, $f_1$=5.0)

Fig. 9.8

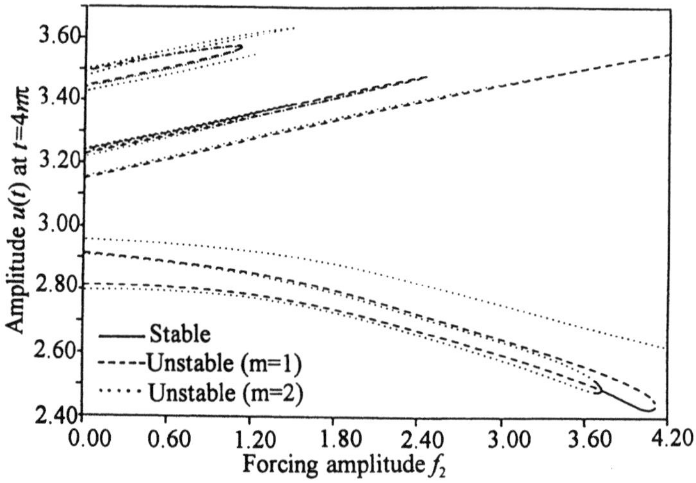


Bifurcation graph of Duffing equation
($\mu$=0.1, $k$=0.0, $\alpha$=1.0, $\omega$=0.5, $f_1$=10.0)

Fig. 9.9(a)

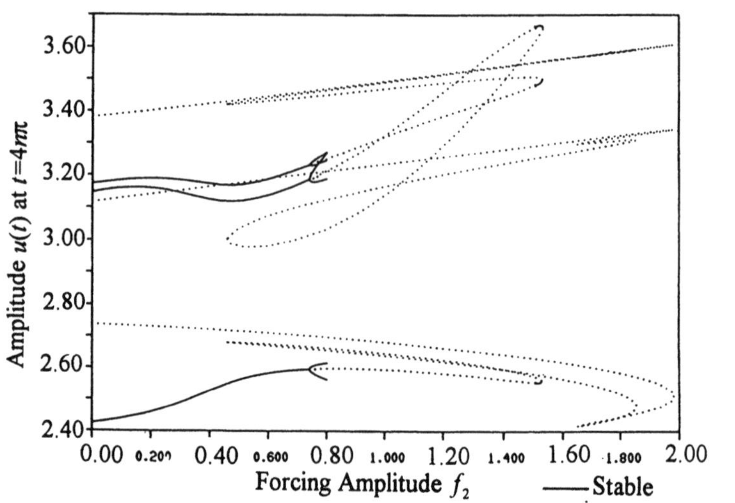


Bifurcation graph of Duffing equation
($\mu$=0.1, $k$=0.0, $\alpha$=1.0, $\omega$=0.5, $f_1$=10.0)

Fig. 9.9(b)

**Table 9.3. Critical points for Fig. 9.8**

$\omega$=0.5, $\mu$=0.1, $k$=0.0, $\alpha$=1.0, $f_1$=5.0

| $f_2$ | |
|---|---|
| 0.0000 | stable 1T1, 2T1, unstable 3T1 |
| 0.4482 | 1T1 to 1T2 |
| 0.5700 | stable 4T2 and unstable 5T2 born |
| 0.5728 | 4T2 to 4T4 |
| 0.5824 | 1T2 to 1T4 |
| 0.9526 | 1T4 to 1T2 |
| 0.9764 | stable 1T2 and unstable 5T2 collide |
| 1.0508 | 2T1 to 2T2 |
| 1.1192 | stable 6T2 and unstable 7T2 born |
| 1.1343 | 6T2 to 6T4 |
| 1.1744 | stable 2T2 and unstable 7T2 collide |
| 1.7520 | 6T4 to 6T2 |
| 1.7910 | 6T2 to 6T1 |
| 1.8346 | stable 6T1 and unstable 3T1 collide |
| 2.1146 | stable 8T2 and unstable 9T2 born |
| 2.1338 | 8T2 to 8T4 |
| 2.1763 | stable 10T2 and unstable 11T2 born |
| 2.2533 | 4T4 to 4T2 |
| 2.2731 | stable 4T2 and unstable 11T2 collide |
| 2.9730 | stable 10T2 and unstable 9T2 collide |
| 3.2431 | stable 12T2 and unstable 13T2 born |
| 3.3313 | 8T4 to 8T2 |
| 3.3519 | 12T2 to 12T1 |
| 3.3534 | stable 8T2 and unstable 13T2 collide |
| 3.9500 | 12T1 to 12T2 |
| 4.0000 | stable 12T2 and unstable 12T1 |

### 9.6.2.3 Bifurcation Graph for $f_2$, with $f_1$=10.0

For larger forcing amplitude, Figs. 9.9a and 9.9b show the bifurcation graphs with $f_1$=10.0. Figures 9.9a shows the period doubling bifurcations for initially unstable solutions. Only those solutions with small periodicity are drawn. The graph shows that only in a narrow range of the parameter, $3.7 \le f_2 \le 4.15$, is a stable periodic solution possible. At the turning points the initially unstable solutions become stable for a very narrow range and lose their stability with period doubling bifurcation. The occurrence of this periodic solution depends on the initial conditions. In general, chaotic solutions are obtained in the region.

It is found that for small $f_2$, third-order subharmonic solutions are co-existing, as shown in Fig. 9.9b. The solutions lose their stability by period doubling bifurcations and tangential bifurcations. These solutions are limited within the range $0.0 \le f_2 \le 2.0$. Stable third order subharmonic solutions exist in the range $0.0 \le f_2 \le 0.74$ and lose their stability by a period doubling bifurcation. The sixth-order subharmonic solutions are stable within the range $0.74 \le f_2 \le 0.81$. Higher order subharmonic bifurcations follow. Stable third order subharmonic solutions also exist within a narrow range around a fold. However, these are hardly observable.

## CONCLUDING REMARKS

By using the incremental harmonic balance method, the steady-state solutions, both stable and unstable, can be determined. Higher order subharmonic solutions can be constructed by adding subharmonic terms. The stability of the solution is determined by examining the amplitude of the eigenvalues of the transformation matrix according to Floquet theory. The period doubling bifurcation graphs are constructed for different combinations of parameters. The boundaries dividing qualitatively different steady-state solutions can be determined iteratively by the method of parametric unfolding. The regions enclosed by higher order subharmonic solution boundaries can be regarded as chaotic. The characteristic diagrams for different combinations of parameters are presented. For the Duffing equation subjected to two-harmonic excitation, the diagram is found to be highly foliated so that many steady-state solutions are co-existing at the same time. Chaotic solutions may be among the members. The folding structure causes sudden disappearance of the originally stable periodic solution and produces a chaotic solution with a small change of parameter.

# Chapter 10

# Computational Methods

We include three computational methods in this chapter, namely normal form theory, symplectic integration and the imbedded partial differential equation method.

Normal form theory is important for studying the qualitative behaviour of the non-linear oscillator. In some cases, higher order normal forms are required to understand the dynamic behaviour near an equilibrium or a periodic orbit. However, the computation of high-order normal forms is usually quite complicated. We provide an explicit formula for the normalisation of non-linear differential equations. The higher order normal form is given explicitly. Illustrative examples include a cubic system, a quadratic system and a Duffing—van der Pol system. We use exact arithmetic and find that the undamped Duffing equation can be represented by an exact polynomial differential amplitude equation in a finite number of terms.

In section 10.2, an accurate beam finite element is used to solve non-linear vibration of arched beams and framed structures. The non-linear governing equations of a skeletal structure are integrated numerically using symplectic integration schemes so that the Poincaré integral invariant of a Hamiltonian flow are preserved during the evolution. The elementary stiffness matrices need not be assembled into global form, because the integration is completed on an element level so that many elements can be handled in core by a small computer. Testing examples include arched beams and frames with and without damping in free and forced vibration. The dynamic symmetry breaking phenomena are noted at the dynamic buckling point.

Invariant torus is a very important case in the study of non-linear autonomous systems governed by ordinary differential equations (ODE). Section 10.3 provides a new numerical approach to approximate the multi-periodic surface formed by an invariant torus by embedding the governing ODE into a set of partial differential equations (PDE). A system with two strongly coupled van der Pol equations is taken as an illustrative example. Both characteristic exponents and rotation numbers are employed to determine the existence of multi-periodic surface.

## 10.1 Normal Form Theory

The invariant manifold of a non-linear oscillator near an equilibrium or a periodic orbit is determined by the structure of its vector field. Two often-used mathematical tools to simplify the original system are centre manifold and normal forms. The normal form theory is a technique of transforming the original non-linear differential equation to a simpler standard form by appropriate changes of co-ordinates, so that the essential features of the manifold become more evident. Basic references on normal forms and their applications may be found in Poincaré [167], Birkhoff [165], Arnold [164], Chow and Hale [115], Guckenheimer and Holmes [36], Ioos and Joseph [52], Sethna and Sell [168], Van der Beek [171], Vakakis and Rand [170] and Leung and

Zhang [166]. In this section a computational approach based on the classical normal form theory of Poincaré and Birkhoff is introduced. The relationship of the coefficients between the original equations and the normal form equations is explicitly constructed. The technique presented in our approach follows the idea of Takens 169]. A linear operator and its adjacent operator are defined with an inner product on the space of homogeneous polynomials. The resultant normal form keeps only the resonant terms, which cannot be eliminated by a non-linear polynomial changes of variables, in the kernel of the adjacent linear operator. The normal form simplifies the original systems so that the dynamic stability and bifurcation can be studied in a standard manner and the classification of manifold in the neighbourhood of an equilibrium or a periodic orbit can be achieved with relatively little effort. The normal forms of several typical non-linear oscillators: the cubic system, the quadratic system, and the Duffing—van der Pol system are provided and the steady-state solutions of the method are compared with existing results.

### 10.1.1 Transformation to Normal Form

Consider the non-linear ordinary differential equations

$$\dot{u} = f(u), \qquad f \in C^r(R^n) \tag{10.1}$$

where $f$ is an $n$-vector function of $u$, differentiable up to order $r$. Suppose eq. (10.1) has a fixed point at $u = u_0$. We first perform a few linear transformations to simplify eq. (10.1). By the variable change $v = u - u_0$, we eliminate the constant terms and shift the fixed point to the origin under which eq. (10.1) becomes

$$\dot{v} = f(v + u_0) \equiv H(v) \tag{10.2}$$

where $H(v)$ is at least linear in $v$. We next split the linear part of the ordinary differential equation and write (10.2) as follows

$$\dot{v} = D_v H(0)v + \overline{H}(v) \tag{10.3}$$

where $\overline{H}(v) \equiv H(v) - D_v H(0)v$ and where, $\overline{H}(v) = O(|v|^2)$ is at least quadratic in $v$, and $D_V$ denotes differentiation with respect to $v$. We further transform $D_V H(0)$ into Jordan canonical form by the canonical matrix $T$, i.e., $v=Tx$, and obtain

$$\dot{x} = T^{-1} D_v H(0) Tx + T^{-1}\overline{H}(Tx) \tag{10.4}$$

which can be written alternatively as

$$\dot{x} = Jx + F(x) \tag{10.5}$$

where $J$ is the Jordan canonical form of $D_V H(0)$ and $F(x)$ is the non-linear part of the equation. In Poincaré's normal form theory, the non-linear function $F(x)$ is expanded by a series of homogeneous polynomials

$$\dot{x} = Jx + F_2(x) + F_3(x) + \cdots + F_r(x) + O(|x|^{r+1}) \tag{10.6}$$

where $F_k \in H_k$, which is the set of homogeneous polynomials of order *k*. To transform eq. (10.6) into its normal form, Poincaré introduces a nearly identity non-linear co-ordinate transformation of the form

$$x = y + h_k(y) \tag{10.7}$$

where $h_k(y)$ is *k*th order in *y*, $h_k(y) \in H_k$, $2 \le k \le r$.

Substituting eq. (10.7) into eq. (10.6) gives

$$\dot{x} = \left(I + D_y h_k(y)\right)\dot{y} = J\left(y + h_k(y)\right) + \cdots + F_k\left(y + h_k(y)\right) + O\left(|y|^{k+1}\right) \tag{10.8}$$

where $I$ denotes the $n \times n$ identity matrix and the term $\left(I + D_y h_k(y)\right)$ is inevitable for sufficiently small $y$ so that

$$\left(I + D_y h_k(y)\right)^{-1} = I - D_y h_k(y) + O\left(|y|^2\right) \tag{10.9}$$

Substituting eq. (10.9) into eq. (10.8) and applying similar nearly identity non-linear transformations up to the *k*th order, we obtain

$$\begin{aligned} \dot{y} &= Jy + F_2(y) + \cdots + F_{k-1}(y) + \overline{F}_k(y) + \left(Jh_k(y) - D_y h_k(y) Jy\right) + O\left(|y|^{k+1}\right) \\ &= Jy + F_2(y) + \cdots + F_{k-1}(y) + \overline{F}_k(y) + ad_J h_k(y) + O\left(|y|^{k+1}\right) \end{aligned} \tag{10.10}$$

where the overbar is used to represent the original polynomial matrices and $ad_J h_k(y)$ denotes an adjacent operator equivalent to the function of the *Lie Bracket*,

$$ad_J h_k(y) \equiv L_J(H_k) = Jh_k(y) - D_y h_k(y) Jy \tag{10.11}$$

To simplify the terms of order *k* as much as possible, we choose a specific form for $h_k(y)$ such that

$$\overline{F}_k(y) = L_J(H_k) \tag{10.12}$$

If $\overline{F}_k(y)$ is the range of $L_J(H_k)$, then all the terms of order *k* can be eliminated completely from eq. (10.10). Otherwise we must find a complementary space $G_k$ to $L_J(H_k)$ and let $H_k = L_J(H_k) \oplus G_k$ so that only terms of order *k* which are in $G_k$ remain in the resultant expression. It is interesting to note that simplifying terms of order *k* would not affect the coefficients of any lower terms. However, terms of order higher than *k* will be changed. Therefore, it is only necessary to keep track of the way that the higher order terms are modified by the successive co-ordinate transformations.

### 10.1.2 Normal Forms of Oscillating Systems

We now specialise the normal form formulation in two dimensions for the Hopf bifurcation below. Consider a system with a small perturbed parameter $\mu$ and an equilibrium point with eigenvalues $\pm i\omega_0, \omega_0 > 0$

$$\dot{x} = f(x, \mu), \qquad x \in R^2, \mu \in R^1 \tag{10.13}$$

where we suppose, after the shifting of origin and canonical transformation,

$$f(0,0) = 0, \quad D_x f(0,0) = \begin{bmatrix} 0 & -\omega_0 \\ \omega_0 & 0 \end{bmatrix}$$

Furthermore, the Taylor expansion of eq. (10.13) gives

$$\dot{x} = Ax + F_2(x) + F_3(x) + \cdots + F_r(x) + O\left(|x|^{r+1}\right) \tag{10.14}$$

$$\text{where} \quad A \cong D_x f(0,0) + D^2_{x\mu} f(0,0)\mu = \begin{bmatrix} \beta\mu & -(\alpha\mu + \omega_0) \\ (\alpha\mu + \omega_0) & \beta\mu \end{bmatrix} \tag{10.15}$$

$$F_2(y) = \begin{bmatrix} F_{21}(y) \\ F_{22}(y) \end{bmatrix} = \begin{bmatrix} a_{20} & a_{11} & a_{02} \\ b_{20} & b_{11} & b_{02} \end{bmatrix} \begin{Bmatrix} y_1^2 \\ y_1 y_2 \\ y_2^2 \end{Bmatrix} \tag{10.16}$$

We would like to find a co-ordinate transformation (10.7) so that the non-linear terms of order $k$+1 in the new system of function $y$ vanish, rather than of order $k$ in the original system of function $x$. Inserting

$y = (y_1, y_2)^T \in R^2$ and $h_k(y) = (h_{k1}(y), h_{k2}(y))^T \in H_k$ into (10.13), we have the Lie bracket,

$$L_J(H_k) = \begin{bmatrix} 0 & -\omega_0 \\ \omega_0 & 0 \end{bmatrix} \begin{bmatrix} h_{k1} \\ h_{k2} \end{bmatrix} - \begin{bmatrix} \partial h_{k1}/\partial y_1 & \partial h_{k1}/\partial y_2 \\ \partial h_{k2}/\partial y_1 & \partial h_{k2}/\partial y_2 \end{bmatrix} \begin{bmatrix} 0 & -\omega_0 \\ \omega_0 & 0 \end{bmatrix} \begin{bmatrix} y_1 \\ y_2 \end{bmatrix} \tag{10.17}$$

### 10.1.3 Second-Order Normal Form

If the smallest order of non-linear terms appearing in (10.13) is two, we try to find a transformation $h_2$ of the form

$$x = y + h_2(y) \tag{10.18}$$

where

$$h_2(y) = \begin{bmatrix} h_{21}(y) \\ h_{22}(y) \end{bmatrix} = \begin{bmatrix} \sum_{j=0}^{2} c_{ij} y_1^i y_2^j \\ \sum_{j=0}^{2} d_{ij} y_1^i y_2^j \end{bmatrix}; \qquad i = 2 - j \tag{10.19}$$

Perform the Lie bracket operation to each basis element on $H_2$

$$\mathrm{H}_2 = \text{span}\left\{ \begin{pmatrix} y_1^2 \\ 0 \end{pmatrix}, \begin{pmatrix} y_1 y_2 \\ 0 \end{pmatrix}, \begin{pmatrix} y_2^2 \\ 0 \end{pmatrix}, \begin{pmatrix} 0 \\ y_1^2 \end{pmatrix}, \begin{pmatrix} 0 \\ y_1 y_2 \end{pmatrix}, \begin{pmatrix} 0 \\ y_2^2 \end{pmatrix} \right\}$$

thus,

$$L_J \begin{pmatrix} y_1^2 \\ 0 \end{pmatrix} = \begin{bmatrix} 0 & -\omega_0 \\ \omega_0 & 0 \end{bmatrix} \begin{bmatrix} y_1^2 \\ 0 \end{bmatrix} - \begin{bmatrix} 2y_1 & 0 \\ 0 & 0 \end{bmatrix} \begin{bmatrix} 0 & -\omega_0 \\ \omega_0 & 0 \end{bmatrix} \begin{bmatrix} y_1 \\ y_2 \end{bmatrix} = \omega_0 \begin{bmatrix} 2y_1 y_2 \\ y_1^2 \end{bmatrix},$$

$$L_J\begin{pmatrix} y_1y_2 \\ 0 \end{pmatrix} = \omega_0\begin{bmatrix} y_2^2 - y_1^2 \\ y_1y_2 \end{bmatrix},\ L_J\begin{pmatrix} y_2^2 \\ 0 \end{pmatrix} = \omega_0\begin{bmatrix} -y_1y_2 \\ y_2^2 \end{bmatrix},\ L_J\begin{pmatrix} 0 \\ y_1^2 \end{pmatrix} = \omega_0\begin{bmatrix} -y_1^2 \\ 2y_1y_2 \end{bmatrix},$$

$$L_J\begin{pmatrix} 0 \\ y_1y_2 \end{pmatrix} = \omega_0\begin{bmatrix} -y_1y_2 \\ y_2^2 - y_1^2 \end{bmatrix}, \text{ and } L_J\begin{pmatrix} 0 \\ y_2^2 \end{pmatrix} = -\omega_0\begin{bmatrix} y_2^2 \\ 2y_1y_2 \end{bmatrix}.$$

The matrix representation of $L_J(H_2)$ can then be written as

$$L_J(H_2) = \text{span}\left\{\begin{bmatrix} y_1^2 & y_1y_2 & y_2^2 & 0 & 0 & 0 \\ 0 & 0 & 0 & y_1^2 & y_1y_2 & y_2^2 \end{bmatrix} \cdot A_J^2\right\} = H_2 \cdot A_J^2 \quad (10.20)$$

where

$$A_J = \omega_0\begin{bmatrix} 0 & -1 & 0 & -1 & 0 & 0 \\ 2 & 0 & -2 & 0 & -1 & 0 \\ 0 & 1 & 0 & 0 & 0 & -1 \\ 1 & 0 & 0 & 0 & -1 & 0 \\ 0 & 1 & 0 & 2 & 0 & -2 \\ 0 & 0 & 1 & 0 & 1 & 0 \end{bmatrix}.$$

Because the $\det\left[A_J^2\right] = 8\omega_0 > 0$, which does not vanish for $\omega_0 > 0$, we obtain the following null complementary space for the homogeneous equations $A_J^2\xi = \{0\}$:

$$G_2 = \xi \equiv \{0\}, \qquad \xi \in R^6 \quad (10.21)$$

Therefore, we can eliminate the quadratic term completely from eq. (10.10) by means of the second-order co-ordinate transformation (10.19) and obtain the second-order normal form

$$\begin{cases} \dot{y}_1 = \left(\beta\mu y_1 - (\alpha\mu + \omega_0)y_2 + O\left(|y_1|^3, |y_2|^3\right)\right) \\ \dot{y}_2 = \left((\alpha\mu + \omega_0)y_1 + \beta\mu y_2 + O\left(|y_1|^3, |y_2|^3\right)\right) \end{cases} \quad (10.22)$$

The coefficients in eq. (10.19) are required for further development and are found from eqs. (10.10)—(10.12). Substituting eq. (10.19) into eq. (10.10) and truncating to the *M*th degree yields

$$\dot{y} = \sum_{N=0}^{M}(-1)^N\left[D_y h_2(y)\right]^N\left\{Ay + Ah_2(y) + \sum_{n=2}^{M}\sum_{m=0}^{n}\frac{D_y^m F_n(y)}{m!}\left[h_2(y)\right]\right\} = Ay + \sum_{n=2}^{M}F_n^{(1)}(y) \quad (10.23)$$

where

$$F_n^{(1)}(y) = \begin{bmatrix} F_{n1}^{(1)}(y) \\ F_{n2}^{(2)}(y) \end{bmatrix} = \begin{bmatrix} \sum_{j=0}^{n} a_{ij}^{(1)} y_1^i y_2^j \\ \sum_{j=0}^{n} b_{ij}^{(1)} y_1^i y_2^j \end{bmatrix}; \quad n = 2, \ldots, M,\ i = 2 - j \quad (10.24)$$

in which the number in the superscript parentheses refers to the index of co-ordinate transformation. From eq. (10.22) we know that the complementary space of second-order normal form, $G_2$, is null so that we have the following equation

$$F_2^{(1)}(y) = F_2(y) + Jh_2(y) - D_y h_2(y) \cdot J \cdot y = 0 \tag{10.25}$$

Then, we find coefficients $c_{ij}, d_{ij}$ in eq. (10.19) by solving the linear algebraic equation, eq. (10.25):

$$\begin{bmatrix} c_{20} \\ c_{11} \\ c_{02} \\ d_{20} \\ d_{11} \\ d_{02} \end{bmatrix} = \frac{-1}{3\omega_0} \begin{bmatrix} b_{20} + a_{11} + 2b_{02} \\ -2a_{20} - b_{11} + 2a_{02} \\ 2b_{20} - a_{11} + b_{02} \\ -a_{20} + b_{11} - 2a_{02} \\ -2b_{20} + a_{11} + 2b_{02} \\ -2a_{20} - b_{11} - 2a_{02} \end{bmatrix} \tag{10.26}$$

### 10.1.4 Third-Order Normal Form

We further perform the nearly identity transformation (10.7) to third order to obtain the more accurate non-linear description of eq. (10.14) in the neighbourhood of the original singular point:

$$x = y + h_3(y) \tag{10.27}$$

where $y$ represents the new co-ordinate and $x$ is its old one.

$$h_3(y) = \begin{bmatrix} h_{31}(y) \\ h_{32}(y) \end{bmatrix} = \begin{bmatrix} \sum_{j=0}^{3} c_{ij} y_1^i y_2^j \\ \sum_{j=0}^{3} d_{ij} y_1^i y_2^j \end{bmatrix}; \qquad i = 3 - j \tag{10.28}$$

The third-order monomial base is given by

$$\mathrm{H}_3 = \mathrm{span}\left\{ \begin{pmatrix} y_1^3 \\ 0 \end{pmatrix}, \begin{pmatrix} y_1^2 y_2 \\ 0 \end{pmatrix}, \begin{pmatrix} y_1 y_2^2 \\ 0 \end{pmatrix}, \begin{pmatrix} y_2^3 \\ 0 \end{pmatrix}, \begin{pmatrix} 0 \\ y_1^3 \end{pmatrix}, \begin{pmatrix} 0 \\ y_1^2 y_2 \end{pmatrix}, \begin{pmatrix} 0 \\ y_1 y_2^2 \end{pmatrix}, \begin{pmatrix} 0 \\ y_2^3 \end{pmatrix} \right\} \tag{10.29}$$

Similar to eq. (10.20), the matrix representation of $L_J(H_3)$ is given by

$$L_J(H_3) = H_3 \cdot A_J^3 \tag{10.30}$$

where

$$A_J^3 = \omega_0 \begin{bmatrix} 0 & -1 & 0 & 0 & -1 & 0 & 0 & 0 \\ 3 & 0 & -2 & 0 & 0 & -1 & 0 & 0 \\ 0 & 2 & 0 & -3 & 0 & 0 & -1 & 0 \\ 0 & 0 & 1 & 0 & 0 & 0 & 0 & -1 \\ 1 & 0 & 0 & 0 & 0 & -1 & 0 & 0 \\ 0 & 1 & 0 & 0 & 3 & 0 & -2 & 0 \\ 0 & 0 & 1 & 0 & 0 & 2 & 0 & -3 \\ 0 & 0 & 0 & 1 & 0 & 0 & 1 & 0 \end{bmatrix}$$

We observe that the homogeneous equations $A_J^3\xi = \{0\}$, $\xi \in R^8$ have two zero eigenvectors for the matrix $A_J^3$ corresponding to its two zero eigenvalues,

$$e^T = \left\{(1,0,1,0,0,1,0,1)^T \; (0,-1,0,-1,1,0,1,0)^T\right\} \tag{10.31}$$

Thus, $A_J^3$ has a complementary space spanned by the monomial basis (10.29):

$$G_3 = H_3 \cdot e^T = \text{span}\left\{\begin{pmatrix} y_1(y_1^2+y_2^2) \\ y_2(y_1^2+y_2^2) \end{pmatrix} \; \begin{pmatrix} -y_2(y_1^2+y_2^2) \\ y_1(y_1^2+y_2^2) \end{pmatrix}\right\} \tag{10.32}$$

Finally, through a proper co-ordinate transformation, we determine the third-order normal form of the original system:

$$\begin{cases} \dot{y}_1 = \left(\beta\mu y_1 - (\alpha\mu+\omega_0)y_2 + a_1 y_1(y_1^2+y_2^2) - b_1 y_2(y_1^2+y_2^2) + O(|y_1|^5,|y_2|^5)\right) \\ \dot{y}_2 = \left((\alpha\mu+\omega_0)y_1 + \beta\mu y_2 + a_1 y_2(y_1^2+y_2^2) + b_1 y_1(y_1^2+y_2^2) + O(|y_1|^5,|y_2|^5)\right) \end{cases} \tag{10.33}$$

where $a_i$ and $b_i$ are to be determined.

We now want to develop a systematic procedure to evaluate the coefficients of normal forms and the normal transformations of third order. We see from eq. (10.11) that

$$\begin{aligned} \dot{x} &= \sum_{N=0}^{M}(-1)^N\left[Dh_3(y)\right]^N\left\{Ay + Ah_3(y) + \sum_{n=2}^{M}\sum_{m=0}^{n}\frac{D^m F_n(y)}{m!}\left[h_3(y)\right]^m\right\} \\ &= Ay + F_2^{(1)}(y) + \sum_{n=3}^{M} F_n^{(2)}(y) \end{aligned} \tag{10.34}$$

where

$$F_n^{(2)}(y) = \begin{bmatrix} F_{n1}^{(2)}(y) \\ F_{n2}^{(2)}(y) \end{bmatrix} = \begin{bmatrix} \sum_{j=0}^{n} a_{ij}^{(2)} y_1^i y_2^j \\ \sum_{j=0}^{n} b_{ij}^{(2)} y_1^i y_2^j \end{bmatrix}; \quad n = 3,\ldots,M,\; i = 2-j \tag{10.35}$$

We know from eq. (10.24) that the third-order normal form is equal to its complement space so that

$$F_3^{(2)}(y) = F_3^{(1)}(y) + J \cdot h_3(y) - Dh_3(y) \cdot J \cdot y = G_3(y) \tag{10.36}$$

The resultant linear eq. is

$$A_J^3 \cdot \xi = \eta \tag{10.37}$$

where $\xi = \{c_{30}, c_{21}, c_{12}, c_{03}, d_{30}, d_{21}, d_{12}, d_{03}\}^T$

$$\eta = \left\{a_{30}^{(1)} - a_1, a_{21}^{(1)} + b_1, a_{12}^{(1)} - a_1, a_{03}^{(1)} + b_1, b_{30}^{(1)} - b_1, b_{21}^{(1)} - a_1, b_{12}^{(1)} - b_1, b_{03}^{(1)} - a_1\right\}^T$$

or
$$\omega_0 \begin{bmatrix} B & -I \\ I & B \end{bmatrix} \begin{bmatrix} \tilde{c} \\ \tilde{d} \end{bmatrix} = \begin{bmatrix} \tilde{a} \\ \tilde{b} \end{bmatrix} \tag{10.38}$$

in which

$$B = \begin{bmatrix} 0 & -1 & 0 & 0 \\ 3 & 0 & -2 & 0 \\ 0 & 2 & 0 & -3 \\ 0 & 0 & 1 & 0 \end{bmatrix}; \quad \tilde{c} = \begin{bmatrix} c_{30} \\ c_{21} \\ c_{12} \\ c_{03} \end{bmatrix}$$

$$\tilde{d} = \begin{bmatrix} d_{30} \\ d_{21} \\ d_{12} \\ d_{03} \end{bmatrix}; \quad \tilde{a} = \begin{bmatrix} a_{30}^{(1)} - a_1 \\ a_{21}^{(1)} + b_1 \\ a_{12}^{(1)} - a_1 \\ a_{03}^{(1)} + b_1 \end{bmatrix}; \quad \tilde{b} = \begin{bmatrix} b_{30}^{(1)} - b_1 \\ b_{21}^{(1)} - a_1 \\ b_{12}^{(1)} - b_1 \\ b_{03}^{(1)} - a_1 \end{bmatrix}$$

Dividing eq. (10.38) into two parts, we obtain

$$\begin{cases} \omega_0\left(B^2 + I\right) \cdot \tilde{c} = B \cdot \tilde{a} - \tilde{b} \\ \omega_0\left(B^2 + I\right) \cdot \tilde{d} = B \cdot \tilde{b} + \tilde{a} \end{cases} \tag{10.39}$$

Note that there are two more unknowns existing in eq. (10.39) than the number of equations. However, the coefficients of the normal form $a_1$, $b_1$ can be evaluated independently by performing an orthogonal linear transformation on one of the equations in eq. (10.39),

$$T = \begin{bmatrix} 0 & -1 & 0 & 1 \\ -3 & 0 & 1 & 0 \\ 0 & 3 & 0 & 1 \\ 1 & 0 & 1 & 0 \end{bmatrix} \tag{10.40}$$

where $T$ is the canonical matrix consisting of the eigenvectors of $B^2+I$. The second eq. in (10.39) becomes

$$LHS = T^{-1}\left(B^2 + I\right)T \cdot \widetilde{d}$$

$$= \omega_0 \begin{bmatrix} 0 & -1/4 & 0 & 1/4 \\ -1/4 & 0 & 1/4 & 0 \\ 0 & 1/4 & 0 & 3/4 \\ 3/4 & 0 & 1/4 & 0 \end{bmatrix} \begin{bmatrix} -2 & 0 & 2 & 0 \\ 0 & -6 & 0 & 6 \\ 6 & 0 & -6 & 0 \\ 0 & 2 & 0 & -2 \end{bmatrix} \begin{bmatrix} 0 & -1 & 0 & 1 \\ -3 & 0 & 1 & 0 \\ 0 & 3 & 0 & 1 \\ 1 & 0 & 1 & 0 \end{bmatrix} \cdot \widetilde{d}$$

$$= \omega_0 \begin{bmatrix} 0 & 2 & 0 & -2 \\ 2 & 0 & -2 & 0 \\ 0 & 0 & 0 & 0 \\ 0 & 0 & 0 & 0 \end{bmatrix} \cdot \widetilde{d} \tag{10.41}$$

$$\text{RHS} = T^{-1}\left(B \cdot b + a\right) = \begin{bmatrix} 0 & -1/4 & 0 & 1/4 \\ -1/4 & 0 & 1/4 & 0 \\ 0 & 1/4 & 0 & 3/4 \\ 3/4 & 0 & 1/4 & 0 \end{bmatrix} \begin{bmatrix} -2a_1 + a_{30}^{(1)} + b_{21}^{(1)} \\ a_{21}^{(1)} + 2b_1 + 2b_{12}^{(1)} - 3b_{30}^{(1)} \\ a_{12}^{(1)} - 2a_1 + 3b_{03}^{(1)} - 2b_{21}^{(1)} \\ a_{03}^{(1)} + 2b_1 - b_{12}^{(1)} \end{bmatrix}$$

$$= \begin{bmatrix} a_{03}^{(1)} - a_{21}^{(1)} - 3b_{12}^{(1)} + 3b_{30}^{(1)} \\ a_{12}^{(1)} - a_{30}^{(1)} + 3b_{03}^{(1)} - 3b_{21}^{(1)} \\ 3a_{03}^{(1)} + a_{21}^{(1)} + 8b_1 - b_{12}^{(1)} - 3b_{30}^{(1)} \\ -8a_1 + a_{12}^{(1)} + 3a_{30}^{(1)} + 3b_{03}^{(1)} + b_{21}^{(1)} \end{bmatrix}$$

By comparing the terms in the last two rows of eq. (10.41), we obtain

$$\begin{cases} a_1 = \dfrac{1}{8}\left(a_{12}^{(1)} + 3a_{03}^{(1)} + b_{21}^{(1)} + 3b_{03}^{(1)}\right) \\ a_1 = \dfrac{-1}{8}\left(a_{21}^{(1)} + 3a_{03}^{(1)} - b_{12}^{(1)} - 3b_{30}^{(1)}\right) \end{cases} \tag{10.42}$$

Substituting eq. (10.42) into eq. (10.39) and solving (10.39), the third-order homogeneous polynomial (10.28) is then determined by:

$$\begin{bmatrix} c_{30} \\ c_{21} \\ c_{12} \\ c_{03} \end{bmatrix} = \frac{1}{8\omega_0} \begin{bmatrix} 0 \\ -3a_{12}^{(1)} - b_{03}^{(1)} + 3a_{30}^{(1)} + b_{21}^{(1)} \\ -3a_{03}^{(1)} + b_{12}^{(1)} + 3a_{21}^{(1)} + b_{30}^{(1)} \\ 0 \end{bmatrix}$$

$$\begin{bmatrix} d_{30} \\ d_{21} \\ d_{12} \\ d_{03} \end{bmatrix} = \frac{1}{4\omega_0} \begin{bmatrix} a_{12}^{(1)} - b_{03}^{(1)} + a_{30}^{(1)} - b_{21}^{(1)} \\ 3a_{03}^{(1)} + 5b_{30}^{(1)} + a_{21}^{(1)} - b_{12}^{(1)} \\ a_{12}^{(1)} + 5b_{03}^{(1)} + 3a_{30}^{(1)} + b_{21}^{(1)} \\ a_{03}^{(1)} + b_{30}^{(1)} + a_{21}^{(1)} + b_{12}^{(1)} \end{bmatrix} \tag{10.43}$$

### 10.1.5 Higher Order Normal Form

The higher order normal form in a rectangular co-ordinate system is derived accordingly and can be written as follows

$$\dot{y} = Ay + \sum_{i=1}^{L} W_{2i+1}(y) + O\left(|y|^{2L+3}\right) \tag{10.44}$$

where

$$A = \begin{bmatrix} \beta\mu & -(\alpha\mu + \omega_0) \\ (\alpha\mu + \omega_0) & \beta\mu \end{bmatrix}, \quad y = \begin{bmatrix} y_1 \\ y_2 \end{bmatrix},$$

$$W_{2i+1}(y) = \left(y_1^2 + y_2^2\right)^i \begin{bmatrix} a_i & -b_i \\ b_i & a_i \end{bmatrix} \begin{bmatrix} y_1 \\ y_2 \end{bmatrix} \qquad W_{2i+1} \in G_{2i+1}$$

and $L$ is a given degree of the normal form equation. The validity of such simplification is guaranteed by the implicit function theorem, as for each $\mu$ near $\mu_0$, there will be an equilibrium $p(\mu)$ near $p(\mu_0)$ that varies smoothly with $\mu$. The normal form of eq. (10.44) in polar co-ordinates can then be written as the form:

$$\begin{cases} y_1 = r\cos\theta, \quad y_2 = r\sin\theta \\ \dot{r} = r\left(\beta\mu + \sum_{i=1}^{L} a_i r^{2i}\right) + \text{hot} \\ \dot{\theta} = \omega_0 + \alpha\mu + \sum_{i=1}^{L} b_i r^{2i} + \text{hot} \end{cases} \tag{10.45}$$

where hot denotes "higher order terms".

A general formula for each $a_i$, $b_i$ is not presently available but we can derive their expressions up to any desired higher order using the algorithm mentioned above recursively, such that

$$\begin{cases} a_3 = \dfrac{1}{128}\left(35a_{70}^{(5)} + 5a_{52}^{(5)} + 3a_{34}^{(5)} + 5a_{16}^{(5)} + 5b_{61}^{(5)} + 3b_{43}^{(5)} + 5b_{25}^{(5)} + 35b_{07}^{(5)}\right) \\ b_3 = \dfrac{-1}{128}\left(5a_{61}^{(5)} + 3a_{43}^{(5)} + 5a_{25}^{(5)} + 35a_{07}^{(5)} - 35b_{70}^{(5)} - 5b_{34}^{(5)} - 3b_{34}^{(5)} - 5b_{16}^{(5)}\right) \end{cases}$$

$$\begin{cases} a_4 = \dfrac{1}{256}\left(63a_{90}^{(7)} + 7a_{72}^{(7)} + 3a_{54}^{(7)} + 3a_{36}^{(7)} + 7a_{18}^{(7)} + 7b_{81}^{(7)} + 3b_{63}^{(7)} + 3b_{45}^{(7)} + 7b_{27}^{(7)} + 63b_{09}^{(7)}\right) \\ b_4 = \dfrac{-1}{256}\left(7a_{81}^{(7)} + 3a_{63}^{(7)} + 3a_{45}^{(7)} + 7a_{27}^{(7)} + 63a_{09}^{(7)} - 63b_{90}^{(7)} - 7b_{72}^{(7)} - 3b_{54}^{(7)} - 3b_{36}^{(7)} - 7b_{18}^{(7)}\right) \end{cases}$$

and for the eleventh order

$$\begin{cases} a_5 = \dfrac{1}{1024}\left(231a_{B0}^{(9)} + 21a_{92}^{(9)} + 7a_{74}^{(9)} + 5a_{56}^{(9)} + 7a_{38}^{(9)} + 21a_{1A}^{(9)} + 21b_{A1}^{(9)}\right. \\ \qquad \left. + 7b_{83}^{(9)} + 5b_{65}^{(9)} + 7b_{47}^{(9)} + 21b_{29}^{(9)} + 231b_{0B}^{(9)}\right) \\ b_5 = \dfrac{-1}{1024}\left(21a_{A1}^{(9)} + 7a_{83}^{(9)} + 5a_{65}^{(9)} + 7a_{47}^{(9)} + 21a_{29}^{(9)} + 231a_{0B}^{(9)} - 231b_{B0}^{(9)}\right. \\ \qquad \left. - 21b_{92}^{(9)} - 7b_{74}^{(9)} - 5b_{56}^{(9)} - 7b_{38}^{(9)} - 21b_{1A}^{(9)}\right) \end{cases}$$

where the subscripts $A$, $B$ refer to the indices 10 and 11, respectively.

The complexity of higher order normal forms rapidly becomes apparent as pointed out by Leung and Zhang [166]. Thus, the algorithm is required to be implemented with the symbolic manipulation in *Mathematica* (Wolfram [172]). In the case of $m>5$, however, it may also cause overflow in computers equipped with conventional memory if we apply the nearly identity normal transformations directly into eqs. (10.23) and (10.34). A general explicit formula representing the homogeneous polynomial terms is therefore very useful to reduce the size of the problem.

If the nearly identical change of co-ordinate of order $k$

$$x = y + h_k(y) \tag{10.46}$$

is applied then the new homogeneous polynomials of order $n$ can be obtained by

$$n<k, \quad F_n^*(y) = F_n(y) \tag{10.47}$$

$$n=k, \quad F_n^*(y) = F_n(y) + J \cdot h_k(y) - Dh_k(y) \cdot J \cdot y \tag{10.48}$$

$$n>k, \quad F_n^*(y) = F_n(y) + \left[DF_{n-k+1}(y) \cdot h_k(y) - Dh_k(y) \cdot F_{n-k+1}(y)\right]_{n \geq 2k-1} + \sum_{m=2}^{k_m} \left[ \frac{D_y^m F_{n+m(1-k)}(y)}{m!} \left[h_k(y)\right]^m \right]_{n+m(1-k) \geq k} \tag{10.49}$$

where $F_n^*(y)$ denotes the resulting homogeneous polynomials, $k_m = \min[k_1, k_2]$, and

$$k_1 = \text{floor}\left[\frac{n-k}{k-1}\right], \quad k_2 = \text{floor}\left[\frac{n}{k}\right]$$

The operator floor $[\cdot]$ gives the greatest integer less than or equal to the variable in the bracket. The result confirms that only odd-order terms would appear in the normal form equation, corresponding to Poincaré resonance (Guckenheimer and Holmes [36]).

### 10.1.6 Oscillators with Odd Non-linearity

Here we consider the non-linear differential equations with $f(u)$ to be an odd function of $u$. To illustrate this, two examples are presented. The first one is the well-known Duffing oscillator, which arises from various physical and engineering problems. The solution of this oscillator has been thoroughly studied so that the accuracy of the results can be compared with the existing methods. In the second example, we will work with an oscillator having a term of fifth power.

Consider the Duffing oscillator

$$\ddot{x} + \omega_0^2 x = -\varepsilon x^3 \tag{10.50}$$

with the initial conditions

$$x(0) = a, \quad \dot{x}(0) = 0 \tag{10.51}$$

where the parameter $\varepsilon>0$ and $\omega_0$ is the linear frequency. In classical perturbation methods, it is usually assumed that $\varepsilon$ is small. In the present case, however, $\varepsilon$ need not be small. With the transformation $x = x_1, \dot{x} = -\omega_0 x_2$, we rewrite eq. (10.50) as

$$\begin{bmatrix} \dot{x}_1 \\ \dot{x}_2 \end{bmatrix} = \begin{bmatrix} 0 & -\omega_0 \\ \omega_0 & 0 \end{bmatrix} \begin{bmatrix} x_1 \\ x_2 \end{bmatrix} + \begin{bmatrix} 0 \\ \dfrac{\varepsilon}{\omega_0} x_1^3 \end{bmatrix} \tag{10.52}$$

Comparing eq. (10.52) with the standard third-order normal form $\{\dot{x}\} = \{x\} + F_3(x)$, we have $b_{30} = (\varepsilon/\omega_0)$ and all the other coefficients are zero. If we substitute these coefficients into eq. (10.42), we obtain $a_1 = 0,\ b_1 = (3\varepsilon/8\omega_0)$.

The third-order non-linear transformation of co-ordinate $\{x\} = [J]\{y\} + h_3(y)$ can be determined by eq. (10.43) and the coefficients of $h_3(y)$ are $c_{30} = c_{03} = c_{21} = d_{30} = d_{12} = 0$,

$$c_{12} = \frac{\varepsilon}{8\omega_0^2},\quad d_{21} = \frac{5\varepsilon}{8\omega_0^2},\quad d_{03} = \frac{\varepsilon}{4\omega_0^2}$$

Here we use $y$ to denote the new co-ordinates and $x$ the old ones. Under the fourth-order non-linear transformation, all the coefficients of the homogeneous polynomial are identically zero:

$$c_{40} = c_{31} = c_{22} = c_{13} = c_{04} = d_{40} = d_{31} = d_{22} = d_{13} = d_{04} = 0$$

According to the method mentioned in the previous section, such computation can be repeated up to the desired order. Finally, we have determined that all the coefficients of the even-order co-ordinate homogeneous polynomials $h_{2n}(y)$ are equal to zero and all the even-order terms are already in their simplest forms. Thus, only the odd-order homogeneous polynomials need to be handled in our computation.

The coefficients of normal form $a_2$ and $b_2$ can be obtained by taking the change of variable

$$\{y\} = \{z\} + h_5(z);\quad a_2 = 0,\ b_2 = \frac{-21\varepsilon^2}{256\omega_0^3}$$

Then, the fifth-order non-linear co-ordinate change $h_5(z)$ is given by

$$c_{50} = c_{05} = c_{23} = c_{41} = d_{50} = d_{14} = d_{32} = 0$$

$$c_{32} = \frac{21\varepsilon^2}{256\omega_0^4},\quad c_{14} = \frac{25\varepsilon^2}{256\omega_0^4},\quad d_{41} = \frac{-39\varepsilon^2}{256\omega_0^4},\quad d_{23} = \frac{-67\varepsilon^2}{256\omega_0^4},\quad d_{05} = \frac{-\varepsilon^2}{32\omega_0^4}$$

Similarly, after taking the seventh-order polynomial change of variables

$$\{z\} = \{w\} + h_7(w)$$

we obtain

$$a_3 = 0,\quad b_3 = \frac{237\varepsilon^3}{4096\omega_0^5}$$

and $h_7(w)$ is given by

$$c_{70}=c_{07}=c_{61}=c_{43}=c_{25}=d_{70}=d_{52}=d_{34}=d_{16}=0$$

$$c_{52}=\frac{-183\varepsilon^3}{4096\omega_0^6},\quad c_{34}=\frac{-91\varepsilon^3}{1024\omega_0^6},\quad c_{16}=\frac{-133\varepsilon^3}{4096\omega_0^6},\quad d_{61}=\frac{363\varepsilon^3}{4096\omega_0^6},$$

$$d_{43}=\frac{217\varepsilon^3}{4096\omega_0^6},\quad d_{25}=\frac{633\varepsilon^3}{4096\omega_0^6},\quad d_{07}=\frac{5\varepsilon^3}{256\omega_0^6}$$

Performing the variable changes further to the ninth order, $\{w\}=\{u\}+h_9(u)$, yields

$$a_4=0,\quad b_4=\frac{-11241\varepsilon^4}{262144\omega_0^7}$$

and $h_9(u)$ is given by

$$c_{90}=c_{09}=c_{81}=c_{63}=c_{45}=c_{27}=d_{90}=d_{72}=d_{54}=d_{36}=d_{18}=0$$

$$c_{72}=\frac{7089\varepsilon^4}{262144\omega_0^7},\ c_{54}=\frac{19215\varepsilon^4}{262144\omega_0^7},\ c_{36}=\frac{19003\varepsilon^4}{262144\omega_0^7},\ c_{18}=\frac{7901\varepsilon^4}{262144\omega_0^7},$$

$$d_{81}=\frac{-16035\varepsilon^4}{262144\omega_0^7},d_{63}=\frac{-46821\varepsilon^4}{262144\omega_0^7},\ d_{45}=\frac{-57969\varepsilon^4}{262144\omega_0^7},\ d_{27}=\frac{-38575\varepsilon^4}{262144\omega_0^7},$$

$$d_{09}=\frac{-29\varepsilon^4}{2048\omega_0^7}$$

If further transformations beyond the ninth order were performed, it is interesting to note that all the following higher order coefficients of normal forms $a_i, b_i (i=5,6,7\ldots)$ as well as the co-ordinate transformation $h_k(y)(k=10,11,12,\ldots)$ vanish. Therefore, the normal form of our cubic is of degree nine. The dynamic system can be represented by an exact polynomial differential amplitude equation in a finite number of terms.

The asymptotic solution of this normal form equation is

$$\begin{cases} u_1=r\cos\theta \\ \dot{r}=0 \\ \dot{\theta}=\omega_0+\dfrac{3\varepsilon r^2}{8\omega_0}-\dfrac{21\varepsilon^2r^4}{256\omega_0^3}+\dfrac{237\varepsilon^3r^6}{4096\omega_0^5}-\dfrac{11241\varepsilon^4r^8}{262144\omega_0^7}\end{cases}\tag{10.53}$$

The steady-state periodic solution of eq. (10.53) is

$$\begin{cases} u_1=a\cos\theta,\quad u_2=a\sin\theta \\ \theta=\left(\omega_0+\dfrac{3\varepsilon a^2}{8\omega_0}-\dfrac{21\varepsilon^2a^4}{256\omega_0^3}+\dfrac{237\varepsilon^3a^6}{4096\omega_0^5}-\dfrac{11241\varepsilon^4a^8}{262144\omega_0^7}\right)t\end{cases}\tag{10.54}$$

To get the steady-state periodic solution in the original co-ordinate, we could trace back all the transformation, for example

$$\begin{aligned}
\{x\} &= \{y\} + h_3(y) \\
&= \{z\} + h_5(z) + h_3\left(z + h_5(z)\right) \\
&= \{w\} + h_7(w) + h_5\left(w + h_7(w)\right) + h_3\left(\{w\} + h_7(w) + h_5\left(u + h_7(w)\right)\right) \\
&= \{u\} + h_9(u) + h_7\left(\{u\} + h_9(u)\right) + h_5\left(\{u\} + h_9(u) + h_7\left(\{u\} + h_9(u)\right)\right) \\
&\quad + h_3\left(\{u\} + h_9(u) + h_7\left(\{u\} + h_9(u)\right) + h_5\left(\{u\} + h_9(u) + h_7\left(\{u\} + h_9(u)\right)\right)\right)
\end{aligned} \tag{10.55}$$

The steady-state periodic solution of the Duffing equation is

$$\begin{cases}
x = \left(a - \dfrac{\varepsilon a^3}{32\omega_0^2} + \dfrac{23\varepsilon^2 a^5}{1024\omega_0^4} - \dfrac{167\varepsilon^3 a^7}{16384\omega_0^6} + \dfrac{3431\varepsilon^4 a^9}{524288\omega_0^4}\right)\cos\theta \\
\quad + \left(\dfrac{\varepsilon a^3}{32\omega_0^2} - \dfrac{3\varepsilon^2 a^5}{128\omega_0^4} + \dfrac{309\varepsilon^3 a^7}{32768\omega_0^6} - \dfrac{7033\varepsilon^4 a^9}{1048576\omega_0^4}\right)\cos 3\theta \\
\quad + \left(\dfrac{\varepsilon^2 a^5}{1024\omega_0^4} + \dfrac{31\varepsilon^3 a^7}{32768\omega_0^6} + \dfrac{191\varepsilon^4 a^9}{524288\omega_0^4}\right)\cos 5\theta \\
\quad - \left(\dfrac{3\varepsilon^3 a^7}{16384\omega_0^6} + \dfrac{227\varepsilon^4 a^9}{1048576\omega_0^4}\right)\cos 7\theta + \left(\dfrac{\varepsilon^4 a^9}{65536\omega_0^4}\right)\cos 9\theta \\
\theta = \left(\omega_0 + \dfrac{3\varepsilon a^2}{8\omega_0} - \dfrac{21\varepsilon^2 a^4}{256\omega_0^3} + \dfrac{237\varepsilon^3 a^6}{4096\omega_0^5} - \dfrac{11241\varepsilon^4 a^8}{262144\omega_0^7}\right)t
\end{cases} \tag{10.56}$$

We compare the solution by the Lindstedt—Poincaré (L-P) method with the same initial conditions below:

$$\begin{cases}
x = \left(a - \dfrac{\varepsilon a^3}{32\omega_0^2} + \dfrac{23\varepsilon^2 a^5}{1024\omega_0^4} - \dfrac{547\varepsilon^3 a^7}{32768\omega_0^6} + \dfrac{6713\varepsilon^4 a^9}{524288\omega_0^4}\right)\cos\theta \\
\quad + \left(\dfrac{\varepsilon a^3}{32\omega_0^2} - \dfrac{3\varepsilon^2 a^5}{128\omega_0^4} + \dfrac{297\varepsilon^3 a^7}{16384\omega_0^6} - \dfrac{15121\varepsilon^4 a^9}{1048576\omega_0^4}\right)\cos 3\theta \\
\quad + \left(\dfrac{\varepsilon^2 a^5}{1024\omega_0^4} - \dfrac{3\varepsilon^3 a^7}{2048\omega_0^6} + \dfrac{883\varepsilon^4 a^9}{524288\omega_0^4}\right)\cos 5\theta \\
\quad + \left(\dfrac{\varepsilon^3 a^7}{32768\omega_0^6} - \dfrac{9\varepsilon^4 a^9}{131072\omega_0^4}\right)\cos 7\theta + \left(\dfrac{\varepsilon^4 a^9}{1048576\omega_0^4}\right)\cos 9\theta \\
\theta = \left(\omega_0 + \dfrac{3\varepsilon a^2}{8\omega_0} - \dfrac{21\varepsilon^2 a^4}{256\omega_0^3} + \dfrac{81\varepsilon^3 a^6}{2048\omega_0^5} - \dfrac{6549\varepsilon^4 a^8}{262144\omega_0^7} + O\left(|a|^{10}\right)\right)t
\end{cases} \tag{10.57}$$

It is clear that the coefficients of the first three orders of $\varepsilon$ in eq. (10.57) are exactly the same as those given in eq. (10.55). Nevertheless, eq. (10.55) is different from eq. (10.57) in the higher order terms. The result derived by normal form theory gives a finite, asymptotic power series, but using the L-P method the solution is represented by an infinite one.

### 10.1.7 Oscillator with Fifth-Power Non-linearity

Consider an oscillator with a fifth-power non-linearity as the second example,

$$\ddot{x}+\omega_0^2 x=-\varepsilon x^5 \tag{10.58}$$

with initial conditions (10.51). Using the normal form method described previously, we obtain the following results

$$a_1=0,\quad b_1=\frac{5\varepsilon}{16\omega_0}$$

and the fifth-order non-linear co-ordinate change is given by

$$c_{50}=c_{05}=c_{23}=c_{41}=d_{50}=d_{14}=d_{32}=0$$

$$c_{32}=\frac{-3\varepsilon}{16\omega_0^2},\quad c_{14}=\frac{-7\varepsilon}{48\omega_0^2},\quad d_{41}=\frac{11\varepsilon}{16\omega_0^2},\quad d_{23}=\frac{31\varepsilon}{48\omega_0^2},\quad d_{05}=\frac{\varepsilon}{6\omega_0^2}$$

Then, taking seventh-order transformation of co-ordinates, we obtain $a_2=0, b_2=0$, with $h_7(w)=\{0\}$. After that, we directly take the ninth-order change of variable which yields $a_3=0,\ b_3=\left(-215\varepsilon^2/3027\omega_0^3\right)$, and

$$c_{90}=c_{09}=c_{81}=c_{63}=c_{45}=c_{27}=d_{90}=d_{72}=d_{54}=d_{36}=d_{18}=0$$

$$c_{72}=\frac{295\varepsilon^2}{3072\omega_0^4},\ c_{54}=\frac{2915\varepsilon^2}{9216\omega_0^4},\ c_{36}=\frac{327\varepsilon^2}{1024\omega_0^4},\ c_{18}=\frac{937\varepsilon^2}{9216\omega_0^4},\ d_{81}=\frac{-445\varepsilon^2}{3072\omega_0^4},$$

$$d_{63}=\frac{-5825\varepsilon^2}{9216\omega_0^4},\ d_{45}=\frac{-6173\varepsilon^2}{9216\omega_0^4},\ d_{27}=\frac{-2027\varepsilon^2}{9216\omega_0^4},\ d_{09}=\frac{-\varepsilon^2}{72\omega_0^4}$$

The steady-state periodic solution of the oscillator with fifth-power non-linearity is

$$\begin{cases} x=\left(a-\dfrac{\varepsilon a^5}{24\omega_0^2}+\dfrac{3791\varepsilon^2 a^9}{147456\omega_0^4}\right)\cos\theta+\left(\dfrac{5\varepsilon a^5}{128\omega_0^2}-\dfrac{5\varepsilon^2 a^9}{192\omega_0^4}\right)\cos 3\theta \\ \quad +\left(\dfrac{\varepsilon a^5}{384\omega_0^2}\right)\cos 5\theta+\left(\dfrac{95\varepsilon^2 a^9}{294912\omega_0^4}\right)\cos 7\theta+\left(\dfrac{\varepsilon^2 a^9}{98304\omega_0^4}\right)\cos 9\theta \\ \theta=\left(\omega_0+\dfrac{5\varepsilon a^4}{16\omega_0}-\dfrac{215\varepsilon^2 a^8}{3072\omega_0^3}\right)t \end{cases} \tag{10.59}$$

The terms of the asymptotic normal form series expansion of eq. (10.59) is also finite with all the higher order normal forms equal to zero.

### 10.1.8 Oscillators with Even Non-linearity

The normal form method can also be easily extended to analyze an oscillator with even non-linearity. A quadratic non-linearity equation is considered in the following calculation.

Consider a system having quadratic non-linearity

$$\ddot{x} + \omega_0^2 x = -\varepsilon x^2 \tag{10.60}$$

with initial conditions (10.51). According to the normal form method, one needs to do a second-order homogeneous polynomial transformation of co-ordinates to simplify the second-order terms in the original equations. The coefficients of this change can be evaluated by eq. (10.26), the results being

$$c_{20} = \frac{-\varepsilon}{3\omega_0^2}, \quad c_{11} = 0, \quad c_{02} = \frac{-2\varepsilon}{3\omega_0^2}, \quad d_{20} = 0, \quad d_{11} = \frac{2\varepsilon}{3\omega_0^2}, \quad d_{02} = 0$$

The subsequent third-order variable change gives

$$a_1 = 0, \quad b_1 = \frac{-5\varepsilon^2}{12\omega_0^3}$$

and

$$c_{30} = c_{03} = c_{21} = d_{30} = d_{12} = 0$$

$$c_{12} = -\frac{\varepsilon^2}{12\omega_0^4}, \quad d_{21} = -\frac{\varepsilon^2}{4\omega_0^4}, \quad d_{03} = -\frac{\varepsilon^2}{2\omega_0^4}$$

The fourth-order homogeneous polynomial of transformation cannot be overlooked in this case and finally its coefficients are found to be

$$c_{31} = c_{13} = d_{50} = d_{22} = 0$$

$$c_{40} = -\frac{\varepsilon^3}{9\omega_0^6}, \quad c_{22} = -\frac{8\varepsilon^3}{9\omega_0^6}, \quad c_{04} = -\frac{28\varepsilon^3}{27\omega_0^6}, \quad d_{31} = \frac{4\varepsilon^3}{9\omega_0^6}, \quad d_{13} = \frac{16\varepsilon^3}{27\omega_0^6}$$

Then normal-form coefficients $a_2$, $b_2$ and the fifth-order homogeneous polynomial of co-ordinate transformation can be evaluated as

$$a_2 = 0, \quad b_2 = \frac{-455\varepsilon^4}{1728\omega_0^7}$$

with

$$c_{50} = c_{05} = c_{23} = c_{41} = d_{50} = d_{14} = d_{32} = 0$$

$$c_{32} = \frac{23\varepsilon^4}{1728\omega_0^8}, \quad c_{14} = \frac{217\varepsilon^4}{5184\omega_0^8}, \quad d_{41} = \frac{-365\varepsilon^4}{1728\omega_0^8}, \quad d_{23} = \frac{-499\varepsilon^4}{5184\omega_0^8}, \quad d_{05} = \frac{263\varepsilon^4}{648\omega_0^8}$$

The sixth-order homogeneous polynomial of the transformation is determined by its coefficients:

$$c_{51} = c_{33} = c_{15} = d_{60} = d_{42} = d_{24} = d_{06} = 0$$

$$c_{60} = -\frac{25\varepsilon^5}{324\omega_0^{10}}, \quad c_{42} = -\frac{23\varepsilon^5}{54\omega_0^{10}}, \quad c_{24} = \frac{5\varepsilon^5}{243\omega_0^{10}}, \quad c_{06} = \frac{182\varepsilon^5}{243\omega_0^{10}},$$

$$d_{51} = \frac{25\varepsilon^5}{81\omega_0^{10}}, \quad d_{33} = \frac{88\varepsilon^5}{243\omega_0^{10}}, \quad d_{15} = -\frac{32\varepsilon^5}{243\omega_0^{10}}$$

The coefficients of the normal form $a_3, b_3$ are

$$a_3 = 0, \quad b_3 = \frac{-28675\varepsilon^6}{12441\omega_0^{11}}$$

The seventh-order co-ordinate transformation is determined by:

$$c_{70} = c_{07} = c_{61} = c_{43} = c_{25} = d_{70} = d_{52} = d_{34} = d_{16} = 0$$

$$c_{52} = -\frac{4967\varepsilon^6}{12441\omega_0^{12}}, \quad c_{34} = \frac{19105\varepsilon^6}{93312\omega_0^{12}}, \quad c_{16} = -\frac{51263\varepsilon^6}{373248\omega_0^{12}}, \quad d_{61} = -\frac{37285\varepsilon^6}{124416\omega_0^{12}},$$

$$d_{43} = \frac{-77285\varepsilon^6}{93312\omega_0^{12}}, \quad d_{25} = -\frac{114743\varepsilon^6}{124416\omega_0^{12}}, \quad d_{07} = -\frac{11201\varepsilon^6}{23328\omega_0^{12}}$$

Substituting the transformation into the original equation, we have the steady-state periodic solution of the oscillator with quadratic non-linearity

$$\begin{cases} x = -\dfrac{\varepsilon a^2}{2\omega_0^2} - \dfrac{13\varepsilon^3 a^4}{24\omega_0^6} + \dfrac{319\varepsilon^5 a^6}{1728\omega_0^{10}} \\ \quad + \left(a - \dfrac{\varepsilon^2 a^3}{48\omega_0^4} + \dfrac{143\varepsilon^4 a^5}{2073\omega_0^8} - \dfrac{35005\varepsilon^6 a^7}{20736\omega_0^{12}}\right)\cos\theta \\ \quad + \left(\dfrac{\varepsilon a^2}{6\omega_0^2} + \dfrac{25\varepsilon^3 a^4}{54\omega_0^6} - \dfrac{12479\varepsilon^5 a^6}{31104\omega_0^{10}}\right)\cos 2\theta \\ \quad + \left(\dfrac{\varepsilon^2 a^3}{48\omega_0^4} - \dfrac{5\varepsilon^4 a^5}{576\omega_0^8} + \dfrac{9799\varepsilon^6 a^7}{33177\omega_0^{12}}\right)\cos 3\theta \\ \quad + \left(-\dfrac{7\varepsilon^3 a^4}{216\omega_0^6} + \dfrac{2353\varepsilon^5 a^6}{15552\omega_0^{10}}\right)\cos 4\theta + \left(\dfrac{37\varepsilon^4 a^5}{20736\omega_0^8} - \dfrac{5633\varepsilon^6 a^7}{995328\omega_0^{12}}\right)\cos 5\theta \\ \quad - \left(\dfrac{41\varepsilon^5 a^6}{3456\omega_0^{10}}\right)\cos 6\theta - \left(\dfrac{641\varepsilon^6 a^7}{1492992\omega_0^{12}}\right)\cos 7\theta \\ \quad \theta = \left(\omega_0 - \dfrac{5\varepsilon^2 a^2}{12\omega_0^3} - \dfrac{455\varepsilon^4 a^4}{1278\omega_0^7} - \dfrac{28675\varepsilon^6 a^6}{124416\omega_0^{11}} + O\left(|a|^8\right)\right)t \end{cases} \tag{10.61}$$

### 10.1.9 Duffing—van der Pol Oscillator

In the problem of flow-induced oscillations, the governing differential equation can be described by the well-known Duffing—van der Pol equation (Blevins, [180])

$$\ddot{x} - 2\varepsilon\lambda\dot{x} + x + \varepsilon x^3 + \varepsilon x^2 \dot{x} = 0 \tag{10.62}$$

with initial conditions (10.50). Here we assume that $\lambda$ is a control variable of the system and $\varepsilon$ is a book-keeping device for small perturbations. Equation (10.59) can further be approximately expressed by a Jordan canonical matrix form

$$\begin{bmatrix} \dot{y}_1 \\ \dot{y}_2 \end{bmatrix} = \begin{bmatrix} \varepsilon\lambda & -1 \\ 1 & \varepsilon\lambda \end{bmatrix} \begin{bmatrix} y_1 \\ y_2 \end{bmatrix} - \varepsilon \begin{bmatrix} y_1 y_2^2 + y_2^3 \\ 0 \end{bmatrix} \tag{10.63}$$

in which the first-order near identity transformation is given by

$$\begin{bmatrix} x_1 \\ x_2 \end{bmatrix} = \begin{bmatrix} \varepsilon\lambda & -1 \\ 1 & 0 \end{bmatrix} \begin{bmatrix} y_1 \\ y_2 \end{bmatrix} \tag{10.64}$$

where we choose $x_1 = x$, $x_2 = -\dot{x}$. Note that the higher order terms in $\varepsilon$ are neglected during the computation. We take transformations up to order five. The resultant normal form of eq. (10.63) in polar co-ordinates is expressed as follows:

$$\begin{cases} y_1 = r\cos\theta, \quad y_2 = r\sin\theta \\ \dot{r} = \varepsilon r\left(\lambda - \dfrac{1}{8}r^2\right) \\ \theta = 1 + \dfrac{3\varepsilon}{8}r^2 \end{cases} \tag{10.65}$$

The coefficients of the intermediate co-ordinate transformations are

$$c_{30} = c_{03} = 0,\ c_{21} = c_{12} = \frac{3\varepsilon}{8},\ d_{30} = d_{03} = \frac{-\varepsilon}{4},\ d_{21} = \frac{-3\varepsilon}{8},\ d_{12} = \frac{-\varepsilon}{8}$$

respectively. The asymptotic solution of eq. (10.65) could also be obtained if we trace back all the nearly identity co-ordinate transformations

$$x = \varepsilon a\left(-\lambda + \frac{7}{32}a^2\right)\cos\theta + \frac{\varepsilon}{32}a^3\cos 3\theta + \left(-a + \frac{9\varepsilon}{32}a^3\right)\sin\theta + \frac{\varepsilon}{32}a^3\sin 3\theta \tag{10.66}$$

where the constant $a$ denotes the amplitude defined by the initial condition. Further discussion of the normal-form amplitude eq. of (11.56) reveals the Hopf bifurcation of the Duffing—van der Pol equation on the parametric plane $r$—$\lambda$. As shown in Fig. 10.1, a spiral sink exists at (0,0)

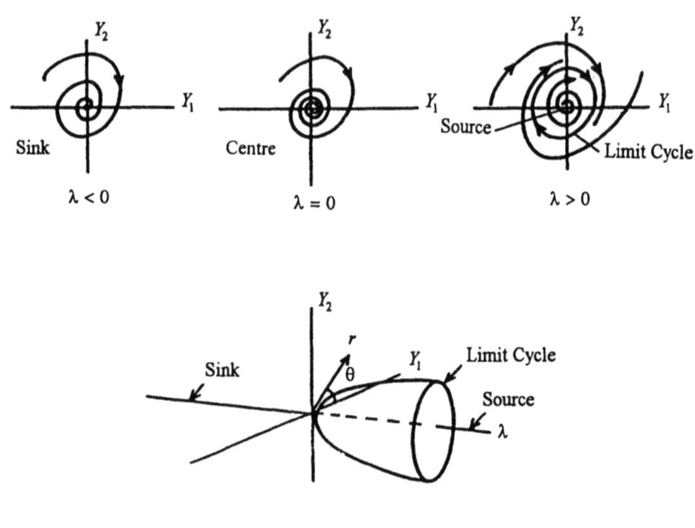


Fig. 10.1

for $l<0$ and a source at (0,0) surrounded by a limit cycle for $\lambda>0$. The limit cycle evolves continuously from the centre at (0,0) for $\lambda=0$. The Hopf bifurcation is of importance in situations where a flow-induced oscillator is subjected to flutters or self-exciting movements. In such circumstances, the orbits of the steady-state periodic solutions stay on the surface of the parabola rotated by $\lambda=(1/8)r^2$.

### 10.1.10 Conclusions

We have presented an arithmetic algorithm to compute the higher order normal forms. By applying the explicit formula proposed in this discussion, we can achieve, in a standard manner, the desired higher order normal forms for non-linear differential polynomial equations. We found that the steady-state solution of the undamped Duffing equation can be represented by a finite cosine series with varying phase in finite polynomial terms. To show the versatility of the algorithm, we illustrated an example in which the order of non-linearity is not restricted to odd numbers. The application to limit cycle bifurcation is also demonstrated by the Duffing—van der Pol oscillator.

## 10.2 Symplectic Integration and Geometric Non-linear Finite Element Method

The geometric non-linear analyses of framed structures for both static and dynamic loading are of considerable interest and practical importance. Chajes and Churchill [238] and Meek and Tan [246] considered the influence of higher order non-linear terms and derived the stiffness matrix expressions accurately. But considerable efforts have been concentrated on the tangent or secant stiffness matrices that consider geometric non-linearity of various degrees. In Leung and Mao [244], the second- and third-order non-linear stiffness matrices were introduced by the order of nodal displacements. The equilibrium equations were accurately discretised and an explicit beam finite element was established. The dynamic problems of beams with large displacement and small strain and rotation have been solved. Using the explicit beam finite element, both programming and computing the strain energy of the discretised systems and the derivatives of the strain energy with respect to the nodal displacements are convenient and more advantageous.

In this section, a beam finite element formulation (Leung and Mao [244]) is extended to framed structures. The symplectic numerical integration schemes (Feng and Qin [241], Ruth [250], Wu [253]) are applied to solve the equations element by element without forming the global matrices. This makes it possible to solve non-linear vibrations of large framed structure using a small computer. Symplectic integration schemes are used. Emphasis is on the asymmetric response of symmetric structures subject to symmetric dynamic loads.

### 10.2.1 Lagrangian Equations

For an initially straight Euler—Bernoulli beam undergoing large displacement with small strain and small rotation, the axial strain $\varepsilon_x$ at the centroid of the cross-section can be expressed as (Crisfield [239])

$$\varepsilon_x = \frac{\partial u}{\partial x} + \frac{1}{2}\left(\frac{\partial w}{\partial x}\right)^2 \tag{10.67}$$

Assuming plane sections remain plane, the displacement in the $x$ direction $u^z$ and the strain $\varepsilon_x^z$ at distance $z$ from the centroid are given by

$$u^z = u - z\frac{dw}{dx} \tag{10.68}$$

$$\varepsilon_x^z = \varepsilon_x + z\chi \tag{10.69}$$

where the curvature $\chi$ is defined as

$$\chi = -\frac{\partial^2 w}{\partial x^2} \tag{10.70}$$

Here $u(x,t)$ and $w(x,t)$ are the axial and transverse displacements of the neutral axis, respectively. The axial resultant force $S$ and the bending moment $M$ are related to $\varepsilon_x$ and $\chi$ and are given by

$$S = EA\varepsilon_x \tag{10.71}$$

$$M = EI\chi \tag{10.72}$$

where $E$ is Young's modulus, $A$ the cross-sectional area, and $I$ the second moment of area of the cross-section. The corresponding Lagrangian for an undamped beam of length $l$ is

$$L = \frac{1}{2}\int_0^l \rho A\left[\left(\frac{\partial u}{\partial t}\right)^2 + \left(\frac{\partial w}{\partial t}\right)^2\right]\mathrm{d}x - \frac{1}{2}\int_0^l \left(EA\varepsilon_x^2 + EI\chi^2\right)\mathrm{d}x + \int_0^l \left(F_u u + F_w w\right)\mathrm{d}x \tag{10.73}$$

where $\rho$ is the mass density of material, $t$ is the time variable, and $F_u(x,t)$ and $F_w(x,t)$ are the axial and transverse loads, respectively.

### 10.2.2 Finite Element Discretisation

Consider an initially straight beam. The displacements $u(x,t)$ and $w(x,t)$ are interpolated by their nodal values $u$(t), $w$(t), respectively, so that

$$\begin{aligned} u(x,t) &= [n(x)]\{u(t)\} \\ w(x,t) &= [N(x)]\{w(t)\} \end{aligned} \tag{10.74}$$

where $[N(x)] = [1 - 3\xi^2 + 2\xi^3, \xi(\xi^2 - 2\xi + 1)l, 3\xi^2 - 2\xi^3, (\xi^3 - \xi^2)l]$, and $[n(x)] = [1 - \xi, \xi]$ are the shape functions and $\xi = x/l$. After integration, eq. (10.73) will be discretized into

$$L = \tfrac{1}{2}\{\dot{u}\}^T[M_u]\{\dot{u}\} + \tfrac{1}{2}\{\dot{w}\}^T[M_w]\{\dot{w}\} - \tfrac{1}{2}\{u\}^T[M_u]\{u\} - \tfrac{1}{2}\{w\}^T[M_w]\{w\}$$
$$- \tfrac{1}{2}\{u\}^T[M_{uq}]\{w_q\} - \tfrac{1}{8}\{w_q\}^T[M_q]\{w_q\} + \{F_u\}^T\{u\} + \{F_w\}^T\{w\} \tag{10.75}$$

where, for the element $e$ with node $i$ and $j$,

$$\{w^e\} = \{w_i, \theta_i, w_j, \theta_j\}, \ \{u^e\} = \{u_i, u_j\}$$

$$[M_u^e] = \int_0^l \rho A[n]^T[n]\mathrm{d}x, \quad [M_w^e] = \int_0^l \rho A[N]^T[N]\mathrm{d}x$$

$$[K_u^e] = \int_0^l EA[n_{,x}]^T[n_{,x}]\mathrm{d}x, \quad [K_w^e] = \int_0^l EI[N_{,xx}]^T[N_{,xx}]\mathrm{d}x$$

$$\{F_u^e\} = \int_0^l [n]^T F_u \mathrm{d}x, \quad \{F_w^e\} = \int_0^l [N]^T F_w \mathrm{d}x$$

The following terms are noted: the quadratic nodal displacement

$$\{w_q^e\} = \{w_i^2, w_i\theta_i, w_i w_j, w_i\theta_j, \theta_i^2, \theta_i w_j, \theta_i\theta_j, w_j^2, w_j\theta_j, \theta_j^2,\}$$

the second-order stiffness matrix

$$[K_{uq}^e] = \int_0^l EA[n_{,x}]^T[Q]\mathrm{d}x$$

and the third-order stiffness matrix

$$[K_q^e] = \int_0^l EA[Q]^T[Q]\mathrm{d}x$$

in which $[Q]$ is defined by the interpolation, $(\partial w/\partial x)^2 = [Q]\{w_q^e\}$ and is given by

$$[Q] = \Big[\frac{36}{l^2}(\xi^2 - 2\xi^3 + \xi^4), \frac{12}{l}(-\xi + 5\xi^2 - 7\xi^3 + 3\xi^4), \frac{72}{l^2}(-\xi^2 + 2\xi^3 - \xi^4),$$
$$\frac{12}{l}(2\xi^2 - 5\xi^3 + 3\xi^4), (1 - 8\xi + 22\xi^2 - 24\xi^3 + 9\xi^4), \frac{12}{l}(\xi - 5\xi^2 + 7\xi^3 - 3\xi^4),$$
$$2(-2\xi + 11\xi^2 - 18\xi^3 + 9\xi^4), \frac{36}{l^2}(\xi^2 - 2\xi^3 + \xi^4), \frac{12}{l}(-2\xi^2 + 5\xi^3 - 3\xi^4),$$
$$(4\xi^2 - 12\xi^3 + 9\xi^4)\Big]$$

An overdot denotes differentiation with respect to $t$. The matrices $[K_{uq}^e]$ and $[K_q^e]$, which are independent of the nodal displacements, are given explicitly in Leung and Mao [244]. The others are available in most of the references quoted in this section.

### 10.2.3 Symplectic Integration Schemes

The Lagrangian formulation is more popular than the Hamiltonian because the former uses physical quantities directly in the configuration space. The Hamiltonian methods are not particularly superior to Lagrangian techniques for the direct solution of mechanical problems. Rather, the usefulness of the Hamiltonian viewpoint lies in the invariant symplectic structure of the resulting equations which are very convenient in theoretical and numerical studies.

All dynamic evolutions of Hamiltonian systems are symplectic (canonical) transformations; the time discretisation algorithm should also be symplectic. Ruth [250], Feng and Qin [241], and Feng *et al.* [242] have all developed symplectic integration methods for Hamiltonian systems. Simo *et al.* [251] gave second-order accurate methods, which preserved both momentum and energy exactly, and a detailed discussion of exact energy—momentum conserving algorithms and symplectic schemes for non-linear dynamics. Robert and Pau [249] discussed the accuracy of the symplectic integrator in the energy error. Wu [253] proposed the time-centred Euler scheme with second-order accuracy to an arbitrary *mth*-order accuracy for ordinary differential equations.

Assuming that the mechanical systems are holonomic and that the forces are monogenic, then the dynamic evolution problems may be expressed in the canonical system of differential equations

$$\begin{aligned}\frac{\partial p}{\partial t} &= -H_q(p,q,t)\\ \frac{\partial q}{\partial t} &= H_p(p,q,t)\end{aligned} \tag{10.76}$$

for a given Hamiltonian $H$. The Hamiltonian is reduced to the energy function $H(p_1, \ldots, p_n, q_1, \ldots, q_n, t)$ for the inertial frame of reference, where $H_p = \partial H/\partial p$, $H_q = \partial H/\partial q$.

Before discussing non-autonomous systems, let us consider the autonomous Hamiltonian system when $H$ is independent of time

$$\dot{z} = J\frac{\partial H}{\partial z}, \quad z \in R^{2n} \tag{10.77}$$

where $J = \begin{bmatrix} 0 & -1 \\ 1 & 0 \end{bmatrix}$. Here 1 is the standard $n \times n$ unit matrix and 0 is the $n \times n$ zero matrix, and $z_i = p_v \quad z_{i+n} = q_i \quad i \le n$, where $n$ is the number of degrees of freedom. Its phase flow is denoted as $g^t(z) = g(z,t) = g_H(z,t)$, being a one-parameter group of canonical maps (Abraham and Marsden [173], Arnold [236]) i.e. $g^0 = \text{identity}.\ g^{t_1+t_2} = g^{t_1}g^{t_2}$, and if $z_0$ is taken as an initial condition, then $z(t) = g^t(z_0)$ is the solution of (10.77) with the initial values $z_0$. Different symplectic integration schemes for the system (10.77) have been constructed (Feng and Qin [241]). Symplectic schemes for Hamiltonian systems preserve all the linear conservative quantities. Moreover, the implicit time-centred symplectic scheme preserves all the linear and quadratic conservative quantities. Let us consider the first-order and the second-order canonical difference schemes for eq. (10.77)

$$\begin{aligned}p_i^{k+1} &= p_i^k - hH_{q_i}(p^{k+1}.q^k)\\ q_i^{k+1} &= q_i^k + hH_p(p^{k+1}.q^k)\\ p_i^{k+1} &= p_i^k - hH_{q_i}(p^{k+1}.q^k) - \frac{h^2}{2}\left(\sum_{j=1}^{n} H_{qj}H_{pj}\right)_{q_i}(p^{k+1}.q^k)\end{aligned} \tag{10.78}$$

$$q_i^{k+1} = q_i^k + hH_{pi}\left(p^{k+1}, q^k\right) + \frac{h^2}{2}\left(\sum_{j=1}^{n} H_{qj} H_{pj}\right)_{p_i} \left(p^{k+1}, q^k\right) \tag{10.79}$$

where $h$ is the time step. These schemes are explicit with respect to the co-ordinate $q$ and are $A$ stable. In general, the Hamiltonian for non-linear vibration problems of structures is separable, i.e. $H = U(p) + V(q)$, where $U(p)$ is quadratic in $p$ representing the kinetic energy. Thus, in this case, eqs. (10.78) and (10.79) are a set of linear equations in $p$. The time-centred Euler scheme (Feng and Qin [241]) for eq. (10.77) is

$$z^{k+1} = z^k + hJH_z\left(\frac{z^{k+1} + z^k}{2}\right) \tag{10.80}$$

which may be written further as

$$\begin{aligned} p_i^{k+1} &= p_i^k - hH_{q_i}\left(\frac{p^{k+1} + p^k}{2}, \frac{q^{k+1} + q^k}{2}\right) \\ q_i^{k+1} &= q_i^k + hH_{p_i}\left(\frac{p^{k+1} + p^k}{2}, \frac{q^{k+1} + q^k}{2}\right) \end{aligned} \tag{10.80a,b}$$

The scheme (10.80) is an implicit second-order scheme. Thus scheme (10.80) requires iterations while schemes (10.78) and (10.79) do not. For non-autonomous Hamiltonian systems, we regard the time $t$ as an additional dependent variable. That is, letting $q_{n+1} = t$, we can choose a parameter $\tau$ as a new independent variable. It is well known (Goldstein [243]) that $p_{n+1} = -H$, which has a unit of energy, and is the generalised momentum conjugate to the time $t$. For this special choice, the function $K(w) = p_{n+1} + H(z)$ with $w = (q_1, \ldots, q_n, t, p_1, \ldots, p_n, -H)^T$ will take the place of the Hamiltonian function $H$.

Consider the ordinary differential equation (ODE)

$$\frac{dx}{dt} = F(x) \tag{10.81}$$

where $F$: $\mathcal{R}^N \rightarrow \mathcal{R}^N$ is an analytic function in $\mathcal{R}^N$, $F = (F_1, F_2, \ldots, F_N)^T$

For time-dependent ODEs, we use a new time variable $\tau$ to make the system time independent. Let $t=\tau$ and $\dfrac{dt}{d\tau} = 1$, then the time-dependent ODEs in $\mathcal{R}^N$ can be rewritten as time-independent equations in $\mathcal{R}^{N+1}$. To make a set of ODEs into a Hamiltonian system, we introduce a matching system (Wu [253]):

$$\begin{aligned} \frac{dX}{dt} &= -\left(X^T F'(x)\right)^T, \\ F'(x) &= \frac{DF}{Dx} \end{aligned} \tag{10.82}$$

Assuming that the Hamiltonian $H = X^T F(x)$, the systems (10.81) and (10.82) together can be rewritten as a Hamiltonian system

$$\frac{dx}{dt} = \frac{\partial H}{\partial X}, \qquad \frac{dX}{dt} = -\frac{\partial H}{\partial x} \tag{10.83}$$

Wu [253] gave a time-centred Euler scheme with second-order accuracy to an arbitrary $m$th-order accuracy for system (10.83). This scheme is $L$ stable. The second time-centred Euler scheme is

$$x_i^{k+1} = x_i^k + hF_i\left(\frac{x^{k+1} + x^k}{2}\right) \tag{10.84}$$

This is an implicit second-order scheme independent of the matching system (10.82).

If the Hamiltonian system (10.76) is separable, then

$$H = U(p) + V(q,t) \tag{10.85}$$

Ruth [250] constructed a third-order explicit symplectic scheme for the separable Hamiltonian (10.85). Qin *et al.* [248] constructed a fourth-order explicit symplectic difference scheme.

We may use the different symplectic integration schemes mentioned above to solve Hamiltonian equations. From the viewpoint of computational stability, implicit schemes are better. For explicit schemes, variable time steps may be used to guarantee the stability of computation. The time step is controlled by the superior and inferior limits of motion, i.e. $\left|q_i^{k+1} - q_i^k\right| \le \varepsilon$ or $\left|x_i^{k+1} - x_i^k\right| \le \varepsilon$. In this discussion we use the first-order scheme (10.78) and the time-centred Euler schemes (10.80) and (10.84) for the numerical examples below.

### 10.2.4 Finite Element Equations of Motion

For an initially straight beam, the finite element equations are obtained from eq. (10.73) by the Hamiltonian principle

$$\begin{aligned} &[M_u]\{\ddot{u}\} + [K_u]\{u\} + \tfrac{1}{2}[K_{uq}]\{w_q\} = \{F_u\}, \\ &[M_w]\{\ddot{w}\} + [K_w]\{w\} + \tfrac{1}{2}[w_{q,w}]^T[K_{uq}]\{u\} + \tfrac{1}{4}[w_{q,w}]^T[K_q]\{w_q\} = \{F_w\} \end{aligned} \tag{10.86}$$

where the subscript *(w)* denotes differentiation with respect to $\{w\}$. For damped forced vibration problems, the finite element equations are obtained by adding the appropriate damping terms

$$\begin{aligned} &[M_u]\{\ddot{u}\} + [C_u]\{\dot{u}\} + [K_u]\{u\} + \tfrac{1}{2}[K_{uq}]\{w_q\} = \{F_u\}, \\ &[M_w]\{\ddot{w}\} + [C_w]\{\dot{w}\} + [K_w]\{w\} + \tfrac{1}{2}[w_{q,w}]^T[K_{uq}]\{u\} + \tfrac{1}{4}[w_{q,w}]^T[K_q]\{w_q\} = \{F_w\} \end{aligned} \tag{10.87}$$

where $[C_u]$, $[C_w]$ are the corresponding damping matrices.

Equations (10.86) and (10.87) are different from some existing references, such as Yang and Saigal [254], in that induced axial force $S$ is a function of displacements and is not averaged and, in finite element discretisation, eqs. (10.73) has not been linearized, i.e. the discretisation is more accurate. Because of the accurate discretisation, iteration procedures are not needed in calculating every incremental step. The accuracy for the unaveraged $S$ is important in the study of non-linear vibrations, chaos and bifurcations.

Now we use the Hamiltonian formalism. For free vibration and undamped forced vibration problems the Hamiltonian of the Lagrangian [10.73] is obtained by the method introduced in Goldstein [243]

$$\begin{aligned} H &= \tfrac{1}{2}\{p_u\}^T[M_u]^{-1}\{p_u\} + \tfrac{1}{2}\{p_w\}^T[M_w]^{-1}\{p_w\} + \tfrac{1}{2}\{u\}^T[K_u]\{u\} \\ &\quad + \tfrac{1}{2}\{w\}^T[K_w]\{w\} + \tfrac{1}{2}\{u\}^T[K_{uq}]\{w_q\} + \tfrac{1}{8}\{w_q\}^T[K_q]\{w_q\} \\ &\quad - \{F_u\}^T\{u\} - \{F_w\}^T\{w\} \end{aligned} \tag{10.88}$$

where $\{p_u\} = [M_u]\{\dot{u}\}$, $\{p_w\} = [M_w]\{\dot{w}\}$ are the momenta. Hamilton's equations corresponding to eq. (10.88) are

$$\begin{aligned} &\frac{\partial\{u\}}{\partial t} = [M_u]^{-1}\{p_u\}, \quad \frac{\partial\{w\}}{\partial t} = [M_w]^{-1}\{p_w\}, \\ &\frac{\partial\{p_u\}}{\partial t} = -[K_u]\{u\} - \tfrac{1}{2}[K_{uq}]\{w_q\} + \{F_u\}, \\ &\frac{\partial\{p_w\}}{\partial t} = -[K_w]\{w\} - \tfrac{1}{2}[w_{q,w}]^T[K_{uq}]^T\{u\} - \tfrac{1}{4}[w_{q,w}]^T[K_q]\{w_q\} + \{F_w\} \end{aligned} \tag{10.89}$$

For damped forced vibration problems, extended Hamiltonian equations are obtained by writing eq. (10.87) in the form of eq. (10.81)

$$\begin{aligned} &\{\dot{U}\} = -[M_u]^{-1}\left([C_u]\{U\} + [K_u]\{u\} + \tfrac{1}{2}[K_{uq}]\{w_q\} - \{F_u\}\right), \\ &\{\dot{W}\} = -[M_w]^{-1}\Big([C_w]\{W\} + [K_w]\{w\} + \tfrac{1}{2}[w_{q,w}]^T[K_{uq}]^T\{u\} \\ &\qquad + \tfrac{1}{4}[w_{q,w}]^T[K_q]\{w_q\} - \{F_w\}\Big), \\ &\{\dot{u}\} = \{U\}, \quad \{\dot{w}\} = \{W\} \end{aligned} \tag{10.90}$$

For an initially straight beam, forming general stiffness matrices $[K_{uq}]$ and $[K_q]$ is not complicated (Leung and Mao [244]). But for framed structures with many elements, computing the general non-linear stiffness matrices is of some complexity. However, when one uses explicit integration schemes, the global stiffness matrices are not needed. Thus we may integrate the Hamiltonian equations element by element.

For an arbitrarily oriented beam element in the plane, the nodal displacements of the $i$th node in terms of local co-ordinates of element $e$, $\{d_i^e\}$, and those in terms of global co-ordinates $\{d_i\}$ are related by

$$\begin{Bmatrix} u_i^e \\ v_i^e \\ \theta_i^e \end{Bmatrix} = \begin{bmatrix} \lambda & \mu & 0 \\ -\mu & \lambda & 0 \\ 0 & 0 & 1 \end{bmatrix} \begin{Bmatrix} u_i \\ v_i \\ \theta_i \end{Bmatrix} \tag{10.91}$$

or symbolically, $\{d_i^e\} = [T_1]\{d_i\}$ (10.92)

where $\lambda = \cos\alpha$ and $\mu = \sin\alpha$ with $\alpha$ being the angle of orientation of the beam. The Hamiltonian may be written on the elementary level,

$$\begin{aligned} H = \tfrac{1}{2}\{p_d\}^T[M_d]^{-1}\{p_d\} + \sum_e\Big(\tfrac{1}{2}\{u^e\}^T[K_u^e]\{u^e\} + \tfrac{1}{2}\{w^e\}^T[K_w^e]\{w^e\} \\ + \tfrac{1}{2}\{u^e\}^T[K_{uq}^e]\{w_q^e\} + \tfrac{1}{8}\{w_q^e\}^T[K_q^e]\{w_q^e\} - \{F_u^e\}^T\{u^e\} - \{F_w^e\}^T\{w^e\}\Big) \end{aligned} \tag{10.93}$$

where $[M_d]$ is the conventional global mass matrix in terms of global co-ordinates and $\{p_d\} = [M_d]\{\dot{d}\}$ are the momenta. Denote the expression after $\sum_e$ in eq. (10.93) by $V_1^e$, then eq. (10.94) may be written as

$$H = \frac{1}{2}\{p\}^T[M_d]^{-1}\{p\} + \sum_e V_1^e \tag{10.94}$$

$V_1^e = U_1^e + \Omega_1^e$ is the total potential energy of element $e$, with $U_1^e$ being the strain energy and $\Omega_1^e$ the potential energy of the external forces. The first- and second-order derivatives of $V_1^e$ with respect to $\{d\}$ are easily computed. Thus, the Hamiltonian equations corresponding to eq. (10.93) may be obtained. For damped forced vibration problems, the finite element equations are obtained by adding the appropriate damping terms,

$$[M_d]\{\ddot{d}\} + [C_d]\{\dot{d}\} + \Big(\sum_e V_1^e\Big)_{,d} = 0 \tag{10.95}$$

where $[C_d]$ are the corresponding damping matrices. Equation (10.95) may be rewritten in the form of eq. (10.90).

### 10.2.5 Numerical Examples

In the case of transient excitation by pulses (i.e. shock load, explosions, etc.), damping has very little effect on the response qualitatively, so an undamped analysis is generally adequate. However in the case of steady-state vibration, damping is relatively important. Several examples on the dynamics of non-linear structures with and without damping are presented. The material of the structures is isotopic linear elastic. Results are compared with those of other investigators when possible. Examples 1 and 2 were calculated previously in Leung and Mao [244] by means of forming all global stiffness matrices that gave the same results, as this study worked on element matrices only.

## Example 1: Undamped Forced Vibration of a Clamped—Clamped Beam by a Sudden Load

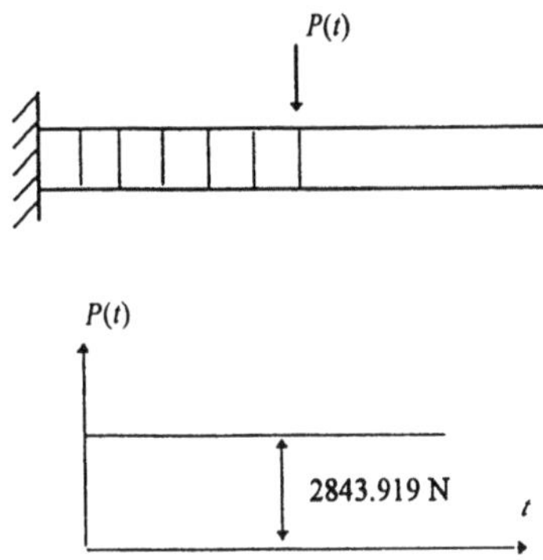


Fig. 10.2

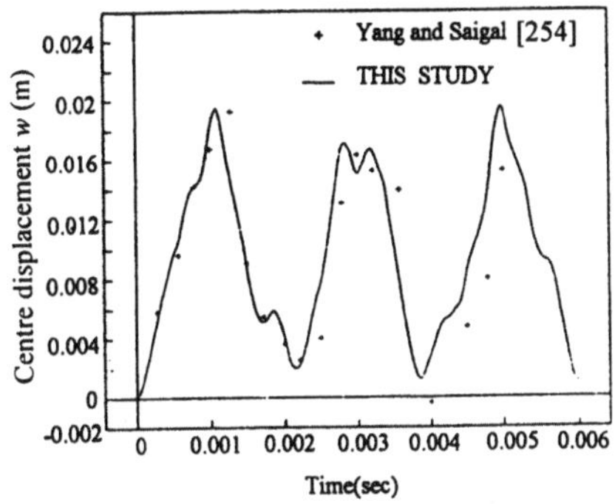


Fig. 10.3

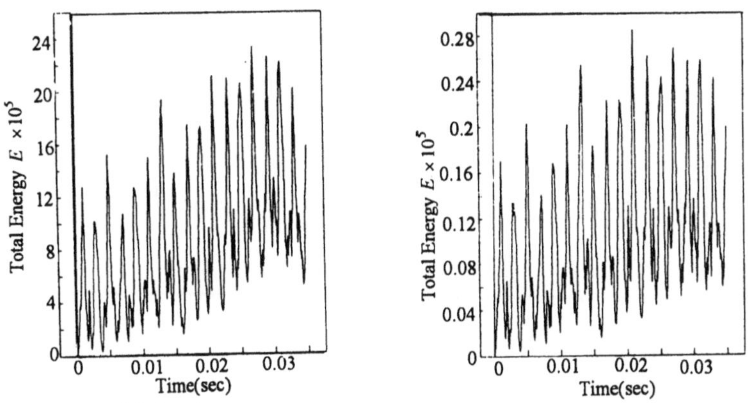


(a) Δt=0.1 μsec

(b) Δt= 1 μsec

Fig.10.4

A clamped—clamped beam under a static concentrated force of 2843.919 N acting at the midspan at time $t$=0 is considered. The modulus of elasticity is $E$=2.07 × $10^8$ kN/m$^2$, the mass density is ρ=2.71 × $10^{-3}$ kg/cm$^3$, the length is $l$=50.8 cm, and the cross section is $2.54 \times 0.3175$ cm. The beam geometry with a concentrated load is shown in Fig. 10.2. Because of symmetry, one-half of the beam is modelled by six finite elements. Figure 10.3 gives the non-linear responses of the midspan deflection obtained using the time-centred Euler scheme with an equal time step, $\Delta t$=1 μs. Figures 10.3 and 10.4 give the total energy $E$= kinetic energy $K$ + potential energy $V$, and the total kinetic energy $K$, respectively. This problem has been studied by many investigators. Mondkar and Powell [247] used five 8-node plane stress elements to model one-half of the beam. Yang and Saigal [254] used six beam elements with $\Delta t$=5 μs and Δ$t$=10 μs. McNamara [245] used five beam bending elements based on a central-difference operator with $\Delta t$=5 μs. The maximum displacement and the period of the first cycle were 0.02286 m and 2884 μs in McNamara [245], 0.019558 m and

2300 μs in Mondkar and Powell [247] and Yang and Saigal [254], and 0.019456 m and 2151 μs in this study.

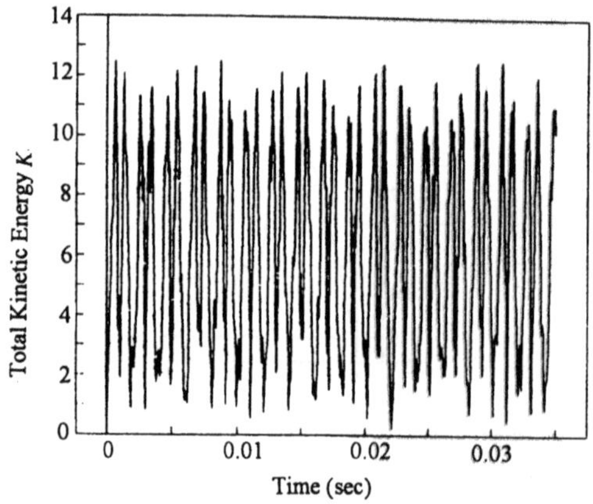


Fig. 10.5

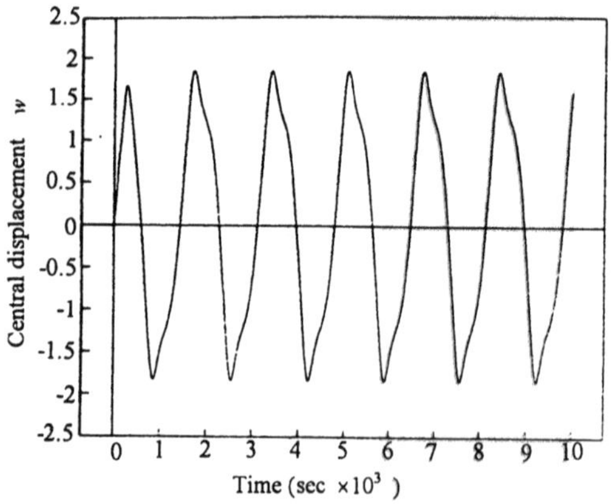


Fig. 10.6

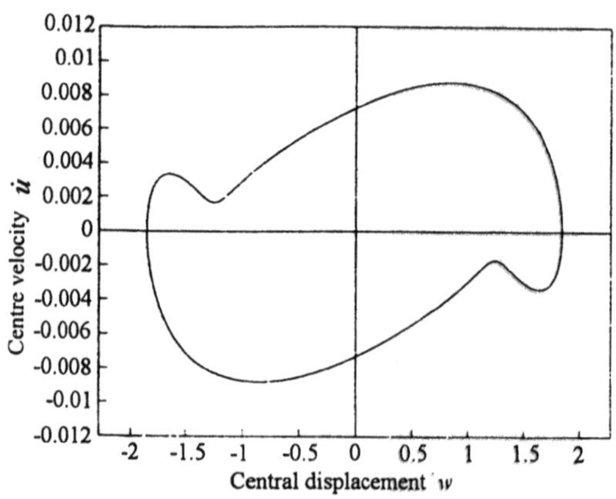


Fig. 10.7

Simo *et al.* [251] and Crisfield and Shi [240] considered that the time-centred Euler scheme is the idea really of "midpoint equilibrium". For non-linear Hamiltonian systems the time-centred Euler scheme fails to conserve the total energy of the system (Simo *et al.* [251]). But from Figs. 10.4 and 10.5 we may know that the amplitude is very small by referring to the total kinetic energy, and that the total energy is almost preserved when using small time steps.

**Example 2 Damped Forced Vibrations of a Hinged—Hinged Beam by Harmonic Force**

Consider a hinged—hinged uniform beam with immovable edges under a concentrated force of $F\cos\omega t$ acting at the midspan, where $F$=2.0 × $10^{-3}$, $\omega$=1.1$\omega_L$, where $\omega_L$ is the linear natural frequency $\omega_L$=3.4189 × $10^{-3}$. The modulus of elasticity, $E$=1; the mass density, $\rho$=1; the length, $l$=28.8675; the second moment of inertia, $I$=0.083333; and the cross-sectional area, $A$=1.0. Assuming the damping forces are proportional to the mass matrix, i.e. $C=\alpha M$, and the damping ratio $\alpha$ =0.01. Because of symmetry, one-half of the beam was modelled by four finite elements. We use the time-centred Euler scheme with a variable time step. The response curve is shown in Fig. 10.6. The phase plane trajectories of steady-state motion are shown in Fig. 10.7.

**Example 3 Elastic Dynamic Snap Buckling of an Arch**

A dynamic buckling analysis of the circular arch shown in Fig. 10.8 was carried out. Owing to symmetry, one-half of the arch is modelled by nine finite elements. Figure 10.9 shows the displacement response predicted in this study using the time-centred Euler scheme. In Fig. 10.9, the deflection $\delta$ is defined as,

$$\delta = \frac{\text{vertical displacement at apex}}{\text{average rise to arch} = H/2} \tag{10.96}$$

We use the following parameters, $R$=170.47 cm, $E$=6.9872 × $10^6$ N/cm$^2$ , $h$=2.54 cm, $\rho$=2.6085 × $10^{-3}$ kg/cm$^2$, $b$=2.54 cm, $C=\sqrt{E/\rho}=5.141\times10^5$cm/s, $\beta$=15°, $P=\left(\frac{R}{h}\right)^2\times\frac{q}{E}$, $\tau=\frac{C}{R}\times t$.

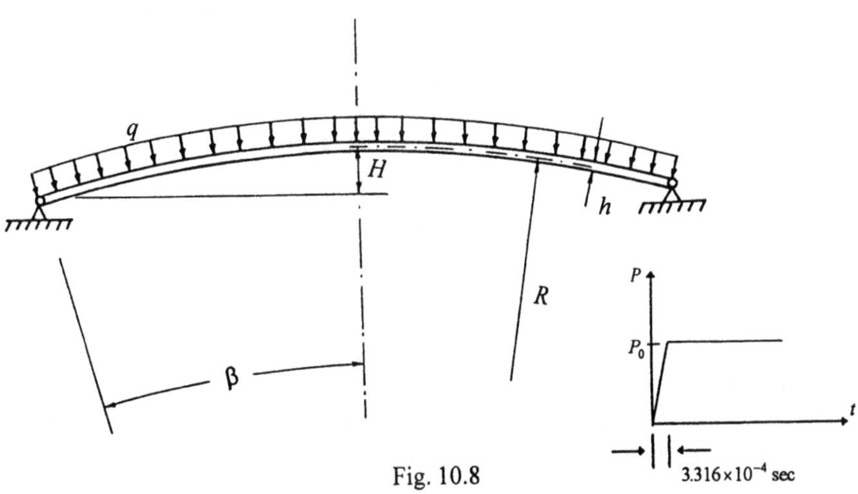


Fig. 10.8

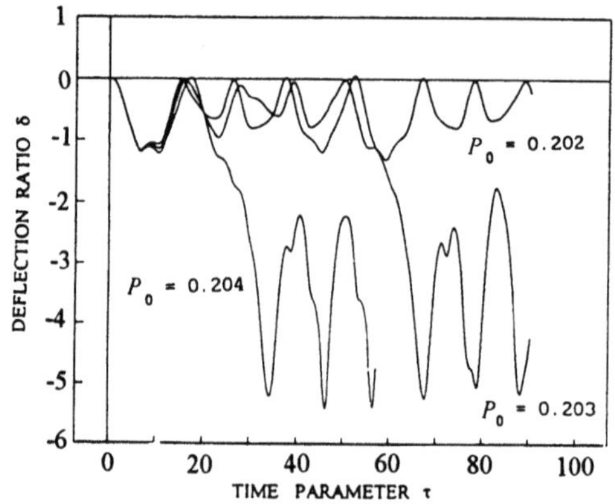


Fig. 10.9

The dynamic buckling of the arch occurs at the load level at which a sudden increase in the deflection ratio is measured. Fig. 10.9 shows that at $P_0 = 0.202$, it oscillates about a position of approximately $\delta = -0.5$, and that at $P_0 = 0.203$ it snaps through at $\tau = 60.0$, and oscillates about a position of approximately $\delta = -4.0$. Therefore, for $\tau$ from 0 to 90.0, the buckling load predicted here lies between $P_0 = 0.202$ and $P_0 = 0.203$. For $\tau$ from 0 to 50, the buckling load predicted here lies between $P_0 = 0.203$ and $P_0 = 0.204$, which is about 2.0% higher than that predicted by Bathe *et al.* [237]. Moreover, results obtained using the whole arch are the same as those using one-half of the arch.

**Example 4 Dynamic Analysis of Plane Frames**

The geometry for the frame is given in Fig. 10.10, where the modulus of elasticity is $E = 2.07 \times 10^8$ kN / m$^2$, the mass density is $\rho$=2.71 × 10$^{-3}$ kg/cm$^3$, and the cross section is 2.54 × 3.175 cm. Every beam member is modelled by four elements.

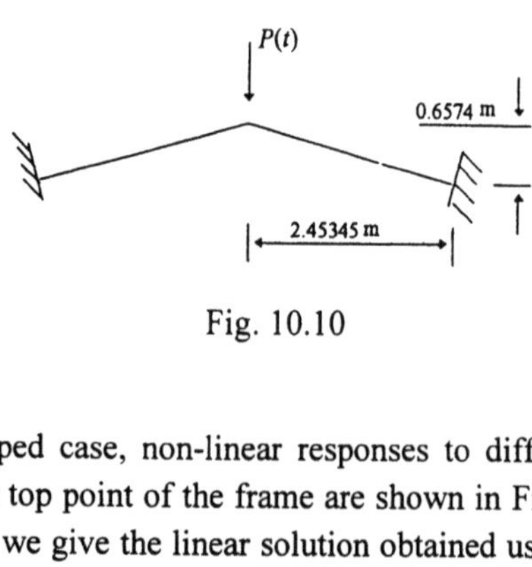


Fig. 10.10

For the undamped case, non-linear responses to different concentrated load $P(t) = F_c$ acting at the top point of the frame are shown in Fig. 10.11(a—e) and Fig. 10.12. In Fig. 10.11(e) we give the linear solution obtained using the whole frame for $F_c = 120$ kN.

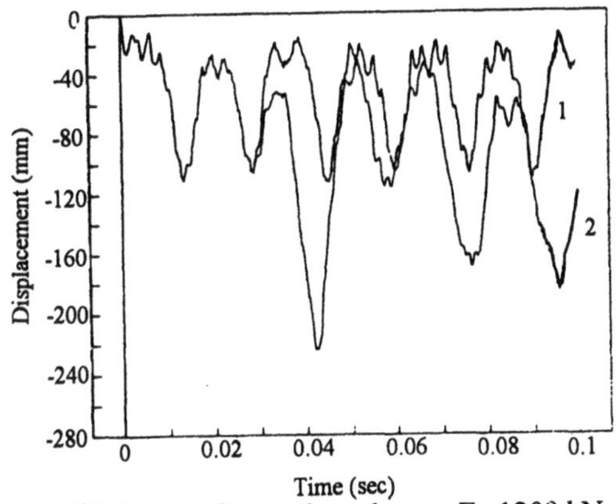


(a) Using the first-order scheme, $F_c$=1200 kN

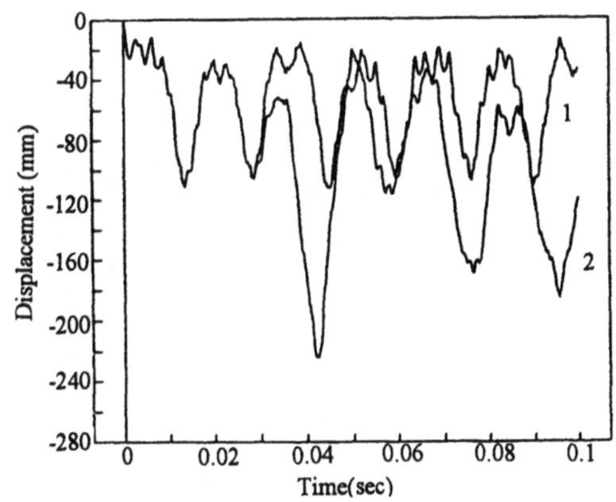


(b) Using time-centred Euler scheme scheme, $F_c$=1200 kN

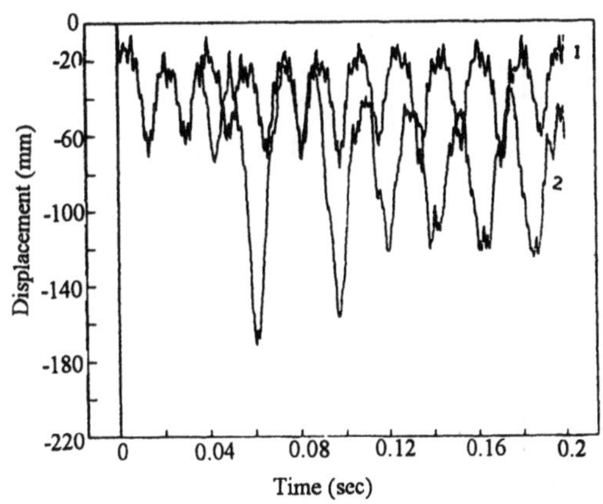


(c) Using time-centred Euler scheme, $F_c$=1000 kN

Fig. 10.11 (a,b,c)

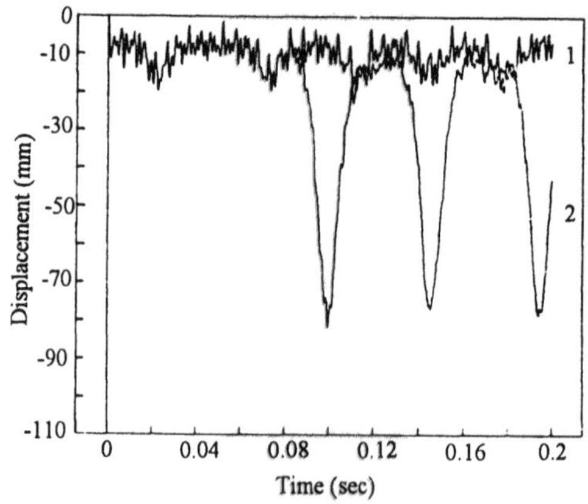


(d) Using time-centred Euler scheme, $F_c$=600 kN

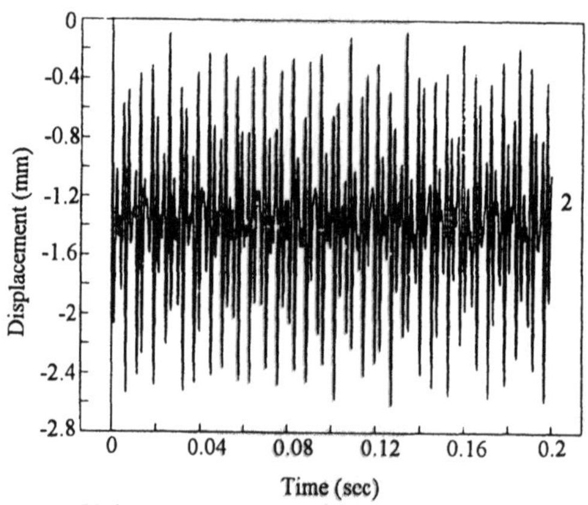


(e) Using time-centred Euler scheme, $F_c$=120 kN

Fig. 10.11 (d,e)

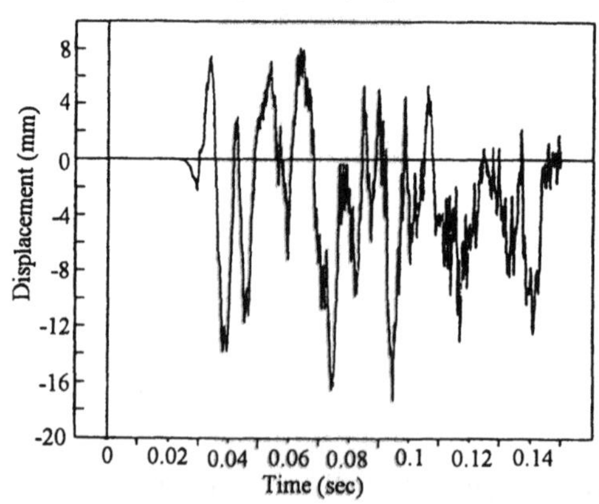


Fig. 10.12 (a) Horizontal Displacement

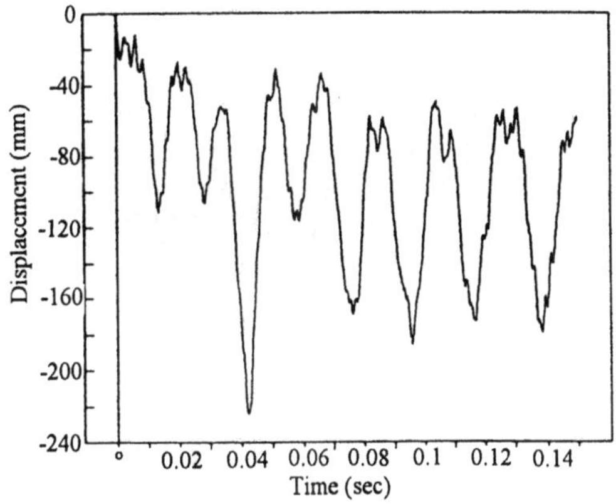


Fig. 10.12 (b) Vertical Displacement

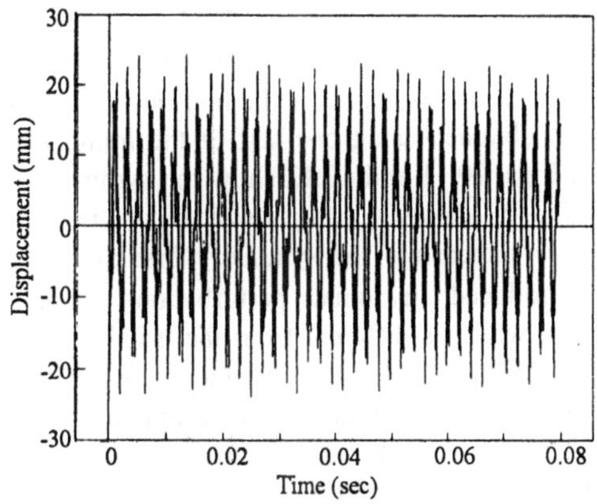


(a) Linear solution using whole frame

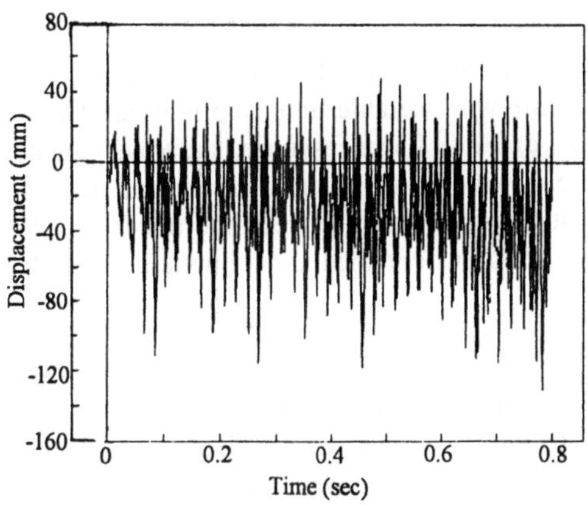


(b) Non-linear solution using one-half of frame

Fig. 10.13 (a,b)

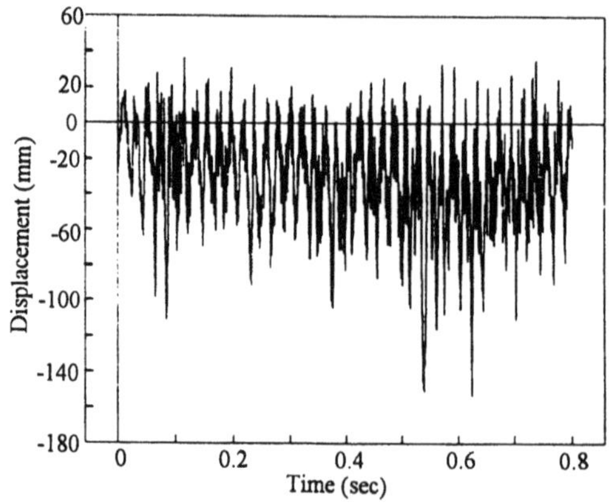


Fig. 10.13 (c) Non-linear solution using whole frame

The non-linear subharmonic solutions are obvious. Intuitively, due to symmetry, results obtained using one-half of the frame should be the same as using the whole frame. However, in numerical calculations, the difference is very large in some cases, when anti-symmetric modes are resonated by the non-linear frequency. In such cases, the symmetric solutions obtained using one-half of the frame are, in fact, unstable. The lower branches of Figs. 10.11 and 10.12 show the symmetry breaking branches. Moreover, Fig. 10.11(a—e) shows a comparison of the non-linear responses for different loads. With $F_c$ increasing, the symmetry breaking emerges more easily. Here, we have not introduced any small perturbations artificially to obtain the asymmetric deformation, which is different from the usual references. For example, Wood and Zienkiewicz [252] calculated the asymmetric deformation of a two-hinged deep arch by imposing a small perturbation to the radius of the arch. Of course, the rounding error of computers may be regarded as small perturbations. Figure 10.11(a—e) also shows a comparison of the linear and non-linear response. The considerable difference in the maximum displacements of the linear and non-linear solutions can also be noted. For proportionally damped vibration $(C = \alpha M)$, where the damping ratio $\alpha$=0.1, $P(t) = F\cos\omega t$ N acting at the top point, $\omega$=1.1$\omega_L$, so $F$=1200 kN in which the linear fundamental frequency of symmetry mode $\omega_L = 275.844\ rad/s$. The time-centred Euler scheme with variable time step is used. The response curves are shown in Fig. 10.13(a—c). For a linear response, solutions obtained using the whole frame are in agreement with those obtained using one-half of the frame. Fig. 10.13(a—c) shows that the differences between the linear and non-linear responses and difference between solutions obtained using one-half of the frame and using the whole frame still exist.

### 10.2.6 Conclusions

Symplectic methods are successful in integrating the Hamiltonian equations. The equations of motion are integrated element by element to avoid the formation of a global stiffness matrix so that complicated framed structures with many elements can be handled in core. The response after dynamic symmetry breaking is an order higher, which cannot be overlooked in a structural design. It is advisable to use the whole structure rather than its symmetric portion when non-linear dynamic response is of interest.

## 10.3. Construction of the Invariant Torus

The existence of the invariant torus is recognized as one of the celebrated scenarios of a non-linear dynamic system with multiple degrees of freedoms. The study of the invariant torus has recently attracted greater interest because of its increasing applicability to mathematical models in diverse disciplines of science and developing research into the manifold of non-linear dynamical systems. The invariant torus can be categorised into two classes according to the nature of the system in question. The first class is conservative in which the total energy of the system is idealised to be constant with no energy losses. A typical example of such a system is a time-periodic Hamiltonian (Arnold [164]). In this case, it is traditional to represent the flow of the system as an attracting manifold with time being a cyclic co-ordinate around the torus. The geometrical characterisation of the manifold of Hamiltonian systems is considered analytically in much of the literature of classical mechanics (Goldstein [243], Arnold [164]). The second class of invariant tori is a non-conservative system where the total energy of the system is a periodic system with a balance of dissipative force and auto-parametric excitation. The dynamic behaviours of such systems become more complicated, therefore the solutions usually need both geometrical techniques and numerical approaches for ordinary differential equations (ODEs). One of the topics of earlier studies on this system is devoted to the modal interaction of coupled van der Pol oscillators (e.g. Minorsky [267], Hall [263], Ioos and Joseph [52], Rand and Holmes [268], Storti and Rand [269], Hall and Iwan [264]). However, most of these analyses presumed some particular time-dependent solutions, although the mutual interaction within the oscillators is known as a time-independent motion associated with a periodic surface. Dieci *et al.* [257] developed a modified leap-frog numerical algorithm using a partial differential equation (PDE) approach to obtain the response of interactive motions between two linearly coupled oscillators. The natural association of a solution to the PDEs and a solution of the original ODEs has been applied in their work in the same way as the centre manifold of Carr [58] and non-linear normal mode of Shaw and Pierre [270]. However, no explicit expression for the periodic surface can be obtained from their algorithm. More recently, Gilsinn [261] presented a symbolic algebraic algorithm in which a periodic surface of two mutually coupled van der Pol oscillators has been constructed by means of a so-called multi-frequency Poisson series using Galerkin's variational method. As to the outcome, the amount of computation in his algorithm is so excessively large that it soon becomes impracticable with increasing strength of non-linearity. The stability of the periodic surface of an invariant torus, on the other hand, has been less emphasised in the past. Hall [263] established a characteristic method to identify the stability properties of an almost periodic system exemplified by coupled van der Pol oscillators. His stability theorem has proved to be useful to identify the convergence of trajectory of almost periodic motions. However, he has also concluded that Hall's theorem cannot detect the disappearance of a multi-periodic surface, i.e. arising from bifurcation of a stable to an unstable torus.

This section is arranged as follows. In 10.3.1, we present an invariant embedding technique that converts an original system of ordinary differential equations (ODEs) to its angular normal form of partial differential equations (PDEs). The characteristics of such system exhibit a close relationship to the behaviour of the Poincaré sections of the original system. In 10.3.2, we develop a new numerical approach, the Topelitz Jacobian Matrix/Fast Fourier Transform (TJM/FFT) algorithm, for the construction of a multi-periodic surface. This method is a quasi-periodic extension of the method of Leung and Ge [266] and Ge and Leung [259]. We suggest a new stability criterion for the two-invariant torus in 10.3.3. The development of such a criterion is stated in detail as well as the analysis of variational techniques in the Poincaré section. It is in fact the only criterion available to identify the existence or non-existence of a periodic surface at present. In 10.3.4 we illustrate the application of our methodology to coupled van der Pol oscillators. A comparison of solutions between the numerical integration method and the TJM/FFT method is given and the validity and reliability of the newly proposed stability criterion for the two-invariant torus are verified.

### 10.3.1 Invariant Embedding

As illustrated previously, the computation of a two-dimensional invariant manifold can be reduced to the computation of an invariant curve for an associated Poincaré map. A PDE approach provides a natural connection between these two solutions. An obvious advantage is that the dimension of the original unknown object has been reduced by one in the new system. The operation is at the core of the invariant embedding techniques. Consider the first-order equations of motion of the form

$$\dot{x}_i = y_i, \tag{10.97}$$

$$\dot{y}_i = f_i(x_1, x_2, \cdots, x_q; y_1, y_2, \cdots, y_q), \quad i \in \mathbf{N}[1, q] \tag{10.98}$$

where $q$ is the total degrees of freedom and $i$ stands for its index, and $(x_i, y_i)$ describes the state, i.e. displacement and velocity of the system in Cartesian co-ordinates. Such a motion can be put into a functional-dependent form in which all displacements and velocities are constrained to a finite dimensional manifold described by polar co-ordinates

$$y_i = \omega_k r_i^{\nu} \cos\omega_k \theta_k, \tag{10.99}$$

$$y_i = \omega_k r_i^{\nu} \cos\omega_k \theta_k, \qquad k \in \mathbf{N}[1, p] \tag{10.100}$$

where each $\omega_k > 0$ is prescribed by the linear part of equation (10.98) and $\theta_k$ stands for the angular variable in a certain polar co-ordinate. The parameter $\nu > 0$ may be introduced in order to further simplify the resulting expressions. A torus $T$ in $p$ space is defined by

$$T^p = \{\Theta = (\omega_1\theta_1, \cdots, \omega_p\theta_p) : \omega_i\theta_i \in \mathcal{R}(\mathrm{mod}2\pi) \tag{10.101}$$

in which the function mod($2\pi$) gives real module $2\pi$. In particular, when $p = 2$, eqs. (10.99) and (10.100) refer to a torus in phase space. As a result, governing eqs. (10.97) and (10.98) become

$$\frac{\mathrm{d}\Theta}{\mathrm{d}t} = \mathbf{f}(\Theta, \mathbf{r}), \quad \frac{\mathrm{d}\mathbf{r}}{\mathrm{d}t} = \mathbf{g}(\Theta, \mathbf{r}) \tag{10.102, 10.103}$$

where $(\Theta, \mathbf{r}) \in T^p \times \mathcal{R}^q$ and functions $f : T^p \times \mathcal{R}^q \to \mathcal{R}^p$, $g : T^p \times \mathcal{R}^q \to \mathcal{R}^p$ are assumed to be smooth. Similar to the conclusions of Shaw and Pierre [270], any motion which satisfies eqs. (10.99), (10.100), (10.102) and (10.103) is defined to be a non-linear normal mode

$$M = \{(\Theta, \mathbf{r}(\Theta)); \Theta \in T^p\} \subset T^p \times \mathcal{R}^q \tag{10.104}$$

We further take the time derivative of constraint eqs. (10.99) and (10.100) and use the chain rule for differentiation to obtain the following PDE

$$\mathbf{w} = \sum_{j=1}^{p} f_j(\Theta, \mathbf{r}(\Theta)) \frac{\partial \mathbf{r}}{\partial \theta_j}(\Theta) - \mathbf{g}(\Theta, \mathbf{r}(\Theta)) \equiv 0, \quad \Theta \in T^p \tag{10.105}$$

where $\mathbf{w}$ is a note of expression and the periodic boundary condition of radii $r_i$ are implemented with

$$r_i(\omega_1\theta_1, \cdots, \omega_{k-1}\theta_{k-1}, 0, \omega_{k+1}\theta_{k+1}, \cdots, \omega_p\theta_p) = r_i(\omega_1\theta_1, \cdots, \omega_{k-1}\theta_{k-1}, 2\pi, \omega_{k+1}\theta_{k+1}, \cdots, \omega_p\theta_p)$$
$$i \in \mathbf{N}[1, q], \quad k \in \mathbf{N}[1, p] \tag{10.106}$$

### 10.3.2 TJM/FFT Algorithm

Compared with the original ODEs (10.98), the most obvious advantage of the PDE expression (10.105) lies in the explicit periodicity of the independent parameter $\theta_k$. Solutions of the non-linear partial differential eq. (10.105) are difficult to express in analytic mathematical terms, and thus apparently preclude spectral analysis where a closed form representation of the manifold is to be sought. The TJM/FFT method for spectral analysis can be used here so that samples of the coefficients are taken over one complete period and are fed to a computer algorithm to find the invariant manifold.

The pair of multi-periodic Fourier Transform for vectors $\{\mathbf{r}\}$ and $\{\mathbf{w}\}$ can be defined in the same way as that in the mono-periodic case (Brigham [255]),

I) Inverse Fourier transform

$$\begin{bmatrix} r_i(\theta_1, \theta_2, \cdots, \theta_p) \\ w_i(\theta_1, \theta_2, \cdots, \theta_p) \end{bmatrix} = \sum_{k_1=-N_1}^{N_1} \sum_{k_2=-N_2}^{N_2} \cdots \sum_{k_p=-N_p}^{N_p} \begin{bmatrix} R_i(k_1, k_2, \cdots, k_p) \\ W_i(k_1, k_2, \cdots, k_p) \end{bmatrix} \cdot e^{i\,(k_1\omega_1\theta_1 + k_2\omega_2\theta_2 + \cdots + k_p\omega_p\theta_p)} + \text{hot} \tag{10.107}$$

II) Fourier transform

$$\begin{bmatrix} R_i(k_1, k_2, \cdots, k_p) \\ W_i(k_1, k_2, \cdots, k_p) \end{bmatrix} = \frac{\prod_{k=1}^{p} \omega_k}{(2\pi)^p} \int_0^{2\pi/\omega_1} \int_0^{2\pi/\omega_2} \cdots \int_0^{2\pi/\omega_p} \begin{bmatrix} r_i(\theta_1, \theta_2, \cdots, \theta_p) \\ w_i(\theta_1, \theta_2, \cdots, \theta_p) \end{bmatrix} \cdot e^{-i\,(k_1\omega_1\theta_1 + k_2\omega_2\theta_2 + \cdots + k_p\omega_p\theta_p)} d\theta_1 d\theta_2 \cdots d\theta_p \tag{10.108}$$

where $k_i$ represents the wave number, $R_i$ and $W_i$ are spectral counterparts of function $r_i$ and $w_i$ respectively, and hot stands for 'higher order terms'. The above mentioned process may be termed as the Galerkin procedure (Urabe and Reiter [271]). The resulting set of simultaneous equations in eq. (10.105) is non-linear, therefore the Newton—Raphson procedure is applied to seek the unknowns in a predictor—

corrector manner, e.g. $\{\mathbf{R}\} = \{\mathbf{R}\}_0 + \{\Delta\mathbf{R}\}$, in which the subscription zero represents the nearby solution vector of the equation, $\{\Delta\mathbf{R}\}$ is the unknown incremental vector given by

$$[\mathbf{J}]\{\Delta\mathbf{R}\} = -\{\mathbf{Z}\} \tag{10.109}$$

where $[\mathbf{J}]$ is the analytical Jacobian matrix

$$J_{i,j} = \left(\frac{\partial W_i}{\partial R_j}\right) \qquad i,j \in \mathbf{N}[-N,N] \tag{10.110}$$

and $\{\mathbf{Z}\}$ is the right-hand residual vector used by the iteration. It should be pointed out that although it was recognized to be difficult in general (Cameron and Griffins [256]), the evaluation of the analytical Jacobian $[\mathbf{J}]$ is particularly desirable. Because we have neither very close initial guesses of the exact solutions nor good estimations of related numerical derivatives to non-linear eq.(10.105). However, as illustrated below, a significant simplification is possible if the TJM/FFT method is applied.

Denote

$$\mathbf{Dr} = \frac{\partial \mathbf{r}}{\partial \Theta} = \left[\frac{\partial r_i}{\partial \theta_k}\right] \tag{10.111}$$

Variations are introduced in order to represent the perturbations of **r**, **Dr** and the strength of non-linearity $\varepsilon$ at a given position $\Theta$,

$$w_i(\mathbf{r}_0 + \Delta\mathbf{r}, \mathbf{Dr}_0 + \Delta\mathbf{Dr}, \varepsilon_0 + \Delta\varepsilon, \Theta) = w_i(\mathbf{r}_0, \mathbf{Dr}_0, \varepsilon_0, \Theta)$$
$$+ \sum_{k=1}^{p} \left.\frac{\partial w_i}{\partial (D_k r_i)}\right|_0 \Delta D_k r_i + \sum_{j=1}^{q} \left.\frac{\partial w_i}{\partial r_j}\right|_0 \Delta r_j + \left.\frac{\partial w_i}{\partial \varepsilon}\right|_0 \Delta\varepsilon + \text{hot} = 0 \tag{10.112}$$

where the parameterized functions $\alpha_k$ and $\beta_{i,j}$ are defined respectively by

$$\alpha_k = \left.\frac{\partial w_i}{\partial (D_k r_i)}\right|_0 = f_k^0, \; \beta_{i,j} = \left.\frac{\partial w_i}{\partial r_j}\right|_0 = \left[\sum_{k=1}^{p} \frac{\partial f_k}{\partial r_j} \cdot \frac{\partial r_i}{\partial \theta_k} - \frac{\partial g_i}{\partial r_j}\right]^0; \tag{10.113}$$

and the perturbed function $\xi_i$ is selected according to its use as predictor or corrector

$$\xi_i = \begin{cases} \left.\dfrac{\partial w_i}{\partial \varepsilon}\right|_0 \Delta\varepsilon & \textit{predictor} \\ w_i(\mathbf{r}_0, \mathbf{Dr}_0, \varepsilon_0, \Theta) & \textit{corrector} \end{cases} \tag{10.114}$$

Sampling each phase plane with $[M_1, M_2, \cdots, M_p]$ interval-equivalent discrete points respectively and substituting the backward Fourier transform (10.107) into eq. (10.112), we obtain

$$w_i(s_1, s_2, \cdots, s_p) =$$
$$\sum_{k_1=-N_1}^{N_1} \sum_{k_2=-N_2}^{N_2} \cdots \sum_{k_p=-N_p}^{N_p} \left[\sum_{j=1}^{q} B_{i,j}(s_1, s_2, \cdots, s_p) + i\sum_{j=1}^{p} k_j \omega_j \alpha_j(s_1, s_2, \cdots, s_p)\right] \times \tag{10.115}$$
$$\Delta R_i(k_1, k_2, \cdots, k_p) \cdot e^{i\sum_{j=1}^{p} \frac{2\pi s_j k_j}{M_j}} + \xi_i(s_1, s_2, \cdots, s_p) + \text{hot} = 0$$

In order to avoid aliasing distortions in the above sampling process, the requisite numbers of retaining point are given by $M_k \geq 2N_k + 1$ (Brigham [255]).

After applying the inverse discrete Fourier transform and interchanging the order of the summations, we obtain

$$W_i(l_1, l_2, \cdots, l_p) =$$
$$\sum_{k_1=-N_1}^{N_1} \sum_{k_2=-N_2}^{N_2} \cdots \sum_{k_p=-N_p}^{N_p} \left[ \sum_{j=1}^{q} B_{i,j}(l_1 - k_1, l_2 - k_2, \cdots, l_p - k_p) + i \sum_{j=1}^{p} k_j \omega_j A_j(l_1 - k_1, l_2 - k_2, \cdots, l_p - k_p) \right]$$
$$\cdot \Delta R_i(k_1, k_2, \cdots, k_p) + Z_i(l_1, l_2, \cdots, l_p) = 0 \tag{10.116}$$

where $A_{i-j}$ and $B_{i-j}$ are the components of the Toeplitz matrices $[\mathbf{A}]$ and $[\mathbf{B}]$ which can be evaluated directly by applying the inverse fast Fourier Transform algorithm to the parameterized functions $\alpha$ and $\beta$ respectively,

$$A_j(\tau_1, \tau_2, \cdots, \tau_p) = \frac{1}{M_1 M_2 \cdots M_p} \sum_{s_1=0}^{M_1-1} \sum_{s_2=0}^{M_2-1} \cdots \sum_{s_p=0}^{M_p-1} \alpha_j(s_1, s_2, \cdots, s_p) \cdot e^{-i \sum_{j=1}^{p} \frac{2\pi s_j \tau_j}{M_j}}$$
$$B_{i,j}(\tau_1, \tau_2, \cdots, \tau_p) = \frac{1}{M_1 M_2 \cdots M_p} \sum_{s_1=0}^{M_1-1} \sum_{s_2=0}^{M_2-1} \cdots \sum_{s_p=0}^{M_p-1} \beta_j(s_1, s_2, \cdots, s_p) \cdot e^{-i \sum_{j=1}^{p} \frac{2\pi s_j \tau_j}{M_j}} \tag{10.117}$$

where $\tau_j$ denotes the index of the convolution sequences,

$$\tau_j = \operatorname{mod}\left( \frac{M_j + l_j - k_j}{M_j} \right) . l_m . k_m \in \mathbf{N}[-N_m . N_m], \; m = 1,2. \tag{10.118}$$

Therefore, the Jacobian matrix used in the Newton—Raphson iteration is

$$\frac{\partial W_i(l_1, l_2, \cdots, l_p)}{\partial R_i(k_1, k_2, \cdots, k_p)} =$$
$$\sum_{j=1}^{q} B_{i,j}(l_1 - k_1, l_2 - k_2, \cdots, l_p - k_p) + i \sum_{j=1}^{p} k_j \omega_j A_j(l_1 - k_1, l_2 - k_2, \cdots, l_p - k_p) \tag{10.119}$$

The right-side vectors are determined directly by the inverse fast Fourier transform of the perturbed function $\xi$,

$$Z_i(l_1, l_2, \cdots, l_p) = \frac{1}{M_1 M_2 \cdots M_p} \sum_{s_1=0}^{M_1-1} \sum_{s_2=0}^{M_2-1} \cdots \sum_{s_p=0}^{M_p-1} \xi_i(s_1, s_2, \cdots, s_p) \cdot e^{-i \sum_{j=1}^{p} \frac{2\pi s_j l_j}{M_j}} \tag{10.120}$$

### 10.3.3 Stability of a Periodic Surface

We know from studies of the periodic Hamiltonian system (Guckenheimer and Holmes [36]), the use of the Poincaré section $\Sigma$ may provide the most direct way to evaluate the character of the solutions along their trajectories. Some of the most important properties of dynamical systems may be better understood from this viewpoint. In this section, we extend the idea to study a system of multi-periodic functions and analyze the characteristic exponents of the linear variation equation associated with the expression of a periodic surface.

The convergence of trajectory of an invariant torus or a so-called periodic surface was studied by Diliberto *et al.* [258], Hall [263] and Gilisinn [260]. In their studies, the left-hand side of the amplitude eqs. (10.102) and (10.103) which is almost periodic in $t$, was assumed to be integrable

$$\mathbf{r}_{n+1}-\mathbf{r}_n=\lim_{T\to\infty}\frac{1}{T}\int_{nT}^{(n+1)T}\mathbf{g}(\Theta(t),\mathbf{r}(t))\cdot dt,\quad \mathbf{r}_n=\mathbf{r}(nT);\ \mathbf{r}_{n+1}=\mathbf{r}(nT+T)\qquad(10.121)$$

A Poincaré mapping gives a good idea of the asymptotic behaviour of the trajectories. A Poincaré section defines a return map in the plane $\Sigma, \mathbf{r}_{n+1}=\mathbf{G}(\mathbf{r}_n)$ in which all trajectories originating from $\Sigma$ return to $\Sigma$ when $T$ is approaching to infinity. The section of a torus can then be regarded as the integrable twist mapping $\Sigma^2$ if the periodic surface of the torus is continuous, smooth and time invariant in $\mathbf{R}^3$. It was concluded in earlier studies (Hall [263]) that the stability of the periodic surface is locally the same as the stability properties of the linear variational equation

$$\delta\dot{\mathbf{r}}=\mathbf{DG}[\mathbf{r}]\cdot\delta\mathbf{r}\qquad(10.122)$$

here $\mathbf{G}[\mathbf{r}]=\dfrac{\omega_1\omega_2}{4\pi^2}\displaystyle\int_0^{2\pi/\omega_1}\int_0^{2\pi/\omega_2}\mathbf{g}(\Theta(t),\mathbf{r}(t))\cdot d\theta_1 d\theta_2$ and $\mathbf{DG}[\mathbf{r}]=[\partial\mathbf{G}/\partial\mathbf{r}]$ (10.123)

in which $\theta_1,\theta_2$ are linearly independent over the integers. The negative real part of the eigenvalues of Jacobian matrix **DG** is $\mathrm{Re}[\lambda_i]<0$, which suggest a stable trajectory. Such a condition is necessary but not sufficient for the existence of a periodic surface because the motion of a mutually non-linear interaction may undergo the route of a 'Devil staircase' (Hao [265]) with increasing strength of non-linearity. In some cases, there exists a periodic rather than a quasi-periodic interaction which cannot form a smooth and continuous periodic surface. In this section, the nature of the transition from the existence of torus to the non-existence of such a manifold is further investigated in the Poincaré section $\Sigma^2$. The result may be helpful to detect the first disappearance of a periodic surface.

Applying the averaging method to the right-hand sides of eq. (10.102), according to Hall [263], we obtain two simplified equations:

$$\begin{aligned}\omega_1\dot{\theta}_1&=\frac{\omega_1\omega_2}{4\pi^2}\int_0^{2\pi/\omega_1}\int_0^{2\pi/\omega_2}f_1(\Theta,\mathbf{r})\cdot d\theta_1 d\theta_2=\eta_1;\\ \omega_1\dot{\theta}_2&=\frac{\omega_1\omega_2}{4\pi^2}\int_0^{2\pi/\omega_1}\int_0^{2\pi/\omega_2}f_2(\Theta,\mathbf{r})\cdot d\theta_1 d\theta_2=\eta_2;\end{aligned}\qquad(10.124)$$

Substituting $\dot{\theta}_1,\dot{\theta}_2$ from (10.123) into eq. (10.105) and averaging the result with respect to $\theta_1,\theta_2$, we obtain

$$\frac{\omega_1\omega_2}{4\pi^2}\int_0^{2\pi/\omega_1}\int_0^{2\pi/\omega_2}\left((\frac{\eta_1}{\omega_1})\frac{\partial\mathbf{r}}{\partial\theta_1}+(\frac{\eta_2}{\omega_2})\frac{\partial\mathbf{r}}{\partial\theta_2}\right)\cdot d\theta_1 d\theta_2=\frac{\omega_1\omega_2}{4\pi^2}\int_0^{2\pi/\omega_1}\int_0^{2\pi/\omega_2}\mathbf{g}(\Theta,\mathbf{r})\cdot d\theta_1 d\theta_2$$

(10.125)

As shown in Fig. 10.14(a), Green's theorem provides the fundamental relationship between the closed line integrals $C$ in the Poincaré section $\sum^2$ and the double integrals in the area $\Omega$ in the left-hand side of eq. (10.125), if $\Omega$ is integrable.

$$\begin{aligned}
&\int_0^{2\pi/\omega_1}\int_0^{2\pi/\omega_2}\left((\eta_1/\omega_1)\frac{\partial r}{\partial\theta_1}+(\eta_2/\omega_2)\frac{\partial r}{\partial\theta_2}\right)\cdot d\theta_1 d\theta_2 \\
&=\oint_C(-r\eta_1)\cdot d(\omega_1\theta_1)+(r\eta_2)\cdot d(\omega_2\theta_2) \\
&=\oint_{C_1}-r(\theta,0)\eta_1\cdot d\theta+\oint_{C_2}r(2\pi,\theta)\eta_2\cdot d\theta-\oint_{C_3}r(\theta,2\pi)\eta_1\cdot d\theta+\oint_{C_4}r(0,\theta)\eta_2\cdot d\theta \\
&=\int_0^{2\pi}\eta_1[r(2\pi,\theta)-r(0,\theta)]\cdot d\theta+\int_0^{2\pi}\eta_2[r(\theta,2\pi)-r(\theta,0)]\cdot d\theta
\end{aligned} \tag{10.126}$$

Furthermore, we have two independent linear variational equations corresponding to the Poincaré section of $\theta_1$ and $\theta_2$ as shown in Fig. 10.14(b).

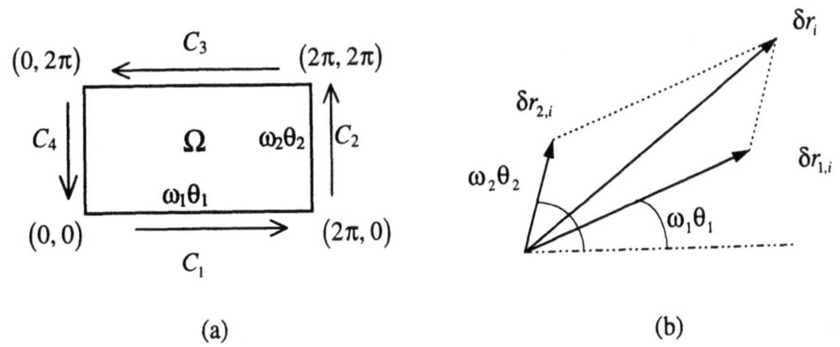


Fig. 10.14

$$\delta r_{n+1.1}=1/\eta_1\, DG[r_n]\cdot\delta r_{n.1},\ \delta r_{n+1.2}=1/\eta_2\, DG[r_n]\cdot\delta r_{n.2} \tag{10.127}$$

Finally, it is easy to find that the asymptotically orbital stability on the Poincaré sections can be ensured if

$$|\lambda_i|\le\max(\eta_1,\eta_2),\qquad i=1,2 \tag{10.128}$$

are satisfied (Hirsch and Smale [40]).

### 10.3.4 Application

For illustration, the method developed in this study is applied to coupled van der Pol oscillators (Hall [263])

$$\ddot{u}_1+\omega_1^2u_1=\varepsilon f_1(u_1,u_2)\dot{u}_1,\ \ddot{u}_2+\omega_2^2u_2=\varepsilon f_2(u_1,u_2)\dot{u}_2 \tag{10.129}$$

where $f_1(u_1,u_2)=1-u_1^2-au_2^2-bu_1^2u_2^2$, $f_2(u_1,u_2)=1-\alpha u_1^2-u_2^2$ and $\varepsilon>0$. Suppose that $\omega_1>0$ and $\omega_2>0$ are linearly independent such that

$\omega_j \neq 0\ (\mathrm{mod}\,\omega_k)$, $j \neq k$, and $k\omega_2 + j\omega_1 \neq 0$ for all integers $k, j$. In this problem $p = q = 2$. Transform eq. (10.129) into a polar co-ordinate

$$u_j = r_j^{\nu}\sin\omega_j\theta_j,\ \dot{u}_j = \omega_j r_j^{\nu}\cos\omega_j\theta_j,\ \omega_j\theta_j \in [0.2\pi]\ j = 1.2; \qquad (10.130)$$

where we can choose $\nu = 1/2$ to obtain a set of much simplified equations of (10.129). Furthermore, we obtain

$$\begin{bmatrix}\dot{\theta}_j \\ \dot{r}_j\end{bmatrix} = \begin{bmatrix} 2\omega_j r_j\cos\omega_j\theta_j & \sin\omega_j\theta_j \\ -2\omega_j r_j\sin\omega_j\theta_j & \cos\omega_j\theta_j\end{bmatrix}^{-1} \cdot \begin{bmatrix} 2\omega_j r_j\cos\omega_j\theta_j \\ 2\varepsilon f_j r_j\cos\omega_j\theta_j - 2\omega_j r_j\sin\omega_j\theta_j\end{bmatrix} \qquad (10.131)$$

so eqs. (10.129) become

$$\dot{\theta}_j = 1 - \frac{\varepsilon}{2\omega_j} f_j(r_1\sin\omega_1\theta_1, r_2\sin\omega_2\theta_2)\sin 2\omega_j\theta_j, \qquad (10.132a)$$

$$\dot{r}_j = \varepsilon r_j f_j(r_1\sin\omega_1\theta_1, r_2\sin\omega_2\theta_2)(1 + \cos 2\omega_j\theta_j), \qquad j = 1,2 \qquad (10.132b)$$

The complete set of double periodic solutions is in the form of

$$\begin{bmatrix} r_1(s_1,s_2) \\ r_2(s_1,s_2)\end{bmatrix} = \sum_{k_1=-N_1}^{N_1}\sum_{k_2=-N_2}^{N_2}\begin{bmatrix} R_1(k_1,k_2) \\ R_2(k_1,k_2)\end{bmatrix} \cdot e^{i[(2\pi s_1 k_1)/M_1 + (2\pi s_2 k_2)/M_2]} \qquad (10.133)$$

However, in order to reduce the amount of computational work, further simplifications to the solution form are introduced,

$$\begin{bmatrix} r_1(s_1,s_2) \\ r_2(s_1,s_2)\end{bmatrix} = \sum_{\substack{k_1=0 \\ |k_1|+|k_2|\le K_m}}^{N_1}\ \sum_{\substack{k_2=-N \\ k_1, k_2 = even}}^{N_2}\begin{bmatrix} R_1(k_1,k_2) \\ R_2(k_1,k_2)\end{bmatrix} \cdot e^{i2\pi(s_1k_1+s_2k_2)/M} + \mathrm{cc} \qquad (10.134)$$

where cc stands for the complex conjugate of the proceeding terms, and from previous experience we know that all periodic components

$$\nu_m = k_{1m}\omega_1 + k_{2m}\omega_2 \qquad (10.135)$$

with $$|k_{1m}| + |k_{2m}| \le K_m \qquad (10.136)$$

are likely to be eminent. Therefore, in this particular case we can choose $K_m = 12$ and only consider the wave whose numbers are included in (10.136). Furthermore, to exhibit the results $\{r_1, \theta_1, r_2, \theta_2\}$ in 3-D space, we may construct the toroidal co-ordinates in the way described by Diliberto *et al.* [258]:

$$\begin{aligned} x &= (r_1 + r_2\cos\omega_2\theta_2)\cdot\cos\omega_1\theta_1 \\ y &= (r_1 + r_2\cos\omega_2\theta_2)\cdot\sin\omega_1\theta_1 \\ z &= r_2\sin\omega_2\theta_2 \qquad & 0 < r_2 < r_1 \end{aligned} \qquad (10.137)$$

The first estimation of the equilibrium points is performed by using the averaged eq. (10.132) with respect to $\theta_1, \theta_2$:

$$\begin{aligned} r_1(1 - r_1/4 - ar_2/2 - br_1r_2/4) &= 0 \\ r_2(1 - \alpha r_1/2 - r_2/4) &= 0 \end{aligned} \qquad (10.138)$$

The solutions of eq. (10.138) $\{r_1, r_2\}^T = \{r_{01}, r_{02}\}^T$ may be almost periodic, periodic and static depending on the stability properties of nearby equilibrium points at the beginning (Hall [263]). It is evident that a torus may occur in the case of $r_{01} \neq 0$ and $r_{02} \neq 0$. We further consider the two cases when $b = 0$ and $b \neq 0$. If $b = 0$ it is easy to find the almost periodic equilibrium point of eq. (10.138)

$$r_{01} = \frac{4-8a}{1-4a\alpha} > 0,\ r_{02} = \frac{4-8\alpha}{1-4a\alpha} > 0 \tag{10.139}$$

If $b \neq 0$ then the periodic equilibrium points are difficult to determine in the system

$$r_{01} = 2 + \left(\Gamma \pm \sqrt{\Delta}\right)/8b > 0, \quad r_{02} = \left[2 - \left(\Gamma \mp \sqrt{\Delta}\right)/8b\right]/2\alpha > 0 \tag{10.140}$$

where

$$\Delta = 16a^2\alpha^2 + 128ab\alpha - 128b\alpha - 8a\alpha + 256b^2 + 32b + 1 \geq 0;\ \Gamma = 4a\alpha - 1 \tag{10.141}$$

Equation (10.140) shows that the solutions of eq. (10.138) may not be completely evaluated if $b \neq 0$. Consequently situations may arise where one can obtain a further set of solutions from eq. (10.140) and eq. (10.141). However, the existence conditions are examined for each invariant torus individually. The asymptotic stability of the equilibrium point is determined by the eigenvalues of the coefficient matrix of the linear variational equation associated with $\{r_{01}, r_{02}\}^T$. The negative real part is achieved when $0 < a < 1/2$ and $0 < \alpha < 1/2$ which represents the equilibrium stable point. The associated coefficient matrix of the linear variation equation is

$$\delta\mathbf{r}_{n+1} = \mathbf{DG}[\mathbf{r}_n] \cdot \delta\mathbf{r}_n \tag{10.142}$$

$$\text{where } \mathbf{DG}[\mathbf{r}_0] = \begin{bmatrix} \partial g_1/\partial r_1 & \partial g_1/\partial r_2 \\ \partial g_2/\partial r_1 & \partial g_2/\partial r_2 \end{bmatrix} = \begin{bmatrix} J_{11} & J_{12} \\ J_{21} & J_{22} \end{bmatrix} \tag{10.143}$$

$J_{11} = \varepsilon(4 - 2r_{10} - 2ar_{20} - br_{10}r_{20})/4$; $J_{12} = -\varepsilon r_{10}(4a + br_{10})/8$; $J_{21} = -\varepsilon\alpha r_{20}/2$; and $J_{22} = \varepsilon(2 - \alpha r_{10} - r_{20})/2$.

The eigenvalues of $\mathbf{DG}[\mathbf{r}_0]$ are

$$\lambda_{1,2} = \left[J_{11} + J_{22} \pm \sqrt{(J_{11} + J_{22})^2 - 4(J_{11}J_{22} - J_{12}J_{21})}\right]/2 \tag{10.144}$$

The boundary values $\eta_1, \eta_2$ of the eigenvalues are given by

$$\begin{aligned} \eta_1 &= \omega_1\dot{\theta}_1 = \omega_1\omega_2 \int_0^{2\pi/\omega_1}\int_0^{2\pi/\omega_2} f_1(\Theta, \mathbf{r}) \cdot d\theta_1 d\theta_2 = \left(1 + \omega_1^2\right)/2 \\ \eta_2 &= \omega_1\dot{\theta}_2 = \omega_1\omega_2 \int_0^{2\pi/\omega_1}\int_0^{2\pi/\omega_2} f_2(\Theta, \mathbf{r}) \cdot d\theta_1 d\theta_2 = \left(1 + \omega_2^2\right)/2 \end{aligned} \tag{10.145}$$

The stability properties of the invariant tori with some values of $\alpha, a, b$ have been investigated by Hall [263]. Here, since the major objective of our work is to develop a constructive procedure to approximate a periodic surface, we leave this topic for future discussion.

### 10.3.5 Numerical Results and Discussion

We present results for the invariant torus obtained by our approach in the following two cases:

(I) $a = 0.2,\ b = 0,\ \alpha = 0.4,\ \omega_1 = 1,\ \omega_2 = 1.414$

(II) $a = 0.2,\ b = 1,\ \alpha = 0.4,\ \omega_1 = 1,\ \omega_2 = 1.414$

and $\eta_1 = 1.0,\ \eta_2 = 1.4996$ are evaluated from eq. (10.145). Definitions of the invariant tori are shown from in Figs. 10.15 to 10.18. For $\varepsilon = 0.1$ indicated in Fig. 10.15(a,b,c), the torus caused by the non-linear interaction of van der Pol oscillators is quite smooth. It should be pointed out that Gilsinn [260] presented the trajectory on the same periodic surface using a long time numerical integration, as shown in Fig. 10.15(d) from a different viewpoint. However, we achieve a similar picture with much less numerical effort by means of the construction of a periodic surface using the TJM/FFT approach. Through the augmentation of the parameter $\varepsilon$, we can easily obtain the torus with $\varepsilon = 0.65$ as shown in Fig. 10.16(a,b,c) in which a blow-up of the periodic surface is observed. Consequently, when the augmentation reaches $\varepsilon = 1.0$, it is shown in Fig. 10.17(a,b,c) that two inner sides of the invariant torus are connected to each other. By further augmentation up to the strongly non-linear interaction with $\varepsilon = 1.20$, the originally smooth surface of the torus turns out to be rough with the appearance of several peaks as depicted in Fig. 10.18(a,b,c). What the graphics seem to imply is that the surface begins to break into several condensed lines in the space with increasing strength of non-linearity. Such a trend can also be observed with changes of symmetry in the vector fields shown in Fig. 10.19(b), Fig. 10.20 and Fig. 10.21. Since the variation of the vector field of $r_1(\theta_1,\theta_2)$ is not so apparent when compared with that of $r_2(\theta_1,\theta_2)$ for the whole process, we present only one graph for illustration. The ordinary portraits of double periodic responses of the invariant torus are also provided in Fig. 10.22. Some typical numerical results for the invariant torus of coupled van der Pol oscillators using the TJM/FFT approach are presented in table 10.1 for $b = 0$ and table 10.2 for $r_i(\theta_1,\theta_2)$ respectively in which $|r_i|_{\max}$ is evaluated by

$$|r_i|_{\max} = \sqrt{\sum_{\substack{k_1=-N \\ |k_1|+|k_2|\le K_m}}^{N} \sum_{\substack{k_2=-N \\ k_1,k_2=\text{even}}}^{N} [R_i(k_1,k_2)]^2} \tag{10.146}$$

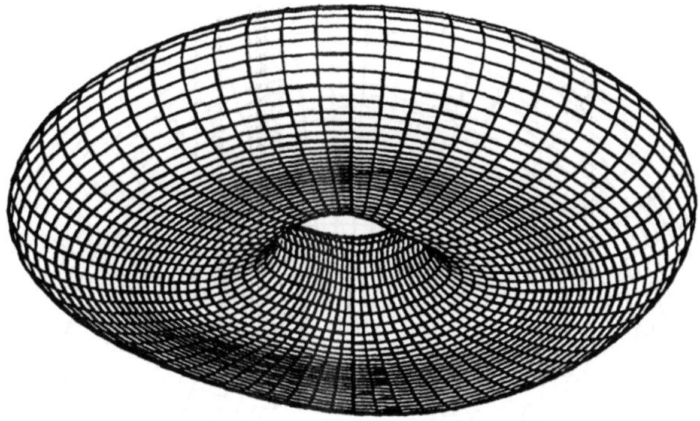

Fig. 10.15(a) Invariant torus of coupled van der Pol oscillators $b$=0, $\varepsilon$ = 0.1
Viewpoint (1.4,-2.6,-1.8)

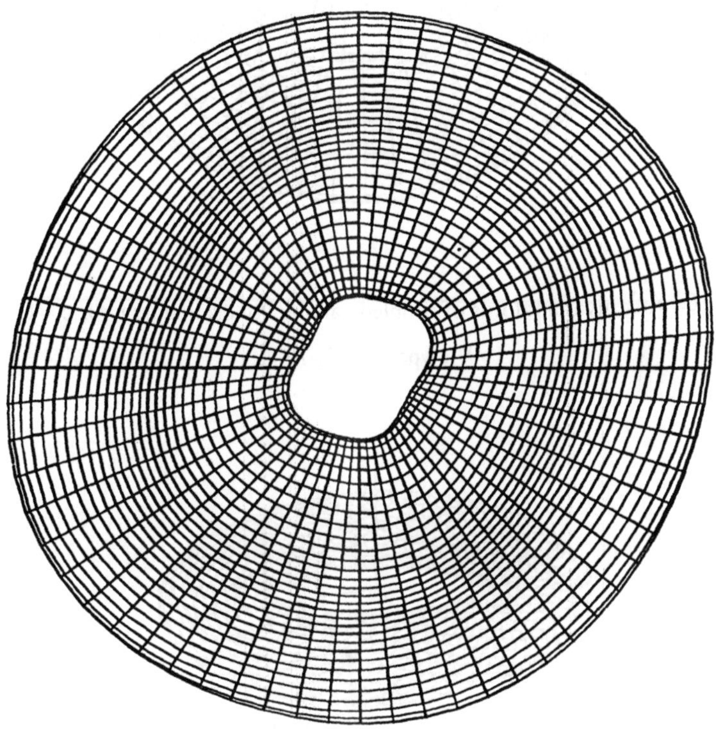

Fig. 10.15(b) Invariant torus of coupled van der Pol oscillators $b$=0, $\varepsilon$ = 0.1

Viewpoint (0,0,1)

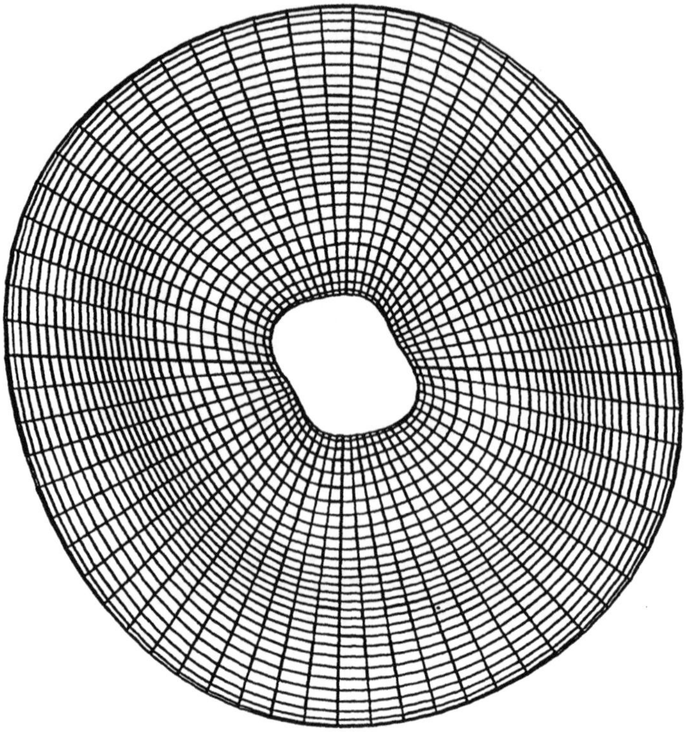

Fig. 10.15(c) Invariant torus of coupled van der Pol oscillators $b=0$, $\varepsilon = 0.1$

Viewpoint (0,0,-1)

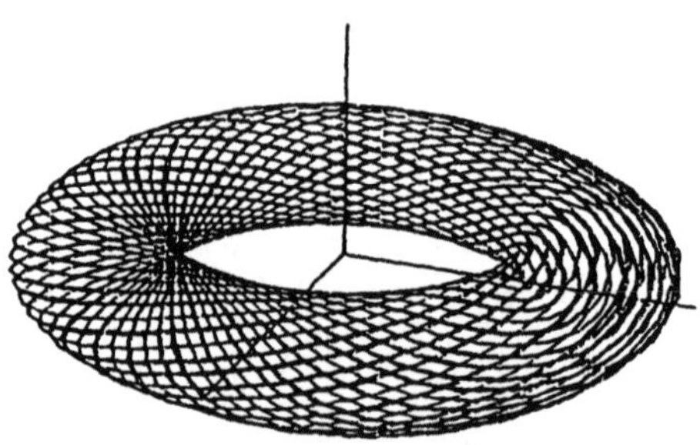

Fig. 10.15(d) Invariant torus of coupled van der Pol oscillators $b=0$, $\varepsilon = 0.1$

Numerical integration result

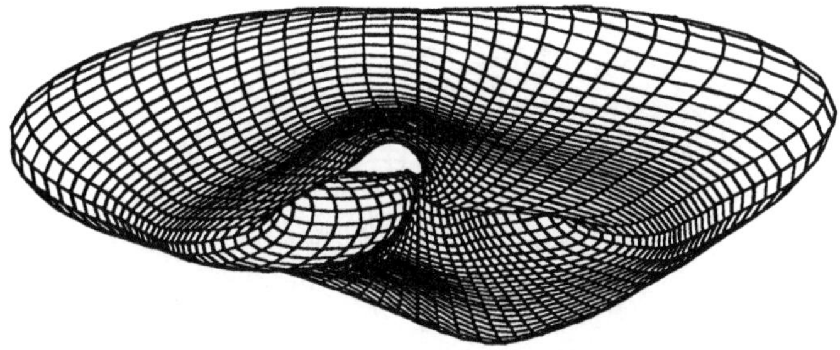

Fig. 10.16(a) Invariant torus of coupled van der Pol oscillators $b$=0, $\varepsilon = 0.65$

Viewpoint (1.4,-2.6,-1.8)

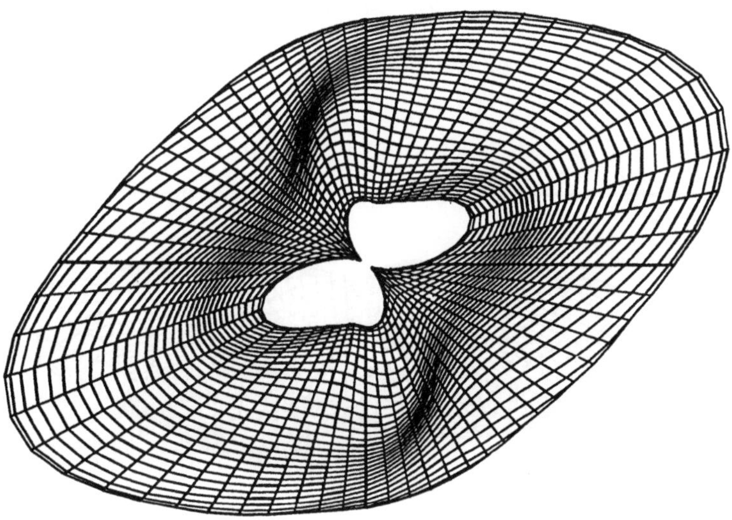

Fig. 10.16(b) Invariant torus of coupled van der Pol oscillators $b$=0, $\varepsilon = 0.65$

Viewpoint (0,0,1)

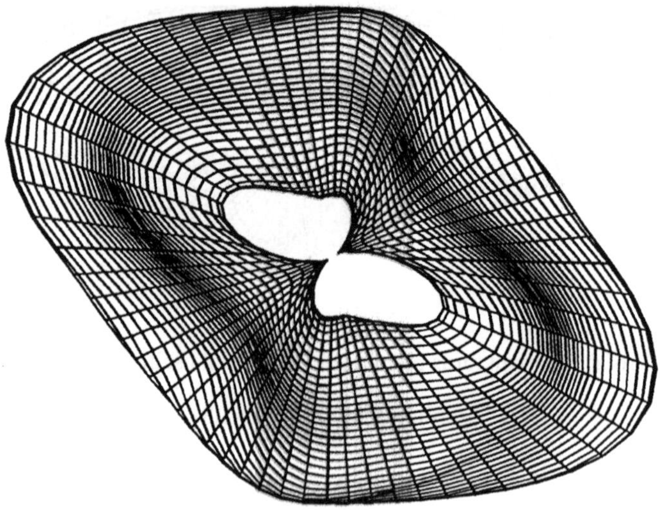

Fig. 10.16(c) Invariant torus of coupled van der Pol oscillators $b=0$, $\varepsilon = 0.65$

Viewpoint (0,0,-1)

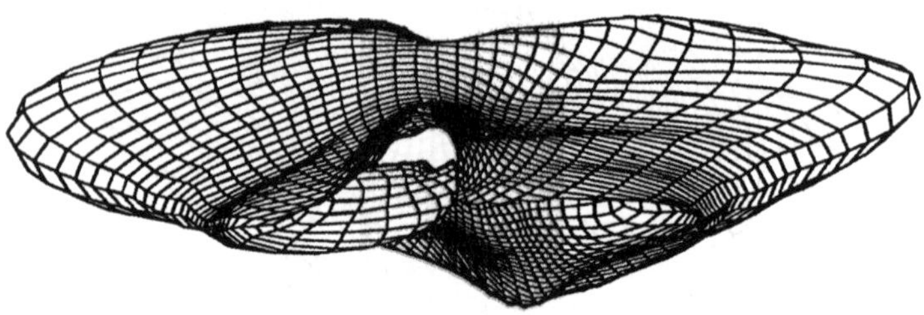

Fig. 10.17(a) Invariant torus of coupled van der Pol oscillators $b=0$, $\varepsilon = 1$

Viewpoint (1.4,-2.6,-1.8)

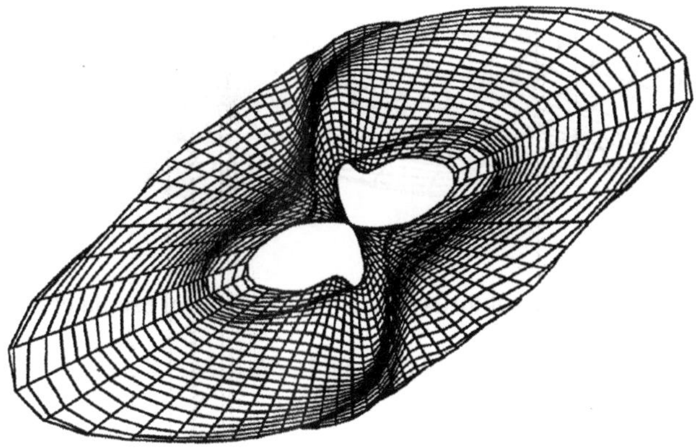

Fig. 10.17(b) Invariant torus of coupled van der Pol oscillators $b$=0, $\varepsilon = 1$

Viewpoint (0,0,1)

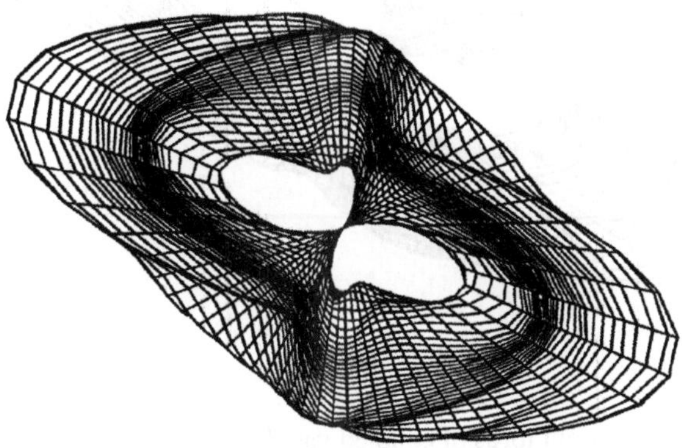

Fig. 10.17(c) Invariant torus of coupled van der Pol oscillators $b$=0, $\varepsilon = 1$

Viewpoint (0,0,-1)

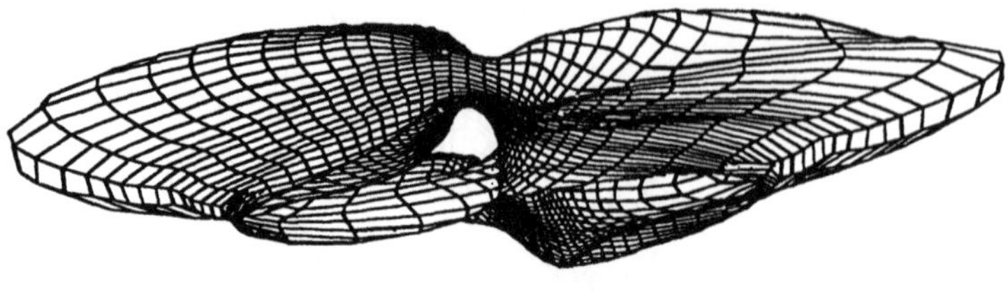

Fig. 10.18(a) Invariant torus of coupled van der Pol oscillators $b$=0, $\varepsilon = 1.2$

Viewpoint (1.4,-2.6,-1.8)

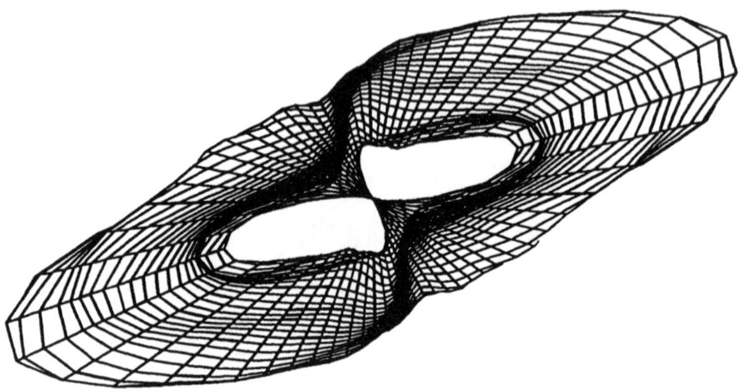

Fig. 10.18(b) Invariant torus of coupled van der Pol oscillators $b$=0, $\varepsilon = 1.2$

Viewpoint (0,0,1)

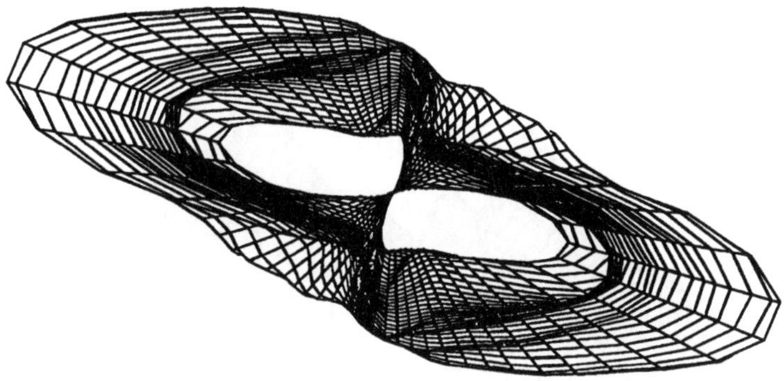

Fig. 10.18(c) Invariant torus of coupled van der Pol oscillators $b$=0, $\varepsilon = 1.2$
Viewpoint (0,0,-1)

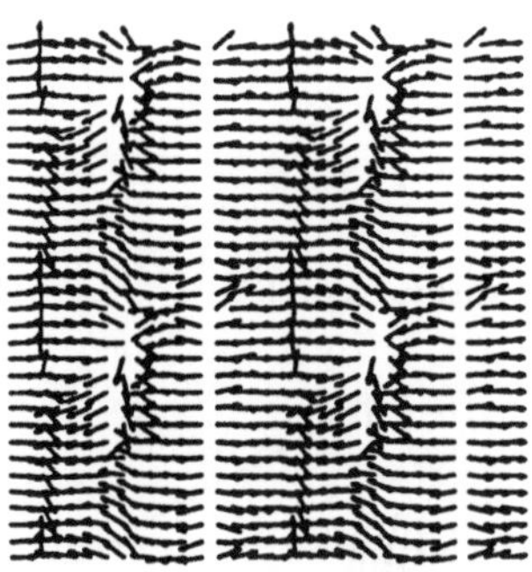

(a) Vector field $r_1(\theta_1,\theta_2)$

(b) Vector field $r_2(\theta_1,\theta_2)$

Fig. 10.19

Vector field $r_2(\theta_1,\theta_2)$

Fig. 10.20

Vector field $r_2(\theta_1,\theta_2)$

Coupled van der Pol oscillatiors. $\varepsilon = 1.2$

Fig. 10.21

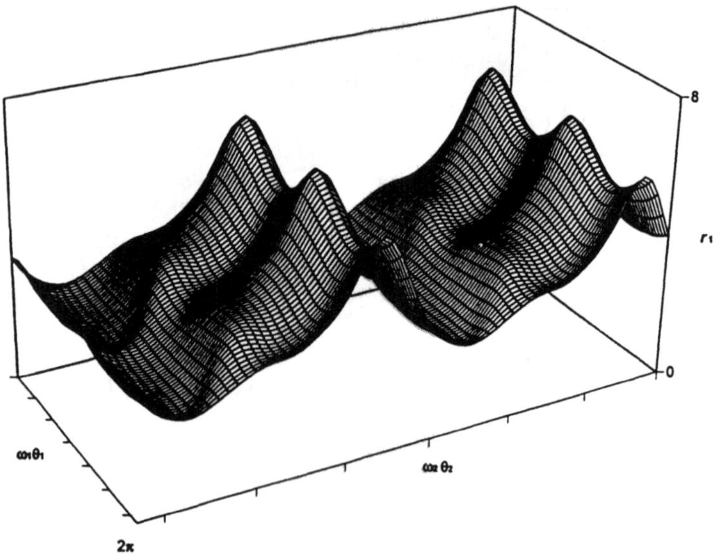


Fig. 10.22 (a) Double periodic surface $\varepsilon=0.85$

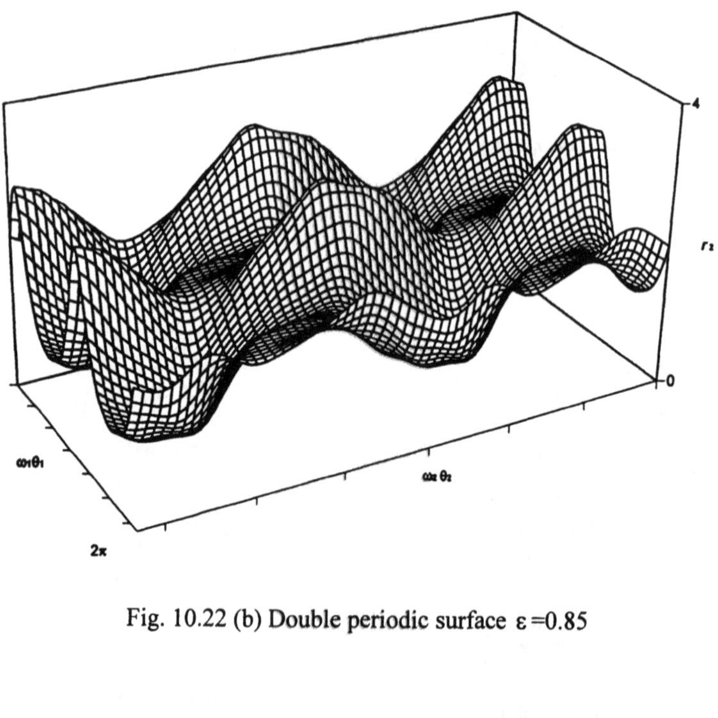


Fig. 10.22 (b) Double periodic surface $\varepsilon=0.85$

**Table 10.1** The invariant torus and its characteristic exponents (b=0)

| $\varepsilon$ | $r_1(0, 0)$ | $r_2(0, 0)$ | $\lvert r_1\rvert_{max}$ | $\lvert r_2\rvert_{max}$ | $\lambda_1$ | $\lambda_2$ |
|---|---|---|---|---|---|---|
| 0.1 | 3.531 | 1.6638 | 3.5333 | 1.7143 | −0.03 | −0.11 |
| 0.2 | 3.5362 | 1.6643 | 3.5453 | 1.7174 | −0.07 | −0.23 |
| 0.3 | 3.5455 | 1.6644 | 3.5658 | 1.7215 | −0.1 | −0.34 |
| 0.4 | 3.5599 | 1.6631 | 3.5956 | 1.7258 | −0.14 | −0.46 |
| 0.5 | 3.5799 | 1.6598 | 3.6352 | 1.7294 | −0.17 | −0.58 |
| 0.6 | 3.6063 | 1.6541 | 3.6852 | 1.7316 | −0.21 | −0.7 |
| 0.7 | 3.639 | 1.6458 | 3.7456 | 1.7323 | −0.25 | −0.82 |
| 0.8 | 3.6781 | 1.6352 | 3.8163 | 1.7314 | −0.29 | −0.95 |
| 0.9 | 3.724 | 1.6205 | 3.898 | 1.7273 | −0.33 | −1.09 |
| 1 | 3.7778 | 1.5991 | 3.9923 | 1.7181 | −0.38 | −1.23 |
| 1.1 | 3.8343 | 1.5808 | 4.0932 | 1.7121 | −0.42 | −1.38 |
| 1.2 | 3.8941 | 1.5624 | 4.2009 | 1.705 | −0.47 | −1.53 |
| 1.3 | 3.9596 | 1.5376 | 4.3186 | 1.6911 | −0.51 | −1.69 |
| 1.4 | 4.0322 | 1.5027 | 4.4483 | 1.667 | −0.56 | −1.85 |
| 1.45 | 4.0719 | 1.4796 | 4.5188 | 1.6497 | −0.58 | −1.94 |
| 1.49 | 4.1057 | 1.4575 | 4.5784 | 1.6329 | −0.6 | −2.01 |

**Table 10.2** The invariant torus and its characteristic exponents (b=1)

| $\varepsilon$ | $r_1(0, 0)$ | $r_2(0, 0)$ | $\lvert r_1\rvert_{max}$ | $\lvert r_2\rvert_{max}$ | $\lambda_1$ | $\lambda_2$ |
|---|---|---|---|---|---|---|
| 0.1 | 1.0656 | 4.4509 | 1.0664 | 4.5857 | −0.08 | −0.18 |
| 0.2 | 1.0683 | 4.4491 | 1.0716 | 4.5895 | −0.16 | −0.36 |
| 0.3 | 1.0726 | 4.4477 | 1.08 | 4.5976 | −0.24 | −0.54 |
| 0.4 | 1.0784 | 4.4486 | 1.0913 | 4.6119 | −0.33 | −0.72 |
| 0.5 | 1.0853 | 4.4542 | 1.1052 | 4.6351 | −0.41 | −0.91 |
| 0.6 | 1.0932 | 4.4661 | 1.1215 | 4.6685 | −0.5 | −1.1 |
| 0.7 | 1.1029 | 4.4825 | 1.1409 | 4.7105 | −0.6 | −1.29 |
| 0.8 | 1.1132 | 4.5056 | 1.1619 | 4.7629 | −0.69 | −1.5 |
| 0.9 | 1.1235 | 4.5361 | 1.1842 | 4.8266 | −0.8 | −1.71 |
| 1 | 1.1346 | 4.5718 | 1.2087 | 4.8988 | −0.91 | −1.93 |
| 1.01 | 1.1358 | 4.5755 | 1.2113 | 4.9064 | −0.92 | −1.95 |
| 1.02 | 1.137 | 4.5793 | 1.2139 | 4.914 | −0.93 | −1.97 |
| 1.03 | 1.1382 | 4.5831 | 1.2166 | 4.9217 | −0.94 | −1.99 |
| 1.04 | 1.1395 | 4.5869 | 1.2193 | 4.9294 | −0.95 | −2.02 |

Tables 10.3 and 10.4 exhibit several typical results given by the TJM/FFT approach. Outputs of such a method, as mentioned before, can be in the form of the coefficients of the responses or the response itself. To check the stability of the invariant torus, we attempted to evaluate the characteristic exponents and rotating numbers with augmentation of parameter $\varepsilon$. Italic texts in tables 10.1 and 10.2 represent the unstable solutions of the invariant torus examined by the characteristic exponents. Beyond the stability boundary of characteristic exponent, it is observed in Figs. 10.23 and 10.24 that the rotation number $\rho$, $\rho = [(\omega_1,\theta_1)/(\omega_2,\theta_2)]$ (Hao [265]), behaves like a Devil's staircase with the exhibition of torching down and launching out domains of rationality where a periodic solution rather than a quasi-periodic solution is found by the test of long-time numerical integration.

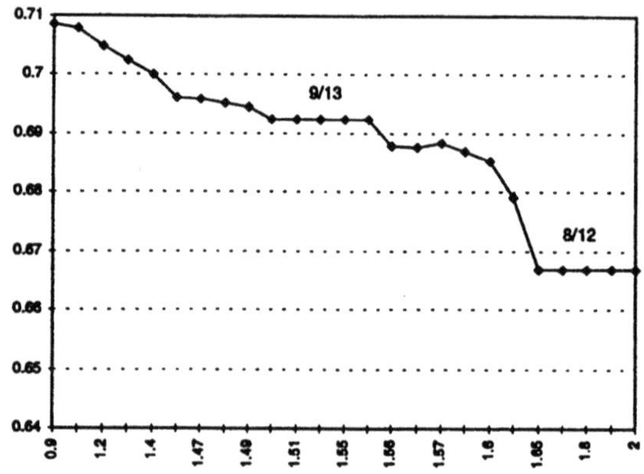


Fig. 10.23 Devil's staircase b=0

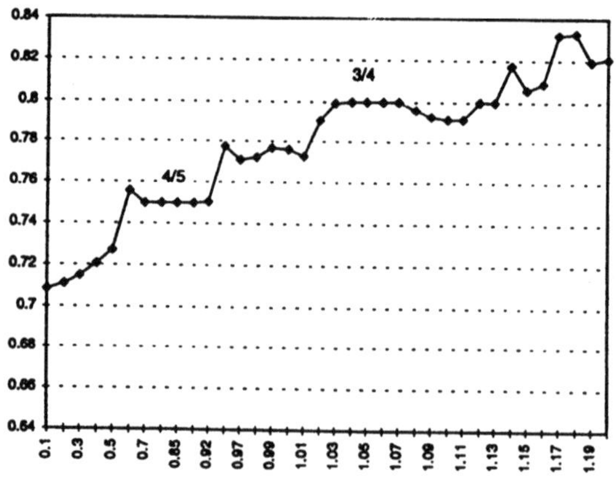


Fig. 10.24 Devil's staircase b=1

**Table 10.3** Dominant coefficients for the invariant torus

| $\omega_1\theta_1$ | $\omega_2\theta_2$ | $b$=0, ε=0.85 | | | | $b$=1, ε=0.95 | | | |
|---|---|---|---|---|---|---|---|---|---|
| | | $R_1$ | | $R_2$ | | $R_1$ | | $R_2$ | |
| $k_1$ | $k_2$ | Re | Im | Re | Im | Re | Im | Re | Im |
| 0 | 0 | 3.7001 | 0 | 1.6287 | 0 | 101289 | 0 | 4.5535 | 0 |
| 0 | 2 | 0.0316 | 0.0844 | 0.2387 | -0.0962 | 0.0325 | –0.1283 | 0.8781 | –0.527 |
| 0 | 4 | 0.0053 | 0.0138 | 0.0325 | -0.0759 | 0.0064 | –0.0225 | 0.2307 | –0.4392 |
| 0 | 6 | –0.0014 | -0.003 | 0.0007 | -0.0261 | –0.0059 | –0.008 | 0.0233 | –0.1949 |
| 0 | 8 | –0.001 | -0.0009 | –0.0048 | -0.0063 | –0.003 | –0.0032 | -0.0363 | –0.0848 |
| 0 | 10 | –0.0004 | -0.0001 | –0.003 | -0.0011 | –0.0016 | –0.0004 | -0.0306 | –0.0311 |
| 0 | 12 | –0.0001 | 0 | –0.001 | 0.0002 | –0.0009 | 0 | -0.0179 | –0.0086 |
| 2 | -10 | –0.0003 | 0.0001 | –0.0033 | -0.0003 | –0.0009 | 0.0012 | -0.007 | 0.0058 |
| 2 | -8 | 0.0002 | 0.001 | –0.0084 | 0.0039 | 0.0004 | 0.003 | -0.0065 | 0.1388 |
| 2 | -6 | –0.0001 | 0.0046 | –0.0129 | 0.0136 | 0.0025 | 0.0077 | -0.0031 | 0.0351 |
| 2 | -4 | 0.0046 | 0.0015 | 0.012 | 0.0515 | 0.0128 | 0.0073 | 0.0444 | 0.0497 |
| 2 | -2 | 0.1079 | 0.0029 | 0.0483 | 0.1657 | 0.0491 | 0.0331 | 0.1318 | 0.0845 |
| 2 | 0 | 0.1939 | -0.6347 | 0.0639 | -0.2196 | 0.0666 | –0.1836 | 0.1353 | –0.1911 |
| 2 | 2 | –0.0036 | -0.031 | 0.0346 | -0.0747 | –0.0197 | –0.0318 | 0.0414 | –0.0807 |
| 2 | 4 | –0.0025 | -0.006 | –0.002 | -0.022 | –0.0048 | –0.0075 | -0.0088 | –0.0471 |
| 2 | 6 | –0.0016 | -0.0012 | –0.006 | -0.008 | –0.0035 | –0.001 | -0.0117 | –0.0246 |
| 2 | 8 | –0.0006 | -0.0001 | –0.0031 | -0.002 | –0.0018 | –0.0001 | -0.0130 | –0.0081 |
| 2 | 10 | –0.0002 | 0.0001 | –0.0014 | 0 | –0.0007 | 0.0003 | -0.0074 | –0.0019 |
| 4 | -8 | –0.0004 | -0.0008 | –0.0033 | -0.0013 | –0.001 | –0.0002 | 0.0013 | –0.0001 |
| 4 | -6 | 0.0027 | -0.0006 | –0.0117 | 0.0021 | –0.0015 | –0.0012 | 0.0062 | –0.0017 |
| 4 | -4 | 0.0003 | 0.0071 | –0.0271 | -0.0009 | –0.0161 | 0.0033 | 0.0059 | –0.0157 |
| 4 | -2 | –0.0036 | -0.0238 | –0.0245 | -0.018 | –0.0175 | 0.0471 | -0.0178 | –0.0654 |
| 4 | 0 | 0.0686 | -0.3259 | 0.0045 | 0.0006 | 0.0324 | –0.0888 | -0.0033 | –0.001 |
| 4 | 2 | –0.0131 | -0.0116 | –0.0047 | -0.0032 | –0.0271 | –0.0042 | -0.0097 | –0.0114 |
| 4 | 4 | –0.003 | -0.0022 | –0.003 | -0.0025 | –0.005 | 0.0008 | -0.0046 | –0.0058 |
| 4 | 6 | –0.001 | -0.0001 | –0.0015 | -0.0003 | –0.0017 | 0.0019 | -0.0045 | –0.0011 |
| 4 | 8 | –0.0003 | 0.0001 | –0.0008 | 0.0003 | –0.0005 | 0.0009 | -0.003 | 0 |
| 6 | -6 | –0.0027 | -0.003 | 0.0084 | -0.0071 | –0.0174 | –0.0008 | 0.0085 | –0.0166 |
| 6 | -4 | –0.001 | -0.0074 | 0.007 | -0.0166 | –0.0153 | 0.0122 | 0.0003 | –0.0231 |
| 6 | -2 | 0.0118 | -0.0222 | 0.0158 | 0.0026 | 0.0258 | 0.0013 | 0.0142 | 0.008 |
| 6 | 0 | –0.0815 | -0.057 | –0.0163 | -0.0114 | –0.0218 | –0.02 | -0.0141 | –0.0159 |
| 6 | 2 | –0.0068 | -0.0025 | –0.0048 | -0.0052 | –0.0061 | 0.0031 | -0.0036 | –0.002 |
| 6 | 4 | –0.0017 | 0 | –0.0019 | -0.0005 | –0.0014 | 0.0017 | -0.004 | 0.0012 |
| 6 | 6 | –0.0005 | 0.0002 | –0.001 | 0.0003 | –0.0001 | 0.001 | -0.0018 | 0.0007 |
| 8 | -4 | 0.0052 | 0.0003 | –0.0018 | 0.0094 | 0.0066 | –0.0024 | -0.0005 | 0.0135 |
| 8 | -2 | –0.0078 | 0.0008 | –0.0053 | 0.003 | 0.0052 | 0.0125 | -0.01 | –0.0046 |
| 8 | 0 | –0.0466 | -0.0314 | 0.0012 | 0.0014 | –0.007 | –0.0134 | 0.0032 | 0.0012 |
| 8 | 2 | –0.0038 | 0.0017 | –0.0006 | 0.0005 | –0.0035 | 0.0054 | -0.0026 | 0.0007 |
| 8 | 4 | –0.0009 | 0.0005 | –0.0006 | 0.0003 | 0.0002 | 0.0009 | -0.0007 | 0.0001 |
| 10 | -2 | –0.004 | -0.0055 | 0.0016 | -0.0026 | 0.0053 | –0.0109 | 0.0055 | 0.0025 |
| 10 | 0 | –0.0142 | 0.0103 | –0.0026 | 0.002 | –0.0061 | 0.0031 | 0.0039 | 0.0011 |
| 10 | 2 | –0.0013 | 0.0012 | –0.0011 | 0.0004 | 0.0003 | 0.0011 | 0.0002 | 0.0001 |
| 12 | 0 | –0.0103 | 0.0061 | –0.0004 | -0.0005 | –0.0041 | –0.0019 | 0.0007 | –0.0008 |

**Table 10.4** Comparison to numerical integration with full scale system ($r_{10}$=3.53, $r_{20}$=1.18, $b$=0, ε=1.2, Tol=$10^{-4}$)

| $\omega_1\theta_1$ | $\omega_2\theta_2$ | $NI_{(5000\ Runs)}$ | | $NI_{(6000\ Runs)}$ | | TJM/FFT | |
|---|---|---|---|---|---|---|---|
| $k_1$ | $k_2$ | $r_{1max}$ | $r_{2max}$ | $r_{1max}$ | $r_{2max}$ | $r_{1max}$ | $r_{2max}$ |
| 0 | 0 | 3.8979 | 1.5713 | 3.8985 | 1.5712 | 3.8941 | 1.5624 |
| 0 | 2 | 0.1382 | 0.2623 | 0.1387 | 0.2622 | 0.1407 | 0.2598 |
| 0 | 4 | 0.0202 | 0.1098 | 0.0203 | 0.1098 | 0.019 | 0.1118 |
| 0 | 6 | 0.0051 | 0.0355 | 0.0055 | 0.0355 | 0.0053 | 0.0361 |
| 0 | 8 | 0.0025 | 0.0147 | 0.0025 | 0.0147 | 0.0025 | 0.0161 |
| 0 | 10 | 0.001 | 0.0059 | 0.0005 | 0.0059 | 0.0006 | 0.007 |
| 0 | 12 | 0.0007 | 0.0025 | 0.0004 | 0.0025 | 0.0004 | 0.0022 |
| 2 | -10 | 0.0007 | 0.0041 | 0.0009 | 0.0042 | 0.0014 | 0.0015 |
| 2 | -8 | 0.002 | 0.0089 | 0.0017 | 0.009 | 0.0006 | 0.012 |
| 2 | -6 | 0.0087 | 0.0238 | 0.0089 | 0.0239 | 0.0088 | 0.0264 |
| 2 | -4 | 0.0015 | 0.0545 | 0.0011 | 0.0546 | 0.0059 | 0.0576 |
| 2 | -2 | 0.1547 | 0.1775 | 0.155 | 0.1776 | 0.1584 | 0.1652 |
| 2 | 0 | 0.9464 | 0.302 | 0.947 | 0.302 | 0.9455 | 0.2964 |
| 2 | 2 | 0.0521 | 0.1046 | 0.0526 | 0.1047 | 0.0527 | 0.1046 |
| 2 | 4 | 0.0107 | 0.0328 | 0.0109 | 0.0329 | 0.0107 | 0.0324 |
| 2 | 6 | 0.0031 | 0.0172 | 0.0034 | 0.0172 | 0.0036 | 0.0174 |
| 2 | 8 | 0.0013 | 0.0064 | 0.0012 | 0.0063 | 0.0013 | 0.0064 |
| 2 | 10 | 0.001 | 0.0031 | 0.0004 | 0.003 | 0.0006 | 0.0031 |
| 4 | -8 | 0.0012 | 0.0056 | 0.0008 | 0.0055 | 0.0033 | 0.0235 |
| 4 | -6 | 0.0068 | 0.0153 | 0.0068 | 0.0152 | 0.0035 | 0.0231 |
| 4 | -4 | 0.0127 | 0.0277 | 0.013 | 0.0276 | 0.0237 | 0.0673 |
| 4 | -2 | 0.0475 | 0.0575 | 0.048 | 0.0575 | 0.0491 | 0.0648 |
| 4 | 0 | 0.4889 | 0.005 | 0.4894 | 0.0049 | 0.4892 | 0.0052 |
| 4 | 2 | 0.0342 | 0.0119 | 0.0348 | 0.012 | 0.0354 | 0.0113 |
| 4 | 4 | 0.0063 | 0.0069 | 0.0062 | 0.007 | 0.0066 | 0.007 |
| 4 | 6 | 0.0015 | 0.0031 | 0.0017 | 0.0031 | 0.0021 | 0.0033 |
| 4 | 8 | 0.0011 | 0.002 | 0.0008 | 0.002 | 0.0009 | 0.0021 |
| 6 | -6 | 0.0055 | 0.0077 | 0.0057 | 0.0077 | 0.0036 | 0.0439 |
| 6 | -4 | 0.0012 | 0.0337 | 0.0016 | 0.0337 | 0.0051 | 0.051 |
| 6 | -2 | 0.0476 | 0.0249 | 0.0479 | 0.0248 | 0.052 | 0.0291 |
| 6 | 0 | 0.1919 | 0.036 | 0.1924 | 0.0361 | 0.1917 | 0.0345 |
| 6 | 2 | 0.014 | 0.0118 | 0.0143 | 0.0118 | 0.0156 | 0.0118 |
| 6 | 4 | 0.0032 | 0.0039 | 0.0033 | 0.004 | 0.0037 | 0.0039 |
| 6 | 6 | 0.0012 | 0.0023 | 0.0009 | 0.0023 | 0.0013 | 0.0023 |
| 8 | -4 | 0.0091 | 0.0052 | 0.0085 | 0.0052 | 0.0117 | 0.0067 |
| 8 | -2 | 0.0171 | 0.0169 | 0.0177 | 0.017 | 0.0175 | 0.0186 |
| 8 | 0 | 0.1071 | 0.0037 | 0.1069 | 0.0036 | 0.11 | 0.0031 |
| 8 | 2 | 0.0097 | 0.0023 | 0.0099 | 0.0025 | 0.0112 | 0.0023 |
| 8 | 4 | 0.0019 | 0.0015 | 0.0021 | 0.0015 | 0.0024 | 0.0016 |
| 10 | -2 | 0.016 | 0.008 | 0.0157 | 0.0078 | 0.0175 | 0.0079 |
| 10 | 0 | 0.0461 | 0.0087 | 0.0464 | 0.0087 | 0.0447 | 0.0074 |
| 10 | 2 | 0.0039 | 0.0027 | 0.0039 | 0.0026 | 0.0048 | 0.0026 |
| 12 | 0 | 0.0281 | 0.0019 | 0.0278 | 0.0018 | 0.0311 | 0.0014 |

There are two types of errors that may corrupt the computations. Aliasing error arises when the number of terms considered is not sufficient or not distributed in the form described in eq. (10.136). The energy of higher order terms is aliased or folded back to the existing coefficients in eq. (10.134). Leakage error occurs when the invariant torus is unstable; for example, when the rotating number of oscillators becomes rational and proceeds with harmonic motions. In this case the terms of the biperiodic solution of eq. (10.134) do not coincide precisely with the components of the single periodic representation. The power at the location which cannot be represented is leaked to the components that are considered. We compared the numerical results obtained by the 5000-run and 6000-run numerical integration to the original full scale. Equation (10.129) with the solutions achieved by our PDE-TJM/FFT approach in the stable area. The error tolerance used in these computations was restricted to $10^{-4}$. As shown in Figs. 10.25(a,b), 10.26(a,b) and table 10.4, the consistent results from these two different approaches demonstrate the success of this time-saving development.

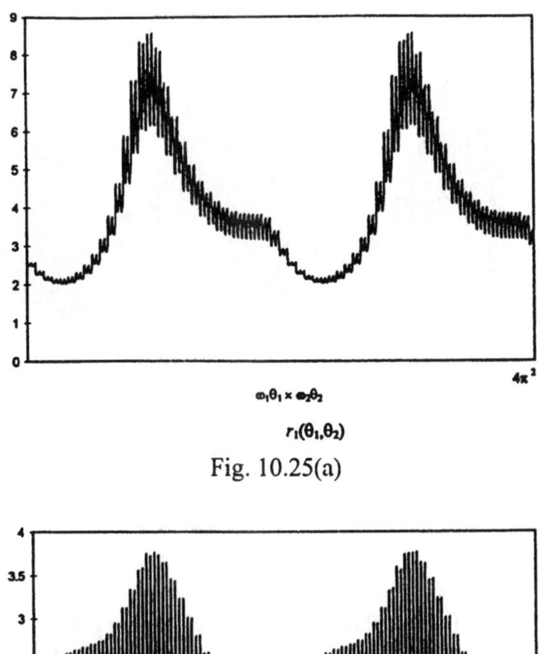


Fig. 10.25(a)

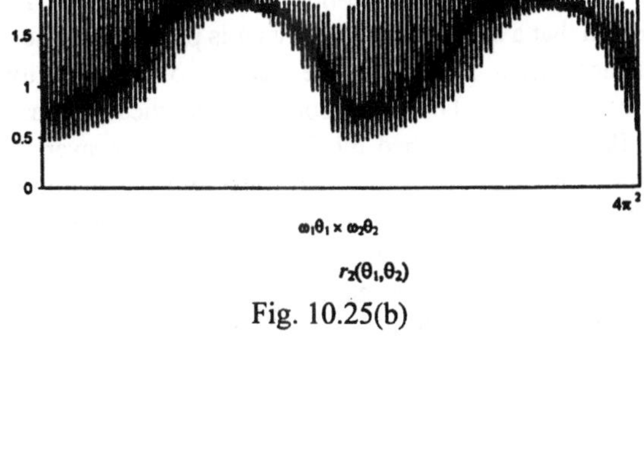


Fig. 10.25(b)

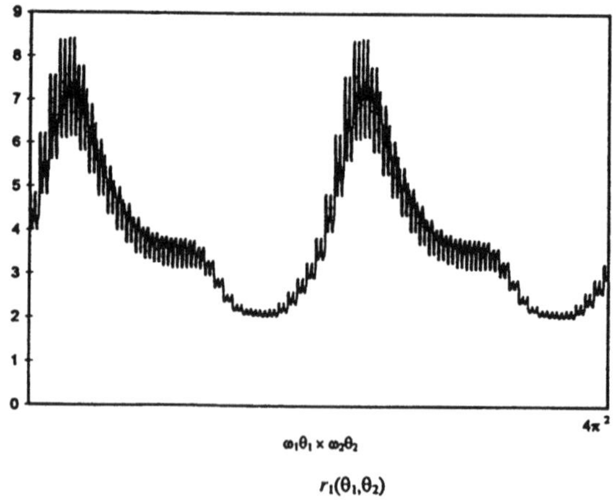


Fig. 10.26(a)

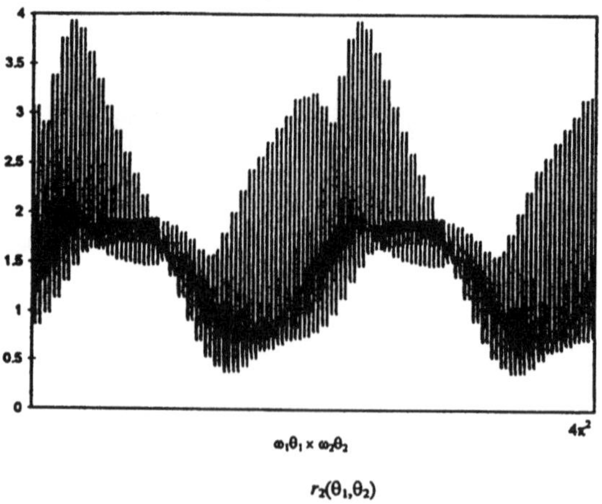


**Responses obtained by TJM/FFT method**
**($r_{01} = 3.53, r_{02} = 1.18, b = 0, \varepsilon = 1.2, Tol = 10^{-4}$)**

Fig. 10.26(b)

### 10.3.6 Conclusion

The present study describes an alternative approach for the prediction of invariant tori for autonomous non-linear oscillators. This method can easily incorporate partial differential equations and perform parameter studies even for the state of strong non-linearity. In the application to two non-linearly coupled van der Pol oscillators, the computation showed that a significant time saving is possible for the study of flows on the surface of a torus using TJM/FFT. The stability of invariant tori are considered along with the computation. The corresponding analytical expression provided by TJM/FFT is similar to that described by Gilsinn [261], however it is shown in this section that the present technique is different from previous methods in its generality and efficiency of implementation. A sufficient—necessary condition for characteristic exponents is also proposed and compared with the devil's staircase of rotation number. A comparison of the present approach with the numerical integration of the full system of coupled oscillators is also carried out.

# Chapter 11

# Non-Linear Structural Dynamics

In previous chapters, some examples were given in order to demonstrate the usefulness and advantages of the modern analytical method. In this chapter, to explain the generality and importance of non-linear dynamic problems, we introduce some engineering examples of bifurcation theory. The examples include the buckling of a bar, the flutter of an elastic structure, the non-linear resonance of an unbalanced rotating shaft with internal damping, galloping vibration analysis for an elastic structure and bifurcation in chemical engineering systems. Though the study is simple from the viewpoint of the modern analytical method, it is illustrative for engineering applications.

## 11.1 Bifurcations in Solid Mechanics

Elastic buckling is a common static bifurcation phenomenon in solid mechanics. For economical purposes, thin-walled and shell structures are used extensively in engineering constructions, e.g. tall buildings and long bridges, cars and ships, aircraft and space shuttles. Elastic buckling of bars, beams, arches, plates and shells is an important topic in solid mechanics [174—176]. Mathematically, elastic buckling is a multi-solution problem of equilibrium. Bifurcation analysis aims to find the critical load when the number of solutions changes and to find the post-buckling states corresponding to all the solutions, stable or unstable. Only a non-linear problem has multiple solutions. The non-linearity of elastic buckling arises from two causes: physical non-linearity of the stress-strain relation and geometric non-linearity of the configuration—displacement relation. When the time factor is important, elastic stability is a dynamic bifurcation problem. In this section we also consider the dynamic bifurcation problem of flutter.

### 11.1.1 The Buckling Problem of a Bar

We shall study the buckling problems of a bar using the Liapunov—Schmidt reduction and singularity theory.

#### 1. Buckling of a Uniform Straight Bar

The changing role of torsional angle $u=u(x)$ of a bar with two hinged ends under axial compression constitutes an initial value problem of the ordinary differential equation

$$F(u,\mu) = -\ddot{u} - \mu \sin u = 0, \qquad 0 \le u \le \pi \tag{11.1}$$

with an initial condition

$$\dot{u}(0) = \dot{u}(\pi) = 0 \tag{11.2}$$

where $\mu = \dfrac{P}{EI}$, $E$=modulus of elasticity and $I$=moment of inertia of cross-section. For simplicity, we suppose the length of bar is $\pi$. We take the non-linear operator

$F: X \times R^+ \to Y$, and the functional space

$$X = \{u \in C^2(0,\pi) | \dot{u}(0) = \dot{u}(\pi) = 0\}$$
$$Y = C^0(0,\pi)$$

Obviously for any $\mu \in R^+$, there exists $F(0,\mu) = 0$, i.e. $u = 0$ is a trivial solution of eqs. (11.1) and (11.2). We are concerned with the problem that from $u = 0$ the solution bifurcates (Fig. 11.1).

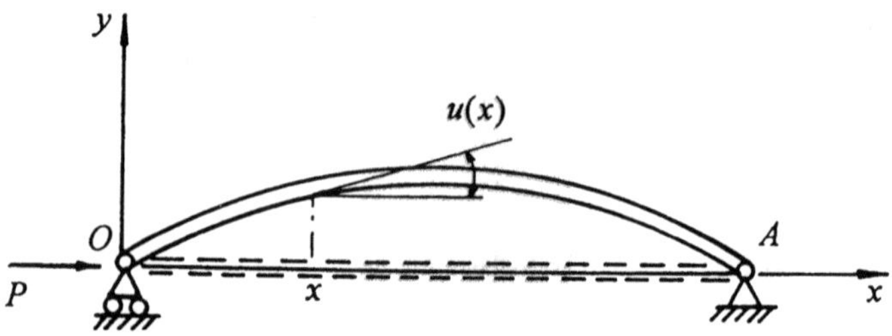


Fig. 11.1

The Linearizing eq. (11.1) with respect to the trivial solution $u = 0$ is

$$F_u(0,\mu) \equiv -\ddot{u} - \mu u = 0 \tag{11.3}$$

Considering the initial condition (11.2), when

$$\mu = \mu_n = n^2, \qquad n = 1,2 \tag{11.4}$$

we have the non-trivial solution of eq. (11.3):

$$u_n = C\cos nx, \qquad n = 1,2,\ldots \tag{11.5}$$

where $C$ is an arbitrary constant. That is, when $\mu=\mu_n$, linear operator $F_u(0,\mu)$ has zero eigenvalue, i.e. $F_u(0,\mu)$ is singular. Hence, $(0,\mu_n)$ ($n$=1,2,...) is a singular point of $F(u,\mu)$. The linear stability depends on the eigenvalue of linear operator $F_u(0,\mu)$. When $\mu=\mu_n$, the trivial solution is critical, and the corresponding critical load is $P_n = \mu_n EI = n^2 EI$. Next we discuss the static bifurcation emanated from the trivial solution at $\mu = \mu_1 = 1$.

Firstly, we reduce the non-linear operator using the Liapunov—Schmidt method.

(i) Decomposition of spaces $X$ and $Y$.

Let $L \equiv F_u(0,1)$. From eq. (11.3), the eigenspace corresponding to the zero eigenvalue of the linear operator $L$ is a one-dimensional subspace (i.e. nullspace of $L$):

$$N = \{u = \beta v_1 | \beta \in R\}$$

where $v_1 = \cos x$. Defining the inner product in spaces *X* and *Y*:

$$\langle u_1, u_2 \rangle = \int_0^\pi u_1(\xi) u_2(\xi) d\xi, \qquad u_1, u_2 \in X \left(\text{or } Y\right)$$

we make the decomposition into a direct sum:

$$X = N \oplus U \tag{11.6}$$

$$Y = N^* \oplus R \tag{11.7}$$

where *R* is the range of *L*.

$$U = N^\perp \equiv \{u \in X | < u, v_1 >= 0\}$$

$$N^* = R^\perp \equiv \{u \in Y | < u, v >= 0, \forall v \in R\}$$

Because *L* is self-adjoint, $N^* = N$, i.e. $N^*$ is the one-dimensional space spanned by $v_1$.

(ii) Equation (11.1) is equivalent to the following equation group:

$$E\, F(u, \mu) = 0 \tag{11.8a}$$

$$(I - E) F(u, \mu) = 0 \tag{11.8b}$$

where *E* is a projective operator from *Y* to *R*, and *I* is the identical operator.

(iii) According to (11.6), *u* can be decomposed into

$$u = v + w, \qquad v \in N, \quad w \in U$$

and substituting into eq. (11.8a), we get

$$E\, F(v + w, \mu) = 0 \tag{11.9}$$

From eq. (11.9), we obtain

$$w = W(v, \mu) \tag{11.10}$$

[*Remark*: In singularity analysis, we do not need the explicit expression of the function *W(v,μ)*.]

(iv) Substituting eq. (11.10) into eq. (11.8b), we obtain

$$\varphi(v, \mu) \equiv (I - E) F(v + W(v, \mu)) = 0 \tag{11.11}$$

Any $v \in N$ is represented by $v = \beta v_1$. Equation (11.11) is equivalent to the bifurcation equation

$$g(\beta, \mu) \equiv \langle v_1, \varphi(\beta v_1, \mu) \rangle = 0 \tag{11.12}$$

where $\beta \in R$ is unknown. The solution of eqs. (11.1) and (11.2) in the neighbourhood of $(\varphi, \mu) = (0,1)$ is one to one corresponding to the solution of bifurcation eq. (11.12) in the neighbourhood of $(\beta, \mu) = (0,1)$. So they have the same bifurcation behaviour. Therefore, we need to study only the bifurcation behaviour in the neighbourhood of $(\beta, \mu) = (0,1)$ of eq. (11.12). Then we use singularity theory to study the reduced equation. Using the formulae given in Chapter 4, we obtain *g* and its derivatives at (0,1):

$$g(0,1) = g_\beta(0,1) = g_{\beta\beta}(0,1) = g_\mu(0,1) = 0$$
$$g_{\beta\beta\beta}(0,1) = \frac{3\pi}{8} > 0, \qquad g_{\beta\mu}(0,1) = -\frac{\pi}{2} < 0 \tag{11.13}$$

According to the identification theory of singularity, $g(\beta,\mu)$ is equivalent to the normal form in the neighbourhood of (0,1)

$$h(\beta,\mu) = \beta^3 - (\mu - 1)\beta$$

The solution of eq. (11.12) yields pitchfork bifurcation. Similarly, the solution of eqs. (11.1) and (11.2) is the same bifurcation behaviour as in $(\varphi,\mu) = (0,1)$. The relationship between $U_{max}$ and $\mu$ is shown in Fig. 11.2. In this figure, the solid line represents the stable solution and the broken line represents the unstable solution. Obviously this coincides with the accurate solution.

Usually, the results are local and qualitative as obtained by singularity theory. But we sometimes get important global information. The asymptotic expression of the buckling solution can be obtained using the singular perturbation method [176].

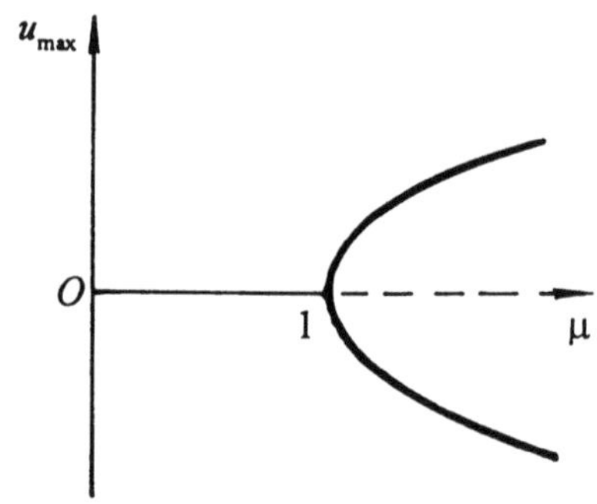


Fig. 11.2

**2. The Buckling of a Bar with an Asymmetric Defect**

In the buckling problem discussed above, suppose the bar is straight and symmetrical. When compression $P$ exceeds the critical load, the bar yields by buckling, but we do not know whether the bar bends up or down. In fact, a real bar has some asymmetric defects, such as some small original bending, asymmetry of the trachoma and so on. These affect the buckling behaviour of the bar. For example, the bar shown in Fig. 11.3, bends up slightly. (i.e. $\delta > 0$) when it is free from a compressive load. In giving the bar a compressive load, the bar bends in continuously towards the direction of the original bending. But if the original bend is small, when the load $P$ increases, the transversal displacement increases very slowly at first until $P$ reaches the critical load of the symmetric bar. Then the transversal displacement increases quickly. This buckling state of an asymmetric bar is shown in Fig. 11.4. Let $\mu = \frac{P}{EI}$ be a horizontal co-ordinate, and the maximum torsion angle (i.e. $u(0)$) be a vertical co-ordinate. In this figure the curve $b$ represents the buckling result of a symmetric bar and the curve $a$ represents the buckling result of an asymmetric bar; they have different qualitative behaviour.

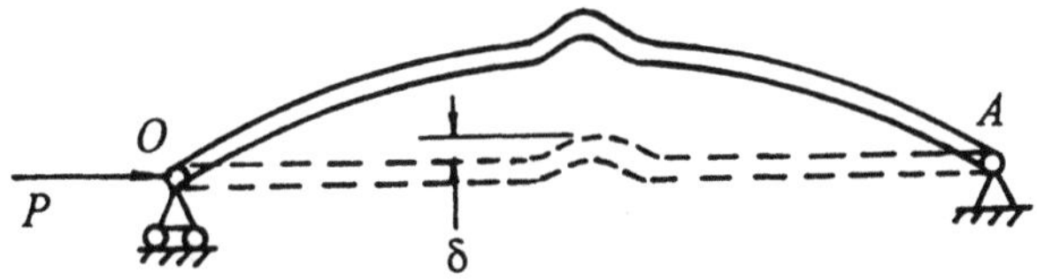


Fig. 11.3

If the original bend is very small (i.e. $|\delta|$ is very small), we can consider it as a small perturbation. So the buckling of the asymmetric bar is the perturbation result of the symmetric bar. Using the universal unfolding theory of bifurcation, we can carry out the classification and qualitative investigation of all the possible cases of the buckling problem of the asymmetric bar [177]. In addition, the solution can be obtained by the matching asymptotic extension method [178].

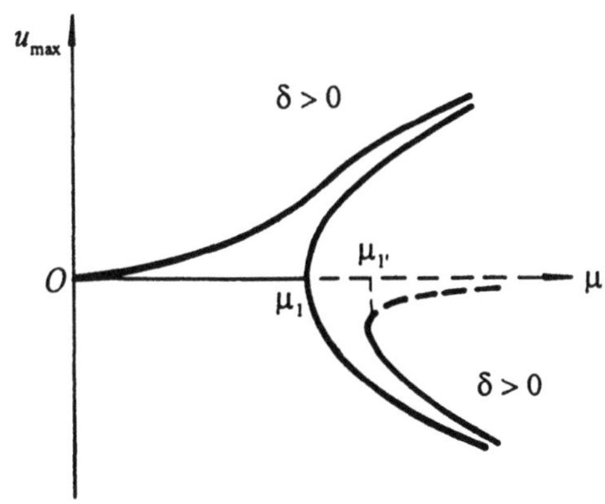


Fig. 11.4

### 11.1.2 Flutter

In flow-induced vibration, flutter is important in industry, especially in the design of aircraft, rockets and jet engines.

When the air current passes over the surface of a thin body (such as the wings of an aircraft or the vanes of an engine), if the velocity of flow is not large, then there is only slight random vibration induced by the air current. By increasing the velocity of flow, the amplitude of the vibration of the body increases, which affects the field of flow and the distribution of the intensity of pressure. The latter will change the motion of the body. This "feedback" effect is the main reason for the loss of stability in aerodynamic elasticity. The experimental result indicates that, when the velocity of flow exceeds a certain value, the vibration of the body changes from random into regular harmonic. The amplitude of the vibration increases rapidly. This phenomenon is called flutter. It is a self-excited vibration phenomenon. Figure 11.5 shows the role of the change of the amplitude $W_{max}$ with dynamical pressure $\mu = \rho_\infty U_\infty^2/2$, where

$\rho_\infty, U_\infty$ are density and velocity of the far-away flow, and $\mu_f$ is the critical value of the arising flutter.

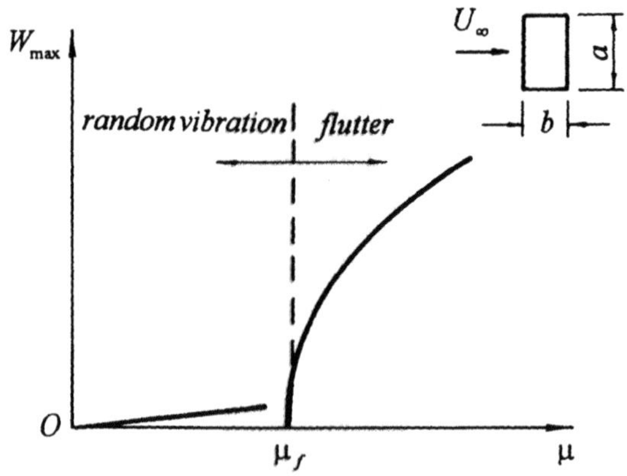


Fig. 11.5

Let us consider the one-dimensional non-linear plate equation as a simple non-linear model of aero-elasticity:

$$DW'''' - \left[\frac{Eh}{2a}\int_0^a (W'')^2 \,\mathrm{d}x\right]W'' + m\ddot{W} + \Delta p = 0 \tag{11.14}$$

where $\cdot = \partial/\partial t, ' = \partial/\partial x, W(x,t)$ is the plate deflection (transversal displacement), $D(x)$ is the bending rigidity, $m(x)$ is the mass of unit area, $x$ is the co-ordinate in the direction of flow, and $\Delta p$ is the aerodynamic intensity of pressure. $\Delta p$ consists of two parts:

$$\Delta p = \Delta p^M + \Delta p^E \tag{11.15}$$

where $\Delta p^M$ is the intensity of pressure produced by plate motion, which is written as follows

$$\Delta p^M = \rho_\infty \alpha_\infty \left(U_\infty W' + \dot{W}\right) \tag{11.16}$$

where $\alpha_\infty$ is the far-away acoustic velocity; $\Delta p^E$ is the intensity of pressure not related to the motion of the plate (i.e. not related to $W$), which can be determined by the experimental method. Suppose the solution $W_0(x,t)$ of eq. (11.14) without flutter is the basic state. Next we discuss the dynamical bifurcation from this basic state to flutter. Let $W = W_0 + w$, where $w$ is the perturbation. Then the linearized equation of eq. (11.14) with respect to $W_0$ is:

$$Dw'''' + m\ddot{w} + \Delta p^M = 0 \tag{11.17}$$

We use the harmonic orthogonal model method to determine the critical state of flutter. Let $\psi_k(x)$ $(k = 1,2,\ldots)$ be the harmonic orthogonal model, when $\Delta p=0$. $\omega_k$ $(k=1,2,\ldots)$ is the corresponding intrinsic frequency, that is,

$$D\psi_k'''' - m\omega_k^2\psi_k = 0, \quad k = 1,2,\ldots \tag{11.18}$$

and $\psi_k$ satisfies the same boundary condition of $w$. According to vibration theory, $\psi_k$ satisfies the following orthogonal conditions:

$$\int_0^a m\psi_k\psi_j \mathrm{d}x = \begin{cases} A_k, & \text{for } k = j \\ 0, & \text{for } k \neq j \end{cases} \tag{11.19}$$

where $A_k$ is a normalisation constant. Expanding $w(x,t)$ to

$$w(x,t) = \sum_{j=1}^{\infty} q_j(t)\psi_j(x) \tag{11.20}$$

where $q_j(t)$ is to be determined. Substituting eq. (11.20) into eq. (11.17), we get

$$\sum_{j=1}^{\infty}\left(Dq_j\psi_j'''' + m\ddot{q}_j\psi_j\right) + \Delta p^M = 0 \tag{11.21}$$

Multiplying eq. (11.21) by $\psi_i$, and integrating it from 0 to $a$, using eq. (11.18) and eq. (11.19), we obtain

$$A_k\left(\ddot{q}_k + \omega_k^2 q_k\right) + \mu Q_k^M = 0, \quad k = 1,2,\ldots \tag{11.22}$$

where $\mu$ is the dynamical pressure, $Q_k^M \equiv \int_0^a \frac{2\Delta p^M}{\rho_\infty U_\infty^2}\psi_k \mathrm{d}x$. From eqs. (11.16) and (11.20) one can obtain further

$$Q_k^M = \sum_{j=1}^{\infty} Q_{kj}^M \tag{11.23}$$

where $$Q_{kj}^M = \frac{2}{M_\infty}\left(q_j \int_0^a \psi_k'\psi_j \mathrm{d}x + \frac{\dot{q}_j}{U_\infty}\int_0^a \psi_k\psi_j \mathrm{d}x\right) \tag{11.24}$$

$M_\infty = U_\infty / \alpha_\infty$ is the Mach number of far-away flow. Equation (11.22) is a second-order differential equation with constant coefficients. Let $q_k = \tilde{q}_k e^{\omega t}$. Substituting into eq. (11.22), we have

$$A_k\left(\omega_k^2 + \omega^2\right)\tilde{q}_k + \mu\sum_{j=1}^{\infty} B_{kj}\tilde{q}_j = 0, \quad k = 1,2,\ldots \tag{11.25}$$

where $B_{kj}$ can be obtained from eq. (11.24), and $\tilde{q}_k$ and $\omega$ are to be determined. Equation (11.25) is the infinite algebraic equation of $\tilde{q}_j$. For the asymmetric result, we take $n$ modes only. So there are $n$ linear homogeneous algebraic equations:

$$A_k\left(\omega_k^2 + \omega^2\right)\tilde{q}_k + \mu\sum_{j=1}^{\infty} B_{kj}\tilde{q}_j = 0, \quad k = 1,2,\ldots,n \tag{11.26}$$

From linear algebra, the necessary and sufficient conditions for eq. (11.26) to have a non-zero solution are that the determinant of the coefficients is equal to zero, i.e.

$$\det\left[A_k\left(\omega_k^2 + \omega^2\right)\delta_{kj} + \mu B_{kj}\right] = 0 \tag{11.27}$$

which is the equation for , the natural frequency of the aerodynamically elastic system. Obviously, depends on the dynamical pressure $\mu$ and Mach number $M_\infty$. Usually is complex; denote it as $= \omega_R + i\omega_I$. For a given $(\mu, M_\infty)$, from

$$q_k(t) = \tilde{g}e = \left(\tilde{q}_k e^{\omega_R t}\right)e^{i\omega_I t} \tag{11.28}$$

and eq. (11.20) we see that if the real part of a certain eigenvalue is bigger than zero, i.e. $\omega_R > 0$, then perturbation $w$ increases with time, so basic state $W_0$ is linearly unstable. Only when the real parts of all eigenvalues $\omega$ are less than zero is the basic state linearly stable. Let $\omega^*$ be the eigenvalue with maximum real part, obviously $\omega_R^* = 0$ corresponds to the critical state of linear stability, and $\omega_R^*$ relates $\mu$ and $M_\infty$. In the $(\mu, M_\infty)$ parameter plane the curve consisting of points that satisfy $\omega_R^* = 0$ is the boundary of linear stability, and we call it a curve. In this case, generally there exists $\omega_I^* \neq 0$, corresponding to Hopf bifurcation (i.e. flutter). So we can use the neutral curve to determine the critical parameter value of the flutter.

To verify the appearance of flutter and study the dynamical behaviour of flutter, we must analyse non-linear eq. (11.14) for bifurcation. The centre manifold method and the Galerkin method are very useful for this purpose. It should be pointed out that eq. (11.14) is only a preliminary model to study flutter, for further study we must consider the more accurate aerodynamically elastic model [179]. The vibration problems induced by flow separation can be found in [180].

## 11.2 Non-Linear Dynamics of an Unbalanced Rotating Shaft

Rotor structure is commonly used in machinery building. With the rapid development of large-scale and high-speed rotors, to guarantee stable motion in different working situations is the most important dynamic problem in machine building. The study of the destabilising problem is very significant in rotor dynamics. There are many factors that affect the stability of motion of a rotor, such as: internal damping of material, damping of sliding bearing, damping produced by flow-solid action, air damping etc., as well as the system parameters: rigidity of rotor and support, and mass (or moment of inertia of cross-section).

The dynamic response of an unbalanced rotating shaft with internal damping is considered in this section. It is well known that internal damping can have a destabilising effect on a rotating shaft. The straight configuration of a perfectly balanced shaft loses stability when the impressed rotational speed exceeds the first critical speed. This critical speed depends on the relative magnitudes of internal and external damping, as well as other system parameters. The corresponding frequency response diagram for this case is shown in Fig. 11.6. The resulting whirling is synchronous only in the case where the external damping is zero; in general it is sub-synchronous.

The familiar amplitude—frequency relationship near the main resonance for a non-linear unbalanced rotating shaft without internal damping is depicted in Fig. 11.7. The lower branch of the steady-state amplitude is stable from the turning point to $\Omega \to \infty$. However, when included in the model, the destabilising effects due to internal damping will cause the lower branch to lose stability at some point. This can be seen by comparing the two cases represented by Fig. 11.6, the perfectly balanced shaft, and Fig. 11.7, the unbalanced shaft: the unstable solution in Fig. 11.6 for $\Omega > \Omega_1^*$ should be perturbed (for a small amount of imbalance in the shaft) to the lower branch solution in Fig. 11.7. As depicted, there is no way to go smoothly from Fig. 11.6 to Fig. 11.7.

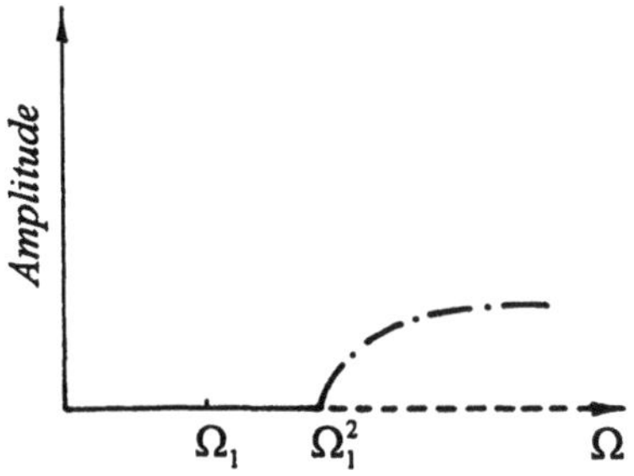

Fig. 11.6

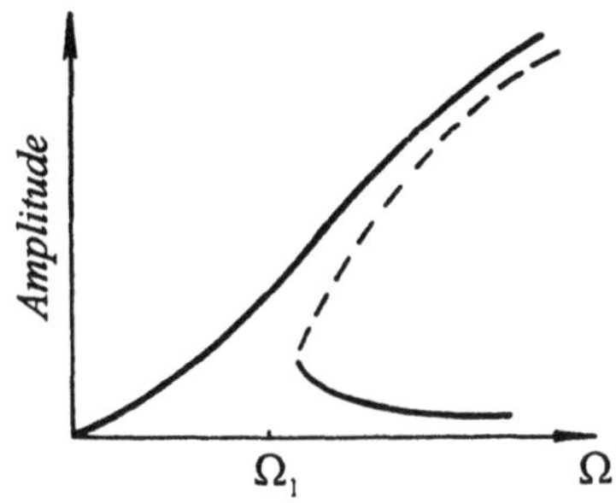

Fig. 11.7

This section provides a qualitative description of the response of the shaft near the points of instability, and presents some global pictures of the response in the neighbourhood of the first critical speed for a small imbalance. In this parameter region, detailed results can be obtained by considering a set of reduced dynamic equations using the centre manifold method.

First of all, we establish the equations of motion and assumptions. Near the primary resonance the first mode dynamics dominates the response. Hence, the equations of motion for the first shaft mode are used. These can be obtained from the full partial differential equations for a rotating shaft with structural non-linearities187]. The model to be considered consists of a straight, slender, uniform, circular cross-sectioned, simply supported shaft mode of a Voigt visco-elastic material rotating at a constant speed (see Fig. 11.8). The corresponding equations of motion are

$$\begin{aligned}
&\ddot{u} + \left(\mu_e + \mu_i \Omega_1^2\right)\dot{u} - 2\Omega\dot{v} + \left(\Omega_1^2 - \Omega^2\right)u - \mu_e \Omega v \\
&\quad + \alpha\Omega_1^2\left(u^2 + v^2\right)u + 2\alpha\mu_i\Omega_1^2\left(u\dot{u} + v\dot{v}\right)u = e\Omega^2, \\
&\ddot{v} + \left(\mu_e + \mu_i \Omega_1^2\right)\dot{v} + 2\Omega\dot{u} + \left(\Omega_1^2 - \Omega^2\right)v + \mu_e \Omega u \\
&\quad + \alpha\Omega_1^2\left(u^2 + v^2\right)v + 2\alpha\mu_i\Omega_1^2\left(u\dot{u} + v\dot{v}\right)v = 0,
\end{aligned} \tag{11.29}$$

where $u$ and $v$ are the transverse displacements in the $x$ and $y$ directions respectively, $\Omega_1$ is the first natural frequency of flexural vibrations, $\mu_e$ and $\mu_i$ are the external and internal damping constants respectively, $\Omega$ is the constant rotational speed of the shaft, $\alpha$ represents the effects of non-linear mid-line stiffening (since $\alpha > 0$, this is a hardening type of non-linearity), and $e$ is the parameter representing the effect of imbalance. The effects of gravity, axial inertia, flexural rotational inertia and transverse shear have been neglected. The equations of motion are presented in terms of a co-ordinate frame $X$—$Y$—$Z$ which is rotating at the rotational speed $\Omega$

about the $Z$—axis; hence a non-zero equilibrium point in these co-ordinates represents a synchronous whirling of the shaft. The rotating co-ordinate system provides, in this case, an autonomous set of governing equations, much like the van der Pol transformation and averaging process used in periodically forced systems. In the present case the transformation works exactly, whereas in more typical cases such transformations are only approximate. The exact transformation removes the explicit time dependence of the equations, facilitating the analysis considerably. In fact, whereas in terms of non-rotating co-ordinates the problem appears as a resonantly perturbed Hopf bifurcation, in terms of rotating co-ordinates it becomes a double-zero bifurcation with symmetry breaking due to imbalance.

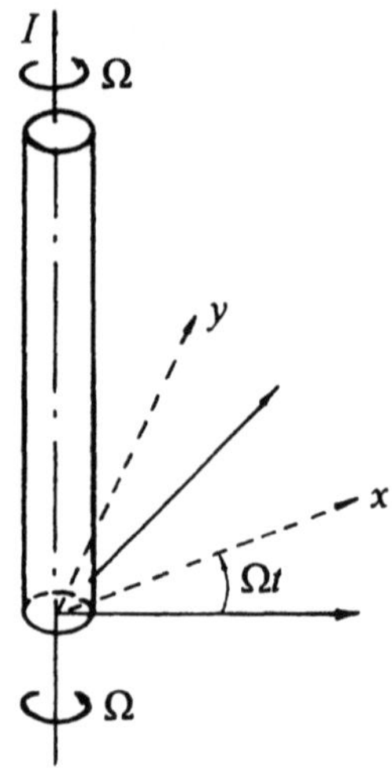


Fig. 11.8

When $e$=0 the amplitude versus rotational speed response diagram for eq. (11.29) is as shown in Fig. 11.6, where

$$\Omega_1^* = \Omega_1 + \frac{\mu_e}{\mu_i}\Omega_1 \tag{11.30}$$

is the first critical speed at which the trivial solution loses its stability and bifurcates into a stable periodic solution. However, for $e \neq 0$ and small, the trivial configuration is no longer a solution of eq. (11.29), but the resulting steady-state synchronous whirling solutions can be obtained by setting all time derivatives equal to zero and solving for the amplitude and phase of the response.

Now we discuss the synchronous steady-state solutions. Synchronous motions of constant amplitude exist in the form

$$\bar{u} = R\cos\theta, \qquad \bar{v} = R\sin\theta \tag{11.31}$$

where the amplitude, $R$, and the phase, $\theta$, of the steady-state solutions, are constants. Substituting eq. (11.31) into eq. (11.29), we obtain two algebraic equations relating $R$ and $\theta$ to the system parameters:

$$\left[\alpha\Omega_1^2 R^3 + \left(\Omega_1^2 - \Omega^2\right)R\right]^2 + \left(\mu_e \Omega R\right)^2 = e^2\Omega^4 \tag{11.32}$$

$$\tan\theta = -\mu_e\Omega \big/ \left[\alpha\Omega_1^2 R^2 + \left(\Omega_1^2 - \Omega^2\right)\right] \tag{11.33}$$

Although the equations of motion (11.29) are non-linear, the steady-state solutions (11.31) obtained by solving eqs. (11.32) and (11.33) are exact and represent the synchronous response of the shaft.

Equation (11.32) is a quadratic equation in $\Omega^2$ in terms of system parameters and $R$. Two solutions of $\Omega^2$ are obtained for every value of R. The requirements that $\Omega^2$ is greater than zero and real lead to the following three cases, depending on the level of external damping, which distinguish three types of possible synchronous response. The stability of the solutions is discussed subsequently. The three cases are as follows:

(1) $0 \le \mu_e^2 \le \alpha e^2 \Omega_1^2$. A frequency response curve for this case is depicted in Fig. 11.9(a). It is a typical non-linear response curve for a hardening non-linearity (provided here by the midline stretching of the shaft). For $\Omega < \Omega_{SN1}$, a single stable synchronous steady-state exists. At $\Omega = \Omega_{SN1}$, a vertical tangency, or in terms of bifurcation analysis a saddle-node bifurcation, occurs which introduces two response amplitude solutions. One of these is unstable; it increases in amplitude as $\Omega$ increases and asymptotically approaches the uppermost stable solution as $\Omega \to \infty$. The lowest solution branch is initially stable but undergoes a flutter instability, i.e. a Hopf bifurcation, at $\Omega_H$ and remains unstable as $\Omega$ is increased further.

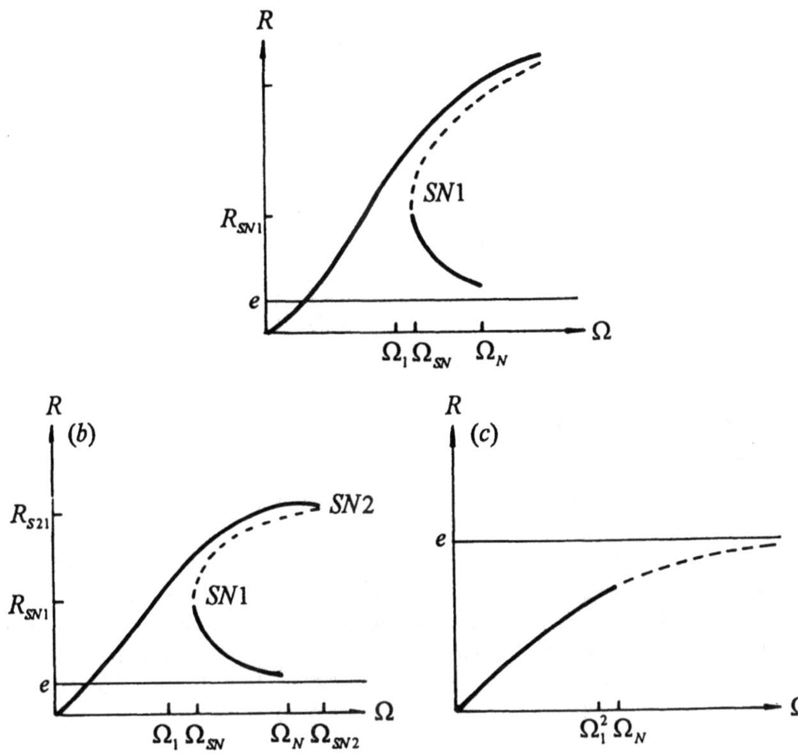


Fig. 11.9

(2) $\alpha e^2 \Omega_1^2 < \mu_e^2 < 2\Omega_1^2(\alpha e^2 + 1)$. The frequency response in this case is shown in Fig. 11.9(b). It is very similar to that of Case (1) except that the two upper solution branches merge and annihilate each other in another vertical tangency at $\Omega = \Omega_{SN2}$, leaving only the unstable lower branch.

(3) $\mu_e^2 \ge 2\Omega_1^2(\alpha e^2 + 1)$. Unlike the two previous cases, no multiple solutions of $R$ exist. The amplitude increases in a monotonous fashion and asymptotes towards $e$ as $\Omega$ increases; this is shown in Fig. 11.9(c). The response undergoes a Hopf bifurcation

at $\Omega_H$. Evidently the different unfolding parameters correspond to the different bifurcation structure diagrams; this is the persistence problem, which we shall not discuss in detail here. linearization is used to determine the stability of the synchronous steady-states. Small perturbations $(\tilde{u},\tilde{v})$ of the steady-state $(\bar{u},\bar{v})$ are introduced as follows: $u=\bar{u}+\tilde{u}$ and $v=\bar{v}+\tilde{v}$. These are substituted into eq. (11.29) which are then expanded in the Taylor series in terms of $(\tilde{u},\tilde{v})$ to yield the following linear equations which govern the dynamics of the perturbations, and thus the stability of $(\bar{u},\bar{v})$:

$$\begin{aligned}\ddot{\tilde{u}}+g_1\dot{\tilde{u}}+g_2\dot{\tilde{v}}+g_3\tilde{u}+g_4\tilde{v}&=0\\ \ddot{\tilde{v}}+g_5\dot{\tilde{v}}+g_6\dot{\tilde{u}}+g_7\tilde{v}+g_8\tilde{u}&=0\end{aligned} \tag{11.34}$$

Here the $g_i$ are coefficients which depend on the system parameters as well $(\bar{u},\bar{v})$ and $\Omega$ (but not on time). The eigenvalues of eqs. (11.34) determine the stability characteristics of $(\bar{u},\bar{v})$. The quadratic equation for the eigenvalues $\lambda$ is:

$$\lambda^4+c_1\lambda^3+c_2\lambda^2+c_3\lambda+c_4=0 \tag{11.35}$$

The coefficients in eq. (11.35) are

$$c_1=2\left(\mu_e+\mu_i\Omega_1^2+\alpha\mu_i\Omega_1^2R^2\right)$$

$$c_2=2\left(\Omega_1^2+\Omega^2+2\alpha\Omega_1^2R^2\right)+\left(\mu_e+\mu_i\Omega_1^2\right)\left(\mu_e+\mu_i\Omega_1^2+2\alpha\mu_i\Omega_1^2R^2\right)$$

$$c_3=2\mu_e\left(\Omega_1^2+\Omega^2+2\alpha\Omega_1^2R^2\right)+2\mu_i\Omega_1^2\left[\Omega_1^2-\Omega^2+\alpha R^2\left(3\Omega_1^2-\Omega^2+\alpha\Omega_1^2R^2\right)\right]$$

$$c_4=\left(\Omega_1^2-\Omega^2\right)\left(\Omega_1^2-\Omega^2+4\alpha\Omega_1^2R^2\right)+3\alpha^2\Omega_1^4R^4+\mu_e^2\Omega^2$$

$\Omega$ is varied and $R$ is evaluated at the roots of eq. (11.32): i.e. at the steady-state amplitude.

Equation (11.35) is solved numerically for the system's linearised eigenvalues along the response branches as $\Omega$ is varied. It is found that the upper branch is always stable, the middle branch is always unstable (of saddle type) and that the lower branch may or may not change stability type depending on the values of $\mu_i$ and $\mu_e$. For the lower branch there exists one pair of complex conjugate eigenvalues with real part which move further into the left half of the complex plane as $\Omega$ increases; these can be ignored as one searches for further instabilities. It is known that at least one eigenvalue is zero at point $SN1$ (since a saddle-node bifurcation occurs), which is readily verified by noting that $C_4(R_{SN1})=0$.

As one starts from point $SN1$ and follows the lower branch as $\Omega$ increases, three cases must be considered. (i) When $\mu_i=0$, the zero eigenvalue becomes negative. It merges with the negative (real) one, and splits into a pair of complex conjugate eigenvalues which move further to the left of the complex plane; this is the case shown in Fig. 11.9 in which the lower branch is stable for all $\Omega$ values for which it exists. (ii) When $0<\mu_i<\mu_i^*$ the zero eigenvalue first moves into the left of the complex plane, then meets the other negative eigenvalue and splits into a pair of complex conjugate eigenvalues with negative real part; the pair of complex conjugate eigenvalues travel rightward as $\Omega$ increases and eventually cross the imaginary axis at a point designated by $\Omega_H$ (determined below); hence the stable lower branch becomes unstable in a Hopf bifurcation at point $\Omega_H$ as shown in Fig. 11.9(a—c). (iii) When $\mu_i \geq \mu_i^*$ there are either two zero eigenvalues ($\mu_i = \mu_i^*$) or one zero and one positive and real ($\mu_i > \mu_i^*$) eigenvalues at point *SN*1 which merge and become complex conjugates with positive real parts. Then move continually rightward in the complex plane as $\Omega$ increases: i.e. the lower branch is unstable at its origin (point *SN*1) and remains unstable as $\Omega$ increases.

An expression for $\mu_i^*$ can be obtained as follows. For $\mu_i = \mu_i^*$ there are two zero eigenvalues at point *SN*1; this implies $c_3=0$ in addition to $c_4=0$. Setting $c_3=0$ and utilising $c_4=0$ to complex $R_{SN1}$ and $\Omega_{SN1}$ yields

$$\mu_i^* = \frac{\mu_e\left(\Omega_1^2 + \Omega_{SN1}^2 + 2\alpha\Omega_1^2 R_{SN1}^2\right)}{\Omega_1^2\left[\Omega_{SN1}^2 - \Omega_1^2 + \alpha R_{SN1}^2\left(\Omega_{SN1}^2 - 3\Omega_1^2 - \alpha\Omega_1^2 R_{SN1}^2\right)\right]} \tag{11.36}$$

To obtain an expression for $\Omega_H$ one can apply the Routh—Hurwitz criterion to eq. (11.35). At $\Omega = \Omega_H$ the following condition must hold (in order to have one pair of purely imaginary eigenvalues):

$$c_1c_2c_3 - c_3^2 - c_1^2c_4 = 0 \tag{11.37}$$

Equation (11.37) (with the $c_i$ from eq. (11.35)) and (11.32) are combined to solve for $\Omega_H$ implicitly.

In order to determine post-critical dynamics for the synchronous solutions and the bifurcating solutions, including periodic, quasi-periodic, etc., the qualitative theory of dynamical systems, in particular the centre manifold method, can be used in a restricted parameter region. The parameter range to be considered is as follows: (i) $\mu_e / \mu_i\Omega_1$, the ratio of external to internal damping, is small; (ii) rotational speeds are close to the first critical speed $\Omega_1^*$; and (iii) the imbalance, $e$, is small. Condition (i) implies that the critical speed is near the first resonance. Condition (ii) restricts attention to rotational speeds near critical (and resonance), the region of interest. And condition (iii) simply states that the effects of a slight imbalance on the instability are of interest (although it could be equally argued that it is the effects of internal damping on resonance that are of interest). In this case the first mode dynamics are dominated by the behaviour associated with the two eigenvalues which have, in this parameter region, nearly zero real part, and the imbalance. The dynamics associated with the remaining two eigenvalues, which lie well into the left half of the complex plane, will rapidly decay. Owing to non-linear coupling effects, the dynamics associated with these latter eigenvalues cannot simply be ignored in general, and the centre manifold technique provides a constructive tool by which the reduced slow dynamics can correctly account for such effects. The treatment begins by splitting the linearised dynamics into slow and fast components. The centre manifold (which is a non-linear invariant subspace tangent to the slow linear eigenspace) is then constructed and finally the dynamic equations on the centre manifold, i.e. the reduced dynamic equations, are then applied and the inferred shaft response can be described.

By introducing new co-ordinates $X = (x_1, x_2, x_3, x_4, e, \varepsilon)^T = (u, v, \dot{u}, \dot{v}, e, \varepsilon)^T$, where $\varepsilon = \Omega_1^* - \Omega_1$, system (11.29) can be written in first-order form as follows:

$$\dot{X}=\begin{Bmatrix}\dot{x}_1\\ \dot{x}_2\\ \dot{x}_3\\ \dot{x}_4\\ e\\ \varepsilon\end{Bmatrix}=\begin{bmatrix}0&0&1&0\\0&0&0&1\\ \Omega^2-\Omega_1^2&\mu_e\Omega&-\mu_e-\mu_i\Omega_1^2&2\Omega\\ -\mu_e\Omega&\Omega^2-\Omega_1^2&-2\Omega&-\mu_e-\mu_i\Omega_1^2\\0&0&0&0\\0&0&0&0\end{bmatrix}\begin{bmatrix}0&0&x_1\\0&0&x_2\\ \Omega^2&0&x_3\\0&0&x_4\\0&0&e\\0&0&\varepsilon\end{bmatrix}$$

$$+\begin{Bmatrix}0\\0\\ -\alpha\Omega_1^2\left(x_1^2+x_2^2\right)x_1-2\alpha\mu_i\Omega_1^2\left(x_1x_3+x_2x_4\right)x_1\\ -\alpha\Omega_1^2\left(x_1^2+x_2^2\right)x_2-2\alpha\mu_i\Omega_1^2\left(x_1x_3+x_2+x_4\right)x_2\\0\\0\end{Bmatrix}=BX+F(X) \qquad (11.38)$$

From this point on $e$ and $\varepsilon$ are treated as variables, rendering terms such as $\varepsilon x_1$ etc. non-linear in the co-ordinates $X\in R^6$, and this permits application of the centre manifold theorem for parameter values away from, but near to, the critical case. The term $BX$ represents the linearized shaft dynamics, and $F(X)$ represents the structural non-linear terms. At $\Omega=\Omega_1^{\bullet}$, the eigenvalues of $B$ are $\pm i\omega_1, \sigma_2\pm i\omega_2$, 0 and 0, where $\omega_1=\frac{\mu_e}{\mu_i}\Omega_1$, $\sigma_2=\mu_e+\mu_i\Omega_1^2$, $\omega_2=2\Omega_1+\frac{\mu_e}{\mu_i}\Omega_1$ and $i^2=-1$. Upon letting $\Omega=\Omega_1^{\bullet}+\varepsilon$, where $\varepsilon$ measures the deviation of the speed $\Omega$ from $\Omega_1^{\bullet}$, and decomposing $B$ into $B_0+B$, eq. (11.38) becomes

$$\dot{X}=B_0X+B_\varepsilon X+F(X) \qquad (11.39)$$

where $B_0$ has the same eigenvalues as $B$ at $\Omega=\Omega_1^{\bullet}$ and $B$ is identical to the zero matrix when $\varepsilon$=0. By employing the change of co-ordinates $X=QZ$, which contains the eigenvectors of $B_0$ in correspondence to the eigenvalues $\pm i\omega_1$, $\sigma_2\pm i\omega_2$, 0 and 0, eq. (11.39) is transformed into the standard form needed in order to apply the centre manifold theorem.

$$\begin{Bmatrix}\dot{Z}_1\\ \dot{Z}_2\\ \dot{Z}_3\\ \dot{Z}_4\\ \dot{q}\\ \dot{\varepsilon}\end{Bmatrix}=\begin{bmatrix}-\sigma_2&-\omega_2&0&0&0&0&Z_1\\ \omega_2&-\sigma_2&0&0&0&0&Z_2\\0&0&0&-\omega_1&0&0&Z_3\\0&0&\omega_1&0&0&0&Z_4\\0&0&0&0&0&0&q\\0&0&0&0&0&0&\varepsilon\end{bmatrix}+\begin{Bmatrix}Q^{-1}B_\varepsilon QZ+Q^{-1}F(QZ)\\0\\0\end{Bmatrix} \qquad (11.40)$$

where $q=\dfrac{e}{\omega_1\left(\omega_2^2+\sigma_2^2\right)}$ is used in place of $e$ and $Z_i$, $i$=1—4, are linear combinations of $u$, $\dot{u}$, $v$ and $\dot{v}$. The centre manifold, given by $(Z_1,Z_2)=h(Z_3,Z_4,q,\varepsilon)$, may be approximated as a power series in its variables. The dynamics on the centre manifold is then governed by the last four of eq. (11.40) after substituting h for $Z_1$ and $Z_2$. The resulting reduced dynamics on the centre manifold is given by

$$\begin{Bmatrix} \dot{Z}_3 \\ \dot{Z}_4 \end{Bmatrix} = \begin{bmatrix} \mu(\varepsilon) & -\omega_1 - \beta(\varepsilon) \\ \omega_1 + \beta(\varepsilon) & \mu(\varepsilon) \end{bmatrix} \begin{Bmatrix} Z_3 \\ Z_4 \end{Bmatrix} + \begin{Bmatrix} a(Z_3^2 + Z_4^2)Z_3 + b(Z_3^2 + Z_4^2)Z_4 \\ -b(Z_3^2 + Z_4^2)Z_3 + a(Z_3^2 + Z_4^2)Z_4 \end{Bmatrix}$$

$$+ \begin{Bmatrix} d_1(\varepsilon)q + d_2 q^3 + (d_3 Z_3 + d_4 Z_4)q^2 + (d_5 Z_3^2 + d_6 Z_3 Z_4 + d_7 Z_4^2)q \\ d_8(\varepsilon)q + d_9 q^3 + (d_{10} Z_3 + d_{11} Z_4)q^2 + (d_{12} Z_3^2 + d_{13} Z_3 Z_4 + d_{14} Z_4^2)q \end{Bmatrix} \qquad (11.41)$$

$$\dot{q} = 0, \qquad \dot{\varepsilon} = 0$$

The constants $\mu, \beta$, $d_1$ and $d_8$ depend on $\varepsilon$ and the other system parameters vanish when $\varepsilon$ vanishes. The remaining constants depend on the system parameters only and not on $\varepsilon$. The terms in the last bracket of eq. (11.41) represent the perturbation to the perfectly balanced shaft since these terms vanish when $q = 0$. The remaining terms describe the Hopf bifurcation of a balanced shaft. A supercritical Hopt bifurcation, corresponding to the birth of a stable limit cycle from the trivial solution, is depicted in Fig. 11.6, and is thus concluded since $a = -\alpha\Omega_1^2\sigma_2 \Big/ \left[\sigma_2^2 + (\omega_2 - \omega_1)^2\right]$ is negative. The dynamics of the full system (eq. (11.29)) with small unbalance and $\Omega$ near $\Omega_1^*$ are fully captured by the dynamics of eq. (11.41), the reduced equations.

Instead of analysing the full system of four first-order non-linear differential equations, the investigation can proceed by studying only two first-order equations. This provides a method for predicting and describing the response without the use of the extensive simulations which would be required in a numerical study as a result of the number of parameters. The analysis utilises standard phase plane techniques: it involves locating the equilibrium points, determining their stability, and checking for the possible occurrence of limit cycles and global bifurcations, such as saddle connections.

In order to interpret the shaft motions from the dynamics on the centre manifold, the facts and correspondences given below should be kept in mind. First, the centre manifold is a two-dimensional surface embedded in the full four-dimensional phase space. It is the invariant solutions that remain on the centre manifold. It is also the proximate locally attractive solutions which start to tend towards it asymptotically (since the dynamics away from the centre manifold will be dominated by the linear terms corresponding to the two eigenvalues that have relatively large negative real parts). Equilibrium points in the centre manifold correspond to constant amplitude and phase synchronous shaft motion, and also correspond to the stability types. A limit cycle in the centre manifold corresponds to a shaft motion which undergoes amplitude and phase oscillations. The shaft will oscillate in a periodic manner in both amplitude and phase with a frequency approximately equal to that associated with a Hopf bifurcate: i.e. $\omega_1$ (recall that $\pm i\omega_1$ are the eigenvalues associated with the instability). In terms of fixed, laboratory co-ordinates, the shaft undergoes a quasi-periodic motion consisting of two frequencies (plus combinations and harmonics); these frequencies correspond to $\approx \Omega_1$ and $\Omega$. For the restricted parameter conditions considered here ($\mu_e/\mu_i\Omega_1$ small), $\omega_1$ is small and thus the shaft will exhibit a motion which looks like synchronous whirling undergoing a low-frequency amplitude modulation (the relative phase is also modulated with the same low frequency). This is not unexpected since, for $\mu_e/\mu_i\Omega_1$ small, the internal damping drives it towards exactly synchronous whirling. Thus, when combinations of these two motions occur, the motion is composed of two oscillations of nearly equal frequency, and a beating, or amplitude modulated, response with a long period envelope exists.

The equilibrium points are obtained by setting $\dot{Z}_3 = \dot{Z}_4 = 0$ in eq. (11.41). The two resulting algebraic equations are cubic in $Z_3$ and $Z_4$ and one, two, or three real solutions are possible for each set of $(e,\varepsilon)$. These correspond to the $(R,\theta)$ determined in eqs. (11.32) and (11.33). The stability of the equilibrium is determined via linearization. Since the balanced shaft ($e = 0$) undergoes a Hopf bifurcation at $\Omega = \Omega_1^*$ ($\varepsilon = 0$), it is expected that at least one equilibrium solution will undergo a Hopf bifurcation at some points with $\varepsilon > 0$. Once the Hopf bifurcation is found, its type (sub-or super-critical) is known to be super-critical from the unperturbed system, since the small perturbations due to $e$ cannot affect the (order-one) coefficient, $a$, in the normal form for the Hopf bifurcation. Thus a stable limit cycle will occur.

Fig. 11.10

Because of the complexity of eq. (11.41), numerical root-solving methods and simulations were employed. Figure 11.10 shows schematic diagrams taken from a series of numerically calculated phase portraits obtained by fixing $e$ and increasing $\varepsilon$ from below zero. These clearly show the occurrences of a saddle-node bifurcation (portrait 2), a Hopf bifurcation (portraits 3 and 4), a saddle connection (portrait 5), and the birth of a limit cycle immediately after the second saddle-node bifurcation (portrait 8). In fact, a saddle-node, saddle connection occurs here. A bifurcation diagram with $e$ and $\varepsilon$ as the unfolding parameters is shown in Fig. 11.11, with the regions of the associated phase portraits from Fig. 11.10 indicated. Bifurcation diagrams for eq. (11.29), based on the dynamic equivalence to eq. (11.41) and on the properties derived in eqs. (11.32) and (11.33) are shown in Fig. 11.12. A discussion of the main results follows.

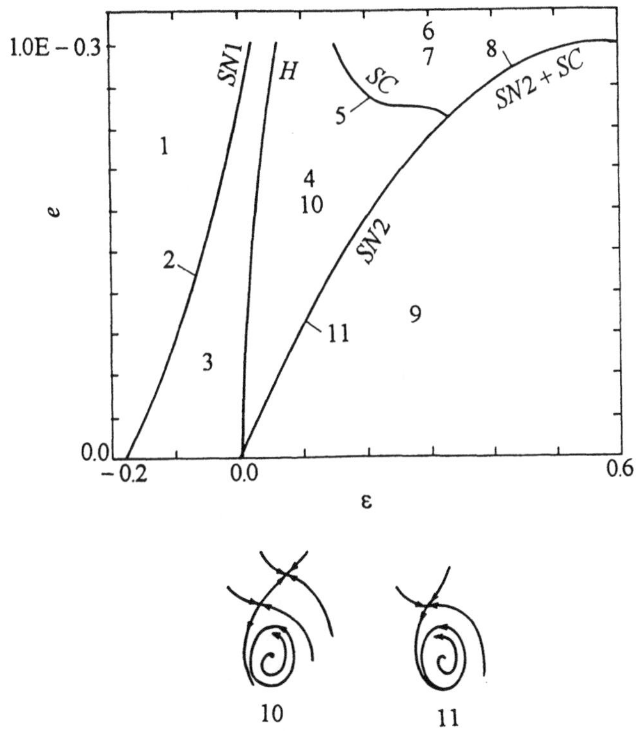


Fig. 11.11

The bifurcation curves for the shaft response are depicted in Fig. 11.12. Each point in the $(e, \Omega)$ space corresponds to a specific magnitude imbalance and rotational speed. The discussion will be focused on frequency response diagrams for the shaft; these display a measure of the shaft's vibrational amplitude versus rotational speed. By including the imbalance as a parameter in Fig. 11.12, the response curves for various levels of imbalance can be obtained. In Fig. 11.12, stable synchronous response amplitudes are indicated by solid lines, unstable synchronous amplitudes are indicated by dashed lines, and quasi-periodic response amplitudes are shown as dash-dotted lines. The amplitudes and stability types for synchronous motions are determined from an analysis in the section of steady-state solutions. The existence and stability of the amplitude-modulated motions, as well as the interaction of these with synchronous motions, are inferred from results presented in the section of application of the centre manifold theorem and some selective simulations. It should be noted that results regarding equilibrium and stability are known rigorously, while results regarding global bifurcations are based on simulations of the centre manifold eq. (11.41) and verified by simulations of the full eq. (11.29).

In Fig. 11.12, the curves marked *SN*1 are associated with conditions for the first saddle-node bifurcation in which the lower two branches of the synchronous response curve are born as $\Omega$ increases. The curves *SN*2 and *SN*2 +*SC* correspond to the saddle-node bifurcation in which the upper and middle synchronous response branches merge and disappear. Note that this curve is asymptotic to a finite value of $e$ ($e$ such that $\mu_e^2 = \alpha e^2 \Omega_1^2$) as $\Omega$ increases, indicating that for sufficiently large imbalance this bifurcation will not occur. The *SN*2 corresponds to a simple saddle-node. The *SN*2 +*SC* corresponds to a more complicated bifurcation in which the saddle-node bifurcation interacts with a saddle connection, resulting in the appearance of a limit cycle corresponding to amplitude-modulated motion (this is depicted in Fig. 11.10, portraits 7—9, with 8 showing the condition at *SN*2+*SC*; see [188]). The curve *H* corresponds to the point at which the lowest synchronous response branch becomes unstable in a Hopf bifurcation, at which point the quasi-periodic response is born. The curve marked *SC* corresponds to the condition in which the limit cycle amplitude has grown to the point where it touches the saddle point corresponding to the unstable middle branch of the synchronous response curve, forming a simple saddle connection (portrait 5 in Fig. 11.10). This bifurcation annihilates the limit cycle. The points where these curves meet correspond to rather complicated co-dimension two bifurcations which are, in themselves, of interest. This is not pursued here, however, since the required information is obtained in a direct fashion using simulations; the interested reader is referred to [205].

Figure. 11.12(a) [or 11.12(b) respectively] is a bifurcation diagram for the case in which the Hopf bifurcation disappears before [or after, respectively] the second saddle-node bifurcation disappears as the eccentricity is increased; the two cases are distinguished by the relative magnitudes of unbalance $e$ which correspond to the conditions $\mu_e^2 = \alpha e^2 \Omega_1^2$ and $\mu_i = \mu_i^*$. These two cases result in minor qualitative differences in the frequency response of the shaft for different levels of imbalance; this is described more fully below. The response of the unbalanced shaft as its rotational speed is varied (near $\Omega_1$) can be inferred from these bifurcation diagrams and the corresponding frequency response diagrams.

The response curves corresponding to the bifurcation diagram of Fig. 11.12(a) are described here. There are four distinct frequency responses possible in Fig. 11.12(a), depending on the level of imbalance. These correspond to cases marked by unbalance levels labelled I, III, IV and V. Cases II represents a special situation which lies on a boundary separating two generic cases. The stable, that is observable, motions are described below.

For very small levels of unbalance, case I, the shaft response consists of a frequency region in which only a single possible synchronous whirling motion exists, $\Omega < \Omega_{SN1}$. Two possible synchronous whirl amplitudes (the observed steady-state depends on initial conditions) exist for $\Omega_{SN1} < \Omega < \Omega_H$, one large and one small. A large-amplitude synchronous whirl and an amplitude-modulated whirl for $\Omega_H < \Omega < \Omega_{SN2}$; and only the amplitude-modulated whirl for $\Omega_{SN2} < \Omega$. In case I the amplitude modulated whirl never interacts with the synchronous whirl. Case II represents the level of imbalance for which this interaction begins; here the amplitude-modulated whirl is just the tangent to the synchronous whirl at the second saddle-node point, $\Omega_{SN_2}$.

Case III is very similar to case I except that the amplitude-modulated whirl is destroyed in a collision with the middle synchronous whirl branch. This results in an additional frequency region, $\Omega_{SC} < \Omega < \Omega_{SN_2}$, in which only the large-amplitude synchronous response is stable. The ranges $\Omega < \Omega_{SN_1}$, $\Omega_{SN_1} < \Omega < \Omega_H$ and $\Omega > \Omega_{SN_2}$ are qualitatively the same as in case I. The range $\Omega_H < \Omega < \Omega_{SC}$ is qualitatively like that for $\Omega_H < \Omega < \Omega_{SN_2}$ in case I.

Case IV, and the borderline level of imbalance given by $\mu_i = \mu_i^*$ between cases III and IV (at which the Hopf bifurcation occurs exactly at the first saddle-node point) are dynamically similar. In this case, and for all larger values of imbalance, i.e. case V, the lowest synchronous response branch is nowhere stable. In case IV the synchronous response grows and becomes amplitude modulated for $\Omega \geq \Omega_{SN_2}$, and at each $\Omega$ there is a unique stable response. In case V there exists a unique stable synchronous response for all values of $\Omega$. Here the imbalance dominates the response, the instability merely renders the lowest synchronous response branch unstable and no amplitude-modulated motion occur.

For Fig. 11.12(b), cases I, II, III and V are the same as those in Fig. 11.12(a). The differences between the two figures are in cases IV. In order to describe these differences the following notation is introduced. Let $e_1$ correspond to the $e$ value such that $\mu_i = \mu_i^*$, above which (in the $e$ direction) the lowest synchronous response branch is unstable for all $\Omega > \Omega_{SN_1}$, and let $e_2$ denote the value corresponding to $\mu_e^2 = \alpha e^2 \Omega_1^2$, above which the uppermost synchronous response branches do not merge. In Fig. 11.12(a), $e_2 > e_1$; in this case $e_1$, that is, the $e$ value above which the lowest response branch is completely unstable (i.e. unstable over $\Omega_{SN1} < \Omega < \infty$) is below the large $\Omega$ asymptote of the *SN2+SC* curve, $e_2$. In this situation, for $e_1 < e < e_2$ the response curves exhibit a completely unstable lower branch, and the upper two branches merge at *SN2+SC* (see Case IV in Fig. 11.12(a)). In Fig. 11.12(b), $e_2 < e_1$; here case IV is different in that the lowest response branch is completely unstable, and the upper two branches do not merge as $\Omega$ is increased.

Simulations of eq. (11.29) verify the bifurcation diagrams of Fig. 11.12. Here we present only simulation results for case III of Fig. 11.12; these are presented in terms of non-rotating, i.e. fixed, co-ordinates in order to aid in visualization of the shaft response. The unique stable synchronous whirl for $\Omega < \Omega_{SN_1}$ is shown in Fig. 11.13(a). Two possible synchronous whirling motions, one large and one small, are shown in Fig. 11.13(b) for an $\Omega$ value in the range $\Omega_{SN_1} < \Omega < \Omega_H$. A large amplitude synchronous whirl and an amplitude-modulated whirl for an $\Omega$ value in the range $\Omega_H < \Omega < \Omega_{SC}$ are shown in Fig. 11.13(c). The actual steady-state solution achieved for the parameter range $\Omega_{SN_1} < \Omega < \Omega_{SC}$ depends on initial conditions. The unique stable large synchronous whirl for an $\Omega$ value in the range $\Omega_{SC} < \Omega < \Omega_{SN_2+SC}$ is shown in Fig. 11.13(d). At this point the previous amplitude-modulated whirl has been annihilated through the saddle connection. The unique stable amplitude-modulated whirl is depicted in Fig. 11.13(e) for an $\Omega > \Omega_{SN_2+SC}$. The long-period envelope is evident in Fig. 11.13(c,e). For $\mu_e/\mu_i\Omega_1$ small, all amplitude-modulated motions which exist near $\Omega_1^*$ are expected to have such long-period envelopes. This period is directly related to the period of the corresponding limit cycles born in Hopf bifurcation at $\Omega=\Omega_H$ with frequencies close to $\omega_1$ ($=\mu_e/\mu_i\Omega_1$) which is small, implying long-period oscillations in amplitude. As $\Omega$ is increased beyond $\Omega_H$, the limit cycle frequency may change, but it is interesting to note that in many cases this limit cycle grows into a saddle connection, which corresponds to a limit cycle with an infinite period. Thus, it is reasonable to expect long-period envelopes for all amplitude-modulated motions near $\Omega_1^*$.

From the above discussion, it is obvious that the shaft response summarized in Fig. 11.12 is much more complicated than that found in either Fig. 11.6 or Fig. 11.7. Several issues remain unexplored. First, singularity analysis is required. The present analysis has dealt only with the response near the first resonance. It is known, however, that rotor-dynamic instability may occur at a much higher value of $\Omega$ if the ratio $\mu_e/\mu_i\Omega_1$ is not small. In these cases envelopes may not have such long-periods since the whirl frequency may not be near to synchronous. However, situations may arise in which the instability interacts with the non-linear resonance of a higher mode, or with a subharmonic or superharmonic resonance.

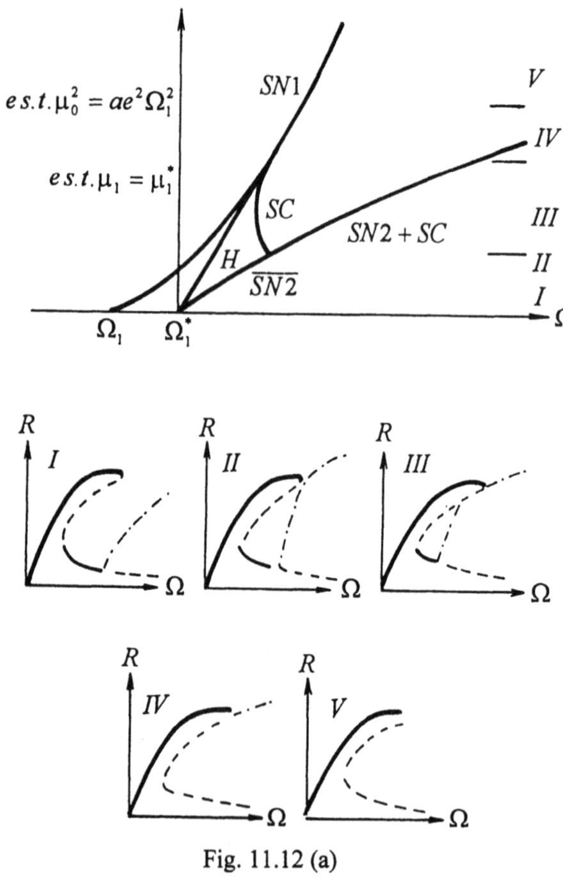


Fig. 11.12 (a)

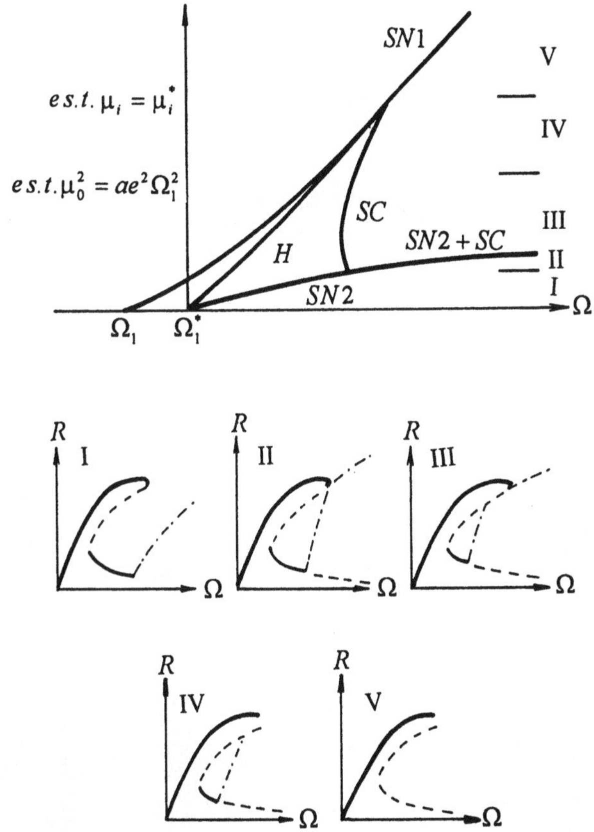


Fig. 11.12 (b)

Also, the dynamic model given in eq. (11.29) is greatly simplified; the types of damping and structural non-linearity used were chosen to illustrate a procedure by which the response may be computed and to provide what are hoped to be qualitatively typical results. It should be noted that if the non-linearity are of the softening type ($\alpha$<0) the response may be quite different.

It may be that irregular (that is, non-periodic) amplitude modulations of the motion are observed in simulation studies of eq. (11.29) or similar models. These can arise from chaotic motions in the full four-dimensional phase space, which cannot be captured by the centre manifold dynamics since they are two-dimensional. These would most likely occur near the saddle connection bifurcations since saddle-connections (that is, homoclinic motions) in a four-dimensional phase space can give rise to chaos in quite general circumstances. Chaotic amplitude modulations have been found to occur in the vibrations of systems with internal resonance; the paper by Johnson and Bajaj [189] provides an example of this.

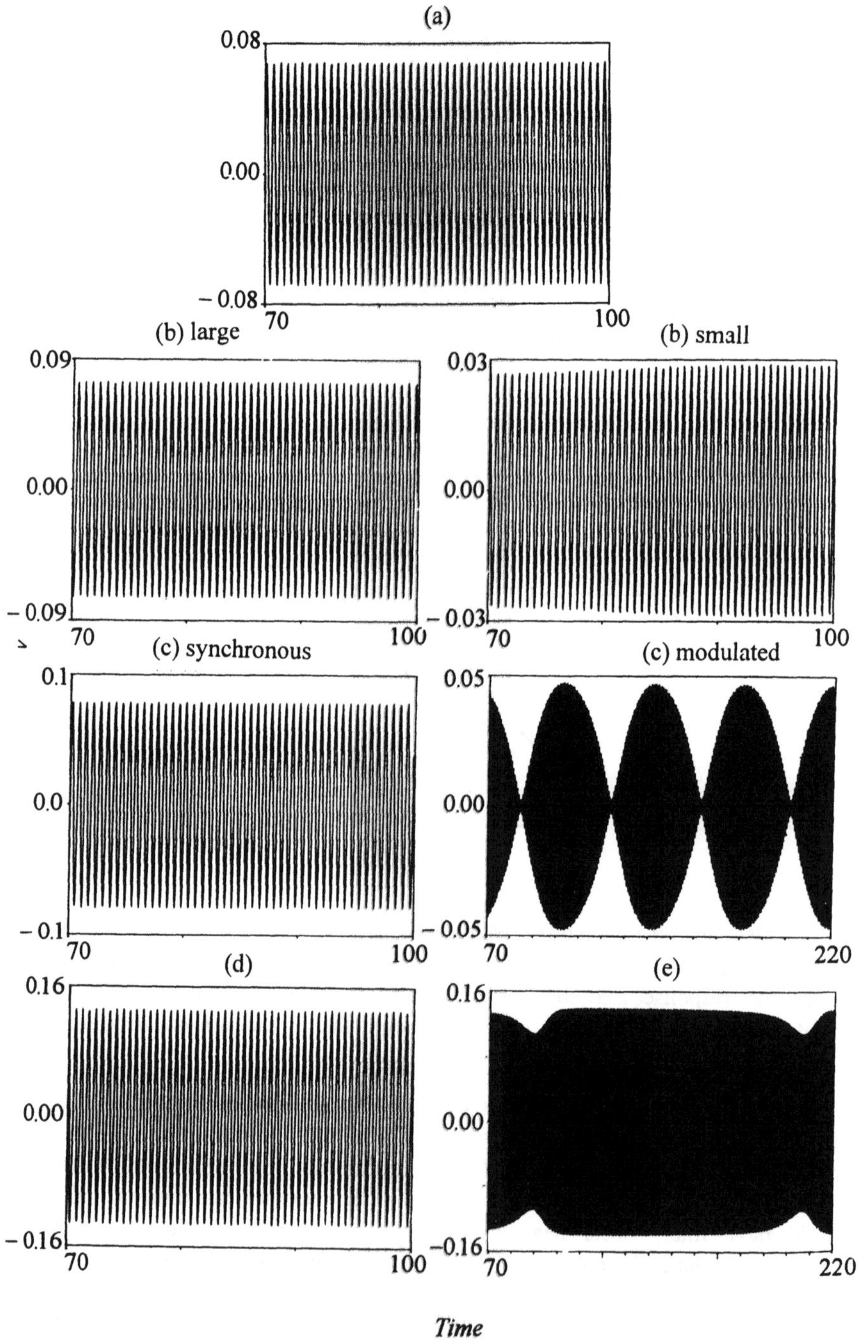
(a)
0.08
0.00
− 0.08
70
100
(b) large
0.09
0.00
− 0.09
70
100
(b) small
0.03
0.00
− 0.03
70
100
(c) synchronous
0.1
0.0
− 0.1
70
100
(c) modulated
0.05
0.00
− 0.05
70
220
(d)
0.16
0.00
− 0.16
70
100
(e)
0.16
0.00
−0.16
70
220
Time

Fig. 11.13

## 11.3 Galloping Vibration Analysis for an Elastic Structure

When an elastic structure oscillates in a steady airflow, the flow relative to the structure also oscillates. The aerodynamic forces produced by this flow can cause a growth in the amplitude of the vibration — large-amplitude, low-frequency plunge motions — a phenomenon called galloping vibration. Such a phenomenon is observed in tall chimneys, a space shuttle on its launching pad, suspension bridges, and overhead power-transmission lines. The galloping of electrical conductors is usually produced by a steady wind flowing across a transmission line having an asymmetrical cross-section. This asymmetry may arise from ice or rime deposits during winter or, more rarely, from the stranding of the cable itself. The galloping phenomenon is highly complex because of the non-linear, geometry-dependent, time-varying aerodynamic loads and the resulting large amplitudes of vibration, which may cause interactions between adjacent spans and their support hardware.

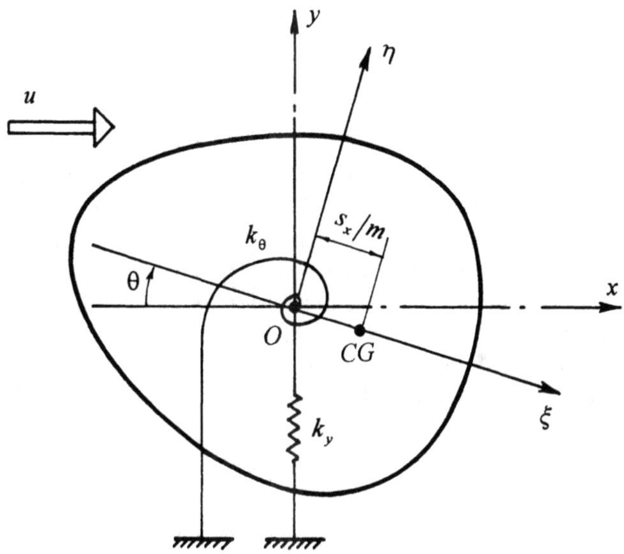


Fig. 11.14

Now we discuss the two-degree-of-freedom autonomous model of Yu [173, 180], which is illustrated in Fig. 11.14. Its equations of motion are described by

$$\begin{aligned} m\ddot{y} + c_y\dot{y} - s_x\ddot{\theta} + k_y y &= F_y \\ I\ddot{\theta} + c_\theta\dot{\theta} - s_x\ddot{y} + k_\theta\theta &= F_M \end{aligned} \tag{11.42}$$

where $y$ is the vertical or plunge displacement, $\theta$ is the angle of twist, and a dot superscript denotes differentiation with respect to time. The $k_y$ and $k_\theta$ represent the stiffness in plunge and torsion, respectively, and the $c_y$ and $c_\theta$ are the corresponding viscous dampers (which are not shown in Fig. 11.14). Furthermore, $m$ and $I$ are the mass and the moment of inertia of the iced conductor's cross-section, A, about its centre of rotation, O. They are defined by

$$m = \int_A \mu d\xi d\eta, \qquad I = \int_A (\xi^2 + \eta^2)\mu d\xi d\eta \tag{11.43}$$

The $s_x$ in eq. (11.42) indicates the eccentricity of the system. It is given by

$$s_x = \int_A \xi\mu d\xi d\eta \tag{11.44}$$

where $\mu$ is the generalized mass density over $A$ and $s_x/m$ is the lateral position of the centre of gravity measured from O.

If the maximum velocity of the structure, $\dot{y}$, is much less than the wind's free-stream velocity $U$, the angle of attack, $\alpha$, is approximately

$$\alpha \approx \theta - \frac{R_1}{U}\dot{\theta} - \frac{1}{U}\dot{y} \tag{11.45}$$

where $R_1$ is the characteristic radius of the section [181]. Then the aerodynamic force and moment acting on $A$, $F_y$ and $F_M$, respectively, can be expressed by

$$\begin{aligned} F_y &= -\frac{1}{2}\rho U^2 d C_y(\alpha) \\ F_M &= -\frac{1}{2}\rho U^2 d^2 C_M(\alpha) \end{aligned} \tag{11.46}$$

where $\rho$ is the density of air. The $\alpha$ is a characteristic structural length which is usually taken as the maximum width of the cross-section normal to the free stream. Coefficients $C_y$ and $C_M$ are relatively smooth, continuous functions of $\alpha$ which may be expressed as experimentally determined polynomials in this variable. The representation of $C_y$ and $C_M$ gives rise to the nature of the non-linearity in the system (and the complexity of the subsequent analysis). Because a simple cubic approximation was shown to capture many of the qualitative features of the galloping of a symmetrically iced conductor [181] a cubic polynomial will be used so that

$$\begin{aligned} C_y &= a_1\alpha + a_3\alpha^3 \\ C_M &= b_1\alpha + b_3\alpha^3 \end{aligned} \tag{11.47}$$

where $a_1$, $a_3$, $b_1$ and $b_3$ are best fit coefficients.

A combination of eqs. (11.42), (11.46) and (11.47) produces

$$\begin{aligned} \ddot{Y} - e_x\ddot{\theta} + Y &= -2\xi_y\dot{Y} - \eta_y U_y^2(a_1\alpha + a_3\alpha^3) \\ \ddot{\theta} - \eta e_x\ddot{Y} + \omega^2\theta &= -2\xi\omega\dot{\theta} - \eta_\theta U_y^2(b_1\alpha + b_3\alpha^3) \end{aligned} \tag{11.48}$$

where

$$\begin{aligned} &Y = \frac{y}{d}, \quad e_x = \frac{s_x}{md}, \quad \omega_y^2 = \frac{k_y}{m}, \quad \omega_\theta^2 = \frac{k_\theta}{I}, \quad \omega = \frac{\omega_\theta}{\omega_y}, \\ &\xi_y = \frac{c_y}{2m\omega_y}, \quad \xi_\theta = \frac{c_\theta}{2I\omega_\theta}, \quad U_y = \frac{U}{\omega_y d}, \quad \eta_y = \frac{\rho d^2}{2m}, \quad \eta_\theta = \frac{\rho d^4}{2I}, \quad \eta = \frac{\eta_\theta}{\eta_y} \end{aligned} \tag{11.49}$$

and $\alpha$ is rewritten in the form

$$\alpha = \theta - \frac{r_1}{U_y}\dot{\theta} - \frac{1}{U_y}\dot{Y} \tag{11.50}$$

where $r_1 = R_1/d$. A dot superscript now indicates a derivative with respect to the non-dimensional time $\tau = \omega_y t$. It can be seen easily from eqs. (11.48) and (11.50) that $Y = \dot{Y} = \theta = \dot{\theta} = 0$ is an equilibrium solution. Suppose the system to be initially in stable equilibrium but a parameter change (e.g. the steady wind speed $U$), increases so that the initial equilibrium becomes unstable. Suppose the damping and aerodynamic forces are small compared to the forces arising from the conductor's stiffness. Then the terms on the right of eq. (11.48) are of order $\varepsilon$, a positive small parameter. $\varepsilon = 0$ denotes a critical point of system (11.48).

Suppose $Y = A_y \cos\omega t$, $\theta = A_\theta \cos\omega t$ then the characteristic polynomial of the homogeneous system (11.48), $G(\omega_c^2)$, can be obtained at the critical point, $c$:

$$G(\omega_c^2) = (1 - \eta e_x^2)\omega_c^4 - (1 + \omega^2)\omega_c^2 + \omega^2 \tag{11.51}$$

which in turn yields the two characteristic frequencies

$$= \omega_{(1,2)c}^2 = \frac{1}{2(1 - \eta e_x^2)}\left[(1 + \omega^2) \pm \sqrt{(1 + \omega^2)^2 - 4(1 - \eta e_x^2)\omega^2}\right] \tag{11.52}$$

Subscript $c$ indicates values associated with the critical point $c$. From eqs. (11.43) and (11.44), it can be shown, by using the Schwartz Inequality [182], that

$$s_x < mI, \quad \text{i.e. } I - \eta e_x^2 > 0 \tag{11.53}$$

which guarantees $\omega_{(1,2)c} > 0$.

The linear transformation

$$\begin{bmatrix} Y \\ \dot{Y} \\ \theta \\ \dot{\theta} \end{bmatrix} = \begin{bmatrix} 1 & 0 & K_2 & 0 \\ 0 & \omega_{1c} & 0 & K_2\omega_{2c} \\ K_1 & 0 & 1 & 0 \\ 0 & K_1\omega_{1c} & 0 & \omega_{2c} \end{bmatrix} \begin{bmatrix} x_1 \\ x_2 \\ x_3 \\ x_4 \end{bmatrix} \tag{11.54}$$

where $$K_1 = \frac{\omega_{1c}^2}{\omega_{1c}^2 - \omega^2}\eta e_x, \qquad K_2 = \frac{\omega_{2c}^2}{\omega_{2c}^2 - 1}e_x \tag{11.55}$$

is introduced into eq. (11.48) to obtain

$$\{\dot{x}\} = [D]\{x\} + \{N\} \tag{11.56}$$

in which $\{x\}^T = \{x_1, x_2, x_3, x_4\}$ and superscript $T$ denotes a transpose. The $[D]$ is a $4 \times 4$ matrix in which non-vanishing components are as follows:

$$D_{12} = \omega_{1c}, \quad D_{21} = -\omega_{1c} - E_1\omega_{1c}K_1\eta_y U_y^3(a_1 + K_1 b_1),$$

$$D_{22} = -E_1 U_y \omega_{1c}^2\left[2\xi_y + 2K_1^2\frac{\omega}{\eta}\xi_\theta - \eta_y U_y(1 + K_1 r_1)(a_1 + K_1 b_1)\right],$$

$$D_{23} = -E_1\omega_{1c}\eta_y U_y^3(a_1 + K_1 b_1),$$

$$D_{24} = -E_1 U_y \omega_{1c}\omega_{2c}\left[2K_2\xi_y + 2K_1\frac{\omega}{\eta}\xi_\theta - \eta_y U_y(K_2 + r_1)(a_1 + K_1 b_1)\right],$$

$$D_{34} = \omega_{2c},\ \ D_{41} = -E_1\frac{\omega_{2c}}{\omega^2}K_1\eta_\theta U_y^3(K_2 a_1 + b_1),$$

$$D_{42} = -E_1 U_y\frac{\omega_{1c}\omega_{2c}}{\omega^2}\left[2K_2\eta\xi_y + 2K_1\omega\xi_\theta - \eta_\theta U_y(1 + K_1 r_1)(K_2 a_1 + b_1)\right],$$

$$D_{43} = -\omega_{2c} - E_1 U_y\frac{\omega_{2c}}{\omega_2}\eta_\theta U_y^3(K_2 a_1 + b_1),$$

$$D_{44} = -E_1 U_y\left(\frac{\omega_{2c}}{\omega}\right)^2\left[2K_2^2\eta\xi_y + 2\omega\xi_\theta - \eta_\theta U_y(K_2 + r_1)(K_2 a_1 + b_1)\right]. \tag{11.57}$$

The non-linear components of $\{N\}$ are

$$\{N\}^T = E_1\left[-K_1 U_y x_1 + \omega_{1c}(1 + K_1 r_1)x_2 - U_y x_3 + \omega_{2c}(K_2 + r_4)x_4\right]^3\{N_1\}'$$

where
$$\{N_1\}' = \left[0, \omega_{1c}\eta_y(a_3 + K_1 b_3), 0, \frac{\omega_{2c}}{\omega^2}\eta_\theta(K_2 a_3 + b_3)\right] \tag{11.58}$$

and
$$E_1 = \frac{1}{(1 - K_1 K_2)U_y} \tag{11.59}$$

The Jacobian matrix of eq. (11.56), evaluated at the critical point for the initial equilibrium solution, $\{x\}=\{0\}$, now takes the canonical form

$$J_c = \begin{bmatrix} 0 & \omega_{1c} & 0 & 0 \\ -\omega_{1c} & 0 & 0 & 0 \\ 0 & 0 & 0 & \omega_{2c} \\ 0 & 0 & -\omega_{2c} & 0 \end{bmatrix} \tag{11.60}$$

Only the non-resonant case (where the ratio $\omega_{1c}/\omega_{2c} \neq m/n$, $m$ *and* $n$ integers) is considered. The periodic and quasi-periodic solutions and the associated stability conditions can be derived from eq. (11.56). This is accomplished by first assuming that the quasi-periodic, as well as the periodic solutions are described in the vicinity of the critical point (where $\varepsilon$=0) by the two-time scale Fourier series

$$x_j(\tau_1,\tau_2,\varepsilon) = \sum_{\substack{m=0 \\ (m_1+m_2=0)}}^{M}\left[p_{j,m_1m_2}(\varepsilon)\cos(m_1\tau_1 + m_2\tau_2) + r_{j,m_1m_2}\sin(m_1\tau_1 + m_2\tau_2)\right] \tag{11.61}$$

here $j$=1,2,3,4; and $m$, $m_1$ and $m_2$ are integers (but $m \geq 0$), whereas $M$ is an arbitrary positive integer. Furthermore,

$$\tau_1 = \omega_1(\varepsilon)\tau,\ \ \tau_2 = \omega_2(\varepsilon)\tau \tag{11.62}$$

The $\omega_1$, $\omega_2$, $p_{j,m_1m_2}$ and $r_{j,m_1m_2}$ are functions of the system's parameters which are associated with the symbolic small parameter, $\varepsilon$. They are determined through a perturbation procedure [183].

Based on eqs. (11.56) and (11.61), the multiple-scale intrinsic harmonic balancing and unification technique can be applied to obtain the rate equations:

$$\begin{aligned}\dot{A}_y &= E_1 U_y \omega_{1c}^2 \eta_y A_y\left[B_1 + E_2\left(B_3 A_y^2 + 2B_4 A_\theta^2\right)\right],\\ \dot{A}_\theta &= E_1 U_y \frac{\omega_{1c}^2}{\omega^2}\eta_\theta A_\theta\left[B_2 + E_3\left(2B_3 A_y^2 + B_4 A_\theta^2\right)\right],\end{aligned} \tag{11.63}$$

$$\dot{\phi}_y = \omega_{1c}\left\{1 + K_1 E_1 U_y^2 \eta_y\left[\frac{1}{2}U_y(a_1 + K_1 b_1) + (a_3 + K_1 b_3)\left(B_3 A_y^2 + 2B_4 A_\theta^2\right)\right]\right\},$$

and

$$\dot{\phi}_\theta = \omega_{2c}\left\{1 + E_1\frac{U_y^2}{\omega^2}\eta_\theta\left[\frac{1}{2}U_y(K_2 a_1 + b_1) + (K_2 a_3 + b_3)\left(2B_3 A_y^2 + B_4 A_\theta^2\right)\right]\right\} \tag{11.64}$$

where
$$\begin{aligned}B_1 &= \frac{1}{2}U_y(1 + K_1 r_1)(a_1 + K_1 b_1) - \left[\frac{\xi_y}{\eta_y} + K_1^2\frac{\omega\xi_\theta}{\eta_\theta}\right],\\ B_2 &= \frac{1}{2}U_y(K_2 + r_1)(K_2 a_1 + b_1) - \left[K_2^2\frac{\xi_y}{\eta_y} + \frac{\omega\xi_\theta}{\eta_\theta}\right],\\ B_3 &= \frac{3}{8U_y}\left[K_1^2 U_y^2 + (1 + K_1 r_1)^2\omega_{1c}^2\right](> 0),\\ B_4 &= \frac{3}{8U_y}\left[U_y^2 + (K_2 + r_1)^2\omega_{2c}^2\right](> 0),\\ E_2 &= (1 + K_1 r_1)(a_3 + K_1 b_3),\\ E_3 &= (K_2 + r_1)(K_2 a_3 + b_3).\end{aligned} \tag{11.65}$$

Setting $\dot{A}_y = \dot{A}_\theta = 0$ yields the steady-state solutions $\overline{A}_y$ and $\overline{A}_\theta$ which are the amplitudes of the periodic solutions having frequencies $\omega_1(\varepsilon) = \dot{\phi}_y$ and $\omega_2(\varepsilon) = \dot{\phi}_\theta$, respectively. The degenerate case of a concentric cross-section, which was studied by Blevins and Iwan [181], can be obtained by setting $e_x = 0$ (*i.e.*, $K_1 = K_2 = 0$) in eqs. (11.63)—(11.65).

Now we study the steady-state solutions from eq. (11.63) by setting $\dot{A}_y = \dot{A}_\theta = 0$. First, it may be noted that

$$A_y = A_\theta = 0 \tag{11.66}$$

is a solution which represents the initial equilibrium solution. Second, there are two distinct families of limit cycles for the periodic solutions which bifurcate from the initial equilibrium solution. These limit cycles are described by

$$A_y^2 = -\frac{1}{E_2}\frac{B_1}{B_3}, \quad A_\theta^2 = 0 \tag{11.67}$$

and $$A_y^2 = 0, \quad A_\theta^2 = -\frac{1}{E_3}\frac{B_2}{B_4} \tag{11.68}$$

Solution eq. (11.67) represents a purely plunge vibration having a frequency

$$\omega_1 = \dot{\phi}_y = \omega_{1c}\left[1 + \frac{K_1\left(\xi_y + K_1^2 \frac{\omega\xi_\theta}{\eta}\right)E_1U_y^2}{(1+K_1r_1)}\right] \tag{11.69}$$

whereas eq. (11.68) corresponds to a solely torsional vibration with frequency

$$\omega_2 = \dot{\phi}_\theta = \omega_{2c}\left[1 + \frac{\left(\xi_\theta + K_2^2 \frac{\eta\xi_y}{\omega}\right)E_1U_y^2}{\omega(K_2+r_1)}\right] \tag{11.70}$$

From eqs. (11.52) and (11.55), it can be demonstrated that

$$\frac{K_1}{K_2} = -\frac{\eta}{\omega^2}, \text{ and, therefore } K_1K_2 < 0 \tag{11.71}$$

(which implies that $E_1>0$). It can be also shown that $K_1 > 0$ (so that $K_2 < 0$) if and only if $\omega < 1$, and that $K_1 < 0$ ($K_2 > 0$) if and only if $\omega > 1$.

A more interesting coupled plunge—torsion steady-state solution can be determined from eq. (11.63) as

$$\begin{aligned} A_y^2 &= \frac{1}{3B_3}\left(\frac{B_1}{E_2} - \frac{2B_2}{E_3}\right), \\ A_\theta^2 &= \frac{1}{3B_4}\left(\frac{B_2}{E_3} - \frac{2B_1}{E_2}\right) \end{aligned} \tag{11.72}$$

if $E_2E_3 \neq 0$. Relation (11.72) represents two distinct amplitudes of a family of quasi-periodic motions which lie on an invariant torus. Such motions, although observed commonly in the galloping of conductors, are not possible for the concentric, $e_x$=0, case. The quasi-periodic motions have frequencies $\omega_1$ and$\omega_2$ which can be found from eq. (11.64). These frequencies are surprisingly identical to those of the uncoupled periodic solutions given by eq. (11.69) and eq. (11.70), respectively. First-order asymptotic expressions for the family of two-dimensional tori are expressed by

$$\begin{aligned} x_1 &= A_y \cos\omega_1\tau, & x_2 &= -A_y \sin\omega_1\tau, \\ x_3 &= A_\theta \cos\omega_2\tau, & x_4 &= -A_\theta \sin\omega_2\tau \end{aligned} \tag{11.73}$$

(More accurate third-order approximations, which may be found by using a conventional averaging approach, can be obtained from [183].)

Next the initiation of galloping and stability analysis are given.

The conditions will be sought for the initiation of galloping (i.e. the instability of the initial equilibrium solution (11.66)) and the instabilities of the three steady-state solutions (11.67), (11.68) and (11.72). In other words, critical stability boundaries will be found where a steady-state solution becomes unstable and jumps to a new state. Stability conditions may be determined on the basis of the Jacobian of eq. (11.63), that is

$$J\left(A_y, A_\theta, \varepsilon\right) = E_1 U_y \begin{bmatrix} J_{11} & J_{12} \\ J_{21} & J_{22} \end{bmatrix} \tag{11.74}$$

where $J_{11} = \omega_{1c}^2 \eta_y \left[B_1 + E_2\left(3B_3 A_y^2 + 2B_4 A_\theta^2\right)\right]$,

$$J_{12} = 4\omega_{1c}^2 \eta_y E_2 B_4 A_y A_\theta,$$

$$J_{21} = 4\frac{\omega_{1c}^2}{\omega^2}\eta_\theta E_3 B_3 A_y A_\theta,$$

$$J_{22} = \frac{\omega_{1c}^2}{\omega^2}\eta_\theta\left[B_2 + E_3\left(2B_3 A_y^2 + 3B_4 A_\theta^2\right)\right] \tag{11.75}$$

The resulting bifurcation diagram, having $B_1$ and $B_2$ as its co-ordinates, and associated stability regions are illustrated in Fig. 11.15. In this figure, I.E.S., H.B.S.(P), and H.B.S.(T) represent the initial equilibrium solution, the Hopf bifurcation solution for a periodic plunge vibration, and the Hopf bifurcation solution for a periodic torsional vibration, respectively. Now these solutions will be detailed.

Substituting eq. (11.66) or (11.67) or (11.68) into eq. (11.74) results in $J_{12}=J_{21}=0$. Thus, the initial equilibrium solution (I.E.S.) or the plunge periodic motion (H.B.S.(P)) or the torsional periodic motion (H.B.S.(T)) is stable if $J_{11}<0$ and $J_{22}<0$. These conditions yield critical boundaries where a new state may bifurcate from the previous steady-state solutions. These results are summarized in table 11.1. It can be observed from the I.E.S. stability condition that the initial equilibrium solution and the plunge (torsional) motion exchange their stability on the critical boundary $C_1$ ($C_2$).

It has been observed in the field that a larger steady wind speed [184] is an important factor in causing instability or the initiation of galloping. This is clear from eq. (11.65) which indicates that at least one of $B_1$ and $B_2$ grow when $U_y$ increases. Table 11.1 suggests that the possibility of the instability of the corresponding initial equilibrium solution also increases.

The stability of the quasi-periodic motions is found by first substituting eq. (11.72) into the Jacobi (11.74) to produce

$$J_{(IV)}\left(A_y, A_\theta, \varepsilon\right) = E_1 U_y \begin{bmatrix} 2\omega_{1c}^2 \eta_y E_2 B_3 A_y^2 & 4\omega_{1c}^2 \eta_y E_2 B_4 A_y A_\theta \\ 4\frac{\omega_{1c}^2}{\omega^2}\eta_\theta E_3 B_3 A_y A_\theta & 2\frac{\omega_{1c}^2}{\omega^2}\eta_\theta E_3 B_4 A_\theta^2 \end{bmatrix} \tag{11.76}$$

The $A_y$ and $A_\theta$ are given of course by solution (11.72). Stability conditions for this family of two-dimensional tori are determined by the trace and determinant of $J_{(IV)}$, that is, trace $J_{(IV)} < 0$ and det $J_{(IV)} > 0$ for which the eigenvalues of the characteristic polynomial of $J_{(IV)}$ have negative real parts. More specifically,

$$\text{trace } J_{(IV)} = 2E_1U_y\left[\omega_{1c}^2\eta_y E_2 B_3 A_y^2 + \left(\frac{\omega_{1c}}{\omega}\right)^2 \eta_\theta E_3 B_4 A_\theta^2\right], \tag{11.77}$$

and
$$\det J_{(IV)} = -12\left(E_1U_y\omega_{1c}\frac{\omega_{2c}}{\omega}\right)^2 \eta_y\eta_\theta E_2E_3B_3B_4A_y^2A_\theta^2 \tag{11.78}$$

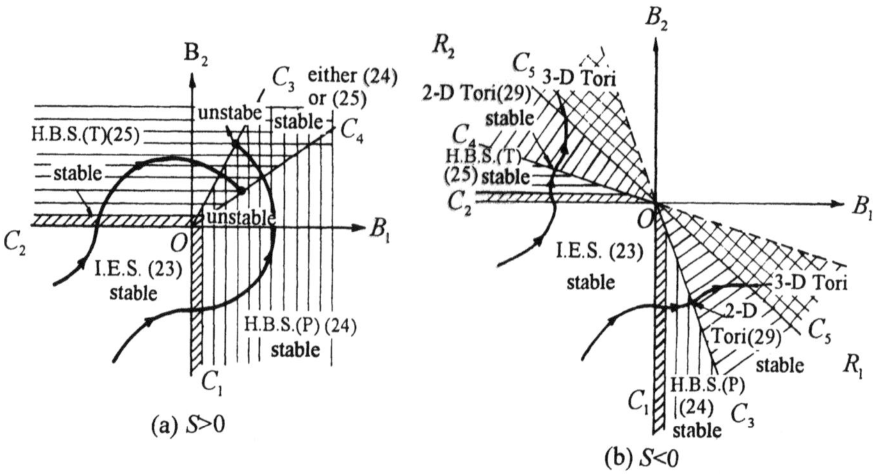


Fig. 11.15

**Table 11.1** Stability conditions and critical boundaries for the initial equilibrium and periodic solutions

| Solution | I.E.S. (23) | | H.B.S.(P) (24), (26) | H.B.S.(T) (25), (27) |
|---|---|---|---|---|
| Existence Condition | | | $\frac{B_1}{E_2} < 0$ | $\frac{B_2}{E_3} < 0$ |
| Stability Condition | $B_1 < 0$, $B_2 < 0$ | | $B_1 > 0$, $B_2 - \frac{2E_3}{E_2}B_1 < 0$ | $B_1 - \frac{2E_2}{E_3}B_2 < 0$, $B_2 > 0$ |
| Critical Boundary | $C_1$: $B_1=0$ | $C_2$: $B_2=0$ | $C_3$: $B_2 - \frac{2E_3}{E_2}B_1 = 0$ | $C_4$: $B_1 - \frac{2E_2}{E_3}B_2 = 0$ |
| Bifurcation Solution | H.B.S.(P) (24), (26) | H.B.S.(T) (25), (27) | Quasi-periodic Motion (26), (27), (29) | |

Substituting eq. (11.72) into eq. (11.77) results in the stability condition

$$\left[\omega_{1c}^2 - \frac{2E_3}{E_2}\left(\frac{\omega_{1c}}{\omega}\right)^2\eta\right]B_1 + \left[\left(\frac{\omega_{1c}}{\omega}\right)^2\eta - \frac{2E_3}{E_2}\omega_{1c}^2\right]B_2 < 0 \tag{11.79}$$

and $\quad E_2E_3 < 0$ (11.80)

Recalling the generic condition $E_2E_3 \neq 0$, it should be noted that the solution of the two-dimensional torus and its stability (see (11.80)) cannot be determined if $E_2E_3 = 0$. Both the coefficients of $B_1$ and $B_2$ in (11.79), on the other hand, are positive due to condition (11.80). Thus, at least one of $B_1$ and $B_2$ must be negative to satisfy condition (11.79) — otherwise the family of two dimensional tori is unstable. Of course, the existence conditions for the family of two-dimensional tori have to be considered. These conditions can be determined readily from eq. (11.72) by requiring $A_y^2 > 0$ and $A_\theta^2 > 0$ so that

$$\frac{B_1}{E_2} - \frac{2B_2}{E_3} > 0 \text{ and } \frac{B_2}{E_3} - \frac{2B_1}{E_2} > 0 \tag{11.81}$$

It is not difficult to see that conditions (11.79) to (11.45) are compatible. In other words, a region does occur in which stable quasi-periodic solutions, represented by eq. (11.72), exist.

Condition (11.79) implies that another stability boundary (see Fig. 11.15(b))

$$C_5 : \left[\omega_{1c}^2 - \frac{2E_3}{E_2}\left(\frac{\omega_{1c}}{\omega}\right)^2 \eta\right]B_1 + \left[\left(\frac{\omega_{1c}}{\omega}\right)^2 \eta - \frac{2E_2}{E_3}\omega_{1c}^2\right]B_2 = 0 \tag{11.82}$$

arises when the quasi-periodic solution becomes unstable and may bifurcate into a three-dimensional torus or chaotic motion. In fact, at the critical boundary defined by eq. (11.82), the Jacobi (11.74) has a pair of complex conjugate eigenvalues on the imaginary axis. This phenomenon gives rise to the well-known classical Hopf bifurcation. Such a bifurcation will create a new amplitude and a third frequency, corresponding to a new family of limit cycles, for the family of two-dimensional tori, eq. (11.72). If this new amplitude is independent of the amplitude of the two-dimensional torus and the third frequency is well separated from the $\omega_1$ and $\omega_2$ of the two-dimensional torus, the combined motion will lie on an invariant three-dimensional torus. However, the critical boundary, $C_5$, should be shown to be located in the region where the quasi-periodic solution (11.72) exists.

For simplicity, let

$$s = \frac{E_3}{E_2} \tag{11.83}$$

The $s$ must be negative for the existence of the stable family of two-dimensional tori, eq. (11.72), due to condition (11.80). With this constraint placed on $s$, the region defined by condition (11.81) can be rewritten as either

$$R_1 = \begin{cases} B_2 - 2sB_1 > 0 \\ B_2 - \frac{s}{2}B_1 < 0 \end{cases} \qquad (\text{for } E_3 > 0) \tag{11.84}$$

or

$$R_2 = \begin{cases} B_2 - 2sB_1 < 0 \\ B_2 - \frac{s}{2}B_1 > 0 \end{cases} \qquad (\text{for } E_3 < 0) \tag{11.85}$$

Equation (11.69) then takes the form

$$B_2 - \frac{\omega_1^2 - \left(\frac{\omega_{2c}}{\omega}\right)^2 \eta(2s)}{\omega_{1c}^2\left(\frac{2}{s}\right) - \left(\frac{\omega_{2c}}{\omega}\right)^2 \eta} B_1 = 0 \tag{11.86}$$

Furthermore, it can be shown straightforwardly that

$$2s < \frac{\omega_{1c}^2 - \left(\frac{\omega_{2c}}{\omega}\right)^2 \eta(2s)}{\omega_{1c}^2\left(\frac{2}{s}\right) - \left(\frac{\omega_{2c}}{\omega}\right)^2 \eta} B_1 < \frac{1}{2}s, \qquad \forall s < 0 \tag{11.87}$$

Thus, it is easy to verify that $C_5$ is located in $R_1$ for $B_1 > 0$ and $B_2 < 0$ whereas $C_5$ is in $R_2$ for $B_1 < 0$ and $B_2 > 0$. Therefore it is possible for a stable family of two-dimensional tori to exist when the critical boundary $C_5$ is intersected.

It should be noted from Fig. 11.15(a), which requires $s > 0$, that the stable region in this figure is similar to that for the concentric case [181]. Also, it is worth noting that parameter $s$ plays a significant role in determining the system's behaviour. No mixed mode exists if $s > 0$. When $s < 0$, on the other hand, Fig. 11.15(b) suggests that quasi-periodic motions (a mixed mode), which lie on a two-dimensional or even a three-dimensional torus, are possible.

The theory has been applied to a 24-Km test span of Ontario Hydro having a single "Drake", 795-Mcm conductor [185]. The spanwise properties of this transmission line and a simulated ice accretion are considered in [186]. A procedure to calculate the numerical properties of an equivalent oscillator is also given. By following this procedure, values of the properties appearing in eqs. (11.47) and (11.49) are:

$a_1 = 261.9478,\ 274.4842;$ $\qquad a_3 = -34.2889,\ -52.1284;$

$b_1 = 28.2562,\ 29.8533;$ $\qquad b_3 = -3.1132,\ -2.9278;$

$\omega_y = 1.9312,\ 3.4312;$ $\qquad \omega_\theta = 15.2413,\ 21.2343;$

$\xi_y = 1.0031 \times 10^{-2},\ 0.5646 \times 10^{-2};$

$\xi_\theta = 0.7936 \times 10^{-2},\ 1.1393 \times 10^{-2};$

$\eta_y = 0.2312 \times 10^{-5},\ 0.2323 \times 10^{-5};$

$\eta_\theta = 1.3849 \times 10^{-5},\ 1.5522 \times 10^{-5};$

$e_x = 0.2823 \times 10^{-1},\ 0.5099 \times 10^{-1};$ $\qquad U_y = 38.9597,\ 20.7119;$

$r_1 = 0.5293,\ 0.50.$

A parameter's first value corresponds to a limit cycle of a one loop per span oscillation whereas the second is the analogous two loops per span case. The corresponding $B_1$ and $B_2$ obtained from eq. (11.65) are

$$B_1 = 755.04,\ 395.90;\ B_2 = -4137.30,\ -4296.16$$

Both these points lie in the fourth quadrant of Fig. 11.15(a).

The study of the galloping of an eccentrically iced conductor indicates that the important feature is associated with a mixed-mode motion which has a two- or three-dimensional torus, respectively. Each of the radii is associated with a distinct frequency. If these frequencies are not resonant or, more generally, if the frequencies are irrational with each other, quasi-periodic motions do not repeat a given orbit. In other words, the torus is covered by quasi-periodic orbits.

From a practical viewpoint, the critical boundaries where bifurcations occur, can be used to assess instability trends with respect to a system's parameters. In particular, the effects of eccentricity may be found and the feasibility of vibration control devices like dampers may be assessed. It has been found that an eccentrically iced conductor may exhibit more complicated dynamic behaviour that needs to be studied more deeply.

## 11.4 Other Applications of Bifurcation Theory

We have given some application examples of bifurcation theory in elastic mechanics, non-linear vibration, structure dynamics and mechanical engineering. It is known that bifurcation theory has a great influence upon physics, chemical engineering, biology, ecology and economics. Because of space limitations, only a brief summary of the application of bifurcation theory to hydrodynamics, continuous stirred tank reactors and reaction—diffusion systems will be given below.

### 11.4.1 Bifurcation of Hydrodynamic Systems

Hydrodynamics is the study of the motion of fluid under the action of external force. For viscous incompressible fluid its motion is described by the Navier—Stokes equations. Let the fluid be full of space region $\Omega$, then dimensionless fluid motion equations are as follows:

$$\bar{v}_t + (\bar{v}\cdot\nabla)\bar{v} = -\nabla p + \frac{1}{R}\nabla^2\bar{v} + \bar{F}(\bar{x},t) \tag{11.88}$$

$$\nabla\cdot\bar{v} = 0 \tag{11.89}$$

where $\bar{v}(\bar{x},t)$ is dimensionless velocity, $\bar{x}\in\Omega$, $t>0$, $p$ is intensity of pressure, and $R$ is Reynolds number, $\bar{F}(\bar{x},t)$ is the known field of external force. Suppose the fluid is without slip and separation on boundary $2\Omega$, and the motion velocity of the boundary is $\bar{v}_B(\bar{x},t)$, so the boundary condition is

$$\bar{v}(\bar{x},t) = \bar{v}_B(\bar{x},t), \qquad \text{for } \bar{x}\in\partial\Omega \tag{11.90}$$

Suppose the velocity is $\bar{v}_0(\bar{x})$ at $t$=0, the initial condition is

$$\bar{v}(\bar{x},0) = \bar{v}_0(\bar{x}), \qquad \text{for } \bar{x}\in\Omega \tag{11.91}$$

If the flow is stationary (not concerned with $t$), the initial condition need not be considered. If there are other fields (such as field of temperature, electromagnetic field, etc.), additional equations must be introduced, and eq. (11.88) is modified necessarily. Obviously, the solution of initial-boundary value problems, (11.88)—

(11.91), depends on Reynolds number $R$ (or the parameters which depend on $R$; other parameters are possible, when other fields can be presented).

For the case of small Reynolds number, it is proved that solutions corresponding to different initial values $\bar{v}_0$ tend to the same solution $\bar{v}^*$, when $t \to \infty$. $\bar{v}^*$ is decided uniquely by $\bar{v}_B$ and $\bar{F}$, and is not dependent on $\bar{v}_0$. It is indicated that in the case of small Reynolds number the flow is fully decided by boundary condition and field of external force. So the problem of hydrodynamics is easily solved. But in the case of larger Reynolds number, the above mentioned uniqueness is usually incorrect. Under the same boundary conditions and fields of external force, it is possible to present different solutions. One needs to completely understand and describe the behaviour of these solutions, such as periodicity, stability and multiplicity etc.; this make the problems of hydrodynamics in the case of larger Reynolds number more complicated and the treatment difficult. Bifurcation theory and stability theory of flow are the vital methods needed to solve these problems. There are some excellent monographs available on this subject [200—205]. It is known that bifurcation of flow closely parallels the study of stability theory. In particular, the critical values of appearance of bifurcation are decided by the eigenvalue problem of the linearized stability of flow.

The study of bifurcation of flow possesses important applications for the search for the mechanism producing turbulence. Since Reynolds in 1883 found the transition phenomenon from layer flow to turbulence occurs at a critical Reynolds number, turbulence has been widely studied and is still a difficult problem. In recent years, experiments, numerical calculations and theoretical results [205] indicate that turbulence possesses complicated chaotic motion behaviour, and particularly its high sensitivity to initial conditions may induce a random dependence on time. Ruelle and Takens [206] have pointed out that, in a fluid system, chaotic motion may appear through finite time bifurcations. This theory was supported by some experiments and research results from low-dimension dynamical systems. Therefore, by deep analysis of different bifurcation processes, we can effectively search every road to turbulence.

The study of the bifurcation problems of flow has produced important developments. For typical problems of hydrodynamics — such as Couette—Taylor flow and Rayleigh—Benard convection, a large number of new results [207—209] are available.

### 11.4.2 Bifurcation in Chemical Reactors

In the last twenty years, have become interested in multiple stationary structures, periodic oscillations and some catastrophe phenomena in chemical reaction systems. These problems closely depend on bifurcation studies. Chemical reaction systems are classified into concentrate variable systems and distribution variable systems. The state variable of a concentrate variable system is independent of the space variable, and is usually described by an ordinary differential equation. The state variable of a distribution variable system is dependent on a space variable, is usually described by a partial differential equation.

Bifurcation phenomena in chemical reactors can be classified into two kinds: stationary and dynamical bifurcations. Stationary bifurcation theory mainly analyses the multiple equilibrium state, dynamical bifurcation theory (particularly Hopf bifurcation theory) studies periodic oscillations, change of stability, etc. These phenomena are decided by complicated mechanisms of chemical reactionary processes. We describe them qualitatively and quantitatively by means of the bifurcation theory, so that designers can make choose the optimum scheme from a larger region of parameters, and at the same time can take effective control of reactors using corresponding bifurcation sets and singular structure. Therefore the study of bifurcation for design and control of chemical reactors has important applications. In addition, the rich bifurcation phenomena appearing in chemical reactors are studied, which not only helps to explain bifurcation problems in reactors, but also promotes development of bifurcation theory.

From a historical viewpoint, the study of bifurcation in chemical reactors began with Liljenroth's work on stationary multiplicity in 1918. Zeldovich and Zisin [41] in 1941 pointed out that there are possibly different multiple stationary structures, when the active energy reaction is high. In 1958, Aris and Amundson [210] first used bifurcation theory to calculate these. Then the treatment of multiplicity and periodicity of the state in chemical reactors by bifurcation theory expanded spectacularly [211-218]. In the study of bifurcation problems of chemical reactors, the application of singularity theory is particularly worth pointing out. It does not merely recognise and classify bifurcation behaviour produced by change of different parameters in chemical reactors, but can also distinguish some parameters that reflect the nature of a stable structure. Thus one can entirely control the stability behaviour of the system. This gives the study of multiplicity and periodicity of chemical reactive systems an effective treatment method. But when one considers more complicated bifurcation problems, one needs to use the Birkhoff normal form etc.

### 11.4.3 Bifurcation of Reaction—Diffusion Systems

In many physical, chemical, biological and bionics systems there occur at the same time both reaction and diffusion. Here reaction is transformation through mutual action of every element of the system (such as chemical reaction, species competition etc.). Diffusion is macroscopic movement formed by spatial random motion of every element (such as matter diffusion, species moves away to another place, etc.). They are many important dissipative systems, and in mathematics they are described by the reaction—diffusion equation.

Consider a reaction—diffusion system consisting of $n$ elements, with $u_k$ $(k=1,2,\ldots,n)$ the density of the $k$th element; such a system is usually a function of time $t$ and space variable $\bar{x}$. Let $\bar{u}=(u_1,\ldots,u_n)$ be the state variable of the system, and $\bar{\mu}=(\mu_1,\ldots,\mu_m)$ be control parameters (such as temperature or intensity of pressure, which are controlled by the external world). In most cases, diffusion defers to Fick's law. The diffusion flow produced by spatial, not well, distribution of every element can be described by the following vector

$$\bar{j}_k=-D_k\nabla u_k, \qquad k=1,\ldots,n \tag{11.92}$$

where $D_k$ $(k=1,2,\ldots,n)$ are diffusion matrices, which are positive definite matrices, and in practice are often diagonal matrices, which can depend on density. Let the product rate produced by reaction of every element be $f_k$ ($k$=1,...,$n$), which are usually non-linear functions of density. According to the mass balance relationship of every element:

$$u_{k,t} = -\nabla \cdot \bar{j}_k + f_k(\bar{u}, \bar{\mu}), \qquad k = 1,\ldots,n \tag{11.93}$$

and from eq. (11.92) one can obtain the following reaction—diffusion equation:

$$u_{k,t} = \nabla \cdot (D_k \nabla u_k) + f(\bar{u}, \bar{\mu}), \qquad k = 1,\ldots,n \tag{11.94}$$

In particular, when $D_k$ ($k$=1,2,...,$n$) are all diagonal matrices and diagonal elements are the same, eq. (11.94) can be written

$$u_{k,t} = D_k \nabla^2 u_k + f_k(\bar{u}, \bar{\mu}), \qquad k = 1,\ldots,n \tag{11.95}$$

where $D_k$ ($k$=1,2,...,$n$) are coefficients of diffusion. The reaction—diffusion equation includes a first-order derivative of time and a second-order partial derivative of space, and $D_k$ is a positive definite matrix. Therefore it is a second-order parabolic partial differential equation. In addition, the proper initial condition and boundary condition must be given.

Reaction—diffusion eq. (11.94) has two important extra cases:

(i) If there is no diffusion, i.e. $D_k$=0 or $u_k$=$u_k(t)$ ($k$=1,...,$n$), eq. (11.94) becomes a first-order ordinary differential equation

$$u_{k,t} = f_k(\bar{u}, \bar{\mu}), \qquad k = 1,\ldots,n \tag{11.96}$$

which is usually called the reaction dynamical equation.

(ii) If there is no reaction, i.e. $f_k$=0 ($k$=1,...,$n$), eq. (11.94) becomes an ordinary diffusion equation

$$u_{k,t} = \nabla \cdot (D_k \nabla u_k), \qquad k = 1,\ldots,n \tag{11.97}$$

For the above ordinary differential equation and diffusion equation, a large number of research works exist. Although both equations reflect the evolution process of time, their research methods and results have basic differences. In reaction-diffusion systems, because reaction and diffusion are coupled, a more complicated dynamical effect must be produced [219—222]. The studies indicate that competition between reaction and diffusion under certain condition can form in system an ordered determinate time—space structure (space pattern, persistent oscillations, etc.), i.e. a dissipative structure. In reaction—diffusion systems, how dissipative structure is produced from time—space in an unordered state, and how transforms occur between these dissipative structures, are the principal bifurcation problems. Diffusion is induced by space non-distribution, so the dissipative structure that we are concerned with in reaction—diffusion systems depends on $\bar{x}$, such as the space periodic stationary state (no dependence on $t$), the time periodic state (periodic change with $t$), the travelling wave and other wave states.

Reaction—diffusion eq. (11.94) can be extended according to the following practical cases:

(i) The reaction term may depend again on $x$, $t$, $\nabla u_1, \ldots, \nabla u_n$ etc., that is

$$f_k = f_k(\bar{x},t,\bar{u},\nabla u_1,\ldots,\nabla u_n,\mu), \qquad k=1,\ldots,n \tag{11.98}$$

(ii) In some cases, such as a fluid of some film or a highly non-linear chemical reaction negative diffusion can appear, i.e. the matter moves in he inverse direction of the density gradient [223]. In this case the diffusion matrix is not a positive definite matrix.

(iii) In some cases, the diffusion term may not follow Fick's diffusion law, but instead the Cahn—Hilliard generalized diffusion law [223,224], i.e.

$$\bar{j}_k = -D_k \nabla u_k + p_k \nabla\left(\nabla^2 u_k\right), \qquad k=1,\ldots,n \tag{11.99}$$

where $p_k$ is a constant. This produces an order-4 reaction—diffusion equation

$$u_{k,t} = \nabla\cdot\left(D_k \nabla u_k\right) - p_k \nabla^4 u_k + f_k, \qquad k=1,\ldots,n \tag{11.100}$$

Stationary bifurcation of the Brusselator system and its Hopf bifurcation, and the time—space pattern of reaction need to be studied more deeply.

# References


1 Chen Yushu, *Nonlinear Vibrations*, Tian Jin Science and Technology Press 1983 (in Chinese)

2 M Golubitsky and WF Langford, Classification and Unfolding of Degenerate Hopf Bifurcations, *Journal Differential Equations*, No 3, 1981

3 Chen Yushu and WF Langford, The Subharmonic Bifurcation Solution of Nonlinear Mathieu's Equation and Euler Dynamic Buckling Problems, *Acta Mechanica Sinica*, No 4, 1988

4 M Globitsky and DG Schaeffer, *Singularities and Groups in Bifurcation Theory*, **1** and **2**, Springer-Verlag, New York, 1985, 1988

5 Zhu Zhaoxuan, Chaos in Nonlinear Dynamics, *Advance of Mechanics*, **14**, No 2, 1984 (in Chinese)

6 Lin Rei and Chen Yushu, A New Method for Multi-degrees of Freedom of Nonlinear Vibration Systems and Application in Analysis of Snake Motion of Rail Vehicles, *Acta Applied Mechanics*, **3**, No 4, 1986 (in Chinese)

7 NK Cooperrider, Nonlinear Behavior in Rail Vechicle Dynamics, *Numerical Methods for Bifurcation Problems*, Proc Conf of UD, August 1984

8 K Zeman, H Troger and R Scheidle, Bifurcation Theory and Vehicle Dynamics with Application to the Tractor Semitrailer, *Proc 7th IAVSD Symposium in Cambridge UK* 1981, Swets and Zeitlinger BV, Lisse, 1982

9 RK Mehra and JV Carroll, Bifurcation Analysis of Aircraft High Angle of Attack Flight Dynamics, *New Approches to Nonlinear Problems in Dynamics SIAM*, Philadelphia, 1980

10 MP Paidoussis and NT Issid, Dynamic Stability of Pipes Conveying Fluid, *Journal of Sound Vibration*, No 33, 1974

11 PR Sethna, Bifurcation Theory and Averaging in Mechanical Systems, *Numerical Method for Bifurcation Problems*, Proc Conf of UD, August 1983, Birkhaüser Verlag, 1984

12 GR Garalas, *Non-linear Differencial Equations of Chemically Reacting ystems*, Springer-Verlag, New York, 1968

13 V Volterra, Variations and Fluetuations of Number of Individuals in Animal Species Living Together, *Journal de Conseil International Pour l'Exploration de la Mer,* **3**, No1,1928

14 EC Zeeman, Catastrophe Theory, *Scientific American*, April 1976

15 JM Greenberg, On the Equilibrium Configurations of Compressible Slender Bars, *Scientific American*, 27, 1967

16 WT Kaiter, Post-Buckling Analysis of a Simple Two-Bar Frame, *Recent Progress in Applied Mechanics*, Almquist and Wiksell, Stockholm, 1967

17 S Alntman, Equilibrium States of Nonlinearly Elastic Rod, *Recent Progress in Applied Mechanics*, Almquist and Wiksell, Stockholm, 23, 1968

18 H Troger, Verzweigungstheorie—Eine Herunsforderung *für* Mathematiker undIngenieure, *GAMM*-Nachrichten, 2, 1982

19 JH Wolkowisky, Existence of Buckled States of Circular Plates, *Communication of Pure and Applied Maths* **20**, 1967

20 D Hui and JS Hansen, Two Mode Buckling of an Elastically Supported Plate and Its Relation to Catastrophe Theory, *Journal Applied Mechanics*, **47**,1980

21 RH Plaute, Buckling of Shallow Elastic Structures, *New Approaches to NonlinearProblems in Dynamics, STAM*, Philadelphia,1980

22 R Scheidl and H Troger, Buckling and Postbuckling of Complete Spherical Shells, *Proc Euromech Coll on Flexible Shells*, Munich, 1983, Springer-Verlag

23 GH Knightly and D Sather, Buckled States of a Spherical Shell under Uniform External Pressure, *Arch Rat Mech Anal*, 72,1980

24 FH Busse, Patterns of Convection in Spherical Shell, *Journal Fluid Mechanics*, **72**,1975

25 VA Zlatoustov, DE Okhotsimsky, VA Sarychev and AP Torzhevsky, Investigation of Satellite Oscillations in the Plane of an Elliptic Orbit, *Proc IUTAM Conf*, Munich 1964

26 J Moser, Is the Solar System Stable? FS *The Mathematical Intelligen*er **1**, 1978

27 DK Campbell, Nonlinear Science---from Normal example to Applications, *Advance of Mechanics*, Translated from Los Alamos Science, No 15 (Special Issue), 1987 by Huang Yongnian (in Chinese)

28 VI Iudovich, Secondary Flows and Fluid Instability Between Rotating Cylinders, Journal Applied Mathematics and Mechanics, **30**, 1966

29 K Kirchgassner, Bifurcation in Nonlinear Hydro-dynamic Stability, *SIAM Review*, **17**,1975

30 JK Hale, *Ordinary Differential Equations*, Wiley, New York, 1969

31 EA Coddington and N Levinson, *Theory of Ordinary Differetial Equations*, McGraw-Hill, New York, 1985

32 JA Sander and F Verhaulst, *Averaging Methods in Nonlinear Dynamical Systems*, Springer-Verlag, New York, 1985

33 Zhu Zhaoxuan, *Lectures in Non-linear Mechanics*, Beijin University, 1988

34 G Duffing, *Erzwungene Schwingungen bei Veranderlicher Eigenfrequenz, Braunschweig*, 1918

35 T Haiason, *Forced Vibrations in Non-linear Systems*, Foresign Lit Press, 1957 (in Russian)

36 J Guckenheimer and P Holmes, *Non-linear Oscillations, Dynamical Systems, and Bifurcations of Vector Fields*, Springer-Verlag, New York, 1983

37 MW Hirsch, CC Pugh and M Shub, *Invariant Manifold, Springer*, Lecture Notes in Mathematics, Vol 583, Springer-Verlag, New York, 1977

38 Chen Yushu and Yang Shaopu, Bifurcation of a Self-excited Vibration Systems with Hysterestic Nonlinearity, *Proc of International Conference on Vibration*

*Problems in Engineering,* Wuhan-Chongqing, PRChina, June 1990

39 IG Martin, *Theory of Motion Stability*, Science Press, 1966 (in Russian)

40 MHirsch and SSmale, *Differential Equations, Dynamical Systems and Linear Algebra,* Academic Press, New York, 1974

41 MM Peixoto, Structural Stability on Two-Dimensional Manifolds,*Topology*, **1**, 1962

42 MM Peixoto, *Dynamical Systems,* Academic Press, New York, 1973

43 J Palis and W Melo, *Geometric Theory of Dynamical Systems, An Introduction,* Springer-Verlag, 1982

44 Zhang Zusheng, *Differential Dynamical Systems*, Science Press, 1987 (in Chinese)

45 Zhang Zhifen, Ding Tongren, Huang Wenzao and Dong Zhenxi, *Qualitative Theory of Differential Equations,* Science Press, 1985 (in Chinese)

46 Ye Yanqian, *Limit Circle Theory*, Shanghai Science and Technology Press, 1984 (in Chinese)

47 W Szlenk, *An Introduction to the Theory of Smooth Dynamical Systems*, Wiley, 1984

48 Lu Qishao, *Qualitative Methods and Bifurcation of Ordinary Differential Equations,* Beijin Aerospace University Press, 1989 (in Chinese)

49 Ye Min, Zhan Kaijun and Chen Yushu, Experimental Study of 1/2 Subharmonic Bifurcation of Nonlinear Mathieu's Equation, *Proceediings of 5th Chinese Nonlinear Vibration Conference*, Dunhuang, Gansu, China, Aug 1989 (in Chinese)

50 Chen Yushu and Zhang Qichang, A New Method for Solving Asymptotical Solutions of Nonlinear Vibration Systems--Computing Coefficients of Normal Form of Vector Fields, *Acta Mech Sinaca*, No 4, 1990 (in Chinese)

51 BD Hassard, ND Kazarinoff and YHWan, *Theory and Applications of Hopf Bifurcation,* Cambridge University Press, Cambridge,1981

52 G Ioos and DD Joseph, *Elementary Stability and Bifurcation Theory*, Springer-Verlag, New York, Heidelberg, Berlin, 1981

53 F Takens, Singularities of Vector Fields, Publ Mathematics IHES, 43,1974

54 FTakens, Forced Oscillations and Bifurcations, *Communication Mathematics Inst Rijkuniversiteit Utrecht*, **3**, 1974

55 RI Bogdanov, Versal Deformations of a Singular Point on the Plane in the Case of Zero Eigenvalues, *Functional Analysis and its Applications*, **9**(2), 1975

56 F Takens, Detecting Strange Attractors in Turbulence, In *Dynamical Systems and Turbulence*, DA Rand and LS Yong(eds) Springer Lecture Notes in Mathmatics, Springer-Verlag, New York, Vol 898, 1980

57 PJ Holmes and DA Rand, *Phase Portraits and Bifurcations of Nonlinear Oscillator* $x'' + (\alpha + \gamma x^2)x' + \beta x + \delta x^3 = 0$, International Journal of Nonlinear Mechanics, 15, 1980

58 J Carr, *Applications of Center Manifold Theory*, Springer-Verlag, New York, 1981

59 Hao Belin, Bifurcation, Chaos, Strange Attractor, Turbulence and Others-Inherit Stochasticy Existed in Determined Systems, *Advance of Physics*, No 3, Sept 1983 (in Chinese)

60 EN Lorenz, Deterministic Nonperiodic Flows, *Journal Atmospheric Science* **20**, 1963

61 MJ Feigenbaum, Quantitative Universality for a Class of Nonlinear Transformations, *Journal of Statistical Physics*, **19**, 1978; **21**, 1979

62 VI Arnold, *Achievement of Mathematics Science*, 18, 1963 (in Russian)

63 J Moser, On Invariant Curves of Aera-preserving Mappings of an Annulus, Nachr Akad, *Wiss Gottingen Mathematics Physics Kl 2*, 1962 64 98 (1954)

64 AN Kolmogorov, *DAN SSSR*, **98**, 1954 (in Russian)

65 M Henon and C Heiles, The Applicability of the Third Integral of Motion: Some Numerical Experiments, *Astron Journal*, 69, 1964

66 D Ruelle and F Takens, On the Nature of Turbulence, *Communication Mathematics Physics*, 20, 1971; 23, 1971

67 S Smale (see SS Cairns' book), *Differential and Combinatorial Topology*, Princeton, 1965

68 S Smale, *Bull Amer Mathematics Society,* **73**, 1967

69 Y Ueda, Steady Motions Exhibited by Duffing's Equation: A Picture Book of Regular and Chaotic Motions, *New Approaches to Nonlinear Problems in Dynamics*, ed PJ Holmes, SIAM, Philadelphia, 1980

70 Y Ueda and N Akamatsu, IEEE Trans Cas 28, 3a, 1981, *New Approaches to Nonlinear Problems in Dynamics*, ed PJ Holmes, SIAM, Philadelphia, 1980

71 JGuckenheimer (see Marsden, McCracken's book), *Hopf Bifurcation and its Applications,* Sec12, 1976

72 K Tomita, *Prog Theor Physics Suppl*, **69**, 1980

73 Y Pomeau, *Communication Mathematics Physics*, **74**, 1980

74 TUeza and Y Aizawa, *Prog Theor Physics*, **68**, 5 (1982); **68**, 6 (1982)

75 P Collet and JP Eckmann, Iterated Maps on the Interval as Dynamical Systems, *Prog on Physics*, Vol I, Birkhauser, Boston, 1980

76 T Giesel and J Nierwetberg, *Physics Rev Lett*, **47**, 14, 1981; **48**, 1, 1982

77 AI Shacovsky, *Journal Ukrainian Mathematics*, **16**, 1964 (in Russian)

78 TY Li and JA Yorke, *Am Mathematics Monthly*, **82**,1975

79 N Metropolis, ML Stein and PR Stein, On Finite Sets for Transformations on the Unit Interval, *Journal Combin Theor*, **A15**, 1973

80 J Guckenheimer, Invent, Mathematics 39, 2 1977

81 B Derrida and Y Pomenau, *Physics Lett*, **80A**, 1980

82 AB Zisook, *Physics Rev*, **A24**, 1981

83 Zhu Zhaoxiun, *Chaos*, Lecture in Beijin University, Oct 1984

84 Zhang Jinyan, *Smale Horseshoe in Henon Mapping*, Science Bull, **24**, 1984

85 F Hausdorff, Dimension and Auberes Mab *Mathematics Ann*, 1918-B,**79**, H2-s

86 Li Jibin, *FK Chaos and Melnicov's Method*, FS Chongqing University Press, 1989 (in Chinese)

87 C Sparrow, *The Lorenz Equations: Bifurcations, Chaos and Strange Attractors*, Springer-Verlag, New York, 1982

88 SE Newhouse, D Ruelle and F Takens, Occurence of Strange Axiom an Attractors near Quasiperiodic Flows on $T^m$, $m \geq 0$, *Communication Mathematics Physics*, **64**, 1978

89 AJ Lichtonberg and MA Liberman, *Regular and Stochastic Motion*, Springer-Verlag, 1982

90 S Wiggins and P Holmes, (i) Periodic Orbits in Slowly Varying Oscillators, and (ii) Homoclinic Orbits in Slowly Varying Oscillators, *SIAM, Journal Mathematics Anal*, **18**, 3, 1987

91 NN Bogoliubov and EA Mitropolsky, *Asympototic Methods in Nonlinear Vibration Theory*, Science Press, 1974 (in Russian)

92 AH Nayfeh and DT Mook, *Nonlinear Oscillations* , Wiley, New York, 1979

93 Zhang Jinyan, *Geometric Theory and Bifurcation Problems of Ordinary Differential Equations* (Revised edition), Beijing University Press, 1987

94 B Hassard and YH Wan, Bifurcation Formulae Derived form Center Manifold Theory, *Journal Mathematics Anal Appl*, **63**, 1978

95 WW Farr and R Aris, Degenerate Hopf Bifurcation in the CSTR with Reactions $A \rightarrow B \rightarrow C$, in *Oscillations, Bifurcations and Chaos*, FVAtkinson, WF Langford and AB Mingarelli (eds), CMS Conference Proceedings, Vol 8, American Mathematical Society, Providence, RI, 1987

96 IS Labouriau, Degenerate Hopf Bifurcation, *SIAM Journal Mathematics Anal*, 16, 1985

97 WW Farr, Chengzhi Li, IS Labouriau and WF Langford, Degenerate Hopf Bifurcation Formulas and Hilbert 16th Problem, *SIAM Journal Mathematics Anal*, **20**, 1, 1989

98 NNBautin, *On the Number of Limit Cycles Which Appear with the Variation of the Coefficients from an Equilibrium Position of Focus or Center Type*, American Mathematics Society Translation No 100, Providende, RI, 1954 (Russian Original in Mat Sb (NS) 30, 1952)

99 Chen Yushu, Zhan Kaijun and WFLangford, Extended Results of Subharmonic Resonance Bifurcation of Nonlinear Mathieu Equation and Some Experimental *Results, Nonlinear Dynamics in Engineering Systems,* IUTAM Symposium Stuttgart/Germany 1989, Springer-Verlag, Berlin, Heidelberg, 1990

100 Chen Yushu and Zhan Kaijun, New Results of Degenerate Bifurcation of Subharmonic Resonance of Parametrically Excited Vibration Systems, *Acta Vibration Engineering,* **3**, 2, 1990 (in Chinese)

101 S Wiggins, *Global Bifurcations and Chaos, Analytical Methods,* Springer-Verlag, New York, 1988

102 VI Arnold, *Dynamical Systems III,* Springer-Verlag, World Publishing Corporation, Beijing, 1989

103 James Gleick, *Chaos---Making a New Science*, Viking Penguin Inc, New York, Twelfth Printing, April 1988

104 Lu Kan et al, *Chaotic Dynamical Systems*, Shanghai Translation Publishing Company, 1990

105 Chen Yushu and Mei Lintao, Bifurcation Solutions of Resonant Cases of Nonlinear Mathieu Equations, *Science in China (Series A)*, No 12, 1990

106 Chen Yu Shu and Zhen Kaijun, A Class of Bifurcation Solutions of Almost-periodic Parametric Vibration Systems, *Acta Mechanica Sinica*, No 3, 1990

107 Chen Yushu, Wu Jianguo and Jin Zhisheng, The Bifurcation Solution of Nonlinear Parametrically Excited Torsion Vibration Problem of Crank, *Journal of Vibration Engineering*, No 1, 1987 (in Chinese)

108 108 Lin Rui and Chen Yushu, Multi-Frequencies Bifurcation of Nonlinear Vibration System with Multi-Degrees of Freedom, *Journal Mathematics and Physics*, 9, 1989

109 Yang Shaopu and Chen Yushu, Bifurcation of a Certain One-Degree of Freedom Hysteretic Solf-Excited Vibration System, *Journal Vibration Engineering*, No 2, 1992 (in Chinese)

110 Chu Jinyun and Chen Yushu, The Periodic Solution of a Nonlinear Dynamic System with Closed Cycle Control, *Journal Applied Mechanics*, No 2, 1991 (in Chinese)

111 Chen Yushu and Cao Qingjie, Random Vibration and Bifurcation under United Parametrically Excited and Noice, *Proceedings of the Third Chinese Conference on Random Vibration*, Taian, Aug 1991 (in Chinese)

112 Yang Shaopu and Chen Yushu,The Bifurcation and Singularities of a parametrically Excited System, *Mechanics Research Communications*, No 4, 1991

113 Chen Yushu and Yang Shaopu, Global and Degenerate Vector Fields Bifurcations of Parametrically Hysteretic Excited Vibration System, *Modern Mathematics and Mechanics, MMM-IV,* Lanzhou University Publishing House, 1991 (in Chinese)

114 Zhang Qichang and Chen Yushu, The Study of Nonlinear Vibration Problems of Multi-Degrees of Freedom by Dynamical System Theory, *Journal of Vibration Engineering,* No 4, 1991 (in Chinese)

115 Chow Shui-Nee and JKHale, *Methods of Bifurcation Theory*, Springer-Verlag, New York, 1982

116 Chow Shui-Nee and Wang Duo, Normal Form of Bifurcating Periodic Orbits, Multiparameters Bifurcation Theory, *Contemporary Mathematics*, **56**, 1986

117 E Knobloch, Normal Form Coefficients for the Nonresonant Double Hopf Bifurcation, *Physics Lett* **116**, 8, 1986

118 Wang Duo, Approximate Calculation of the Coefficients of Bogdanov Normal Form, *Proceedings of the Fifth National Conference on Nonlinear Vibration*, Dunhuang, 1989

119 Wang Duo, An Introduction to the Normal Form Theory of Ordinary Differential Equations, *Advances in Mathematics*, 19, 1, 1990

120 DK Arrosmith and CM Place, Bifurcations of a Cusp Singularity with Applications, *Acta Applicadae Mathematicae*, **2**, 1984

121 AK Bajaj, Resonant Parametric Perturbations of the Hopf Bifurcation, *Journal Mathematics Anal Applications*, **115,** 1984

122 AK Bajaj, Bifurcations in a Parametrically Excited Nonlinear Oscillator, *International Journal of Nonlinear Mechanics*, **23,** 1, 1987

123 M Golubisky and I Stewart, Hopf Bifurcation in the Presence of Symmetry, *Arch Rat Mech Anal*, **87**, 1985

124 JD Jin and Y Matsuzaki, Bifurcations in a Two-Degree-of-Freedom System with Follower Forces, *Journal Sound Vibration*, **126**(2), 1988

125 V Kacani et al, Dynamics and Bifurcations in the Motion of Tractor-Semitrialler Vehicles, *Oscillations, Bifurcations and Chaos*, 1987

126 RW Kolkka, Singular Perturbations of Bifurcation with Multiple Independent Bifurcation Parameters, *SIAM Journal Applied Mathematics*, 44, 1984

127 FL Sanders, *Averaging Methods in Nonlinearity Dynamical Systems*, Springer-Verlag, New York, 1985

128 S Yano, Parametric Excitation in the Self-excited Vibration System with Dry Friction (1,2*), Bull JSNE*, **27** (224), 1984

129 ST Ariartnam and Wei Chau Xie, Lyapunov Exponents and Stochastic Bifurcations, *Nonlinear Dynamics in Engineering Systems*, Springer-Verlag, 1989

130 VV Boltin, *Random Vibration of Elastic Systems*, Martinus Nijboff Publishers, The Hague, 1984

131 GQ Cai, A New Approximate Solution Technique for Randomly Excited Nonlinear Oscilltors, *International Journal Nonlinear Mechnics*, **23**, 5/6, 1988

132 SH Crandall and WQ Zhu, Random Vibration; a Survey of Recent Developments, *ASME Journal Applied Mechanics*, 50, 1983

133 R Graham, Forward Hopf Bifurcation with Multiplicative Gaussian White Noise: Exact Fokker Planck Solution, *Physics Lett*, 80A, 5/6,1980

134 RA Ibrahim and H Heo, Autoparametric Vibration of Coupled Beams under Random Support Motion, *ASME* Paper No 85-DET-184, 1985

135 RN Iyengar, Higher Order Linearization in Nonlinear Random Vibration, *International Journal Nonlinear Mechanics,* **23,** 5/6, 1988

136 S Janeczko and EWajnryb, Bifurcations in Stochastic Dynamical Systems with Simple Singularities, *Stochastic Process and Their Applications*, **31**,1989, North-Holland

137 J Scheurle, Chaotic Solutions of Systems with Almost Periodic Forcing, *Journal Applied Mathematics Physics (ZAMP)*, **37**,1986

138 T Kapitaniak, *Chaos in Systems with Noise*, World Scientific Publishing Co, Pteltd, 1988

139 A Lemarchand, H Lemarchand and ESulpice, Interaction of a Hopf Bifurcation and Symmetry-Breaking Bifurcation: Stochastic Potential and Spatial Correlations, *Journal Statistic Physics*, **53**, 3/4, 1988

140 YK Lin and GQ Cai, Exact Stationary Response Solutions for Second Order Nonlinear Systems Under Parametric and External White Noise Excitations; Part II, *ASME Journal Applied Mechanics*, **55**,1988

141 N Sri Namachchivaga, Hopf Bifurcation in the Presence of Both Parametric and External Stochastic Excitations, *Journal Applied Mechanics*, **55,** December 1988

142 AH Nayfey and SJ Serhen, Response Statistics of Nonlinear Systems to Combined Deterministic and Random Excitations, *International Journal Nonlinear Mechanics*, **25**, 5, 1990

143 CWS To, On Dynamics Systems Disturbed by Random Parametric Excitation, *Journal Sound Vibratiob*, **123**, 2, 1988

144 Van der Beek, Normal Form and Periodic Solutions in the Theory of Nonlinear Oscillation, Existence and Asymptotic Theory, *International Journal Nonlinear Mechanics*, **24**, 4, 1989

145 GEYoung and RJ Chang, Prediction of the Response of Nonlinear Oscillators under Stochastic Parametric and External Excitations, *International Journal Nonlinear Mechanics*, **22**, 2, 1987

146 MX Jinmintberg, *Nonlinear Random Problems in Mechanical Vibration Systems*, Science Press, 1980 (in Russian)

147 HI Neimark and PS Langa, *Random Vibrations*, Sxience Press, Moscow, 1987

148 SL Lau and YK Cheung, Amplitude Incremental Variational Principle for Nonlinear Vibration of Elastic System, *ASME Journal of Applied Mechanics*, **48**, 1981

149 AYT Leung and TC Fung, Construction of Chaotic Regions, *Journal of Sound and Vibration*, **131**(3), 1989

150 Chen Yushu, Tang Yun, et al *Modern Analytical Method of Nonlinear Dynamics*, Science Press, 1992 (in Chinese)

151 PJ Holmes and FC Moon, Strange Attractors and Chaos in Nonlinear Mechanics, *ASME Journal of Applied Mechanics*, **50**, 1983

152 BH Tongue, Existence of Chaos in a One Degree of Freedom System, *Journal of Sound and Vibration*, **110**, 1986

153 EH Dowell and C Pezeshki, On the Understanding of Chaos in Duffing's Equation Including a Comparison with Experiment, *ASME Journal of Applied Mechanics*, **53**, 1986

154 BH Tongue, Characteristics of Numerical Simulations of Chaotic Systems, *ASME Journal of Applied Mechanics*, **54,** 1987

155 EH Dowell, Chaotic Oscillations in Mechanical Systems, *Computational Mechanics,* **3**, 1988

156 AA Ferri, On the Equivalence of the Incremental Harmonic Balance Methodand the Harmonic Balance—Newton—Raphson Method, *Journal of Applied Mechanics*, **53,** 1986

157 M Urabe, Galerkin's Procedure for Nonlinear Periodic Systems, *Arch Rat Mech Anal,* **20,** 1965

158 M Urabe and A Reiter, A Numerical Computation of Nonlinear Forced Oscillations by Galerkin Procedure, *Journal of Mathematical Analysis and Application,* **14,** 1966

159 RE Mickens, Bounds on the Fourier Coefficients for the Periodic Solutions of Nonlinear Oscillation Equations, *Journal of Sound and Vibration*, **124**, 1988

160 R van Dooren, On the Transition from Regular to Chaotic Behavior in the Duffing Oscillator, *Journal of Sound and Vibration*, **123**, 1988

161 P Friedmann, CE Hammaond and THWoo, Efficient Numerical Treatment of Periodic Systems with Application to Stability Problem, *International Journal for Numerical Methods in Engineering,* **11**, 1977

162 AYT Leung and TC Fung, Phase Increment Analysis of Damped Duffing Oscillators, *International Journal for Numerical Methods in Engineering*, **28**, 1989

163 BA Huberman and JP Crutchfield, Chaotic State of Harmonic Systems in Periodic Field, *Physics Rew Lett*, **43**, 1979

164 VI Arnold, *Geometrical Methods in the Theory of Ordinary Differential Equations*, Springer-Verlag, New York, 1983

165 GD Birkhoff, *Dynamical Systems*, Vol 9, AMS Collection Publications, 1972

166 AYT Leung and QC Zhang, Normal Form Analysis of Hopf Bifurcation Exemplified by Duffing's Equation, *Shock and Vibration*, **1**, 1994

167 H Poincaré, Les Méthods Nouvelles de la Mécanique Céleste, Gauthier-Villars, Paris, 1889

168 PR Sethna and GR Sell, Review of the Hopf Bifurcation and its Applications, *Journal of Applied Mechanics*, **45**, 1978

169 F Takens, Normal Form for Certain Singularities of Vector Fields, *Annals of the Institute of Fourier*, **23**, 1973

170 AF Vakakis and RH Rand, Normal Mode and Global Dynamics of a Two-Degree-of-Freedom Nonlinear System — I Low Energies, *International Journal Non-Linear Mechanics*, **27**, 1992

171 CGA van der Beek, Normal Form and Periodic Solutions in the Theory of Nonlinear Oscillation — Existence and Asymptotic Theory, *International Journal Non-Linear Mechanics,* **24**, 1989

172 S Wolfram, *Mathematica: A System for Doing Mathematics by Computer*, Addison-Wesley, Redwood City, CA, 1991

173 R Abraham and JE Marsden, *Foundations of Mechanics*, 2nd edn, Benjamin/Cummings, Reading, MA, 1978

174 JJ Stoker, *Nonlinear Elasticity*, Gordon and Breach, 1968

175 JMT Thompson and GW Hunt, *A General Theory of Elasticity Stability*,Wiley,1973

176 DO Brush and BO Almroth, *Buckling of Bars, Plates and Shells*, Mc Graw-Hill, New York, 1975

177 M Golubitsky and D Schaeffer, A Theory for Imperfect Bifurcation Theory via Singularity Theory, *Communication Pure Applied Mathematics*, **32**, 1979

178 EL Reiss, Imperfect Bifurcations, *Applications of Bifurcation Theory* (ed PH Rabinowitz), Academic Press, 1977

179 EH Dowell, *Aeroelasticity of Plates and Shells*, Noordhoff, 1975

180 RD Blevins, *Flow-Induced Vibration*, Krieger, 1986

181 RD Blevins and WD Iwan, The Galloping Response of a Two-Degree-of-Freedom System, *ASME Journal of Applied Mechanics*, 41, 1974

182 W Rudin, Real and Complex Analysis, 3rd edn, McGraw-Hill, New York, 1987

183 K Huseyin and P Yu, On Bifurcations into Non-resonant Quasi-Periodic Motions, *Applied Mathematics Modelling*, **12**, 1988

184 AT Edwards and A Madeyski, Progress Report on the Investigation of Galloping of Transmission Line Conductors, *Trans AIEE*, **75,** 1956

185 O Nigol and GJ Clarke, Conductor Galloping and Control Based on Torsional Mechanism, *presented at IEEE Power Engineering Society Meeting,* New York, 1974

186 CY Ip, YM Desai, P Yu, AH Shah and N Popplewell, An Oscillator Model for the Galloping of an Electrical Transmission Line, *Proceedings of the 4th International Conference on Recent Advances in Structural Dynamics*, 1991

187 J Shaw and SW Shaw, Instabilities and Bifurcations in a Rotating Shaft, *Journal of Sound and Vibration*, **132**(2), 1989

188 S Schecter, The Saddle-node Separatrix-loop Bifurcation, *Journal of Mathematical Analysis,* **18,** 1987

189 JM Johnson and AK Bajaj, Amplitude Modulated and Chaotic Dynamics in Resonant Motions of Strings, *Journal of Sound and Vibration*, **128**, 1989

190 MO Hongler and L Streit, On the Origin of Chaos in Gearbox Models, *Physica* D, **29**, 1988

191 PL Ko, Heat Exchanger Tube Fretting Wear: Review of an Application to Design, *ASME Journal of Tribology*, **107**, 1985

192 TW Lee and AC Wang, On the Dynamics of Intermittent-Motion Mechanisms: Part I - Dynamics Model and Response, *ASME Paper 82-DET-64*, 1982

193 SF Masri, *Analytical and Experimental Studies of Impact Dampers*, PhD Dissertation, California Institute of Technology, 1965

194 MT Bengisu, T Hidayetogiu and A Akay, A Theoretical and Experimental Investigation of Contact Loss in the Clearances of a Four-Bar Mechanism,*ASME Journal of Mechanism, Transmission and Automation in Design,* **108**, 1986

195 MD Bryant, Nonlinear Forced Oscillation of a Beam Coupled to an Actuator via Hertzian Contact, *Journal of Sound and Vibration*, **99**, 1985

196 S Satio, Calculation of Nonlinear Unbalanced Response of Horizontal Jeffcott Rotors Supported by Ball Bearings with Radial Clearances, *ASME Paper No 85-DET-33*, 1985

197 SW Shaw and JP Holmes, A Periodically Forced Piecewise-Linear Oscillator, *Journal of Sound and Vibration*, **108,** 1983

198 YB Kim and ST Noah, Stability and Bifurcation Analysis of Oscillators with Piecewise-Linear Characteristics: A General Approach, *Journal of Applied Mechanics,* **58,** June, 1991

199 Chen Yu Shu and Jin Zhisheng, Subharmonic Solution of a Piecewise-Linear Oscillator with Two Degrees of Freedom, *Applied Mathematics Mechanics*, 7(3), 1986

200 CC Lin, *The Theory of Hydrodynamic Stability*, Cambridge, 1955

201 S Chandrasekhar, *Hydrodynamic and Hydromagnetic Stability*, Oxford, 1961

202 DD Joseph, *Stability of Fluid Motions I and II,* Springer-Verlag, 1976

203 A Georgescu, *Hydrodynamic Stability Theory*, Martinus Nijhoff, 1985

204 DH Sattinger, *Topics in Stability and Bifurcation Theory,* Springer-Verlag, 1973

205 HL Swinney and JP Gollub (eds*), Hydrodynamics Instabilities and the Transition to Turbulence,* 2nd edn, Springer-Verlag, 1985

206 D Ruelle and F Takens, On the Nature of Turbulence, *Communication Mathematics Physics,***20**, 1971

207 CD Andereck, SS Liu and HL Swinney, Flow Regimes in a Circular Couette System with Independently Rotating Cylinders, *Journal Fluid Mechanics*, **164**, 1986

208 WF Langford, R Tagg, E Kostelich, HL Swinney and M Golubitsky, Primary Instability and Bicriticality in Flow Between Counterrotating Cylinders, *Physics Fluids,* **31**, 1988

209 M Golubitsky, JW Swift and E Knobloch, Symmetries and Pattern Selection in Rayleigh—Ben$\bar{a}$rd Convection, Physics, 10D, 1984

210 R Aris and N Amundson, An Analysis of Chemical Reactor Stability and Control, *Chemical Engineering Science*, **7**, 1958

211 A Uppal, WH Ray and A Poore, On the Dynamic Behavior of Continuous Stirred Tank Reactors, *Chemical Engineering Science*, **29**, 1974

212 A Uppal, WH Ray and A Poore, The Classification of the Dynamic Behavior of Continuous Stirred Tank Reactors — Influence of Reactor Residence Time, *Chemical Engineering Science,* **31**, 1976

213 M Golubitsky and BL Keyfitz, A Qualitative Study of the Steady State Solutions for a Continuous Flow Stirred Tank Chemical Reactor, *SIAM, Journal Mathematics Anal*, **11,** 1980

214 M Golubitsky, BL Keyfitz and DSchaeffer, A Singularity Theory Analysis of a Thermal-Chain Branching Model for the Explosion Peninsula, *Communication Pure Applied Mathematics,* **34**, 1981

215 V Balakotaiah and D Luss, Analysis of Multiplicity Patterns of a CSTR, *Chemical Engineering Comm*, **13**, 1981

216 V Balakotaiah and D Luss, Structure of the Steady-State Solutions of Lumped-Parameter Chemically Reacting Systems, *Chemical Engineering Science*, **37**, 1982

217 V Balakotaiah and D Luss, Global Analysis of the Multiplicity Features of Multi-Reaction Lumped Parameter Systems, *Chemical Engineering Science*, **39**, 1984

218 Tang Yun, Bifurcation Theory Method and its Application to the Study of Chemical Reactors, *Practice and Recognition of Mathematics*, **3**, 1992

219 PC Fife, *Mathematical Aspects of Reaction—Diffusion Systems*, Springer-Verlag, 1979

220 O Dickmann, and NM Temme, Nonlinear Diffusion Problems, *Mathematisch Centrum,* 1976

221 PC Fife, Asymptotic States for Equations of Reaction and Diffusion, *Bull Amer Mathematics Society*, **84,** 1978

222 G Nicolis, Bifurcations, Fluctuations and Dissipative Structures, in *Nonlinear Phenomena in Physics and Biology* (ed by RH Enns et al), Plenum, 1981

223 G Nicolis and I Prigogine, *Self-Organization in Nonequilibrium Systems*, Wiley, 1977

224 JW Cahn and JE Hillard, Free Energy of a Nonuniform Systems I, *Journal Chem Physics,* **28,** 1958

225 YS Chen and J Xu, Periodic Responses and Bifurcations in Nonlinear Hill's System (I*),* *Journal of Nonlinear Dynamics in Science Tech*, **1**(1), 1993 (in Chinese)

226 YS Chen and J Xu, Periodic Responses and Bifurcations in Nonlinear Hill's System (II), *Journal of Nonlinear Dynamics in Science Tech,* **1**(3), 1994 (in Chinese)

227 YS Chen and J Xu, Nonlinear Dynamical Bifurcations of a Thin Circular Plate Subjected to the Radial Disturbing Force, *Proceedings of ICNM II*, Beijing University Press, Beijing, 1993

228 YS Chen, DS Wang and M Ye, Chaotic Motion of Beam Under Longitudinal Excitation, *Journal of Nonlinear Dynamics in Science Tech*, **1**(2), 1993 (in Chinese)

229 YS Chen and J Xu, Universal Classification of Bifurcating Solutions to a Primary Parametric Resonance in van der Pol—Duffing—Mathieu's Systems, *Science in China (Series A),* **25**(12), 1995

230 YS Chen and J Xu, Global Bifurcations and Chaos in a Van der Pol-Duffing-Mathieu System with a Single-Well Potential Oscillator, *Proc International Conf Structural Dynamics, Vibration, Noise and Control, Hong Kong, Dec 5—7,* 1995

231 YS Chen and J Xu, Global Bifurcations and Chaos in a Van der Pol—Duffing—Mathieu System with a Three-Well Potential Oscillator, *Acta Mechanica Sinica*, **11**(4), 1995

232 YS Chen and J Xu, Local Bifurcation Theory of Nonlinear Systems with Parametric Excitation, *Nonlinear Dynamics*, 1996

233 YS Chen, *Bifurcation and Chaos Theory of Nonlinear Vibration Systems*, High Education Press, Beijin, 1993 (in Chinese)

234 Q Meng and YS Chen, A Study on the New Mechanism of Oil Film Unstability of a Nonlinear Rotor-Bearing System, *Journal Nonlinear Dynamics in Science Tech*, **2**, 3, 1995 (in Chinese)

235 P Yu, AH Shan and N Popplewell, Inertially Coupled Galloping of Iced Conductors, *ASME, Journal Applied Mechanics*, 59, 1992

236 VI Arnold, *Mathematical Method of Classical Mechanics*, Springer-Verlag, New York, 1978

237 KJ Bathe, E Ramm and EL Wilson, Finite Element Formulations for Large Deformation Dynamic Analysis, *International Journal of Numerical Methods in Engineering,* **9**, 1975

238 A Chajes and JE Churchill, Nonlinear Frame Analysis by Finite Element Methods, *Journal of Structural Engineering, ASCE*, **113,** 1987

239 MA Crisfield, *Nonlinear Finite Element Analysis of Solids and Structures,* Wiley, New York, 1991

240 MA Crisfield and J Shi, A Co-Rotational Element/Time-Integration Strategy for Non-Linear Dynamics , *International Journal of Numerical Methods in Engineering*, **37**, 1994

241 K Feng and MZ Qin, Hamiltonian Algorithms for Hamiltonian Systems and a Comparative Numerical Study, *Computer Physics Communications*, **65,** 1991

242 K Feng, HM Wu , MZ Qin, and DL Wang, Construction of Canonical Difference Schemes for Hamiltonian Formalism via Generating Functions, *Journal of Computational Meathematics,* **7**, 1989

243 H Goldstein, *Classical Mechanics*, Addison-Wesley, Reading, MA, 1980

244 AYT Leung and SG Mao, Symplectic Integration of an Accurate Beam Finite Element in Nonlinear Vibration, *Computers & Structures*, **54**, 1995

245 JE McNamara, Solutions Schemes for Problems of Nonlinear Structural Dynamics, *Journal of Pressure Vessel Technology, ASME*, **96**, 1974

246 JL Meek and HS Tan, Geometrically Nonlinear Analysis of Space Frames by Incremental Iterative Techniques, *Computer Methods in Applied Mechanics and Engineering,* **47,** 1984

247 DP Mondkar and GH Powell, Finite Element Analysis of Nonlinear Static and Dynamic Response, *International Journal of Numerical Methods in Engineering,* **11**, 1977

248 MZ Qin, DL Wang and MQ Zhang, Explicit Symplectic Difference Schemes for Separable Hamiltonian Systems, *Journal of Computational Mathematics*, **9**, 1991

249 IM Robert and A Pau, The Accuracy of Symplectic Integrators, *Nonlinearity*, **5**, 1992

250 RD Ruth, A Canonical Integration Technique, *IEEE Transactions on Nuclear Science, NS-30,* 1983

251 JC Simo, N Tarnow and KK Wong, Exact Energy-Momentum Conserving Algorithms and Symplectic Schemes for Nonlinear Dynamics, *Computer Methods in Applied Mechanics and Engineering*, **100**, 1992

252 RD Wood and OC Zienkiewicz, Geometrically Nonlinear Finite Element Analysis of Beams, Frames, Arches and Axisymmetric Shells, *Computers & Structures*, **7**, 1977

253 Y Wu, The Generating Functions for the Solution of ODEs and its Discrete Methods, *Computational Mathematics and Applications*, **15**, 1988

254 TY Yang and S Saigal, A Simple Element for Static and Dynamic Response of Beams with Material and Geometric Nonlinearities, *International Journal of Numerical Methods in Engineering,* **20**, 1984

255 EO Brigham, *The Fast Fourier Transform,* Prentice-Hall, Englewood Cliffs, NJ, 1974

256 TM Cameron and JH Griffin, An Alternating Frequency/Time Domain Methods for Calculating the Steady-State Response of Nonlinear Dynamic System, *ASME Journal of Applied Mechanics*, **56**, 1989

257 L Dieci, J Lorenz and RD Russell,Numerical Calculation of Invariant Tori, *SIAM Journal on Scientific and Statistical Computing*, **12**, 1991

258 SP Diliberto, WT Kyner and RB Freund, The Application of Periodic Surface Theory to the Study of Satellite Orbits, *Astronomical Journal*, **66**, 1961

259 T Ge and AYT Leung, A Toeplitz Jacobian Matrix Method for the Steady State Analysis of Discontinuous Oscillators, *Shock and Vibration*, **2**, 1995

260 DE Gilsinn, Constructing Invariant Tori for Two Weakly Coupled van der Pol Oscillators, *Nonlinear Dynamics*, **4**, 1993

261 DE Gilsinn, Constructing Galerkin's Approximations Invariant Tori Using MACSYMA, *Nonlinear Dynamics*, 1995

262 RD Blevins, *Flow Induced Vibration,* PhD Thesis, California Institute of Technology, Pasadena, Calif, 1974

263 JK Hall, *Oscillations in Nonlinear Systems*, McGraw-Hill, New York, 1963

264 SA Hall and WD Iwan, Oscillation of a Self-Excited Nonlinear System, *ASME Journal of Applied Mechanics*, 51, 1984

265 B Hao, Elementary Symbolic Dynamics and Chaos in Dissipative Systems, World Scientific, Singapore, 1989

266 AYT Leung and T Ge, Toeplitz Jacobian Matrix Method ASME Journal of Applied Mechanics, **117**, 1995

267 N Minorsky, *Nonlinear Oscillators*, Van Nostrand, New York, 1962

268 RH Rand and PJ Holmes, Bifurcation of Periodic Motions in Two Weakly Coupled van der Pol Oscillators, *International Journal of Non-linear Mechanics*, **15,** 1980

269 DW Storti and RH Rand, Dynamics of Two Strongly Coupled van der Pol Oscillators, *International Journal of Non-linear Mechanics*, **17,** 1982

270 SW Shaw and C Pierre, Normal Modes for Nonlinear Vibratory Systems, *Journal of Sound and Vibration*, **164,** 1992

271 M Urabe and A Reiter, Numerical Computation of Nonlinear Forced Oscillations by Galerkin's Procedure, *Journal of Mathematical Analysis and Applications*, **14,** 1966

# INDEX